INTRODUCTION TO MOTOR MECHANICS

SECOND EDITION

LES STACKPOOLE • MAL MORRISON • ALAN GREGORY

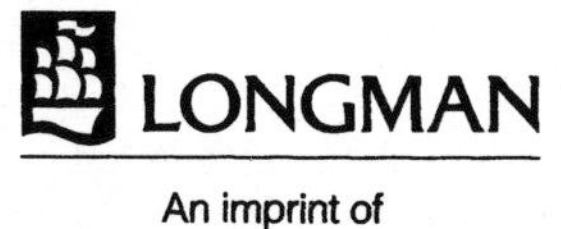

An imprint of
Addison Wesley Longman

Addison Wesley Longman Australia Pty Limited
95 Coventry Street
South Melbourne 3205 Australia

Offices in Sydney, Brisbane and Perth, and associated companies throughout the world.

First published as Introduction to Motor Mechanics 1989
This edition published 1993
Reprinted 1994, 1995, 1996, 1997, (twice), 1999
2 3 4 01 00 99

Edited by Adrienne Linley
Designed by Stanley Wong
Set in 10/11pt Century Schoolbook
Produced by Addison Wesley Longman Australia Pty Ltd

National Library of Australia
Cataloguing-in-Publication data

Stackpoole, Les.
Introduction to motor mechanics.

2nd ed.
Includes index.
ISBN 0 582 909856.

1. Motor vehicles — Maintenance and repair. I. Morrison, Mal. II. Gregory, A. (Alan). III. Title.

629.287

The publisher's policy is to use **paper manufactured from sustainable forests**

CONTENTS

PREFACE

The methods of training automotive mechanic apprentices have undergone major changes in the last eight to ten years. The requirements of the automotive industry for flexible starting dates for their employees, and the recognition by education authorities that students have individual differences and learn at different rates (with the course ending when the student has completed the prescribed topics instead of at the end of the year) have transformed the apprentice mechanic's classroom. The familiar image of a teacher-centred class with all students listening to a lecture, or all working on the same component, and all at the same progress point in the course, has virtually disappeared from motor mechanic apprentice training in technical colleges in Australia.

The absence of the structured lecture and supplementary class notes made it necessary for each student to be good at finding material in text books and manuals. Deficiencies in otherwise excellent technical books and manuals quickly became obvious to teachers working in the field. Many single-volume books attempt to cover the entire automotive area or assume previous knowledge by the reader. This approach means introductory material containing vital learning steps is either bypassed altogether, or compressed to a minimum explanation due to space limitations in the book.

This book will now form the first part of a set of books that will cover the entire content of an apprenticeship course. Volumes 2 and 3 are in preparation. Combined with *Electronics for Motor Mechanics* by the same authors, these books will provide the most comprehensive collection of materials available to motor mechanics.

As with the previous edition, this first volume has been written specifically to assist the beginner in automotive training. No chapter presupposes previous automotive knowledge and all chapters are written in a simple, easy-to-follow style with maximum use of graphics to assist the reader. Each area is expanded to provide a comprehensive base on which students can build as they continue into advanced automotive training.

The content of *Introduction to Motor Mechanics* was designed to assist both teachers and students by reducing the need for supplementary notes and the extensive reference libraries currently required to support the course. This policy has been maintained in the second edition of *Introduction to Motor Mechanics* and so the book has been brought up-to-date to cover technological changes; a new section on metal inert gas (MIG) welding has been included and a new chapter on air conditioning introduced. New illustrations aid comprehension.

Although this first volume is part of a set to service primarily the training of professional motor mechanics, it will be useful on its own as a text book for owner-maintenance courses and secondary school automotive classes.

1

PERSONAL DEVELOPMENT

INTRODUCTION

Many people want to be good at their trade, and skilled in diagnosing and repairing faults in motor vehicles. However, a good trade person must also be skilled in other areas to be a competent worker. Some of these skills include being:

1 able to communicate effectively by speech, by telephone and by writing;
2 good at reading, as you will be required to read and understand service instructions and safety precautions;
3 able to carry out calculations involving money, time and measurements;
4 able to understand the readings on various types of mechanical and electrical measuring and testing instruments.

This chapter will help you in improving these skills.

COMMUNICATION

Communication plays a very important part in our lives. We start communicating at a very early age. For example, a baby cries when it is hungry and yawns when it is tired. This is the baby's way of communicating. There are many ways of communicating with each other. A nod of the head or a wink of an eye may be all that is required to convey a message.

We use one or more of our senses to assist us in communicating:

- **Speech**, for verbal communication;
- **Sight**, for written communication;
- **Hearing**, for audio communication;
- **Touch**, to communicate temperature and the shape and texture of objects;
- **Smell**, to communicate condition of the air.

Governments and business organisations have spent billions of dollars on improving communications. It is possible to receive 'live' television pictures and sound from all around the world. Satellite communication systems, global telephone networks, sophisticated computers and business machines send, receive and print information from around the world.

TELEPHONE USE

One of the most common methods of communication is by using the telephone. It is quick, easy and direct. The telephone allows you to locate and talk to other people, and for them to locate and talk to you. To benefit from every telephone call you make or receive, we offer these suggestions.

1 Before making a telephone call, make sure you have the right telephone number, and know exactly why you are making the call.
2 Have a pen and paper beside the telephone. If you take notes during a conversation, you can check on things later, rather than rely on your memory.
3 Speak directly into the mouthpiece. You do not need to raise your voice, just speak naturally and clearly.
4 When receiving a business call pick up the receiver straight away. Answer the call with

Customer Name .... Telephone Numbers

Address .... Business ....

.... Home ....

Service/Repair Required Order Number ....

....

....

....

....

Date and Time Vehicle in to Workshop.

Date .... Time ....

Date and Time Vehicle Required by the Customer

Date .... Time ....

Estimated Cost of Repair to the Customer $ ....

a pleasant 'Good morning' or 'Good afternoon' followed by the company name and your own name, for example 'Good morning, Joe Blow Smash Repairs, Nick speaking.'

5 Note down the name of the person you are talking to, and use it during the call. People feel important when you are friendly and remember their name.
6 Make sure you fully understand the details of the call. If you are not sure, or not clear on any part of the conversation then politely ask the caller to repeat the details.
7 When finishing a telephone conversation do not just say 'goodbye' and hang up. Be polite, and always remember that telephone politeness may be an effort, but it gets easier with practice and will help you in your job.

To make sure that you get all the information required, when answering a business telephone call, there are information forms available to help you. There is a sample form above.

Emergency telephone call procedure

1 Remain calm and collect all the necessary information regarding the accident, for example the exact location.
2 Dial the emergency number (as listed for your area).
3 When the operator answers he/she will ask you which emergency service you need.
4 You will have to tell the emergency service what sort of emergency it is and exactly where it is.
5 You will also have to give information about anybody who is injured.

Telephone directories

Telephone directories are a good place to find information. Within the major metropolitan areas there are several different types of directories. The two main types are:

- the White Pages; and
- the Yellow Pages

THE WHITE PAGES

The information contained in two books, A–K and L–Z, is:

- emergency services;
- operator assisted services;
- dial-it information services;
- international direct dialling (I.D.D.);
- subscriber trunk dialling (S.T.D.);
- S.T.D. area codes;
- interpreter service;
- health emergency;

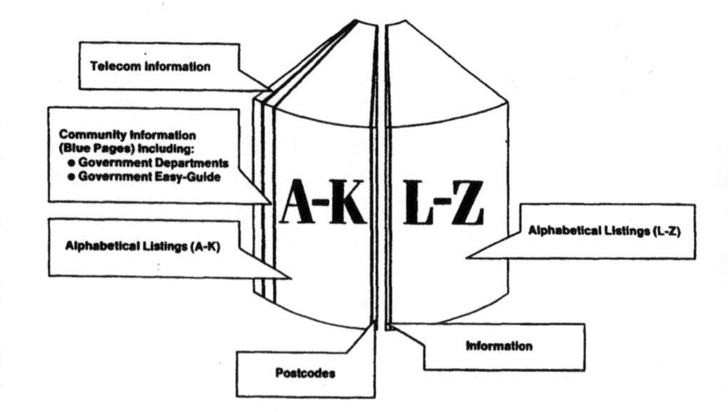

- Australia Post services;
- Commonwealth government;
- local government;
- postcodes;
- alphabetical listings of all telephone subscribers.

THE YELLOW PAGES

The information on products and services is:
- emergency services;
- fast find index (for products and services);
- guide to better buying;
- consumer rights;
- railway information;
- Melbourne telephone district map;
- city locations;
- World Trade Centre;
- dial-it information services;
- alphabetical listing of products and services.

WHITE PAGES USE

This chapter will not explain the many services in the telephone directory, as Telecom has provided a colour-coded reference section at the front of the directory.

The most common use of this directory is to locate the telephone numbers and addresses of private and business subscribers. The guide below will help you with the conventions used in the alphabetical listing.

How To Find Names In This Directory

Names are divided into two parts for sorting. The first part, or the first word, determines the place to find the name. The second part, all the initials or remaining words, determines the order within that group.

Alternative Names
Names such as "John Smith Motors" would most likely appear as "Smith John Motors" but could appear in both places. Check for alternatives.

Ampersand (&), Apostrophes, etc.
Disregard these characters.
Day A B
Day & Graham Pty Ltd
Day's Chemical Co
Dayze Shirt Co

Hyphens
Names which contain a hyphen are treated as two words and are sorted according to the *first* name. The second name is treated as initials.
For example:
A-Grade
Agar L F
Agar-Lyons B
Agar P M
AGRA Products

Capital Letters
Names beginning with a single capital letter, or a group of capital letters, appear as though the capital letters are one word, regardless of whether they are separated by an & (and), full stops, spaces, apostrophes etc.
For example:
A Great impression
Abbott L
A & F Products
African Food
A J's Auto
A J's Bricklaying
Apps D M
ARC Engineering
Archer D F
A S J Transport

Initials
Remember that the second part of a name can be words or initials, so initials are treated as a word.
Jones A
Jones A L
Jones Alfred
Jones Art Supplies

Mt and Mount
Names beginning with "Mt" are treated as though spelt "Mount". Names such as "Mount" appear first, followed by names which have "Mt" or "Mount" as the first part of the name.
For example:
Mount A F
Mount W
Mountain S
Mt Abercrombie
Mountford G N
Mt Hutt Ski Resourt
Mount Macedon

Numbers
Numbers are found under the spelt out word.
Thompson A
Three Angels Cafe
3D Signs
Thursday Travel Agency

Prefix Words
Prefixes are included as part of the first word.
Leather Goods Ltd
Le Cafe Francais
Ledger Supply Co
Le Guin A
Leigh A B

St and Saint
Names beginning with "St" are treated as though spelt "Saint". Names such as "Saint" appear first, followed by names which have "St" or

"Saint" as the first part of the name.
For example:
Saint A
Saint W
St James Church
Saintford L
Saint John Council
St Paul's School
Sainty A

Spelling variations
Names can be spelt in different ways yet sound the same.
Spelling variations are shown at the end of these names.
Reed W
Reed-see also
Read, Reede, Reid

Surnames
The whole surname is included as the first word. If there are several surnames in a business name look under the first one.
Van Aston B
Vance A
Vanden-Berg A
Van der bilt H
Vandyke W
Van Dyke & Wilson Ltd.

The Word "The"
Some business names which begin with "The" are sorted under the next word by request.
Check both places.
China Sales Ltd
China Shop The
China Toy Co.

YELLOW PAGES USE

The Yellow Pages directory, similar to the White Pages, contains information on a number of services. One of the more common uses of this directory can be explained by referring to this example.

To insure a motor vehicle: consult the Yellow Pages and the 'fast find index' section. Select the pages in which insurance companies, dealing in motor vehicle insurance, are listed. Obtain the information required by phoning some of these companies. The name and

FAST-FIND
INDEX
PRODUCT OR SERVICE ALPHABETICALLY LISTED - THEN SELECT HEADING AND PAGE NUMBER

MAIN CLASSIFICATION HEADINGS ARE SET **IN BOLD TYPE**

ACOUSTIC SCREENS See
Office Furniture 1472
Partitions 1552
ACOUSTICAL CONSULTANTS ... 97
ACRYLIC FABRICATORS—See
Plastics - Fabricators
ADVERTISING — MATCHES- See
Match Mfrs &/or W salers 1273
ADVERTISING MEDIA REPRESENTATIVES 106
ADVERTISING
AGENTS See
Advertising Agencies 100
Agents - General 109
AIKIDO INSTRUCTORS- See
Martial Arts & Self Defence
Instruction &/or Supplies

Electrical Insulators &/or Insulating Materials 772
INSURANCE AGENTS 1216
INSURANCE ASSESSORS & LOSS ADJUSTERS 1219
INSURANCE BROKERS 1220
INSURANCE BROKERS INFORMATION SERVICES—See
Information Services 1203
INSURANCE CONSULTANTS—See
Insurance Agents 1216
Insurance Brokers 1220
INSURANCE—CREDIT 1224
INSURANCE—FIRE, MARINE, ACCIDENT & GENERAL 1224
INSURANCE—HEALTH—See
Health Insurance Funds 1117
INSURANCE—LIFE 1230
INSURANCE—LIFE UNDERWRITERS 1234
INSURANCE—MOTOR CYCLE—See
Insurance—Motor Vehicle 1234
INSURANCE—MOTOR VEHICLE 1234

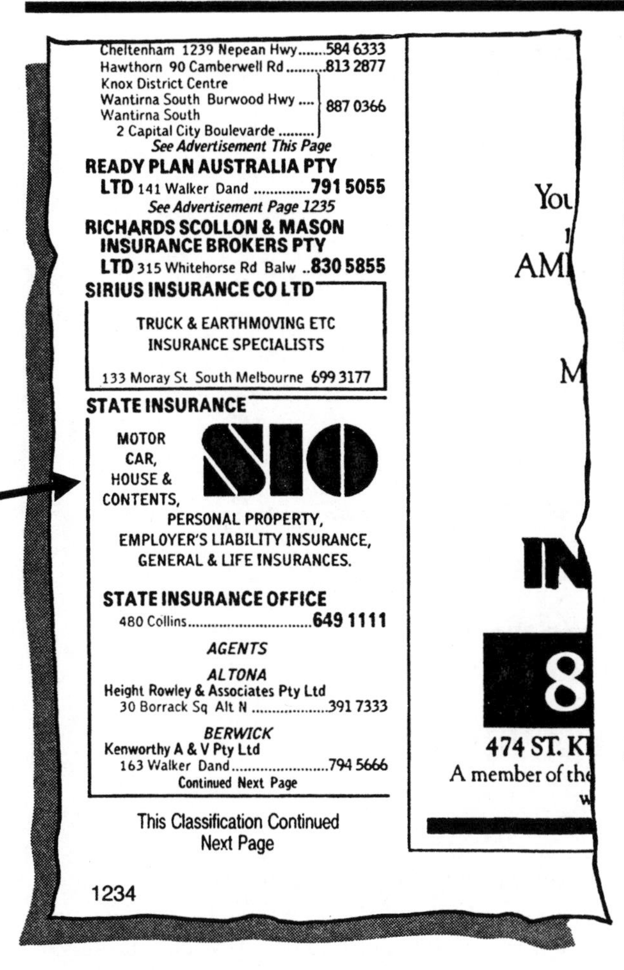
Cheltenham 1239 Nepean Hwy.......584 6333
Hawthorn 90 Camberwell Rd..........813 2877
Knox District Centre
Wantirna South Burwood Hwy.... } 887 0366
Wantirna South
2 Capital City Boulevarde......... }
See Advertisement This Page
READY PLAN AUSTRALIA PTY LTD 141 Walker Dand **791 5055**
See Advertisement Page 1235
RICHARDS SCOLLON & MASON INSURANCE BROKERS PTY LTD 315 Whitehorse Rd Balw ..**830 5855**
SIRIUS INSURANCE CO LTD
TRUCK & EARTHMOVING ETC
INSURANCE SPECIALISTS
133 Moray St South Melbourne 699 3177
STATE INSURANCE
MOTOR CAR, HOUSE & CONTENTS,
SIO
PERSONAL PROPERTY,
EMPLOYER'S LIABILITY INSURANCE,
GENERAL & LIFE INSURANCES.
STATE INSURANCE OFFICE
480 Collins..............................**649 1111**
AGENTS
ALTONA
Height Rowley & Associates Pty Ltd
30 Borrack Sq Alt N 391 7333
BERWICK
Kenworthy A & V Pty Ltd
163 Walker Dand........................794 5666
Continued Next Page
This Classification Continued Next Page
1234

You
AM
M
IN
8
474 ST. K
A member of th

address of the company you have selected may be the:

State Insurance Office
480 Collins St
Melbourne 3000

You may choose to visit this insurance company. Get its address and check a street directory for the location.

STREET DIRECTORY USE

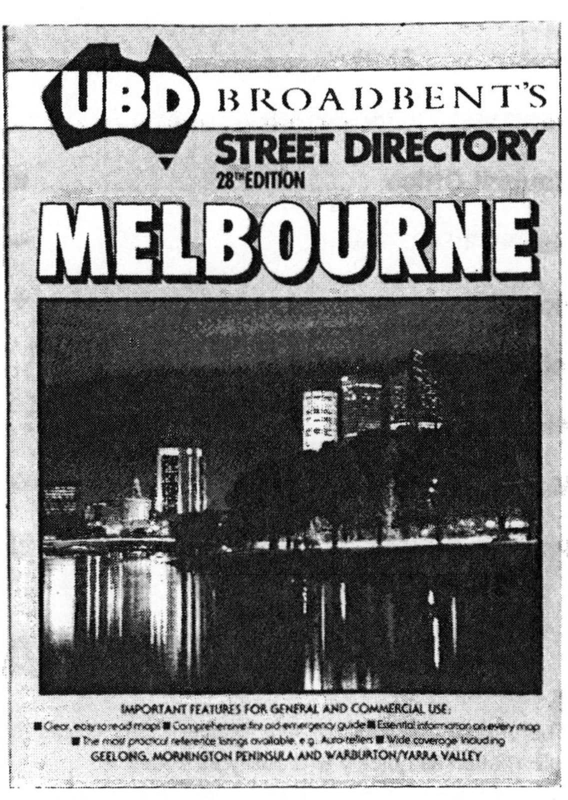

The street directory contains information on:

- road maps;
- tour maps
- public transport;
- index of street names;
- service information;
- index of localities

All street directories contain explanatory notes and legends depicting the symbols used on the maps. A number of typical examples are shown on the following pages.

LEGEND

Ambulance Station................................... ⊕

Automatic Telling Machine, maps 1 & 2.... $

Council Office.. ■

Guide Hall.. ♣

Hospital.. ✚

Infant Welfare Centre.............................. I.W.C. ▪

Kindergarten... K'garten ▪

Masonic Temple... ⚜

Parking Station, maps 1 & 2....................... P

Place of Worship...................................... †

Police Station.. ◉

Post Office, most have public telephones....... ★

School, State... ▲

School, Private.. △

Scout Hall.. ⚜

Telephone.. ☎

Toilets, maps 1 & 2.................................... T

Weighbridge.. Ⓦ

EXPLANATION

The maps and indexes are divided into colour coded sections to help the user to locate the required information as quickly and easily as possible.

The Key Maps, inside the front cover, show how the street maps relate to the major roads and suburbs.

The street maps are divided into two sections, those with blue borders 19 to 416, at a scale of 1:18,450, make up the bulk of the directory. Each map covers 3.5km east to west and 4km north to south. Maps 3 to 18, with orange borders, at a scale of 1:10,200, cover the congested Inner City areas in greater detail. These maps cover 2km east to west and 2.25km north to south. All the street maps are aligned to magnetic north, with a scale and key to symbols appearing at the top of each pair of maps. All the maps have an overlap with the adjoining maps in each direction.

The red bordered City Centre maps 1 & 2 cover the Central Business District at a scale of 1:5,300, showing detail of major buildings, retailers, hotels, theatres, nightclubs, parking stations, etc. with comprehensive indexes.

The special maps 425 to 435, with green borders, show details of Melbourne, Monash, La Trobe & Deakin Universities, RMIT, Swinburne, VFL Park, MCG & National Tennis Centre, the Victorian Arts Centre, Calder Park Thunderdome and Phillip Island Grand Prix Circuit.

For trips outside the area covered by the Street Maps, turn to the Victorian Road Map section comprising maps 450 to 462, with red borders, covering Victoria and southern New South Wales with a Key Map and Index to Major Towns.

In the event of an accident or emergency, the First Aid Guide for Motorists offers five pages of invaluable information for the care and treatment of various injuries. Phone numbers for all Emergency Services on page 610.

For Information about the public transport system, turn to The Met. Four pages, with yellow borders, giving details on ticketing, tram and rail routes.

The Index to Streets, 132 pages in all, contains over 45,000 separate entries.

The Information Section, with grey borders, contains 45 pages of comprehensive index listings with 46 different categories, ranging from Airfields and Automatic Telling Machines thru Libraries and Markets to VFL Clubs and Wharves Refer to the list of Contents on page VIII for complete details.

LEGEND

Freeway/Arterial Link, with Emergency Telephones	FRANKSTON FREEWAY
Proposed Freeway	PROPOSED FREEWAY
Primary Arterial Road, with service road	NEPEAN HIGHWAY
Secondary Arterial Road, with service road	TOORAK ROAD
Collector Road, with occasional building numbers	DOMAIN RD.
Local Access Street	THE BOULEVARD
Non-trafficable Street	
Tramway	
Traffic Light, One Way Street and Roundabout	
Railway Line and Station, with distance from Flinders St. station	road over; road under; level crossing
Rivers and Creeks	Merri Ck.
Bicycle Track	Bicycle Track
Walking Track	Walking Track
Postcode Number and Boundary	3931
Municipal Name and Boundary	Mornington
Suburb Name	MORNINGTON
Locality Name	Parkdale
Direction to City of Melbourne	CITY
Direction to City of Geelong	GEELONG
Park, Reserve, Golf Course, etc.	
School, Hospital, Cemetery, etc.	
Distance, by road from GPO	59
National Route Number	
Freeway Route Number	
Metropolitan Route Number	

The Metropolitan Route Numbering system is being progressively updated during 1989, resulting in many changes to the old system including the abolition of separate Freeway Route Numbers

To locate Collins St, Melbourne in the street directory, check the index of street names. The street names are listed in alphabetical order. Under the listing for Collins we find Collins St, Melbourne. This will give the roadmap number and the grid reference. For example, in a street directory the information for Collins Street, Melbourne could be roadmap 1 and grid reference C6.

COLLINS

Av	Altona North	192	A2
Cl	Scoresby	202	A7
Cr	Berwick	268	E3
Ct	Balwyn	153	H5
Ct	Chelsea	275	D2
Gr	Croydon	137	A1
Pde	Sorrento	356	D3
Pl	Geelong	414	D6
Pl	Kilsyth	137	G7
Rd	Dromana	338	C8
Rd	Melton	394	F1
St	Belmont	410	B5
St	Bentleigh	221	D5
St	Box Hill	176	G2
St	Brighton	219	H7
St	Bulleen	131	H5
St	Chadstone	198	C6
St	Coburg	128	A1
St	Diamond Creek	60	G6
St	Essendon	126	G5
St	Footscray	170	F2
St	Geelong West	408	A6
St	Heidelberg Heights	102	E7
St	Melbourne	1	C6
St	Melbourne	172	D4
St	Melbourne	8	B6
St	Mentone	248	F5
St	Mentone	249	B5
St	Preston	129	A1

Grid references

- Along the top and bottom of the roadmap there are alphabetically marked sections.
- Down the sides of the roadmap there are numerically marked sections.

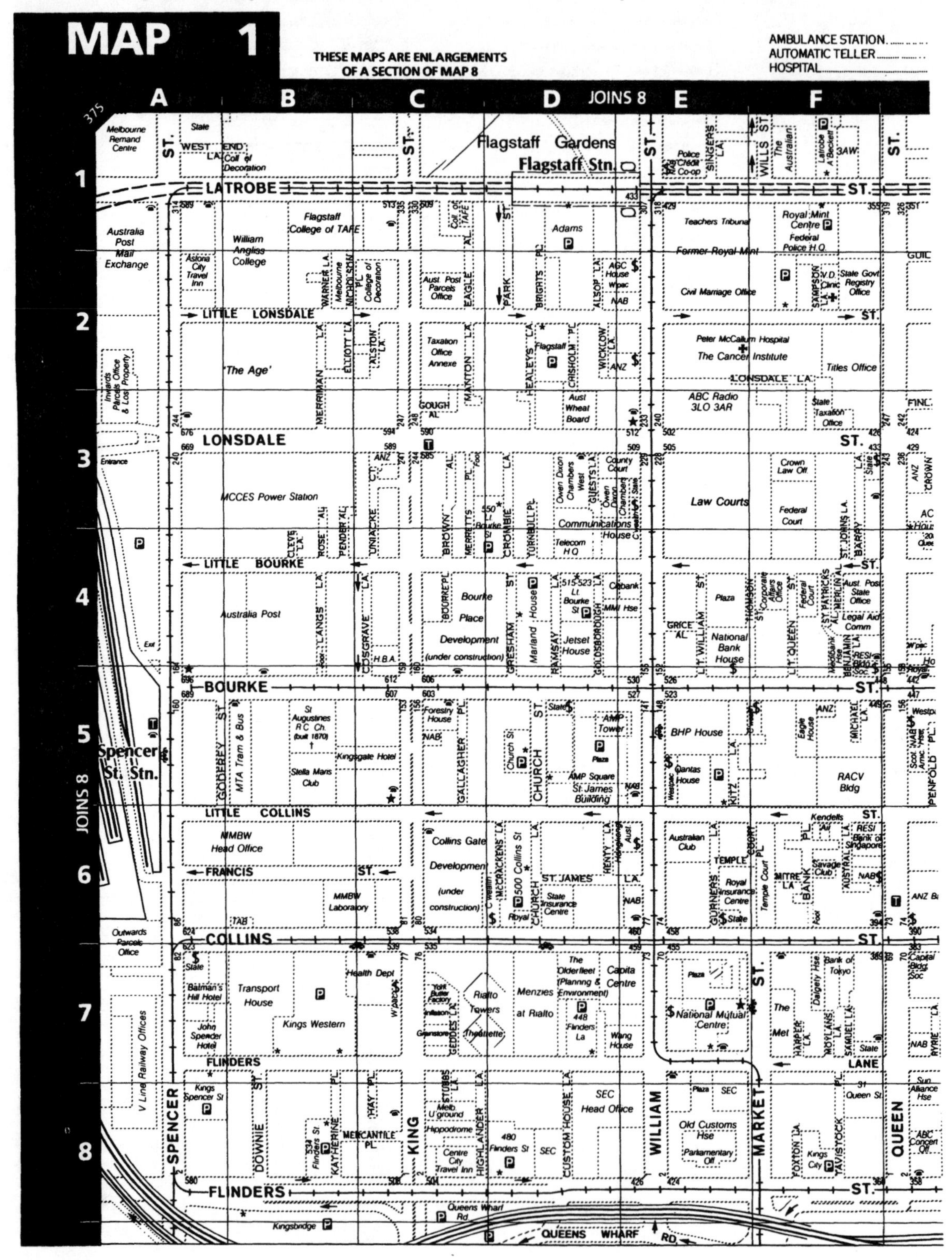

To locate grid reference C6 on roadmap 1:
1 place a finger on the section marked 'C' at the top of roadmap 1.
2 Place a finger of the other hand on the section marked 6 at the side of roadmap 1.

By running your finger down the C section and your other finger along the 6 section, the section in which both fingers meet is referred to as grid reference C6, and the name Collins St will be located in this section.

Collins St is quite a long street. To find the street number of 480, a further reference is given to help you.

Run a finger along the 6 section, until the street number close to 480 is located. This will place the building in that section of Collins St, between William St and King St.

The above example shows how to find house numbers. Some directories also provide information on tram and bus route numbers.

LITERACY

Descriptive writing

Another common method of communication is by writing. Newspapers, magazines and books are good examples of this method of communication. Most of the writing we do can be classed as either personal writing or formal writing. Personal writing is concerned with your personal experiences, hobbies, sporting interests, or thoughts on particular issues. Formal writing may be a business letter, a job application, a service report, or an assignment for education purposes.

Writing demands concentrated thinking. At times we know what we want to say, but find it difficult to 'put it in writing'. When you are given a writing project it is good practice to:
1 organise and plan what you are going to write;
2 prepare by thinking about, talking over and finding out about the subject;
3 write a first draft, which will probably contain bad grammar and long sentences. However, this first draft must contain the main idea of the final written work;
4 check the draft by rewriting the clumsy passages, making better use of words, joining some sentences and making long, rambling text into short sentences that show your meaning;
5 check the final draft. Between your first hazy thoughts and this final draft, your ideas will have become clearer, so that the reader will, hopefully, instantly understand your meaning.

Component part names

It is important to know the correct names of automotive component parts to assist in:
- understanding the way a motor vehicle is built;
- understanding the construction and operation of the 'major' assemblies (engine, transmission, suspension etc.);
- ordering and purchasing spare parts for a motor vehicle.

By looking closely when working on a motor vehicle and in the workshop, you will be able to recognise the many component parts which make up the modern motor vehicle. These component parts are represented in service and spare parts manuals as 'line' drawings and 'exploded' view drawings. Frequent reference to these books and manuals will help you to remember and identify parts.

After ensuring you have the correct component part name, it is essential that the make, model, year of manufacture and engine numbers are quoted when ordering replacement parts.

If you neglect to quote the engine number when ordering a replacement air cleaner element you could receive the wrong type. As different types of engines may be fitted to the same make, model, year of a vehicle, for example EFI and carburettor models.

Service manuals

Vehicle manufacturers spend millions of dollars in producing reliable, economical and comfortable vehicles which are acceptable to their many customers. Service manuals are written for each vehicle model produced. These manuals contain repair procedures, service information, and specifications for the vehicle. To find specific information in a service manual:
1 refer to the index at the front of the service manual for the required section;
2 open the service manual at the required section. A sub-index at the front of this section shows the page number;

3 turn to the page number;
4 search this page for the required information.

Note: It is recommended that, before starting any repairs to a motor vehicle, the service manual be checked for the correct repair procedures and specifications.

Text books

Text books contain general information on the operation and service procedures for automotive assemblies, tools and equipment, test instruments and other related workshop equipment. To assist the reader this information must be presented in an orderly manner, therefore most textbooks are structured with:

- a contents page, usually located at the front of the book;
- an alphabetically listed index, located at the back of the book; and
- a glossary of terms, which explains the meanings of terms used in the automotive trade.

Information within a text book can be found by two methods:

1 contents section for the page location of a major assembly;
2 alphabetical index for the page location of a specific component or items.

Vocabulary

In this book, other books and the automotive trade, you may meet a number of unfamiliar words and terms. It is a good idea to use a glossary or technical dictionary to assist you with the meaning of the unfamiliar words and terms. If, after reading the explanations, you are still unsure of their meaning, ask your teacher, workmates or suitably qualified people to explain to you the meaning of the words or terms. The results of not fully understanding the meaning of words or terms can be dangerous. For example: A sign on a container '**Caution. The contents of this container are highly flammable**'. If you do not know the meaning of the word 'flammable' and were foolish enough to open the container when there was a flame (such as a lighted cigarette or oxy welding equipment) nearby then the situation could be dangerous.

DICTIONARY USE

The words in a dictionary are listed in alphabetical order from A at the beginning to Z at the end of the book. For example, to find the word 'easy' in a dictionary:

1 find the section containing the words starting with the letter 'E';
2 start at the beginning of the 'E' section and locate the first word that has 'A' as its second letter;
3 check the 'EA' group of words until the first word that has 'S' as its third letter is located (this will be near the end of the 'EA' group);
4 check the 'EAS' group of words until the fourth letter 'Y' is located. This is the last letter of the word 'EASY' so its meaning can now be read.

Note: This procedure can be used to find any length word in a dictionary or glossary.

COMPREHENSION AND PRECIS

Comprehension, another word for understanding, and precis (pray-see) aims at reducing the original material but still keeping all the important details. The following examples show two passages, one of which is a precis of the other. In many service manuals the writer takes it for granted that the reader is a trained mechanic, and so does not provide a full statement of the task to be performed. This approach is good as unnecessary words can be annoying to the reader, but trainees must be careful that they fully understand the meaning of the instructions before proceeding.

These *comprehension and precis* examples demonstrate a method of understanding passages from a service manual.

Example 1

Part 1 shows a paragraph intended for a trainee. By picking out the essential parts (underlined), the writer can shorten (precis) the paragraph and still, as shown in part 2, retain the same meaning for a trained person.

1 Clean the distributor cap, with a soft bristle brush and mild cleaning solvent or alcohol. Dry the cap with compressed air. Inspect the cap for cracks, burned contacts, permanent carbon tracks or dirt or corrosion in the sockets. Replace the cap if it is defective.

2 Clean the distributor cap, with a soft brush and alcohol. Dry and inspect the cap for defects and replace if faulty.

Example 2

Part 1 is an acceptable instruction for a trained person, but may be inadequate for a trainee. Part 2 expands the original instruction, by providing the additional information needed to complete the task.

1 Clean spark plugs.
2 Clean the spark plugs on a sand blast cleaner, following the manufacturer's instructions. Do not prolong the use of the abrasive blast as it will erode the insulator and the electrodes. Clean the electrode surfaces with a small file. Dress the electrode to obtain flat parallel surfaces on both the centre and side electrodes.

These examples were provided to highlight the care which must be taken by the trainee when reading technical instructions. Understand the *real* meaning of the instruction before starting any task. It would have been possible to clean the spark plugs by washing them in solvent, but this was not what the instruction meant.

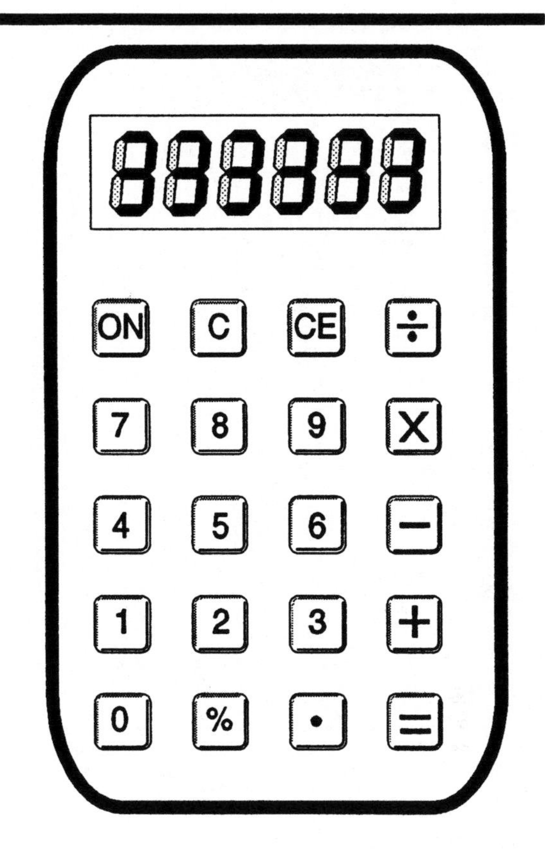

BASIC ACCOUNTING

To be successful in business, it is very important to maintain accurate records of all business transactions, accounts, receipts, salaries, banking and taxes. In our everyday life, we need to keep accurate records and understand our business transactions and finances. If we consistently spend more money than we earn, do not pay our bills, or are unable to calculate sales invoices, time and labour charges, then we will have to accept the consequences.

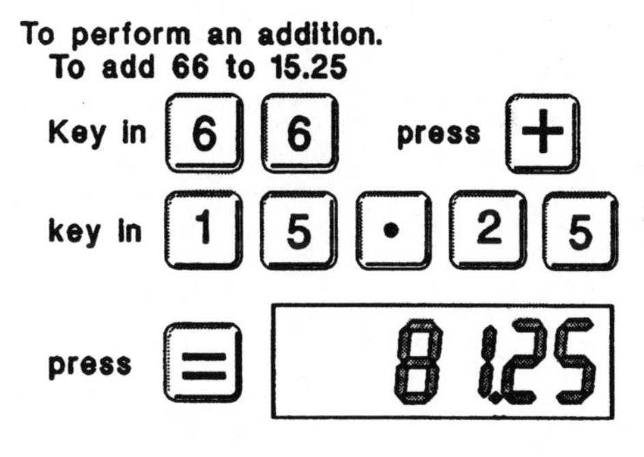

CALCULATOR USE

It is not necessary to have significant knowledge of mathematics to calculate area, volume, salary or invoice totals. Cheap calculators are readily available and are easy to operate. The functions most commonly used are addition, subtraction, multiplication, division and percentage; these are present on very simple calculators. If you are unsure of how to use a calculator, the following hints should be of assistance.

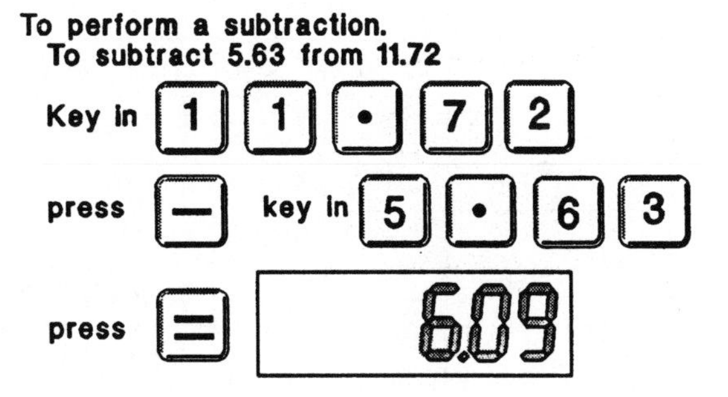

To perform a multiplication.
To multiply 3.142 by 2.6

Key in 3 • 1 4 2

press X key in 2 • 6

press = 8.1692

To perform a division.
To divide 1126 by 36.3

Key in 1 1 2 6

press ÷ key in 3 6 • 3

press = 31.0192

To perform a percentage.
To calculate 12.5% of 32

Key in 3 2 press X

key in 1 2 • 5

press % 4

At times it is possible to make mistakes when entering the numbers on the keypad. Most calculators contain a CE key. The letters CE denotes CLEAR ENTRY. This key is very useful to correct mistakes. Try this example.

Add 66 to 15.25.

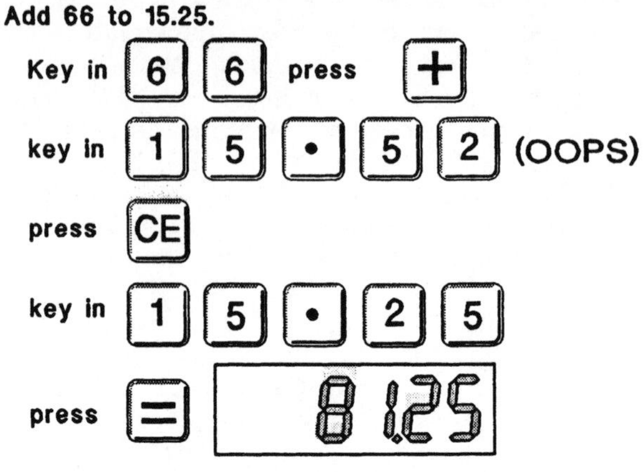

The CE key removed your mistake and allowed the calculation to proceed correctly.

This section will assist you to calculate the total costs of a customer sales docket. Refer to the sample sales docket.

- The unit price refers to the price of one item.
- Sales tax of 20 per cent is added to the unit price.
- Sub-total is calculated by: unit price plus sales tax times quantity

Spark Plugs ($1.10 + 22c) × 6
= $1.32 × 6
= $7.92

Parts and accessories					
Quantity	Part Number	Description	Unit Price	Sales Tax	Sub Total
6	23156	Spark plugs	1.10	20%	$7.92
1	31256	Air cleaner element	7.40	20%	$8.51
2	52316	Fan belts	6.20	20%	$14.80
1	63521	Oil filter	4.60	20%	$5.52
4 1	—	Engine oil	9.60	—	$9.60
		Total			$46.35

Quotations

When a fault has been diagnosed in any section of a motor vehicle, a quotation should be prepared for the customer. A quotation is an accurate list of all parts and labour that would be used in the repair of a vehicle. The information that is needed to prepare a quotation is as follows:

- a full list of all the parts and consumables including their current retail prices;
- a standard times chart which will allow the total time for the repair to be calculated;
- the hourly rate set by the proprietor/s of the workshop;
- a quotation sheet, see the example;
- the name and address of the customer;
- the make and registration number of the vehicle;
- the percentage of discount that the proprietor/s of the workshop has set for cash payments.

Once a quotation has been accepted by the customer, the cost of the repair is fixed and it cannot be changed without an agreement from the customer.

Preparing a quotation:

1 Enter the name and address of the customer.
2 Enter the make and registration number of the vehicle.
3 Enter the date.
4 List all the parts and consumables in the column under the heading 'PARTS'.
5 Insert the current retail price for each item in the respective columns.
6 Add the prices of all the parts and enter the result in the columns next to 'SUB-TOTAL'.
7 Carefully list all the major steps of the repair in the column under the heading 'REPAIRS'.
8 Using the standard times chart, calculate the hours that will take to complete the repair.
9 Calculate the cost for the repair by multiplying the hours for the repair by the set hourly rate.
10 Enter the repair cost in the columns next to the number of hours and the hourly rate.
11 Transfer the sub-totals to the respective columns next to 'REPAIRS — SUB-TOTAL' and 'PARTS — SUB-TOTAL'.
12 Add these two sub-totals and write the result in the columns next to 'TOTAL'.
13 Calculate the discount for a cash payment and enter it in the columns next to 'DISCOUNT'.
14 Subtract the discount from the previous total and enter the result in the columns next to 'TOTAL'. See the example quotation sheet.

Note: Some proprietors do not give discount to customers for cash payments.

Parts ordering

When a fault has been diagnosed in any section of a motor vehicle and the customer has agreed to have the repair carried out, an order for replacement parts will have to be prepared for the parts supplier.

An efficient workshop will have order forms printed in a style that will allow sufficient information to be noted. This will ensure that the parts supplier can fill the order promptly without ringing you for further information. See the example. The following information should appear on an order form:

- the vehicle's identification;
  —make, model and year;
  —serial number and/or chassis number;
- the date of the order;
- the parts supplier and the salesperson;
- your name;
  —the salesperson may need to contact you
  —about the availability of the parts;
- a complete list of the parts;
  —parts must be correctly named;
  —details must be clearly legible.

Note: The salesperson will not be familiar with the particular repair so do not assume that this person will supply parts that you have omitted from the list.

Once the order form has been completed, it may be delivered, posted or phoned through to the parts supplier and a copy should be kept in a prominent place for future reference.

Upon receiving the parts, each part must be checked off against the copy of the order form. Omitted or incorrect replacement parts must be brought to the attention of the parts supplier as soon as possible.

Preparing an order form:

1 Obtain a pen and a piece of 'scrap' paper.
2 Raise the bonnet of the vehicle.
3 Locate the ID plate of the vehicle which is generally on the firewall behind the engine.
4 Note the make, model and year of the vehicle.

GREAT SERVICE MOTORS

PROPRIETORS: Gregory, Stackpoole and Morrison

AUTOMOTIVE REPAIR SPECIALISTS
1432 AUTOMOBILE ROAD, VEHICLEVILLE, VICTORIA, 3999
TELEPHONE: (03) 153 6241 FAX: (03) 153 6242

QUOTATION 1342

NAME .... MAKE ....

ADDRESS .... REG. NO. ....

....

.... DATE ....

QTY	PARTS			REPAIRS		
1	Cylinder head gasket set	67	45	Remove and clean cylinder head		
1	Exhaust valve	15	95	Remove, clean, reface, seat and replace		
1	Inlet valve	12	60	valves		
1	Top radiator hose	5	80	Clean cylinder block face		
1	Radiator coolant additive, 250lt	2	50	Replace cylinder head		
				Adjust valve clearance		
				4.5 Hours @ $35 per hour	157	50
				REPAIRS – SUB-TOTAL	157	50
				PARTS – SUB-TOTAL	104	30
				TOTAL	261	80
				DISCOUNT	26	10
	SUB-TOTAL	104	30	**TOTAL**	235	70

THIS QUOTE IS VALID FOR 30 DAYS

PROPRIETORS: **Gregory, Stackpoole and Morrison**

AUTOMOTIVE REPAIR SPECIALISTS
1432 AUTOMOBILE ROAD, VEHICLEVILLE, VICTORIA, 3999
TELEPHONE: (03) 153 6241 FAX: (03) 153 6242

ORDER FORM **2801**

PARTS SUPPLIER: ....

....

SALES PERSON: .... PARTS ORDER BY: ....

MAKE .... MODEL ....

SERIAL NO. .... CHASSIS NO. ....

YEAR .... DATE ....

QTY	PARTS DESCRIPTION
1	Cylinder head gasket set including valve stem seals and
	rocker cover gasket
1	Exhaust valve
1	Inlet valve
1	Top radiator hose
1	Radiator coolant additive. 250lt.

5 Note the serial and/or chassis numbers.
6 Clean your hands; grease marks on an order form makes it impossible to read or overwrite.
7 Enter the name of the part supplier.
8 Enter the name of the salesperson, if known.
9 Transfer the information from the piece of 'scrap' paper to the order form and ensure it is legible and entered in the correct place.
10 List the parts.
—Enter the quantity of the part in the 'QTY' column.
—Enter the correct name of the part, on the same line as the quantity, in the 'DESCRIPTION' column.
11 Check that the list is complete.
12 Copy the order form or ensure that the copy is legible.
Note: The order forms are generally of a self-copying type.
13 Deliver the form to the parts supplier.
14 Use the copy of the order form as a check-list when the parts have been delivered from the parts supplier.

Pay calculations

Your weekly wage assists you to pay bills, buy clothes, food and other products, and pay for entertainment. Most wage earners are paid an agreed amount of money for doing specific jobs, usually on a weekly basis. A pay slip usually accompanies your weekly wage. On this pay slip is information on:

- personal identification;
- number of hours worked — normal and penalty rates;
- gross wages;
- income tax;
- other deductions;
- net wages.

Personal identification is usually your name and perhaps a pay number.
Number of hours worked — Normal hours are the number of hours you usually work, excluding any overtime.
Penalty rates — are rates paid for overtime worked. Penalty rates differ, depending on the time in which you work the overtime. Time and a half rate means that you receive one and a half times your normal hourly rate. Double time rate means that you receive twice your normal hourly rate.
Gross wage — refers to the total wage you are entitled to, before any deductions are taken from it.
Income tax — is a deduction determined by the Taxation Department.
Other deductions — may be for union dues, superannuation or for a savings account.
Net wage — refers to the actual amount of money you receive after all deductions are taken from your gross wage.

Example
Suppose you are earning a gross weekly wage of $200.00 for forty hours work. To calculate the gross hourly rate, divide 200 by 40, showing your gross rate for one hour would be $5.00. Suppose, during one week you work for forty-five hours:
40 hours are calculated at the normal rate
3 hours may be calculated at the 'time and a half' overtime rate
2 hours may be calculated at the 'double time' overtime rate
Your 'gross' weekly wage would be:

40 × $5.00	= $200.00
3 × $7.50	= $ 22.50
2 × $10.00	= $ 20.00
Total	= $242.50

Income tax deduction. For the gross wage of $242.50 refer to the income tax deduction scale.

If you are seventeen years of age, you are not married, and have lodged a general exemption form, then by checking the $242.50 per week column you will see the income tax deduction is $40.00.
Other deductions. You may have a savings account and have agreed to have $50.00 deducted from your weekly wage and paid directly into the bank.
Net wages. This is the amount of money you actually receive after all the deductions are taken from your gross wage.

Gross wage			$242.50
Deductions —	Income tax	$40.00	
	savings account	$50.00	
Total deductions		$90.00	

Your net wage is calculated by subtracting the total deductions from the gross wage.

Gross wage	$242.50
Subtract deductions	$ 90.00
	$152.50

Your net wage would be $152.50

DEPENDANT AND ZONE REBATE READY RECKONER

Rebate claimed	Instalment value	Rebate claimed	Instalment value	Rebate claimed	Instalment value	Rebate claimed	Instalment value	Rebate claimed	Instalment value	Rebate claimed	Instalment value	Rebate claimed	Instalment value	Rebate claimed	Instalment value
$	$	$	$	$	$	$	$	$	$	$	$	$	$	$	$
1	0.05	7	0.15	40	0.75	90	1.70	376	7.15	780	14.80	1030	19.55	1600	30.40
2	0.05	8	0.15	45	0.85	100	1.90	400	7.60	800	15.20	1100	20.90	1700	32.30
3	0.05	9	0.15	50	0.95	200	3.80	500	9.50	830	15.75	1200	22.80	1800	34.20
4	0.10	10	0.20	60	1.15	270	5.15	600	11.40	900	17.10	1300	24.70	1900	36.10
5	0.10	20	0.40	70	1.35	282	5.35	700	13.30	938	17.80	1400	26.60	2000	38.00
6	0.10	30	0.55	80	1.50	300	5.70	749	14.25	1000	19.00	1500	28.50	3000	57.00

INSTALMENT SCHEDULE

Weekly earnings	Instalment			Weekly earnings	Instalment			Weekly earnings	Instalment			Weekly earnings	Instalment		
	No tax file number	No general exemption	With general exemption		No tax file number	No general exemption	With general exemption		No tax file number	No general exemption	With general exemption		No tax file number	No general exemption	With general exemption
1	2	3	4	1	2	3	4	1	2	3	4	1	2	3	4
$	$	$	$	$	$	$	$	$	$	$	$	$	$	$	$
1	0.50	0.30	-	66	33.15	19.95	-	131	65.80	39.60	9.25	196	98.50	66.35	27.65
2	1.00	0.60	-	67	33.65	20.25	-	132	66.30	39.95	9.50	197	99.00	66.75	27.90
3	1.50	0.90	-	68	34.15	20.55	-	133	66.85	40.35	9.75	198	99.50	67.20	28.15
4	2.00	1.20	-	69	34.65	20.85	-	134	67.35	40.80	10.00	199	100.00	67.60	28.40
5	2.50	1.50	-	70	35.15	21.15	-	135	67.85	41.20	10.25	200	100.50	68.00	28.65
6	3.00	1.80	-	71	35.65	21.45	-	136	68.35	41.60	10.50	201	101.00	68.40	28.90
7	3.50	2.10	-	72	36.15	21.80	-	137	68.85	42.00	10.70	202	101.50	68.85	29.15
8	4.00	2.40	-	73	36.70	22.10	-	138	69.35	42.45	10.95	203	102.00	69.25	29.45
9	4.50	2.70	-	74	37.20	22.40	-	139	69.85	42.85	11.20	204	102.50	69.65	29.70
10	5.00	3.00	-	75	37.70	22.70	-	140	70.35	43.25	11.45	205	103.00	70.05	29.95
11	5.50	3.30	-	76	38.20	23.00	-	141	70.85	43.65	11.70	206	103.50	70.50	30.20
12	6.00	3.65	-	77	38.70	23.30	-	142	71.35	44.10	11.95	207	104.00	70.90	30.45
13	6.55	3.95	-	78	39.20	23.60	-	143	71.85	44.50	12.20	208	104.50	71.30	30.70
14	7.05	4.25	-	79	39.70	23.90	-	144	72.35	44.90	12.45	209	105.00	71.70	30.95
15	7.55	4.55	-	80	40.20	24.20	-	145	72.85	45.30	12.70	210	105.50	72.15	31.25
16	8.05	4.85	-	81	40.70	24.50	-	146	73.35	45.75	12.90	211	106.00	72.55	31.50
17	8.55	5.15	-	82	41.20	24.80	-	147	73.85	46.15	13.15	212	106.50	72.95	31.75
18	9.05	5.45	-	83	41.70	25.10	-	148	74.35	46.55	13.40	213	107.05	73.35	32.00
19	9.55	5.75	-	84	42.20	25.40	-	149	74.85	46.95	13.65	214	107.55	73.80	32.25
20	10.05	6.05	-	85	42.70	25.70	-	150	75.35	47.40	13.90	215	108.05	74.20	32.50
21	10.55	6.35	-	86	43.20	26.00	-	151	75.85	47.80	14.15	216	108.55	74.60	32.75
22	11.05	6.65	-	87	43.70	26.30	-	152	76.35	48.20	14.40	217	109.05	75.00	33.05
23	11.55	6.95	-	88	44.20	26.60	-	153	76.90	48.60	14.65	218	109.55	75.45	33.30
24	12.05	7.25	-	89	44.70	26.90	-	154	77.40	49.05	14.85	219	110.05	75.85	33.55
25	12.55	7.55	-	90	45.20	27.20	-	155	77.90	49.45	15.10	220	110.55	76.25	33.80
26	13.05	7.85	-	91	45.70	27.50	-	156	78.40	49.85	15.35	221	111.05	76.65	34.05
27	13.55	8.15	-	92	46.20	27.85	-	157	78.90	50.25	15.60	222	111.55	77.10	34.30
28	14.05	8.45	-	93	46.75	28.15	-	158	79.40	50.70	15.85	223	112.05	77.50	34.55
29	14.55	8.75	-	94	47.25	28.45	0.25	159	79.90	51.10	16.10	224	112.55	77.90	34.85
30	15.05	9.05	-	95	47.75	28.75	0.45	160	80.40	51.50	16.35	225	113.05	78.30	35.10
31	15.55	9.35	-	96	48.25	29.05	0.70	161	80.90	51.90	16.60	226	113.55	78.75	35.35
32	16.05	9.70	-	97	48.75	29.35	0.95	162	81.40	52.35	16.85	227	114.05	79.15	35.60
33	16.60	10.00	-	98	49.25	29.65	1.20	163	81.90	52.75	17.05	228	114.55	79.55	35.85
34	17.10	10.30	-	99	49.75	29.95	1.45	164	82.40	53.15	17.30	229	115.05	79.95	36.10
35	17.60	10.60	-	100	50.25	30.25	1.70	165	82.90	53.55	17.55	230	115.55	80.40	36.35
36	18.10	10.90	-	101	50.75	30.55	1.95	166	83.40	54.00	17.80	231	116.05	80.80	36.60
37	18.60	11.20	-	102	51.25	30.85	2.20	167	83.90	54.40	18.05	232	116.55	81.20	36.90
38	19.10	11.50	-	103	51.75	31.15	2.40	168	84.40	54.80	18.30	233	117.10	81.60	37.15
39	19.60	11.80	-	104	52.25	31.45	2.65	169	84.90	55.20	18.55	234	117.60	82.05	37.40
40	20.10	12.10	-	105	52.75	31.75	2.90	170	85.40	55.65	18.80	235	118.10	82.45	37.65
41	20.60	12.40	-	106	53.25	32.05	3.15	171	85.90	56.05	19.00	236	118.60	82.85	37.90
42	21.10	12.70	-	107	53.75	32.35	3.40	172	86.40	56.45	19.25	237	119.10	83.25	38.15
43	21.60	13.00	-	108	54.25	32.65	3.65	173	86.95	56.85	19.50	238	119.60	83.70	38.45
44	22.10	13.30	-	109	54.75	32.95	3.90	174	87.45	57.30	19.75	239	120.10	84.10	38.75
45	22.60	13.60	-	110	55.25	33.25	4.15	175	87.95	57.70	20.00	240	120.60	84.50	39.10
46	23.10	13.90	-	111	55.75	33.55	4.40	176	88.45	58.10	20.25	241	121.10	84.90	39.40
47	23.60	14.20	-	112	56.25	33.90	4.60	177	88.95	58.50	20.50	242	121.60	85.35	39.70
48	24.10	14.50	-	113	56.80	34.20	4.85	178	89.45	58.95	20.75	243	122.10	85.75	40.00
49	24.60	14.80	-	114	57.30	34.50	5.10	179	89.95	59.35	21.00	244	122.60	86.15	40.30
50	25.10	15.10	-	115	57.80	34.80	5.35	180	90.45	59.75	21.25	245	123.10	86.55	40.60
51	25.60	15.40	-	116	58.30	35.10	5.60	181	90.95	60.15	21.70	246	123.60	87.00	40.95
52	26.10	15.75	-	117	58.80	35.40	5.85	182	91.45	60.60	22.15	247	124.10	87.40	41.25
53	26.65	16.05	-	118	59.30	35.70	6.10	183	91.95	61.00	22.60	248	124.60	87.80	41.55
54	27.15	16.35	-	119	59.80	36.00	6.35	184	92.45	61.40	23.05	249	125.10	88.20	41.85
55	27.65	16.65	-	120	60.30	36.30	6.55	185	92.95	61.80	23.50	250	125.60	88.65	42.15
56	28.15	16.95	-	121	60.80	36.60	6.80	186	93.45	62.25	23.95	251	126.10	89.05	42.45
57	28.65	17.25	-	122	61.30	36.90	7.05	187	93.95	62.65	24.40	252	126.60	89.45	42.75
58	29.15	17.55	-	123	61.80	37.20	7.30	188	94.45	63.05	24.85	253	127.15	89.85	43.10
59	29.65	17.85	-	124	62.30	37.50	7.55	189	94.95	63.45	25.30	254	127.65	90.30	43.40
60	30.15	18.15	-	125	62.80	37.80	7.80	190	95.45	63.90	25.75	255	128.15	90.70	43.70
61	30.65	18.45	-	126	63.30	38.10	8.05	191	95.95	64.30	26.20	256	128.65	91.10	44.00
62	31.15	18.75	-	127	63.80	38.40	8.30	192	96.45	64.70	26.60	257	129.15	91.50	44.30
63	31.65	19.05	-	128	64.30	38.70	8.55	193	97.00	65.10	26.85	258	129.65	91.95	44.60
64	32.15	19.35	-	129	64.80	39.00	8.75	194	97.50	65.55	27.10	259	130.15	92.35	44.95
65	32.65	19.65	-	130	65.30	39.30	9.00	195	98.00	65.95	27.40	260	130.65	92.75	45.25

Source: AGPS

Time and labour calculations

All trades have a recommended hourly charge rate for their tradespersons' time. This hourly charge rate takes into account many factors. In the automotive trade the factors include:

- the cost or rental of the workshop;
- rates, water, gas and electricity charges;
- the cost of tools and equipment;
- wages of tradespersons and office staff, including holiday pay;
- the cost and the running of service vehicles;
- other incidental costs, such as advertising, accountants' fees, office furniture and equipment etc.

Every service business must be profitable and competitive. There is no future for a business which spends more than it earns and a business service which is too expensive, and therefore not competitive. Hourly labour charges are set after much thought and discussion. When costing and calculating the time for a job, ensure your costing and calculations are accurate.

If the hourly charge rate is $30.00 per hour, the charge for half an hour's work will be half of $30.00, which is $15.00.
The charge for 4.5 hours will be $30 × 4.5 hours = $135.00
The charge for 11 hours will be $30 × 11 hours = $330.00

AREA AND VOLUME CALCULATIONS

The modern motor vehicle is a very well designed and manufactured machine. All the component parts are made to exact specifications and they are precisely fitted together. When repairing and overhauling automotive components, measuring and specific calculations play an important part in the success of the repairs.

To measure the **length of an object**, place a ruler or tape measure across the object, make sure the 'zero' mark on the ruler or tape measure is lined up with one end of the object to be measured. Read the measurement on the ruler or tape measure at the other end of the object. The measured distance across the object is shown to be 30 mm (millimetres). This measurement can also be expressed as 3 cm (centimetres).

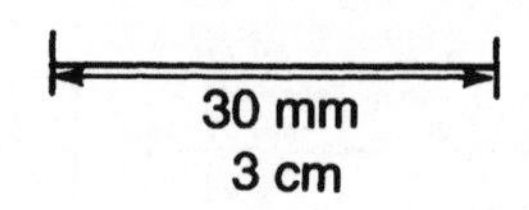

To calculate the **area of an object:**
1. measure the length of the object, as described previously;
2. measure the width of the object, as shown below.

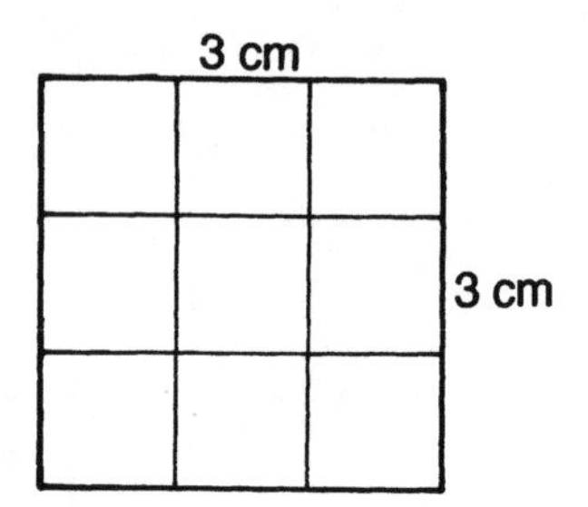

The area is calculated by multiplying
length × width
= 3 cm × 3 cm
= 9 square centimeters (9 cm^2).

To calculate the **volume of a cube:**
1. measure the length of the cube;
2. measure the width of the cube;
3. measure the depth of the cube;

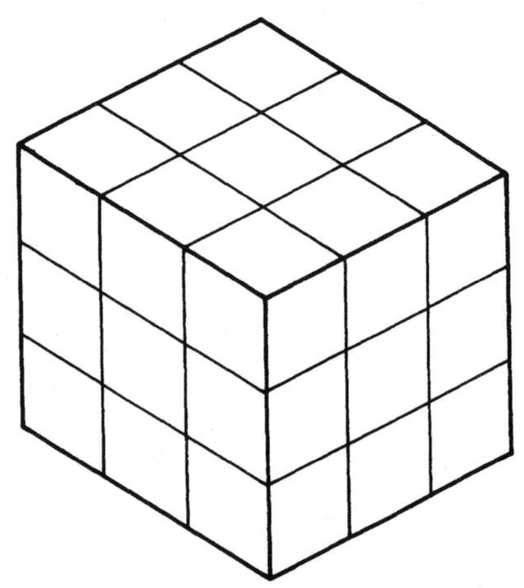

The volume is calculated by multiplying
length × width × depth
3 cm × 3 cm × 3 cm
$3 \times 3 \times 3 = 27\ cm^3$

To calculate the **area of a circle**, a formula is used.
πr^2 π(pi) $= 3.14$ or $\frac{22}{7}$
r^2 = radius multiplied by radius
Note: The radius is the distance from the centre of a circle to the outside (circumference). The radius is equal to half the diameter. For

example, calculate the area of a circle which has a radius of 2 cm.

$\pi r^2 = 3.14 \times 2 \times 2$
$= 12.56\ cm^2$.

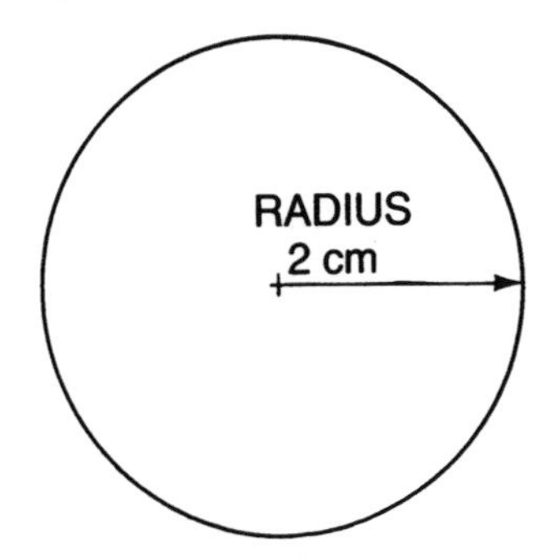

To calculate the **volume of a cylinder**, a formula is used: πr^2 multiplied by the height. For example, calculate the volume of a cylinder having a radius of 2 cm and a height of 3 cm.

$= 3.14 \times 2 \times 2 \times 3$
$= 37.68\ cm^3$.

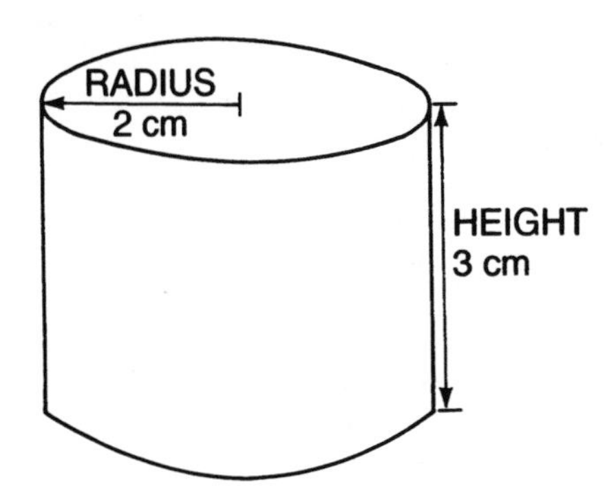

CHAPTER 1 REVISION

1 **Telephone directory**. Locate and record the directory page number, the telephone number, and the address of:
- your place of employment;
- your school/college;
- the nearest Commonwealth Employment Service office.

2 **Street directory**. Locate and record the street directory page number, and grid reference of:
- your place of employment;
- your school/college;
- the nearest Commonwealth Employment Service office.

3 **Telephone use**
1) Describe how you would 'book in' a vehicle for repairs, and make sure you receive all the relevant details from the customer.
2) If you have to telephone emergency services, describe the procedure for obtaining assistance from the police, and ambulance services.

4 **Descriptive writing**. Write a passage of 100–150 words on one of the following topics:

Personal— My hobby.
My favourite sport.
For relaxation I like to . . .

Formal— Describe the rules for your favourite sport.
Describe the propelling, stopping and steering of a vehicle.

5 **Service manuals**. Select a service manual, locate and record the page number in which you would find information on the following:
- engine specifications;
- transmission specifications;
- wheel alignment specifications.

6 **Text books**. Using this text book, locate and record the page numbers on which you would find information on the following:
- the petrol pump;
- a brake disc;
- air cleaner service;
- oil seals;
- spark plugs.

7 **Part names**. From a service manual, select an illustration showing and naming the component parts of an automotive sub-assembly (suspension unit, engine, or brakes). Proceed to the sub-assembly, point out and name the component parts as shown in the illustration.

8 **Vocabulary.** With reference to a dictionary and a technical dictionary, briefly describe the meaning of the following:

• friction	• discernible
• capacity	• response
• force	• serviceability
• volume	• visual
• vacuum	• prerequisite
• torque	• assessment
• pressure	• determine

9 **Comprehension.** With reference to the following passage:
Pry off the hub cap or wheel cover. Take

Parts and accessories					
Quantity	**Part Number**	**Description**	**Unit Price**	**Sales Tax**	**Sub Total**
4	23156	Spark plugs	$1.20	20%	
2	31256	A/C element	7.60	20%	
1	52316	Fan belts	8.20	20%	
2	63521	Oil filter	5.60	20%	
4 l	—	Engine oil	9.60	—	
		Total			$

care you do not damage the hub cap or wheel cover. Remove the bearing dust cap, straighten the cotter pin, remove the cotter pin. Unscrew the adjusting nut, remove the adjusting nut and washer. Shake the wheel from side to side, or pull the wheel out, to assist in the removal of the outer bearing. Remove the outer bearing and place in a clean container.

1 Underline the words which refer to the task to be carried out.
2 In your own words, condense and retain the meaning of the instructions as stated in the passage.

10 **Basic accounting**. With reference to the sales docket, calculate the total cost of all items including sales tax.

11 **Pay calculations**. With reference to the tax scale, and knowledge of your gross weekly wage:
- calculate your net hourly rate;
- calculate your net weekly wage, plus four hours overtime at the penalty rate of time and a half;
- calculate your net holiday wage for four weeks annual leave.

12 **Time and labour calculations**. With reference to the standard hourly rate charged by your employer, calculate:
- the labour charge for 0.75 of an hour;
- the labour charge for 3.5 hours;
- the labour charge for 11.25 hours.

13 **Area and volume calculations.**
- Calculate the volume of a cylinder having a radius of 3 cm and a height of 7 cm.
- Calculate the area of a circle having a radius of 2 cm.
- Calculate the area of a circle having a diameter of 5 cm.
- Calculate the volume of a cylinder having a radius of 3 cm and a height of 7 cm.

2

SAFETY

INTRODUCTION

The purpose of this chapter is to:
- identify some of the safety hazards and dangerous work practices that occur in automotive workshops;
- show that safety must be thought of at all times;
- encourage safe work practices to make sure that nobody is injured and machines, tools or test equipment are not damaged;
- emphasise the awareness of personal safety, by wearing protective clothing and approved personal safety equipment;
- provide information on basic first aid in case of any accidents in the workshop.

WORKSHOP EQUIPMENT

Automotive workshops will generally have the following equipment:
- vehicle hoist;
- tyre servicing equipment;
- grinder;
- safety stands;
- running engines (in operable vehicles).

Caution: before operating any of these items of equipment it is essential to read the manufacturer's operating instructions and safety precautions. Particularly observe the safe working limit (S.W.L.) of the hoist.

Vehicle hoist

The vehicle hoist is used to raise the vehicle clear of the floor and therefore allow the trainee/tradesperson to stand up, rather than lie down, to carry out underbody service/repair operations. A vehicle hoist is a strong, safe item of equipment which is unlikely to fail. However, operators must always be aware that they will be working under a vehicle of 1 to 1.5 tonnes mass.

Types of hoists installed in automotive workshops include the following:
- **two post**, which lifts the vehicle at its jacking points, leaving the wheels free;

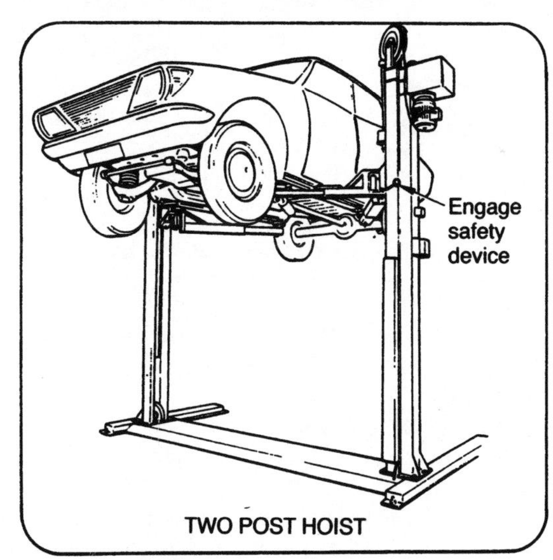

Source: AGPS

- **four post**, which lifts the vehicle at its wheels;
- **rail** (single or double ram) which lifts the vehicle at its jacking points or at structural members.

Precautions when using a hoist:
1 Only use the hoist after you have been instructed in its correct operation.

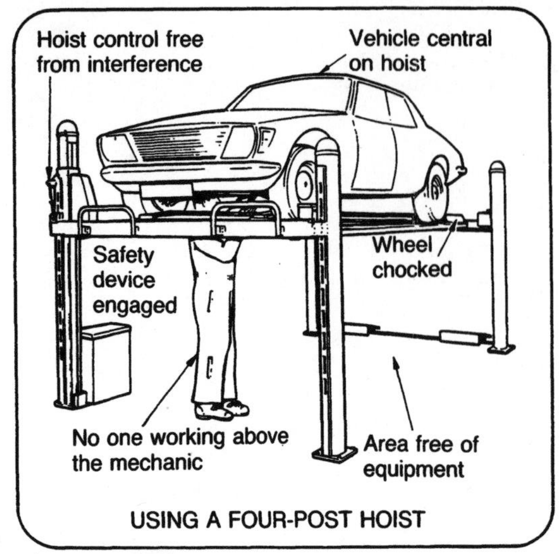

USING A FOUR-POST HOIST

Source: AGPS

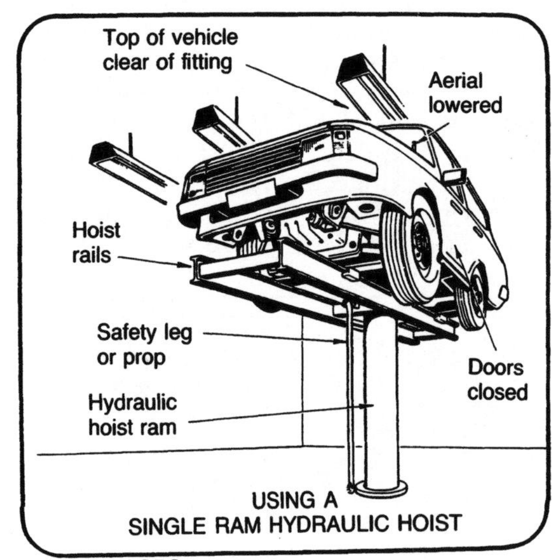

USING A SINGLE RAM HYDRAULIC HOIST

Source: AGPS

2 Drive the vehicle slowly, onto or over the hoist (make sure the vehicle is in a central position).
3 When guiding the driver, do not stand in the direct path of the vehicle being positioned on the hoist.
4 Do not lift a vehicle on the hoist when it is carrying an unstable load, which may shift during the time it is on the hoist.
5 Close vehicle doors, lower or remove radio aerials (C.B. type). Make sure there is enough clearance above the vehicle to prevent damage to air, power or light fittings when it is raised.
6 Once the vehicle is raised make sure the safety system is engaged and operating before you go under the hoist.

OBTAIN ADVICE BEFORE OPERATING UNFAMILIAR EQUIPMENT

Source: AGPS

7 Be careful when removing heavy components from the rear of a front wheel drive vehicle raised on a two post hoist as it could overbalance and tilt forward.

Hoists should never be used when they have one or more of the following faults:

- jerky when it is raised;
- lowers itself from a raised position;
- slowly rises when in use or not;
- lowers very slowly;
- leaks oil out of any seal or hose.

Your supervisor must be notified and a 'DO NOT USE' sign should be attached to the hoist.

Note: Further details of hoist use are contained in chapter 6.

Tyre servicing equipment

A large range of equipment is available to make the task of removing, inspecting and fitting tyres easier. When it is used incorrectly it can be dangerous, even deadly.

A standard ER 78 S14 tyre inflated to 280 kPa will cause a load on the inside of the tyre of approximately 15 to 17 tonnes. This load will force the tyre on to the rim and because of the rubber type used on the bead it will tend to stick to the rim, after deflating the tyre. The force needed to cause the tyre bead to break from the rim and remove or refit the tyre requires the use of a variety of equipment, including:

- tyre levers;
- air compressors;
- bead breakers;
- pneumatic equipment to mount and demount tyres;
- tyre inflation cages and restraining frames;
- tyre inspection tools.

The complete operation of mounting and demounting tyres is explained in chapter 15. However, as this chapter is about safety the following information on precautions and procedures relating to tyres and wheels is appropriate at this point.

1 Deflate the tyre. DO NOT PROCEED until the valve core is removed.
2 Where possible, use an automated tyre changing machine (see below.)
3 Bead breakers and tyre levers, if not used correctly, may cause damage to hands, feet, wheel rim and the tyre side wall or bead.
4 Automated tyre changers make a difficult task easy by reducing the physical force required of the operator. However, the force applied to the wheel and tyre is enough to cause severe harm to the operator. Keep clear of the moving parts of the machine while it is in operation. Observe the manufacturer's operating instructions, which may vary depending on the brand of machine in use.
5 Tyre inflation (especially truck type rims) can also be a hazardous task. Tyre pressures range from 180 kPa for light car applications to 900 kPa for large road transport applications. All types of tyres should be inflated using a suitable cage or restraining device/clamps. There have been many recorded cases of serious injury or death caused by tyre/wheel/rim assemblies 'blowing' apart, particularly split rim truck units. USE A SAFETY CAGE.

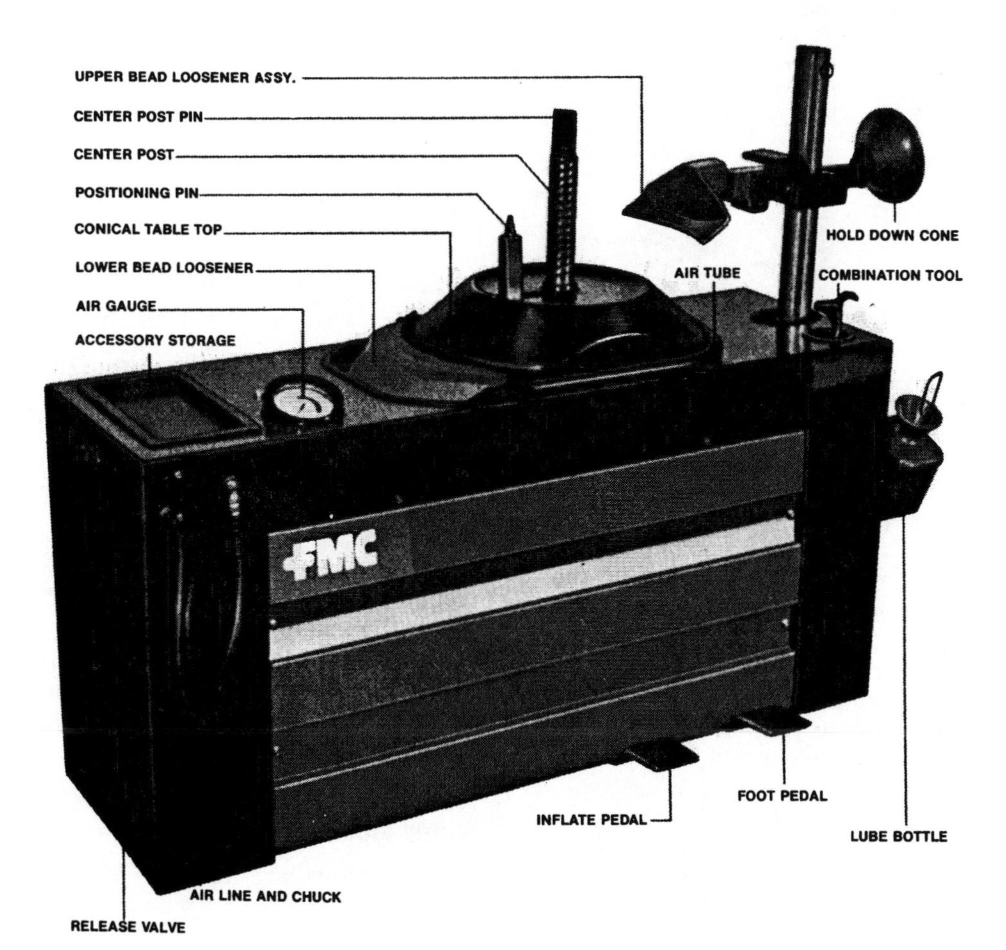

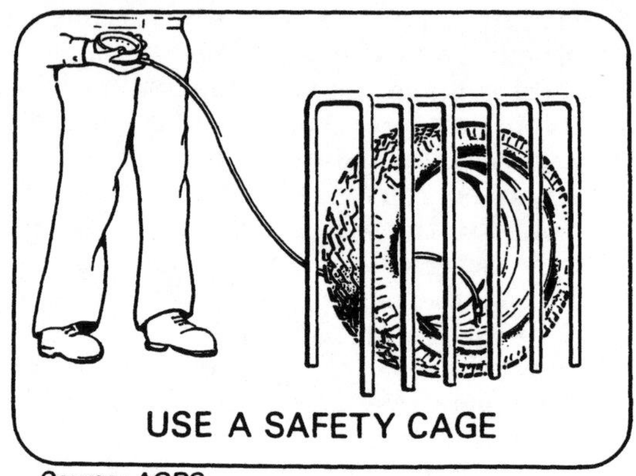

Source: AGPS

Bench grinder/pedestal grinder

The pedestal or bench grinder (same unit but one is bolted to a bench and the other is bolted to a pedestal or stand) runs on electricity, and is designed to remove metallic material by abrasion. To do this a wheel made of an abrasive material is spun at

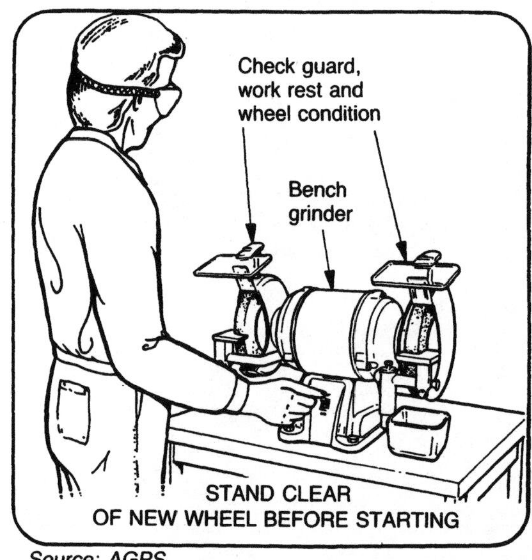

Source: AGPS

high speed. The object to be ground sits on the grinder work rest and is moved into contact with the wheel.

While it can remove material faster than filing or hacksawing, it can also be a safety hazard. A few simple precautions will make sure the grinder remains a safe and useful tool in the workshop.

1 Locate the 'Off' switch.
2 ALWAYS wear safety goggles.
3 Check the grinder wheel regularly for chips, damage or wear. Damaged wheels could

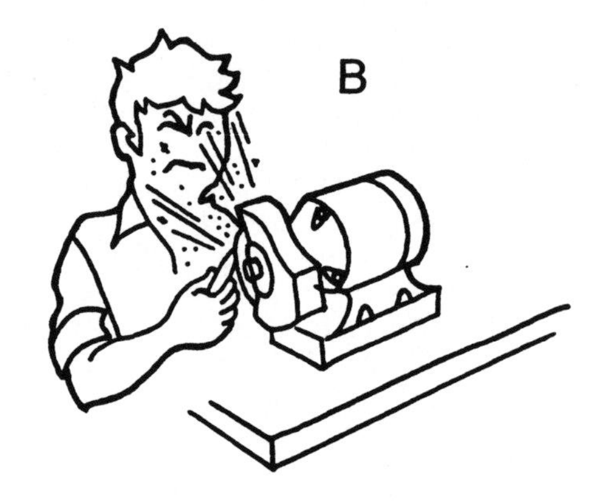

shatter, so do not use a wheel with visible defects.
4 Make sure there is not too much clearance between the rest and the grinding wheel.
5 Check all safety guards are complete and securely attached.
6 Allow the machine to attain its working r.p.m. before starting to grind the workpiece.
7 Make sure your fingers do not come into contact with the rotating wheel. This can easily occur when objects which are too small, curved or of a hoop shape are incorrectly held against the wheel.
8 Do not wear loose clothing, especially unbuttoned sleeve cuffs.
9 Prevent the workpiece from overheating by regularly cooling it in water.
10 Turn the machine off when not in use.
11 Stand clear of the wheel on initial start-up after the machine has been serviced, as incorrect fitting or adjustments could cause the wheel to shatter.

Safety stands

Safety stands are usually made from steel and will include an adjusting mechanism

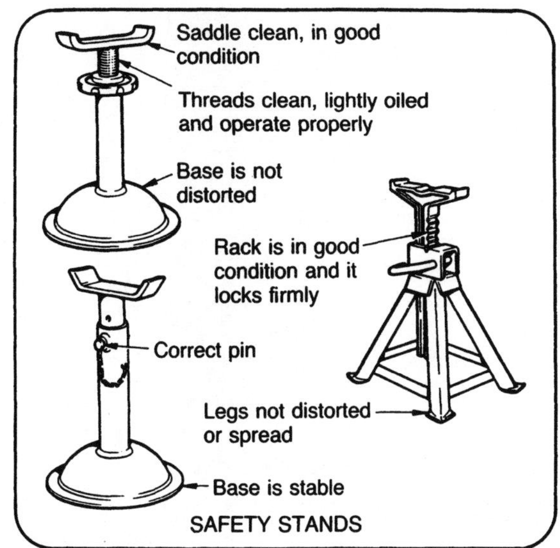

SAFETY STANDS

Source: AGPS

such as screw, hole and pin or ratchet. The purpose of safety stands is to support the vehicle once it has been lifted clear of the floor by a mechanical or hydraulic jacking device. Their use is similar to those of the vehicle hoist.

A vehicle of a 1 to 1.5 tonne mass would cause death or serious injury if it fell on anyone. Damage to the vehicle and equipment will also result from incorrect use of safety stands.

Before using safety stands:

1 Park the vehicle on a flat and level floor.
2 Ensure the saddles at the top of the safety stands are not broken or spread.
3 Wipe the saddles to remove grease, oil or other materials.
4 Make sure the bases or footings of the stands are not bent or distorted.
5 Check the components and operation of the adjuster/locking devices (screws, holes and pins or ratchets).
6 Make sure the stands can support weight of the vehicle.

When using safety stands:

1 Jack up the vehicle to the desired height.
2 Place stands under the axle or frame members. Make sure that the stand saddles make contact on a horizontal member of the vehicle. If this is not done the stand saddle could slip as the vehicle load is applied to it.
3 Make sure the stand is located correctly to prevent damage to pipes, hoses, wires, cables and vehicle floor pan structure.

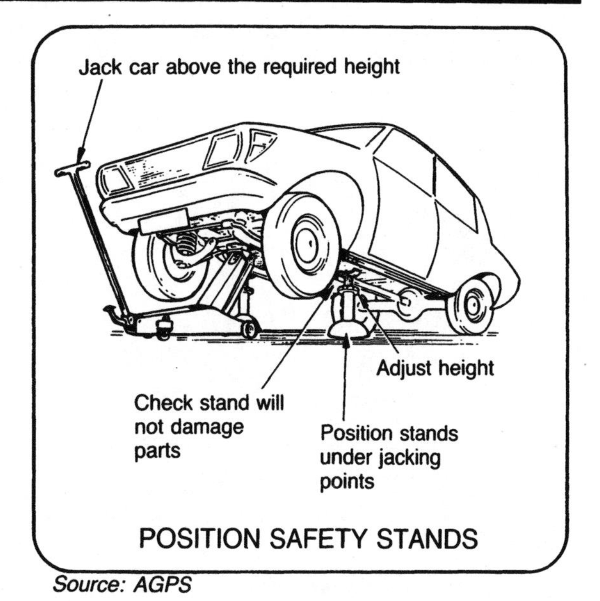

POSITION SAFETY STANDS

Source: AGPS

4 Lower jacking device slowly to ensure that the vehicle load is applied gently to the stands.

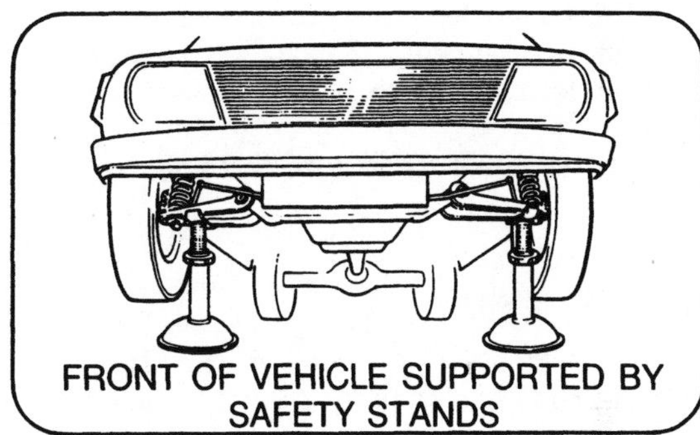
FRONT OF VEHICLE SUPPORTED BY SAFETY STANDS

Source: AGPS

5 Inspect the base of all the stands to see if they are still sitting squarely on the floor. If not, raise the vehicle and reposition stands.
6 Set all the stands to make sure the vehicle is level when the jacking device is removed.

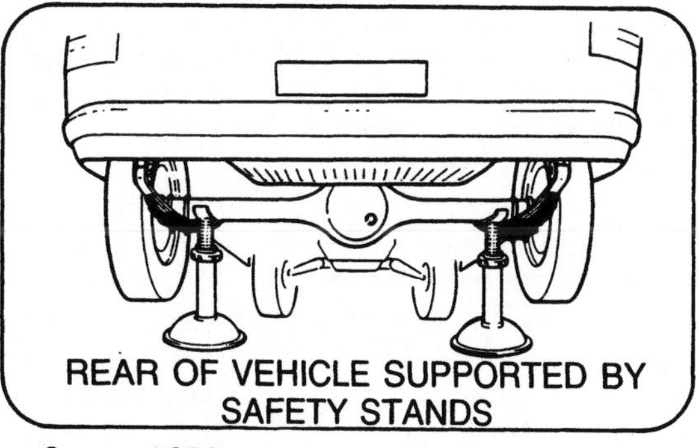
REAR OF VEHICLE SUPPORTED BY SAFETY STANDS

Source: AGPS

7 Before removing the jacking device, check to see that the vehicle is stable and secure on the stands.

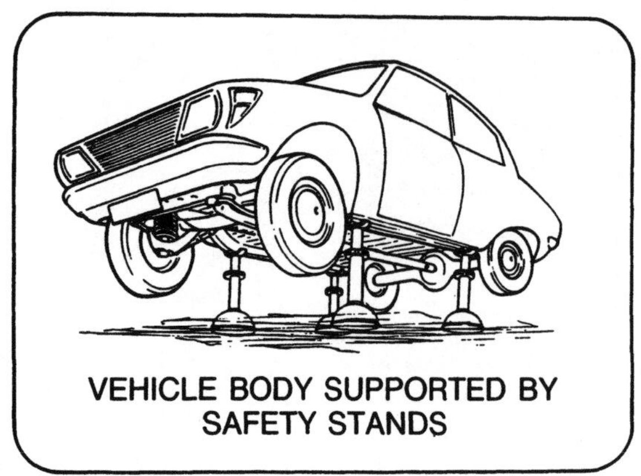
VEHICLE BODY SUPPORTED BY SAFETY STANDS

Source: AGPS

To remove safety stands:

1 Jack up the vehicle to a height which will remove the vehicle's mass from the stands.
2 Move the stands away from the vehicle.
3 Lower the vehicle gently to prevent it slipping off the remaining stands.
4 Jack up the remaining raised section of the vehicle and move stands away from the vehicle.
5 Gently, lower vehicle to the floor.

Operable engines

A number of tasks involve working on an engine that is either running or ready to run. These tasks include:

- checking and adjusting ignition timing;
- checking and adjusting engine idle speed and idle mixtures;
- inspecting for engine squeaks and other noises.

All of these and other checks, adjustments and inspections will be carried out in an area where fans, pulleys and belts are constantly turning. Exhaust manifolds and engine components are running at high to very high pressures and temperatures. Ignition system components and connectors will have voltages ranging from 12 V to 40 000 V and the engine cooling system will be operating at pressures ranging from 90 kPa to 150 kPa.

It is evident that running engines can be safety hazards and all safety precautions must be taken when working on these engines to minimise the chances of personal injuries.

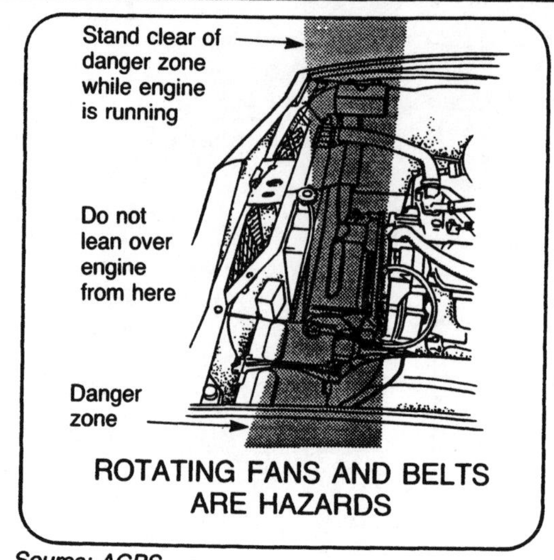

ROTATING FANS AND BELTS ARE HAZARDS

Source: AGPS

Electrical power supply points

Generally three electrical power supply points will be found in most automotive workshops.

- 32 V AC used for portable lighting equipment, such as lead lights.
- 240 V AC (domestic equipment supply) generally used for power tools, test equipment, small electric welding units, air compressors etc.

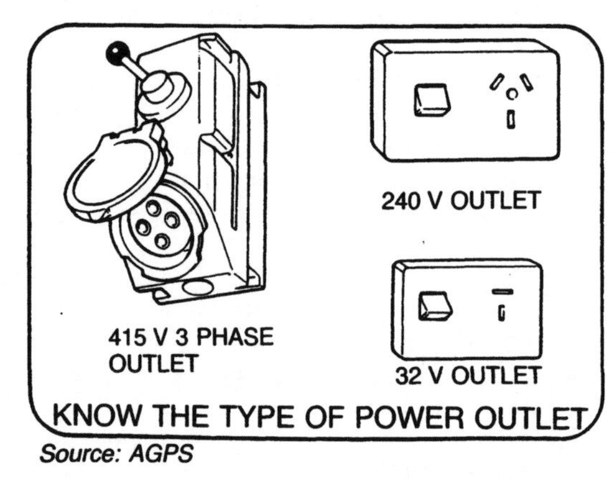

KNOW THE TYPE OF POWER OUTLET

Source: AGPS

- 415 V AC, sometimes called 'three phase' power. It is generally used on such heavy duty equipment as drills, grinders, welding units, wheel balancers and air compressors.

Incorrect voltage supply:

Make sure all electrical equipment is connected to the right voltage supply as a connection to the wrong voltage supply will have serious results. For example:

- connecting a 32 V lead light to a 240 V supply will cause the globe filament to burn out.

- connecting a 240 V device to 415 V supply will cause the device to be burnt out in a fraction of a second.

Normally these connections would not be possible because of different socket/plug fittings, however, it is a good idea to check the voltage requirement stamped on all electrical equipment.

ELECTRICAL EQUIPMENT

Electrical equipment, including power tools, lead lights and test equipment, is designed and manufactured to assist the tradesperson to be more accurate, work faster with less effort, be more efficient and more productive. The equipment can only do its job if it is in correct working order, and is used according to the maker's instructions and within the safety limitations. To make sure that all electrical equipment is used correctly, refer to the following guidelines.

1. Make sure the maker's operating instructions and safety precautions are read, understood, and put into practice when using the equipment.
2. Check that all electrical equipment, including plugs and leads, are in good condition before they are used. If in doubt do not use. Remember, electricity can kill.
3. Place a warning sign on any faulty or suspected faulty electrical equipment.
4. Make sure the area in which the electrical equipment is being used is well ventilated, and the equipment and leads are clear of any water.
5. Be especially careful when using extension leads. When using a 'wind up' extension lead, always unwind it completely, as using it when it is wound up can cause it to overheat, damage the insulation and may cause a fire.

Do not leave an extension lead switched on when the appliance is not in use. Do not place it near water, any sharp object or in a place where people can trip over it.

Main switches and safety switches

The electrical supply system to all workshops must pass into a switchboard. This switchboard is usually located on the wall, inside the building, close to where the electricity supply enters the building. If it is a large building or the building has a number of work areas then more than one switchboard will be used.

For the following safety reasons it is advisable to correctly identify the **main supply switch** in the switchboard:

- if a circuit becomes faulty then the main supply switch has to be turned off before the circuit can be repaired;
- if a person is being electrocuted or is being dragged into a machine, the safest thing to do is to switch off the main supply switch, before attempting to touch the person.

The safety cut-out switch is another safety device used in most workshops. Its function is to switch off supply to a number of wired-in electrical units. These switches are generally located at convenient places around the workshop. To identify the safety cut-out

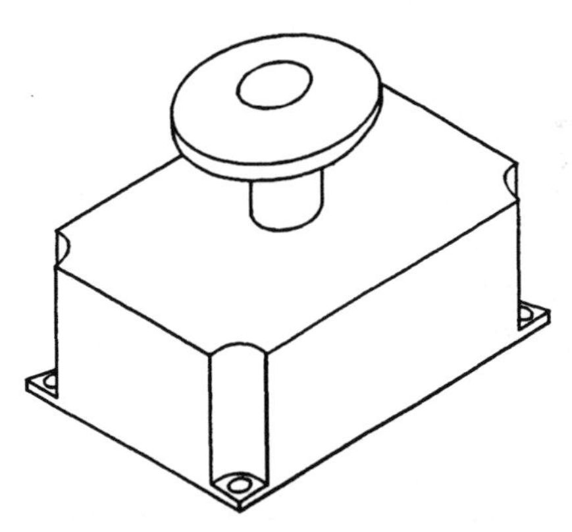

switches look around the walls of the workshop for a blue/grey box (80 mm × 80 mm × 50 mm) with a red button in the centre of the front face. When located, ask the supervisor to turn off the safety cut-out switch to identify the electrical units which it controls.

Booster batteries (jump starting)

If a vehicle engine will not start because of a discharged or flat battery, it may be started by using electrical power from another battery. This procedure is described as 'slave' or 'jump' starting.

Note: On vehicles equipped with any electronic control units, such as electronic ignition modules, engine management microprocessors etc., slave or jump starting is NOT recommended, as damage to these electronic units may occur. It is recommended that the discharged battery be removed and a fully charged battery be fitted.

For other types of vehicles the following procedure for jump or slave starting may be used.

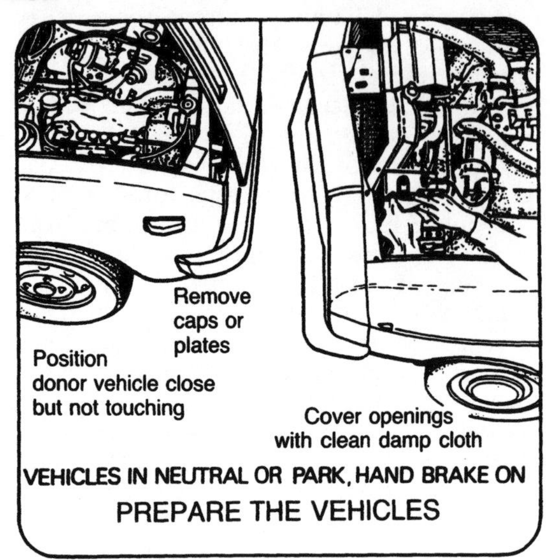

Source: AGPS

1 Make sure both batteries are of the same voltage, for example 12 V.
2 Place the two vehicles close to, but not touching, each other.
3 If vehicles are:
   - manual transmission, apply parking brake and put into neutral gear.
   - automatic transmission, apply parking brake and select 'park' range.
4 Turn off ignition, lights and accessories in both vehicles. (If vehicle is parked in a dangerous position leave hazard warning lights on.)

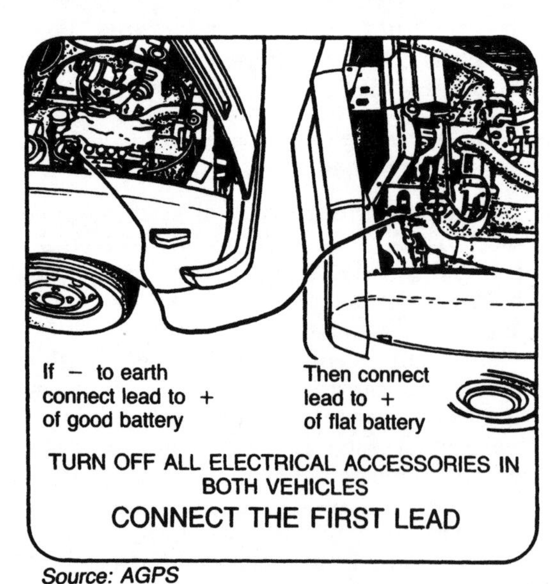

Source: AGPS

5 Connect one end of the first jumper cable to the positive terminal of the good (donor) battery. The positive terminal can be identified by a plus sign '+' on the battery case, and the cable is usually red in colour.
6 Connect the other end of this cable to the positive terminal of the flat battery in the vehicle.
7 Connect one end of the second jumper cable to the negative terminal of the good battery. The negative terminal can be identified by a minus sign '–' on the battery case, and the cable is usually black in colour.

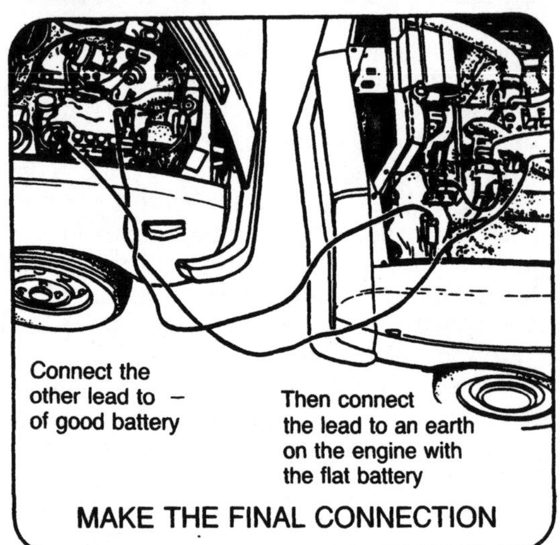

Source: AGPS

8 Connect the other end of this cable to the engine mounting bracket or another good 'earth' connection. Do NOT connect the negative cable to the discharged battery.
9 Make sure that the cables are connected correctly and that they will be clear of any moving parts when the engine starts.
10 Start the engine in the vehicle with the good battery and run it at a fast idle for two to four minutes.
11 Start the engine which has the flat battery and run it at a fast idle for two to four minutes.
12 Disconnect the jumper leads in the reverse order of connecting them.

FIRE EXTINGUISHERS

The possibility of accidentally starting a fire in an automotive workshop is very high. It is important to be able to:

FIRE CLASSIFICATION CHART

FIRES IN A SERVICE AREA GENERALLY FALL INTO THREE CLASSIFICATIONS

CLASS A FIRES

Ordinary Combustible Materials such as Wood, Paper, Textiles and So Forth.
REQUIRES . . . Cooling-Quenching

CLASS B FIRES

Flammable Liquids, Gasoline, Oils, Paints and so Forth.
REQUIRES . . . Blanketing or Smothering

CLASS C FIRES

Electrical Equipment, Motors, Switches and So Forth.
REQUIRES . . . A Non-Conducting Agent

HERE'S HOW TO OPERATE THE PORTABLE FIRE EXTINGUISHER

SODA-ACID: Direct Stream At Base of Flame

PUMP TANK: Place Foot on Footrest and Direct Stream At Base of Flames

CARBON DIOXIDE: Direct Discharge As Close to Fire As Possible. First At Edge of Flames and Gradually Forward and Upward

FOAM: Don't Play Stream Into the Burning Liquid. Allow Foam to Fall Lightly on Fire

	FOAM Solution of Aluminum Sulphate and Bicarbonate of Soda	OK FOR A–B
		NOT FOR C
	**CARBON DIOXIDE** Carbon Dioxide Gas Under Pressure	NOT FOR A
		OK FOR B–C
	**DRY CHEMICAL**	MULTI PURPOSE TYPE: OK FOR A B C
		ORDINARY B–C TYPE: NOT FOR A; OK FOR B C
	**PUMP TANK** Plain Water	OK FOR A
		NOT FOR B–C
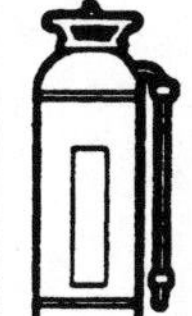	**GAS CARTRIDGE** Water Expelled By Carbon Dioxide Gas	OK FOR A
		NOT FOR B–C
	SODA-ACID Bicarbonate of Soda Solution and Sulphuric Acid	OK FOR A
		NOT FOR B–C

- locate fire-fighting equipment and alarms in your workshop;
- identify the most suitable fire extinguisher for each type of fire;
- correctly operate each type of available fire extinguisher.

If you can do this the possibility of fire damage will be greatly reduced.

Location

The location of a fire extinguisher should be indicated by a sign above the extinguisher. The sign displays the type of fire extinguisher.

The location of the fire alarm should also be indicated by a sign above it.

Identification and operation

Refer to the fire extinguisher chart.

Fire fighting procedure

To minimise the extent of a fire in the workshop, you must be ready to take action by:

1 understanding the 'What to do in the case of fire' procedures at your place of work;
2 knowing the location and the correct operation of all the fire fighting equipment;
3 learning the location of the fire alarms and the emergency exits.

If a fire starts, remember and follow these steps:

1 activate the fire alarm;
2 advise the fire brigade;
3 warn everyone in the area of the fire;
4 fight the fire with the available fire fighting equipment;
5 evacuate the workshop if necessary.

Quick, decisive action during the first few minutes of a fire can prevent a disaster, so be prepared.

PRESSURISED UNITS

There are a number of pressurised units used in the workshop, and some are fitted to motor vehicles. The following are examples of these pressurised systems:

- workshop **compressed air supply** (870 kPa) and the tools connected to this air supply;
- vehicle **engine cooling systems**. As well as being under pressure (90 to 110 kPa), the water in the system is at a high temperature (95°C);
- vehicle **air conditioning systems**. While operating during summer conditions, system pressure may go as high as 2000 kPa.

Now let's look at each system to identify the hazards and the precautions when operating these units.

Workshop compressed air supply

Compressed air is used in workshops to reduce the time and effort necessary in carrying out tasks, for example air powered tools, pneumatic tyre removal/refitting machines, lubrication equipment, cleaning parts, pumping up tyres etc. When the compressed air supply is used correctly it is a labour-saving service, if it is used incorrectly it can be a **safety hazard**.

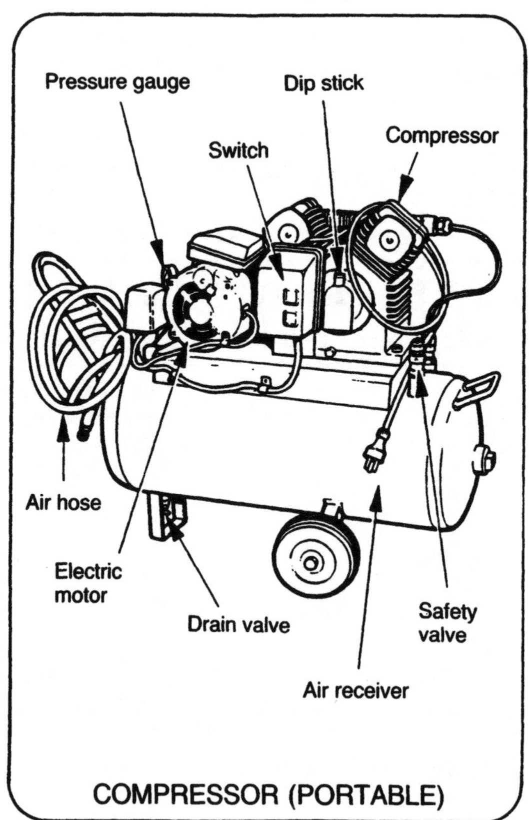

COMPRESSOR (PORTABLE)

Source: AGPS

- If air enters the blood stream it can cause heart failure.

- If air hoses/fittings are allowed to 'whip' around on a free air line, they can cause injuries.
- If compressed air is accidentally directed at eyes or ears, permanent damage to sight and hearing could occur.

Precautions when using compressed air:

1 Check hoses and connectors for possible damage before connection to air supply.
2 Hold the working end of the hose when connecting other end to supply, to prevent hose 'whip'.
3 Wear rubber gloves and safety glasses when cleaning or drying parts with compressed air.
4 Never use compressed air to blow dust from clothes or hair.
5 If ball or roller bearings are to be cleaned or dried with compressed air, do not allow the bearing to spin, because if a bearing is spun at high speed it can shatter, causing serious injury.

Vehicle cooling system

The vehicle cooling system's pressure may be as high as 112 kPa (16 psi) above atmospheric conditions. This can be a safety hazard. If the cooling system's pressure should suddenly be released, by removal of the radiator cap or any other system component, when the system is at operating temperature, the coolant will boil immediately. If the 'boiling' coolant contacts the skin or eyes, severe scalding will result.

Precautions when working on a hot cooling system:
Allow the system to cool down before attempting to remove the radiator cap. If this is not possible the *chances of possible scalding can be reduced* if the following precautions are taken:

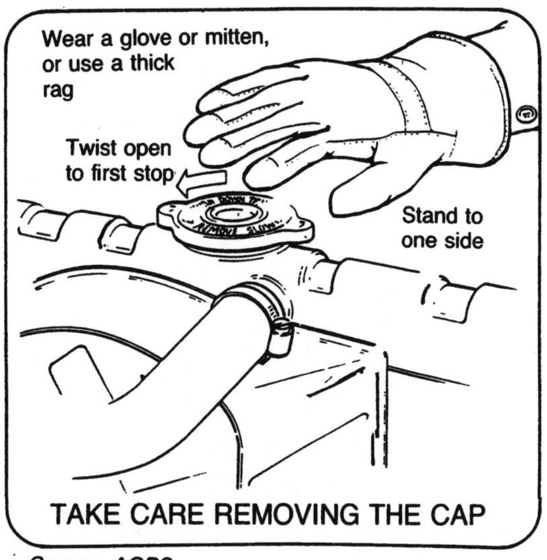

Source: AGPS

1 Protect your hand with a thick glove or rag.
2 Stand to one side of the radiator.
3 Twist cap anti-clockwise until the first stop is felt.
4 You will hear steam escaping out of the radiator overflow hose. Wait until all steam has escaped.
5 Continue twisting cap in an anti-clockwise direction until the second stop is felt.
6 Remove the cap.
7 Remember, even though you have released the system pressure the coolant will still be hot enough to cause scalding.

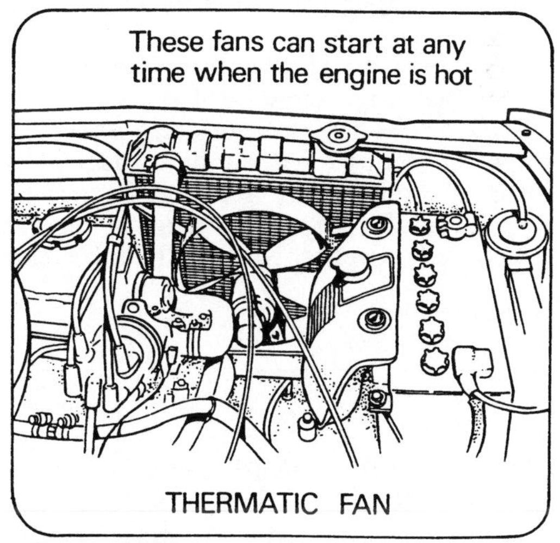

Source: AGPS

Vehicle airconditioning system

The refrigerant used in automotive airconditioning systems is R12. This

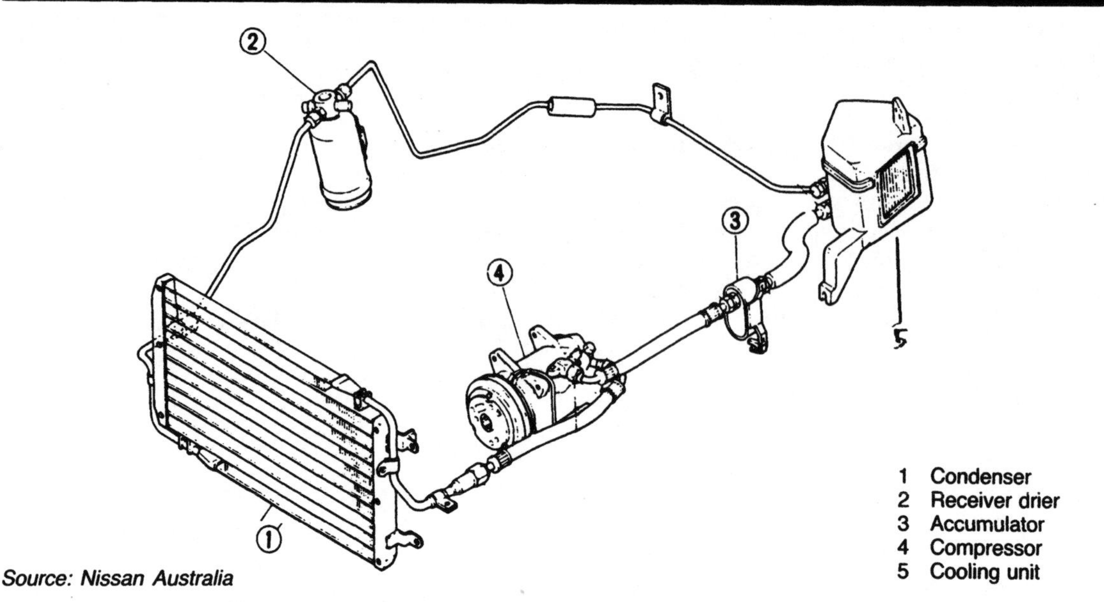

Source: Nissan Australia

refrigerant will be subjected to pressures ranging from 200 kPa to 2000 kPa, depending on whether the system is operating or not. R12 refrigerant boils at –30°C, so if it is accidentally sprayed on to you, it will absorb the heat from that part of your body to become a vapour. In doing so it will leave that part of your body very cold, cold enough to freeze it and cause frostbite.

Precautions to follow:

1 When working on a vehicle with airconditioning fitted, wear safety glasses and keep collar and cuffs buttoned up on overalls.
2 Don't weld or steam clean on or near airconditioning components.
3 If R12 is released into workshop area and it comes in contact with a flame it will not burn or explode but a poisonous gas will be formed. Do not breath these fumes, and evacuate the workshop until the air is clear.

HEALTH HAZARDS

There are many health hazards in an automotive workshop. Being aware of these hazards, adopting good safe work practices and using common sense will be good for everyone. The following list identifies some of the potential health hazards in an automotive workshop, and suggests ways to minimise these potential health hazards.

Source: AGPS

- Oil, petrol and grease spills should be cleaned immediately.

Source: AGPS

- Floors and workbenches should be clean and tidy.

- All tools and test equipment should be stored carefully in boxes, on toolboards or in cupboards.

STORE PARTS AND MATERIAL SAFETY

Source: AGPS

- Pipes, angle iron, metal rods, flat steel and vehicle parts should be stored neatly on racks.
- Always store oils, petrol, cleaning fluids, solvents, acids, adhesives etc. in a cool, safe and well-ventilated area.
- Wearing long hair, metal watches, rings, bracelets, and necklaces can be health hazards as long hair can get caught in machinery, and rings, bracelets, watches and necklaces, being made of metal, conduct electricity.

CARBON MONOXIDE IS DANGEROUS

Source: AGPS

- Make sure the workshop is well ventilated, as exhaust fumes and other air pollutants are health hazards.
- When working in a noisy environment always wear approved hearing protection muffs, to prevent the possibility of loss of hearing.

PROTECTIVE CLOTHING

Over the years there has been a lot of research and development on the design and testing of protective clothing. All protective clothing has to withstand rigorous tests to

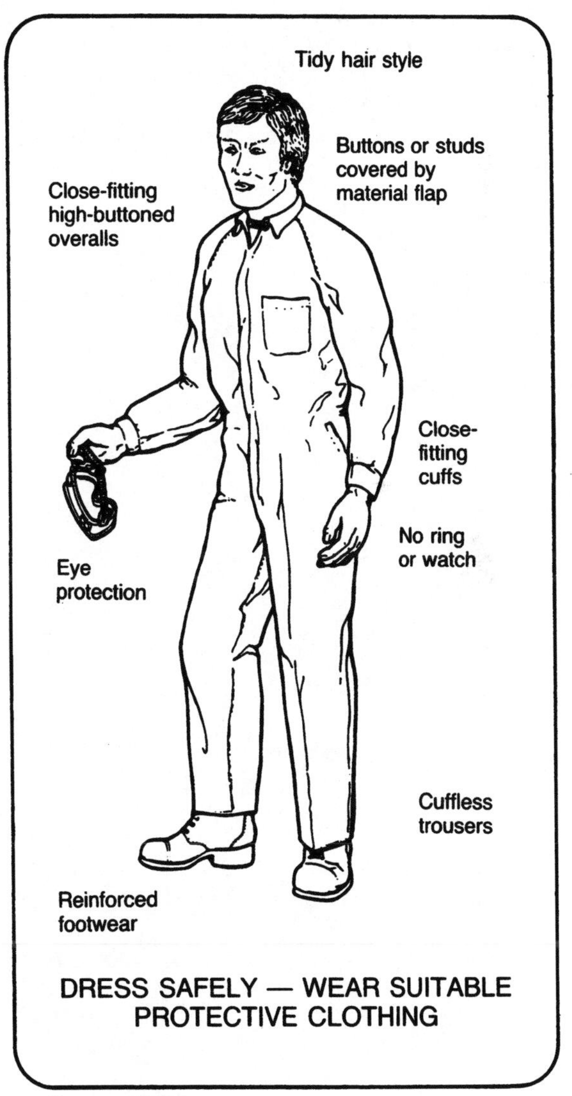

DRESS SAFELY — WEAR SUITABLE PROTECTIVE CLOTHING

Source: AGPS

gain the Australian Safety Standard Association stamp of approval. When buying any type of protective clothing make sure it has been tested to the Australian Safety Standards Association recommendations, then you can be certain that you have purchased a quality item.

Always wear protective clothing and use safety equipment for your protection, when working in the workshop. The following are examples of protective clothing and safety equipment recommended for use in an automotive workshop.

Overalls, coveralls and boilersuit refer to the one-piece garment which covers the body, arms and legs. The overalls should be the correct size, close-fitting, comfortable and made from a flame-resistant, hard-wearing material. Keep all buttons fastened on the overalls and do not roll up the trouser cuffs. Overalls can protect your:

- ordinary clothes from dirt and damage;
- body, arms and legs from heat and hot metal splatter when welding;
- body, arms and legs from grease, oil, petrol, solvents and acids;
- arms and legs from minor cuts, abrasions, and bruising.

Eye protection is vital to everyone, especially in the workshop. There are many types of eye protection devices, from safety spectacles to full-face protection masks. Always wear the correct type of eye protection for each specific task, as follows:

- using the grindstone, wear grinding goggles;
- using gas welding equipment, wear gas welding goggles;
- using arc welding equipment, wear a full-face type arc welding mask;
- using any power tools, air operated or electric, wear grinding goggles;
- using a pedestal drill or a lathe, wear grinding goggles;
- when working on an air conditioning unit, use chemical type goggles.

Head protection. The types of head protection recommended in a workshop are as follows:

- a cotton or wool cap which will protect the head from dirt, grease and minor cuts and bruising;
- a 'bump' cap, which is a lightweight plastic cap which protects the head from dirt, cuts, abrasions, bruises, bumps, oils, greases, solvents etc.;
- a hair net protects the long-haired worker from having his or her hair being caught up in any revolving machinery.

Source: AGPS

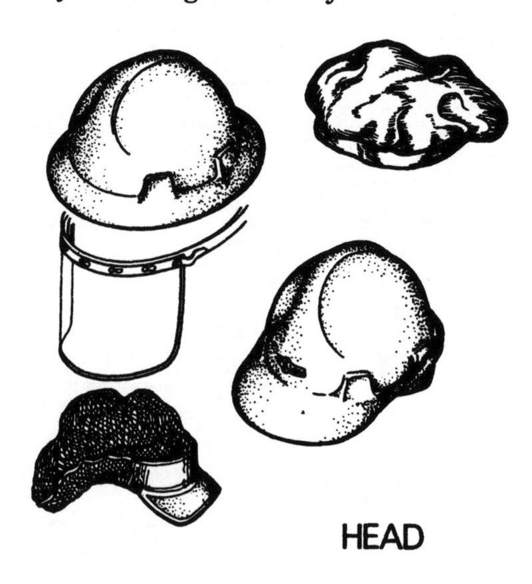

Hearing protection. If people work in a very noisy area for long periods of time, they will gradually lose some of their hearing. This loss of hearing may occur slowly over a period

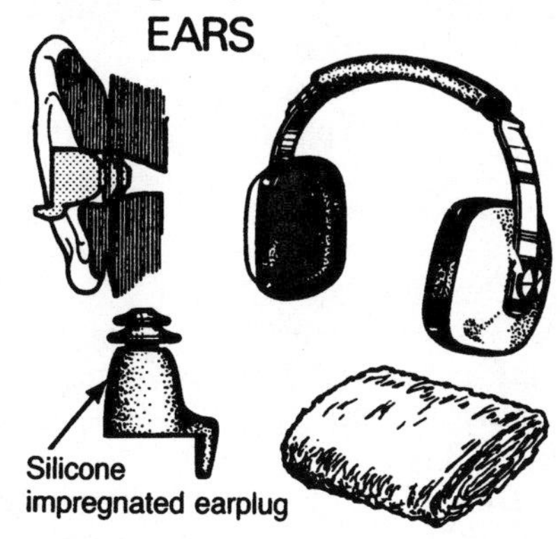

Source: AGPS

of years, so you must wear hearing protection whenever you are working in an noisy environment. To identify a noisy environment a noise meter, measuring decibels, is used. A normal speaking voice is approximately 60 db, and a six cylinder petrol engine running at 3000 r.p.m. produces an under-bonnet noise of approximately 115–120 db. Constantly working in an area which has a decibel reading of 90+ db can cause gradual loss of hearing. The types of hearing protection most commonly used are ear muffs and ear plugs.

Foot protection. Always wear strong, comfortable, leather shoes or boots in the

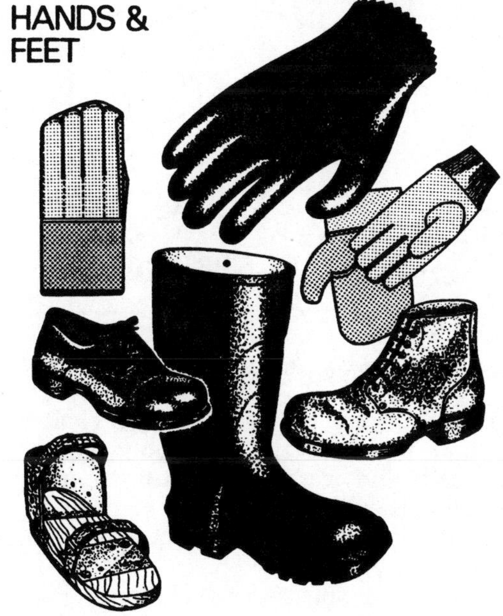

Source: AGPS

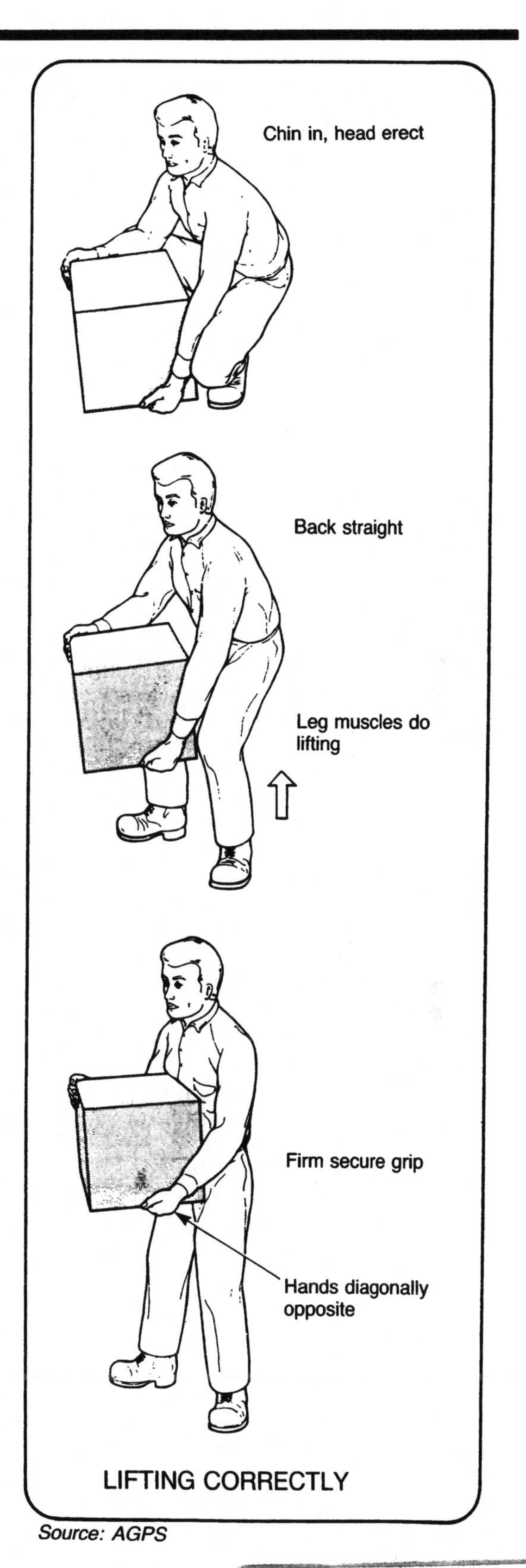

Source: AGPS

workshop. Safety boots with steel toecaps and reinforced soles are strongly recommended, for the following reasons:

- the reinforced soles will give protection from sharp pieces of metal and nails;
- the steel toecaps will protect the toes from heavy falling objects;
- the leather uppers will protect the feet from acids, solvents or hot metals from welding and grinding.

LIFTING

There are a number of heavy components in a vehicle, and there are times when these heavy components have to be removed and replaced. It is good practice to always use mechanical assistance, such as a hydraulic jack, transmission jack, overhead crane, gantry crane, fork lift or block and tackle when lifting any of these heavy components.

Manual lifting should be restricted to components which are not too bulky and having a mass of less than 15 kg. When manually lifting a component, the following procedure is recommended:

1. Place your feet in a secure position, close to the component.
2. Adopt a balanced position, with your knees bent.
3. Get a good, firm, secure grip of the component, using both hands.
4. Keep your back straight.
5. Keep your head straight, your chin tucked in, and take a deep breath before starting to lift the component.
6. The load can now be lifted by straightening your legs and allowing the strong thigh muscles to take most of the effort.

Caution: do not bend your back when lifting any load, as this can result in damage to your spine and back.

FIRST AID

The purpose of this section on first aid is not to give precise medical instructions on treating industrial accident patients. This section offers suggestions, which hopefully will assist the trainee/tradesperson to deal with accidents in the workshop.

Note: 'first aid' refers to someone being 'first' at the scene of an accident, and giving 'aid' (assistance), to the accident victims, then arranging qualified, professional medical assistance from a hospital, a doctor, a nurse or other qualified person.

The automotive workshop is a place in which many and varied types of accidents can occur. They can include:

- abrasions and cuts, resulting in a loss of blood;
- cracked or broken bones;
- shocks from electricity, or from witnessing or being involved in an accident;
- eye injuries from grit, solvents, or from a welding 'flash';
- burns from welding or hot exhaust systems, and scalds from hot coolant from a cooling system.

To minimise the extent of any accident, it is good practice to be aware of the following:

- know where the first aid kit is located;
- know where to locate the telephone numbers and addresses of the nearest doctor and hospital;
- act promptly.

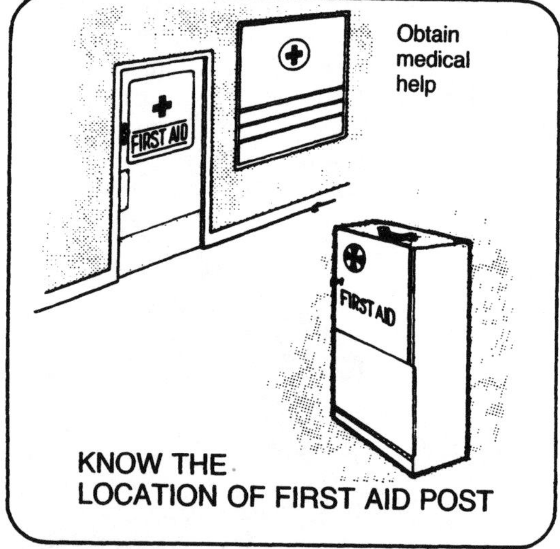

Source: AGPS

First aid treatment suggestions

Cuts. If a cut is bleeding, apply a clean towel or thick pad over the cut, apply pressure on the pad and raise the limb to assist in stopping the bleeding. Then arrange for the patient to be checked by a qualified medical person.

Breaks. If you suspect that a person has broken a bone in a limb, for example a leg

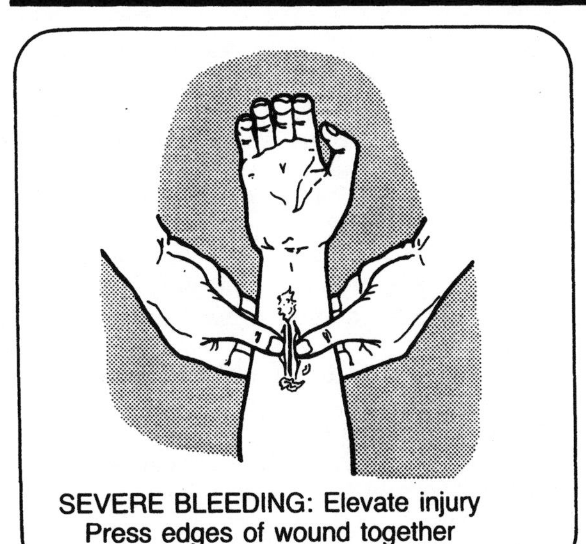
SEVERE BLEEDING: Elevate injury
Press edges of wound together

Source: AGPS

bone, then comfort the patient and immediately send for an ambulance. Move the patient only if she or he is in a dangerous position. Otherwise, make the patient comfortable and try not to move the suspected broken limb until an ambulance, doctor, nurse or other qualified medical person arrives.

Shocks. Every person who is involved in an accident or is a witness to an accident is subject to varying forms of shock. When someone is in a shock condition, then it is advisable to take the patient to a quiet, warm place. Encourage the patient to lie down and elevate his or her legs, loosen any tight clothing and comfort the patient. Then arrange for the patient to be checked by a qualified, professional medical person.

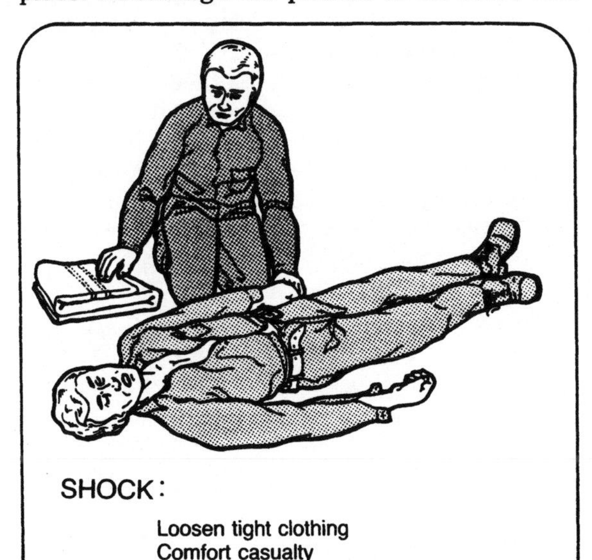
SHOCK:

Loosen tight clothing
Comfort casualty

Source: AGPS

Electrical shock. When required to give first aid to a person who has had an electrical shock then remember to,

1 switch off the source of electrical power, before attempting to touch the person.
2 If the source of electrical power cannot be switched off then move the person away from the 'Live' electricity by using a piece of dry wood or other non conductive material.
3 Treat the patient for shock as previously described, then arrange for the person to be checked by a qualified professional medical person.

Eye injuries. As eyes are very sensitive and precious, it is wise to treat every eye injury as serious. If a person has an eye injury, then the following is recommended.

EYE INJURIES: Bandage the eye loosely
Guide victim to medical attention
Never touch eye surface

Source: AGPS

- Never rub the injured eye.
- Tell the person to hold the eye still.
- Never touch the eye surface with anything.
- Loosely bandage both eyes.
- Take the person to where she or he can be checked by a qualified medical person.

Burns and scalds. The treatment suggested for burns and scalds is to hold the affected area under clean, cold, running water, to remove the heat from the burn or scald.

- Do NOT apply creams or ointments to a burn.
- Do NOT tear or pull clothing away from a burn.
- Do NOT touch a burn where the skin has been broken.
- Do apply plenty of clean, cold, running water over the burn.
- Do arrange to have the injured person checked by a qualified medical person.

Note: All injuries, no matter how minor, should be reported and a record of the time, date, conditions, extent of the injury and method of treatment should be noted. The injury report will assist the patient with any medical compensation claims.

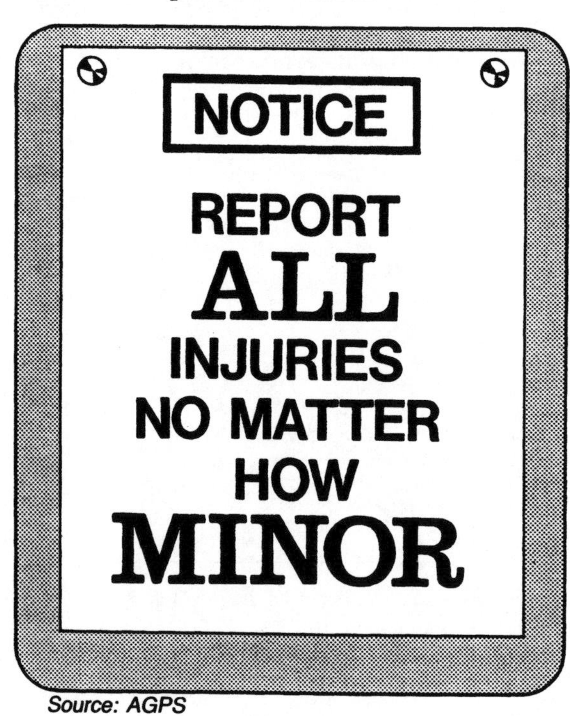

Source: AGPS

CHAPTER 2 REVISION

1 Describe four safety precautions to be observed when using a hoist.

2 Identify five safety precautions to be observed when using a bench grinder.

3 Explain the recommended procedures and safety precautions when placing a vehicle onto safety stands.

4 Give three reasons why running engines can be safety hazards.

5 Show, by drawings, three types of electrical power supply points and give examples of equipment using these power supply points.

6 Name four precautions to be taken before using any electrical appliance.

7 Describe a method of jump or slave starting an engine, using a booster battery.

8 What procedures should be followed if a fire starts in the workshop?

9 Identify and explain the safety hazards and precautions to be observed when working on the following units:
- an air compressor, or supplied compressed air;
- engine's cooling system;
- vehicle's air conditioning unit.

10 List at least five potential health hazards found in an automotive workshop.

11 List the recommended types of safety clothing and safety equipment required when you are:
- using a grindstone;
- working in a noisy area;
- electric welding.

Explain the personal dangers that may occur if the safety clothing and safety equipment is not used.

12 Describe the correct method of lifting an oil drum which weighs less than 15 kg.

13 Describe the action to be taken if any of the following accidents occur in the workshop:
- a deep cut on the forearm;
- electric shock;
- a broken leg;
- eye injury.

3

HAND TOOLS AND EQUIPMENT

INTRODUCTION

An important part of the automotive repair industry is the type, quality and selection of hand tools and equipment. Hand tools are classed as those tools which can be held or supported by the operator's hands while they are being used. Equipment is defined as being those devices which are controlled by an operator but supported by the floor, building or vehicle.

When using all tools and equipment, the safety aspect is extremely important. Tools and equipment used for the purpose for which they were intended will not be a risk to the operator, his or her workmates or the component parts or the vehicle. It is the responsibility of the operator and the workshop manager to service and maintain all tools and equipment.

There are three groups related to the types of hand tools, which include electrical, pneumatic and mechanical. These groups are:

- general
  - — used on a wide range of jobs
  - — they are the responsibility of the tradesperson
  - — their length is limited to approximately 0.5 m;
- specialised
  - — designed for a particular job
  - — they are the responsibility of the workshop manager,
  - — their length may be greater than 0.5 m;
- measuring instruments
  - — these are very accurate measuring devices
  - — they are the responsibility of the workshop manager, although some tradespeople do supply their own
  - — they are generally shorter than 0.5 m.

There is a wide range of workshop equipment available, so this chapter will deal with those items which are considered necessary for the operation of an efficient and effective workshop.

Good quality hand tools and equipment will make sure that the jobs in the workshop can be completed without delay and to a high standard. Before an item is purchased, care must be taken that it will:

- do the required job (suitability);
- last for a long time (durability);
- operate correctly each time it is used (reliability).

The ability to quickly select the correct hand tool and/or equipment to complete a job successfully is achieved through practice. The aim of tool selection is to make ready all the tools needed for the job before work is started. This reduces the time spent 'looking for tools'.

IDENTIFICATION

Hand tools — general

Name and types	Description	Application
Pry bar Figure 3-1	Up to 400 mm long. Blade at one end, point at other end.	Parting component. Removing seals. Aligning holes.
Brush • hair Figure 3-2	Wooden handle. Coarse hair. Up to 250 mm in length.	Washing components with solvent. Removing dirt from parts.
• wire Figure 3-3	Wooden or plastic 300 mm handle. Several rows of wire bristles.	Removing carbon, rust, gasket material, dirt from parts.
Chisel • cold Figure 3-4	Made from special steel alloy. Cutting edge at one end and head on the other end. Up to 300 mm long and 30 mm wide.	Cutting rivets, bolts nuts and sheet steel.
• diamond-point Figure 3-5	Smaller than cold chisel.	Cleaning keyways, threads and corners.
Die Figure 3-6	Made from special steel alloy. Round or hexagonal shape. Used with stock. Many sizes and thread types.	Cutting external threads on rod or bolts.
Die stock Figure 3-7	Two handles attached to a stock. Locking device on stock.	Holding and turning a die during threading procedure.
Drift Figure 3-8	Made from 'soft' metal rod. Tip at one end and head at other end. Up to 300 mm long.	Removing sleeves and bushes. Fitting bushes and seals.

Figure 3-1

This figures are not to scale and are for identification purposes only.

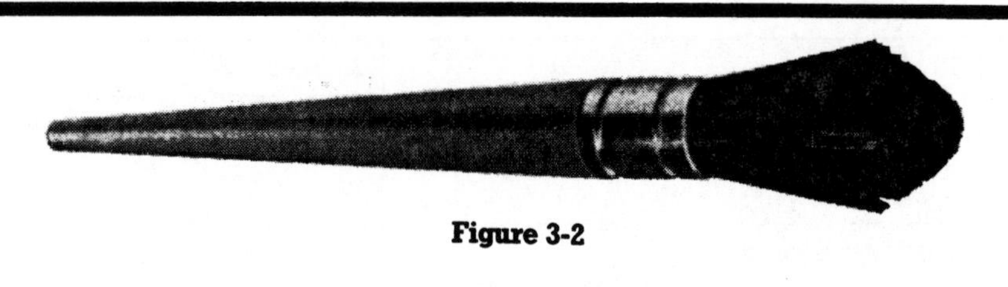

Figure 3-2

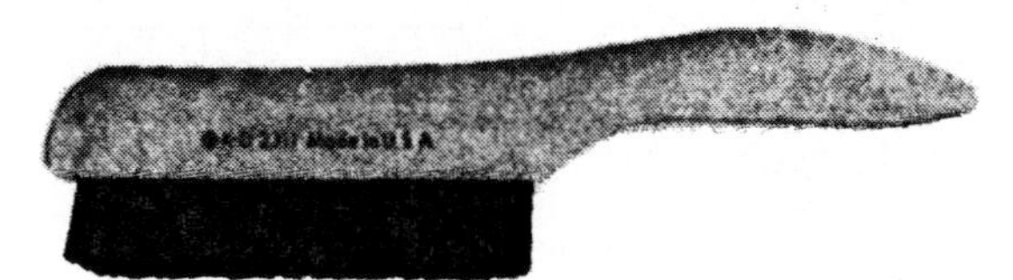

Figure 3-3

Figure 3-4

Figure 3-5

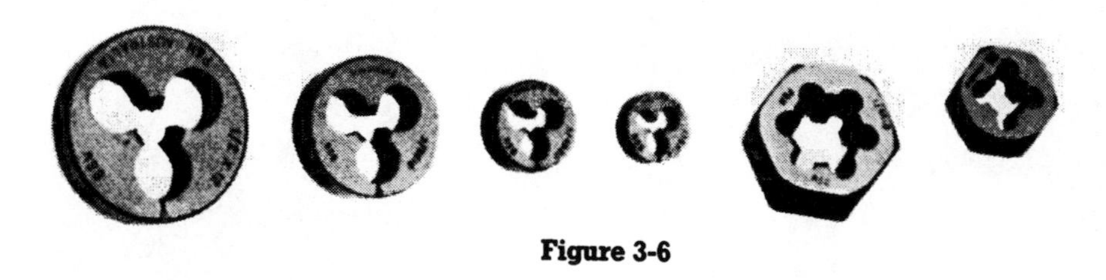

Figure 3-6

Figure 3-7

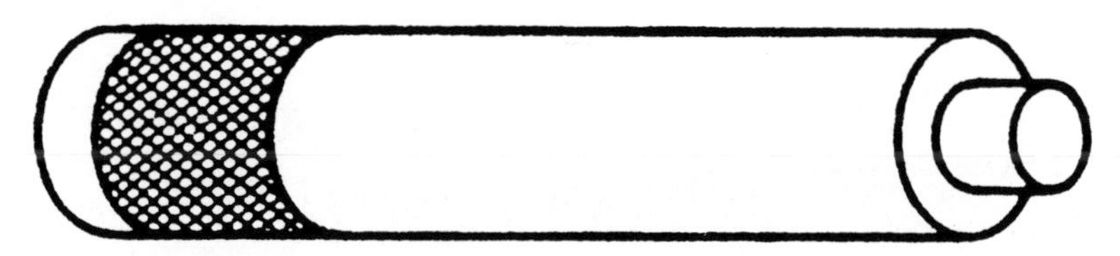

Figure 3-8

These figures are not to scale and are for identification purposes only.

Name and types	Description	Application
Drill • electric Figure 3-9	240 V, multi-speed and/or impact device. Holds up to 13 mm twist drill.	Gripping and turning a twist drill.
• hand Figure 3-10	Mechanical device. Holds up to 13 mm twist drill.	As above.
• twist Figure 3-11	Made from high-speed alloy steel rod. Fluted and pointed at one end. Straight or tapered at other end. Many sizes	Boring holes through metal, wood and plastic.
Extractor — stud • fluted and tapered Figure 3-12	Made from high-speed alloy. Several sizes.	Removing broken studs (stud must be drilled through its centre).
• fluted and straight Figure 3-13	As above.	
• cam Figure 3-14	Made from carbon steel. Serrated cam with a drilled plate on each side.	Removing broken but protruding stud.

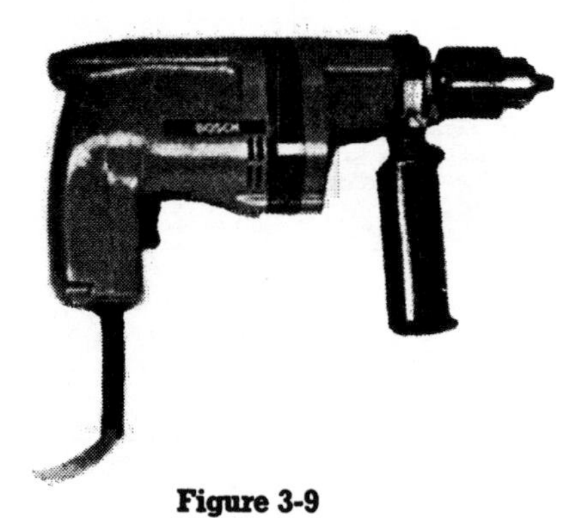

Figure 3-9

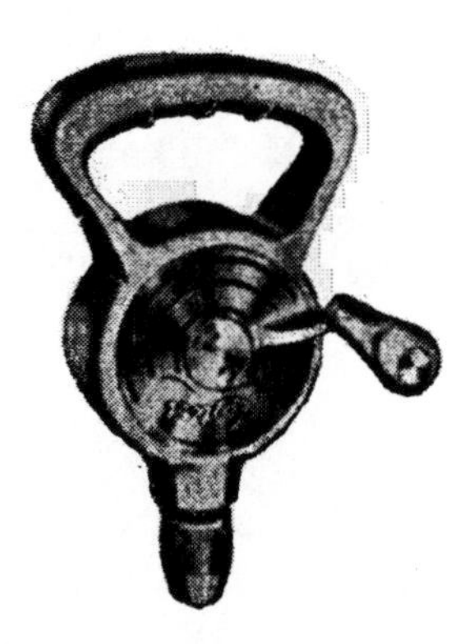

Figure 3-10

Figure 3-11

These figures are not to scale and are for identification purposes only.

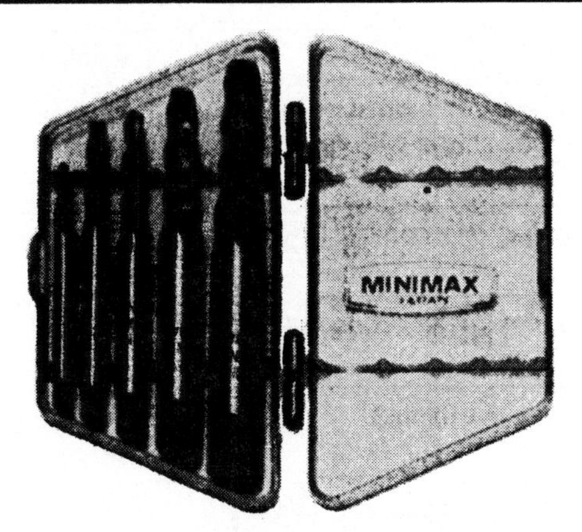

Figure 3-12

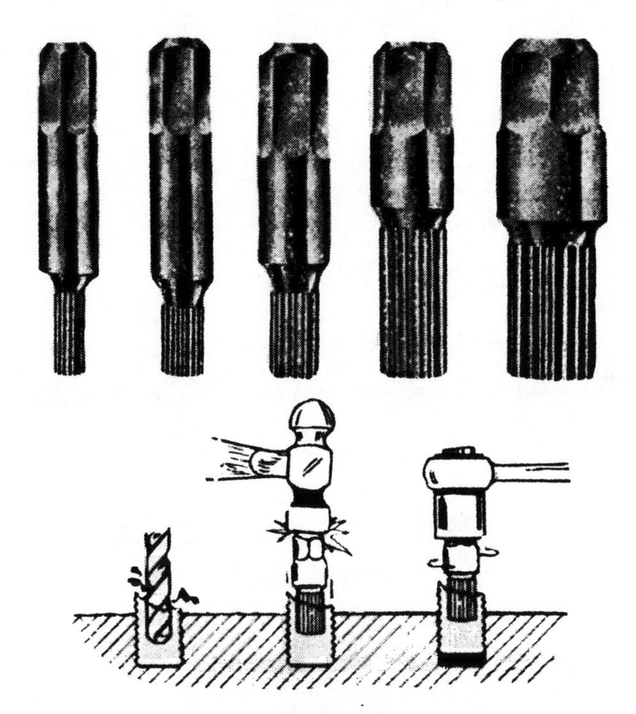

Figure 3-13

Figure 3-14

These figures are not to scale and are for identification purposes only.

Name and types	Description	Application
File • flat and mill Figure 3-15	Made from carbon or high-speed steel. Thin, rectangular shape with teeth cut across its faces.	Restoring surface to a flat and smooth finish.
• half-round Figure 3-16	As above but has one convex (domed) face.	Enlarging holes and rough finishing concave surfaces.
• round (rat tail) Figure 3-17	As above but completely round in section and tapered to hip.	Enlarging small holes or filing fillets.
• points Figure 3-18	Same as flat file but thinner.	Restoring ignition point surfaces to serviceable condition.
• square Figure 3-19	Same as flat file but square in section.	Forming and finishing elongated holes or slots.
• triangle Figure 3-20	Same as flat file but has only three equal faces.	Restoring damaged threads. Sharpening saw teeth.
Note: All these files are available in various sizes, lengths and tooth patterns.		

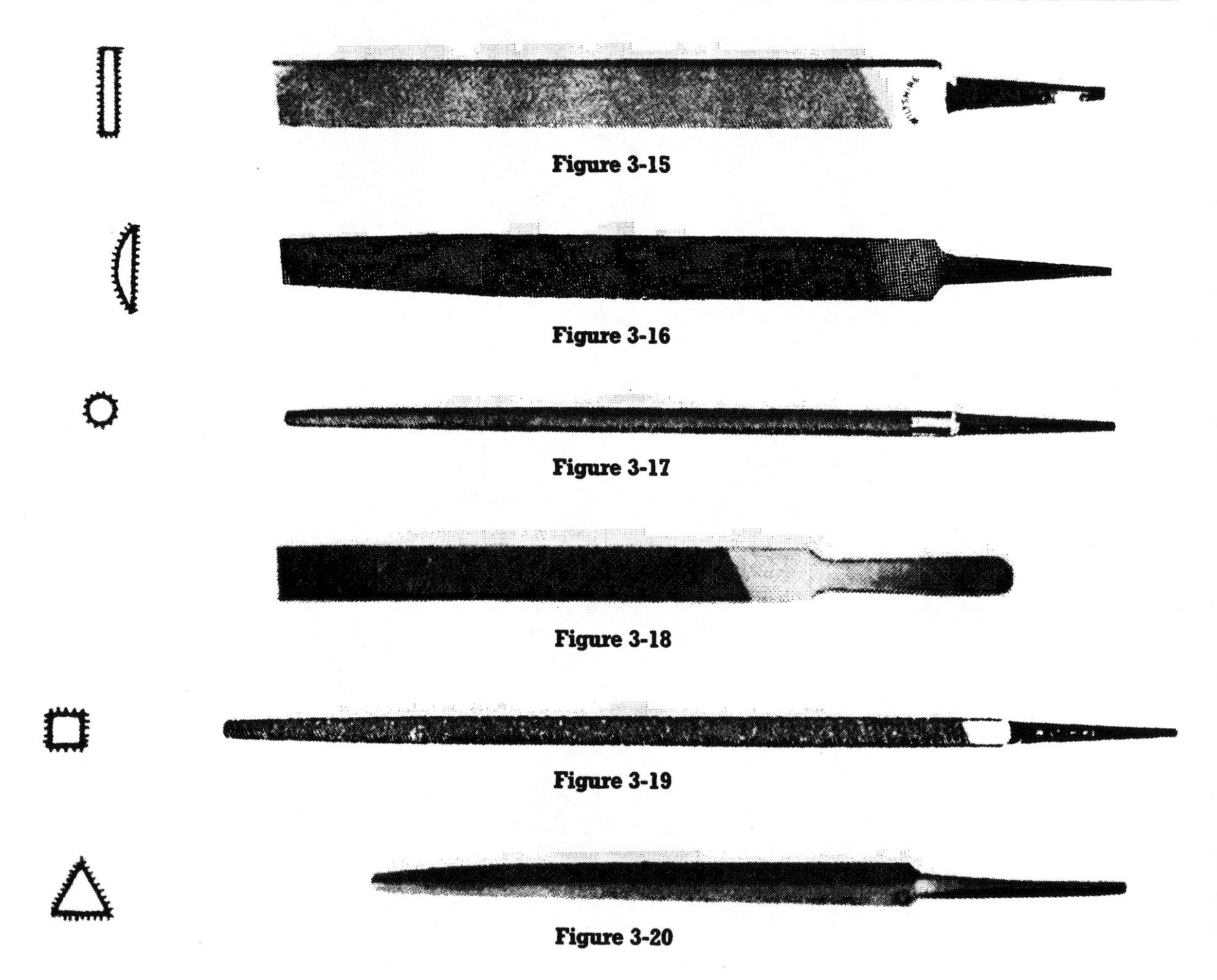

Figure 3-15

Figure 3-16

Figure 3-17

Figure 3-18

Figure 3-19

Figure 3-20

These figures are not to scale and are for identification purposes only.

Name and types	Description	Application
Grinder • portable Figure 3-21	240 V or pneumatic (air) driven. Detachable disc or stone.	Reducing the size of metal parts. Cutting metal parts. Preparing metal for welding.
• angle Figure 3-22	As above.	
Hacksaw Figure 3-23	Frame is adjustable. Tension on blade is adjustable.	Used for cutting most metal objects.

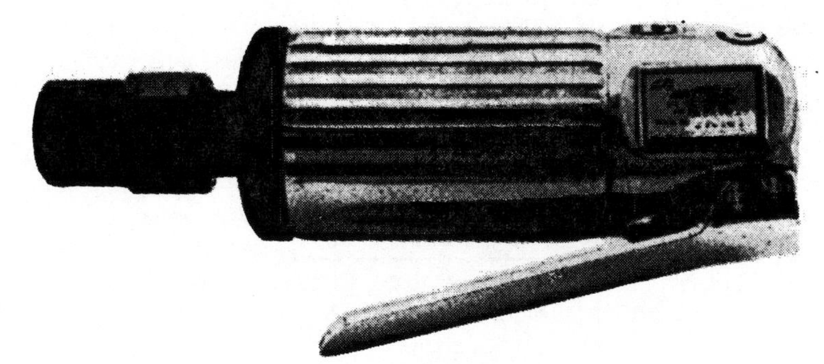

Figure 3-21

Figure 3-22

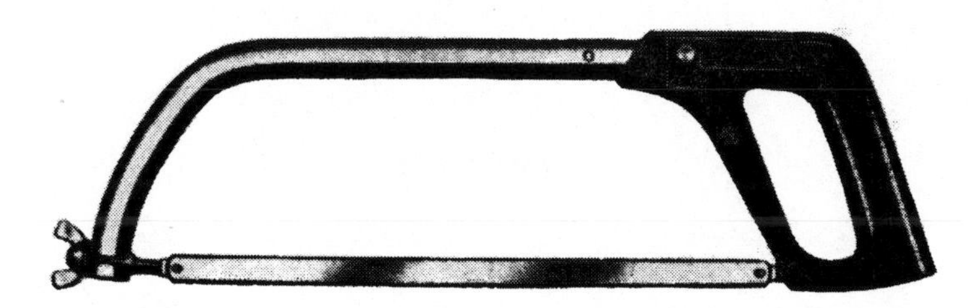

Figure 3-23

These figures are not to scale and are for identification purposes only.

Name and types	Description	Application
Hammer • ball peen Figure 3-24	Made from alloy steel which has been drop-forged and heat treated. Ball on one end and a post with flat surface on other end. An eye, centrally located, is fitted with a hickory (wooden) handle. Up to 1.5 kg in weight.	Rounding off rivets and bolts. Forming metal. Driving drifts, pins, punches and chisels.
• sledge Figure 3-25	Same as ball peen with the ball replaced by another post. Up to 6.5 kg in weight.	Driving large drifts, pins, collars and shafts.
• soft face Figure 3-26	Faces made from brass, raw hide and plastic are fitted to a formed head. The head is cast from steel or aluminium alloy and attached to a handle.	Hammering parts made from materials that bruise easily or have a very hard surface.
Iron-soldering • copper Figure 3-27	Copper head is cast on the end of a steel shaft which is fitted with a wooden handle.	Joining steel, copper or brass parts with solder.
• electric Figure 3-28	240 or 12 V. Copper tip heated by an electric element.	Joining steel or copper wire with resin cored solder.
Keys — allen Figure 3-29	Made from hexagonal, high tensile steel. Bent at right angles near one end. Many different sizes are available.	Loosening or tightening grub or set screws.
Lead light Figure 3-30	32 or 240 V. Fitted with a protected globe.	Illuminating the immediate work area.

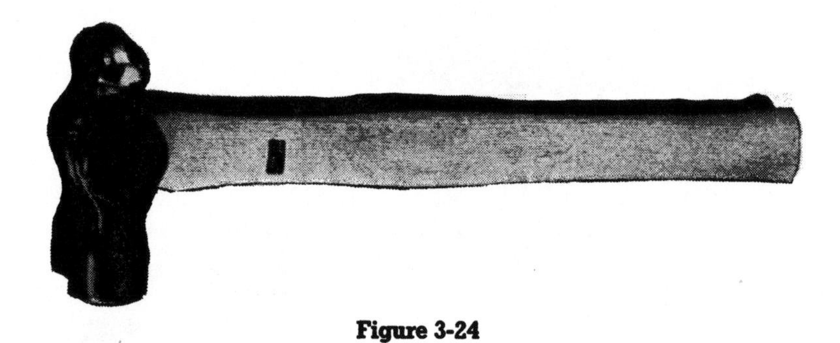

Figure 3-24

Source: Sykes-Pickavant

These figures are not to scale and are for identification purposes only.

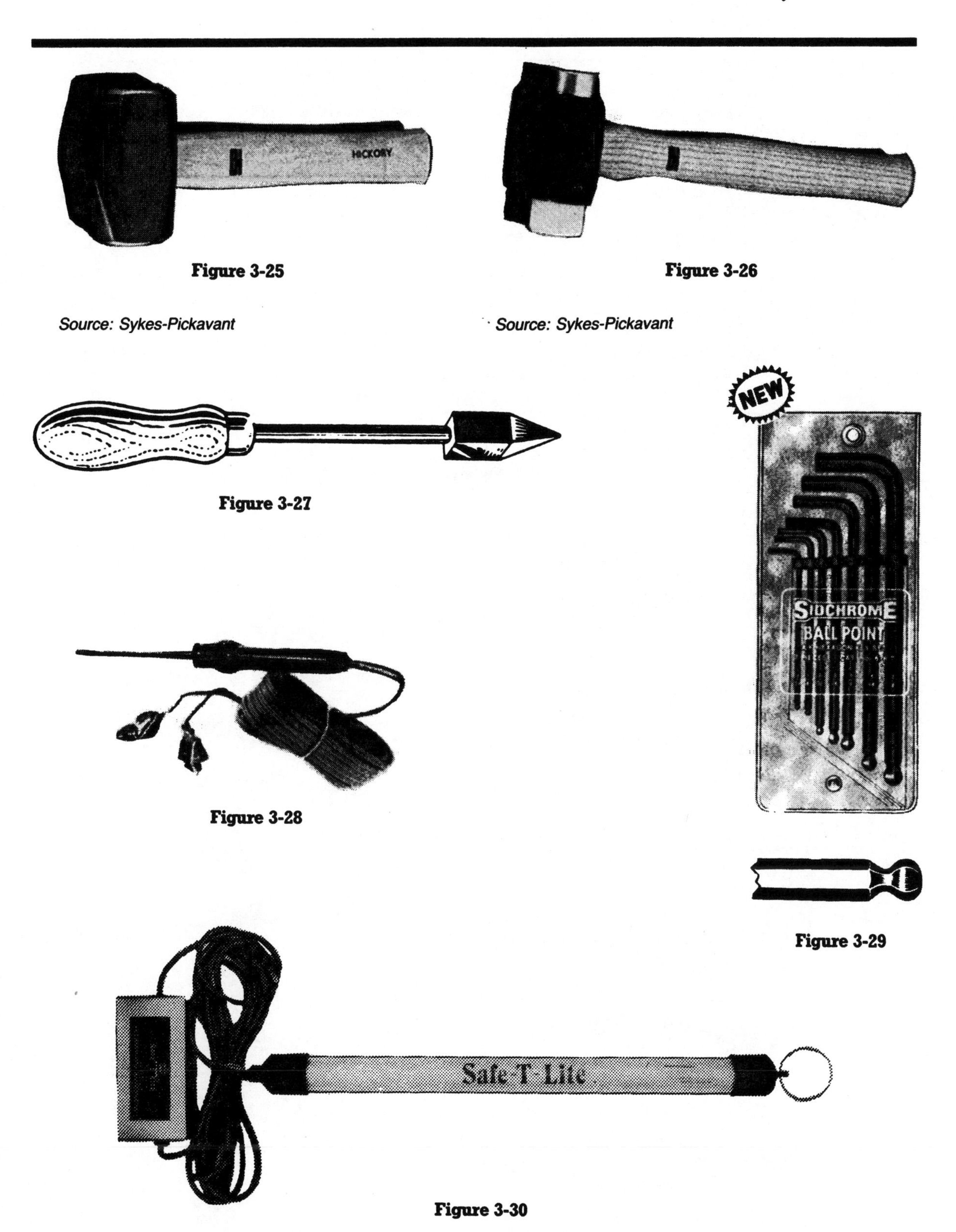

Figure 3-25

Source: Sykes-Pickavant

Figure 3-26

Source: Sykes-Pickavant

Figure 3-27

Figure 3-28

Figure 3-29

Figure 3-30

These figures are not to scale and are for identification purposes only.

Name and types	Description	Application
Pliers • combination Figure 3-31	Forged from steel and ground to produce the various jaws.	Twisting, gripping, bending and cutting most thin sheet metals or wires.
• long nose Figure 3-32	Forged from steel to give long tapered jaws. Machined to produce serrated surface on inside of jaws.	Gripping and holding very small parts or fine wire.
• multi-grip Figure 3-33	Similar to combination type except the opening of the jaws is adjustable.	Gripping and holding round parts.
• side cutting Figure 3-34	Similar to combination type except the jaws are ground to sharp cutting edges.	Snipping wire, small bolts or rods and split pins. Removing and replacing split pins.
• snap ring (circlip) Figure 3-35	Made in the same way as other pliers except each jaw is shaped to fit a particular circlip. Several types are available.	Removing and replacing internal and external circlips.
• vice grip Figure 3-36	Jaws are drop forged. Handles are pressed from steel. Locking mechanism inside handles.	Holding or clamping components.
Note: Pliers are available in various lengths and sizes.		

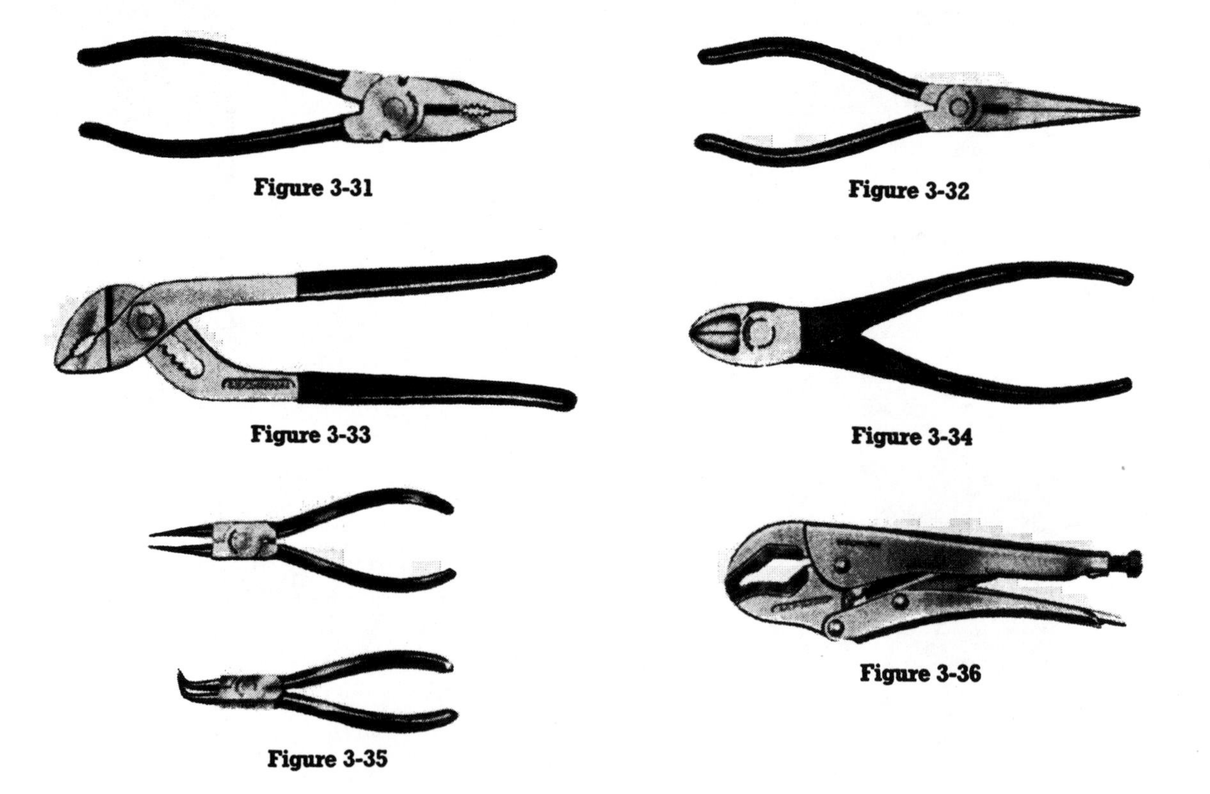

Figure 3-31

Figure 3-32

Figure 3-33

Figure 3-34

Figure 3-35

Figure 3-36

These figures are not to scale and are for identification purposes only.

Name and types	Description	Application
Punch • aligning Figure 3-37	Made from tool steel. Ground to a long taper and pointed at one end.	Aligning mating holes in two or more parts.
• centre Figure 3-38	A smaller type of aligning punch.	Indenting metal surface to locate a twist drill.
• pin Figure 3-39	Made from tool steel. Ground to a specific diameter at one end. Several diameters are available.	Removing tapered or straight pins.
• starter Figure 3-40	Similar to aligning punch except its tip is ground flat. Several tip sizes are available.	Breaking the initial grip of a pin. Used before a pin punch.
Note: Punches are available in many different lengths and sizes		

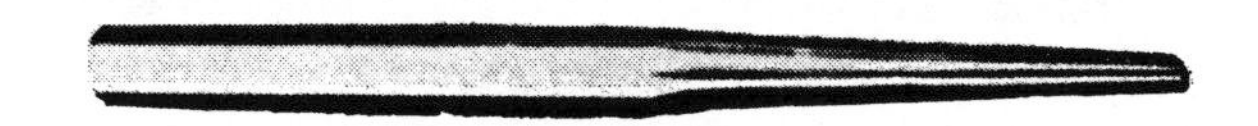

Figure 3-37

Source: Sykes-Pickavant

Figure 3-38

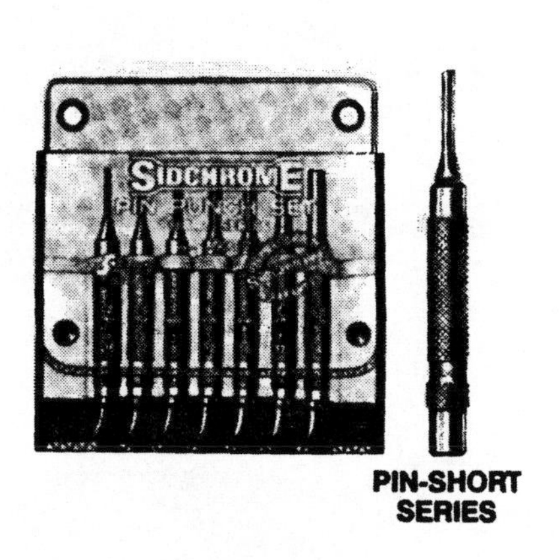

Figure 3-39

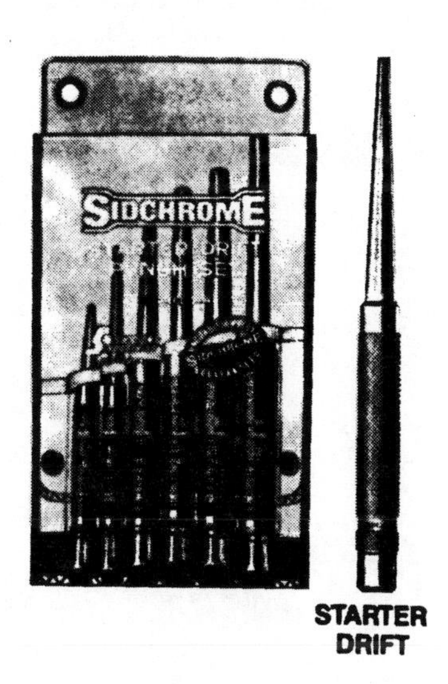

Figure 3-40

These figures are not to scale and are for identification purposes only.

Name and types	Description	Application
Scraper Figure 3-41	Made from spring or carbon steel. Similar in shape to a file but has a cutting edge at its tip.	Removing carbon, rust, and stubborn dirt from flat surfaces.
Screwdriver • standard Figure 3-42	Flat blade formed on the end of a tempered steel bar. Plastic handle formed on other end of bar.	Loosening and tightening screws with a standard slot in their heads.
• phillips Figure 3-43	Same as above but the tip is ground tapered and formed into a cross.	Loosening and tightening screws with a recessed cross in their heads.
Note: Screwdrivers are available in many different lengths and sizes.		

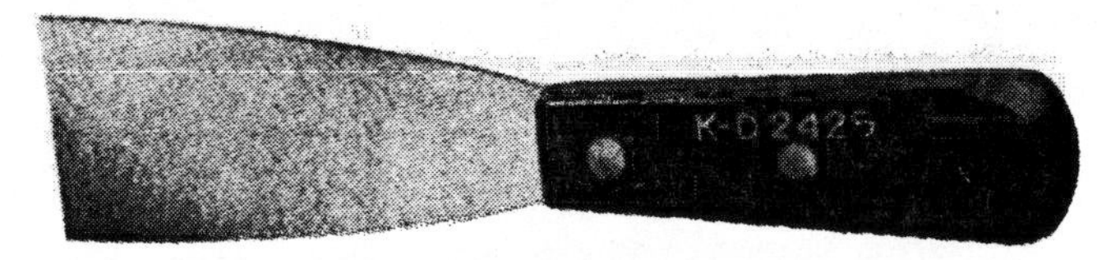

Figure 3-41

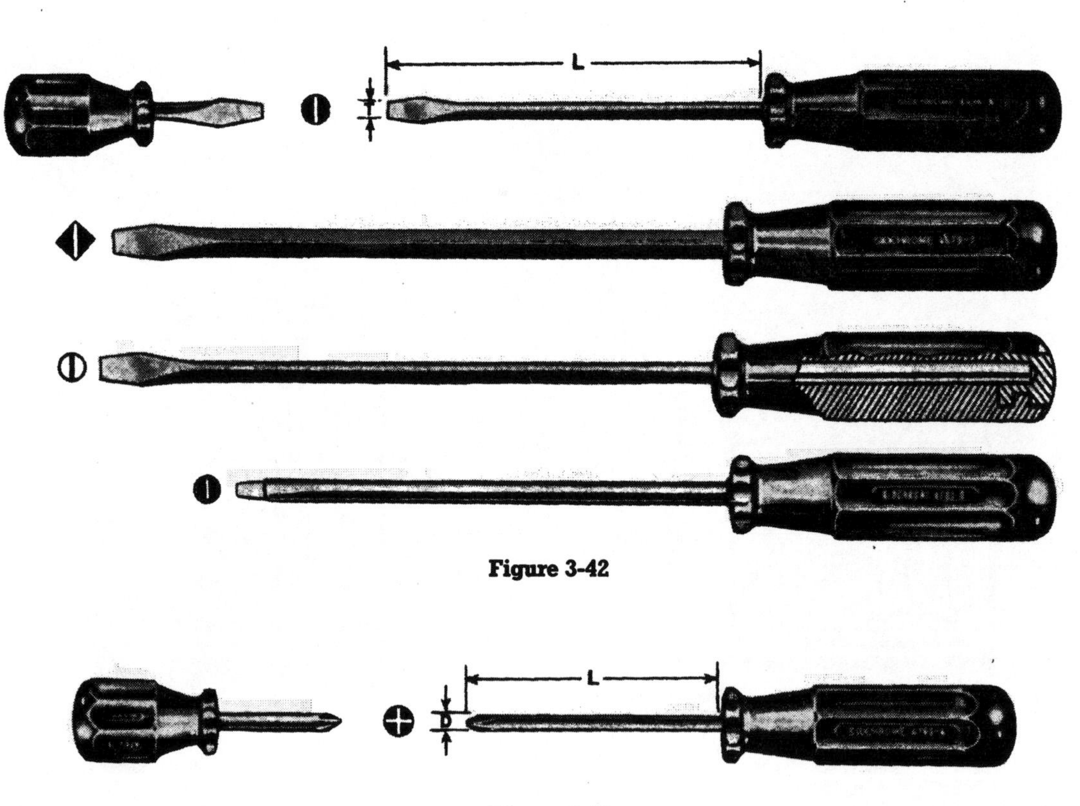

Figure 3-42

Figure 3-43

These figures are not to scale and are for identification purposes only.

Name and types	Description	Application
Socket accessories • brace (speed brace) Figure 3-44	Forged from special steel alloy. Ground to shape and chrome plated for long life. Crank shape fitted with a handle at one end and square drive at the other end.	Unscrewing loose bolts or nuts. Screwing bolts or nuts until they are just tight.
• extension Figure 3-45	Same as above except it is a straight bar with a male drive at one end and female drive at the other end.	Locating the handle or speed brace in a position that clears other components or body work.
• handle-breaker bar Figure 3-46	Plastic handle fitted to one end of a steel shaft and an adjustable drive at the other end.	Initially loosening extremely tight bolts or nuts.
— 'L' bar Figure 3-47	Bar bent close to one end to form an 'L' shape. A square drive is formed at both ends.	Initially loosening bolts and nuts.
— sliding tee Figure 3-48	Bar fitted with a sliding head which has a square drive.	As above.
• ratchet Figure 3-49	Similar to a breaker bar except the drive is fitted with a ratchet.	Unscrewing bolts or nuts when space around them is limited.

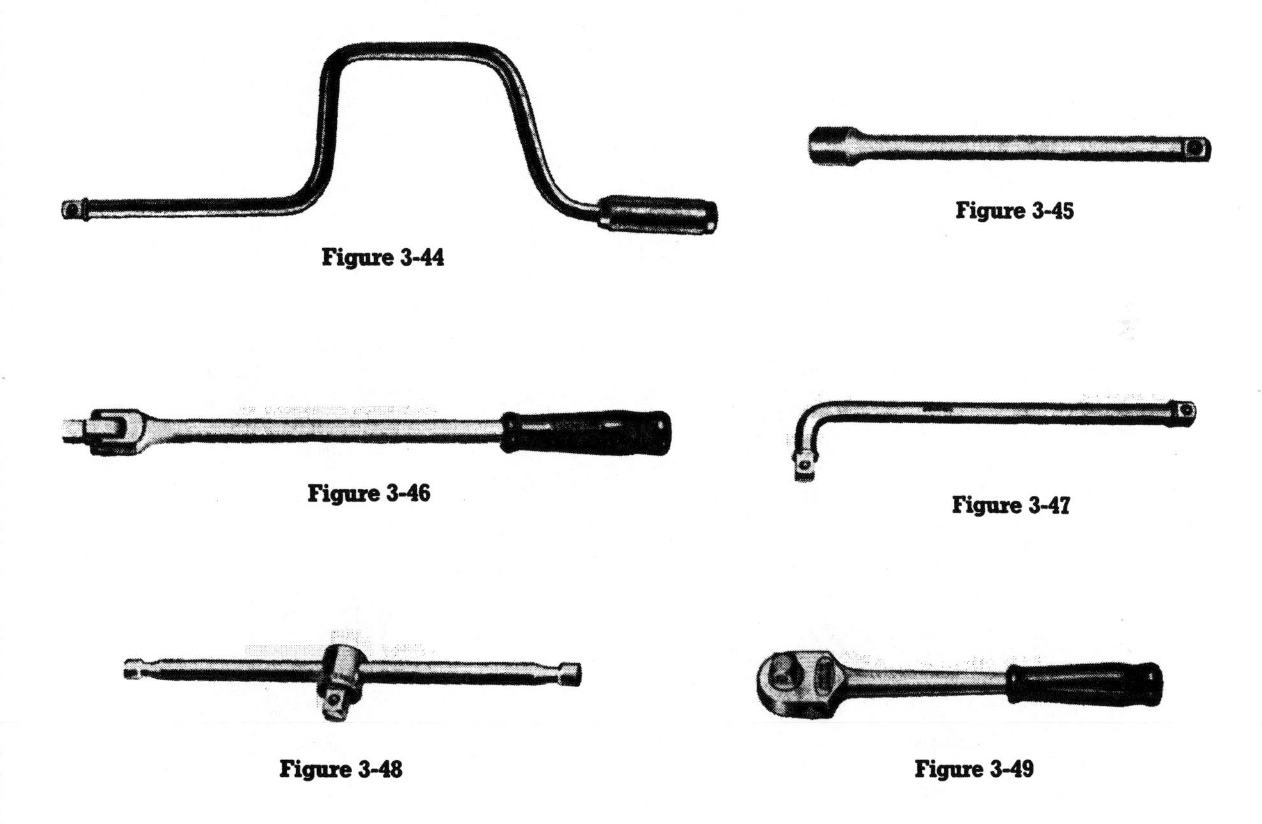

Figure 3-44

Figure 3-45

Figure 3-46

Figure 3-47

Figure 3-48

Figure 3-49

These figures are not to scale and are for identification purposes only.

Name and types	Description	Application
• universal joint Figure 3-50	Made so that the drive can move in small arcs to its body. A spring centralises the drive with its body.	Allows a slight misalignment between a handle and a socket.
Socket • standard Figure 3-51	Made from chrome plated alloy steel. Has a square drive recess at one end and a six or twelve pointed recess at its other end. Its size is stamped on its body.	Used with a handle to loosen or tighten bolts or nuts.
• extra deep Figure 3-52	As above except it is longer.	Loosening or tightening components or nuts fitted some distance from the end of their studs.
• spark plug Figure 3-53	As above except it is fitted with a rubberised tubular insert.	Loosening or tightening spark plugs.
Note: These sockets are classed as thin walled sockets. They must not be used with an impact wrench because they shatter.		
• impact Figure 3-54	Made the same as standard socket except it is a thick walled type and has six points.	Used with an impact wrench to loosen or tighten bolts or nuts.
Note: 'Six points' refers to a single hexagonal shape. 'Twelve points' refers to a double hexagonal shape.		

Figure 3-50

Figure 3-51

Figure 3-52

Figure 3-53

Figure 3-54

These figures are not to scale and are for identification purposes only.

Name and types	Description	Application
Spanner	Forged from a special alloy steel. Generally, chrome plated for long life. Designed with a head at each end of a straight or offset shank. The size of each head is stamped on its shank.	
• open end (flat) Figure 3-55	Each head is machined to form two parallel jaws of a given size.	Loosening or tightening bolts or nuts located in a confined space.
• ring Figure 3-56	Each head is a shallow cylindrical shape with its internal surface formed to a 'single' or 'double' hexagon of a given size.	Loosening or tightening bolts or nuts that have confined space above them.
• combination Figure 3-57	One head is an open type and the other is a ring type of the same size.	Has the advantages of both types of spanners.
Tap (hand) Figure 3-58	Made from carbon or high-speed steel. Cutting teeth are separated by flutes which form the cutting edges and act as channels for the waste. The distance between each tooth is the thread pitch. The thread size and type is stamped on its shank.	Using with a handle to cut a thread in a hole boring through or into most materials.
Note: Available in many different thread types and sizes.		

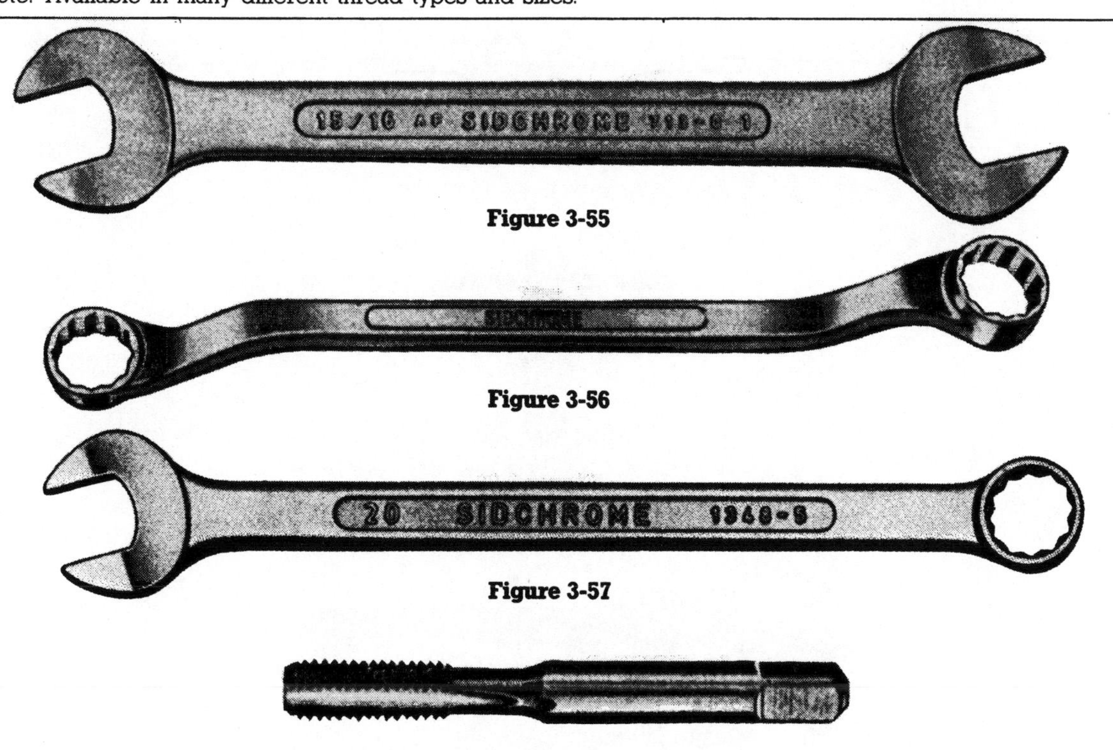

Figure 3-55

Figure 3-56

Figure 3-57

Figure 3-58

These figures are not to scale and are for identification purposes only.

Name and types	Description	Application
Tap handle • stock Figure 3-59	Stock consists of a fixed jaw and a moveable jaw. One handle adjusts the moveable jaw onto the end of the tap.	Holding and turning a tap during a threading process.
• 'T' Figure 3-60	Made with a four jaw chuck at one end of its body and a handle at the other end.	As above.
Tinsnips Figure 3-61	A cutting edge is formed on each of the jaws which are designed to slide past one another.	Shearing (cutting) most sheet metals or thin materials.
Tool box Figure 3-62	Made from sheet steel. Designed with several special compartments. Generally, portable.	Holding, separating and transporting hand tools.
Wrench • adjustable Figure 3-63	Forged from high-quality steel and chrome plated or black finished for long life. Available in several lengths.	Substituting for the correct spanner.
• pipe Figure 3-64	Forged from high-quality steel and fitted with serrated jaws. Available in several sizes.	Gripping and turning cylindrical shapes.

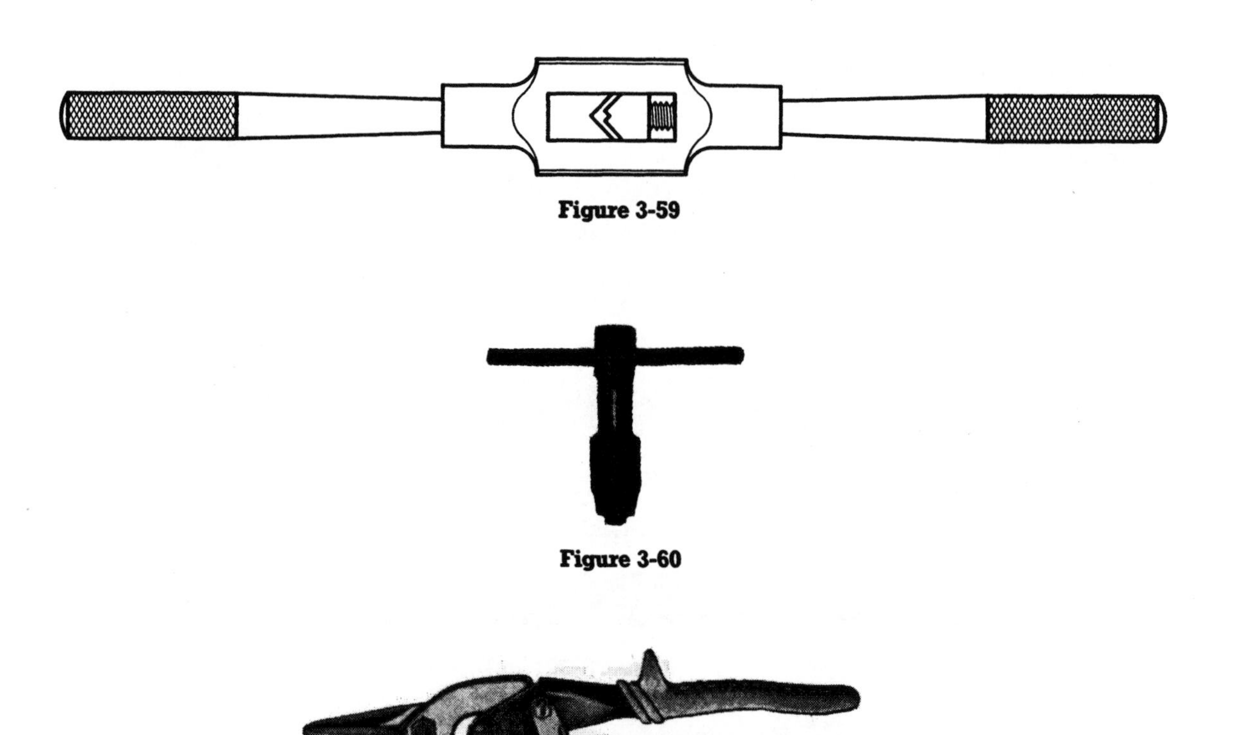

Figure 3-59

Figure 3-60

Figure 3-61

These figures are not to scale and are for identification purposes only.

Figure 3-62

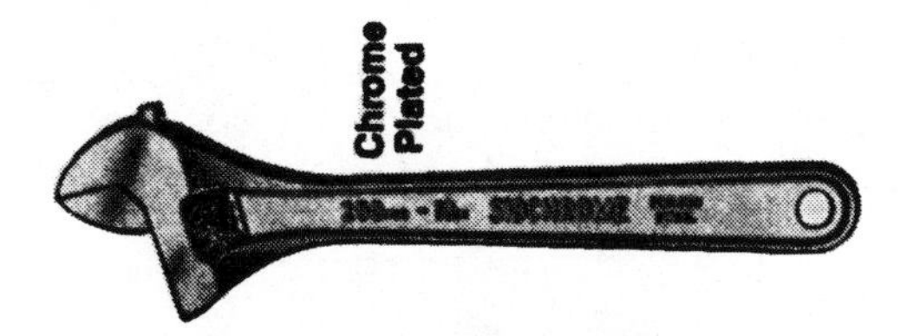

Figure 3-63

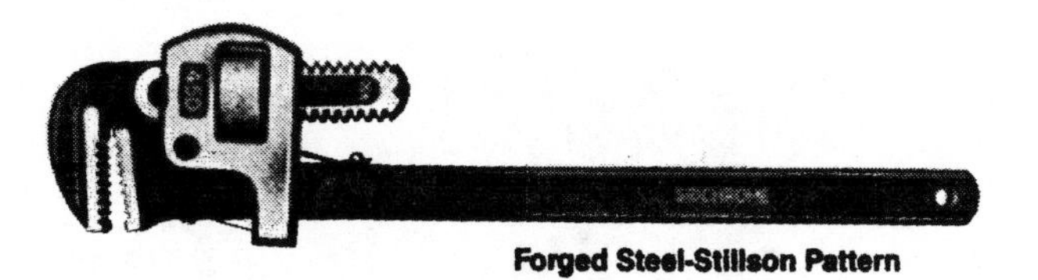

Figure 3-64

These figures are not to scale and are for identification purposes only.

Hand tools — specialised

Name and types	Description	Application
Battery terminal cleaner Figure 3-65	Made with a spring steel cutter inside a cylindrical head and a reamer type cutter on another head. Wire brush types are available.	Cleaning the inside of battery terminals and the outside of battery posts.
Battery terminal puller Figure 3-66	A cross arm supports two spring loaded legs and the pressure screw.	Pulling both battery terminals from their posts.
Brake adjuster Figure 3-67	A small lever with a flat blade formed at each end.	Adjusting the brake shoe to brake drum clearance.
Brake spring pliers Figure 3-68	A large version of long nose pliers which has specially formed tips or jaws to grip the brake shoe and spring.	Removing or replacing the brake return springs.
Clutch aligning tool Figure 3-69	A tapered and stepped sleeve slides on a shaft that has removeable spigots.	Aligning the clutch plate centre with the spigot bearing or bush in the crankshaft.
Drain plug spanner Figure 3-70	• Formed with different sized squares, tapers and a hexagon at each end. • Formed with different sized single hexagonal and square recesses.	Loosening or tightening special recessed drain plugs. Loosening or tightening hexagonal or square headed drain plugs.
Engine sling Figure 3-71	A short wire cable or rope fitted with special eyes and lugs at each end.	Supporting an engine during its removal or replacement.

Figure 3-65

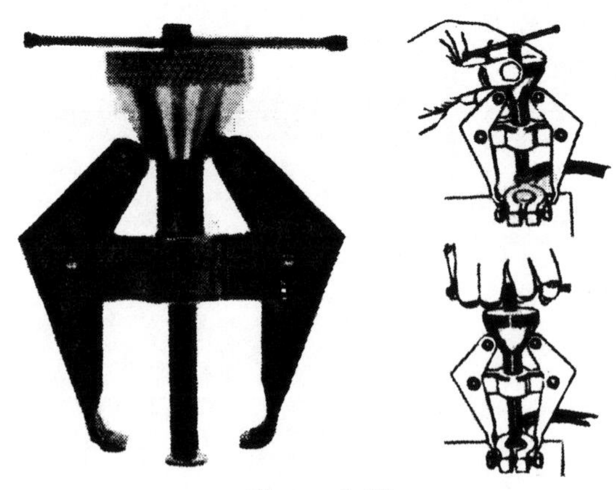

Figure 3-66

These figures are not to scale and are for identification purposes only.

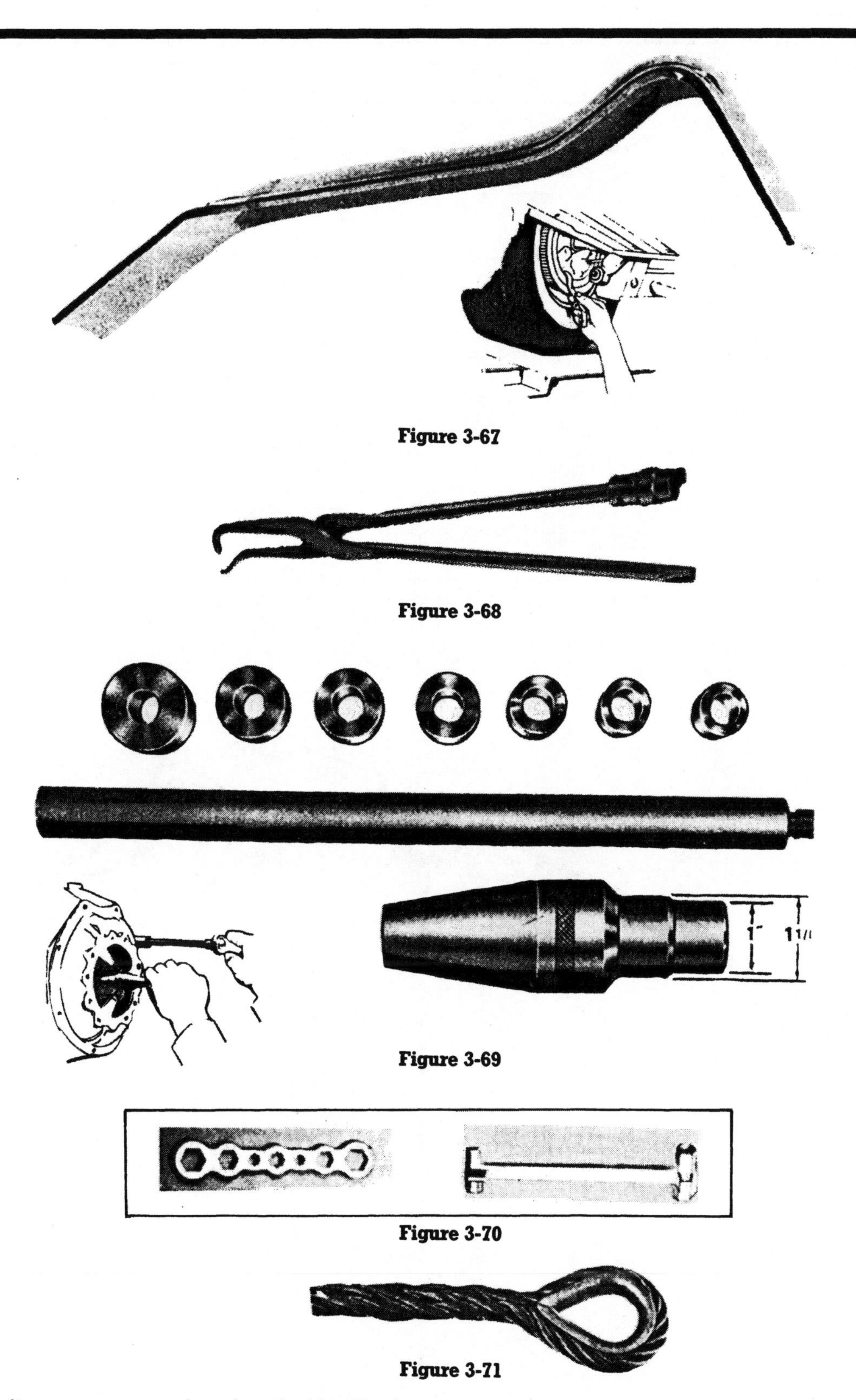

Figure 3-67

Figure 3-68

Figure 3-69

Figure 3-70

Figure 3-71

These figures are not to scale and are for identification purposes only.

Name and types	Description	Application
Grease gun (hand) Figure 3-72	A spring loaded plunger housed inside a tubular body which is fitted with a cast head. The head contains a lever action pump fitted with a flexible outlet hose and nozzle.	Lubricating ball joints with grease.
Impact screwdriver Figure 3-73	A hollow body contains a spring loaded ramp device which slightly turns the drive end. A drive end is recessed to allow different blades to be fitted.	Loosening extremely tight screws.
Impact wrench Figure 3-74	A pneumatically or electrically driven ratchet device that is fitted with a drive square and has both forward and reverse directions.	Removing and replacing bolts or nuts at a fast rate. *Note*: Some allow their tension setting to be changed.
Oil filter remover Figure 3-75	A spring steel strap attached by both ends to a lever (handle). Different sizes are available.	Removing an oil filter.
Oil gun (hand) Figure 3-76	A spring loaded plunger housed inside a tubular body which is fitted with a cast head. The head is fitted with a rigid or flexible outlet pipe.	Withdrawing oil from a component or inserting oil into a component.

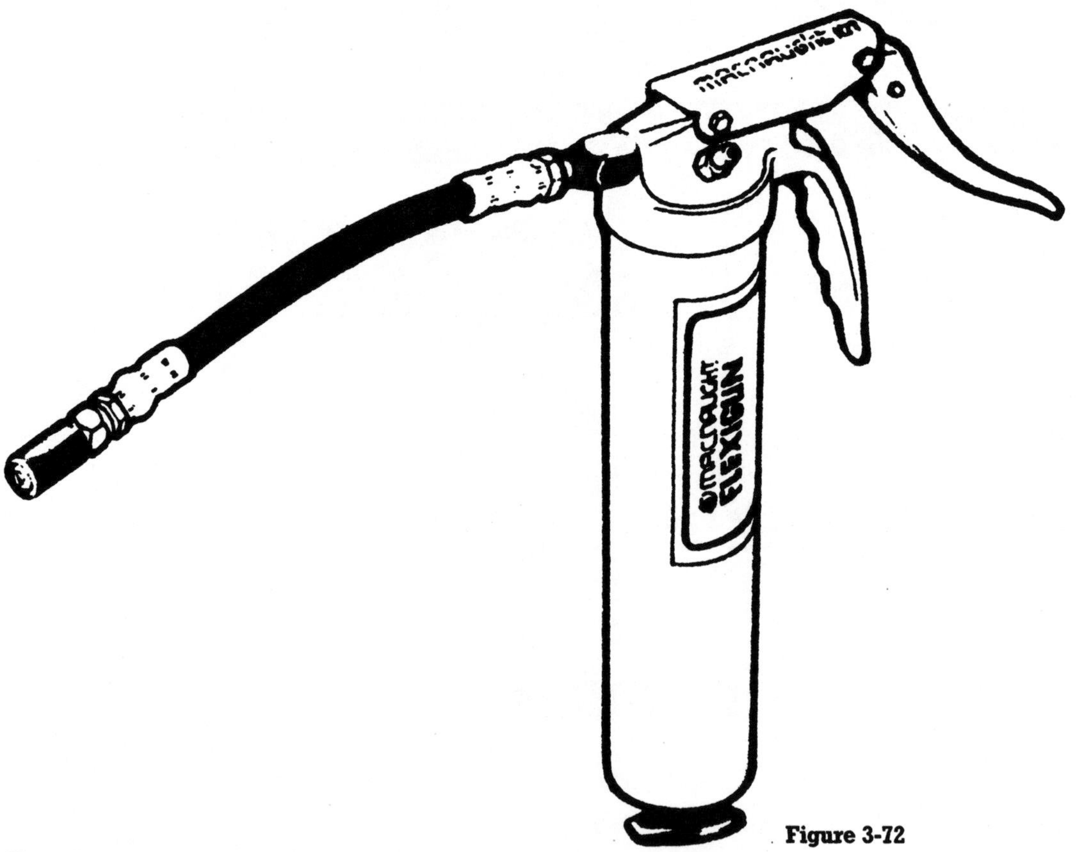

Figure 3-72

These figures are not to scale and are for identification purposes only.

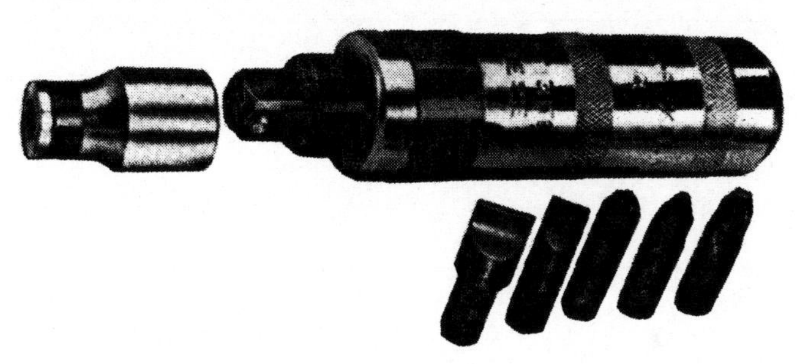

Figure 3-73

Source: Sykes-Pickavant

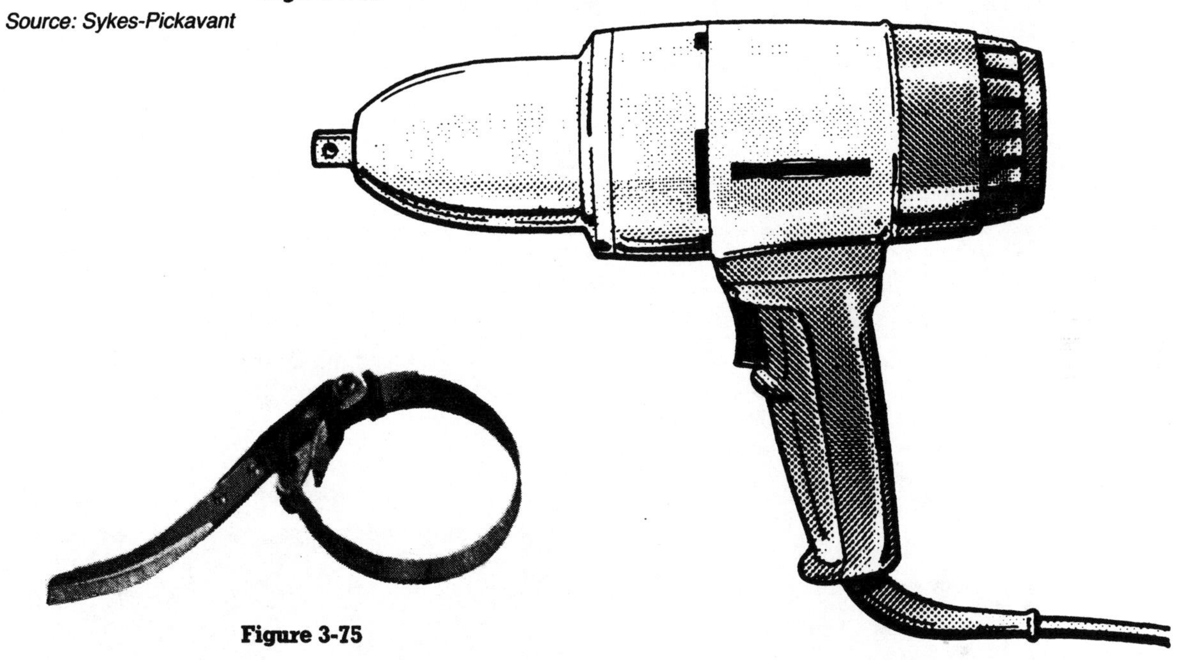

Figure 3-75

Figure 3-74

Figure 3-76

These figures are not to scale and are for identification purposes only.

Name and types	Description	Application
Pipe cutter Figure 3-77	A rotary cutter made from special steel is attached to a pressure screw fitted to one end of a 'C' shaped frame. The other end of the frame is fitted with two small rollers.	Cutting the end of steel or copper tubing so that it is square.
Pipe flaring tool • single Figure 3-78	This unit consists of: • clamp and die; • cutter; • swage on the end of pressure screw.	Forming a uniform lip (flare) on the end of a steel or copper tube (pipe).
• double Figure 3-79	This unit consists of: • forged steel body; • several die; • several special punches; • pressure screw; • tapered swage; • locking plug.	Forming a double thickness lip (flare) on the end of a steel or copper tube (pipe).
Piston ring compressor Figure 3-80	Made from spring steel sheet and fitted with several straps and a ratchet device. Several sizes are available.	Clamping piston rings firmly into their piston grooves during the installation of the piston into its bore.
Ridge remover Figure 3-81	This unit consists of: • two cast end plates; • two cast centralising carriers; • blade carrier and special steel cutter; • adjusting screw.	Cutting the lip (ridge) that is formed on the top of a worn cylinder bore.

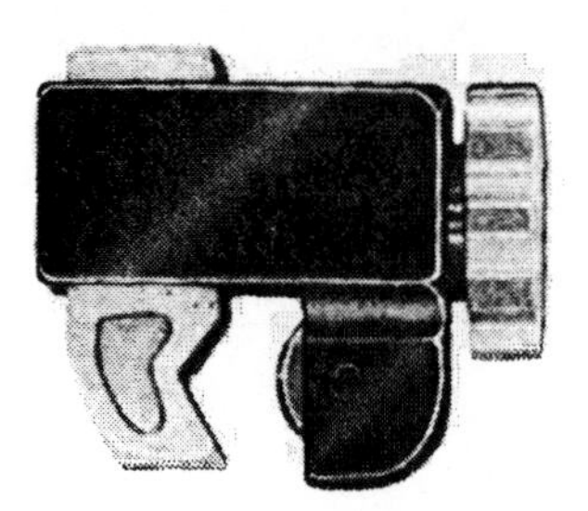

Figure 3-77

Figure 3-78

Source: Sykes-Pickavant

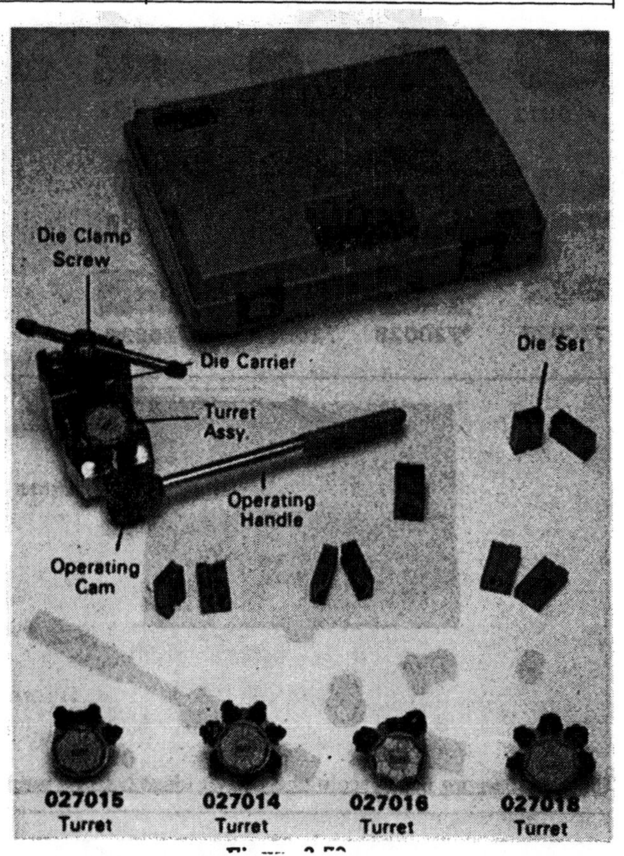

Figure 3-80

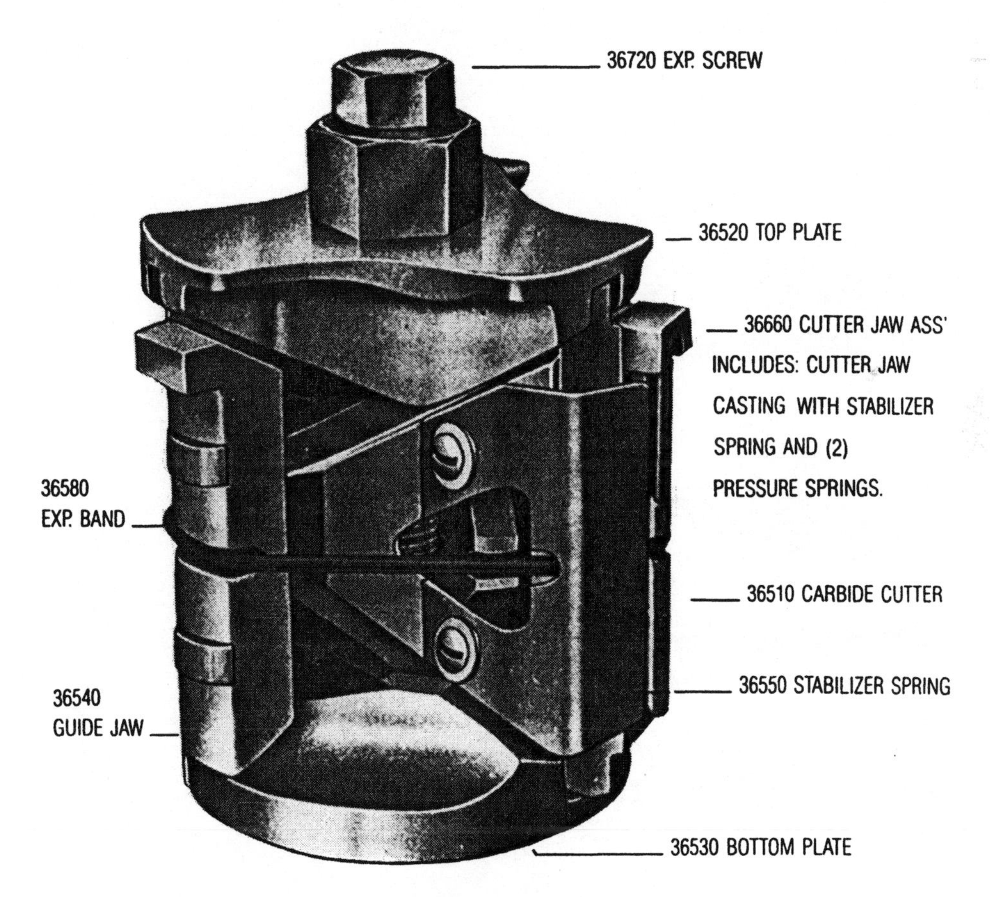

Figure 3-81

These figures are not to scale and are for identification purposes only.

Name and types	Description	Application
Rotary brush Figure 3-82	A small wire brush is formed on the end of a hub that is fitted with a small shank. Several different types are available.	Used with an electric drill to remove carbon from the combustion chambers and ports of a cylinder head.
Seal puller Figure 3-83	A tapered threaded cylindrical body fitted with a special steel pressure screw.	Removing oil seals.
Steering wheel puller Figure 3-84	A high-quality steel plate fitted with a pressure screw and two draw bolts.	Removing steering wheels.
Valve lapper Figure 3-85	A moulded rubberised sucker and handle fitted to a piece of dowelling (wooden rod).	Gripping and turning a valve during a valve grinding on a cylinder head.
Valve seat cutter Figure 3-86	Made from a tool steel disc with cutter blades ground at a given angle to its rim. Several sizes and angles are a part of a set.	Restoring valve seats in the combustion chambers of a cylinder head.
Valve seat hone Figure 3-87	Made from a special carborundum mixture which is moulded onto a lead core. The core is drilled and threaded to suit a drive carrier. It is a part of a synchro seating kit.	Used with a synchro-seating unit which is driven by an electric drill for cutting and honing valve seats in a cylinder head to the desired shape.

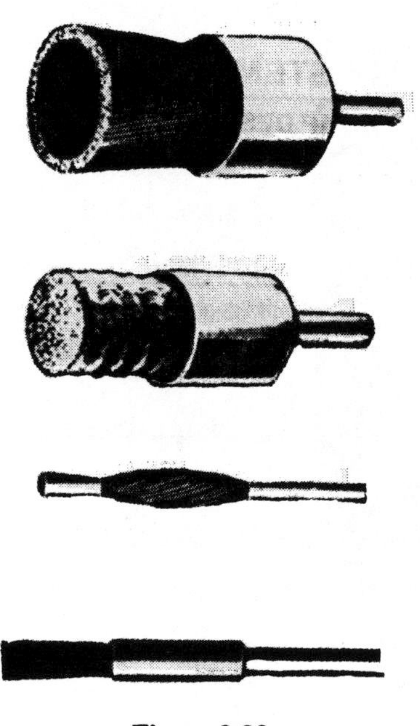

Figure 3-82

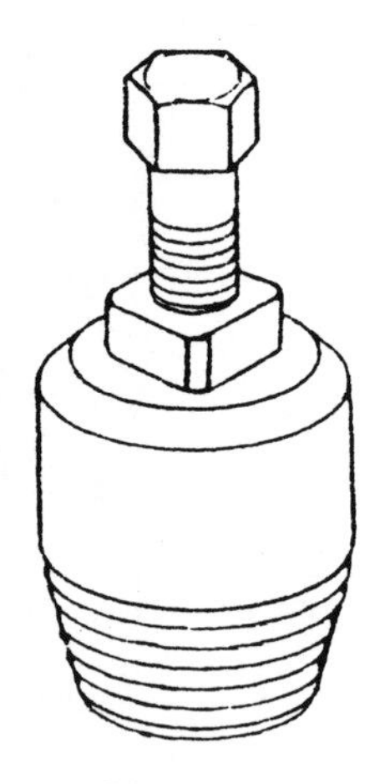

Figure 3-83

These figures are not to scale and are for identification purposes only.

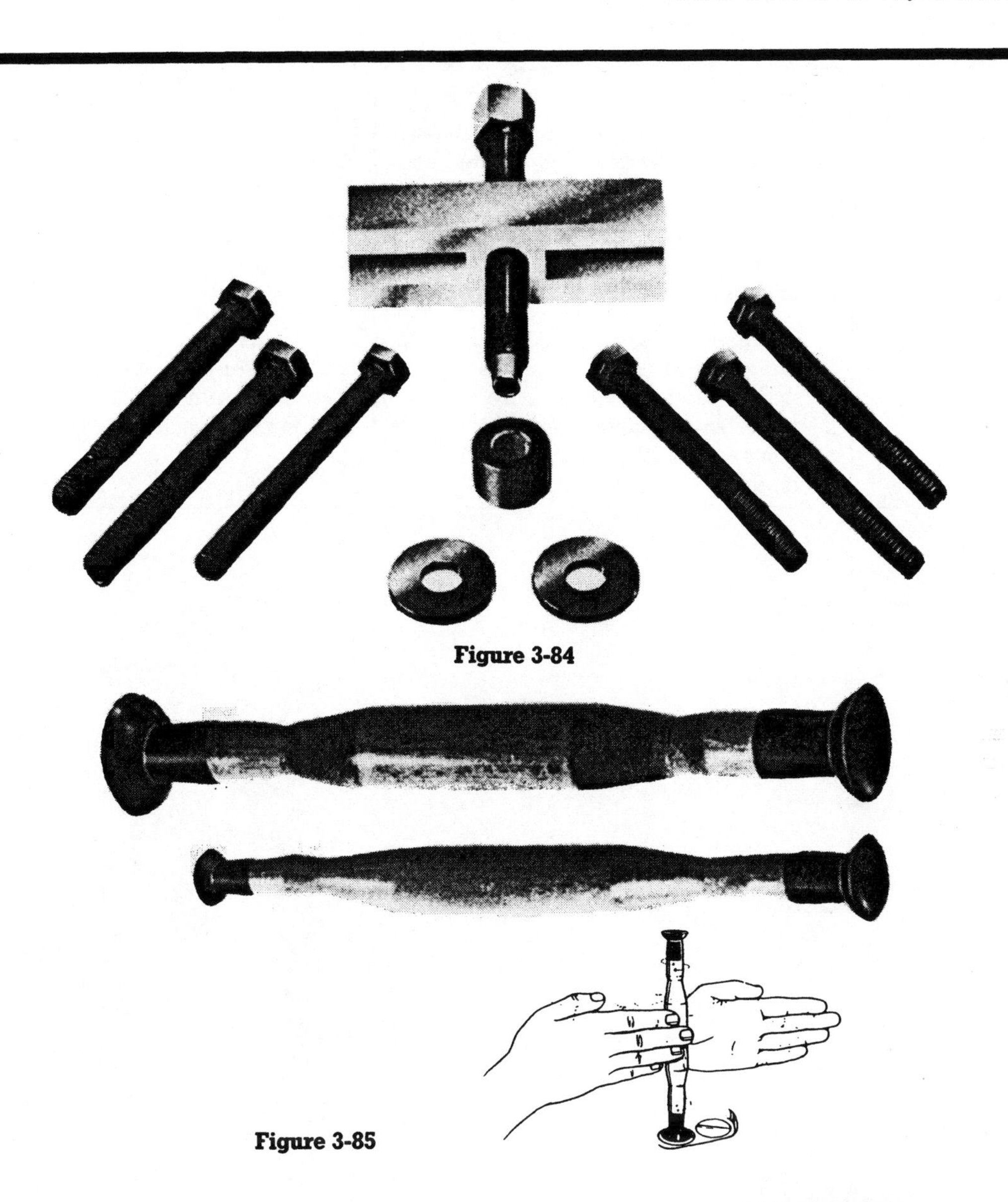

Figure 3-84

Figure 3-85

Figure 3-86

Source: Sykes-Pickavant

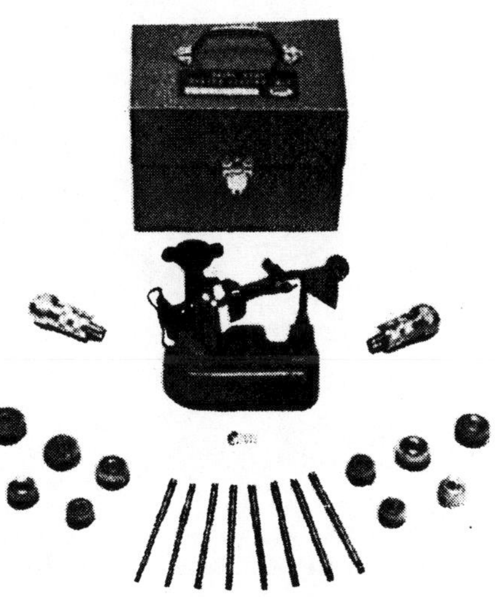

Figure 3-87

These figures are not to scale and are for identification purposes only.

Name and types	Description	Application
Wheel cylinder clamp Figure 3-88	Made from high tensile wire. Formed to the shape of an alligator clip. A 'G' clamp type is also available.	Holding wheel cylinder pistons in their bores while the brake shoes and return springs have been removed.
Wheel (hub) puller Figure 3-89	Its arms and legs are cast from steel. A pressure screw is located in the arm.	Removing hubs from splined or tapered axle shaft ends.
Wheel spanner Figure 3-90	Drop forged from steel. Made in an 'L' or '+' shape. May have one or more single hexagonal sockets of different sizes. Some types have a flat blade formed on the end of one arm.	Loosening or tightening wheel nuts or screws. Those fitted with a blade are used to pry off hub caps.

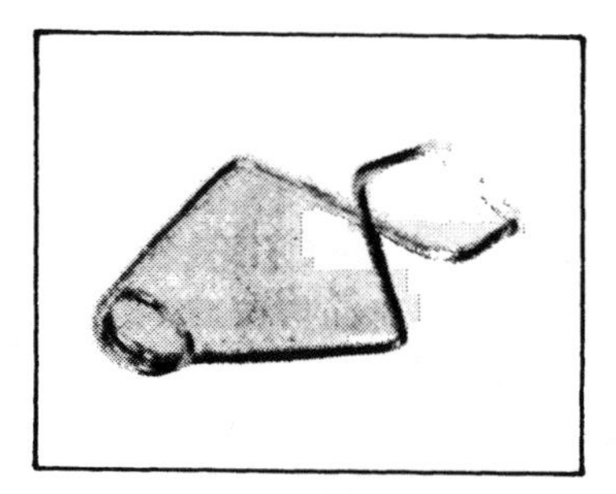

Figure 3-88

Figure 3-89
Source: Sykes-Pickavant

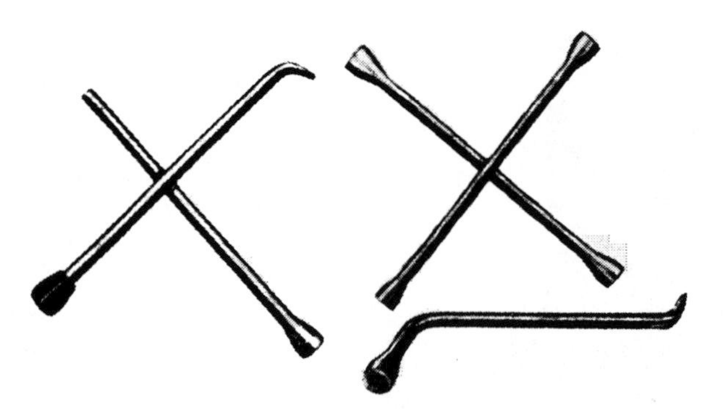

Figure 3-90

These figures are not to scale and are for identification purposes only.

Measuring instruments — mechanical

Name and types	Description	Application
Ball gauge Figure 3-91	Made from high quality steel. Formed with a hollow handle and a split shank and ball. A pin, threaded at one end and flared at the other end, passes through the ball and screws into an adjusting knob. They are available in many different ball sizes.	Used with an outside micrometer for measuring the inside diameters of small holes or bores.
Cooling system tester Figure 3-92	Consists of a pressure gauge, hand pump, release valve, extension piece and several adaptors.	Testing a cooling system for leaks. Testing a radiator cap pressure setting.
Compression gauge Figure 3-93	Consists of a pressure gauge, release valve, solid or flexible tube and a spark plug adaptor.	Testing the pressure created during the compression phase in an engine.
Dial indicator Figure 3-94	Consists of a mechanical gauge fitted with two needles, a moveable dial, a spindle and a stand.	Measuring small distances between 0.01 mm and 10 mm, for example a bend in a crankshaft (very accurate).

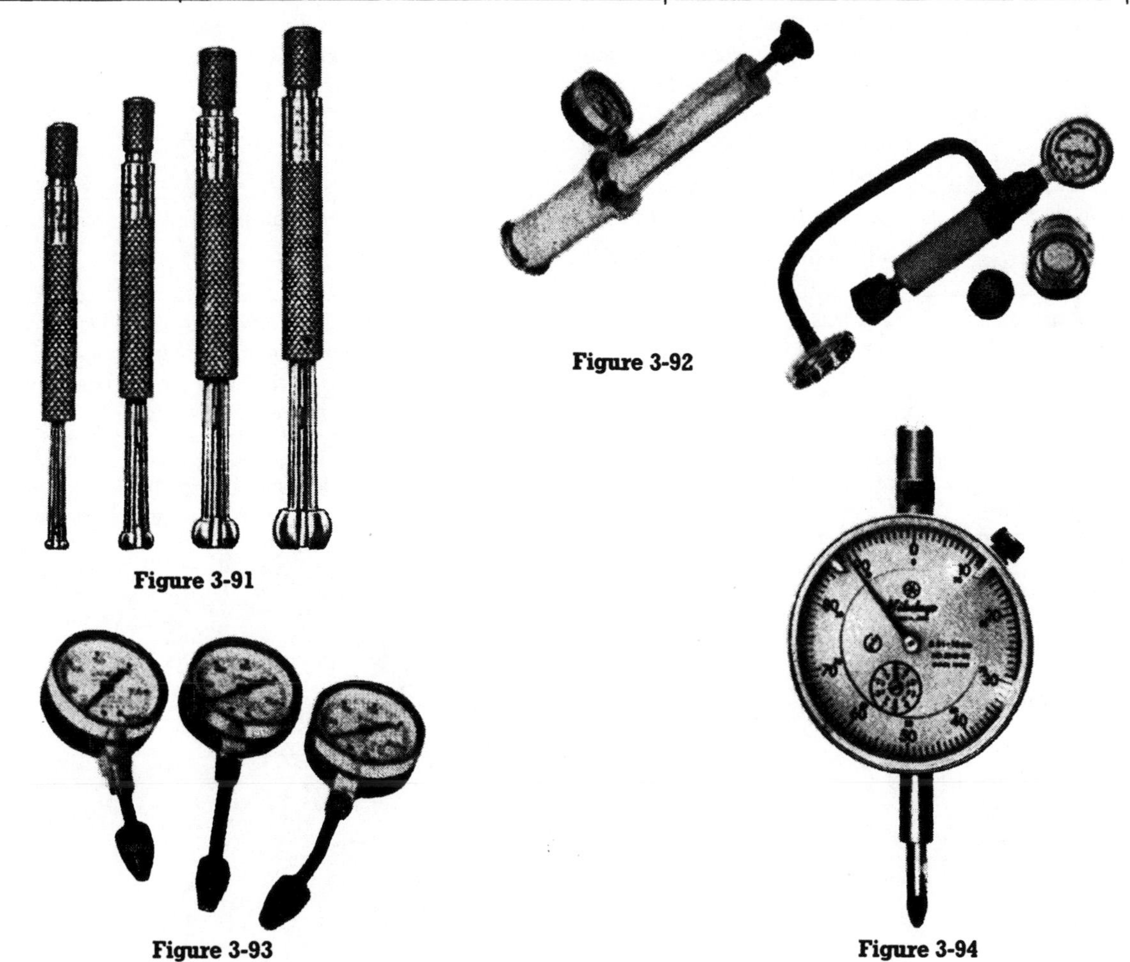

Figure 3-91

Figure 3-92

Figure 3-93

Figure 3-94

These figures are not to scale and are for identification purposes only.

Name and types	Description	Application
External micrometer Figure 3-95	A drop forged 'C' shape frame is fitted with a fixed anvil, a hard faced spindle and a main nut assembly. A graduated thimble is attached to the end of the spindle. A graduated sleeve is positioned on the main nut. Several different sizes are available.	Measuring external diameters or lengths from 0.01 mm to 100 mm, for example the diameter of a piston skirt (very accurate).
Feeler gauge Figure 3-96	Made from a strip of spring steel of constant thickness for its full length. Its size (thickness) is marked on one of its surfaces. A wide range of sizes is available. A set of blades is housed in a shield.	Measuring small clearances or gaps that exist between two components, for example, used to measure tappet clearance.
Inside caliper Figure 3-97	Consists of two legs, a pivot pin, circular spring and an adjusting device.	Used with a steel rule for measuring internal dimensions of components, for example to measure the inside diameter of exhaust tubing.
Inside micrometer Figure 3-98	Consists of head which is fitted with a main nut and curved tipped anvil. A spindle, which is attached to a graduated thimble and anvil screws into the main nut. A graduated sleeve is positioned on the main nut. Its range is varied by placing a distance piece on the curved tipped anvil.	Used with an external micrometer for measuring the internal dimensions of components, for example to measure the diameter of a cylinder bore (very accurate).
Oil pressure gauge Figure 3-99	Consists of a pressure gauge, braided flexible hose and adaptor.	Testing the oil pressure of a running engine.
Outside caliper Figure 3-100	Consists of two legs, a pivot pin, circular spring and an adjusting device.	Used with a steel rule for measuring the external dimensions of components, for example to measure the outside diameter of exhaust tubing.
Pressure and vacuum gauge Figure 3-101	Consists of a pressure gauge, flexible hose and adaptor.	Testing the pressure and vacuum of a mechanical fuel pump. Testing the inlet manifold vacuum of a running engine.

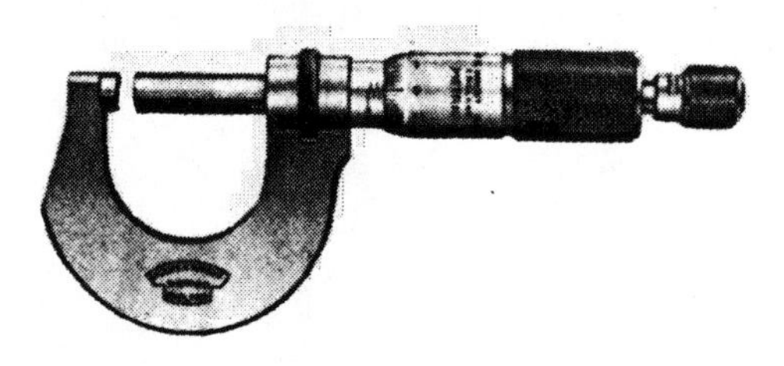

Figure 3-95

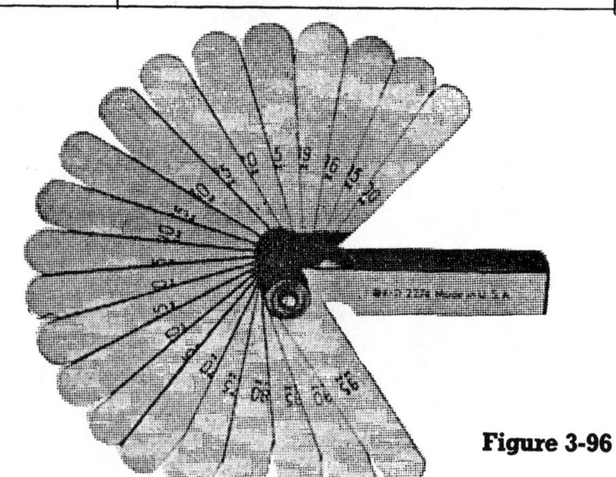

Figure 3-96

These figures are not to scale and are for identification purposes only.

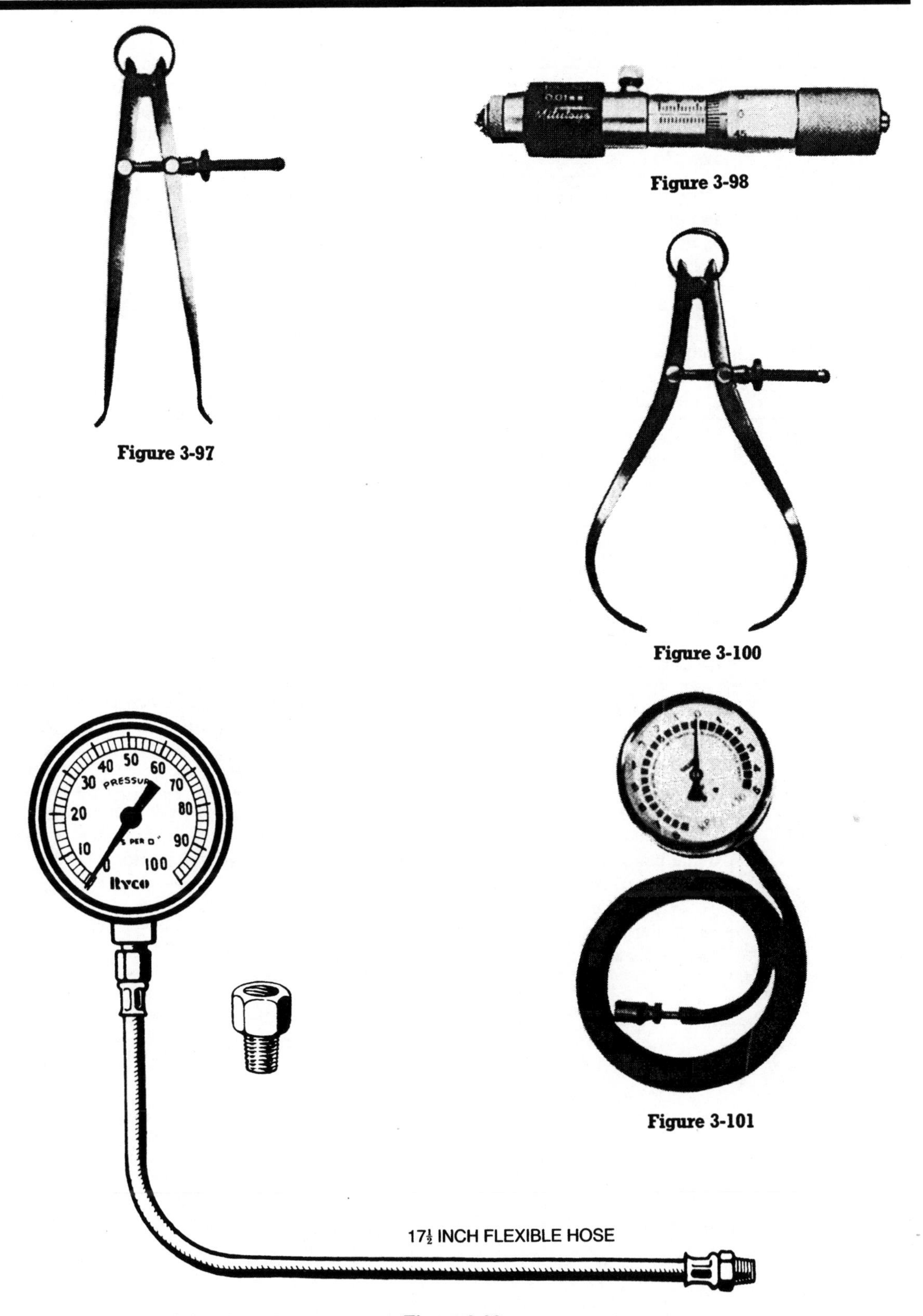

Figure 3-97

Figure 3-98

Figure 3-100

Figure 3-101

Figure 3-99

These figures are not to scale and are for identification purposes only.

Name and types	Description	Application
Spark-plug gauge Figure 3-102	Consists of spring steel wire bent and attached to a blade or body. Several different sized wires form a set. A set includes a gap adjuster.	Measuring the gap that exists between the electrodes of a spark plug. Adjusting the gap of a spark plug.
Steel rule Figure 3-103	Made from alloy steel which is hardened and ground. A graduated scale is etched on one or both surfaces. A common length is 300 mm.	Measuring the dimensions of a component (least accurate of measuring devices).
Steel tape Figure 3-104	Made from a strip of thin spring steel which is slightly curved for its entire length. A graduated scale is marked on its concave side. Many different lengths are available.	Measuring distances that are greater than one metre.
Straight-edge Figure 3-105	Made from a thick strip of alloy steel or aluminium alloy. One surface is broadly chamfered to one edge. A common length is 1000 mm.	Used with a feeler gauge for measuring the twist or buckle (distortion) of a flat surface.
Tension wrench Figure 3-106	Made from a special steel rod that is attached at one end to a handle and at the other end to a drive head. A scale is fitted to its handle and a signal device is located in its drive head arm or its handle. Several sizes are available.	Setting the tension of bolts or nuts to a given value.
Vee block Figure 3-107	Cast and machined from a thick plate of alloy steel.	Used with a flat surface for supporting a round or cylindrical component during a measuring task, for example checking a camshaft for bend.
Vernier caliper Figure 3-107A	Made from a special alloy steel and formed into two pieces. The T-shaped frame has an internal and external jaw. Its body is marked with a millimetre scale. The slide has a vernier scale and an internal and external jaw. Some calipers have a depth gauge.	Measuring dimensions of most types of shapes and depth of holes, to within 0.02 mm (very accurate).

Figure 3-102

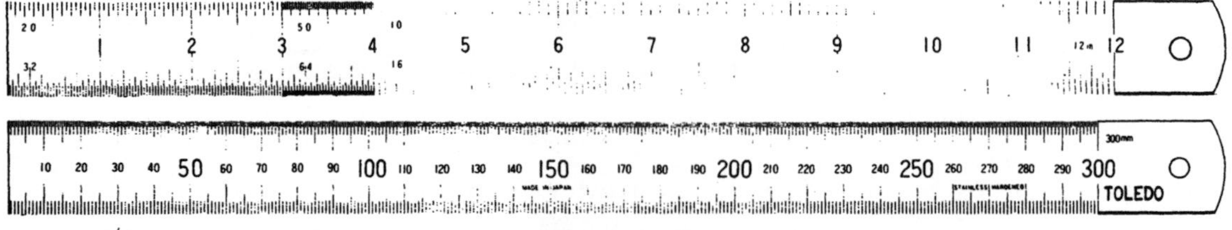

Figure 3-103

These figures are not to scale and are for identification purposes only.

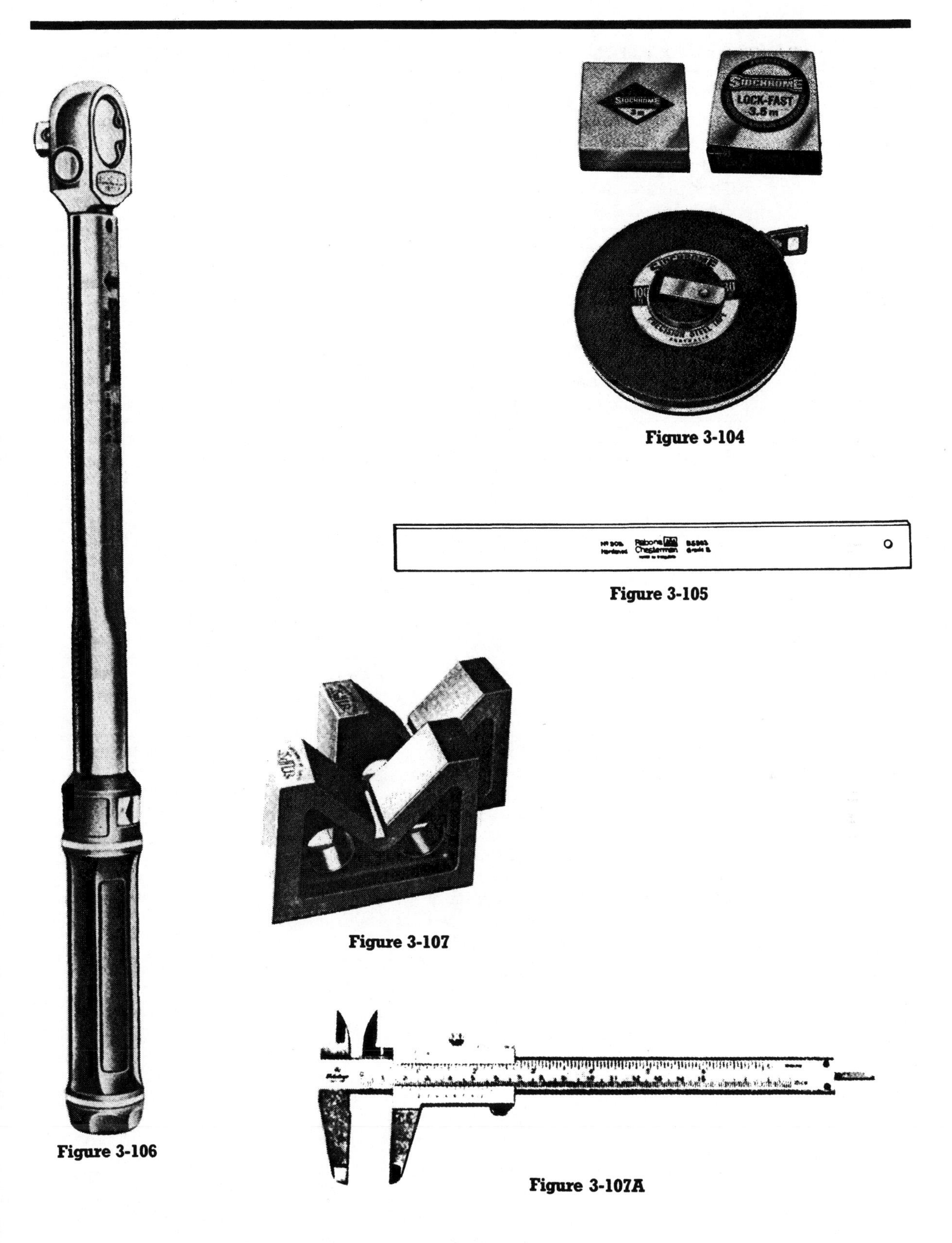

Figure 3-104

Figure 3-105

Figure 3-106

Figure 3-107

Figure 3-107A

These figures are not to scale and are for identification purposes only.

Measuring instruments — electrical

Name and types	Description	Application
Ammeter Figure 3-108	Consists of a moving coil meter, dial, scale selector switch or several terminals, at least two sockets or terminals (red and black), a case and two test leads fitted with insulated clips (red and black).	Measuring current flow in a given electrical circuit.
Caution: connect an ammeter into the electrical circuit.		
Battery discharge tester Figure 3-109	Consists of a moving coil meter, heavy duty push button switch, two heavy duty leads (red and black) fitted with insulated battery clips and a ventilated case.	Testing the condition of a battery by placing it under load.
Hydrometer Figure 3-110	Consists of a graduated float, glass tube, rubberised bulb and pick-up tube.	Testing the specific gravity (S.G.) of the electrolyte in various cells of a battery.
Low voltage test light Figure 3-111	Consists of a transparent body fitted with a pointed metal probe, a low wattage globe, an end cap and a test lead fitted with a clip.	Checking the existence of a voltage supply to an electrical component.
Ohmmeter Figure 3-112	Consists of a moving coil meter, dial, scale selector switch, two test leads with probes (red and black) and a small battery inside its case.	Testing the resistance of an electrical component that has been disconnected or removed from its circuit.
Caution: ensure that the component is disconnected from its battery before the ohmmeter is connected.		
Multi-meter Figure 3-113	Made by including an ammeter, ohmmeter and voltmeter in one case.	Testing amps, ohms and volts related to the components in a given electrical circuit.
Timing light Figure 3-114	Consists of a plastic body (pistol shape) which houses a light, lens and electronic components which are connected to two low voltage (12 V) leads and a high voltage pick-up lead.	Observing the timing marks of a running engine.
Voltmeter Figure 3-115	Consists of a moving coil meter, dial, scale selector switch or several terminals, at least two sockets (red and black), two test leads fitted with insulated clips and case.	Measuring the voltage drop across an electrical component when the circuit is turned on.

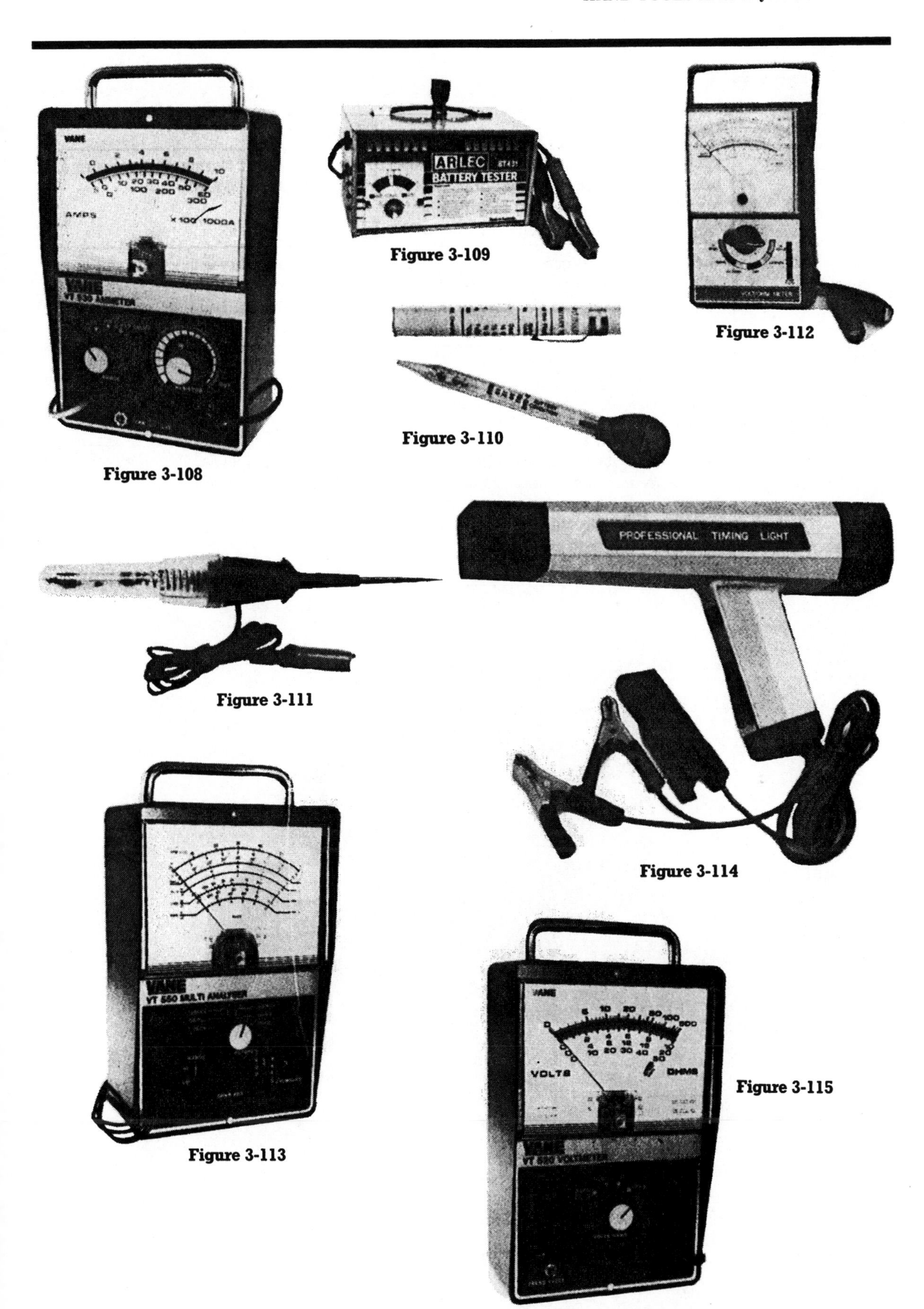

Figure 3-108

Figure 3-109

Figure 3-110

Figure 3-111

Figure 3-112

Figure 3-113

Figure 3-114

Figure 3-115

These figures are not to scale and are for identification purposes only.

Equipment — electrical

Name and types	Description	Application
Air compressor unit Figure 3-116	Consists of a receiver (high pressure tank), safety valve, pressure gauge, drain valve, heavy duty switch, electric motor (240 or 415 V), Vee belts and compressor.	Pressurising and storing air to ensure a steady supply at 550-850 kPa.
Battery charger Figure 3-117	Consists of a metal case with ventilators, voltage selection switch, mains power lead, an ammeter and two battery leads (red and black) with clips.	Charging a battery either in or out of the vehicle. Charging several batteries out of their vehicles.
Bench grinder Figure 3-118	240 V motor fitted with two 150 × 25 mm grinding wheels, metal guards, rests and viewing shields. Bolted to a suitable bench.	Removing material from small metal components. Sharpening hand tools.
Electric welder Figure 3-119	240 or 415 V units. Consists of an amperage indicator, amperage setting device, two high current terminals, one high current cable fitted with an earth clamp and one high current cable fitted with an insulated hand piece.	Welding most metals with an electrode held in its hand piece.

Figure 3-116

These figures are not to scale and are for identification purposes only.

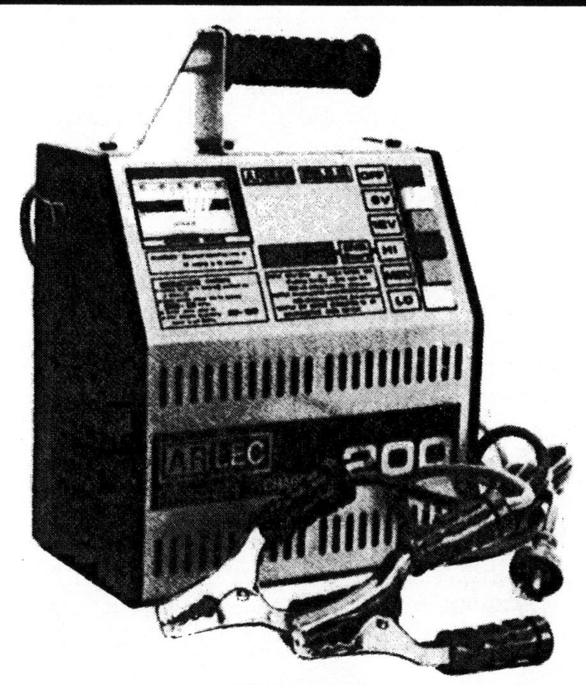

Figure 3-117

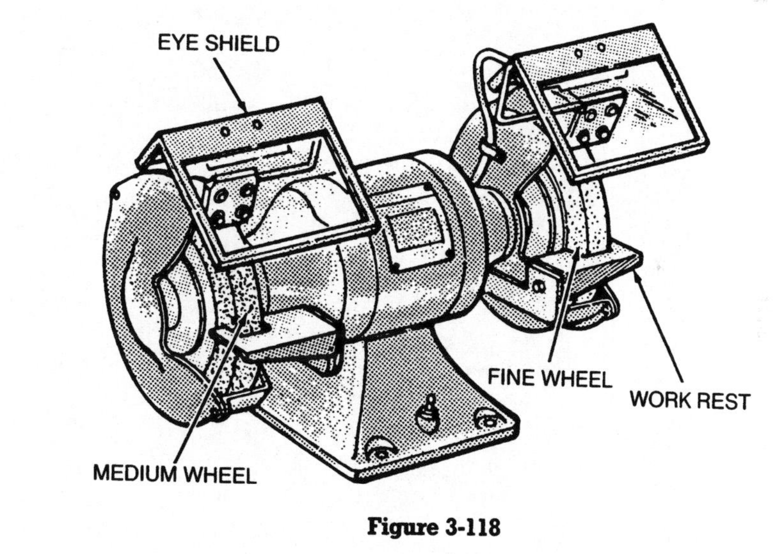

Figure 3-118

Figure 3-119

These figures are not to scale and are for identification purposes only.

Name and types	Description	Application
Hoist Figure 3-120	415 V hydraulic/electrical device. May have two or four posts. A 'two poster' consists of two adjustable arms on a carrier attached to each post, a control switch and locking device. A 'four poster' consists of a ramp supported at each corner by a carrier attached to each post, a control switch and safety lock-device.	Lifting and supporting heavy cars or light commercial vehicles.
Parts cleaner Figure 3-121	Consists of a solvent reservoir, cleaning tank, 240 V solvent pump, flexible hose fitted with a bristle brush, and light.	Removing grease and dirt from small components.
Steam cleaner Figure 3-122	240 V device. Consists of a fuel tank, small furnace, pressure gauge, water inlet hose, high pressure outlet hose fitted with a long insulated nozzle.	Removing stubborn dirt and grease from components or body work.
Wheel balancer Figure 3-123	415 V device. Consists of a control panel control switch, hub adaptor, spindle and safety guard.	Checking and balancing a road wheel when it has been removed from the vehicle.

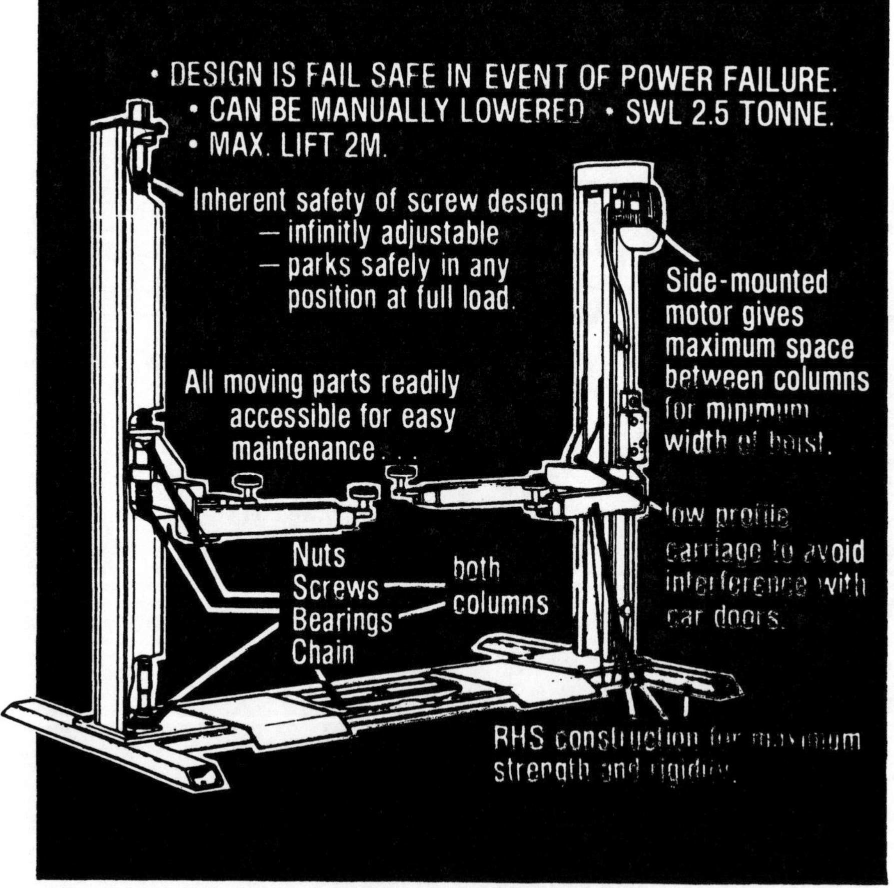

Figure 3-120

These figures are not to scale and are for identification purposes only.

Figure 3-121

Figure 3-122

Figure 3-123

These figures are not to scale and are for identification purposes only.

Equipment — mechanical

Name and types	Description	Application
Bead breaker Figure 3-124	Consists of a pedestal supporting a table and pneumatic ram, a breaker plate attached to the ram, air control valve, wheel lock pin and nut, and a bead lever.	Removing and refitting a tyre to a rim.
Block and tackle Figure 3-125	Consists of a chain block, load chain, endless chain and a tackle head fitted with a hook.	Used with a gantry or frame for lifting an engine from a vehicle.
Creeper Figure 3-126	Made from pressed and welding sheet steel. A small platform with a caster wheel fitted to each corner. A head rest is fitted at one end.	Supporting and protecting the tradesperson from the floor. Allows free movement of the tradesperson while he or she is under a raised vehicle.
Fire extinguisher Figure 3-127	Consists of a cylinder, control valve assembly, hose fitted with a nozzle and a handle. Several types are available.	Extinguishing most types of fires.
Hydraulic press Figure 3-128	Consists of a reinforced frame, press table, table winch, hydraulic ram, control valve and pressure gauge.	Forcing components off or onto shafts, rods, pins or other components, for example a bearing off a shaft.

Figure 3-124

These figures are not to scale and are for identification purposes only.

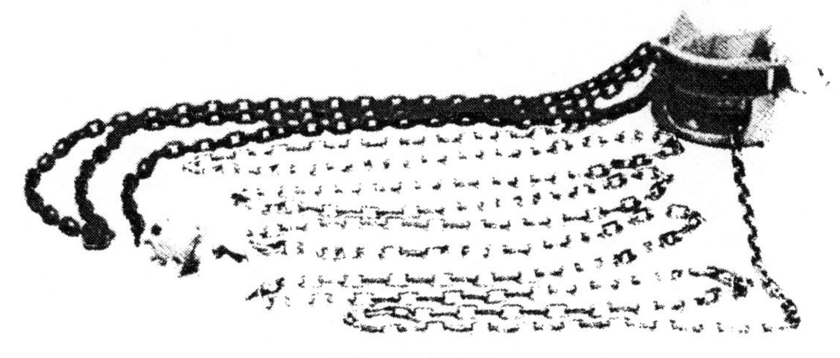

Figure 3-125

Figure 3-126

Figure 3-127

Figure 3-128

These figures are not to scale and are for identification purposes only.

Name and types	Description	Application
Oxygen and acetylene welder Figure 3-129	Consists of two high pressure cylinders (red and black), two pressure regulators and gauge sets, two hoses (red and black), hand piece and trolley.	Welding, cutting and brazing of most metals.
Safety stand Figure 3-130	Made from steel. Consists of a short extendible post, adjustment device, saddle and a wide base.	Supporting a vehicle after it has been raised to a suitable height.

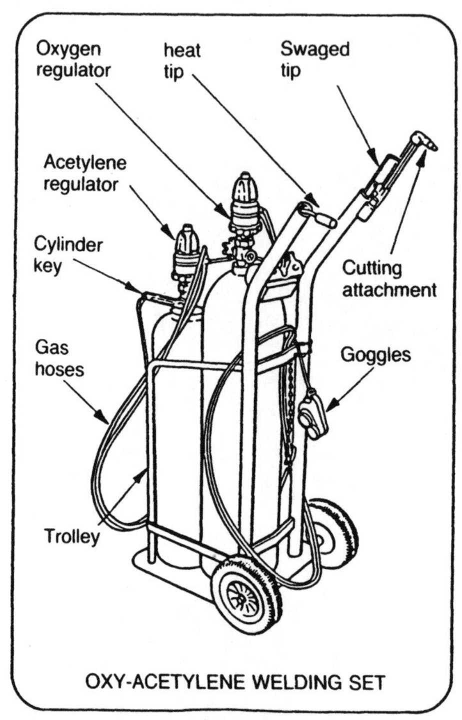

Source: AGPS **Figure 3-129**

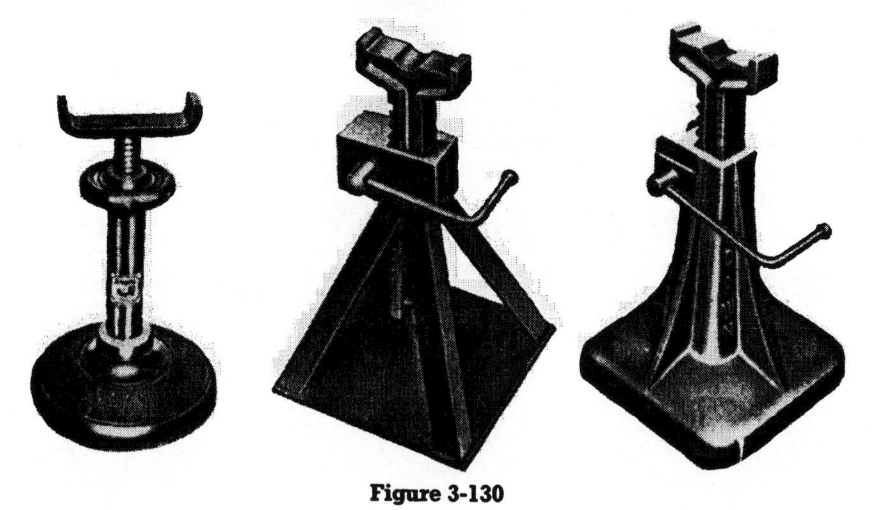

Figure 3-130

These figures are not to scale and are for identification purposes only.

Name and types	Description	Application
Trolley jack Figure 3-131	A sturdy steel chassis which houses a hydraulic ram, pump and control valve. Chassis is fitted with two wide steel wheels at one end and two casters at the other end. A handle is use to control the ram and to tow the jack over the floor. A small hoisting device is connected to the ram and is fitted with a saddle.	Lifting a vehicle off the floor and lowering a vehicle to the floor.
Vice Figure 3-132	Both the body and slide are cast from iron. The body is formed with a base and machined to suit the slide. The slide is fitted with a clamping screw. Both castings are machined to allow a hardened steel jaw to be attached.	Clamping a component or a piece of material during a re-work process, for example filing a carburettor base.
Work bench (portable) Figure 3-133	A tall and narrow steel frame fitted with press steel top and shelves.	Storing and transporting components and tools during a job.

Figure 3-131

Figure 3-132

Figure 3-133

These figures are not to scale and are for identification purposes only.

SELECTION AND USE

The points that must be considered are:
1 planning the job;
2 selecting equipment;
3 selecting hand tools;
4 selecting measuring instruments.

Planning the job

Planning the job is important for several reasons. First, the time that the vehicle is going to be 'out of action' must be considered. When the job is going to take several days or more, the vehicle must be placed in the workshop so that normal operations can continue without it being in the way. Secondly, the nature of the job must be considered. Will special equipment such as a hoist or a pit be needed to gain access to the components? When a large job is undertaken, it may be necessary to use this special equipment to start with and then store the vehicle while the repairs are being done, for example an engine overhaul. Thirdly, an area on the workshop floor and/or a bench may be needed while the component is being dismantled and assembled, and for storing other components that are not being repaired, for example a bonnet or a transmission during an engine overhaul. This area must be prepared before the components are removed from the vehicle. Lastly, the availability of parts will have an influence on the time taken for the job. In the cases of large jobs, it may be necessary to pre-order the parts before the job is scheduled. This will require an excellent diagnosis of the problem so the correct parts can be ordered. Diagnosing is a skill that is based on a sound knowledge of mechanical principles and will not be covered in this book. For our purpose, it has been assumed that the diagnosis is correct and that the planning stage can take place.

Selecting equipment

Once the job has been planned, the type and number of pieces of equipment will be easily identified. Where equipment is going to be used for long periods, it is essential that other jobs will not be affected due to the equipment not being available. In some cases equipment will have to be duplicated for this reason, for example safety stands and floor creepers. The location of the pieces of equipment must be known before the job is started so that time is not wasted looking for them. It is best that the equipment be brought to the work area before starting the job.

The types of equipment that may be needed for a job are:
- seat and guard covers;
- tool and parts trays;
- liquid containers;
- safety stands and jacks;
- special jig or fixtures;
- cleaning materials and devices;
- device for dispensing liquids such as water, oil and petrol;
- various electrical or electronic devices.

Note: The selection of correct safety equipment is important.

Selecting hand tools

The correct selection of hand tools will allow the job to be performed safely, quickly and without damage to the parts or vehicle. The fit of the tool in or on the fastener is the most important aspect of tool selection. The four main tools that fit in or on a fastener are:
- allen keys;
- screwdrivers;
- sockets;
- spanners.

Allen keys fit into single or double hexagonal recesses formed in set or grub screws. The size of an allen key is marked on its shank. To select an allen key:
1 Measure the distance between two opposite flats of the recess in the set screw.
2 Select the same sized allen key.
3 Try the fit of the allen key in the recess. It must be a snug fit.

Note: When a grub screw is below the surface of the component, estimate the size of the recess and try an allen key. Continue until the correct allen key is found.

Screwdrivers fit into slots or recesses formed in the head of screws. The size of a screwdriver blade is selected by its width and thickness. To select a screwdriver:
1 Its blade width must extend almost the full length of the slot.
2 Its thickness must fill the width of the slot.

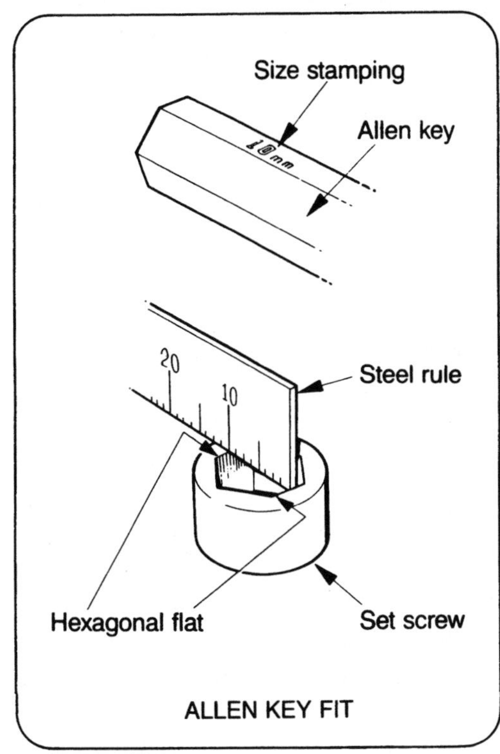

ALLEN KEY FIT

Source: AGPS

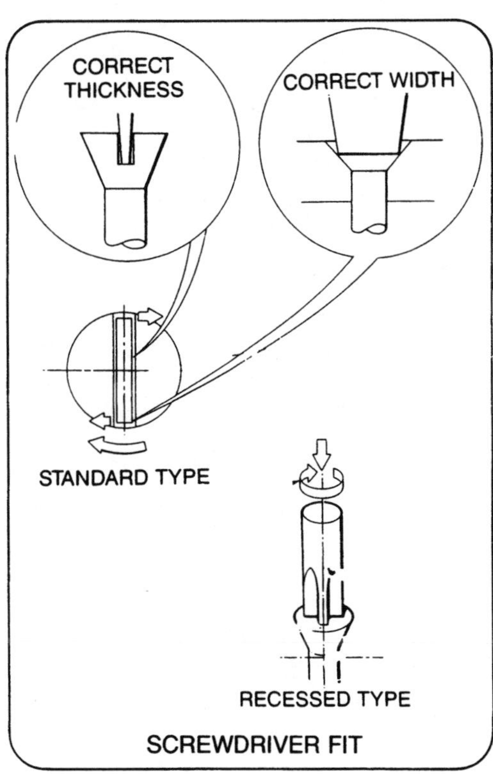

SCREWDRIVER FIT

Source: AGPS

Note: This also applies to a phillips screwdriver.

Sockets fit onto the head of bolts or nuts. Their size and thread type are marked on their curved surface. To select a socket:

1 Measure the distance between two opposite flats on the head of the bolt or nut.
2 Select a socket of the same size.
3 Try the fit of the socket on the bolt or nut. It must be a snug fit.

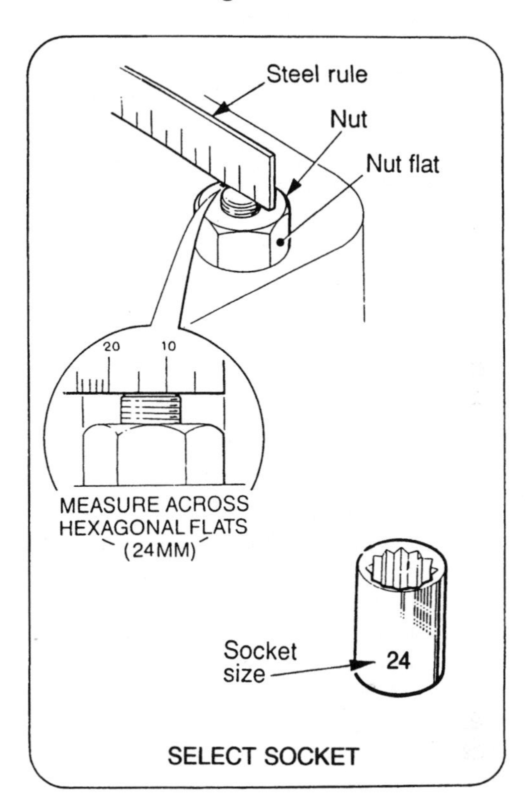

SELECT SOCKET

Note: The space above the head of a bolt or nut will determine whether a socket or a spanner can be used.

Spanners fit onto the heads of bolts or nuts. Their size and thread type are marked on their shanks. To select a flat or ring spanner:

1 Measure the distance between two opposite flats on the head of the bolt or nut.
2 Select a spanner of the same size.
3 Try the fit of the spanner on the bolt or nut. It must be a snug fit.

Note: The space above and around the bolt or nut will determine whether a ring or flat spanner can be used.

The speed of screwing a bolt or nut in until it is tight, or out after is has been initially

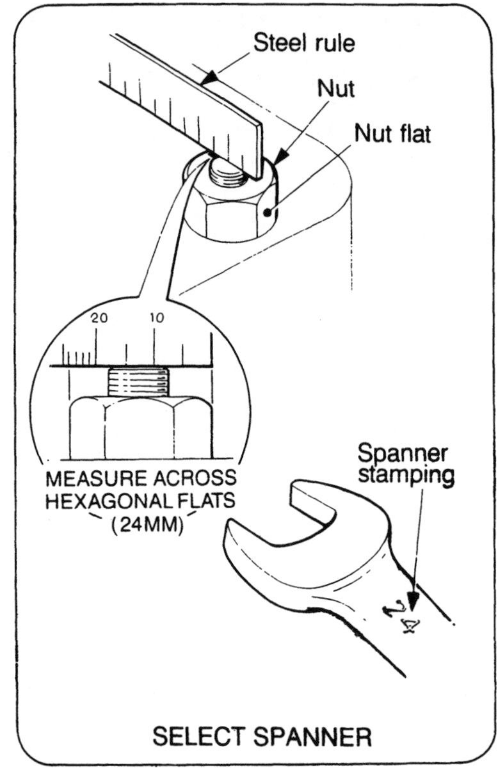

SELECT SPANNER

Source: AGPS

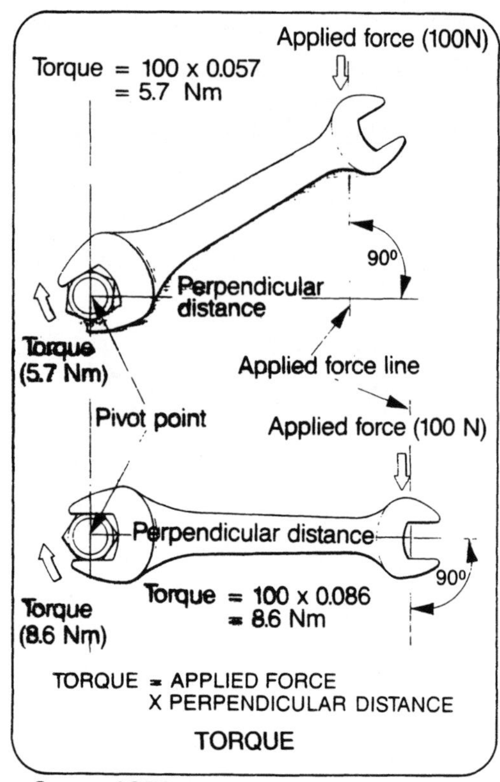

TORQUE

Source: AGPS

loosened, is influenced by the space around and above it. The order of selection is:

- socket fitted to an air ratchet;
- socket fitted to a speed brace;
- socket fitted to a ratchet;
- socket fitted to a handle or breaker bar;
- ring spanner;
- flat spanner.

Using hand tools

After selecting the type and size of a hand tool, its length should be considered, as this reduces the effort (pull) needed to loosen or tighten a bolt or nut. Where possible, the longest spanner or handle should be used as it allows more torque to be applied. A short spanner requires more effort to achieve the same torque. Torque is the turning or twisting effort applied to a bolt or nut when it is being loosened or tightened. Torque (newton metres) is calculated by multiplying the effort (force) in newtons by the length in metres of the spanner. The effort (pull) should be applied 90° to a spanner's shank or a handle's length away from the bolt or nut and it must increase evenly.

Note: More control of a hand tool is gained by pulling it than pushing on it.

The length of a screwdriver shank has an effect on loosening or tightening a screw. A long shank screwdriver is better than a short shank screwdriver.

Selecting mechanical measuring instruments

Factors governing the selection of mechanical measuring instruments are related to the accuracy required and the dimension to be measured. The order of accuracy is:

1. micrometers — to a one-hundredth of a millimetre;
2. dial indicators — to a one-hundredth of a millimetre;
3. vernier calipers — to a one-fiftieth of a millimetre;
4. steel rule — to a half of a millimetre;
5. steel tape — to 1 mm.

The dimensions that can be measured are:

- lengths;
- widths;
- heights;
- inside diameters;

- outside diameters;
- depths;
- radial and axial run-out.

A vernier caliper is used to measure depths and a dial indicator is used to measure radial and axial run-out.

Using mechanical measuring instruments

The skills of using measuring instruments not only relate to handling and adjusting them, but to reading their scales. One method is to first become familiar with the scales on an instrument and then practise using it to measure a dimension. The following guidelines will assist in developing these skills.

OUTSIDE MICROMETER

To use and read an outside micrometer, the equipment needed is:

- pencil;
- sheet of paper;
- 0 to 25 mm outside micrometer;
- piece of rod or a bolts approximately 12 mm in diameter;
- diagram of the parts of a micrometer. See below.

To become familiar with its scale:

1 Slowly turn the ratchet so that the faces move together until a clicking sound is heard.
2 Note that the zero on the thimble is aligned with the zero on the sleeve.
3 Turn the ratchet in the opposite direction until the thimble has rotated exactly one turn. The zero on the thimble is aligned with the centre line of the sleeve's scale.
4 Observe the gap between the faces. It is exactly half a millimetre and is written as 0.50 mm.
5 Write the following on the piece of paper. 'One turn of the thimble moves the face on the spindle exactly half a millimetre (0.50 mm)'.

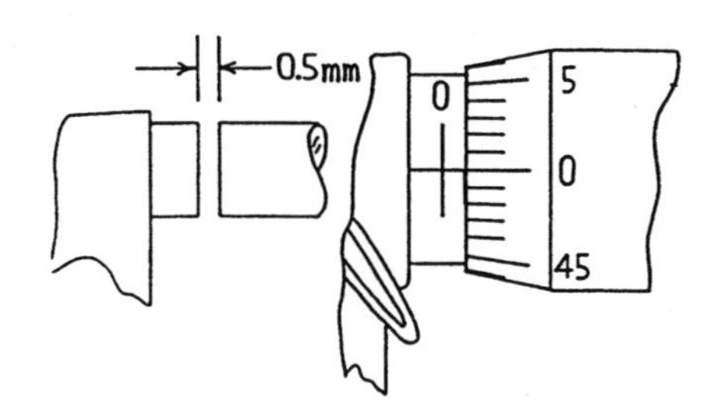

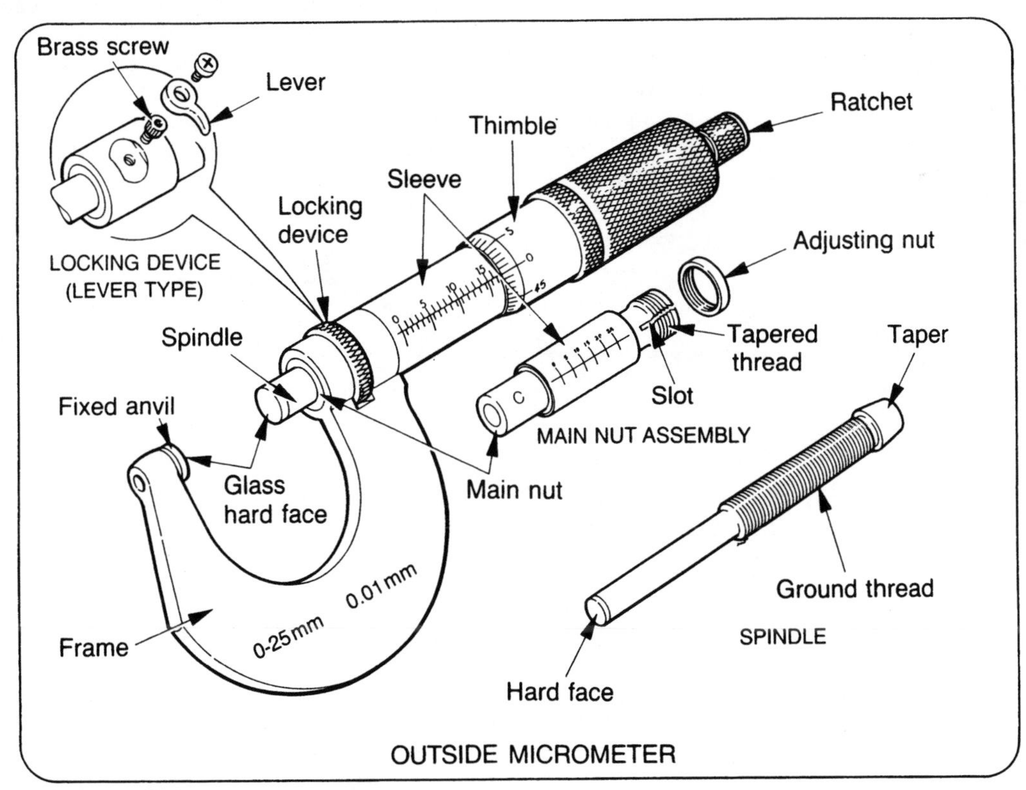

OUTSIDE MICROMETER

Source: AGPS

6 Practice steps 1 to 4 and read step 5, several times.
7 Rotate the ratchet until the thimble has completed another turn.
8 Observe the gap between the faces. It is exactly one millimetre and it is written as 1.00 mm.
9 Write the following on the piece of paper. 'Two turns of the thimble moves the face on the spindle exactly one millimetre (1.00 mm)'.

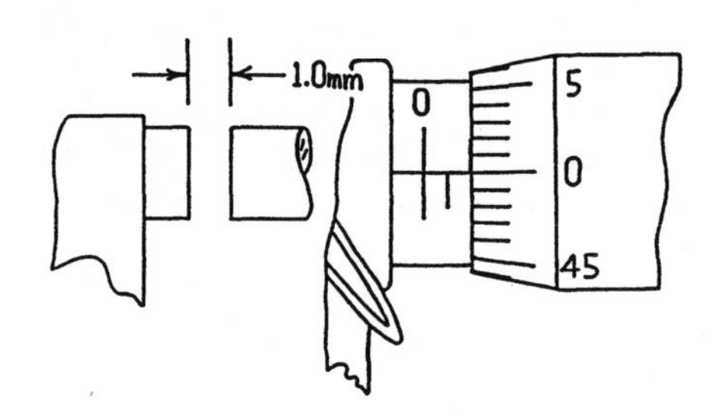

10 Practice steps 1, 2, 3, 7, 8, and read step 9, several times.
11 Rotate the ratchet so that the thimble turns about twenty to thirty times.
12 Observe the scale that is visible on the sleeve.
13 Note that the small marks (graduations) above the centre line are exactly one millimetre (1.00 mm) apart.
14 Note that the small marks below the centre line divide the distance between two marks above the centre line exactly in half. Each lower mark is exactly half a millimetre (0.50 mm) from one of the upper marks.
15 Observe the marks on the end of the thimble. There are fifty of them evenly spaced around the thimble's end.
16 Turn the ratchet until the zero on the thimble aligns with the centre line on the sleeve.
17 Turn the thimble very slowly with the ratchet until its next mark aligns with the centre line.
18 Note that the thimble has moved only one fiftieth of a turn. This means that the thimble and face have moved one fiftieth of a half a millimetre, which is a one hundredth of a millimetre (0.5 divided by 50 equals 0.01 mm.)
19 Summarise the steps by writing the following on the paper.
'One thimble turn equals half a millimetre (0.50 mm)'.
'Two thimble turns equals one millimetre (1.00 mm)'.
'One fiftieth of a thimble turn equals one hundredth of a millimetre (0.01 mm)'.

To read its scale:

1 Turn the thimble until there is a large gap between the faces.
2 Count the number of marks above the centre line that have been uncovered by the end of the thimble (see below).
3 Write the number on the paper. For example, when twelve marks are uncovered, write 12.00 mm.
4 Look at the marks below the centre line. When one of them is closer to the end of the thimble than the mark above the line,

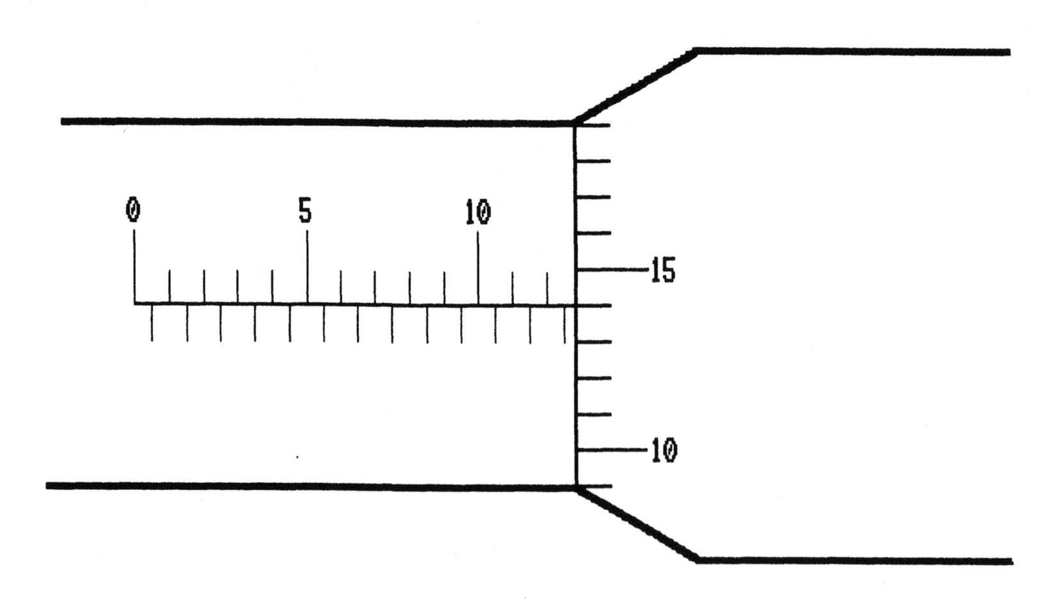

write 0.50 mm directly below the first number.
For example 12.00 mm
0.50 mm

5 Look at the marks on the end of the thimble.
6 Select the thimble mark nearest the centre line and note its value. It may the fourteenth mark, which is 0.14 mm.
7 Write its value directly below the number on the paper.
For example, when the fourteenth mark is nearest the centre line then write 0.14 mm. So,
12.00 mm
0.50 mm
0.14 mm
8 Add the numbers to obtain the reading.
For the above example the reading is:
12.00 mm
0.50 mm
0.14 mm
12.64 mm
9 Repeat the above steps several times.

To use a micrometer to measure the diameter of the piece of rod:

1 Turn the ratchet on the micrometer until the gap between the faces will allow the piece of rod to be inserted.
2 Hold the rod firmly with the left hand.
3 Grip the micrometer with the right hand as shown in the diagram.
4 Turn the ratchet in the direction that moves the face towards the rod.
5 Continue to turn the ratchet until a clicking sound is heard.
6 Slowly pull the micrometer away from the rod. A slight drag should be felt.

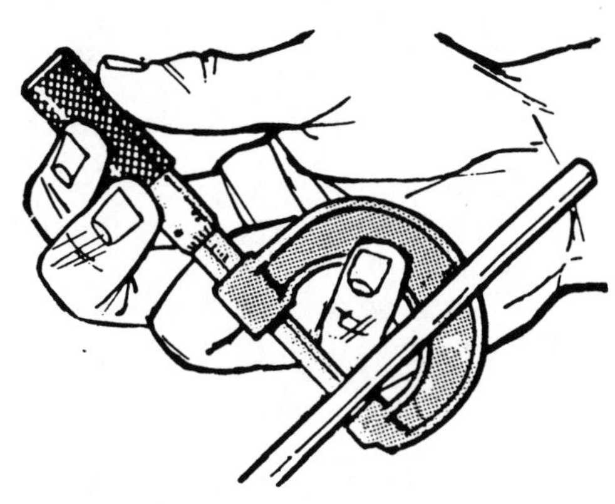

Holding and Operating a Micrometer

Note: Do not force the micrometer when it is set too tightly. Turn the thimble slightly backwards.

7 Lock the thimble and record the reading.
8 Repeat all the steps several times but in different places along the rod.

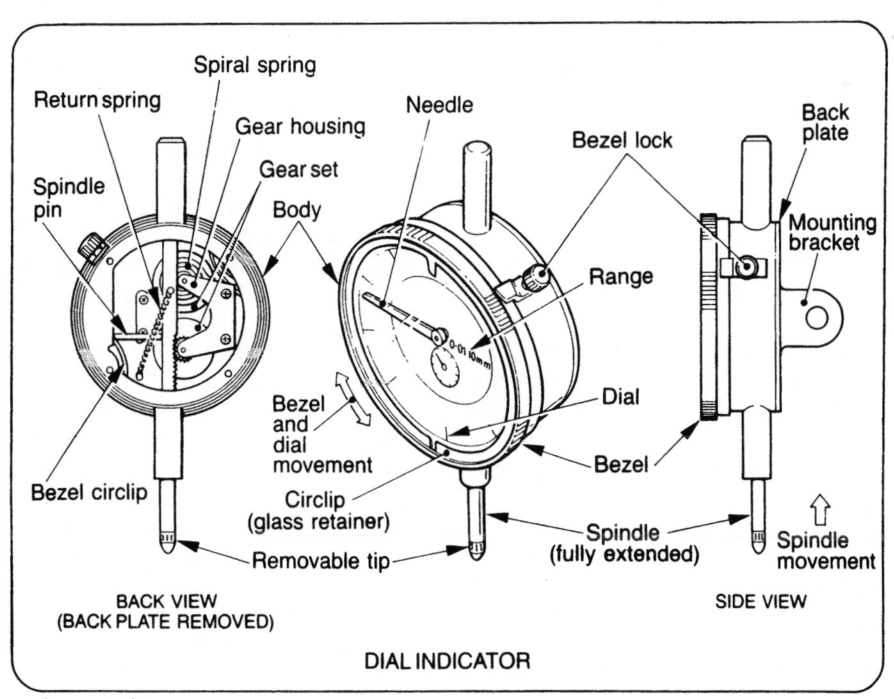

DIAL INDICATOR

Source: AGPS

DIAL INDICATOR

To use and read a dial indicator, the equipment needed is:

- pencil;
- sheet of paper;
- dial indicator;
- camshaft and two suitable Vee blocks;
- diagram of the parts of a dial indicator. See previous illustration.

To become familiar with a dial indicator scale:

1 Observe the face and, with the aid of the diagram, locate its small inner scale, outer scale, bezel and bezel lock.
2 Using the left hand, hold the dial indicator with its face up.
3 Note exactly where the small needle is pointing on the small inner scale.
4 Unlock the bezel.

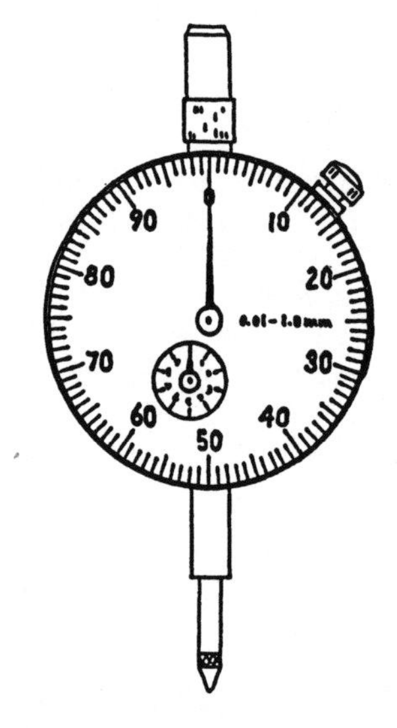

5 Adjust the outer scale so that its zero (0) is below the large needle.
6 Lock the bezel.
7 Using the right thumb, slowly depress the spindle tip until the large needle has completed exactly one turn (once around the outer scale).
8 Note the new position of the small needle. It has moved to the next mark on the small inner scale.
9 Write the following on a piece of paper. 'One complete turn of the large needle is the same as the small needle moving to the next mark.'
'One millimetre (1.00 mm) movement of the spindle causes the large needle to rotate one complete turn.'

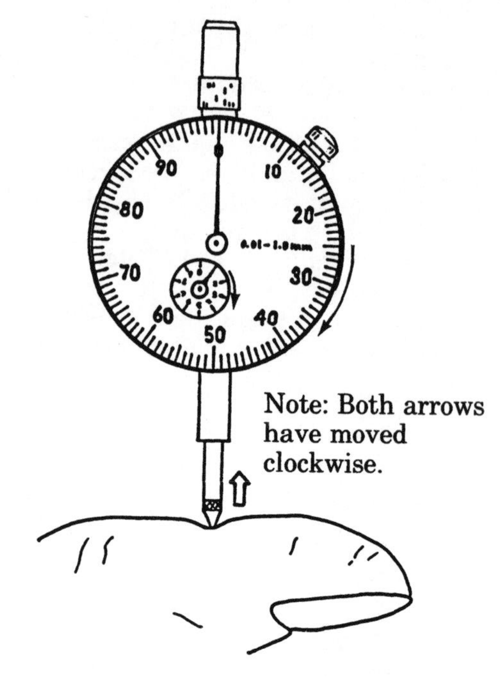

10 Note that the outer scale is divided into one hundred equal parts. This means that each mark (graduation) is a one hundredth of a millimetre (0.01 mm) from the next mark. For example, when the large needle is above the seventh mark from zero (0.00 mm), the reading would be 0.07 mm or seven one-hundredths of a millimetre.

Note: Some dial indicators do not have a small inner scale and needle.

To read a dial indicator scale:

1 Note the reading on the small inner scale or count the number of times the large needle passes zero (0.00 mm).
2 Record the number on a piece of paper. For example, when the small needle has moved two marks away from the initial setting, write 2.00 mm.

Note: When the small needle does not pass a mark or the large needle does not pass zero, the reading is less than one millimetre (1.00 mm).

3 Note the position of the large needle on the outer scale.
4 Record the reading on the piece of paper directly beneath the first number. For example, when the large needle is above the forty-seventh mark and the small needle has moved past two marks, the record would be:

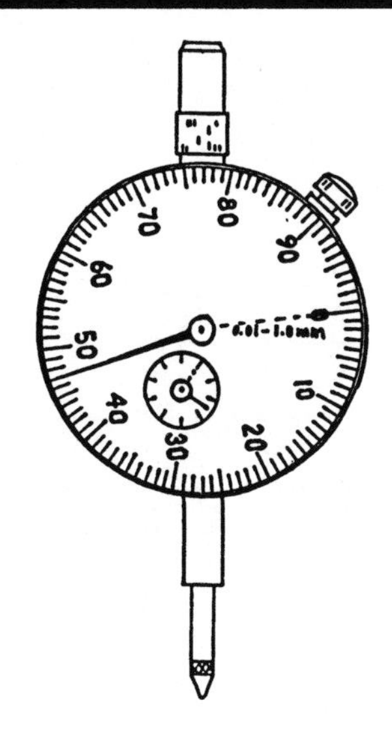

2.00 mm
0.47 mm

Note: Some dial indicators are marked from zero to fifty and then back to zero on their outer scales.

5 Add the numbers to obtain the reading. For the above example the reading is:
2.00 mm
0.47 mm
2.47 mm

To use a dial indicator to measure the radial run-out of a camshaft bearing:

1 Place the Vee blocks onto a flat metal surface (surface plate).
2 Oil two strips of paper and locate them in the Vees of the blocks.
3 Carefully, position the camshaft on the Vee blocks so that it is supported by each end bearing.
4 Place the magnetic stand onto the flat surface near the centre camshaft bearing.
5 Adjust the arm lengths so that the dial indicator spindle is:
 a directly above the bearing;
 b at 90° to the camshaft's axis;
 c slightly depressed.
6 Slowly turn the camshaft to a point where the large needle stops moving to the left (anti-clockwise).
7 Zero the large scale.
8 Slowly turn the camshaft to a point where the large needle stops moving to the right (clockwise).
9 Record the reading on a piece of paper.
10 Repeat steps 4 to 9 several times.

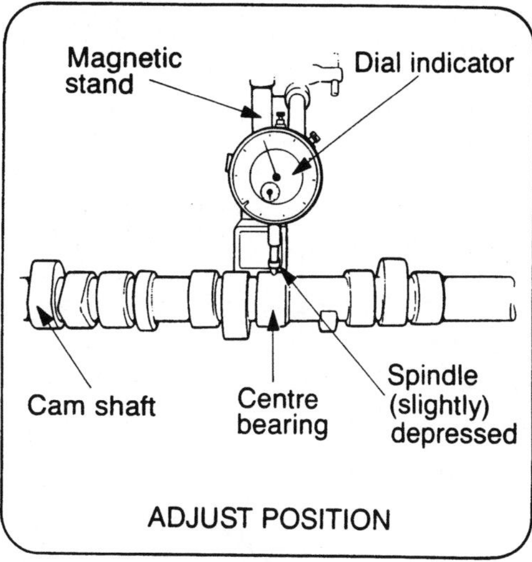

ADJUST POSITION

Source: AGPS

Note: To measure the axial run-out of a flange, the dial indicator spindle is positioned parallel to the parts axis.

VERNIER CALIPER

To use and read a vernier caliper, the equipment needed is:

- pencil;

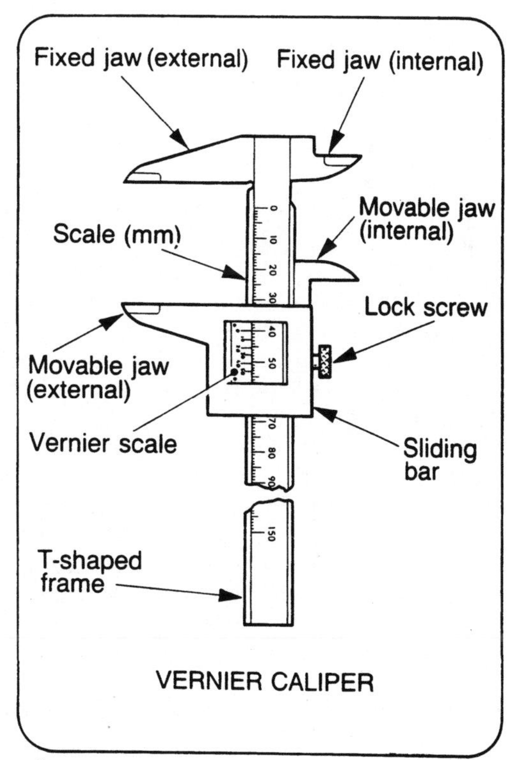

VERNIER CALIPER

Source: AGPS

- sheet of paper;
- vernier caliper;
- short piece of exhaust pipe;
- diagram of the parts of a vernier caliper. See previous illustration.

To become familiar with a vernier scale:

1. Move the sliding bar towards the fixed end until the movable jaw contacts the fixed jaw.
2. Observe and note that the zero (0) on the vernier scale is aligned with the zero on the fixed scale.

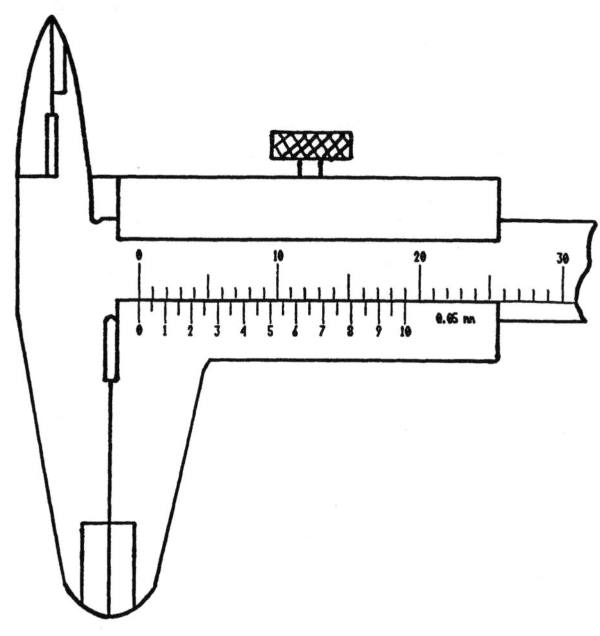

3. Locate the accuracy marking on the vernier scale. For example, when it is marked 0.05, the accuracy or the difference between each mark (graduation) is 0.05 mm (five one-hundredths of a millimetre).
4. Observe the fixed scale and note that the marks are one millimetre (1.00 mm) apart.

To read a vernier scale:

1. Move the sliding bar away from the fixed end about 50 mm.
2. Observe the fixed scale and record the number of millimetres between its zero and the zero on the vernier scale. For example, when the zero on the vernier scale is between forty-eight (48) and forty-nine (49), write 48.00 mm on the piece of paper.
3. Carefully observe the vernier scale and locate the mark that is exactly aligned with a mark on the fixed scale.

Note: Some practice is needed to exactly locate the aligned marks.

4. Note the closest number on the vernier scale between this mark and the zero.

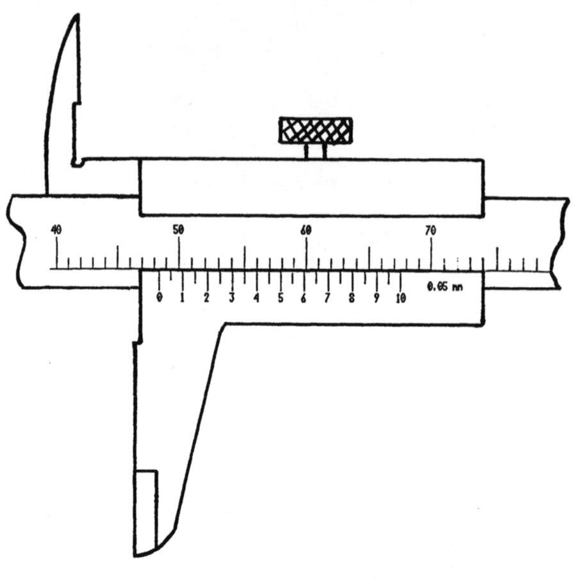

5. Write this number below the first number on the piece of paper. For example, when the number is three, write 0.30 mm below 48.00 mm, so the record would be:
   48.00 mm
   0.30 mm
6. Count the marks between this number and the aligned mark. Include the aligned mark.
7. Multiply the vernier scale accuracy by the number of marks. For example, when the number is 1 and the accuracy is 0.05, the result is 0.05 mm (0.05 × 1 in this case).
8. Write it below the last number on the piece of paper and add the numbers. For the above example the reading is:
   48.00 mm
   0.30 mm
   0.05 mm
   48.35 mm

To use a vernier caliper to measure the outside diameter of a piece of exhaust pipe:

1. Open the jaws on the vernier caliper so that the gap is larger than the outside diameter of the exhaust pipe.
2. Place the exhaust pipe between the jaws and rest it against the fixed jaw.
3. Slide the movable jaw towards the exhaust pipe until it just contacts.
4. Lock the slide (when a lock is fitted).

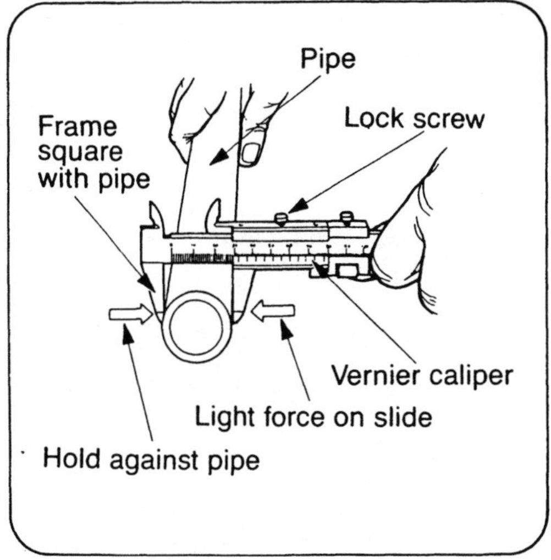

Source: AGPS

5 Carefully, pull the exhaust pipe out of the jaws.
6 Read and record the measurement.
7 Repeat all the steps several times at other points on the exhaust pipe.

Note: To measure the inside diameter, use the other jaws.

OUTSIDE CALIPERS

To use and read an outside caliper, the equipment needed is:

- pencil;
- sheet of paper;
- outside caliper;
- steel rule;
- short piece of exhaust pipe;
- diagram of the parts of an outside caliper. See above right.

To use an outside caliper and a steel rule to measure the outside diameter of a piece of exhaust pipe:

1 Open the legs of the caliper so that the exhaust pipe will pass between their tips.
2 Place the exhaust pipe between the tips of the legs.
3 Bring one of the tips into contact with the pipe.
4 Using the adjusting nut, move the other tip until it contacts the side of the pipe directly opposite the other tip.
5 Slide the tip back and forth across the pipe and adjust the leg position so that a slight drag is felt.
6 Carefully pull the pipe away from the caliper.

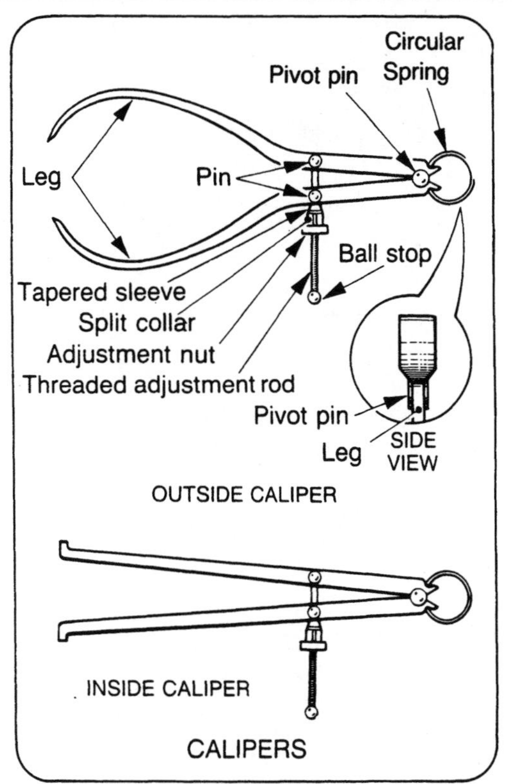

Source: AGPS

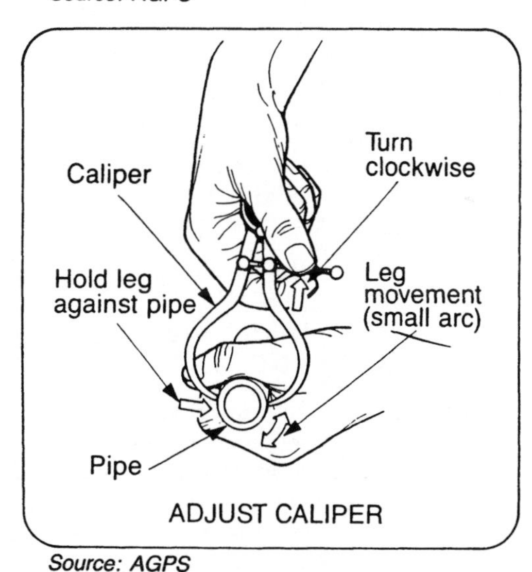

Source: AGPS

To read an outside caliper:

1 Place the tip of one leg against the end of the steel rule.

Caution: Ensure that the end of the steel rule has not been damaged or shortened.

2 Move the other tip over the scale.
3 Select the nearest millimetre to the tip.
4 Record the reading on a piece of paper.

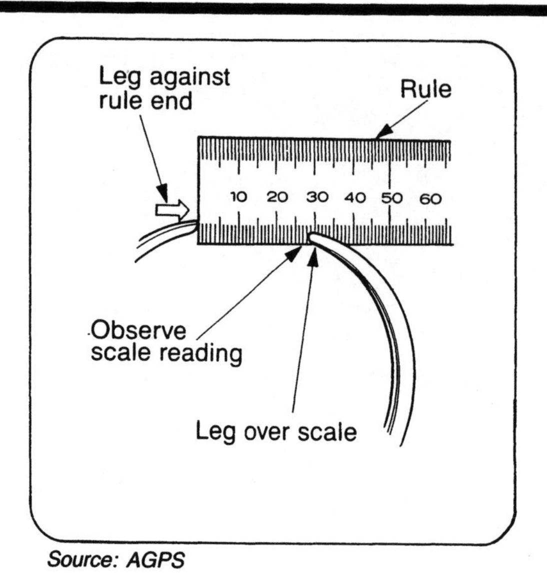

Source: AGPS

Note: To increase the accuracy of the measurement, use a steel rule with half-millimetre graduations.

Inside calipers are used in a similar manner to outside calipers.

Using workshop equipment

The best approach to using workshop equipment is to:

- carefully read the operator's manual;
- operate it according to the manual;
- observe all safety precautions;
- regularly maintain and check the equipment.

This section will concentrate on the safety aspects and the maintenance of the commonly used equipment in a workshop.

MECHANICAL EQUIPMENT

Equipment	Safety precautions	Maintenance
Bead breaker	Fully deflate the tyre. Clamp the tyre firmly to the table. Lubricate beads, lever and rim with soap.	Remove all dirt and dust from table, ram and threads.
Block and tackle	Work within the stated loads. Ensure it is hooked correctly. Ensure that chains are straight.	Inspect hooks for stress or cracks. Keep chains clean. Lubricate at points indicated.
Creeper	Make sure all wheels turn freely. Repair loose or broken metal work.	Clean and lubricate all the wheels. Clean the head rest and the platform.
Fire extinguisher	Display all signs. Use the extinguisher for the fire listed on label. Ensure that extinguishers are in a suitable location.	Check and service them on a regular basis.
Hydraulic press	Fully insert table pins. Secure pressing plates. Work within the rated limits. Shield the parts being pressed apart. Keep the pressing plates as close together as possible.	Check cables and pins regularly. Clean table and frame. Repair fluid leaks.
Oxygen and acetylene welder	Always use the correct gauge pressures. Make sure that the bottles' necks are above their bases. Wear welding goggles. Keep bottles away from hot areas. Never oil the threads on the gauges or hoses.	Keep the tips clean. Check for damage to hoses and clean them regularly. Check for leaks at each fitting.

Safety stand	Use on a level floor. Place under correct points on the vehicle.	Check locking device. Inspect pad and base for cracks.
Trolley jack	Position the jack under correct lifting point. Use on flat surface. Never use a 'creeping' jack. Always remove the jack after the vehicle has been lowered onto safety stands.	Lubricate the wheels once a month. Check the saddle for damage or looseness. Check the hydraulic system for leaks, and repair when necessary.

ELECTRICAL EQUIPMENT

Equipment	**Safety precautions**	**Maintenance**
Air compressor unit	Wear safety glasses. Never blow compressed air directly onto your skin. Never spin a bearing with compressed air.	Drain water from tank or receiver each week. Check hoses and fittings monthly. Check and add oil to compressor.
Battery charger	Use in a ventilated area. Remove battery caps and cover with a damp cloth. Never disconnect the leads with the charger switched on as the battery will explode.	Clean spilt acid from all surfaces. Check leads and clips regularly.
Bench grinder	Wear safety glasses. Never use a grinder that vibrates because the grinding wheel may break. Never stall the wheel with excess pressure.	Dress the wheel when it becomes grooved. Replace the wheel when it becomes too small. Reset the rest after dressing the wheel.
Electric welder	Use the correct welding shield or mask. Wear protective clothing. Select the correct rods and current settings. Use in ventilated area.	Keep cables clean. Repair damaged cables. Replace faulty clamps.
Hoist	Position the vehicle correctly. Adjust arms to contact the lifting points on the vehicle. Always check arm position once the vehicle leaves the floor. Always engage the safety lock. Never lift a vehicle that is over the safe working load. Never use a hoist that 'creeps' or 'drops'.	Check cables, chains, gears or screws monthly. Keep pad and arms clean. Check safety lock monthly.
Steam cleaner	Wear safety goggles and protective clothing. Use in ventilated area. Keep steam pressure within safety limits.	Check hose regularly.
Wheel balancer	Make sure wheel is firmly clamped to spindle. Make sure shield is correctly closed before starting the balancer.	Keep the balancer clean and remove used wheel weights from the trays.

CHAPTER 3 REVISION

1 On a separate sheet of paper, briefly describe the following hand tools and state an application for each.
   a Wire brush
   b Half-round file
   c Hacksaw
   d Allen key
   e Centre punch
   f Phillips screwdriver
   g Breaker bar
   h Impact socket
   i Tap
   j Combination spanner

2 On a separate sheet of paper, briefly describe the following special hand tools and state their application.
   a Battery terminal cleaner
   b Brake adjuster
   c Impact wrench
   d Engine sling
   e Oil filter remover
   f Pipe cutter
   g Ridge remover
   h Valve seat cutter
   i Wheel cylinder clamp
   j Piston ring compressor

3 On a separate sheet of paper, briefly describe the following mechanical measuring instruments and state an application for each.
   a Ball gauge
   b Dial indicator
   c Feeler gauge
   d External micrometer
   e Outside caliper
   f Steel rule
   g Spark-plug gauge
   h Tension wrench
   i Vee block

4 On a separate sheet of paper, briefly describe the following electrical measuring instruments and state their application.
   a Ammeter
   b Battery discharge tester
   c Hydrometer
   d Test light
   e Ohmmeter
   f Multi-meter
   g Timing light
   h Voltmeter

5 On a separate sheet of paper, briefly describe the following electrical equipment and state their application.
   a Battery charger
   b Electric welder
   c Hoist
   d Steam cleaner
   e Wheel balancer

6 On a separate sheet of paper, briefly describe the following mechanical equipment and state their application.
   a Bead breaker
   b Fire extinguisher
   c Hydraulic press
   d Safety stand
   e Trolley jack

7 What are the four reasons for planning a job?

8 List the equipment that would be needed to allow a wheel to be removed from a vehicle.

9 Briefy explain when the following hand tools should be used.
   a Socket
   b Ring spanner
   c Flat spanner
   d Adjustable wrench

10 Name the most accurate mechanical measuring instrument and list the dimensions that it can be used to measure.

11 What are the readings on the external micrometer scales?

a

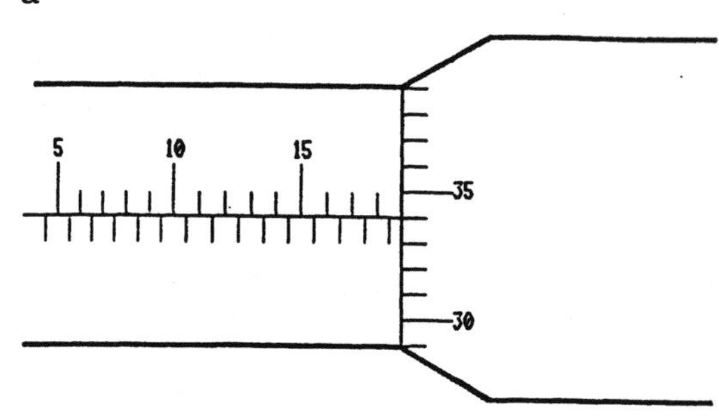

b

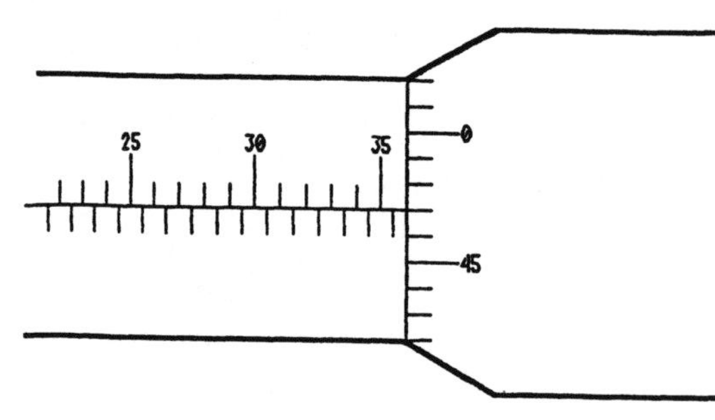

c

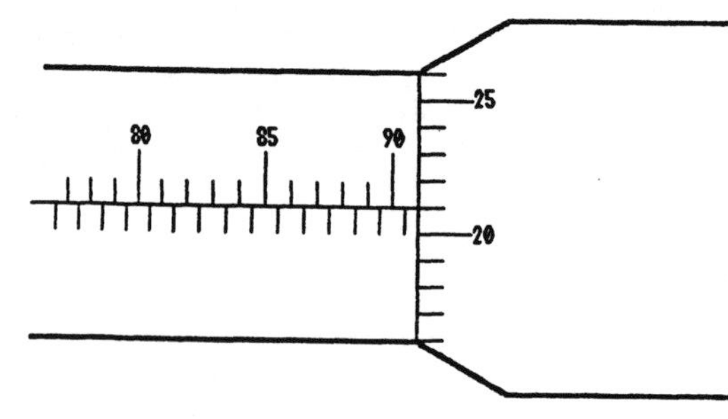

d

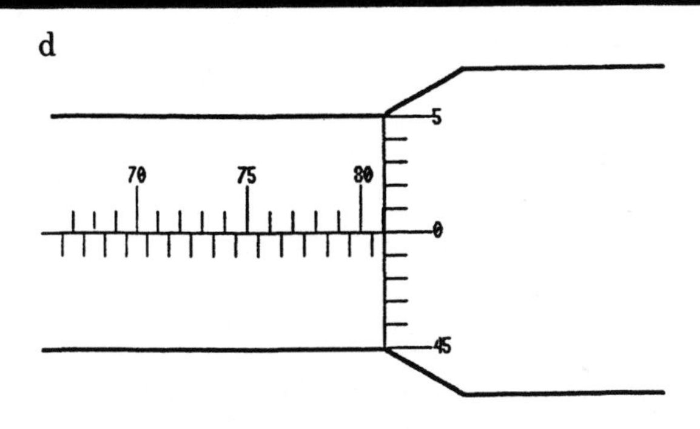

12 What are the readings on the dial indicator scales?
Note: Dotted lines are the starting points on both scales.

a

b

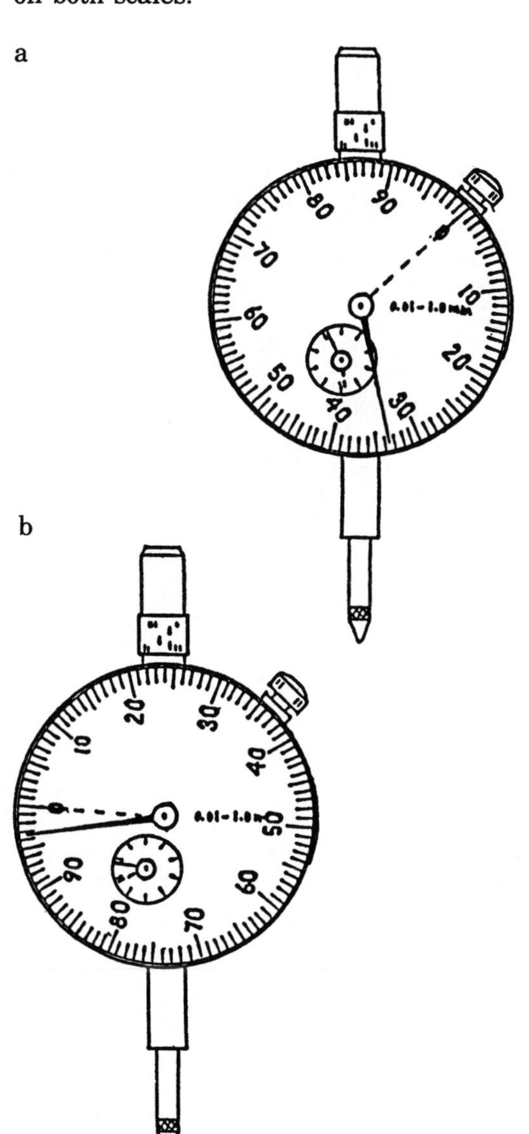

c

d

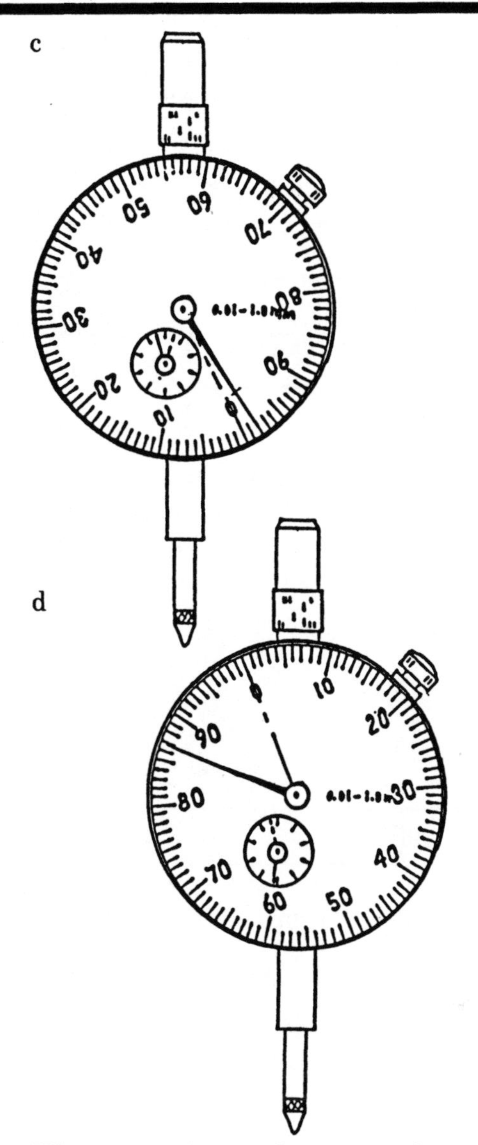

13 What are the readings on the vernier caliper scales?

a

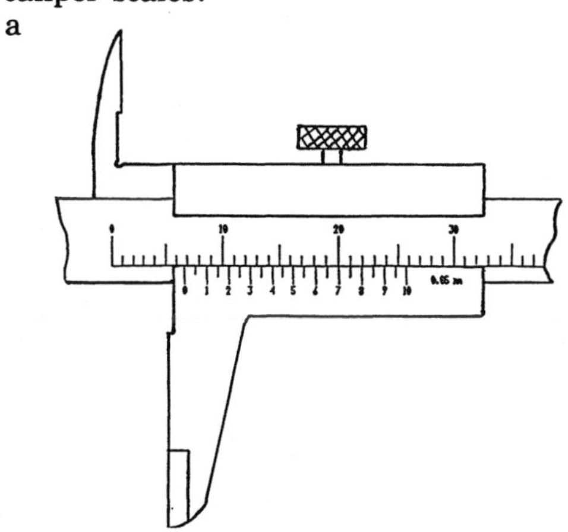

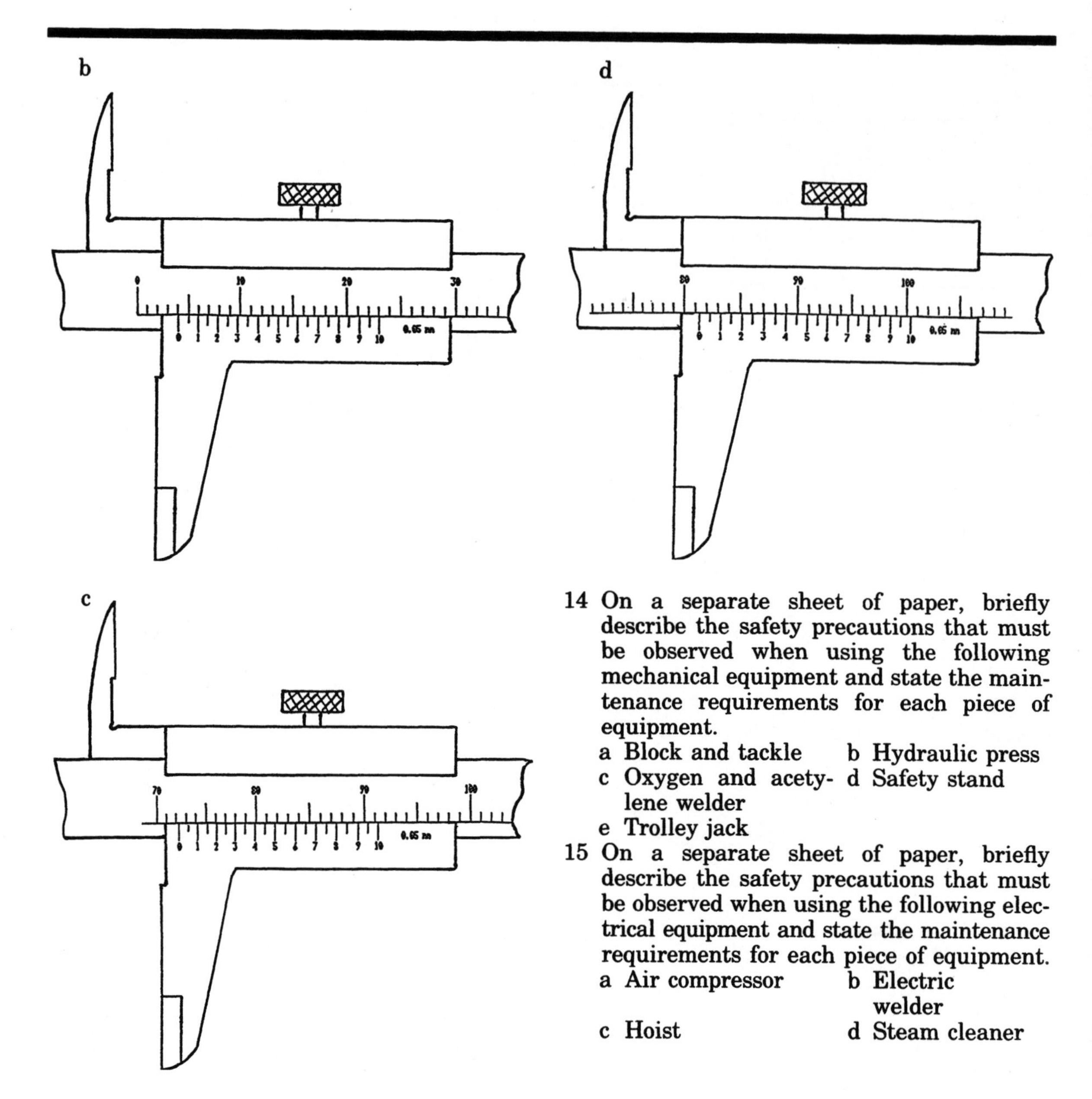

14 On a separate sheet of paper, briefly describe the safety precautions that must be observed when using the following mechanical equipment and state the maintenance requirements for each piece of equipment.

a Block and tackle
b Hydraulic press
c Oxygen and acetylene welder
d Safety stand
e Trolley jack

15 On a separate sheet of paper, briefly describe the safety precautions that must be observed when using the following electrical equipment and state the maintenance requirements for each piece of equipment.

a Air compressor
b Electric welder
c Hoist
d Steam cleaner

4

COMPONENT RECLAMATION

Most of the modern mechanic's work is involved with diagnosis, replacement of defective components and routine maintenance of vehicles. He or she is no longer involved in the making of components nor the detailed repair of major assemblies. Work of this type is normally sub-let to specialist repair machine shops. The result of this trend has been the loss of skills in the mechanic's trade, with less emphasis on fitting and machining skills.

Despite this change in work practices there remains a core of fitting tasks which the mechanic must be capable of performing:

- identification and repair of threads. This skill has become more important with the increasing use of metric threads.
- removal of broken studs or bolts.
- repair of brake or fuel lines, flares and flare nuts.
- resurfacing of manifold or carburettor flanges that have warped in service.
- maintenance of hand tools, hacksaw, centre punches, cold chisels, screwdrivers and drill sharpening.

This chapter will describe the various thread forms in common usage and provide the mechanic with sufficient information to undertake the tasks listed above.

NUTS, BOLTS AND STUDS

Thread types

The vehicle market is a world market, and most technically advanced nations manufacture vehicles for export. At the same time, the car-exporting nations also import cars from many other countries. One of the problems that emerges from this type of trading is compatibility of thread forms. In a similar way to the use of different languages and systems of measurement, vehicles were manufactured using different thread forms. Many of the thread forms are similar in appearance, and the mechanic must make sure the correct thread type has been selected when replacing nuts, bolts or studs. Ten or fifteen years ago it was possible to identify the threads within a vehicle based on country of manufacture, for example European—metric, U.K.—Whitworth or B.S.F., U.S.A. — U.N.F. or N.C., Japan — metric. This type of assumption is no longer valid, as a modern vehicle can be a 'world' car with engine, transmission, suspension, brake system and body panels being manufactured in different countries for final assembly as a badge-engineered product in a country which uses a different system. If the mechanic is not familiar with the product and is unaware of the thread type in use, he or she must identify the thread before 'attempted' replacement.

Terminology

Before ordering replacement bolts or nuts the mechanic should become familiar with the terminology used for descriptions.

- type — bolt, capscrew, machine screw etc.
- length — length is measured from under the head of the bolt to the tip
- major diameter — the major diameter of the threaded section.

- thread length — the length of the threaded portion only.
- pitch — pitch is the distance between a point on a thread and a corresponding point on the adjoining thread. Pitch is expressed in number of threads per inch, for example 20 t.p.i. for British or United States threads. However, the metric system prefers to use a single thread measurement, for example 1.5 mm.
- head size — the distance across the flats of the bolt head, for example 9/16″ or 14 mm. A variation to this measurement is used with B.S.F. and Whitworth, but the use of these threads is declining.
- tensile strength — the strength of a bolt will vary depending on material, method of manufacture and diameter. The tensile strength of a bolt is coded onto the bolt head and a replacement bolt must always be of an equal or superior grading. British and United States coding is by the use of a number of small lines arranged in a radial pattern on the bolt head. An increase in the number of marks indicates an increase in tensile strength.

  Metric system bolt strength is signified by a numeral stamped on the head of the bolt. An increase in the number value indicates an increase in tensile strength.

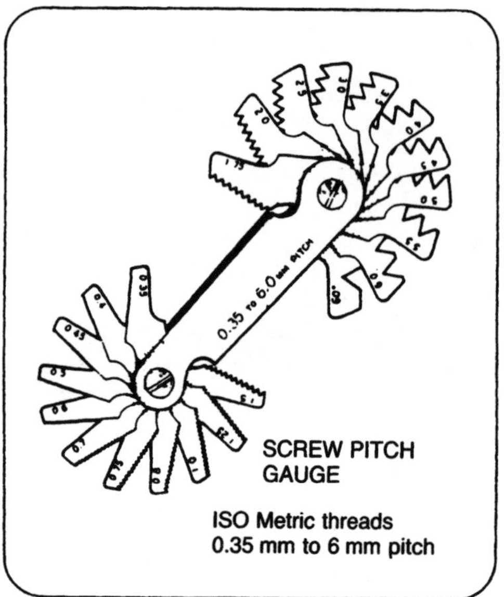

Source: AGPS

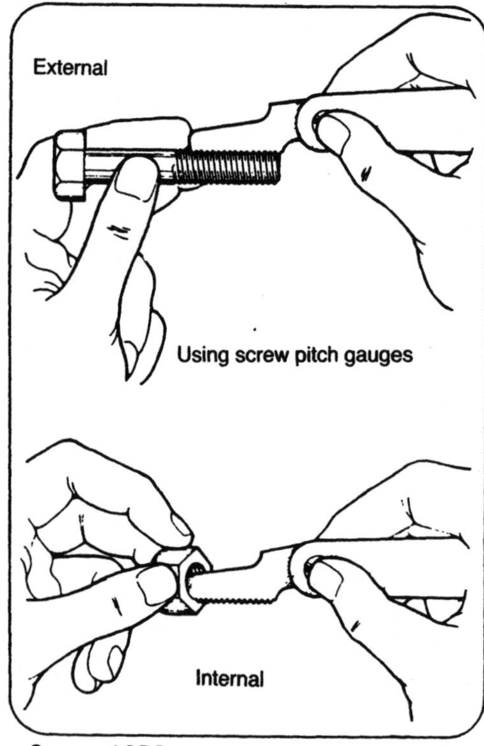

Source: AGPS

Bolt identification

The identification of a bolt requires several steps:

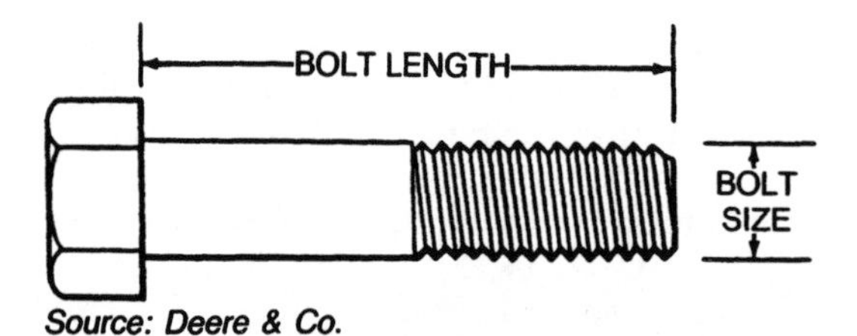

Source: Deere & Co.

1. Measure the length and major diameter (include the thread length in the description if the bolt is not threaded over the full length).
2. Use a thread pitch gauge to measure the pitch.
3. Select a gauge of the correct thread form and insert the teeth of the gauge into the thread of the bolt. Hold the gauge aligned to the bolt and ensure the thread and gauge teeth are an accurate fit. Read the number stamped on the gauge and this will provide a direct reading of the thread pitch (direct readings are for metric threads only).

Note: The numbers stamped on imperial thread gauges are expressed as threads per inch (t.p.i.). To identify the thread type it will be necessary to cross-reference the major diameter and pitch (t.p.i.) on a thread chart.

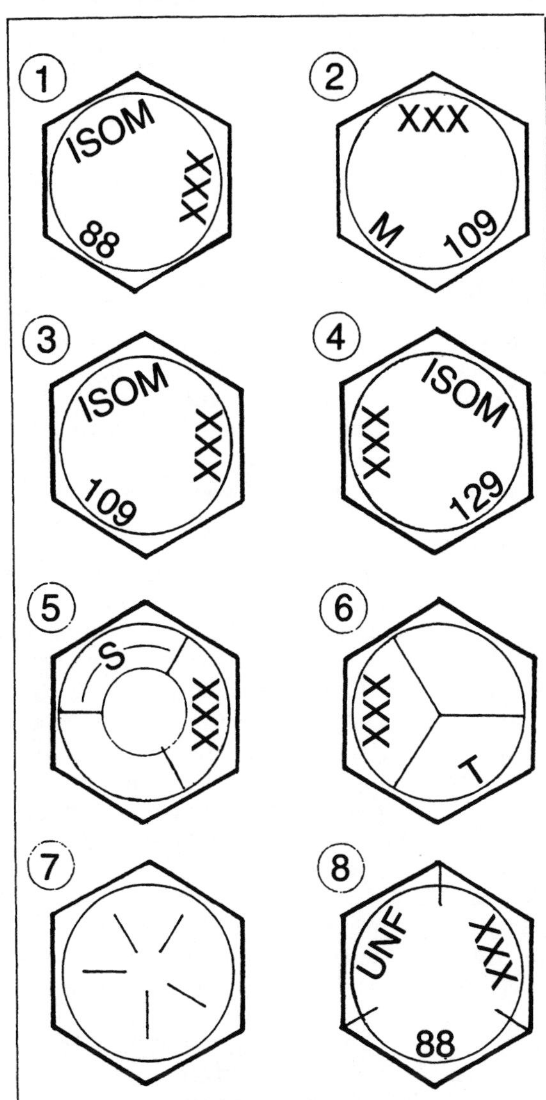

Typical bolt identification marks
(XXX denotes manufacturer's identification mark)

1. ISO metric bolt standard marking
2. ISO metric bolt alternative marking
3. ISO metric bolt strength grade 10.9 — coloured blue
4. ISO metric bolt standard grade 12.9 — coloured red
5. UNF/UNC bolt standard marking — strength grade may be S, T or V
6. UNF/UNC bolt alternative marking
7. UNF/UNC bolt American marking — number of lines denotes strength grade
8. UNF/UNC bolt European manufacturer's marking

There are three main types of bolt head — hexagon, square and countersunk. Hexagon heads are the most common, square heads are used basically for structural or heavy engineering work and countersunk heads — which are slotted to take a screwdriver — are used where it is important that the surface has a flush finish.

4 Determine the tensile strength by checking the coding on the bolt head.

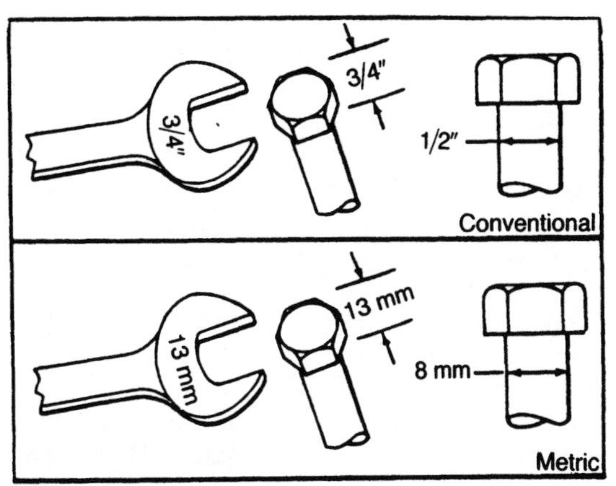

Source: Deere & Co.

5 Spanner size or 'across the flats' of the bolt head is usually standardised and need not be included unless the bolt is for a special application requiring a specific non-standard bolt head.

6 Identify the bolt fully by adding the thread length and overall length to the description. For example:
7/16″ U.N.C., 3″ length 2 1/2″ threaded, grade 5.
12 mm (dia.); 1.75 mm (pitch) x 25 mm (length), 8.8 grade (tensile strength).

Bolt tensioning

Vehicle manufacturers' service manuals recommend torque/tension specifications for the majority of nuts and bolts used in the construction of vehicles.

If bolts or nuts are not tensioned to the correct specifications then damage to components, loss of performance and/or a vehicle breakdown may occur. For example, when a cylinder head bolt is tensioned the mating threads on the bolt and the cylinder block are stretched and become a forced, tight fit between the threads. If there was insufficient tensioning then the bolt could vibrate loose and cause damage to the engine. By over tensioning, the bolt could be weakened by excess stretching and a possible breakage of the bolt may result.

Use of tension/torque wrenches

The tension/torque wrench is the tool most commonly used in automotive servicing to correctly check the tension of bolts and nuts. There are various designs of tension wrenches and all of them provide a method of measuring and checking the torque setting of a bolt and/or nut.

The normal measurement for determining the tension of a bolt or nut is Newton metres (Nm) (metric) or Pound Foot (lbf ft) (imperial).

A torque of 1 Nm is equivalent to having a spanner lever of 1 m length and a force of 1 N applied at the outer end of the lever. Similarly, the application of a 2 N force to the outer end of a 0.5 m long spanner would produce a torque of 1 Nm.

A torque of 1 lbf ft is equivalent to having a spanner lever of 1 ft length with a force of 1 lb applied to the outer end of the lever. Similarly, the application of a 0.5 lb force to the outer end of a 2 ft long spanner would produce a torque of 1 lbf ft.

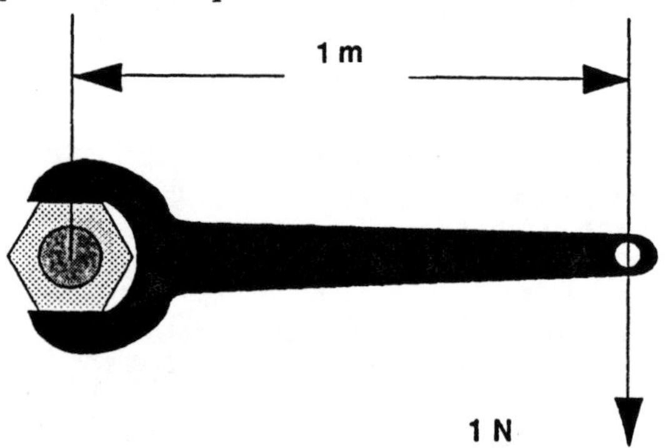

A force of 1 N applied in the direction of the arrow will produce a twisting effort (torque) of 1 Nm about the centre of the bolt.

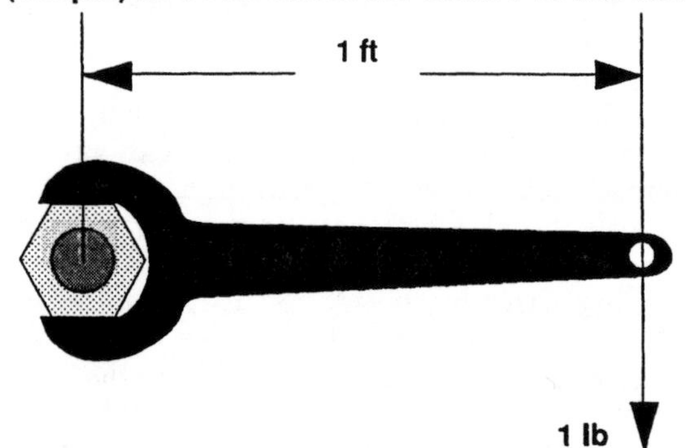

A force of 1 lb applied in the direction of the arrow will produce a twisting effort (torque) of 1 lb ft about the centre of the bolt.

CONVERSION

To convert pound foot to Newton metres multiply by 1.356.

—1 lbf ft = 1 multiplied by 1.356 = 1.356 Nm
—10 lbf ft = 10 multiplied by 1.356 = 13.56 Nm

To convert Newton metres to pound foot divide by 1.356

—1 Nm = 1 divided by 1.356 = 0.737 lbf ft
—10 Nm = 10 divided by 1.356 = 7.37 lbf ft

Bolt/stud thread damage: repair

Broken bolts/studs and stripped threads are common repairs which the mechanic will encounter. A understanding of the repair methods available and the precautions necessary will prevent a minor repair developing into a time-wasting exercise with the possibility of severe damage to the component under repair.

Many bolt breakages are caused by the application of incorrect torque to the bolt. In a similar way many 'stripped' thread problems would not have occurred if the bolt had been 'hand started' into the thread before applying spanner force. A short review of the use of torque settings as a preventive measure would be appropriate before understanding the methods of repairing the results of incorrectly torquing a bolt. The manufacturer will specify torque limits for each bolt application within the vehicle. These settings are specific to the particular application of the bolt/stud and include the effects on the components being attached and the type of gasket material in use.

Avoid warped housing, stripped threads and gasket failures — obtain and USE the recommended torque settings.

Stripped threads: repair

GENERAL PRECAUTIONS

To prevent metal chips falling into a closed end hole or into a component under repair it is advisable to coat the drill/tap with a sticky grease. The metal particles can be cleaned from the tap after the task is completed.

Correctly identify the thread tap/die nut before starting work. The use of an incorrect tap/die nut may completely destroy the damaged thread.

REPAIR

Partially damaged threads can often be restored by the use of a thread tap. Minor damage will 'clean up' as the tap progresses down the thread. If the thread is external, that

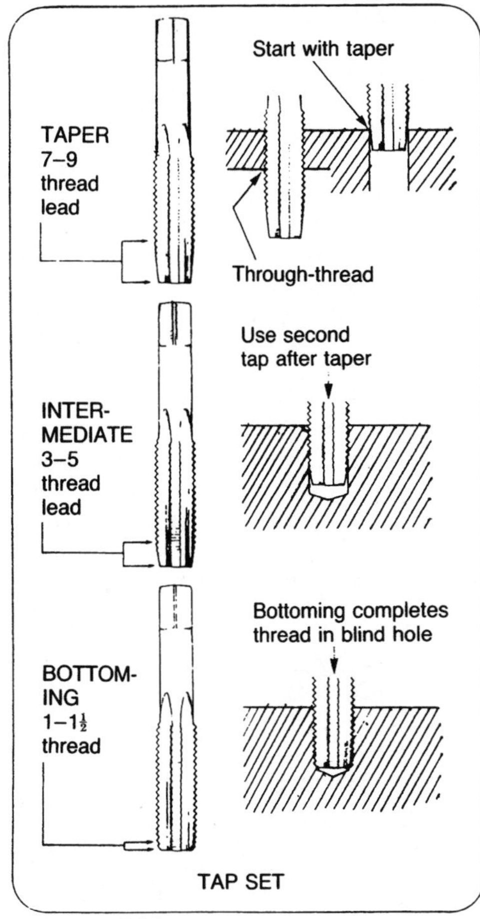

Source: AGPS

is, a bolt or stud, a die nut can be screwed down the thread to recover the damaged area. However, major thread damage requires a more extensive repair. Three of the repair techniques available are detailed below, but the method used for the repair will depend on the location and severity of the damage which has occurred.

Method A. Drill the hole oversize and tap to the next larger bolt size.

1 Select the oversize bolt available.
2 Select the correct tapping size drill from a chart.
3 Drill the hole with the tapping drill.
4 Cut the new thread in the drill hole with a taper tap.
5 Select a bottoming tap and repeat the thread cutting process.
6 Test the selected bolt in the new thread.

Caution: The hole in the component which is to be attached by this bolt must be drilled to a size which will allow clearance for the major bolt diameter.

Method B. Drill the hole oversize, fit a threaded plug, redrill and tap the plug to the original bolt size.

1 Select a suitable oversize threaded plug.
2 Select the correct tapping size drill to suit the plug from a chart.
3 Drill the hole with the tapping drill.
4 Cut a thread into the drilled hole with a taper tap. Clean thread with a bottoming tap if required.
5 Select a tapping size drill (to suit the original bolt size) from a chart.
6 Drill a tapping size hole in the threaded plug (if the plug is not pre-threaded).
7 Cut a thread in the plug with a taper tap. Repeat the thread cutting with a bottoming tap.
8 Fit a bolt and lock nut to the threaded plug.
9 Screw the plug into the component until the plug is firmly locked in place. (It may be necessary to coat the plug threads with a adhesive or 'stake' the plug with a chisel. Follow the plug manufacturer's advice).
10 Loosen the lock nut and remove the bolt.
11 Test the thread with an original size bolt.

Caution: The threaded plug must be flush to the surface of the component. File the surface flat if any protrusion exists.

STANDARD SCREW FITS IN HELI-COIL INSERT

HELI-COIL
TAPPED HOLE

Method C. 'Replace' the thread. Several companies market thread repair kits which consist of a threaded coil, a tapping drill and a tap (and a special thread loading tool if required). Each bolt size requires a kit to suit the thread size and type.

1 Drill the stripped thread from the hole with the drill supplied.
2 Tap the thread into the drilled hole with the special tap.
3 Insert the thread coil into the tapped hole with a loading tool.
4 Depending on the repair kit used, it may be necessary to 'break off' a thread loading tag and/or stake the thread coil to ensure retention of the thread in the tapped hole.
5 Test the thread with an original size bolt.

External thread cutting: repair

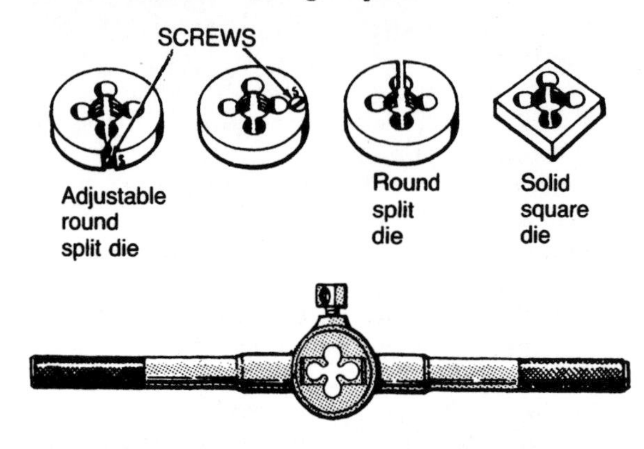

As described earlier in this chapter, minor thread damage can be repaired by the use of a die nut.

NEW THREADS

New threads can be cut onto round metal stock of suitable size.

1 Ensure the metal stock is equal in diameter to the major diameter of the required thread.
2 Select the die and fit the die into a holder.
3 Check the tapered teeth section of the die will be facing the metal stock when thread cutting starts. The internal diameter of the die is slightly larger at this end and the thread will be much easier to 'start' for cutting.
4 Lubricate the thread cutting area and commence cutting the thread. The die must be kept flat and square to the work or an 'out of square' thread start may occur, and the material may have to be scrapped.
5 Test the thread with a nut.

Broken bolts/studs: removal

Several methods are used to remove broken bolts and/or studs. The choice of technique will depend on location, material and the length of the broken part protruding (if any).

LARGE PORTION PROTRUDING

Grip the protruding section with vice grips and loosen the broken stud.

SMALL PORTION PROTRUDING

Two simple extraction methods are used when an adequate section of the broken stud is protruding from the hole.

1 File flats onto the protruding section and loosen the stud with a spanner.
2 Cut a screwdriver slot into the broken section with a hacksaw and loosen the stud with a screwdriver.

NO PART PROTRUDING

Three different repair techniques are available when the stud is broken below the surface of the component.

1 **Tap with pin punch.** Use a sharp pointed chisel or a pin punch and a hammer if the breakage is flush or slightly below the component surface. Tap the broken bolt gently in a counter-clockwise direction. A loose bolt will respond to this treatment and the broken section removal will be a relatively simple task.

Caution: Do not damage the threaded hole in this process. If the broken section does not move readily do not continue with this method.

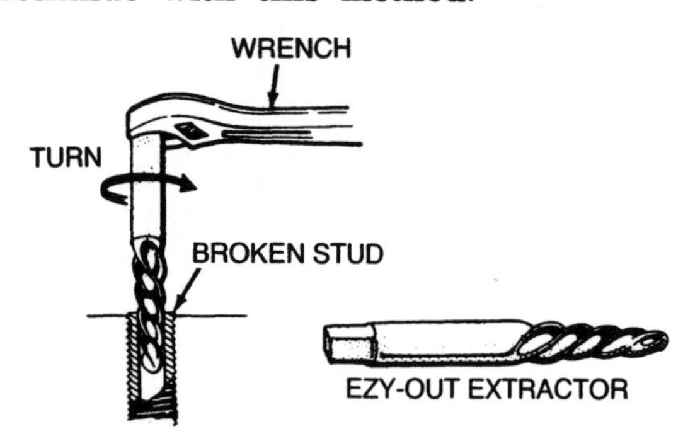

2 **Stud extractor.** Centre punch the EXACT centre of the broken section and drill a small diameter hole through the stud. Select a drill slightly smaller than the thread minor diameter and drill through the broken section using the small drill hole as a 'pilot' to stay on centre. Insert a stud extractor into the drilled hole and wind the extractor holder in a counter-clockwise direction to remove the broken bolt.

Caution: The extractor is hardened and undue force will cause it to break. If excessive force is required, do not continue. Remove the extractor by twisting the holder in a clockwise direction.

3 **Drill, chisel and tap.** Select a drill slightly smaller than the bolt, centre punch the EXACT centre of the broken section and drill through the bolt. If a stud extractor is not available it is sometimes possible to tap a diamond point chisel into the hole and remove the broken section by winding the chisel in a counter-clockwise direction. If

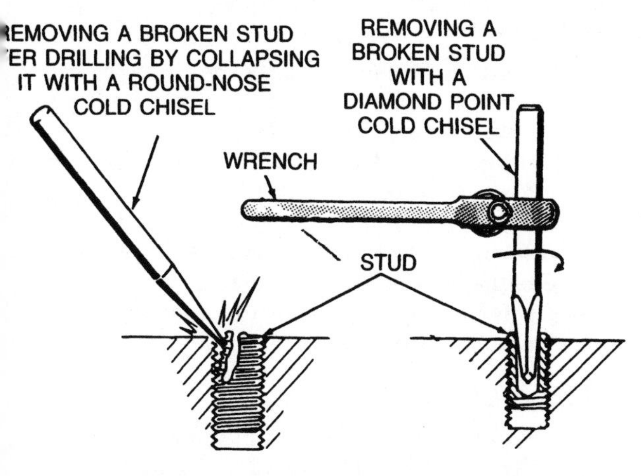

this method is not successful, collapse the shell of the bolt with a cold chisel and remove the damaged bolt.

Remove the remaining thread parts with a taper tap and 'clean' the threaded hole with a bottoming tap.

Caution: The tapping drill must be exactly on centre. If it is not, part of the threaded hole will be destroyed.

FASTENERS

In addition to the standard range of bolts and studs, the automobile contains many different types of fasteners. The range includes special types of bolts, self-tapping screws, metal thread screws, set screws, cotter pins, rivets and circlips. Each fastener is chosen by the manufacturer because of suitability for the application. The mechanic should always replace any damaged or missing fasteners with the same type.

LOCKING DEVICES

The primary purpose of a locking device is to prevent a fastener from loosening in service. The range of these devices includes several types of lock nuts and washers. Locking devices can also be lock wires, split pins or tags. Do not forget to fit locking devices when assembling components. A simple error of omitting a lock tag can result in very expensive damage, such as the total loss of an engine assembly caused by a connecting rod nut failure.

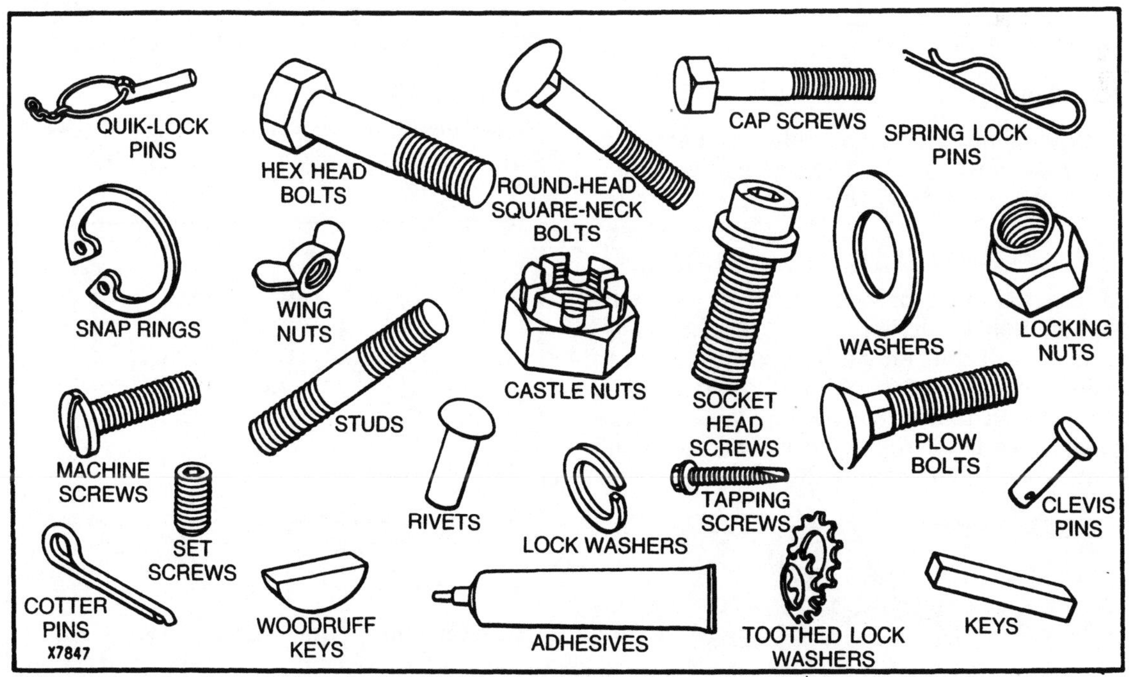

Source: Deere & Co.

FASTENERS

PIPE FLARING

Mechanics are frequently required to repair metal pipe lines on a vehicle. These lines are usually of seamless steel tubing and are flared at each end and retained in position by a flare nut.
Three common types of pipe flare are:
- single lap — low pressure lines only.
- double lap — brake lines and hydraulic fittings.
- 'ball' flare — brake lines and hydraulic fittings.

The single lap flare is rarely used, as most automotive applications specify the use of a double lap or ball (inverted) flare.

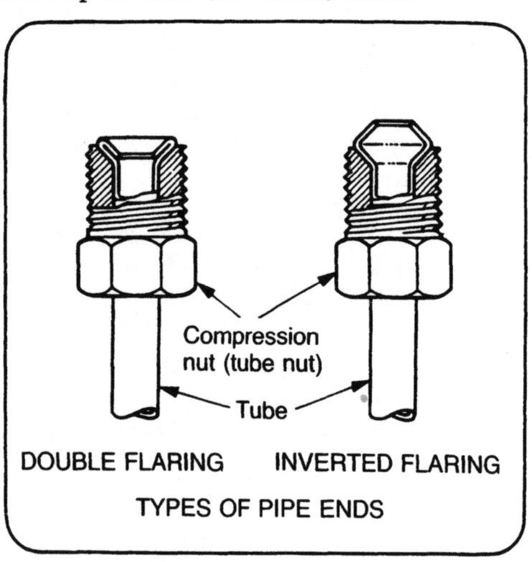

Source: AGPS

Preparation

The task of preparing the tube for flaring is equally as important as the flaring operation, so take careful note of the preparation steps.

REMOVAL OF TUBING FROM A ROLL

Careless handling of bulk tubing can cause twisting and distortion in the finished pipe. The most suitable unwinding technique is to place the tubing roll vertical on a flat surface. Hold one end to the flat surface and roll the tubing away from the secured end, ensuring the unrolled tube remains flat to the surface.

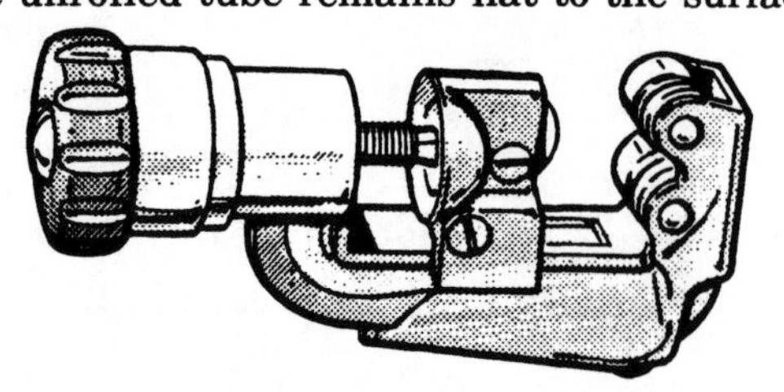

CUTTING

It is possible to use a fine-toothed hacksaw to cut the tube, but a more suitable method is to use a tube cutter. The cutter is a simple tool containing a cutting wheel which can be tensioned against the tube. The tube is supported by a double roller. By rotating the tool around the tube and progressively increasing tension on the cutting wheel after each rotation it is possible to produce a neat square cut end on the tube.

DEBURRING

Remove burrs from the inside diameter of the tube by rotating the reamer (part of the cutting tool) blade in the tube end. Check the pipe end is square to the tube: use a smooth file to align the end if necessary and repeat the deburring operation.

MEASURE PIPE

Compare the length of the old tube to the tubing to be cut. Remember to include sufficient length for curves and bends. If the shape of the tube makes measurement difficult an easy solution is to obtain a piece of string. Starting at one end of the old tube align the piece of string with the tube. By holding the string in contact with the tubing round each of the bends and curves it is possible to obtain an accurate tubing length.

Transfer the tube measurement to the new tube, cut to length and repeat the cutting and deburring steps as previously described.

FLARE NUTS

Select new flare nuts, ensuring the thread and flare type are identical to the old fittings.

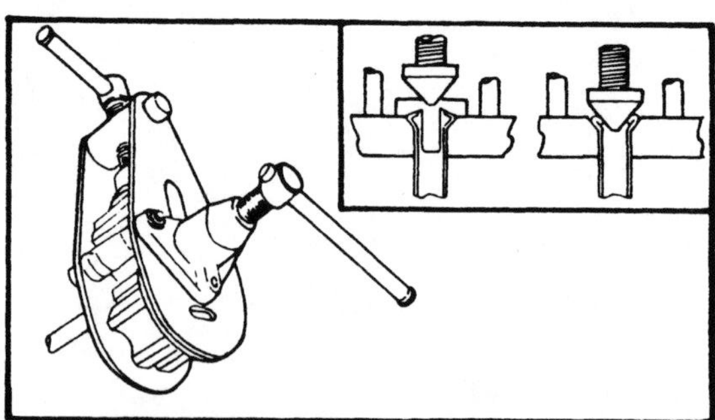

Source: Mitsubishi Sigma GJ series manual

FLARING

1 Determine the flare type to be manufactured on the tube ends (single flare; double lap flare; ball flare).

2 Check the contents of the flaring kit to ensure suitable tube gripping dies and flare-forming dies are present.
3 Slide the flare nut over the end of the tubing until the nut is clear of the work area.
4 Select and position the appropriate tube gripping dies within the tool.
5 Insert the tube through the gripping dies and lightly grip the tube by tightening the tool clamping handle.
6 Select the correct size forming die and push the tubing through the gripping dies until the tube end is level with the shoulder on the forming die.
7 Tighten the clamping handle firmly.

The first seven steps of the flaring process are common to the three basic flare forms. However, the process differs at this stage. Each of the flare types will be described individually, starting in each case at point 8.

SINGLE LAP FLARE

8 Wind the cone-shaped flare die in contact with the tube end.
9 Continue winding the cone inwards until the tube is forced into the required shape.
10 Allow the cone to 'bottom', unwind and remove the cone. If the protrusion depth was correctly set the flare size will be correct.
11 Remove the tube from the tool by releasing the clamping handle.
12 Inspect the flare to ensure the shape and size are correct and the flare is square to the tube.

BALL/DOUBLE LAP FLARE

8 Select and insert a forming die into the tube. Wind the cone in contact with the forming die.
9 Continue winding the cone inwards against the die and observe the tube forming into a bell shape.
10 Allow the die to 'bottom', completing the first stage of the double lap flare.
11 Unwind the cone and remove the forming die from the tube.

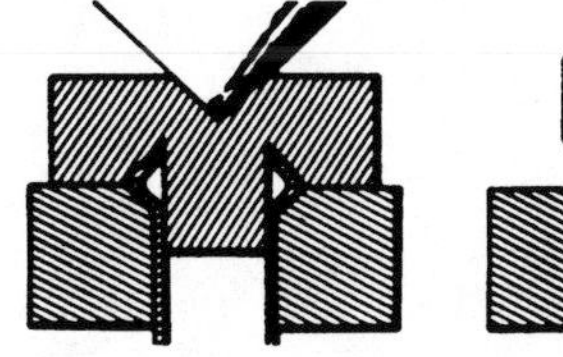

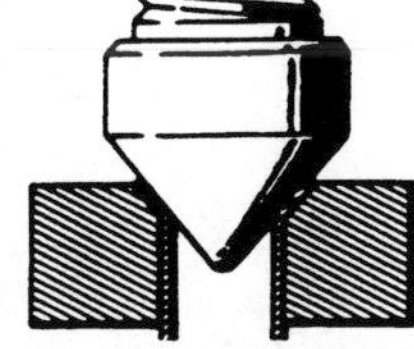

The first stage of the double lap flare, when formed using the tool shown in the illustration, is the correct shape for a 'ball' flare. Therefore if this was the type of flare required the student should go to step 13.

12 Wind the cone inward onto the end of the partially formed tube end until the cone 'bottoms'.
13 Unwind the cone, release the clamping handle and remove the tube from the tool.
14 Inspect the flare for uniformity, splits, size and shape.

The flaring process will be repeated at the other end of the tube, but consideration should now be given to the method of pipe bending that will be used to shape the finished tube. If a spring type bending tool is to be used, the tube must be formed into shape before the second flare nut is fitted, as this will prevent the bending tool passing over the tubing.

Complete the second flaring operation, blow out any particles from the tube with compressed air and the completed pipe is ready for fitting.

HACKSAW

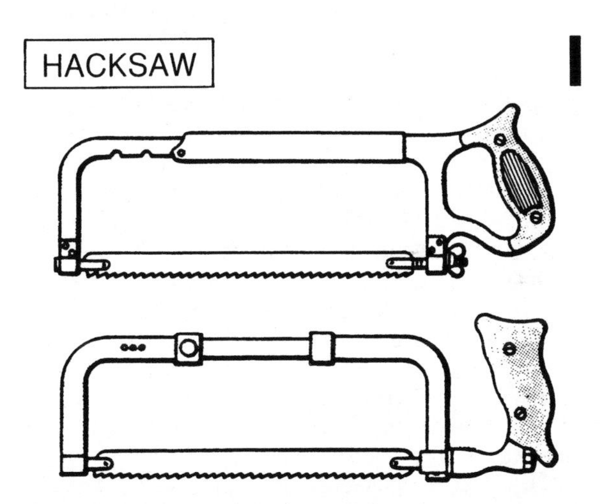

A mechanic frequently needs to cut exhaust pipe, rods, pieces of steel etc., and so the hacksaw is a necessary tool to include on the workshop tool board. Unfortunately this is the only attention many mechanics give to the hacksaw, and the same blade is used (incorrectly) on a wide variety of tasks. The choice of a blade unsuited for the task frequently leads to breakage of the blade, and the difficulty of the task is increased.

Blades are available in a range of teeth sizes. A blade with many small teeth can be described as a fine blade, and a coarse blade would have fewer, but larger, teeth. Each blade is suitable for specific tasks. For example:

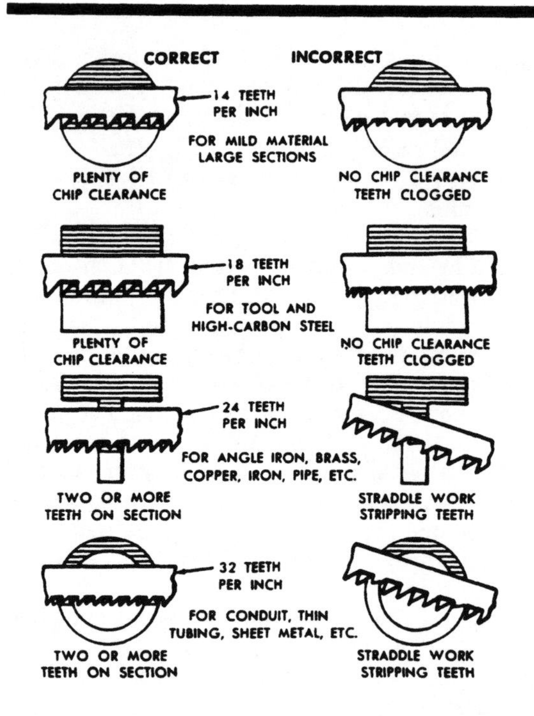

- thin wall pipe would require the use of a fine blade with two or more teeth on the wall section of the pipe at any point of the blade stroke.
- solid round or block would have substantial quantities of metal removed at each blade stroke, therefore the teeth should be larger and fewer in number to ensure the metal chips are cleared from the cut. The chip clearing action will allow maximum cutting action from the teeth.

Between the extremes of fine and coarse there are a number of blades, with each type particularly suited to a certain type of material or cross-section. Correct selection of the blade will reduce blade jamming and breakage, and increase cutting action.

The metal, particularly thin sheets, should be securely braced to prevent movement or flexing during the cutting process to prevent breakage of the cutting blade.

Fitting a blade

1 Select the correct blade for the task.
2 Fit the blade with the cutting teeth facing away from the handle.
3 Tension the blade until the blade resists flexing or twisting.

Using the hacksaw

1 Apply light downward pressure as the blade is pushed across the work on the forward cutting stroke.
2 Release the downward pressure on the return stroke, allowing an easy return.
3 Continue the strokes at a steady rate which allows easy movement of the blade and continuous chip clearance.
4 Observe progress of the cut to ensure the cut continues at the required angle to the work.
5 Prevent metal distortion and a ragged edge to the cut face (particularly on thin metal pipe) by decreasing downward pressure as the blade approaches the end of the cut.

SURFACING

Many automotive components are attached by a flange and a sealing gasket is inserted between the mating faces. Over-torquing of the retaining bolts can cause the gasket material next to the bolt holes to become more compressed than most of the gasket. This causes a bending force to be exerted on the flange, which can become permanently distorted.

To prevent leakage from the flange gasket on reassembly it is necessary to file the flange face flat and smooth.

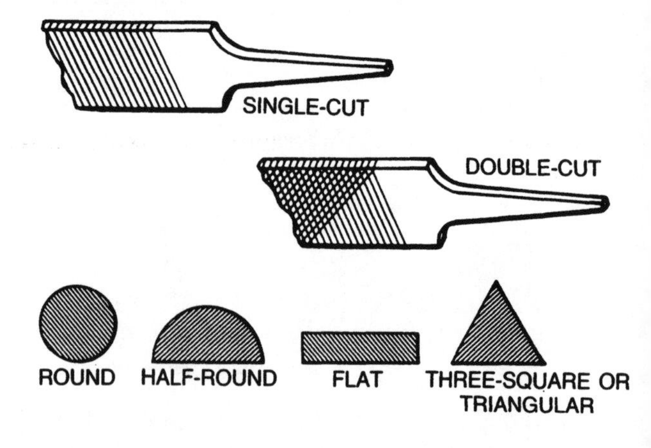

1 Secure the component in a vice. Flange facing up and horizontal, with the component protected by soft vice jaws if necessary.
2 Select a suitable file for the task. Many different file types, shapes and teeth are available but a general purpose flat, medium, double cut file of sufficient length for the task will be adequate for most steel or cast iron flange resurfacing jobs.

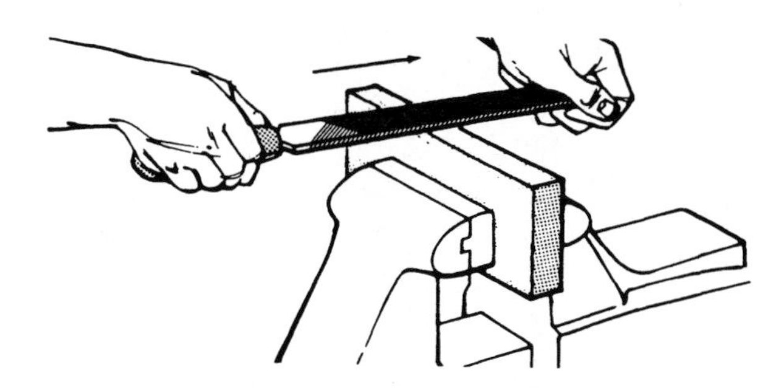

3 Hold the file firmly, flat and square to the flange, and with light downward pressure push the file across the work face. The file should move smoothly and a definite cutting action should be felt.
4 Lift the file slightly on the return stroke. (Soft metals require the file to be dragged slightly on the return stroke to assist with cleaning the file teeth.)
5 Observe the file pattern (scratches) on the work piece to ensure the full surface is covered.
6 Check the work piece for flat and square when the file marks completely cover the surface.

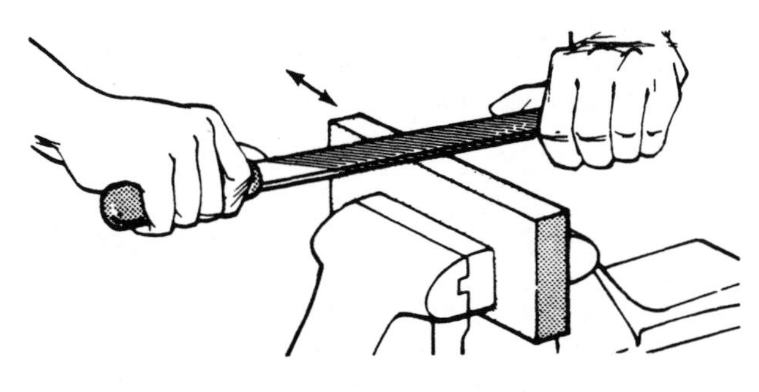

7 Draw-file the surface by moving the file back and forth sideways across the flange.
8 Detect any high and low spots by checking the scratch pattern after draw-filling.
9 Repeat the filing operation until the draw-file test is satisfactory.
10 Final·test of the filed face is by the use of a straight edge and a light source. Place the straight edge on the face and observe the light passing between the faces. Depressions on the filed surface will allow the light to show readily under the straight edge.

TOOL MAINTENANCE

Drill sharpening

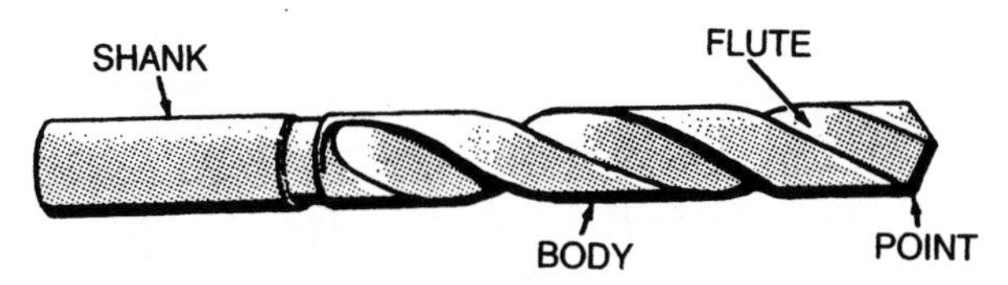

The technique of sharpening a drill requires keen eyesight and practice to produce an acceptable result. Before attempting to sharpen a drill it is necessary to know the angles and clearances required.

A new drill has two cutting lips. When viewed from the sides each lip is angled at 59° to the vertical axis. Each lip or cutting edge must be of equal length so the tip of the drill is central. Unequal length cutting edges will cause the drill to run 'off centre' and the drilled hole will be larger than the drill size.

Rotate the drill slightly and observe the top face as the cutting lip moves away. The top face angles downward at 12–15° from the lip towards the heel. This angle must be present or the drill will not cut.

PRACTICE

1 Obtain a relatively large drill (it will be easier to observe the required action if the drill is 12 mm or greater in diameter).
2 DO NOT START THE GRINDING WHEEL. This is a practice exercise to develop manipulative skill.
3 Hold the drill with both hands — thumb and forefinger grip.
4 Align one cutting lip parallel and horizontally to the wheel. Rest the back of the hand holding the forward section of the drill onto the grinder rest for support. DO NOT PLACE THE DRILL ON THE GRINDER REST.

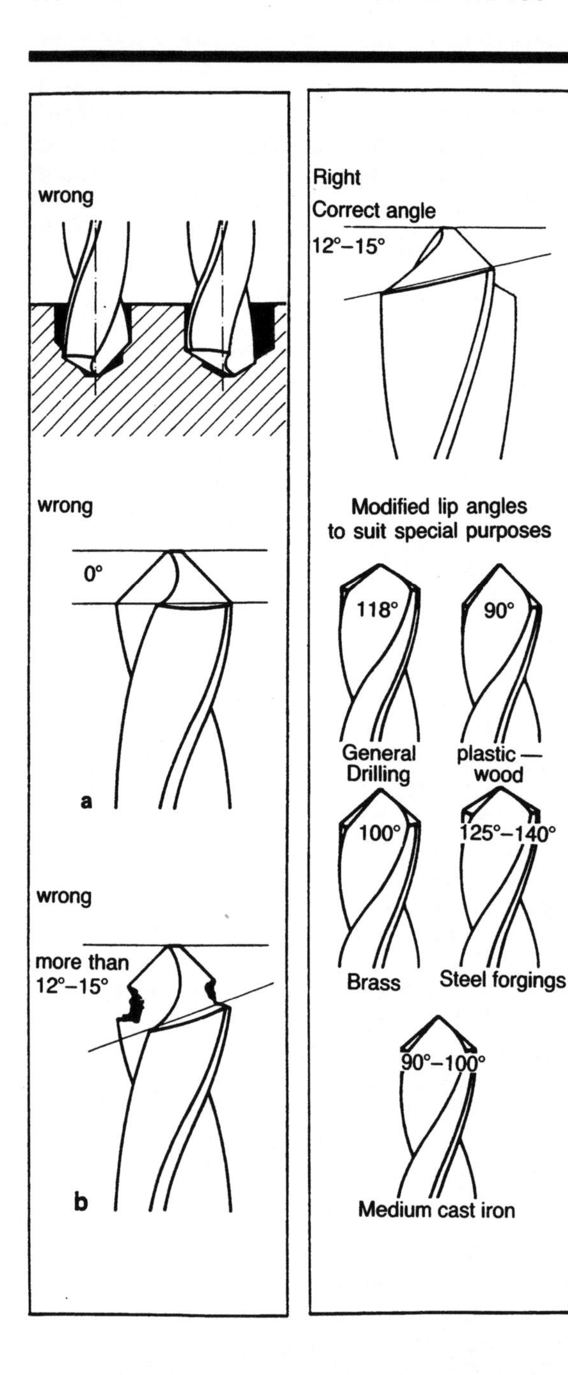

5 Check the drill angle to the wheel face. The drill should be horizontal to the wheel, 3 to 5 cm above the rest and angled at approximately 60°.
6 Starting at the cutting lip, rotate the drill tip to produce the ground surface of the point. *Note:* The shank of the drill, held in the right hand, must be lowered during the rotation to cause the 12–15° clearance angle to be formed as the drill moves toward the heel of the cutting face.
7 Repeat this exercise several times until the cutting face of the new practice drill remains in contact with the wheel during the rotation and lowering manipulation.

SHARPENING

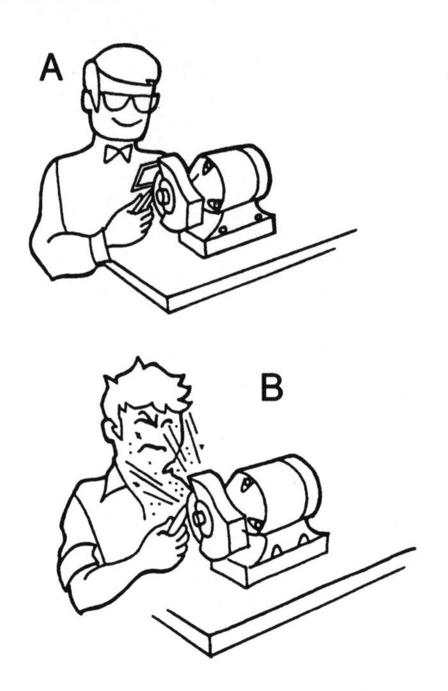

1 WEAR SAFETY GLASSES.
2 Start grinding wheel. DO YOU KNOW HOW TO TURN OFF THE GRINDER?
3 Using a blunted drill of 12 mm or greater size, repeat the practice steps with the grinder operating.
4 Grind each cutting face alternately until a visual inspection indicates the drill is correctly ground.

TESTING

1 Obtain a soft metal block.
2 Fit the sharpened drill into a fixed pedestal drill.
3 Drill a hole in the soft metal block.
4 Observe the drilling operation and watch the chip clearance. A correctly sharpened drill will produce similar chips from each flute.
5 Remove the drill, invert and place the drill shank in the drilled hole. The drill should be a neat fit, indicating the drill is running on centre and the cutting lips are of equal length.

DRILLING 'SOFT' MATERIALS

The angles previously described have been chosen for general purpose use on mild steel. A useful modification to the drill angles is to reduce the cutting edge angles so the drill tip becomes a sharper point. As the material chosen becomes 'softer' the drill cutting angles reduce (tip shape becomes more of a spear shape). For example, mild steel 59°; aluminium 45°; wood 30°.

Grinding wheels

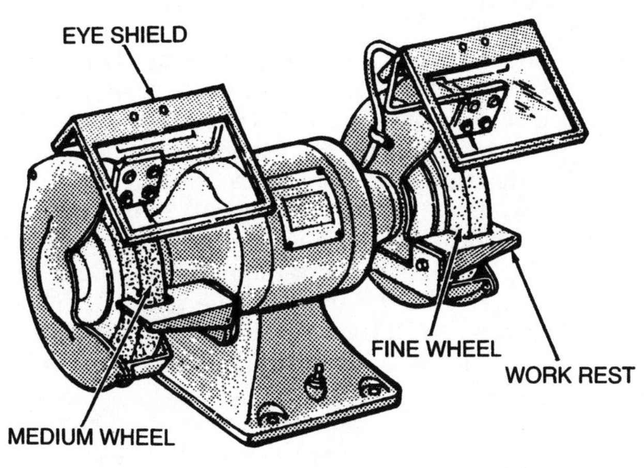

The wheels of a grinder must be maintained so the grinding face is uniform, free of grooves and not clogged by soft metal. The wheel can be 'dressed' by the use of a wheel dressing tool. With the grinder operating the dressing tool is steadied on the rest and a dressing cut is made by moving the tool slowly across the rotating wheel face. Repeat the dressing cuts until the grinding face is clean and free from grooves.

During this operation the wheel will reduce slightly in diameter, opening a gap between the tool rest and the wheel. Adjust the gap to the minimum clearance possible. A large gap may result in a tool jamming between the wheel and the rest.

Do not sharpen tools such as screwdrivers and chisels by angling the tool tip down towards the wheel. Most tools should be sharpened by using the rest to support the tool and angling the face to be ground upward towards the grinding wheel (drill sharpening is described in a separate paragraph).

Screwdrivers

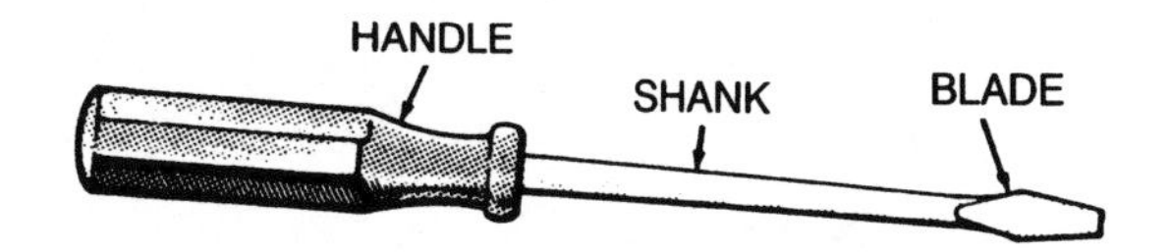

Screwdrivers require tip maintenance after prolonged use. Rounded edges and tip damage can be repaired by grinding the blade to the original shape and size. Do not angle the flats of the blade too sharply as the screwdriver will rise out of the screw head slot when force is applied, damaging the slot. Ensure the flats of the blade are ground close to parallel at the blade tip.

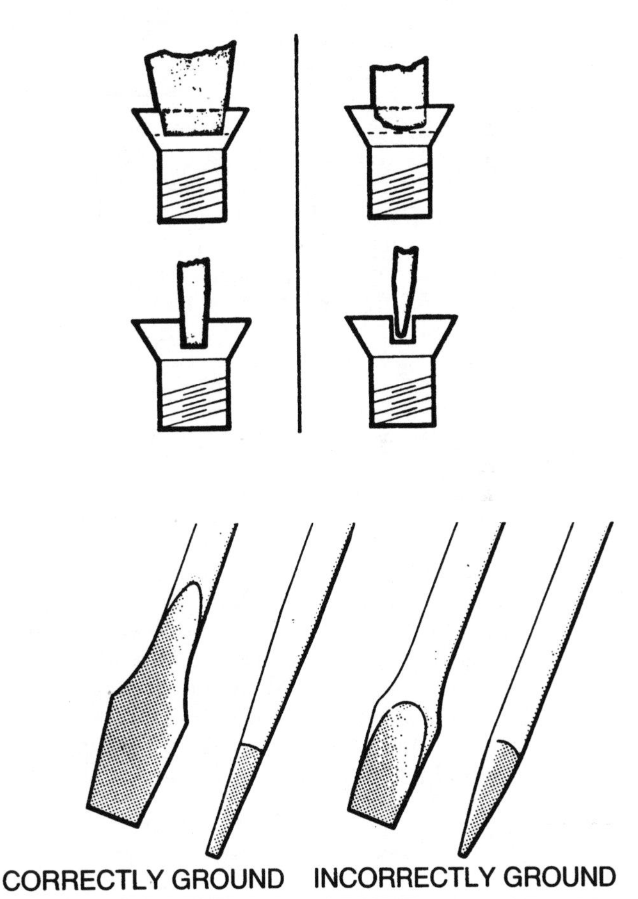

Chisels

Chisels receive continuous impact damage in use, and damage is inevitable. Two areas of the chisel require grinding:

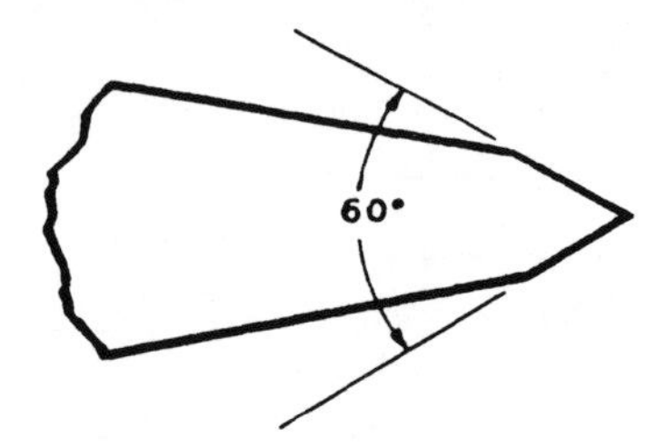

1 the tip must be sharp and should be ground to an included angle of 60°.

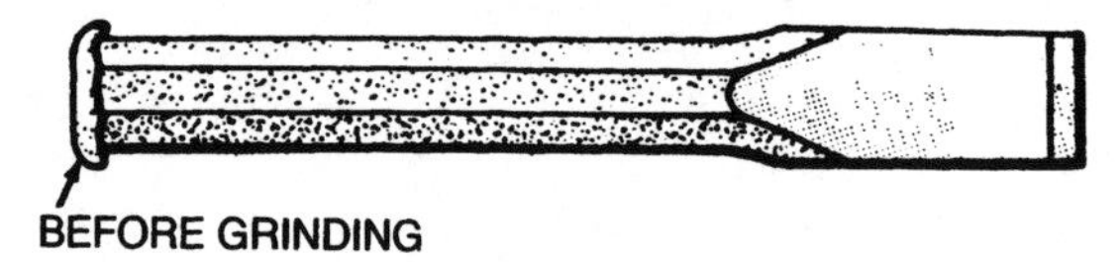

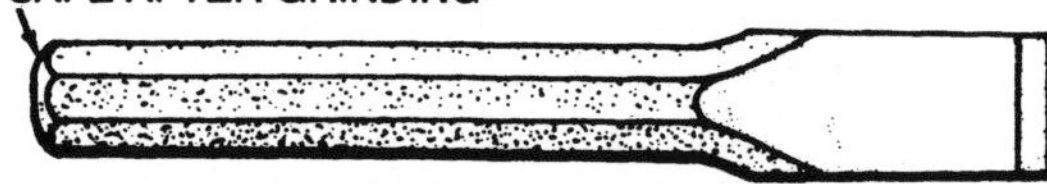

2 the head tends to 'mushroom' from the repeated hammer blows. This is a dangerous condition as the edges could break off and inflict damage to a person. It is also very possible that the user will receive a cut to the hand from the sharp edges. All traces of 'mushrooming' should be removed by grinding a chamfer on the chisel head.

CHAPTER 4 REVISION

1 Name three thread types used on current vehicles.
2 Write down a definition for each of the following terms used to describe a bolt: length, major diameter, thread length, pitch, head size and tensile strength.
3 Obtain a selection of different type and thread form bolts and use a thread gauge and chart to identify each bolt.
4 Why is it necessary to obtain the vehicle manufacturer's bolt torque specifications before tightening a bolt?
5 List three methods of repairing damaged threads in a component and provide a detailed explanation of one of these methods.
6 List the steps required to cut a new thread onto round metal stock.
7 List three methods of removing a broken bolt or stud from a housing.
8 Describe a method for removing a stud which has broken below the surface of the component, and no part of the stud is protruding from the threaded hole.
9 Obtain a selection of automotive fasteners and correctly identify each by type: set screw, cotter pin, circlip etc.
10 Obtain a selection of locking devices, for example split pins, various lock washers, lock tags, and identify each by type.
11 List three types of pipe flares in common use and state an application for each type.
12 Explain a method of removing tubing from a roll without damaging the tube.
13 Obtain the instructions for a pipe flaring tool and write down the method of making a double lap flare.
14 Use a pipe flaring tool and prepare one flare of each of the three common types.
15 What type of hacksaw blade would be suitable for cutting thin wall pipe?
16 Which direction should the teeth face when fitting a hacksaw blade?
17 What does the term 'draw filing' mean?
18 Why, when sharpening a drill, must the two cutting edges be equal in angle and length?
19 Obtain a number of blunt drills and, using the method described in this chapter, sharpen each drill.
20 What type of drill cutting edge angle is suitable for drilling 'soft' materials?
21 Obtain a blunt chisel or screwdriver and use a grindstone to return the tool to an acceptable condition for use.

5

WELDING

SAFETY

1

Oxy-acetylene gas welding equipment is an essential part of every automotive workshop. Because of the wide range of tasks to which the equipment can be applied almost every tradesperson will need to understand the operating techniques and safety precautions associated with welding equipment use. An oxy-acetylene welding plant is a powerful and versatile tool, but it must be handled carefully to avoid the possibility of personal, property or equipment damage. Safety rules for operating gas welding equipment are very simple and easy to learn. The problems which occur with welding equipment are unlikely to be caused by lack of knowledge, but by a failure to use the information already known. Minimise the **DANGER** to you and the work area by **LEARNING** and **OBSERVING** the **SAFETY** rules listed below.

1 **Use a cylinder trolley.** Cylinders must be mounted upright and secured.
2 **Keep cylinders away from heat sources.** Full and empty cylinders should not located near a heat source.
3 **Protect cylinder valves.** Protect the cylinder valves by storing the cylinders in a manner which protects the valves from knocks, falls or weather.
4 ***NEVER*** **strike an electric arc on a cylinder.**
5 **Do not place electric leads across cylinders.**
6 **Oil or grease must not be used on cylinder or regulator threads.** Oxygen under pressure can cause oil or grease to ignite violently.
7 **Leave the key in place on the acetylene cylinder.** In case of emergency the cylinder can be turned off quickly.
8 **Do not use unserviceable equipment.** Leak test the system to check cylinders, regulators, gauges, hoses and the hand-piece are in serviceable condition.
9 **Light the blowpipe with a spark lighter.** Never use gas type cigarette lighters or matches to light the blowpipe.
10 **Safety check the surrounding area.** Clear the surrounding area of all flammable materials. Where this is not practical, an observer should watch the surrounding area for sparks. Avoid welding or cutting on wooden floors or using wooden work supports. Recheck the area for smouldering material after the welding process has finished.
11 **Locate the fire extinguisher.** After a fire is in progress is not the time to check the location of a suitable type of extinguisher. Locate and if necessary position a fire extinguisher near to the work area.
12 **Wear safety clothing.** Observe personal safety by wearing the correct equipment for the task.

Safety clothing

Eye protection. Welding goggles for gas welding and helmets suitable for 'arc' welding are designed to protect the eyes from sparks and harmful rays. Welding goggles have specially tinted lenses fitted into a frame and secured with spring side clips. The frame is formed to fit firmly against the operator's

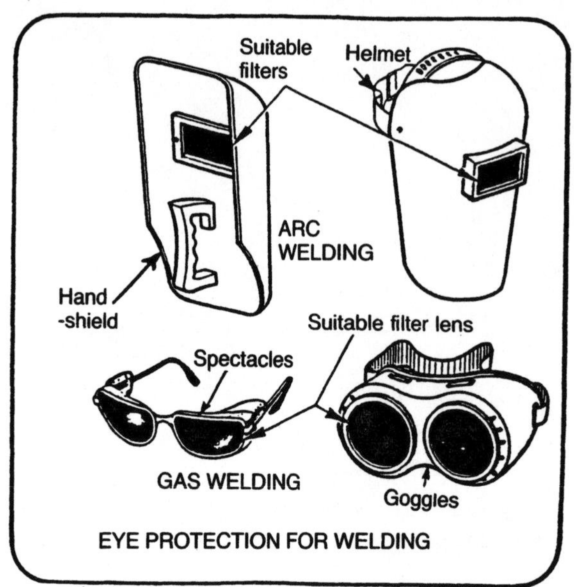

EYE PROTECTION FOR WELDING

Source: AGPS

face, and is held in place by an elasticised head band. A variation of these goggles is to have a clear lens in the frame and the tinted lens in a 'flip type' visor which is lowered into place as welding starts. These goggles provide additional safety for the operator as his or her eyes are shielded from sparks when clear vision is required, such as when adjusting the flame.

Gloves. Full length leather gloves are worn to protect the operator's hands and arms from sparks and harmful rays.

Apron and spats. A full length leather apron and spats over work boots are worn to protect the operator's clothing from sparks, particularly when a cutting torch is in use.

Clothing. The normal workshop clothing consisting of overalls, a cap and steel-capped work boots are recommended for any type of welding operation. For electric 'arc' welding a leather jacket, leather apron, leather spats and gloves are recommended.

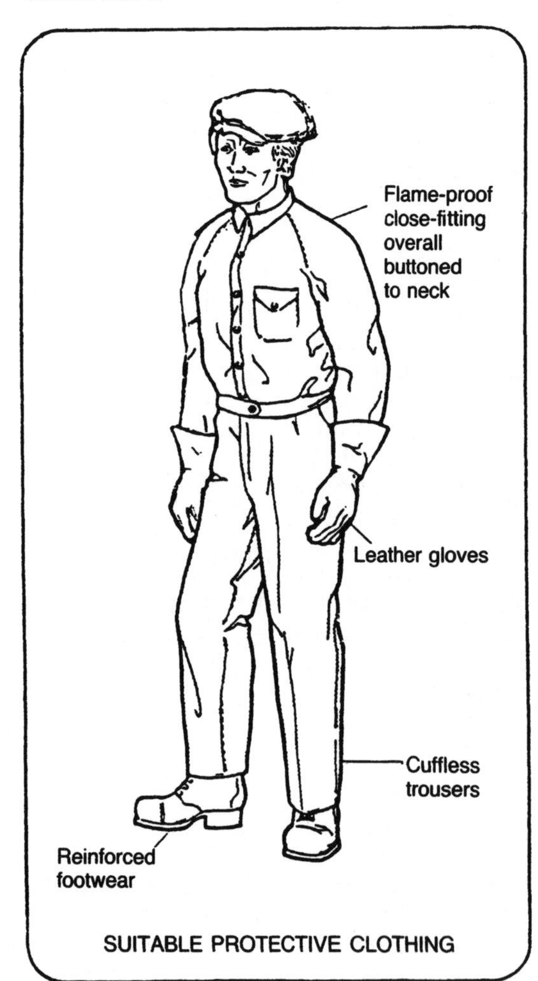

SUITABLE PROTECTIVE CLOTHING

Source: AGPS

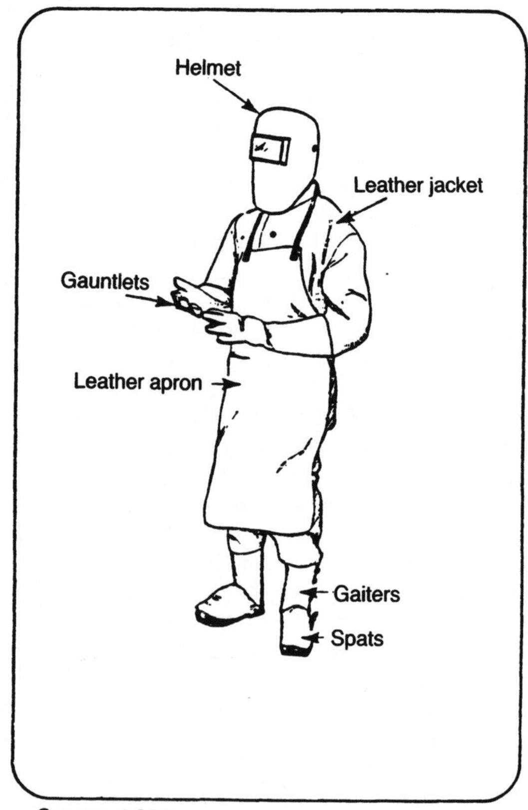

Source: AGPS

COMPONENTS

The standard oxy-acetylene welding equipment used in the automotive repair outlet consists of:

- oxygen cylinder;
- acetylene cylinder;
- regulators;
- gauges;
- hoses;
- handpiece;
- welding tips;
- cutting nozzle;
- accessories — tools.

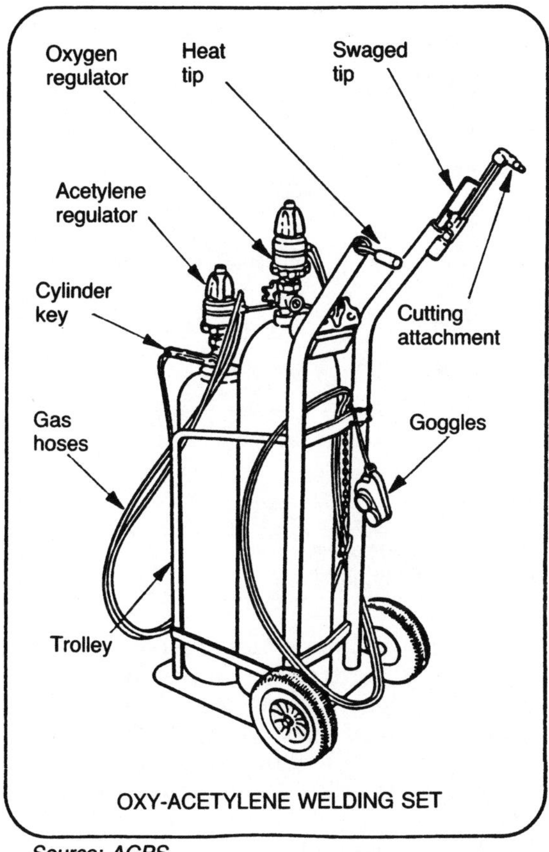

OXY-ACETYLENE WELDING SET

Source: AGPS

Oxygen cylinder. The cylinder is coloured black and the gas outlet is fitted with a right-hand internal thread. The cylinder when fully charged can have a pressure of 13 000 kPa at room temperature.

Acetylene cylinder. Red in colour and fitted with left-hand internal thread at the gas outlet. The fully charged cylinder has a pressure of 1500 kPa at room temperature.

Regulators. A regulator is a pressure-reducing device that can reduce the cylinder pressure to a suitable constant working pressure at the hand-piece. The working pressure is controlled by an adjusting knob on the regulator.

Gauges. Each cylinder has a specially calibrated gauge set. The gauges are coloured red (acetylene) and black (oxygen) and are fitted with spigot nuts to suit the thread in the gas cylinder outlets. Each gauge set consists of a pair of gauges; one gauge indicates the cylinder pressure and the second gauge reads the working pressure being supplied to the hand-piece.

Hoses. The gauge set and the hand-piece are linked by a pair of rubberised high pressure hoses. The hoses are coloured red (acetylene) and black (oxygen) and the end fittings are L.H. thread (acetylene) and R.H. thread (oxygen) to prevent incorrect connections. The length of the hoses can vary, but generally a length of approximately 2 m is adequate.

Hand-piece. The hand-piece consists of the blowpipe and the welding tips. Control valves are fitted to the hand-piece to control the quantity of each gas entering the hand-piece mixing chamber. The welding tips are attached to the hand-piece by a threaded barrel connection.

Welding tips. A range of tips is available for general purpose welding. These tips are identified by a number stamped into the material of the tip. Low numbers mean a small flame for light duty welding, and as the number increases the size of flame available is greater and heavier welding tasks can be undertaken.

Cutting nozzles. Special nozzle adapters must be fitted to the blowpipe when metal cutting is required. These nozzles have an additional oxygen valve to set the pre-heating flame, and a lever which, when depressed, controls additional oxygen to provide the cutting action. The tip has an outer ring of holes for the pre-heating flame and a central hole which operates when cutting is in progress.

Accessories

The standard accessories required to service and maintain an oxy-acetylene welding plant are:

- Spark lighter — a flint gun used to ignite the gas.

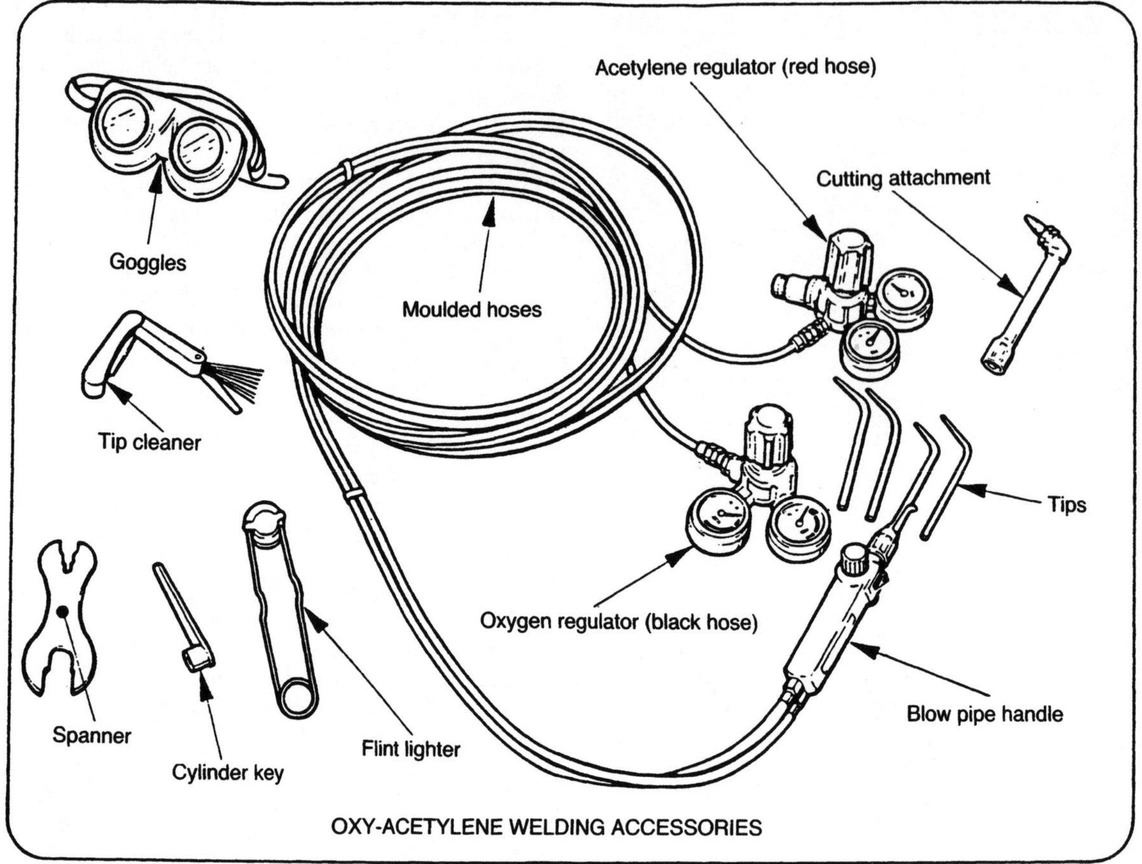

Source: AGPS

- Combination spanner — several width jaws are provided on a single spanner allowing all standard fittings on the plant to be tightened or loosened with one tool.
- Tip cleaners — a set of different sized reams used to clean the carbon from the holes in the end of the tips.
- Valve key — a specially designed key which is used to open or close the valve on the neck of the gas cylinders.

  Note: There are a number of cylinders which are fitted with an integral control knob. These cylinders do not require the use of a valve key.
- Roller guide — consists of two small guide wheels located on an adjustable leg which is attached to the cutting torch head. The guide is used to position and steady the cutting tip during a cutting process.
- Trolley — designed to hold the two gas cylinders in an upright position and to provide an easy method of moving the welding plant to new locations.

ASSEMBLY AND SET UP OF EQUIPMENT

1. Move the oxygen cylinder into position adjacent to the trolley. Do not attempt to carry the cylinder. Grasp the cylinder at the neck and upper girth section. Rotate the cylinder, while holding it at a angle to the ground, so that it rolls along its lower edge.
2. Lift the oxygen cylinder into position on the trolley.
3. Repeat steps 1 and 2 with the acetylene cylinder.
4. Secure both cylinders to the mobile trolley with a chain.
5. Remove the cylinder safety caps. Refilled cylinders are supplied with a plastic safety cap inserted into the valve outlet. These caps must be removed by hand (screwdriver or pliers if necessary).

Oxy-acetylene cylinders are available in several different sizes but all will eventually require refilling by the supplier. The mechanic must therefore be capable of installing a pair of cylinders into a mobile trolley and preparing the equipment for welding. This exercise assumes the trolley is empty and fully charged cylinders are available.

Caution: Do not use compressed gas to eject the cap from the cylinder

6 'Crack' the cylinder valves. Before connecting the regulators to the cylinders, it is essential to clear dirt or foreign material from the valve area. Opening and closing the cylinder valve quickly is known as cracking the valve, allowing the escaping gas to blow out any debris. Do not perform this operation near open flames or other persons. Wear eye protection.

7 Inspect the cylinder valve seat, and the regulator attaching nut and seat.

8 Fit the regulators to the cylinders. Fit the black (oxygen) regulator to the black (oxygen) cylinder. The regulator is fitted with a right hand thread. The nut should be tightened firmly, with the combination spanner. Repeat this step with the red (acetylene) cylinder and the regulator. Remember, the acetylene units are fitted with left hand threads.

9 Release the regulator valves. Cylinder pressure will enter the regulator during the next step and the regulator should be unloaded by winding the adjusting knob on both the oxygen and acetylene regulators, until the knobs are fully released.

10 Open the cylinder valves slowly. Do not open the cylinder valves more than one and a half turns. Cylinder pressure will be indicated on the high pressure gauges.

11 Attach the hoses. Fit the hoses to the regulator. The red hose (L.H. threaded nut) is for acetylene. The black hose (R.H. threaded nut) is for oxygen. Hand start the nuts, then tighten firmly with the combination spanner.

12 Blow out the lines. Wind the adjusting knobs of both regulators inward until gas is flowing from each hose. This will clear any minor foreign material present in the hoses. Release the adjusting knobs.

13 Fit the hoses to the blowpipe. Fit the black hose to the blowpipe fitting marked 'O', (R.H. thread) and the red hose to the fitting marked 'A', (L.H. thread). Hand start the nuts, then tighten firmly with the combination spanner.

14 Check blowpipe flow. Wind the adjusting nuts of the regulators until the gauge pressures indicate approximately 100 kPA. Slowly 'crack' open the blowpipe valves one at a time and ensure there is a gas flow from both valves.

15 Leak test the equipment. Ensure the regulators are set to 100 kPa, then close the cylinder valves. Watch the low pressure gauge; a drop in gauge pressure will indicate leaks in the system. To locate the leaks, coat all the joints with soapy water. Bubbles rising through the soapy water will indicate the location of the leaks.

Caution: Equipment which fails the leak test must not be used.

16 Fit the welding tip. Select and fit the welding tip to the blowpipe. The tip may not align itself into a direction which will be comfortable for welding. The direction the tip faces can be altered by loosening the fluted sleeve nut, twisting the tip to the required position, then retightening the sleeve nut.

OXYGEN & ACETYLENE REGULATOR PRESSURES

Size Tip	Cleaning Drill	Mixer	Regulator Pressures kPa (psi)	
			Oxygen	Fuel Gas
8	8	304002	50 (5–10)	50 (5–10)
10	10	304002	50 (5–10)	50 (5–10)
12	12	304002	50 (5–10)	50 (5–10)
15	15	304002	50 (5–10)	50 (5–10)
20	20	304002	50 (5–10)	50 (5–10)
26	26	304002	50 (5–10)	50 (5–10)
32	32	304002	100 (15)	100 (10–10)
40	38	304003	200 (30)	100 (10–15)
4 × 24HT	24	304003	200 (30)	100 (10–15)

17 Set the working pressures (refer to illustration). Open the blowpipe oxygen valve, adjust the regulator until the required working pressure is indicated on the low pressure gauge. Close the blowpipe oxygen valve. Repeat this procedure with the blowpipe acetylene valve and regulator.

LIGHTING THE FLAME

The welding equipment must be prepared for use and operating pressures set as described in the previous paragraph before attempting to light the flame.

1 Open the blowpipe acetylene valve slightly (less than 3/4 turn).
2 Light the gas with a flint lighter.
3 Adjust the blowpipe acetylene valve until the flame is burning slightly away from the tip and not producing soot.
4 Open the blowpipe oxygen valve SLOWLY. Continue opening the valve until the flame changes from orange into two sections or cones. The section nearest the tip (inner cone) will be white and ragged, the outer cone will be blue and feathery. The setting will be correct for a NEUTRAL flame when the inner cone becomes clearly defined.
5 Recheck the working pressures.
6 The welding unit is now ready for use on general purpose welding.

The neutral flame is widely used in welding processes. When directed onto steel, the

Neutral Flame

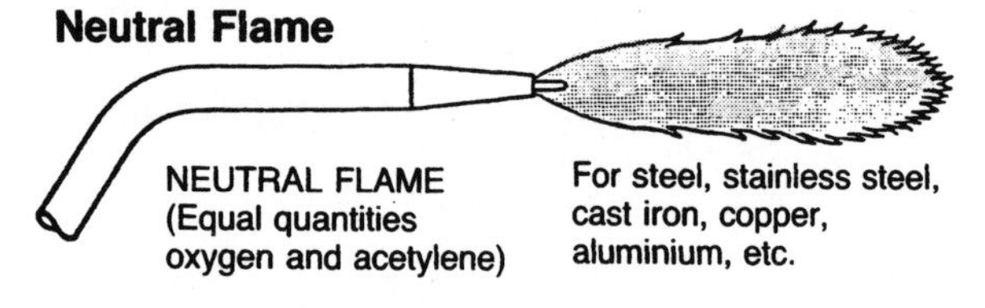

metal melts cleanly without sparking. No carbon is being transferred into or out of the metal, therefore the chemical characteristics of the weld and base metal remain the same. This type of flame produces clean strong welds.

Cutting torches may vary in design and the manufacturer's instructions must be followed at all times. However, the torch will contain a preheating section and a cutting section within the tip. With the high pressure oxygen (cutting) lever released, the preheat flame is adjusted to a NEUTRAL flame. Depress the cutting lever and observe the flame. The preheat flame should still be a neutral flame: adjust the valve as required until the preheat flame remains neutral whether the cutting lever is released or depressed. The oxygen pressure setting for cutting torch applications will be many times that of the acetylene pressure, therefore consult a pressure chart before using a cutting torch.

Two other flame types are possible, but have limited applications.

1 Oxidising flame. When the blowpipe oxygen valve is opened excessively, the flame

Oxidising Flame

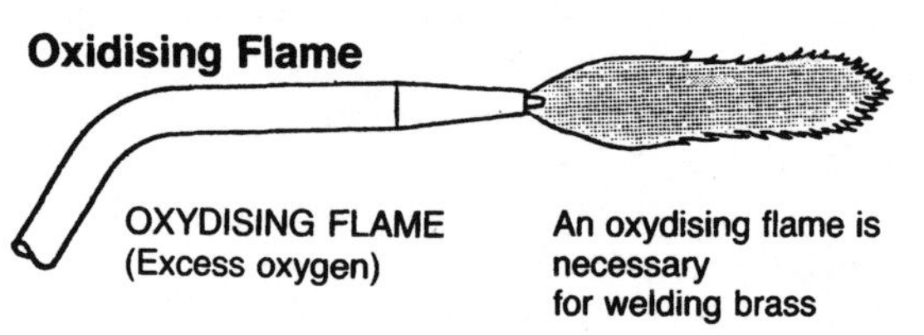

colour in the inner cone tends toward a purple shade and becomes shorter. The outer cone flares at the ends. This type of flame is suitable when braze welding.
2 Carburising flame. A deficiency of oxygen will cause a carburising flame. This flame

Carburising Or Reducing Flame

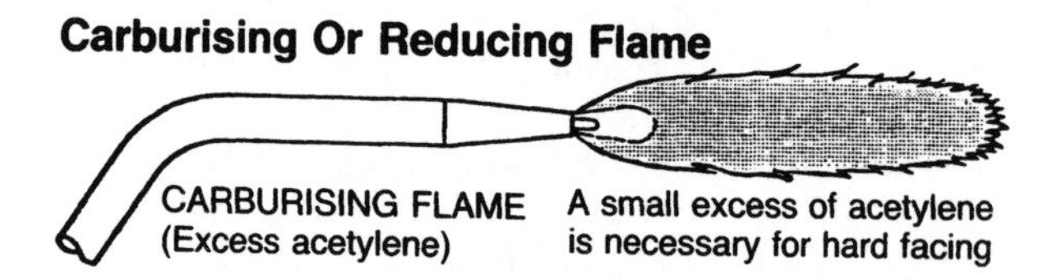

is recognisable by a short white inner cone enclosed by a light blue acetylene feather and a deeper blue outer envelope. This flame can be used for aluminium welding and hard facing processes.

Shut down

CLOSE DOWN (NORMAL)

1 Close the blowpipe acetylene valve. The flame will be extinguished.
2 Close the blowpipe oxygen valve.

Note: Steps 1 and 2 are adequate if the shutdown is for a short period.

3 Close both cylinder valves.
4 Open the blowpipe oxygen valve and allow the gas to flow until both the oxygen gauges read zero. Close the blowpipe oxygen valve.

5 Unwind the oxygen regulator adjusting knob.
6 Repeat steps 4 and 5 for the acetylene valve and regulator.

CLOSE DOWN (EMERGENCY)

Emergency shutdown of the plant is required if a condition known as FLASHBACK occurs. The gas can ignite and burn back into the blowpipe or tubing. If the flame is in the blowpipe (a high pitched hissing sound at the mixer) close the oxygen blowpipe valve then the acetylene valve. The flame should burn out in a few moments.

Gas ignition in the tubing is more serious and the cylinder valves must be closed immediately. It is possible for an acetylene cylinder to become hot if severe flashback has occurred.

There is an approved handling procedure if this condition occurs, however, at this stage of your studies you should not be welding without an instructor present. Discuss the emergency procedure with the instructor.

Flashbacks are a warning that a fault exists. Do not relight the blowpipe until the pressures, tubing, fittings and tips are inspected for faults.

A lesser problem known as a BACKFIRE is more common. The flame may temporarily be extinguished by weld particles, tip overheating or by touching the work with the tip. This problem may clear by itself and if the work is still hot enough the flame may relight. When the flame does not relight, close both blowpipe valves, allow the tip to cool, then relight in a normal manner.

WELDING PROCESSES

An oxy-acetylene plant can be used for:

- braze welding;
- brazing;
- cutting;
- fusion welding;
- heating.

The operating techniques are different in these processes and require continuous practice to maintain high skill levels. This section of the chapter is a general guide to the methods used, however, it is recommended that the equipment manufacturer's procedures are observed at all times.

Braze welding is a method of joining two pieces of metal together, without melting the parent metal, by applying a layer of dissimilar metal (manganese bronze) between each piece. A flux is used during the process to reduce oxidation. The basic steps to be followed are:

1 Inspect and set up the welding equipment.
2 Select the welding tip to suit the work. Braze welding usually requires a tip size smaller than when fusion welding on the same size task.
3 Set the regulators to the pressures required.
4 Prepare the work-pieces. Remove all rust, dirt or grease. Grind a bevel edge along the area to be welded. When repairing a crack, a stronger repair will be obtained by enlarging the crack to a 'V' form. This will provide greater surface area to which the bronze can be bonded.
5 Light the blowpipe.
6 Adjust the control valves to provide a slightly oxidising flame.
7 Preheat the work-piece around the weld area to a dull red.
8 Preheat the tip of the brazing rod
9 Plunge the hot tip of the rod into the braze welding flux.
10 Heat the welding area and, at the same time, melt the tip of the 'fluxed' rod so that it drops into the 'V'.

Note: When the weld area is at the correct temperature the bead will spread out over a wide area.

11 Continue to melt the 'fluxed' rod to form a build-up to the correct level.
12 Move along the 'V' until more flux is required on the rod.
13 Repeat steps 8 to 11 until the 'V' is completely full.

Brazing is a method of joining two pieces of thin metal together by a thin layer of filler metal (similar to soldering and silver soldering). The procedure is:

1 Thoroughly clean the both pieces of metal, especially at the areas to be joined.
2 Fit a small tip to the blowpipe.

Note: Select a tip which is two sizes smaller than one used for a fusion weld on the same thickness material.

3 Light the blowpipe and adjust to a neutral

flame, and set the working pressures on both gauges to 70 kPa.
4 Apply a thin coat of flux paste to the brazing area and the tip of the filler rod.
5 Place the filler rod tip at the start of the brazing area.
6 Using the outer section of the flame, heat around the general area of the filler rod until the rod melts.
7 Lift the rod away from the brazing area while continuing to heat the area.
8 When the filler metal has completely flowed between the two pieces of metal, add more filler rod.
9 Continue steps 6 to 9 until the two pieces of the metal are bonded together.
10 Allow the metal to cool, then remove any excess flux.

Oxy-acetylene cutting of iron or steel is achieved by raising the temperature of the metal to a white heat, and then using an extra stream of oxygen to rapidly burn the metal. The procedure for a straight line cut on metal to 6 mm thickness is:

1 Thoroughly clean around the cut line with a wire brush.
2 Mark the cut line with a piece of chalk.
3 Clamp a piece of scrap metal, having a straight side or edge onto the metal being cut, the straight edge running along the inside of the chalk line.
4 Select the cutting nozzle to suit the thickness of the metal to be cut. (A number 8 nozzle is suitable for metal thicknesses up to 6 mm.)
5 Fit the cutting nozzle to the blowpipe.
6 Practise a cutting run along the chalk line, by resting the nozzle on the edge of the scrap metal and moving slowly from one end to the other.
7 Adjust the position of the scrap metal to align the nozzle with the chalk line.
8 Wear the recommended safety clothing and observe all safety precautions.
9 Light the blowpipe and adjust the pressures to oxygen 200 kPa and acetylene 100 kPa.
10 Hold the nozzle so that the cone of the flame is about 2 mm above the edge of the metal to be cut.
11 Ensure that the nozzle is at right angles to the surface of the metal.
12 When the metal glows 'white hot' tilt the nozzle slightly away from the edge and depress the cutting lever.

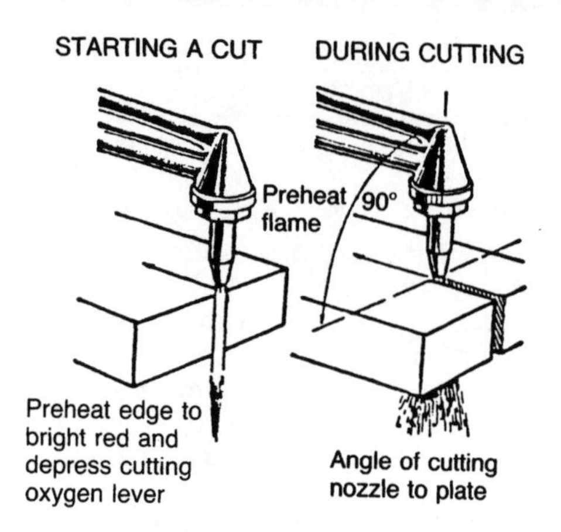

13 When the cutting starts, straighten the nozzle until the edge of the metal has been cut.
14 Tip the nozzle slightly in the direction of the cut and move the nozzle, at a steady rate, along the chalk line.

Note: When the cutting action is lost due to moving the nozzle too fast, release the lever and restart the cut.

15 When the cut is completed, the scrap section may still be attached by the slag. To release the scrap metal, tap it gently with a hammer.

Fusion welding

During this process the joining edges of both pieces of metal are melted. The molten metal then flows together to form a permanent bond. There are various techniques used to produce a fusion weld. The forehand process is suitable for welding metals of up to 5 mm in thickness, and has wide application for automotive repairs.

The technique of fusion welding is complex for a beginner, so a valuable exercise is to practise forming lines of fusion on a piece of sheet metal. This method is called puddling.

PUDDLING (LINES OF FUSION)

The technique of puddling is to carry a molten pool of metal across the surface without the use of a filler rod.

1 Clean the surface of a piece of 1 mm thick sheet steel.
2 Place the sheet steel on a fire brick.
3 Select a number 8 tip.
4 Set the regulator pressures to 50 kPa.

5 Light the blowpipe and adjust to a neutral flame.
6 Beginning at the right hand edge of the sheet, position the blowpipe with the tip angled at 45° to the work and the inner cone just above the surface.
7 Move the tip in a small circle until a pool of molten metal appears.
8 Slowly move the tip to the left so that the small circles overlap and the pool remains a uniform size.

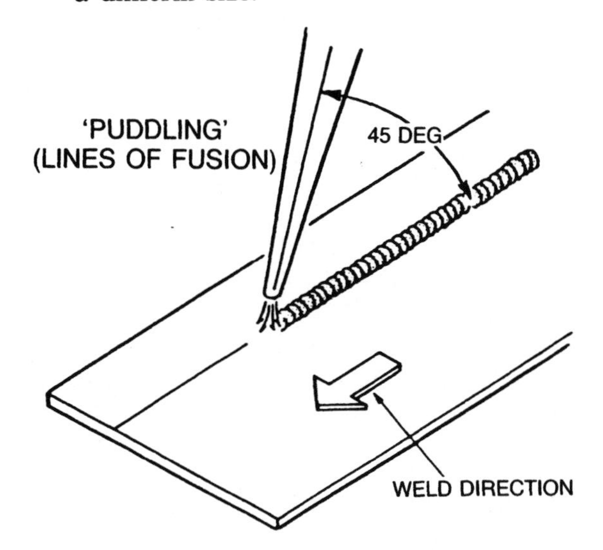

9 Continue moving the tip to the left at a constant rate to ensure the width of the puddle remains the same and no blow holes appear in the metal.
10 Stop about 10 mm from the edge of the metal.
11 After mastering this process, obtain a clean piece of metal and draw several straight lines across one surface.
12 Repeat the puddling process along each of these lines.

FUSION WELDING (NO FILLER ROD)

The procedure for fusion welding two pieces of panel steel 100 mm long, 50 mm wide and 1 mm thick is as follows.

1 Thoroughly clean both pieces of panel steel.
2 Butt the longer edges of the panel steel together.
3 Select a number 8 tip and fit it to the blowpipe.
4 Adjust the regulators to 70 kPa.
5 Light the blowpipe and set to a neutral flame.
6 Tack weld each end of the join. This will assist in controlling distortion.
Use the puddling technique to produce tacks of about 5 mm.
7 Hold the blowpipe at about 60° to the work, and the inner cone of the flame about 2 mm from the surface.
8 Move the flame to the right hand edge of the work.
9 Warm the area along the butt for approximately 20 mm.
10 When the area is red hot, move the cone of the flame to the start of the butt.
11 Using the the puddling technique, fusion weld the join along the full length of the metal.

FUSION WELDING (WITH FILLER ROD)

The procedure for fusion welding two pieces of panel steel 100 mm long, 50 mm wide and 1 mm thick is as follows.

1 Thoroughly clean both pieces of panel steel.
2 Bring the longer edges of the panel steel together so a 0.5 mm gap exists.
3 Select a number 8 tip and fit it to the blowpipe.
4 Adjust the regulators to 70 kPa.

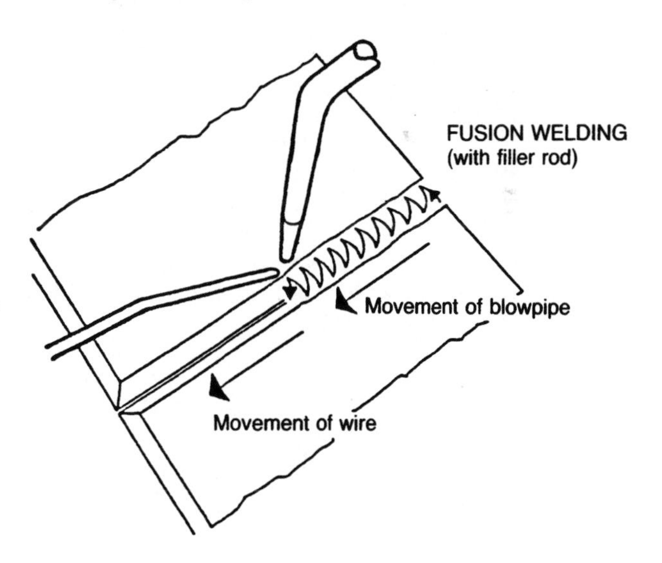

5 Light the blowpipe and set to a neutral flame.
6 Tack weld each end of the join. This will assist in controlling distortion.
Use the puddling technique to produce tacks of about 5 mm.

7 Hold the blowpipe at about 60° to the work, and the inner cone of the flame about 2 mm from the surface.
8 Move the flame to the right hand edge of the work.

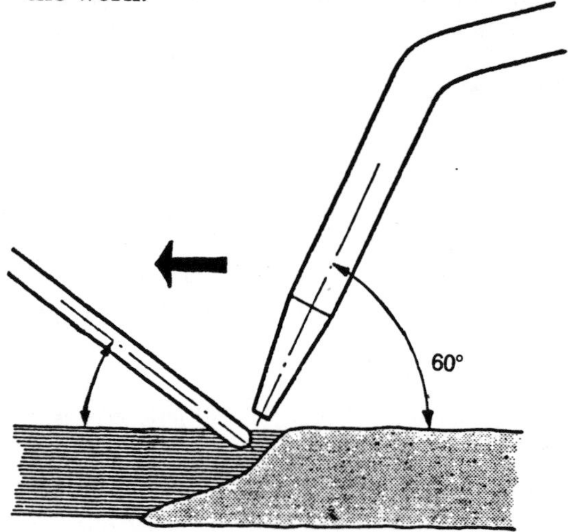

9 Warm the area along the butt for approximately 20 mm.
10 When the area is red hot, move the cone of the flame to the start of the butt.
11 Using a 1.5 mm mild steel filler rod and the puddling technique, add filler metal until the weld bead is slightly above the surface. When metal is needed in the weld, the filler rod is dipped into the puddle. When sufficient metal has been deposited, the tip of the filler rod is withdrawn to the outer envelope of the flame.
12 Continue this process until a fusion weld bead of the join has been formed along the full length of the metal.
13 Inspect the weld bead for consistent width, height and penetration.

LAP AND FILLET WELDING

The welding procedure is similar to butt welding and the differences will be overcome with practice. When lap welding the flame must be concentrated more on the lower plate, as the upper plate edge will melt more readily. An additional problem appears when fillet or 'T' welding. The vertical plate tends to be undercut by the flame but the problem can be minimised by adding the filler rod to the top of the weld pool and maintaining steady progress along the weld line.

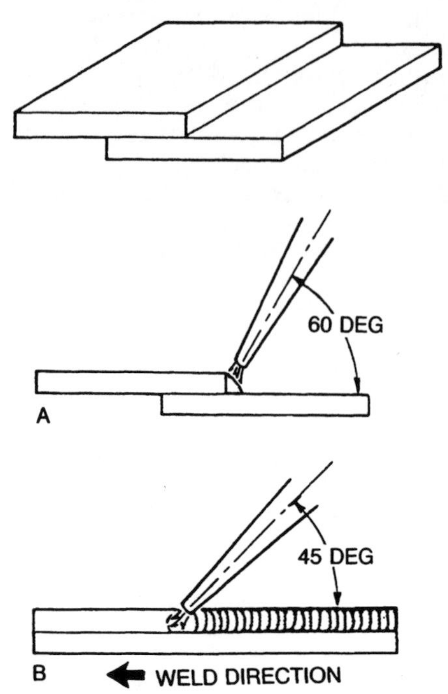

Torch position for the lap weld. A — Torch at 60 degree with the horizontal. B — Torch at 45 degree in the direction of welding.

ARC WELDING

The illustration on the following page shows an electric welder and accessories.

Safety in welding is a matter of applied common sense. The process of arc welding can produce:

- electric shocks;
- burns;
- arc burns and eye damage;
- fumes.

Precautions must be taken to avoid personal injury.

Electric shocks can be avoided by:

- ensuring all the electrical cables are insulated and free of any internal and external defects;
- switching off all electrical appliances when not in use;
- wearing approved safety clothing and equipment;
- ensuring the work area is dry.

Burns can be avoided by:

- wearing the approved safety clothing and equipment;

- correctly disposing of the hot ends of the electrodes;
- marking all completed 'hot' work;
- wearing goggles to protect eyes and face when chipping the hot slag from the weld;
- ensuring the surrounding area is clear of flammable material and liquids;
- having the recommended type of fire extinguisher nearby.

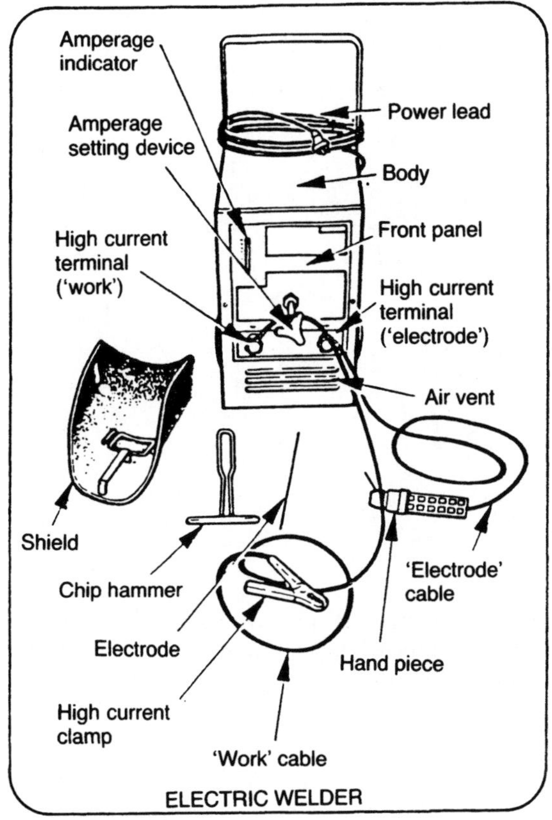

ELECTRIC WELDER

Source: AGPS

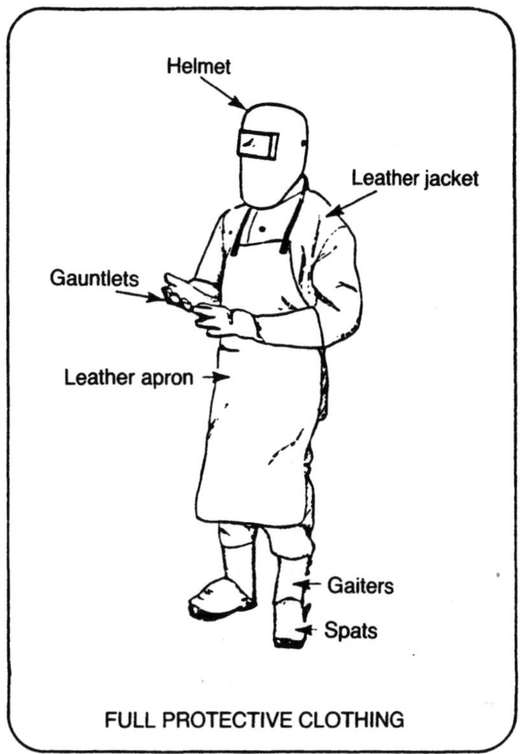

FULL PROTECTIVE CLOTHING

Source: AGPS

Fumes and the harmful effects of fumes can be avoided by:

- ensuring adequate ventilation is available within the work area;
- using an extraction fan;
- wearing approved breathing apparatus when welding galvanised or similar coated metals.

Arc burns and eye damage can be avoided by:

- using the approved welding masks and hand shields to protect the head and neck;
- wearing the recommended safety clothing;
- taking care when striking an arc;
- warning assistants when you are about to strike an arc;
- using adequate screening to protect other workers.

Note: Eye damage can be caused by infra-red and ultra-violet rays coming from the arc welder. To protect eyes from damage, always wear an approved arc welding mask, or hand shield.

Preparation for arc welding

1. Ensure the welding machine is placed in a clean, clear, dry position in the workshop.
2. Remove all flammable materials and liquids from the area.
3. Switch on extraction fans and check ventilation of the workshop.
4. Thoroughly check the welding machine and the leads to ensure they are in a safe, working condition.
5. Check and secure the welding machine's 'earth' clamp.
6. Inspect the electrode holder for burns or cracks.

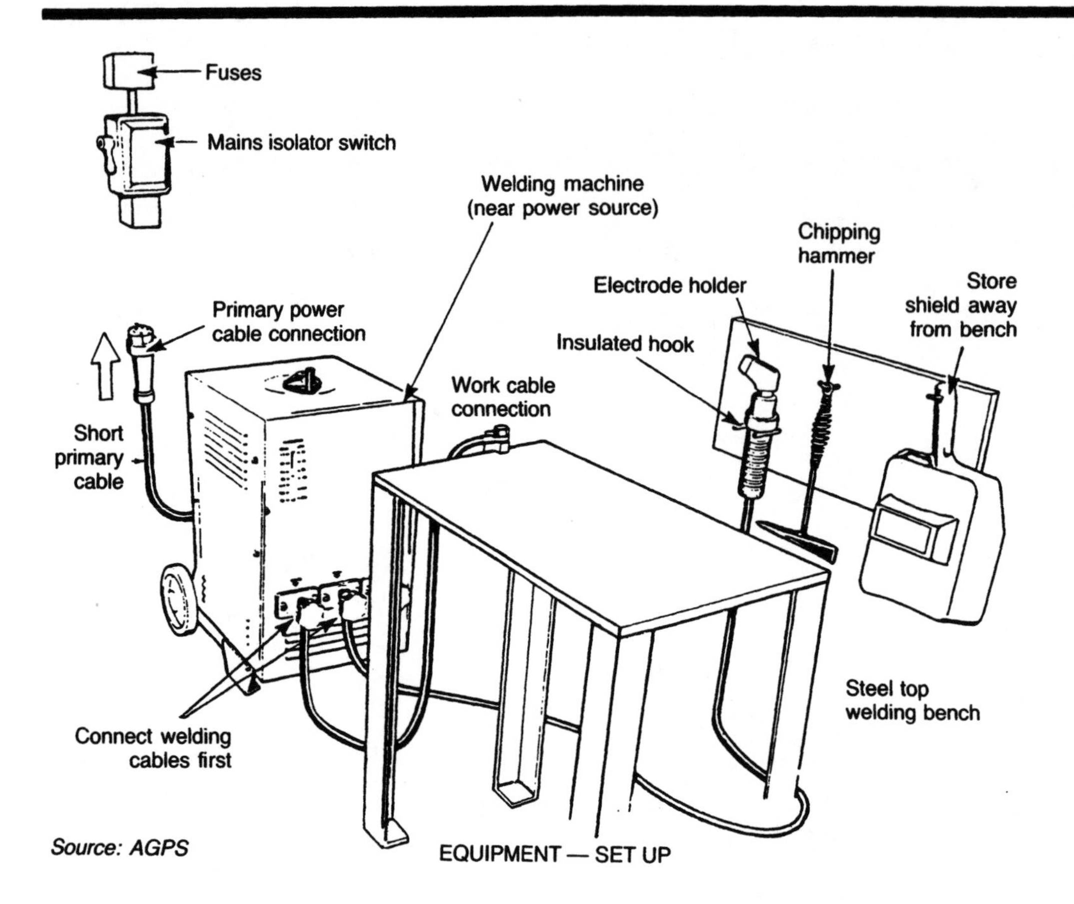

Source: AGPS

EQUIPMENT — SET UP

7 Set the amperage to suit the selected electrode and metal thickness of the material to be welded.
8 Erect screens to protect other workers and onlookers from arc flashes.
9 Have the correct type of fire extinguisher nearby.
10 Inspect the condition of all safety clothing and equipment.
11 Put on the safety clothing and equipment.
12 Ensure the metal to be welded is secure, before starting to weld.

Arc welding procedure

During the process of arc welding, the temperature at the arc can exceed 5000°C. This very high temperature melts the end of the electrode into a molten pool, formed by the fusion of the metals being welded. The flux on the electrode assists the arc welding process by providing a gas shield around the arc and depositing a layer of protective slag around the weld.

Bead runs procedure

1 Select a piece of clean, mild steel plate, approximately 6 mm thick, 100 mm long and 50 mm wide, having square sheared edges.
2 Set up the welding equipment, and wear the necessary safety clothing (refer to 'Preparation for arc welding').
3 Place the piece of steel plate onto the welding bench.
4 Select the recommended electrode, suitable for flat position welding of 6 mm mild plate.
5 Set the welding current to the recommended amperage.
6 Strike an arc (similar action as striking a match) onto one end of the plate.

7 Position the electrode as shown in the illustration below.

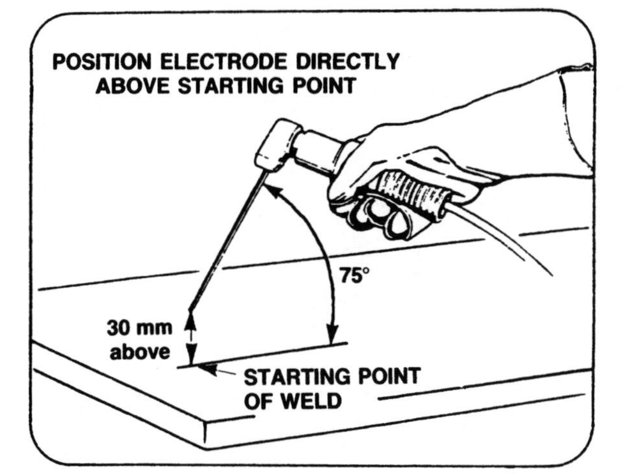

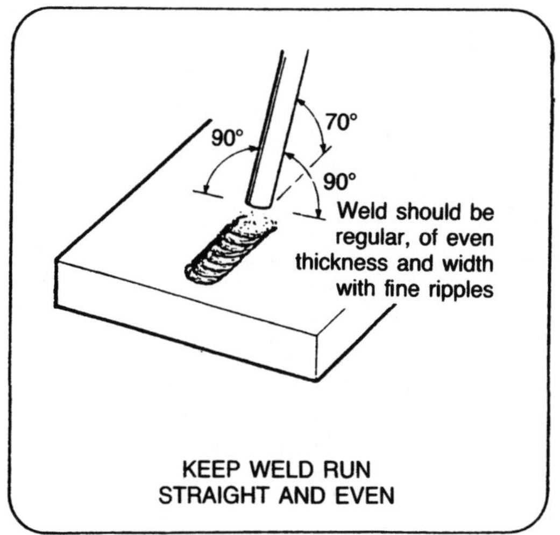

Source: AGPS

8 Once the arc has been established, reduce the arc length to approximately 3 mm by pushing the electrode towards the molten pool. Move the electrode slowly along the weld path, keeping the molten pool at a constant width.

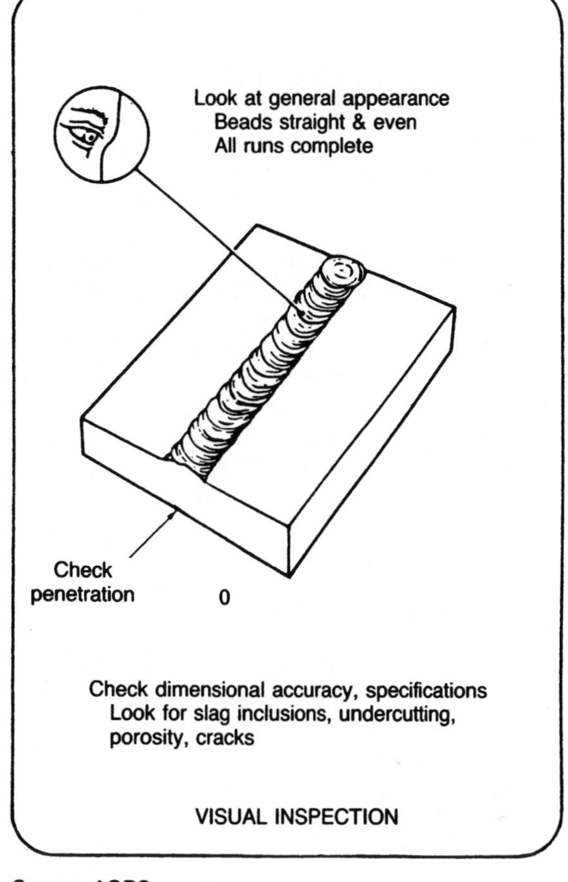

Source: AGPS

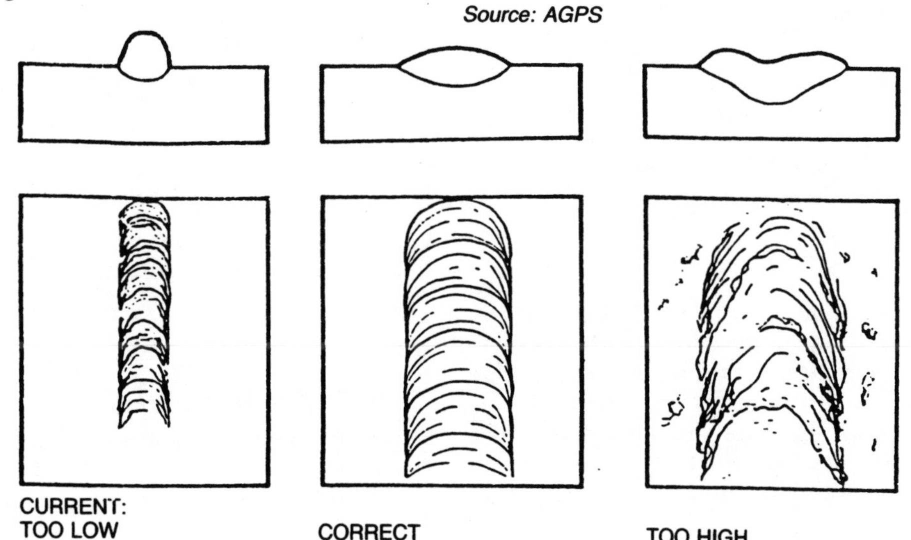

Source: AGPS

9 By ensuring that the arc length and the molten pool are kept constant, the weld will proceed at a steady rate.

10 On completion of the bead run weld, allow the plate to cool, chip off the slag (wear goggles), clean the weld area using a wire brush, and inspect the weld.

BUTT WELD PROCEDURE

1 Select two pieces of clean, mild steel plate, approximately 6 mm thick, 100 mm long and 50 mm wide, having square sheared edges.

2 Set up the welding equipment, and wear the necessary safety clothing.

3 Place the two pieces of plate onto the welding bench, the edges of the plates butted together along the length of the joint.

Note: Leave a gap of approximately 2 mm between the plates to assist in weld penetration and minimise distortion.

4 Select the recommended electrode and check the amperage setting.

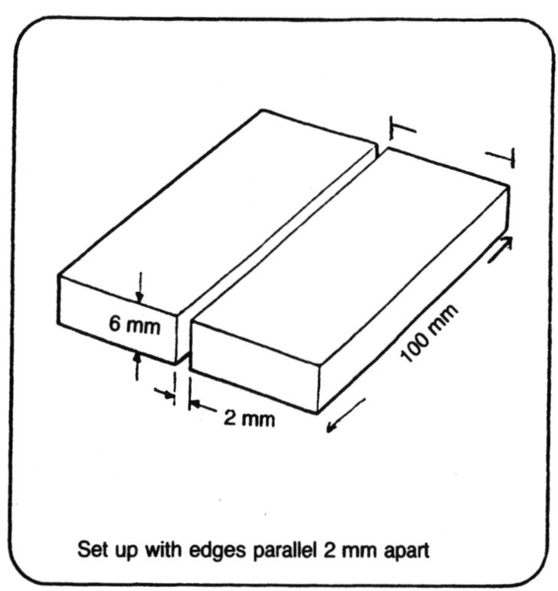

Set up with edges parallel 2 mm apart

Source: AGPS

5 Strike an arc and 'tack' weld the 'butt' at both ends. Refer to lower right illustration.

6 Weld the two pieces of mild steel together as shown in the illustration opposite.

7 By ensuring that the arc length and the molten pool are kept constant, the weld will proceed at a steady rate.

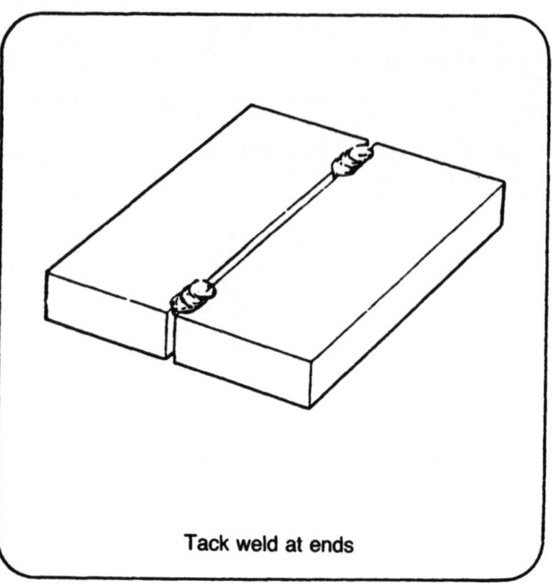

Tack weld at ends

Source: AGPS

8 On completion of the butt weld, allow the plates to cool, chip off the slag (wear goggles), then clean the weld area and inspect the weld.

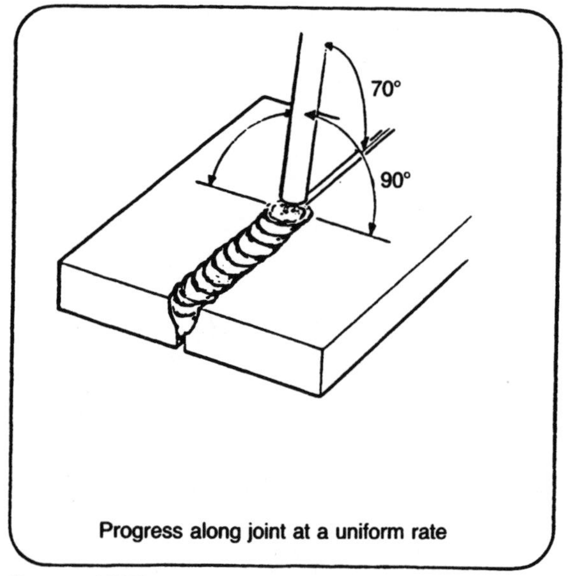

Progress along joint at a uniform rate

Source: AGPS

LAP WELD PROCEDURE

1 Select two pieces of clean mild steel plate, approximately 6 mm thick, 100 mm long and 50 mm wide, having square sheared edges.

2 Set up the welding equipment and wear all the necessary safety clothing.
3 Place the plates onto the welding bench and position them as shown in this illustration.

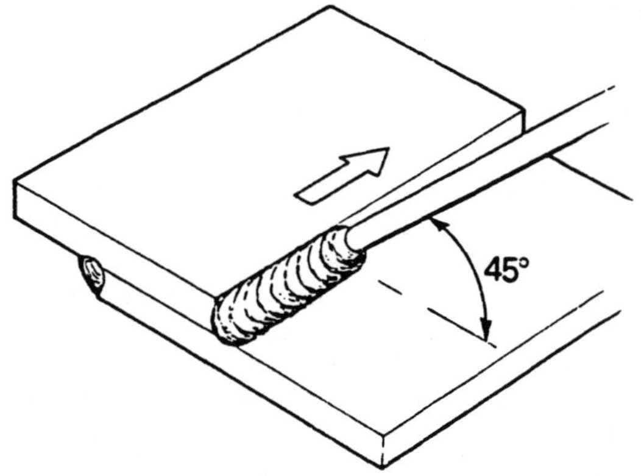

Source: AGPS

4 Select the recommended electrode and check the amperage setting.
5 Tack weld both ends of the plates.
6 Lap weld both plates as shown in the illustration above.
7 By ensuring that the arc length and the molten pool are kept constant, the weld will proceed at a steady rate.
8 On completion, allow the plates to cool, chip off the slag (wear goggles), then clean and inspect the weld.

METAL INERT GAS (MIG) WELDING

The illustration below shows the major components of a MIG welding plant.

Safety in MIG welding is a matter of applied common sense. The process of MIG welding can produce:
- electric shocks;
- burns;
- arc burns and eye damage;
- fumes.

Electric shocks can be avoided by:
- ensuring all the electric cables are insulated and free of internal and external defects;
- switching off all electrical appliances when not in use;
- wearing the approved safety clothing and equipment;
- ensuring the work area is dry.

Burns can be avoided by:
- wearing the approved clothing and equipment;
- marking all completed 'hot' work;
- ensuring the surrounding area is clear of flammable material and liquids;
- having the recommended type of fire extinguisher nearby.

Arc burns and eye damage can be avoided by:
- using the approved welding masks and hand shields to protect the eyes, head and neck;
- wearing the recommended safety clothing;
- warning assistants when you are about to strike an arc;
- using adequate screening to protect other workers.

Note: Eye damage can be caused by infra-red and ultra-violet rays coming from the welding process.

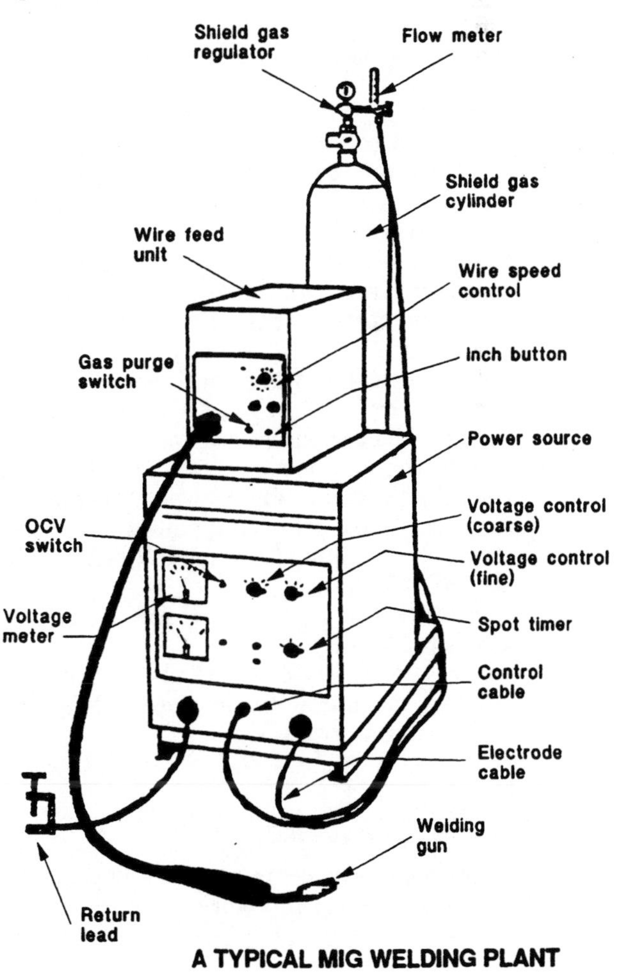

A TYPICAL MIG WELDING PLANT

Fumes. Unlike arc welding, the amount of fumes produced during the MIG process is low. However, the effects of fumes can be avoided by:

- ensuring adequate ventilation is available within the work area;
- using an extraction fan when working in a confined area;
- wearing approved breathing apparatus when welding galvanised or similar coated metals.

Preparation for MIG welding

1. Ensure the welding machine is placed in a clean, dry position in the workshop.
2. Remove all flammable materials and liquids from the area.
3. Switch on extraction fans and check ventilation of the workshop.
4. Thoroughly check the welding machine and the leads to ensure they are in a safe condition.
5. Check and secure the welding machine's 'earth' clamp.
6. Inspect and clean the torch nozzle and liner, feed rollers and wire guides.
7. Ensure the correct size and type of electrode wire has been installed into the wire drive unit.
8. Erect screens to protect other workers and onlookers from arc flashes.
9. Have the correct type of fire extinguisher nearby.
10. Set the voltage to the specified value.
11. Set the wire-feed speed to a value that will produce the recommended welding current.
12. Adjust the gas flow control valve to a setting of approximately 10 l per minute.

Note: The values for setting points 10, 11 and 12 should be obtained from a 'MIG WELDING VARIABLES' chart.

13. Inspect the condition of all safety clothing and equipment.
14. Put on the safety clothing and equipment.
15. Ensure the metal to be welded is secure before starting to weld.

MIG welding procedure

One of two methods is used to deposit the electrode wire onto the work during the MIG welding process. The two methods of MIG welding are:

- spray arc method;
- short arc method.

The method used is automatically selected when the MIG welding plant's settings are adjusted for a particular welding task.

Spray arc method

This is a high current range method which is effective for welding thick metals. During the spray arc welding process, a very high temperature is produced at the end of the electrode wire and on a small area of the metal. This very high temperature melts the end of the electrode wire into a molten pool, formed by the fusion of the metals being welded. The gas flow, from the nozzle, forms a protective shield around the arc and the molten pool, to ensure a clean slag-free weld.

Short arc method

This is a low current range method which is effective for welding thin metals such as those used in the automotive industry. During the short arc welding process, a high temperature is produced at the end of the electrode wire and on a small area of the metal. This high temperature melts the end of the electrode wire and a molten 'blob' of metal stretches out (like chewing gum) until it enters a molten pool formed by the fusion of the metals being welded. The instant the blob hits the pool, the arc stops, causing the string of molten metal to break. Once the string is broken, the arc restarts in order to repeat the cycle. The gas flow, from the nozzle, forms a protective shield around the arc and the molten pool, to ensure a clean slag-free weld.

BEAD WELD PROCEDURE

1. Select a piece of clean, mild steel plate, 3 mm thick, 100 mm long and 50 mm wide.
2. Place the mild steel plate onto the welding bench.
3. Check the setting on the welding plant.
4. Inspect and adjust the stick-out of the electrode wire to about 7 mm.

Note: Stick-out is the distance that the electrode wire protrudes beyond the end of the nozzle tip.

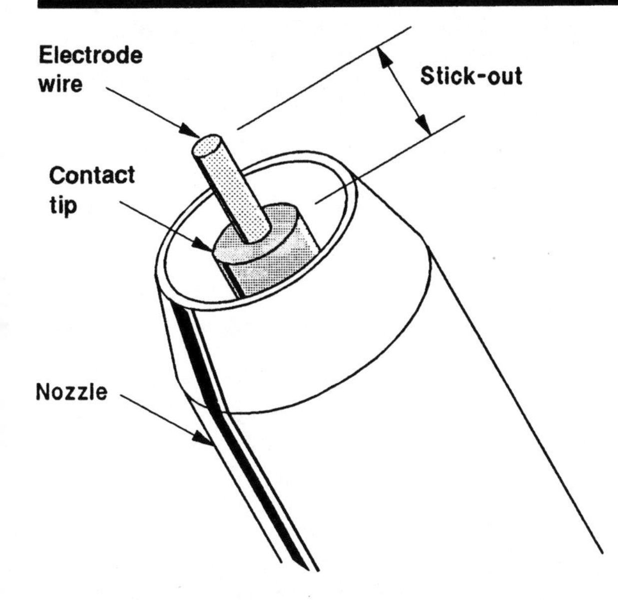

Adjust the stick-out of the electrode wire.

5 Put on the safety clothing and equipment.
6 Position the nozzle tip a distance of 10 mm above the piece of metal.
7 Hold the torch perpendicular to the work, then tilt the torch 30° towards you while preventing it from tipping to one side.

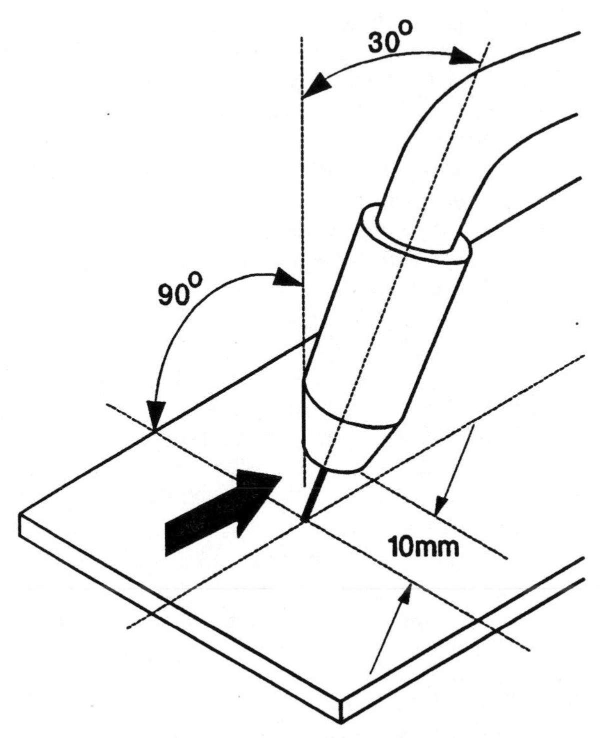

Hold the torch in the correct position.

8 Lower the welding mask by shaking your head.
9 Squeeze the trigger on the torch to establish the arc.
10 Maintain a nozzle tip distance of 10 mm as you move the torch towards you at a constant rate.

Note: The rate of movement of the torch is determined by watching the size of the molten pool as it forms in the metal. When the rate is too slow, the bead will be too high. When the rate is too fast, the bead will be too low, too thin and may even be broken.

11 At the end of the run, release the trigger on the torch, reduce the torch movement, but do not lift the torch away from the work for a few seconds.
12 Examine the bead. It should be a same width and height along the run.
13 When the bead has defects, readjust the voltage and wire feed setting.
14 Using a pair of cutting pliers, remove the blob from the end of the electrode wire and adjust the stick-out.
15 Repeat steps 6 to 14 until a correct bead has been formed on the metal plate.

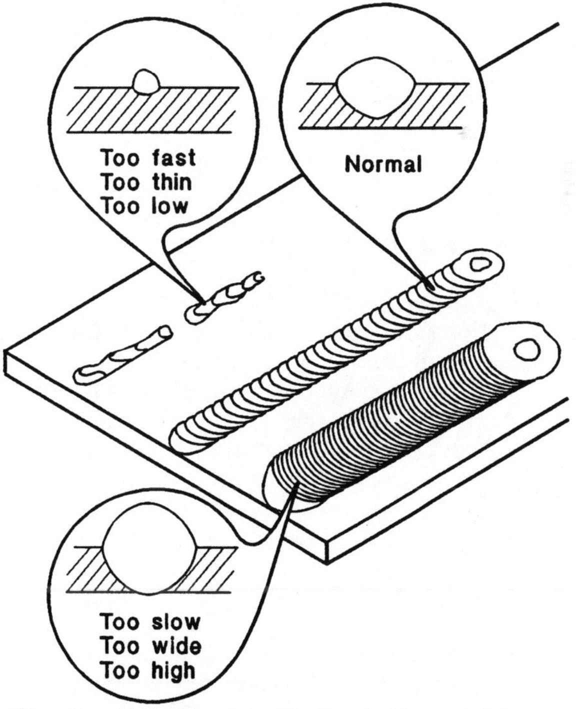

The bead pattern will show the problems with the welding process.

16 Turn off the gas and electrical supplies to the welder.
17 Clean the equipment, particularly the torch nozzle.

Note: Use an anti-spatter spray on the torch nozzle to help keep it clean.

18 Clean the work area and store the protective clothing.

BUTT WELD PROCEDURE

1 Select two pieces of clean, mild steel plate, 3 mm thick, 100 mm long and 50 mm wide.
2 Place the mild steel plates side by side onto the welding bench with a 1.5 mm gap between them.
3 Check the setting on the welding plant.
4 Inspect and adjust the stick-out of the electrode wire to about 7 mm.

Note: Stick-out is the distance that the electrode wire protrudes beyond the end of the nozzle tip.

5 Put on the safety clothing and equipment.

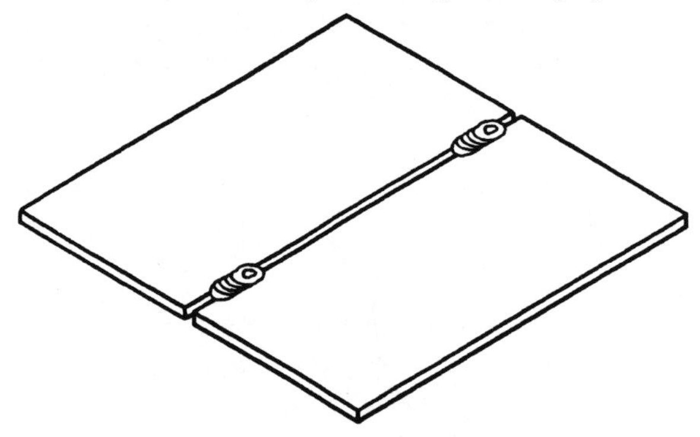

Tack weld the two plates.

6 Run a tack-weld.
  - Position the nozzle tip a distance of 10 mm above and 10 mm from one end of the gap between the pieces of metal.
  - Hold the torch perpendicular to the work, then tilt the torch 30° towards you while preventing it from tipping to one side.
  - Lower the welding mask by shaking your head.
  - Squeeze the trigger on the torch to establish the arc.
  - Maintain a nozzle tip distance of 10 mm as you move the torch towards you, at a constant rate, for about 5 mm.
7 Repeat the tack-weld at the other end of the gap in a similar manner as described in step 6.
8 Run a weld the full length of the gap.
9 Examine the weld. It should be of the same width and height along the full length of the gap.
10 Check the weld penetration.
  - Place the work in a vice so the middle of the weld is positioned just above the jaws.
  - Bend the protruding plate towards the welded face.
  - Examine the under-side of the weld for cracks or flaws.

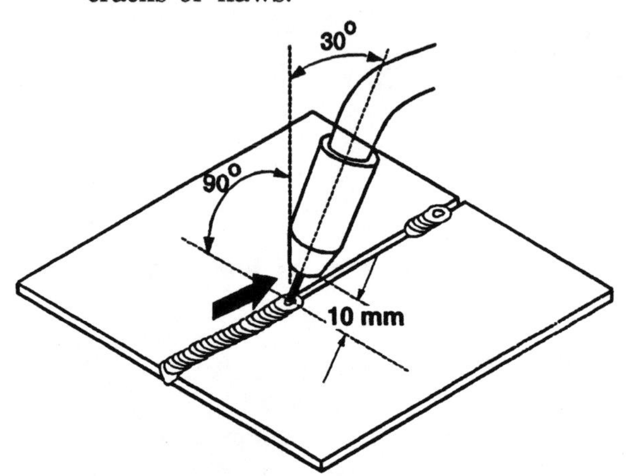

Weld the two plates.

Note: Too much weld penetration is indicated by the weld being below the top face of the plate and by weld also protruding below the bottom face of the plate. Not enough weld penetration is indicated by a mound of weld above the top face of the plate and by a large crack along the bottom face of the plate.

11 When the weld has defects, readjust the voltage and wire feed setting.
12 Using a pair of cutting pliers, remove the blob form the end of the electrode wire and adjust the stick-out.
13 Repeat the above steps until the plates have been correctly welded.
14 Turn off the gas and electrical supplies to the welder.
15 Clean the equipment, particularly the torch nozzle.

Note: Use an anti-spatter spray on the torch nozzle to help keep it clean.

16 Clean the work area and store the protective clothing.

LAP WELD PROCEDURE

1 Select two pieces of clean, mild steel plate, 3 mm thick, 100 mm long and 50 mm wide.

2 Place one of the mild steel plates onto the work bench and then place the other plate on top of the first plate as shown in the illustration.

Note: A small piece of 3 mm plate should be placed between the outer edge of the top plate and the work bench to support the plate until it has been tack welded.

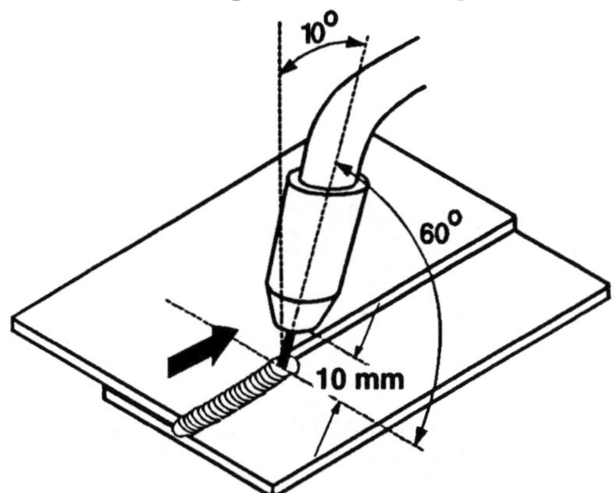

Lap weld the two plates.

3 Check the setting on the welding plant.

4 Inspect and adjust the stick-out of the electrode wire to about 7 mm.

Note: Stick-out is the distance that the electrode wire protrudes beyond the end of the nozzle tip.

5 Put on the safety clothing and equipment.

6 Run a tack-weld.
- Position the nozzle tip a distance of 10 mm above and 10 mm from one end of the 'Vee' formed by the two pieces of metal.
- Hold the torch 60° to the work, then tilt the torch 10° towards you while maintaining the 60° angle. See the illustration.
- Lower the welding mask by shaking your head.
- Squeeze the trigger on the torch to establish the arc.
- Maintain a nozzle tip distance of 10 mm as you move the torch towards you, at a constant rate, for about 5 mm.

7 Repeat the tack-weld at the other end of the 'vee' in a similar manner, as described in step 6.

8 Tip the plates over to show the other 'Vee' formed by the plates.

9 Run a weld the full length of the 'Vee'.

10 Examine the weld. It should be of the same width and height along the full length of the 'Vee'.

11 Tip the plates over to show the 'Vee' which had been previously tack welded.

12 Run a weld the full length of the 'Vee'.

13 Examine the weld. It should be of the same width and height along the full length of the 'Vee'.

14 When the weld has defects, readjust the voltage and wire feed setting.

15 Using a pair of cutting pliers, remove the blob from the end of the electrode wire and adjust the stick-out.

16 Repeat the above steps until the plates have been correctly welded.

17 Turn off the gas and electrical supplies to the welder.

18 Clean the equipment, particularly the torch nozzle.

Note: Use an anti-spatter spray on the torch nozzle to help keep it clean.

19 Clean the work area and store the protective clothing.

CHAPTER 5 REVISION

1 List a minimum of seven safety precautions to be observed when using gas welding equipment.

2 Name the components and accessories required for oxy-acetylene welding.

3 Describe the recommended method of setting up the equipment for gas welding. Include leak testing and the regulator pressures for butt welding of 1 mm panel steel.

4 Sketch three types of flames used in the process of oxy-welding.

5 Describe the processes in which each of these flames is used.

6 What are the oxygen and acetylene regulator pressures set to when cutting 6 mm thick mild steel plate?

7 Describe the procedure for 'forehand' welding of 1 mm panel steel.

8 Give three reasons for using welding goggles when oxy-welding.

9 Name at least three precautions for EACH of the following electric welding hazards.
- Electric shocks.
- Burns.
- Arc flashes.
- Fumes.

10 Describe a method of electric 'butt' welding two pieces of 6 mm thick mild steel plate. Include the amperage setting and the type of electrode to be used.

11 List the personal safety equipment to be used when electric welding and the possible personal dangers if this safety equipment is not used.

6

LUBRICANTS AND VEHICLE SERVICING

LUBRICANTS

Engine oil

VISCOSITY/GRADING

Oil companies conduct exhaustive tests to ensure the lubricants specified for a vehicle application will meet the standards imposed by the vehicle manufacturer. To identify an oil as engine oil is not sufficient, as each engine has requirements based on type, application and climatic conditions. The servicing mechanic should have a working knowledge of the terminology associated with oils, be capable of correctly identifying the oil from information on the manufacturer's package codings and use only those gradings specified by the oil company and/or vehicle manufacturer. Being aware of the specialised nature of lubricants can assist the servicing mechanic to avoid the problems caused by the use of lubricants unsuited for the task.

The most obvious difference in oils, apart from colour, is the flow rate when poured from a container. This flow rate (fluidity) and body of the oil is known as VISCOSITY. A number grading is applied to oils to indicate the viscosity rating; the lower the number the greater the fluidity (thinner) of the oil. As the number grading becomes higher the fluidity of the oil becomes less and the body of the oil increases (oil becomes thicker). Many oils have two numbers and a letter to denote the grading, for example 15W-40. This indicates the oil has the fluidity of a '15' oil, the body and load carrying capability of a '40' oil and the 'W' means the oil was cold weather tested and is suitable for cold weather driving. Oils designated in this manner are known as MULTI-VISCOSITY oils.

The American Petroleum Institute applies a service rating to oils. The blend of additives determines the service rating and as vehicle manufacturers change engines and lubricant requirement A.P.I. conducts tests and issues a standard to which oils must conform if they are to be marketed as meeting the service rating. Generally the new rating replaces the old standard and the earlier rating becomes discontinued. Service ratings for petrol engines have progressed from SA through SB, SC, SD, SE and SF to the current rating of SG. Compression ignition (diesel) engines use different rating codes, for example CA, CB and CC, with CD applicable to engines in severe operating conditions. The packaging of the oil should be checked for this information as use of the wrong oil could invalidate the engine warranty if failure occurs.

ADDITIVES

The allowable distance travelled between oil changes seems to vary considerably and a vehicle owner can be excused for becoming confused by conflicting claims relating to the 'correct' oil change period. Knowledge of the additive action in oils can assist the servicing mechanic to explain to owners why the oil change period for the same vehicle can vary depending on the usage of the vehicle.

Anti-scuff additives protect camshafts, pistons and cylinder walls by polishing the moving parts. High temperature engine operation reduces the life of this type of additive, therefore the oil change period is reduced.

Corrosion inhibitors attack the byproducts of combustion, reducing the formation of harmful acids. Winter, short trips, cold engine and rich mixtures increase the amount of acids produced, therefore the oil change period can be altered by seasonal factors.

Oxidation inhibitors are sludge and varnish fighters. An engine used on cold start, short trips promotes the formation of sludge. This engine may then be subjected to high speed, high temperature driving which will convert part of the sludge to varnish. These deposits affect the operation of hydraulic lifters, piston rings, valves etc. An effective oxidation inhibitor chemically attacks sludge and varnish to lessen the effect of these deposits.

Detergent dispersants work on the concept of keeping contaminants in suspension in the oil. An effective filter will trap the contaminants and keep the engine clean.

Foam inhibitors lessen the tendency for the oil to aerate when agitated by the moving engine parts. If oil foam is developed, vital lubrication surfaces in the engine may be starved of oil and engine failure will occur.

Viscosity index improvers lessen the thinning effect that increases in temperature have on oil. The use of viscosity index improvers enable the oil to flow readily when cold but provide the oil with greater body as the operating temperature of the engine increases.

Pour point depressants reduce cold start cranking effort by increasing the fluidity of the oil.

Caution: Use approved oils only. Oils which are not approved by the vehicle manufacturer may not provide the required level of engine protection.

Gear oils

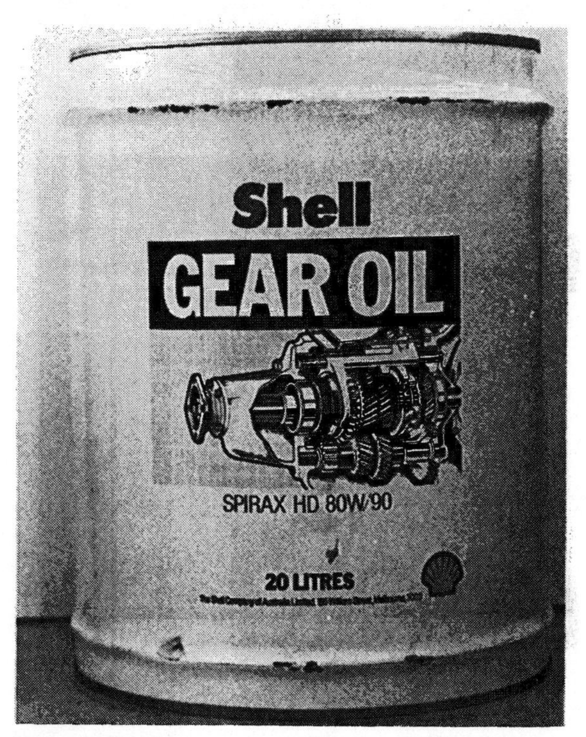

Typical gear oil.

Gear boxes and final drive units contain a variety of gear types, including special requirements such as limited slip differentials. The meshing of gear sets such as spiral bevel, spur bevel, worm drive and hypoid can cause extremely high wiping forces across the gear faces, also the agitating action of the gears in the oil attempts to create oil foam. To minimise these problems and the substantial heating effects produced by transmissions operating under heavy load, it is essential to select the correct oil for the transmission and final drive units.

The American Petroleum Institute issues a classification for transmission oils based on the design of the gear unit and the anticipated service loadings. Gear oils also have an S.A.E. viscosity number, but there is no connection between crankcase oils and the gear lubricants series of numbers.

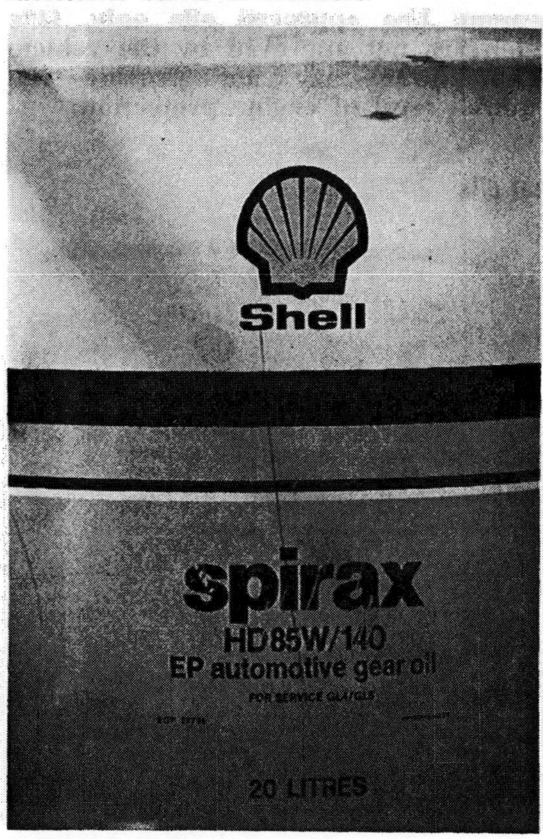

The gear lubricant service designation (eg GL4/GL5) must match the gear requirements.

A.P.I. GEAR LUBRICANT SERVICE DESIGNATION

- GL 1 — Straight mineral gear lubricant.
- GL 2 — Worm drive gear lubricant.
- GL 3 — Extreme pressure gear lubricant for manual transmissions and spiral bevel gear sets operating under moderate service conditions.
- GL 4 — Extreme pressure lubricant suitable for hypoid final drive units in most automotive applications.
- GL 5 — Extreme pressure multipurpose gear lubricant for high speed/low torque low speed/high torque and high speed shock load service requirements of hypoid gear and some limited slip differentials.

Caution: Use of oil gradings heavier or lighter than specified by the transmission manufacturer may result in increased bearing and synchroniser wear.

Auto-transmission fluids

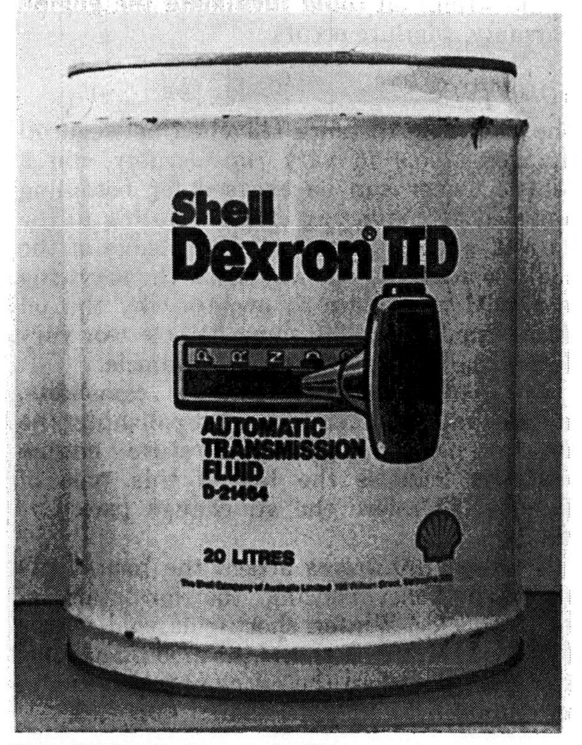

Typical automatic transmission fluid.

Automatic transmissions require a fluid that can resist heat (high oxidation stability), have resistance to foaming, corrosion, and possess good detergency and anti-wear properties. Transmission fluid is red in colour, however, the servicing mechanic must be careful when checking oil colour. Some fluids will, after a short time in service, darken perceptibly but this change in colour does not indicate a transmission malfunction.

The choice of fluids available for passenger vehicles is limited to three types with many sales outlets stocking only one, perhaps two, of the fluids. Most vehicles tend to use Dexron 11, however, Ford prefers the use of a fluid possessing a slightly different frictional characteristic. Mercedes-Benz (and Renault for a number of its transmissions) recommends the use of a fluid which is slightly different from both the previous types. Use of a non-recommended fluid could cause a change in the 'shift quality' of the transmission.

Grease

Follow the oil companies' recommendations and use different grease types where required.

The type of loads, temperature and exposure to water which will be applied to the vehicle component generally determine the type of grease to be used. Each oil company markets a range of greases to suit the requirements of the vehicle manufacturer. Unfortunately, the package coding relates to the marketing effort by the company and the relationship between symbols such as 77, Retinax A, EP grease 2, MP, No. 3L, Graphited No. 3 is difficult to establish. The oil companies' recommendation charts will specify the grease type suitable for the required task. Do not use grease which is not specifically recommended.

Greases are manufactured from different bases. A general outline of the grease composition and the type of automotive application in which the grease is normally used will alert the service mechanic to the need for care in lubricant selection.

Multi-purpose does not mean the grease is suitable for every application. Multi-purpose grease can be soap or lithium based. Soap base grease can be manufactured as a heavy duty grease suitable for extreme pressure applications with good resistance to water and salt attack. However, this type of grease has a low melting point and may fail in high temperature applications (disc brake hub wheel bearings).

Lithium based greases can be blended with molybdenum disulphide to provide extra heavy duty protection (or increase the length of the service period). Other lithium base greases are available for multi-purpose use such as chassis wheel bearings (except where dics brakes are fitted), universal joints and water pumps.

Wheel bearings fitted to a vehicle equipped with disc brakes are subjected to extreme temperatures and a clay base grease, capable of withstanding temperatures up to 260°C, is normally used.

In addition to commonly used types of grease a number of special greases are manufactured. By using a semi-fluid lime-lead based grease as a lubricant it is possible to provide for applications such as gear sets which require the grease to 'self feed' into the moving parts as the grease is consumed, such as some chain saw gearboxes. A lime based grease containing graphite applied to truck spring saddles will withstand the washing action of water from the road in this exposed location and still protect the spring saddle parts.

Brake fluid

Keep container tightly capped when not dispensing fluid.

Brake fluid is the working hydraulic fluid for automotive braking systems. Fluid manufacturers observe strict quality standards, ensuring the fluid is compatible with other brands and will not affect the sealing cups and rubbers within the braking system. The fluid is dyed (usually blue or green) for ease of identification. The major problem in service is not in identification of type (low boiling point fluids are not marketed for automotive use) but in service procedure, handling and contamination. Follow the manufacturer's instructions on storage and spillage problems.

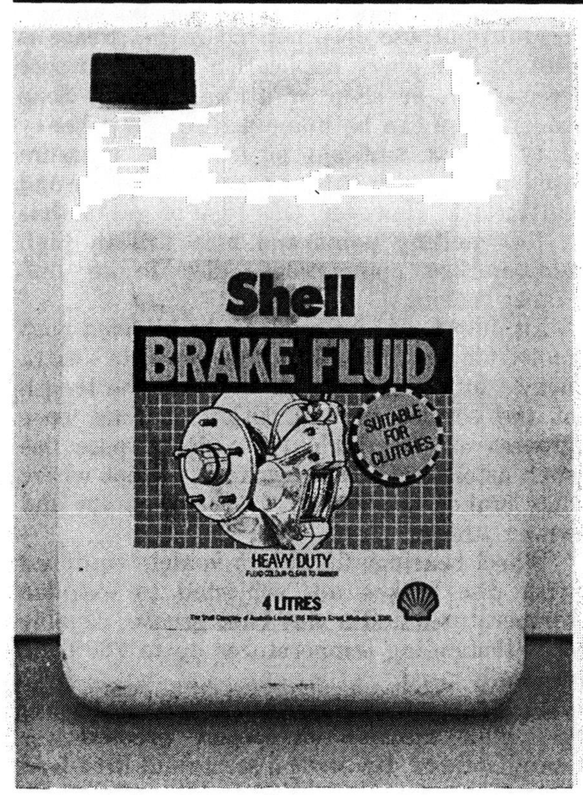

Keep container tightly capped when not dispensing fluid.

Cooling system conditioners

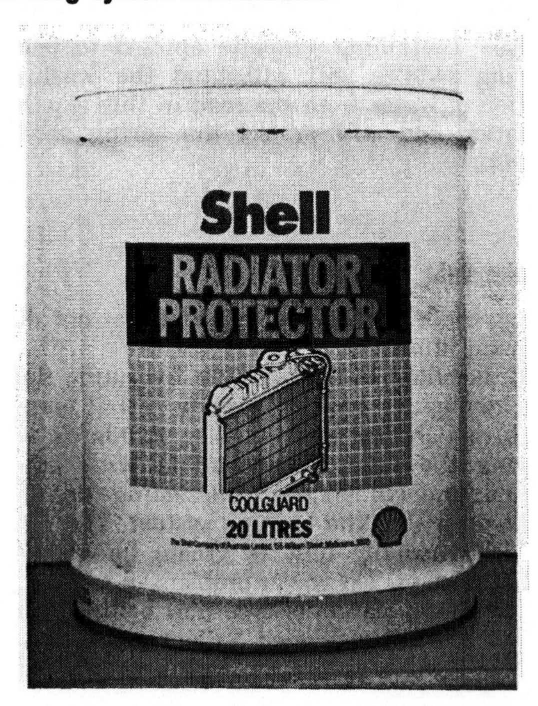

Protect vehicle engines by using cooling system additives.

Many different brands of conditioners are available and care must be taken if mixing different types. Not all conditioners are intended for the same purpose, and they may not be compatible. Unless the conditioner type is known the vehicle cooling system (including the heater) should be drained and flushed before adding conditioner. A conditioner is essential in alloy based engines and the manufacturer's specification should be consulted for the type and the concentration required to combat corrosion.

Specialty products

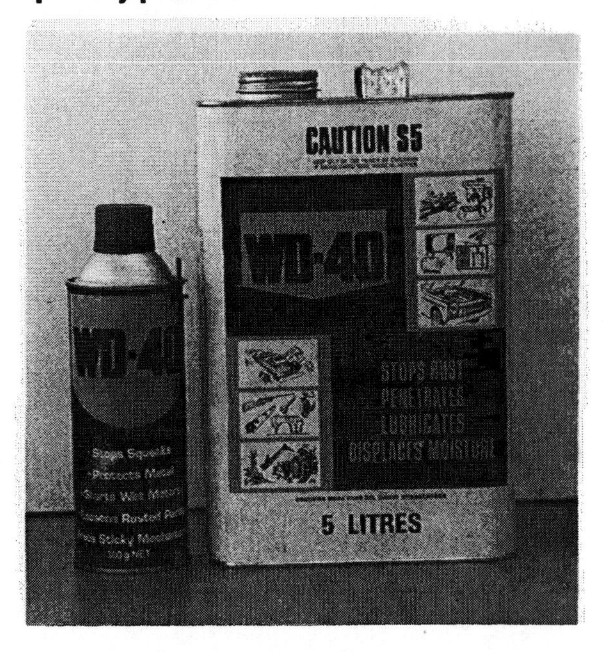

Many other servicing aids are available, such as Drilube for door locks, rubber lubricant for door seals, penetrating oil for freeing rusted components, blended oils for two-stroke engines, powdered graphite for door lock tumblers etc. For all applications, follow the instructions on the product package.

EQUIPMENT IDENTIFICATION

Before servicing a vehicle the operator should have available, and understand the use of, each of the units listed below.

Modern service hoists.

Vehicle hoist

Hoists are available in several types and lift capacities. They are of rugged construction and contain substantial reinforcement sections. Because of the safety considerations, all hoists have a safety lock built in to the design. Locate and learn to use the safety lock before operating the hoist.

The hoist can be a:

- single post frame lift;
- double post frame lift (operating in a similar manner to the single post type and usually only found in heavy vehicle servicing areas);
- double post body lift;
- four post platform lift.

Many hoists use hydraulic fluid for the lift but the method of supplying force to the fluid varies. Frame contact types (single or double) generally use compressed air to force fluid into a lift ram, raising the vehicle. The same objective is achieved by the body/platform type hoists but the hydraulic fluid is pressurised by an electric motor driving a high pressure pump. It is also possible to use an electric motor to drive a screw thread located in the lift posts to provide the lifting action.

Jacks (hand operated)

Universal type garage jack.

A hydraulic jack is a relatively simple device consisting of a pumping chamber, a lifting ram, a one-way valve and a control valve which can be used by the operator. The jack parts are enclosed in a single strong steel container.

The basic jack (with suitable rearrangement of the component parts) is fitted into a steel trolley with four steel wheels, a lifting arm and saddle and becomes a standard garage floor jack. This type of jack,is one of the most versatile tools available to a mechanic. Learn to use the garage floor jack correctly and safely, as it can be used in many lifting applications.

Specialised variations of the jack are available where the amount of work or difficulty of the task justifies the expense of these units. Two examples of specialised jack applications are:

1 Transmission jacks. These are available as attachments to a floor jack for low-level work or as a complete unit for removing transmissions when the vehicle is on a vehicle hoist.
2 Porta-power. Panel beaters use this tool extensively when straightening frames and panels. The pumping chamber and the ram are in separate steel housings joined by a high pressure flexible hose.

Safety stands

Safety stands can carry the load on a single, three or four leg base. Height adjustment can be by screw thread or by locating a pin in a series of spaced holes. The common factor with each of the types is they are all robustly constructed in steel and provide a stable base to support a vehicle load.

Screw thread type safety stands.

Grease guns

The grease gun ranges from a very simple hand lever operated type to a compressed air powered unit operating from a large drum of grease, which can be located in another room, with grease being supplied to the lubrication gun via a high pressure flexible

Air operated mobile grease gun.

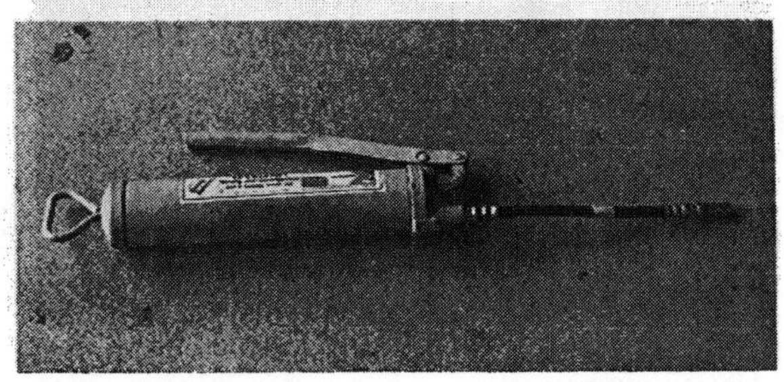

Hand operated grease gun.

hose. Regardless of the supply method the gun head consists of a fitting suitable for attaching to the vehicle greasing points and a trigger or lever to control the flow of grease into the grease nipple on the vehicle. The grease fitting can be attached to the gun body by either a flexible hose or a solid tube.

Servicing areas have a combination of grease gun types — generally the air powered units operating from a roll of hose (concealed in an overhead or cabinet dispenser for neatness) supply the commonly used multi-purpose chassis grease. A number of hand operated guns containing special purpose greases are kept on a rack for use as required.

Oil/fluid dispensers

The simplest method of dispensing oils and fluids is from a container directly into the vehicle through a funnel to prevent spillage. This technique, while suitable for a single application, would be expensive to use in a high volume business. The cost penalties in packaging, storage and disposal of hundreds of containers cause the large operator to install dispensing guns. The dispensers are similar in appearance to an air powered grease gun. Each fluid has a nozzle fitted with a flow control valve and is connected by a flexible hose to a bulk container. The container can be subjected to air pressure to provide a continuous flow of fluid. This type of supply is suitable for engine oil, automatic transmission fluid, gear box and rear axle oils. Water can also be supplied in a similar manner, but the hose would be connected to the mains supply.

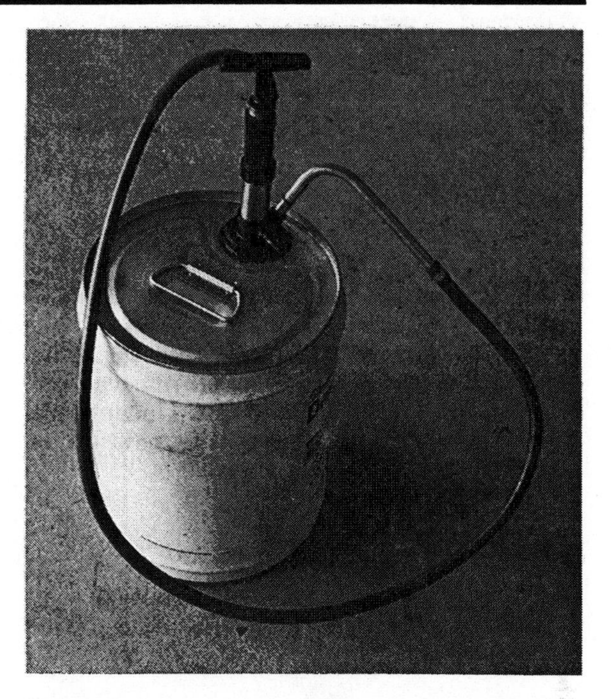

Hand operated pump fitted with flow control valve and inserted into an oil drum.

Hand operated fluid pump.

For the workshop requiring a more cost-effective method than single-use containers but where the business volume does not warrant a full lubritorium it is possible to obtain hand operated pumps which can be inserted into a drum of fluid. The pump has attached to it a length of hose and a flow valve. By opening the valve and pushing down on the pump the operator can supply fluids in a similar way to the powered systems.

Suction guns

As the name implies, this gun removes fluids from vehicle units. The gun is fitted with a flexible hose and is constructed as a hollow tube containing a sliding plunger which can be withdrawn to create a suction. Suction guns are particularly useful for withdrawing oil from a unit not equipped with drain plugs

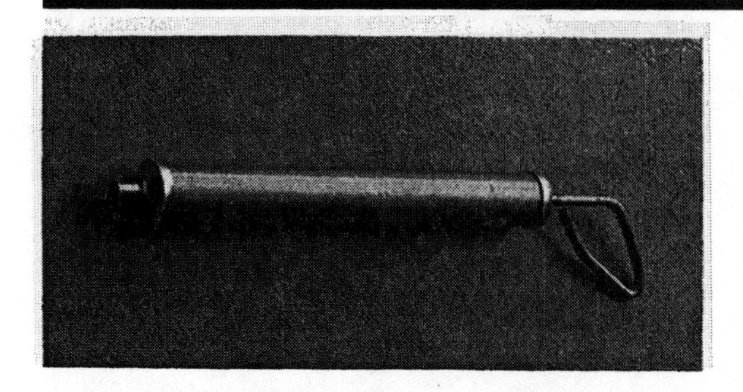

(some differentials). Take care that the same gun is not used for both dirty oil removal and the addition of new oil to a unit. Under these conditions contamination of the new oil is inevitable.

Hydraulic fluid dispensers

Brake fluid has a particular characteristic — it will attract moisture from the atmosphere.

This feature discourages the use of unsealed bulk dispensers. Many workshops use containers less than 1 l in size, ensuring the container is capped when not in use. The maximum container size in general use is a sealed 20 l drum fitted with a hand pump and flow control valve.

Distilled water dispenser

Distilled water should be dispensed from a plastic container fitted with a 'no drip' nozzle. This type of nozzle allows the container nozzle to be positioned into the battery cap opening before flow starts. When the dispenser is lifted from the opening, flow will stop, preventing spillage on the battery top.

APPLICATION OF FLOOR JACKS AND HOISTS

A large percentage of maintenance work on vehicles requires the vehicle to be raised from the floor for under-body access. Mechanics must understand, with absolute certainty, the safety requirements to be observed when using floor jacks and vehicle hoists. A standard passenger vehicle has a mass of between one and two tonnes which, if dislodged from a jack or hoist, can crush the unwary servicing mechanic.

Most modern workshops are equipped with a number of floor jacks and a vehicle hoist. The hoist may be either a drive-on platform type fitted with a pantograph (scissors) vehicle lift or a frame contact type fitted with movable

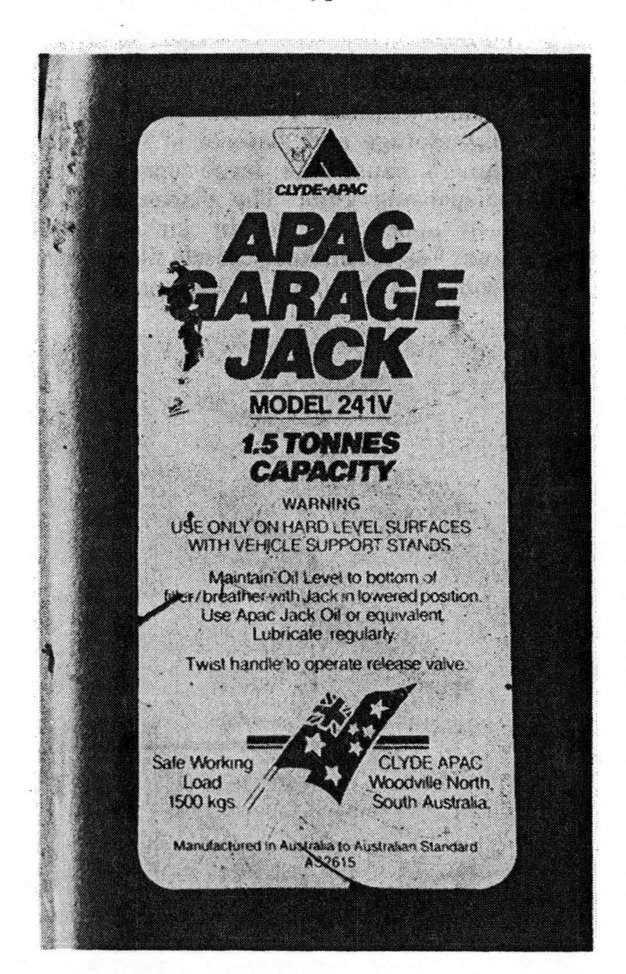

Check capacity of jack before use.

arms and contact pads. Other types of specialised jacks and hoist types are in use, but an explanation of these basic types will be sufficient to alert a mechanic to the basic safety requirements.

Floor jacks

CAPACITY

Floor jacks are available in many sizes and lift capacities. Almost all the four wheel mobile jacks are capable of lifting a standard passenger vehicle but a check of the lifting capacity, if this is the first time you are using the unit, is a wise safety hint. The maximum lift capacity is listed on a decal (sticker) attached to the jack. Do not attempt to raise a vehicle if the vehicle mass is greater than the jack capacity.

OPERATION

Jack operating controls are relatively simple. A hydraulic fluid control valve operated by

To raise, close valve and pump handle. To lower, twist handle counter-clockwise SLOWLY.

the handle opens and closes a passage to the lift ram. When the valve is closed fluid cannot return from the lift ram, therefore the jack lifting saddle will remain in the up position. The lift height can be increased, with the valve closed, by pumping the handle or depressing the foot pedal (if fitted). The valve is placed in the closed position by rotating the handle in a clockwise direction until handle movement stops. The jack lifting saddle is lowered by rotating the handle in a counter-clockwise direction. This action opens the hydraulic valve and fluid in the lift ram returns to the reservoir.

USE

To raise the vehicle, either at the front, rear or side, it is obvious the jack lifting saddle must

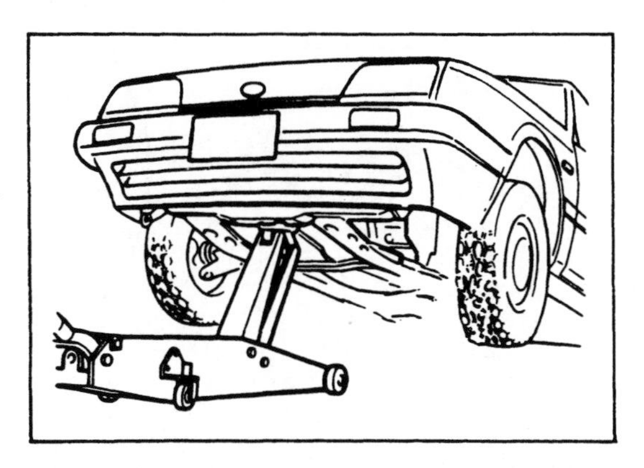

be placed under a solidly constructed vehicle component capable of withstanding the load without damage. This necessity is often forgotten, as large numbers of vehicles have sustained damage to body members, engine sumps, tie-rods and steering linkage because of carelessly placed jacks. It is also essential to place the jack under components that do not lean at an angle as the vehicle is lifted. An example of this problem is lifting by the lower suspension arms which, on some vehicles, develop acute angles as the lift progresses. Under these circumstances the vehicle may slip and fall while being raised. Consult the owner's handbook or the vehicle service manual for the correct vehicle lifting positions.

Caution: Do not remove wheels or work under a vehicle while the vehicle is supported on a hydraulic jack. Steel safety stands must be placed to support the vehicle mass before work commences.

Safety stands

Positioning of safety stands under the body or chassis members is subject to the same

Rear body supporting points.
Source: Mitsubishi Sigma GJ series manual

Rear supporting points.
Source: Mitsubishi Sigma GJ series manual

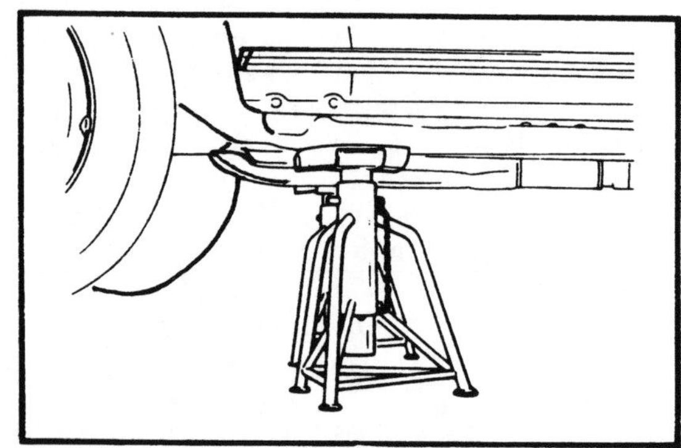

Place stands under reinforced chassis rails.
Source: Mitsubishi Sigma GJ series manual

requirements as those for the jack lifting saddle — the body or chassis member must be capable of holding the load without damage and the point of contact must not be angled or of a shape which will allow the stand to slip.

Vehicle hoists

Any workshop equipped with a hoist should also have a lubrication and servicing guide published by an oil company. These guides are invaluable for providing the location of suitable lift points on an extensive range of vehicles. Consult the guide before lifting an unknown vehicle model on a body contact type hoist.

Body/frame contact hoist

The most common type has two lift posts, one on each side of the vehicle. Each post is fitted with a lifting bar to which two movable arms are attached. Each arm has a sliding extension with a swivel pad attached.

The photo shows the driver's side post and arms. A similar arrangement is present on the passenger side of the vehicle.

1 Before driving into this type of hoist the operator must ensure:
  - the hoist is fully down;
  - the extension arms and swivel pads are

in a position which will not foul the vehicle underbody.

2 Drive the vehicle between the lift posts into a central position to ensure the vehicle will be balanced when raised. The driver should leave the transmission in neutral and the parking brake in the 'off' position.

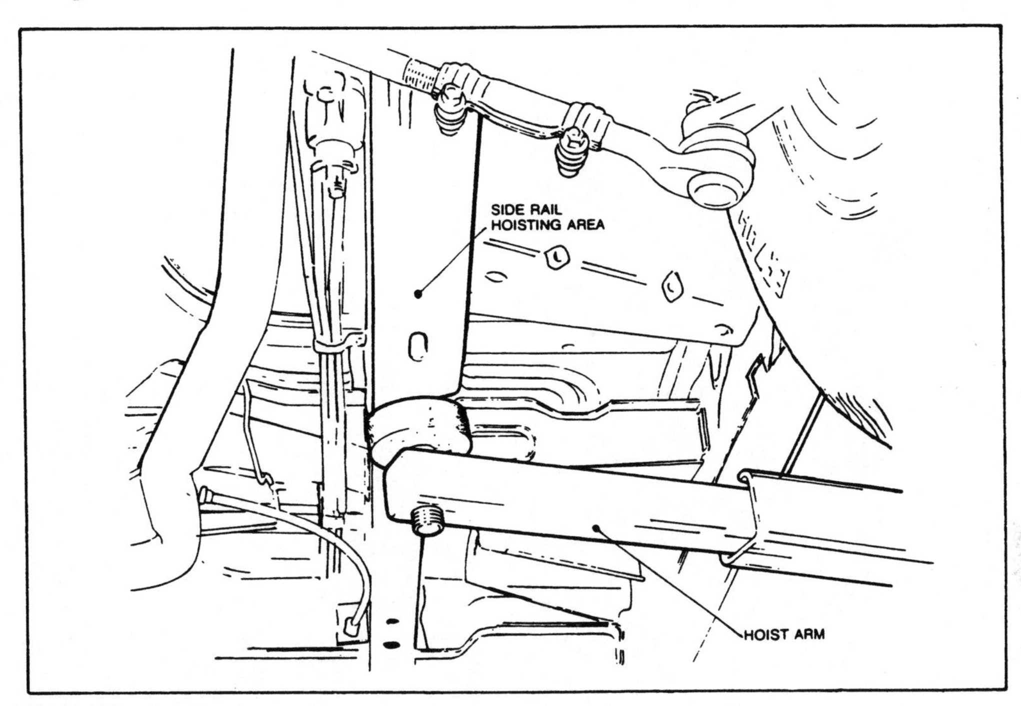

Typical lift points (front).

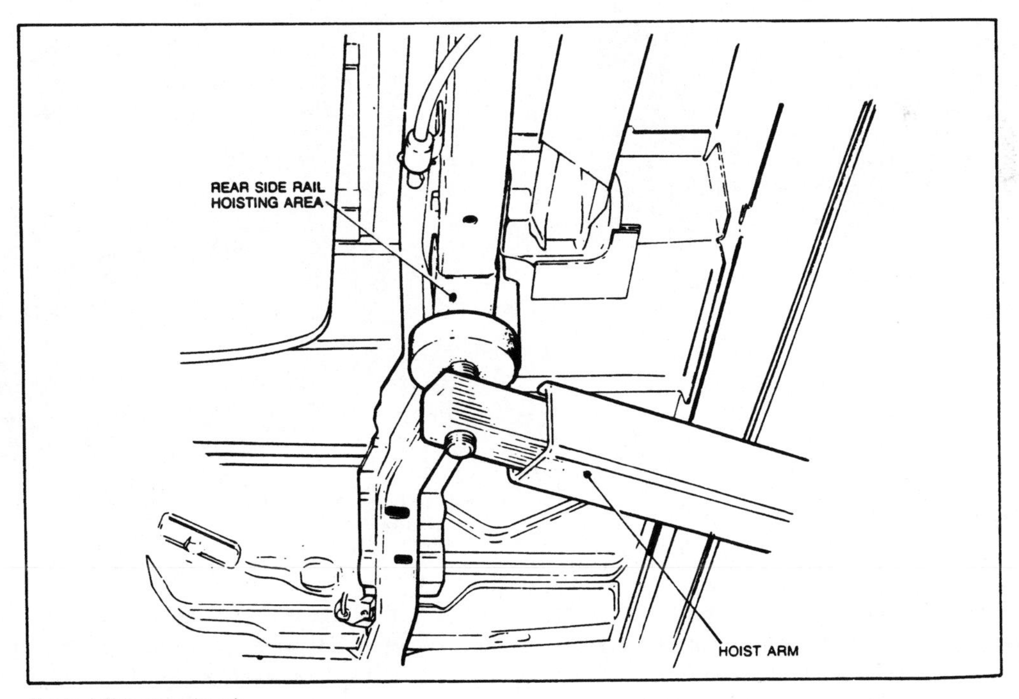

Typical lift points (rear).

3 Position the hoist swivel pads under the vehicle lift points (frame or chassis). Check the lubrication guide if doubt exists regarding the correct location of the swivel pads. CHECK to ensure the hoist will not contact or crush any under-body fittings as the pads raise into contact with the vehicle.

OPERATION

The hoist will have a control section consisting of an:

- UP switch;
- DOWN switch;
- EMERGENCY STOP.

RAISING

1 Check the hoist area is clear.
2 Press the UP button.

Caution: Release the up button the moment the swivel pads make contact with the vehicle and recheck the under-body components for clearance.

3 Recheck the positioning of the swivel pads to ensure no under-body damage can occur and the vehicle is central on the hoist.
4 Press the UP button and the hoist will lift the vehicle.
5 Observe the area above and around the rising vehicle for possible obstructions.
6 When the working height is reached, release the UP button.
7 Check the safety mechanism is latched. A number of hoists have automatic safety mechanisms which do not require operator intervention.

LOWERING

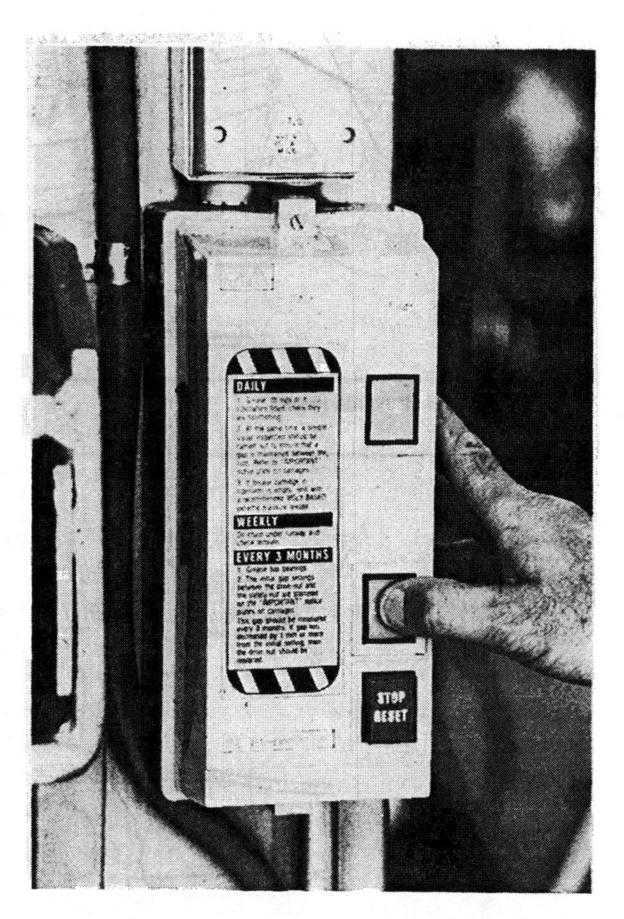

Before pressing the down button, you must check the hoist area is clear.

1 Ensure the vehicle doors are closed and the under-hoist area is clear of obstructions.
2 Release the safety latch (check specific hoist operating instructions).

3 Press the DOWN button and the hoist will start to descend.
4 After the hoist is fully down, fold back the swivel pads and arms to clear the vehicle under-body.
5 Drive the vehicle off the hoist SLOWLY, listening for any under-body contact.

Platform (drive on) hoist

This type of hoist is not favoured by lubritorium operators because the wide platform tends to restrict access to suspension components when lubrication of grease points is required. However, as the lifting components are contained in the four corner posts, the entire under-body section is free of posts. This feature makes the hoist extremely suitable for transmission servicing and/or removal. A platform hoist is generally not manufactured with a central lift post as most servicing advantages would be lost. By equipping the platform with a scissors jack it is possible to raise the vehicle wheels clear of the platform regardless of the height to which the hoist has been raised. This feature allows full access to the wheels and brake units in a similar manner to the frame contact hoist.

Prior to driving on to the hoist a visual inspection will confirm the platform is fully down (in contact with the ground) and the scissors jack is fully retracted.

The vehicle should be driven on to the platform at a slow speed and positioned centrally on the hoist. The driver should leave the transmission in neutral and the parking brake in the 'off' position.

LIFTING

The hoist lift control (electric-hydraulic types) will consist of a single UP/DOWN control. Moving the control to the UP position will cause the hoist to start lifting the vehicle.

The precautions to be observed when operating this type of hoist are similar to the frame contact units. However, two additional safety checks must be observed as the platform rises.

1 The two vehicle wheel chocks must rise into place at the rear of the platform. (The drive-on ramps tilt upwards as the platform rises and provide an effective barrier to the vehicle rolling backwards off the hoist.)
2 The automatic safety feature must be operating. This usually consists of a number of large latching hook, one for each post, that can engage in slots cut into the corner posts. Each post has slots spaced vertically at intervals of 20 to 25 cm, and the latching hooks pass over the slots as the platform rises.

The scissors jack (if fitted) consists of two platforms approximately 1 m in length, which can rise under the door sills, lifting the vehicle clear of the hoist platform. This feature is normally used AFTER the platform had been raised to the required height and the safety latches are locked. The scissors jack is fitted with safety latches which engage at FULL extension.

Other hoist types

There are many other variations of hoists available. They may be powered by compressed air or electric motors and may be single or double ram. Before operating any hoist, ensure you are aware of the safety requirement (not all hoists have automatic safety latches) and the operating technique.

VEHICLE SERVICING

Most owners refer to the maintenance schedule required for a motor vehicle as 'The car needs a service'. The scheduled servicing requirements of a modern vehicle are very different from the earlier concepts of a grease and oil change each 1500 km. Modern vehicle maintenance is commonly based on 10 000 km service intervals, with a number of items at 40 000 km or greater distances. To enable a vehicle to have such extended operational periods before servicing has meant that lubricants, oils filters and sealing methods have undergone remarkable improvements. This approach also highlights the necessity for the correct APPROVED lubricants and premium oils to be used. Vehicles operated on extended schedules and serviced with incorrect or non-manufacturer approved materials may suffer rapid wear due to breakdown of the lubricants.

Oil companies and vehicle manufacturers provide lubrication application charts and

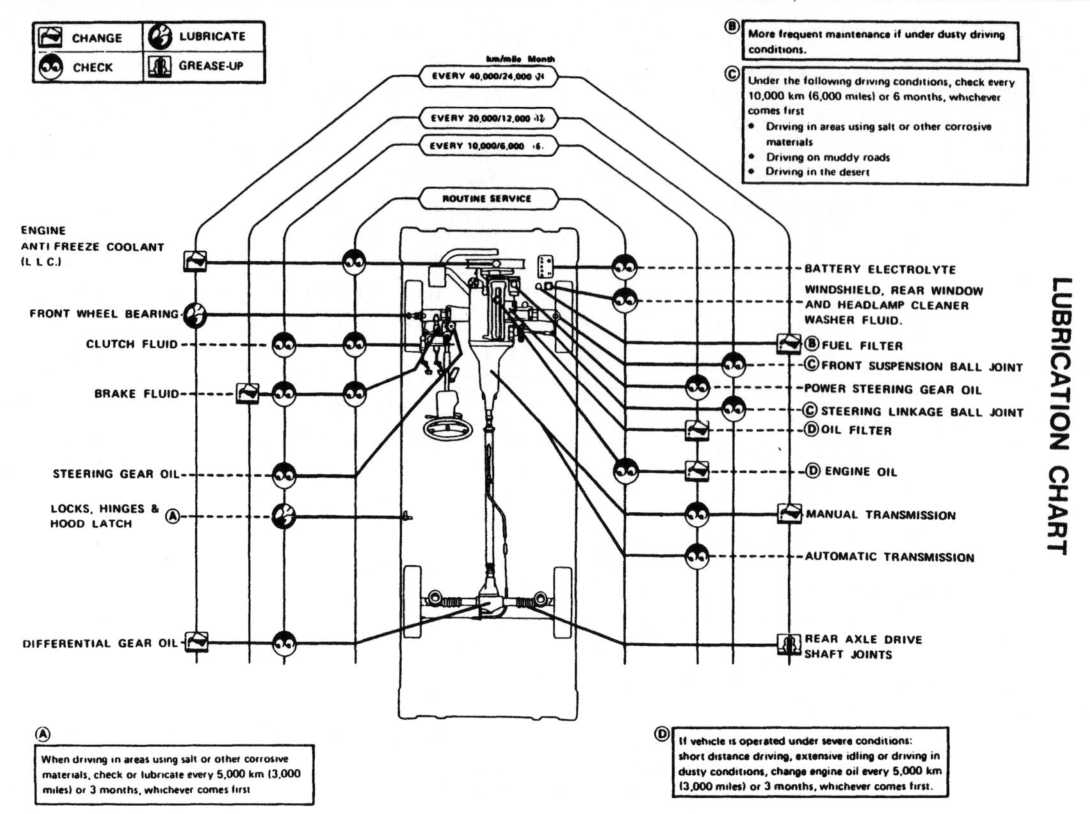

Source Nissan Australia

vehicle servicing schedules for vehicles operating in both 'normal' and 'severe' conditions. Generally severe conditions relate to dust, high speed, prolonged idle, towing or low temperature/short trip operations. The effect of these conditions varies but dust, low temperature/short trip or extended idle shorten engine oil change periods, towing reduces brake life, etc.

Vehicle maintenance on a modern vehicle means maintenance and inspection of engines, transmissions, final drives, brakes, suspension, wheel bearings, emission control systems, cooling systems and drive belts, body parts (door locks etc.) and vehicle electrics. Many of these areas are the subject of separate chapters within this book, therefore the type of service described will be a regular lubrication service not requiring specific or additional servicing work.

Lubrication service

PREPARATION

1 Obtain vehicle and job card. Ensure you are aware of the service work authorised.
2 Clean hands to prevent grease marks on the steering wheel and door locks. Use a seat cover to protect the unholstery.
3 Drive vehicle on to the hoist.

Caution: Drive slowly to prevent the possibility of under-body damage if contact with the hoist occurs.

4 Release bonnet catch, put handbrake in off position and gear lever (manual or automatic) in neutral position.
5 Obtain oil company/vehicle manufacturer's guide for maintenance points and lubricant type.
6 Raise vehicle on hoist.

Caution: Check guide for hoist pad (if applicable to hoist type in use) contact positions and/or lifting precautions.

7 Check hoist safety mechanism is in place and has positively latched.

Under-body servicing

1 Drain the engine oil. Position the oil drain bowl under the sump and remove the sump plug. Allow engine oil to drain while hot.

2 Refit sump plug. Allow engine oil to drain completely. Fit the sump plug after checking plug threads and the sealing washer.

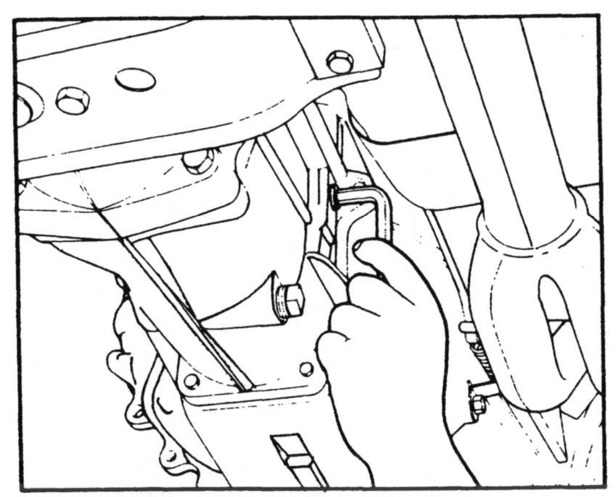

Source: Mitsubishi Sigma GJ series manual

3 Manual transmission only. Remove the fill plug and check the oil level; top up with correct gear oil if required.
4 Final drive. Remove the fill plug and check level; top up with correct gear oil if required.

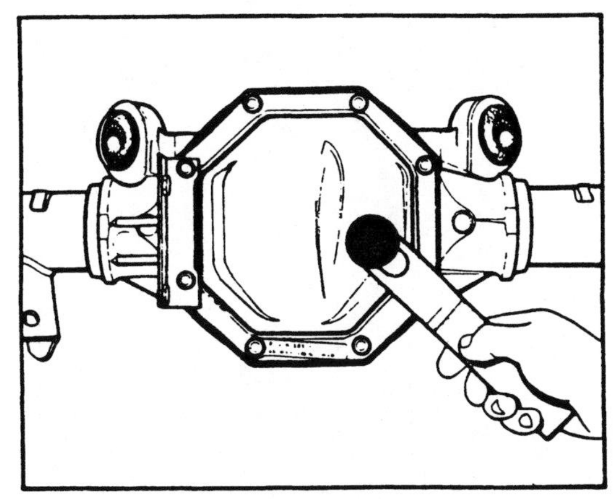

Removing rubber level plug.
Source: Mitsubishi Sigma GJ series manual

5 Oil filter. Unscrew the engine oil filter (if in a position which permits removal from under the vehicle). If necessary use a filter removal tool to loosen a tight filter.

Oil filter removal.

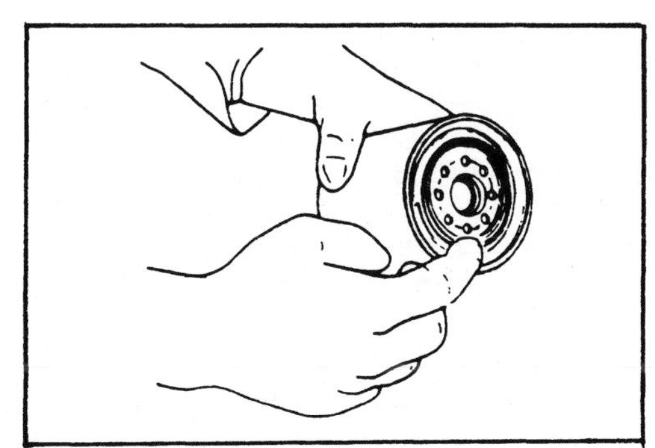

Source: Nissan Australia

Fit a new filter — the seal on 'spin on' one piece filters must be lubricated prior to fitting.

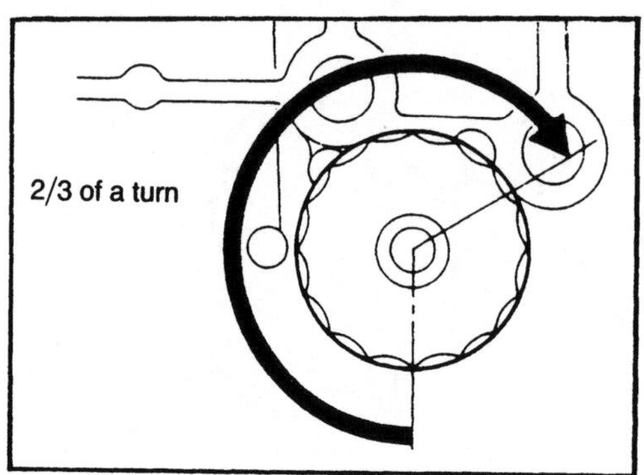

Source: Nissan Australia

Follow the fitting directions printed on the case of many filters. For example, hand tighten 2/3 of a turn after initial contact of seal on seat.

Caution: Shape and size are not an accurate guide to filter application. The internal construction of similar appearance filters can be different. Use filters with the correct part number or listed equivalent number.

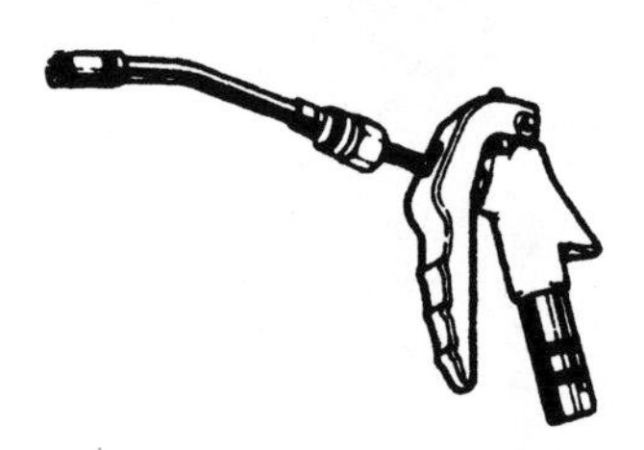

6 Grease points. Use a service guide and locate each lubrication (grease) point under the vehicle. Each fitting must be wiped clean prior to the grease gun being fitted to the greasing nipple. Keep count of the points serviced and compare the total to the service guide to ensure no points are overlooked.

Caution: Seals can be damaged by excessive grease application therefore universal joints, centre bearings etc. should be serviced with a hand operated grease gun filled with the specific grease type required.

UNDER-BODY INSPECTION

The modern vehicle is a reliable unit and the extended service periods mean the vehicle will rarely be raised on a hoist. It is essential that a thorough inspection is performed while the vehicle is being serviced. The items listed below should be considered as a minimum standard. Remember, items detected during inspection do not only alert the vehicle owner to possible hazards but represent a potential revenue source for your garage. Check:

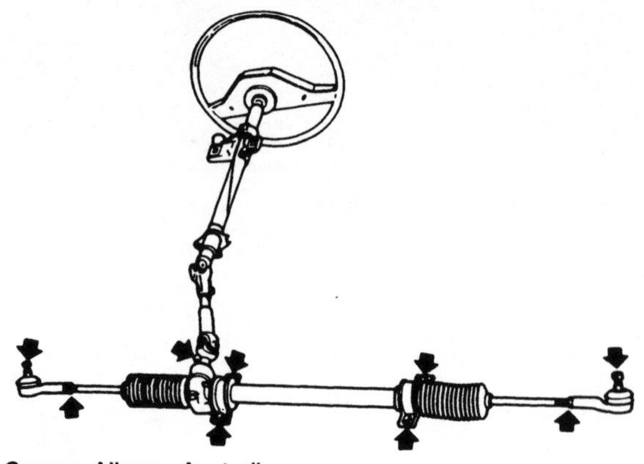
Source: Nissan Australia

1 steering linkage for wear, looseness or split seals;

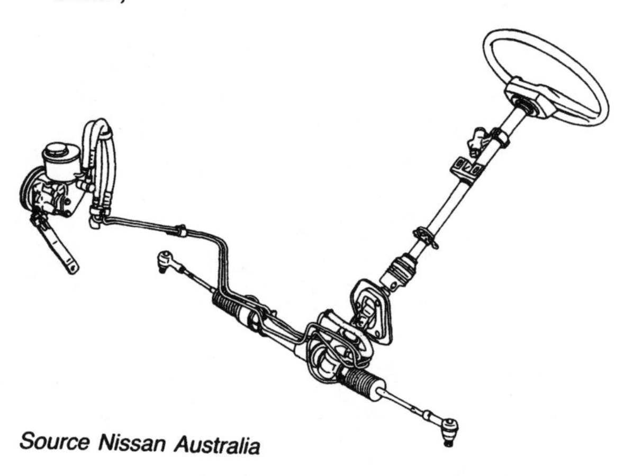
Source Nissan Australia

2 steering box (particularly power assisted types) or fittings for leaks;

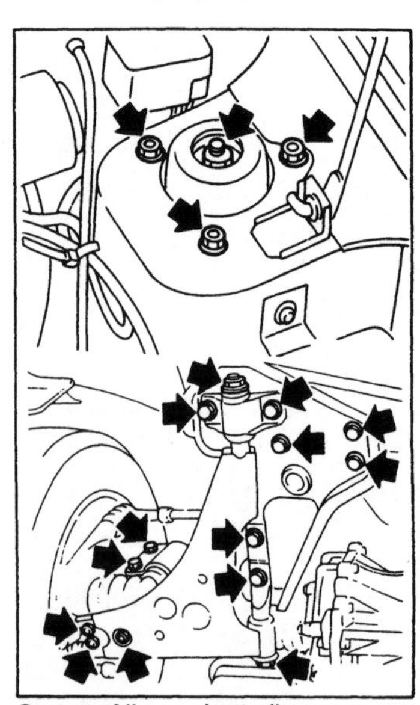
Source: Nissan Australia

3 front suspension pivots and shock absorber mountings for wear or looseness;

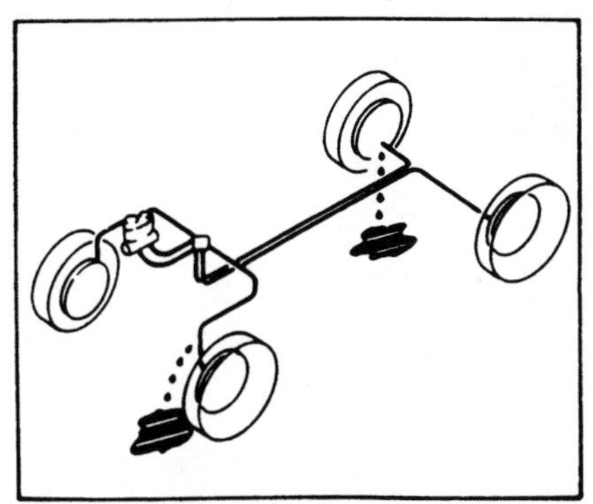

Source: Nissan Australia

4 brake hoses for tyre/frame contact, leaks or deterioration;
5 brake caliper mountings for looseness;
6 rear brake hose/pipes for exhaust pipe contact, leaks or wear;

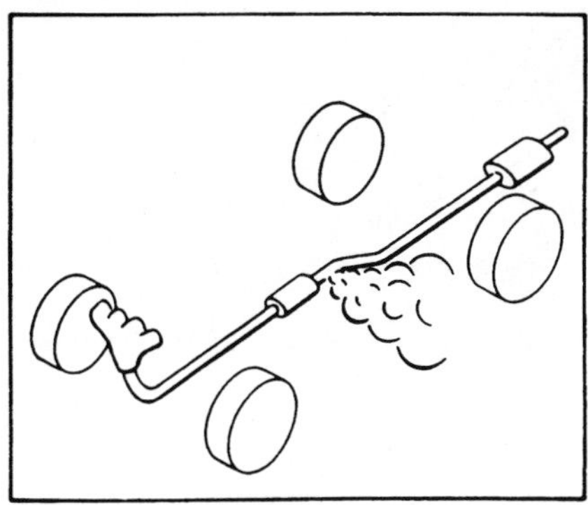

Source: Nissan Australia

7 exhaust pipes/muffler for obvious leakage or mounting breakage;

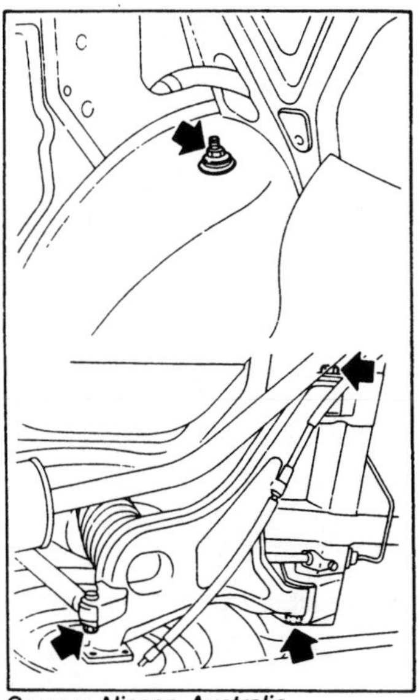

Source: Nissan Australia

8 rear spring/suspension mounting points for looseness or wear.

The next series of checks requires the vehicle wheels to be free to move. Platform (drive on) hoists will require the use of a hoist 'scissors' or jacking system to unload the vehicle wheels.

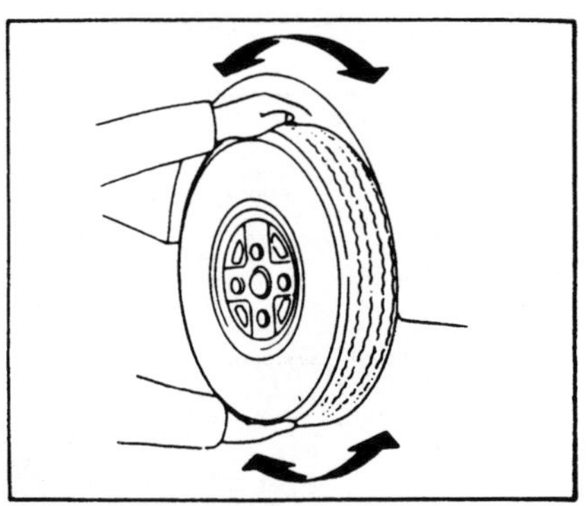

Source: Nissan Australia

9 Grasp front wheels and rock the wheel in both a vertical and horizontal direction to detect loose wheel bearings and worn suspension components.

Caution: This check will not be effective in detecting wear in certain types of ball joint suspension. For specific front suspension component testing refer to the vehicle manufacturer's service instructions.

10 Revolve each wheel and check the tyre for excessive run out, splits, wear, deformity or incorrect tyre 'mix' on the vehicle;

Source: Nissan Australia

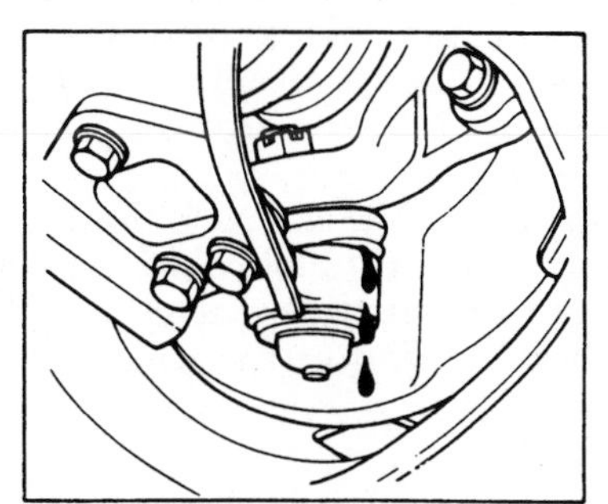

Source: Nissan Australia

11 grease or oil leaks from any under-body assembly or component;
12 record all faults detected on the vehicle service sheet;
13 lower vehicle to ground.

Caution: Check area for obstructions or people before lowering hoist.

UNDER BONNET SERVICING

1 Prepare vehicle. Raise the bonnet. Place guard covers on both guards to protect and keep the paintwork clean.
2 Engine oil filter. Change the engine oil filter

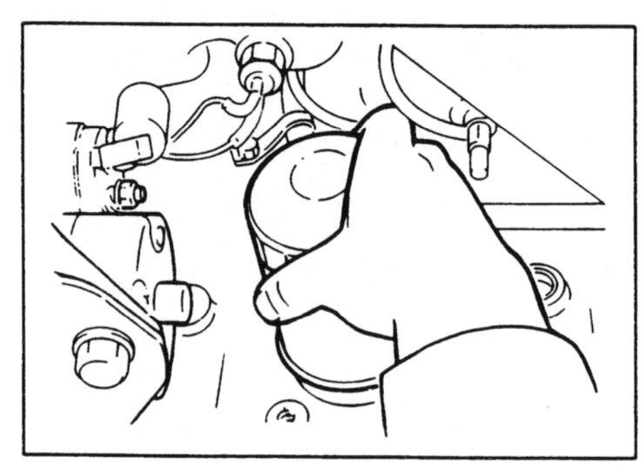

Oil filter installation.

if filter was not changed from under the vehicle.
3 Engine oil. Fill sump with the correct grade, type and quantity of oil, refit the filler cap.

Caution: Additional oil is required when the filter is changed. Recheck oil level after engine has been running and top up if necessary.

4 Brake and clutch fluid. Check fluid level of brake and clutch (if applicable) reservoirs. Many reservoirs are transparent and do not require the removal of the cover for level checking.

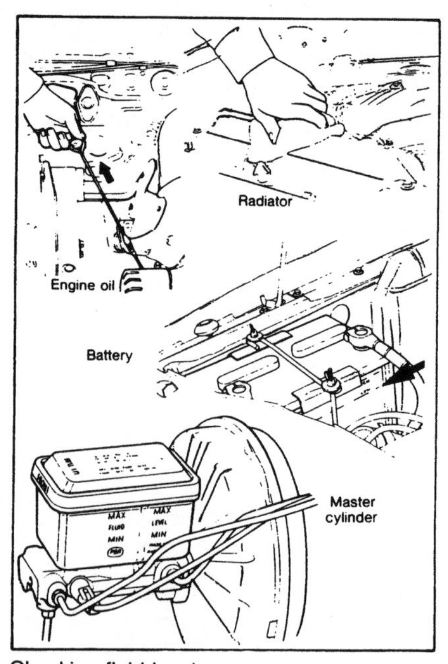

Checking fluid levels.
Source: Mitsubishi Sigma GJ series manual

5 Battery level. Check battery electrolyte level — top up with DISTILLED water if required.

6 Radiator. Check radiator coolant level. Gently squeeze each hose and check for cracks.

Caution: Do not remove radiator cap if engine is at operating temperature. There is extreme danger of steam burns.

7 Transmission oil level check. Automatic transmissions require the transmission to have been placed in each range with the engine running to allow oil to be drawn into the torque convertor and hydraulic system before a dipstick fluid level check. The check should be taken with the engine

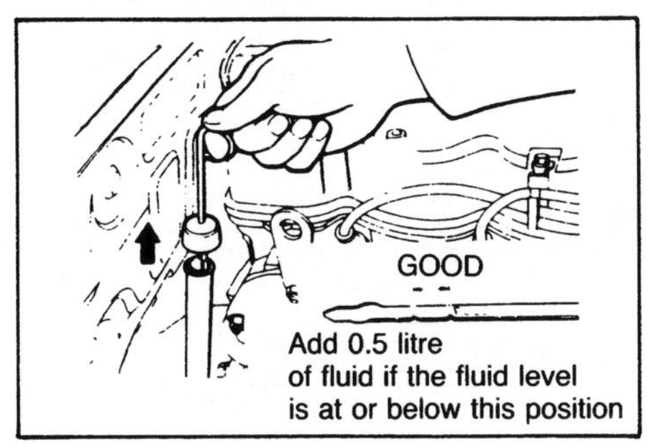

Automatic transmission dipstick markings.
Source: Mitsubishi Sigma GJ series manual

idling and the transmission in 'park'. Failure to follow this procedure will give a false fluid level reading. Check the service guide for any special procedures relevant to the serviced vehicle. Top up with correct fluid if required.

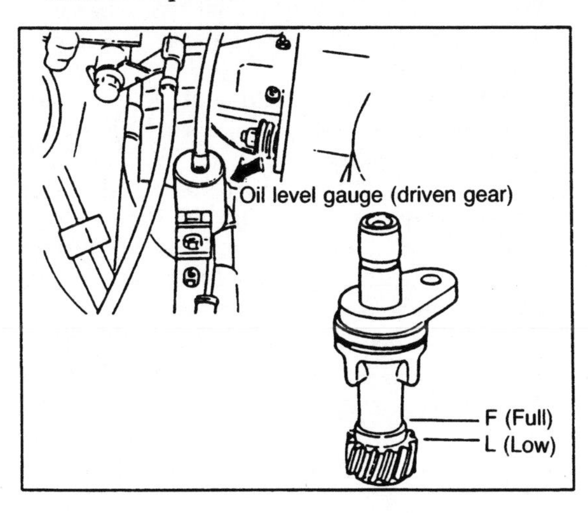

Manual transmissions (front wheel drive) are checked from under the bonnet. Remove the filler plug (a number of vehicles use the speedo drive opening as a fill point), check the oil level and top up with correct gear oil if required.

8 Power steering (if fitted). Check reservoir fluid level.

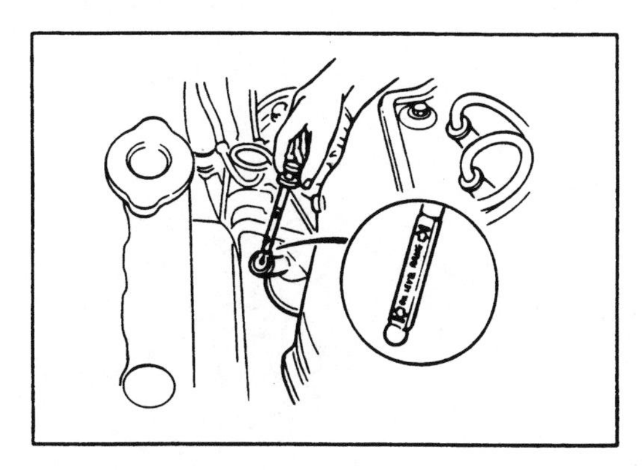

9 Windscreen washer. Check water level in windscreen washer bottle.

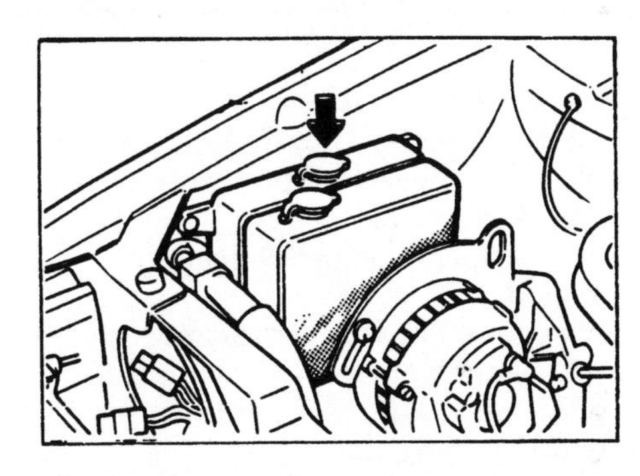

10 Service points. Use an oil can (or lubricant stick if specified) and lubricate wear spots as indicated on the vehicle service guide (bonnet hinges, links, door catches or carburettor dashpots (C.D./S.U.)).

11 Air cleaner service. Refer to the chapter 'Minor servicing' for detailed service procedure.

12 Drive belt inspection. Refer to the chapter 'Minor servicing' for detailed service procedure.

13 Idle and emission system checks. Refer to the chapter 'Fuel systems' for detailed service procedure.

14 Clean up. Close bonnet after removing guard covers and checking for tools left under bonnet. Remember, clean your hands to prevent grease marks on paintwork.

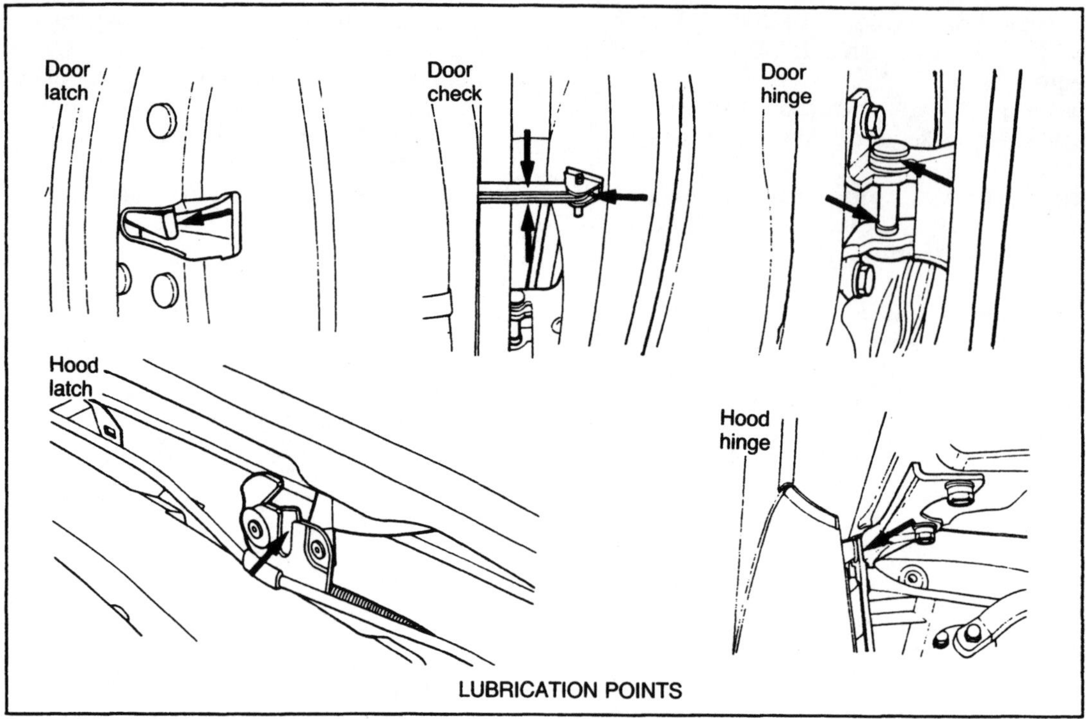

Source: Mitsubishi Sigma TM series manual

BODY CHECKS

Refer to the chapter 'Roadworthy awareness' for detailed checking procedures for the following items.

1 Check wiper arms/blades for correct operation.
2 Check operation of all external lighting.
3 Check instrument panel for malfunctions.
4 Check hand and foot brakes for correct adjustment heights.
5 Bounce test car to check for poor shock absorber performance.
6 Inflate tyres to recommended pressures.
7 Record all faults detected.

VEHICLE DELIVERY

1 Change lubrication stickers, and ensure distance entered is correct.
2 Record all material usage on job card.
3 Check all faults noted are entered on owner's copy of job card.
4 Clean windscreen.
5 Check steering wheel, door locks, trim, guards and bonnet for grease smears. Clean if required.
6 Remove seat cover.

CHAPTER 6 REVISION

1 Explain what the symbols 10W-50 means when applied to an engine oil.
2 What is the service rating requirement of engine oil for the majority of modern engines?
3 Name six additives used in engine oils and give a short description of each additive.
4 Name one automotive assembly for which GL 4 rated gear oil would be suitable.
5 Why is multi-purpose grease unsuitable for wheel bearing use on vehicles fitted with disc brakes?
6 What particular feature of modern car engines makes the use of a cooling system conditioner essential?
7 What is a safety stand and what precautions are necessary in its use?
8 Prepare a list of the servicing equipment required when undertaking a routine service on a passenger car.
9 Explain a basic precaution which should be observed when using suction guns for servicing.

10 What type of fluid is used to top up batteries?
11 What does the term 'jack capacity' refer to?
12 Write a list of precautions to be observed when operating a vehicle hoist.
13 What type of operating conditions may change the distance allowed by the manufacturer between servicing intervals?
14 Prepare a list of the items to be serviced under the headings, under-body, under-bonnet and body checks.
15 Why should a hand operated grease gun be used on some under-body components?
16 What precaution should be observed prior to fitting a new 'spin on' oil filter?
17 Why is it dangerous to remove a radiator cap when the engine is at operating temperature?
18 Consult the servicing guide for the correct method of checking the manual transmission oil level for three different front wheel drive vehicles.
19 Consult the servicing guide for the procedure to be followed when checking the automatic transmission fluid level on at least two popular vehicles.
20 Why is it necessary to supply additional oil when the engine oil filter is changed during the service?

MINOR SERVICING

This chapter refers to minor servicing procedures on the following systems and units:

1 **Cooling system** — operation, inspection and testing;
2 **Drive belts** — inspection and adjustment;
3 **Air cleaners** — identification and servicing.

Always consult the vehicle manufacturer's service instructions and safety precautions before commencing any servicing of a system or unit. Incorrect servicing procedures could result in personal injury, premature wear and damage to components and costly vehicle breakdowns.

COOLING SYSTEMS

In Australia the temperatures can vary from minus ten degrees (–10°C) to plus forty-five degrees (+45°C). For engines to operate effectively throughout this temperature range an efficient cooling system is essential. If a vehicle has a faulty cooling system and is running in a high temperature area, the engine can overheat, causing serious damage to its internal components. The results could be a very expensive breakdown. If an engine is operating in very low temperatures and the cooling system does not have the correct additive or the correct amount of additive in it, the water can freeze and cause cracks to occur in the cylinder head and the cylinder block.

Methods of cooling

There are two methods used in the cooling of engines.

Direct method, used mainly on small engines having cooling fins around the cylinder head and the cylinder block. The heat from the engine is transferred to the cooling fins and to the air passing through and over the fins.

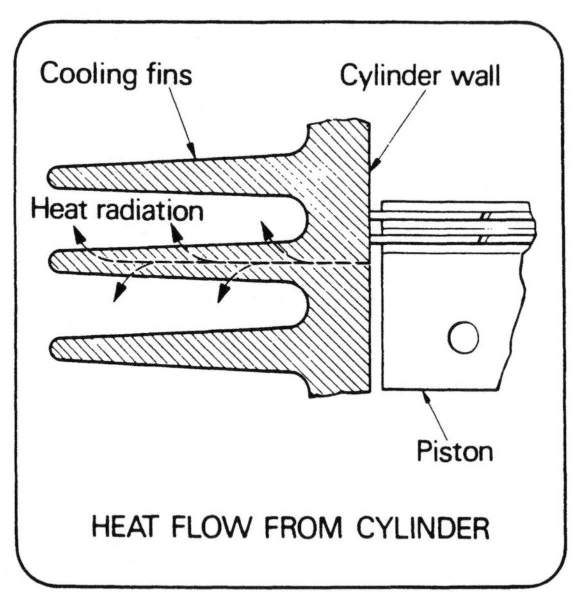

Source: AGPS

Indirect method uses a combination of water (coolant) and air cooling and is the most common type of cooling system used on motor vehicles. In this chapter we will be concentrating on the indirect or water (coolant) system.

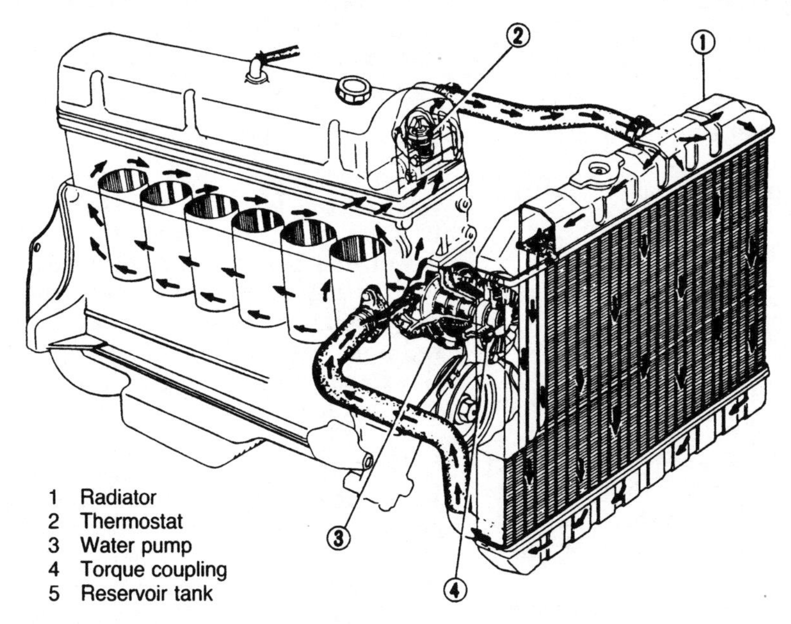

Source: Nissan Australia

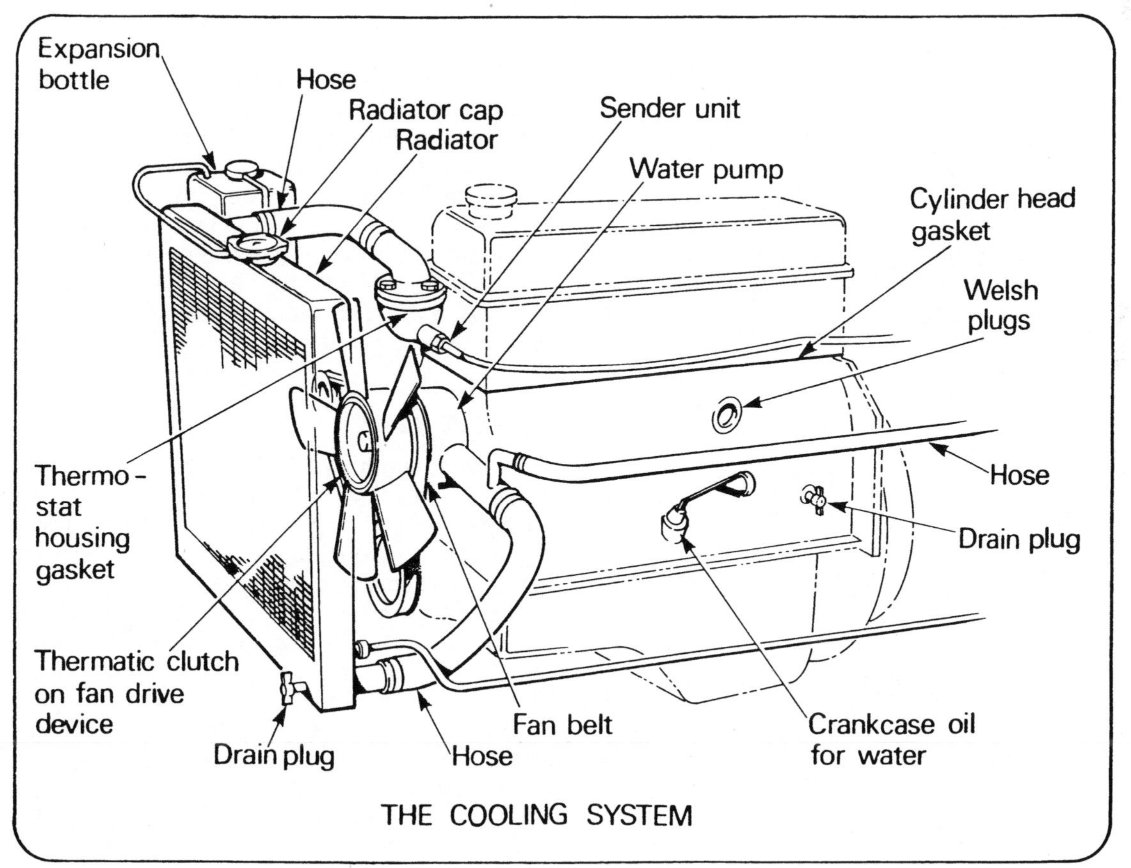

THE COOLING SYSTEM

Source: AGPS

Identification of components

It is necessary to identify correctly the components of a cooling system for the following reasons:

1 to understand the operation of the cooling system;
2 to assist in diagnosing cooling system problems;
3 for ordering spare parts.

The radiator

The radiator is manufactured from a variety of materials, including copper, tin, aluminium and plastic. It consists of two tanks, connected by a 'core' containing a series of tubes and cooling fins. Side plates, connected between the tanks, support, strengthen and provide mounting brackets for the radiator.

Each tank has provision for the connection of a rubber hose to the engine. The top tank has a fitting for the filler cap, and the bottom tank may have a drain tap or plug.

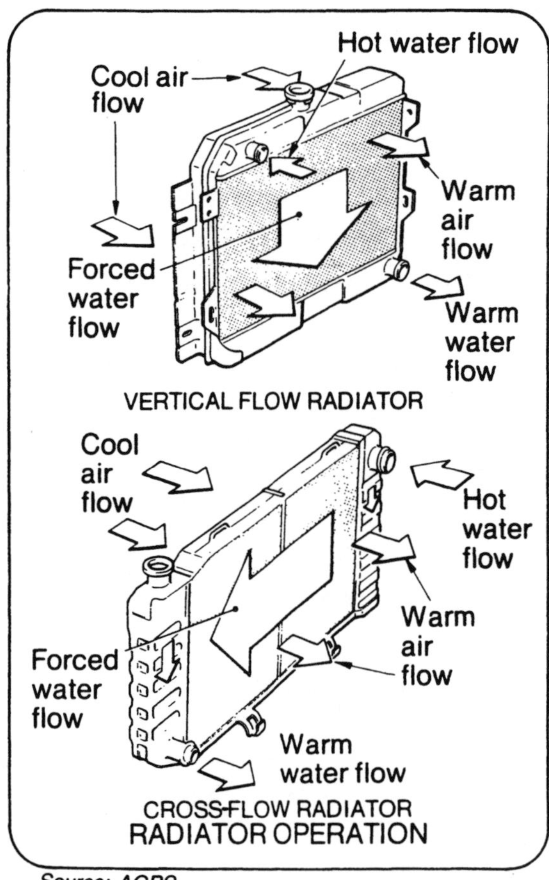

RADIATOR OPERATION

Source: AGPS

The radiator is generally located at the front of the vehicle, as an adequate air flow through the radiator is essential for the efficient operation of the cooling system.

TYPES OF RADIATORS

Generally, there are two types of radiators, the vertical flow type and the cross flow type. In the vertical flow type, the water (coolant) flows from the top tank through the 'core' to the bottom tank, and in the cross flow type the coolant flows from one side of the radiator through the 'core' to the other side of the radiator. This type of radiator is used in vehicles with low bonnets.

OPERATION

In a vertical flow radiator, the water pump circulates the coolant through the cooling system. The heat from the engine is transferred to the coolant. The 'hot' coolant flows through the top tank of the radiator, down through the many tubes of the core, dissipating its heat to the air passing through the fins of the 'core'. The bottom tank receives the 'cooled' coolant and directs it through the bottom radiator hose to the inlet of the water pump.

Some radiators have a drain pipe fitted to the filler neck, to drain off excess coolant caused by expansion. Other types of systems have expansion 'tanks', made from plastic or metal, which collect the excess coolant during expansion and return the coolant to the radiator when the temperature of the engine cools down.

The water pump

The water pump is generally manufactured from cast iron or an aluminium alloy, and mounted onto the front of the engine block. It consists of a housing, in which a shaft and bearing assembly is located. The shaft has a seal and pump impeller fitted to the 'wet' end, and a fan hub fitted to the other end. A fan and belt pulley are usually bolted to the fan hub, and the water pump is belt driven by the engine's crankshaft. The centrifugal pump allows the water to be circulated without requiring excessive power to drive it. The inlet of the pump is connected to the bottom radiator hose and the outlet of the pump transfers the water through the engine block and cylinder head.

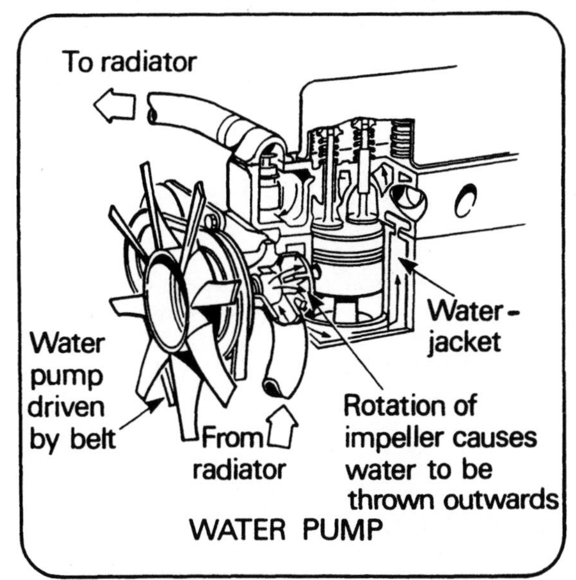

Source: AGPS

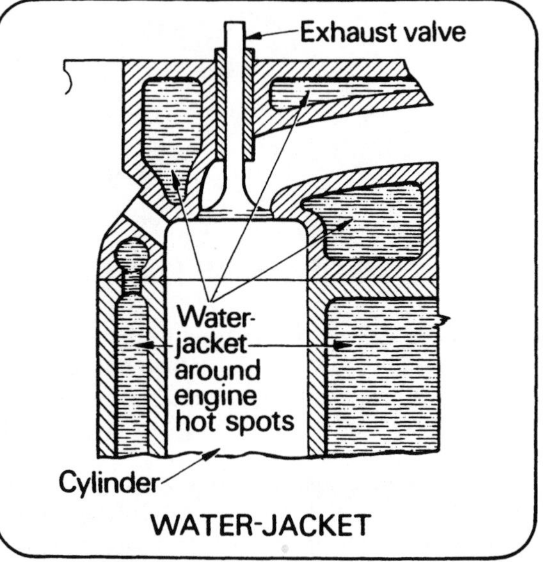

Source: AGPS

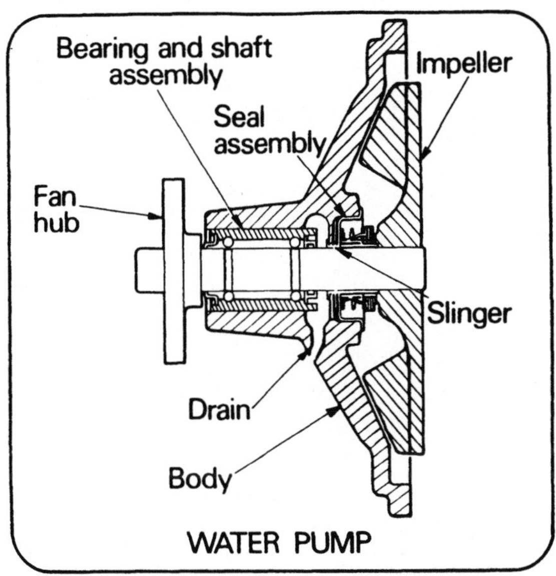

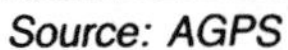
Source: AGPS

The function of the water pump is to circulate the water through the engine block, cylinder head and the radiator.

The water jacket

The water jacket is the space between the 'hot' cylinder walls in the engine's block, and in the cylinder head, around the combustion chambers and valve areas. These spaces are filled with coolant. The heat caused by combustion is transferred into the water jacket.

Welch plugs

Welch plugs (expansion plugs) are fitted to the engine block and cylinder head. During

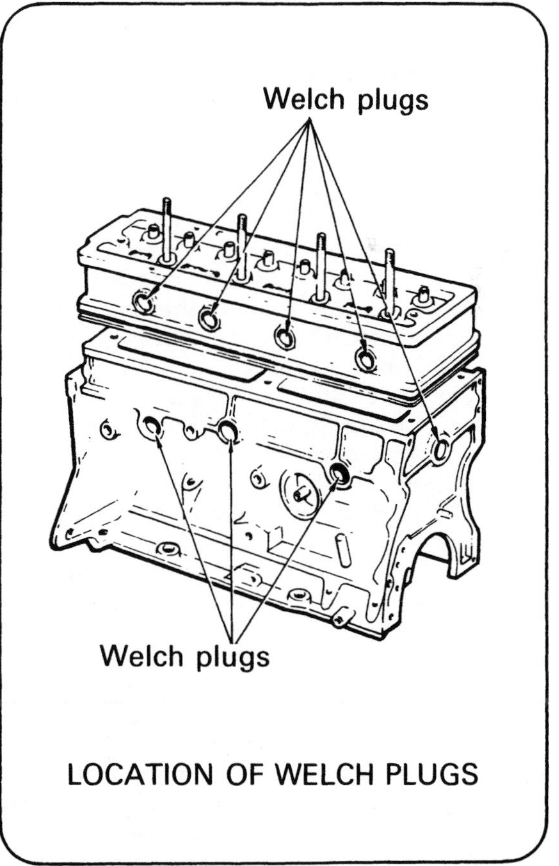

Source: AGPS

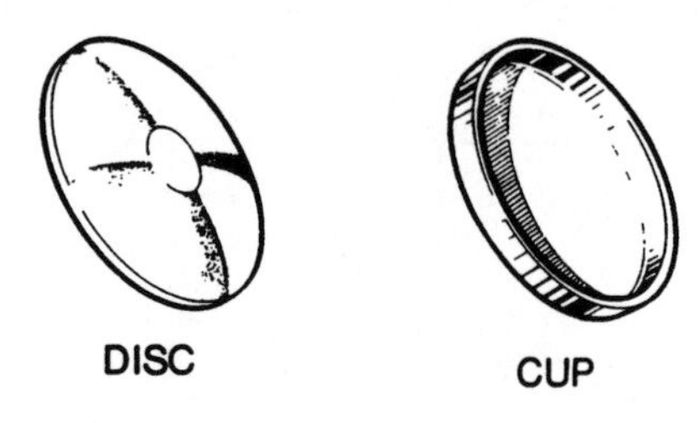

TYPES OF WELCH PLUGS
Source: AGPS

the cylinder block and cylinder head casting processes, holes are required for the removal of the moulding sand from the water jacket areas. These holes are plugged by welch plugs.

The common types are 'cup' and 'disc'. The function of welch plugs is to provide a safety valve for the cylinder block and the cylinder head. If the water in the water jacket freezes, the plugs are forced out. This action assists in saving the cylinder block and cylinder head from cracking.

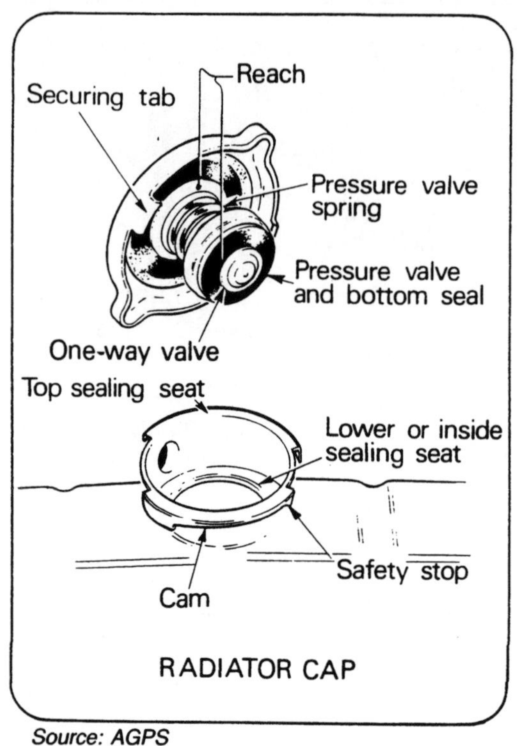

RADIATOR CAP
Source: AGPS

The radiator cap

This seals and pressurises the radiator and cooling system. Each cap has its pressure rating marked on it — normal range from 28 kPa (4 p.s.i) to 147 kPa (21 p.s.i.). Pressurising the cooling system has the following advantages.

1 The boiling point of the coolant is raised, which increases cooling system's efficiency. The greater the temperature difference between the inside of the radiator and the temperature of the air outside the radiator, the more cooling will take place.
2 Smaller radiators can be used.
3 Water loss due to evaporation is reduced.
4 Water surge is reduced.

The radiator cap has a spring loaded seal. When the pressure inside the cooling system reaches the pressure rating of the cap, the spring loaded seal is unseated and the pressure is released. This reduced pressure causes the spring loaded seal to again seal the cooling system.

A small one-way valve is fitted to some radiator caps, its function being to prevent low pressure being created in the system as the water cools.

Caution: Always fit the recommended 'pressure and reach' radiator cap, as failure to do this can cause cooling system problems and damage to components.

The thermostat

This is normally situated in the cylinder head at the engine's water outlet (the radiator's top hose inlet connection). It is a temperature

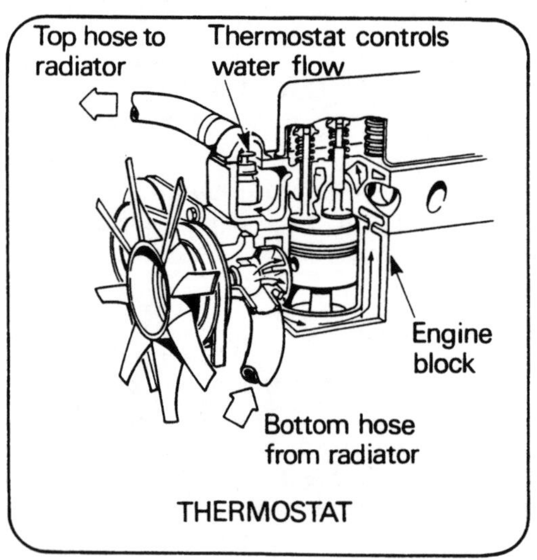

THERMOSTAT
Source: AGPS

sensitive valve: when the engine is cold the thermostat is closed, restricting the water flow to the radiator; as the engine warms, the thermostat opens. The warmer the engine gets the more the thermostat opens, allowing more water to circulate through the system. The thermostat maintains an efficient engine temperature during all climatic (weather) conditions.

There are two types of thermostats — 'bellows' and 'wax'. The bellows thermostat is used in the older, low pressure cooling systems, and in some air cooled engines, to operate the air flow valve.

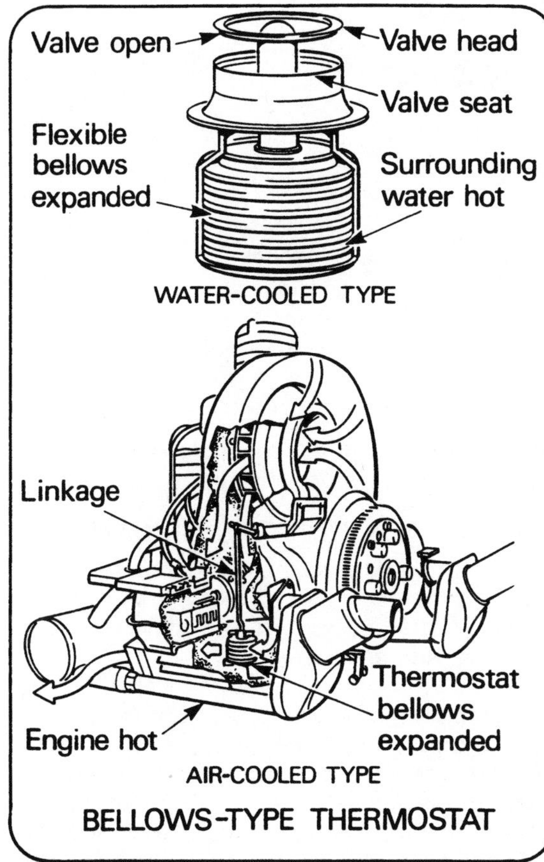

BELLOWS-TYPE THERMOSTAT

Source: AGPS

The wax type is used in the higher pressurised cooling systems of today's modern vehicles.

The opening temperature of a thermostat is marked on it. The temperature ratings range between 76°C and 94°C.

Caution: Always refer to the engine manufacturer's specifications. When replacing a thermostat, ensure the replacement thermostat has the recommended opening temperature marked on it.

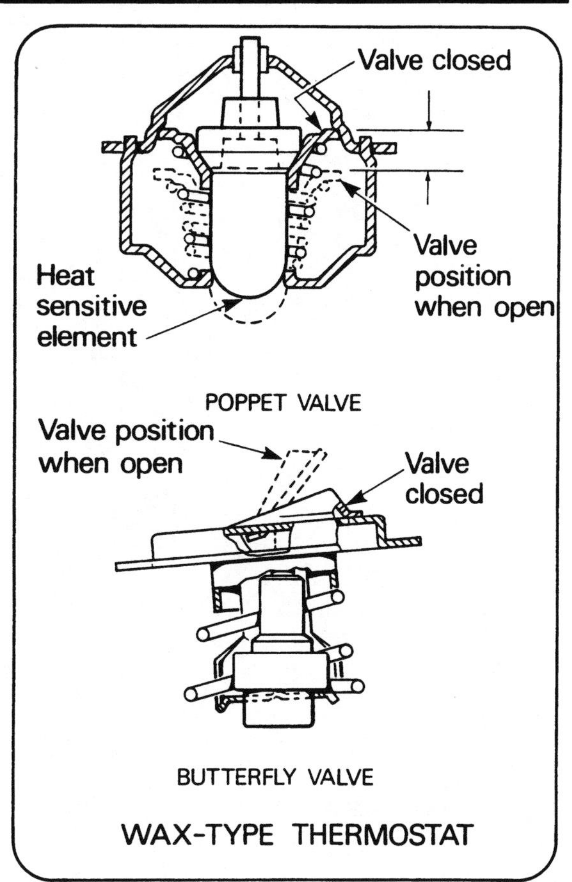

WAX-TYPE THERMOSTAT

Source: AGPS

Cooling system inspection should be carried out thoroughly, and all safety precautions be observed.

Warning: Severe burns to the skin can result from removing the pressure cap from a hot radiator. The increased pressure inside the radiator forces boiling coolant to gush out, and on contact with skin this will cause severe scalding.

Coolant inspection

The coolant level must be checked regularly, and topped up if required. On cooling systems with expansion tanks the coolant level should be maintained between the add and full marks. On other systems, it is necessary to remove the radiator cap to check the coolant.

1 Turn the cap anti-clockwise one quarter of a turn.

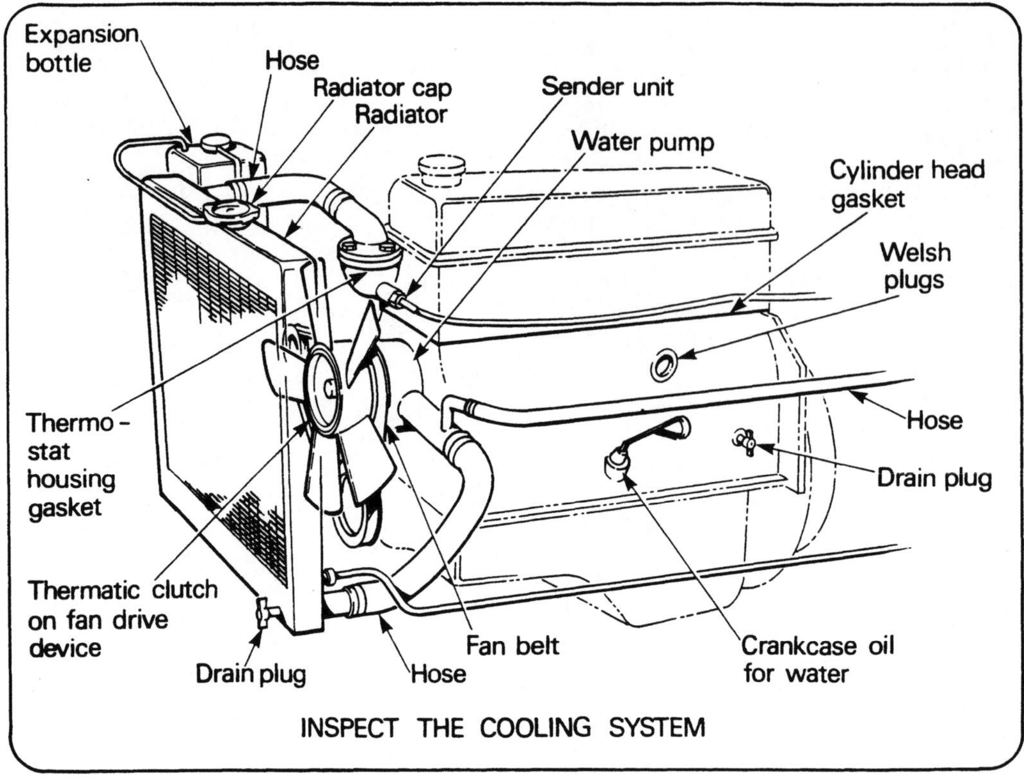

Source: AGPS

2 Press the cap firmly down and turn it a further quarter of a turn.
3 Remove the cap.
4 Note the release pressure marked on the top of the cap and check the engine specifications to ensure that the correct cap has been fitted.

Inspect the coolant for:
- The correct level.
- Cleanliness — ensure that there is no rust, oil or discoloration of the coolant.
- The addition of an inhibitor. The function of a coolant additive is to supply a rust inhibitor to the cooling system. Most additives raise the boiling point and lower the freezing point of the coolant.
- Specific Gravity (S.G.) to ensure the mixture of additive and water is correct.

VISUAL INSPECTION FOR LEAKS

The following areas should be inspected:
1 The radiator core, tank joints and drain plug.
2 The cylinder head gasket.
3 The hoses, and the joints between the hoses and their components.
4 The water pump seal.
5 The temperature gauge sender unit.
6 The welsh plugs and engine drain plug.

VISUAL INSPECTION OF COMPONENTS

Note: Engine at operating temperature.
1 Inspect the radiator core for obstructions such as dirt, insects, oil and leaves, and for bent or damaged fins.
2 Check all hoses for swelling, oil contamination, hardness, cracking and perishing.
3 Check the crankcase oil for water contamination by observing the oil is not grey or white, and that the dipstick does not show any sludge, water drops or excessive rust.
4 Check the fan for looseness, and for cracked or bent blades.
5 Check fan belt for tension, cracks, wear, and oil contamination.

PRESSURE TESTING

To pressure test the cooling system, a cooling system analyser is used. The analyser consists of a hand operated pump, gauge assembly, and a clamping device which fits to the radiator filler neck.

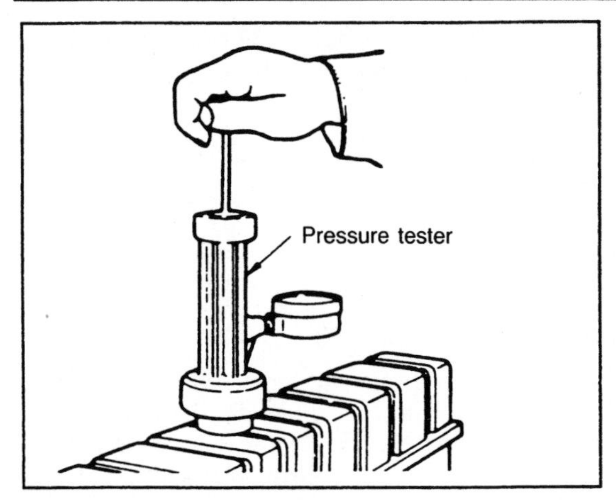

Pressure testing cooling system
Source: Mitsubishi Magna TN series manual

TESTING FOR EXTERNAL LEAKS

1 Remove the radiator cap.
2 Top up the coolant level, if required.
3 Clamp the analyser to the radiator filler neck.
4 Pump the analyser until a pressure equal to the pressure rating of the cap plus 25 per cent, is shown on the gauge.
5 Inspect the radiator and all connections for leaks.
6 The gauge reading should remain steady. A drop in gauge pressure indicates a leak in the system. If no coolant leaks can be found and the system continues to lose pressure, an internal leak from the water jacket or the cylinder head gasket must be suspected.

TESTING FOR INTERNAL LEAKS

1 Remove the radiator cap.
2 Top up the coolant level (leave an air space).
3 Start the engine and allow it to reach operating temperature.
4 Clamp the analyser to the radiator filler neck.
5 Pump the analyser to slightly pressurise the system (7–14 kPa).
6 Observe the analyser gauge while the engine is running. A sudden increase in the pressure reading indicates a fault in the cylinder head or the head gasket.

Caution: Do not allow the pressure to rise above the normal testing pressure, as this could cause damage to the analyser and to the components of the cooling system.

TESTING THE RADIATOR CAP

The pressure in the cooling system is controlled by the spring loaded valve in the radiator cap. To test the radiator cap:

1 Attach the cap to the analyser.
2 Check the release pressure, which is marked on the top of the cap, and check the engine manufacturer's specifications to ensure the correct cap is fitted.

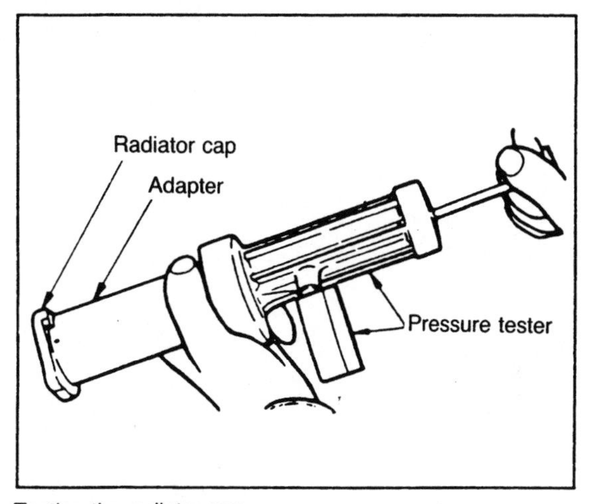

Testing the radiator cap
Source: Mitsubishi Magna TN series manual

3 Slowly pump the analyser and check the release pressure on the gauge. The gauge will show a distinct drop and then hold steady, as the valve releases.
4 If the valve release pressure is incorrect, or the cap pressure does not hold steady on the gauge, the cap is faulty and should be replaced.

Drive belts

A drive belt transfers power between the shafts of two or more components which are in the same axis, parallel and a distance apart.

Correct operation of the drive belt is dependent on the condition and the correct alignment of the pulleys. When a correctly tensioned 'V' belt is driven by a crankshaft, friction takes place between the sides of the belt and the sides of the pulley. It is important that the 'V' belt is the correct length, width and 'V' angle. Ensure the 'V' of the pulley is always deeper than the thickness of the belt so that the belt does not 'bottom' in the 'V' groove of the pulley.

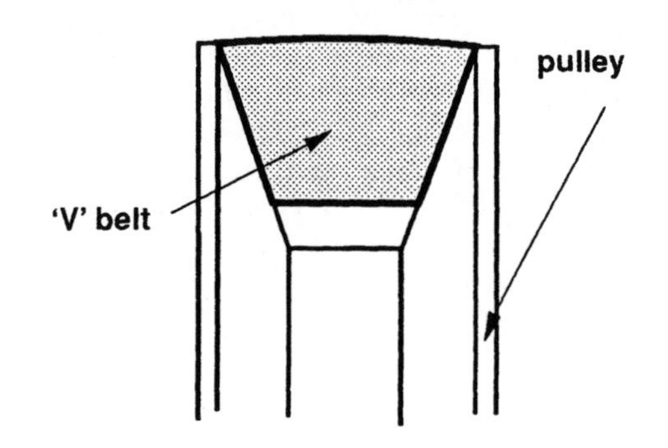

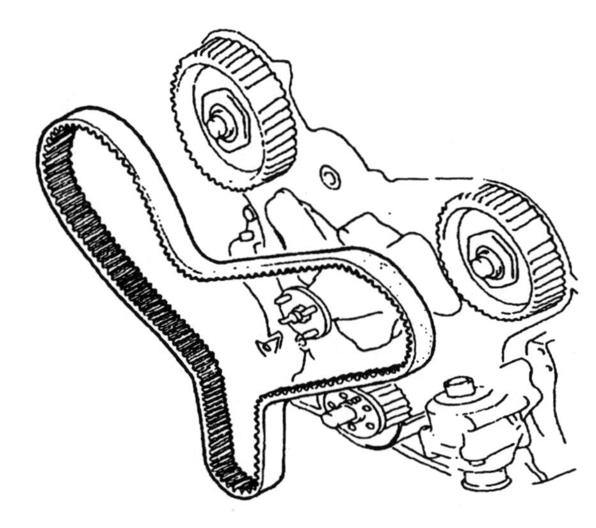

Internally toothed camshaft drive belt.

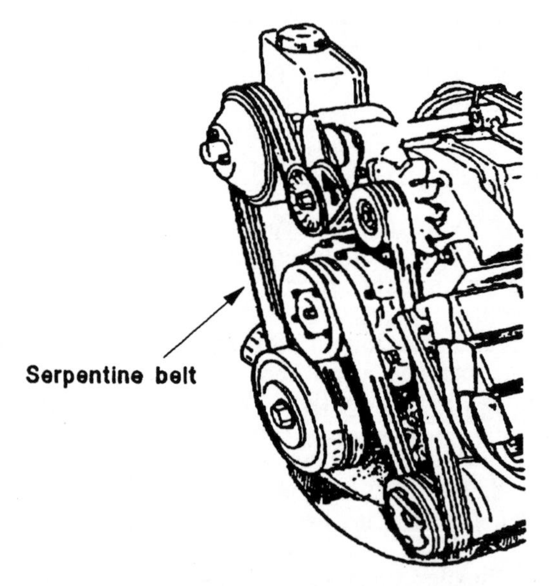

Single, internally grooved, serpentine belt used to drive accessories.

Types of drive belts commonly used on automotive engines include:

1 'V' type;
2 internal tooth type;
3 internal groove (serpentine) type.

DRIVE BELT AND PULLEY SERVICE

To remove a fan belt from an engine, first consult the engine manufacturer's service manual. However, in the absence of service instructions the following procedure may be used.

'V' type fan belt removal

1 Locate and loosen the alternator pivot bolts and adjusting clamp bolt.
2 Push the alternator towards the engine block. This will slacken the fan belt.
3 Roll the fan belt off the alternator pulley, the crankshaft pulley and the water pump pulley.
4 The fan belt can now be removed.

Inspect a fan belt for:

1 Glazing. This is indicated by a hard shiny surface on the side of the belt, caused by the belt continuously slipping.
2 Incorrect seating in the pulleys. This is caused by wear or the wrong sized belt being fitted.
3 Damaged belts, cracked, fraying, torn or oil contaminated belts.

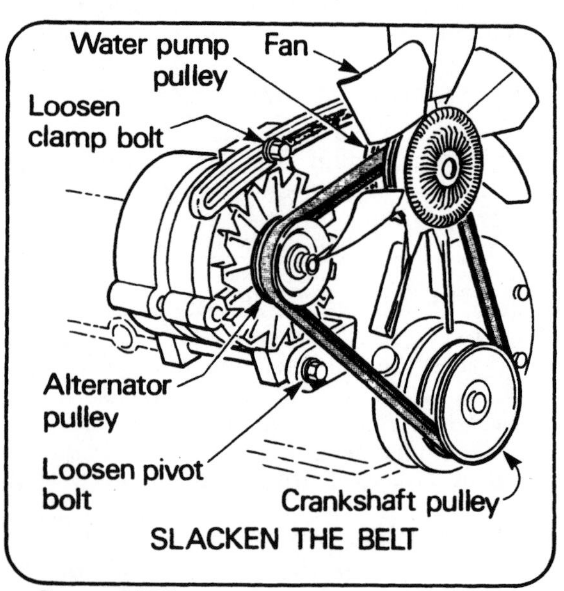

Source: AGPS

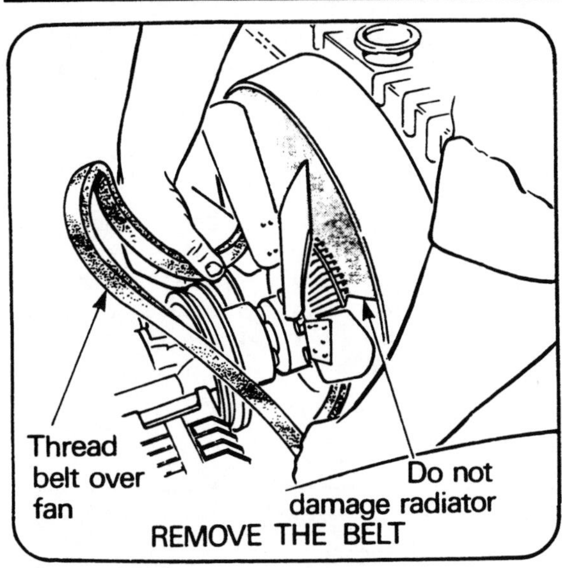

Source: AGPS

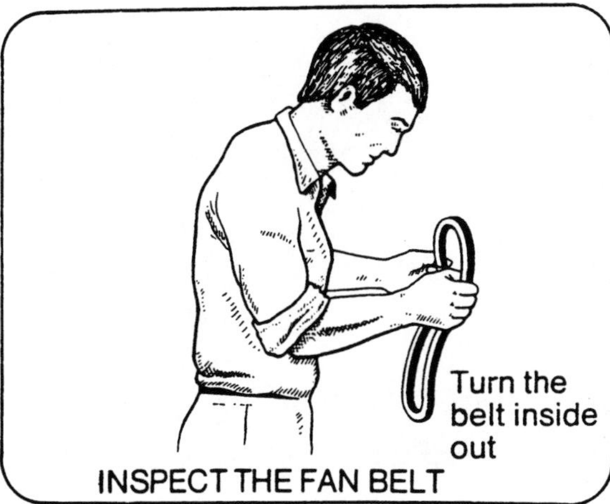

Source: AGPS

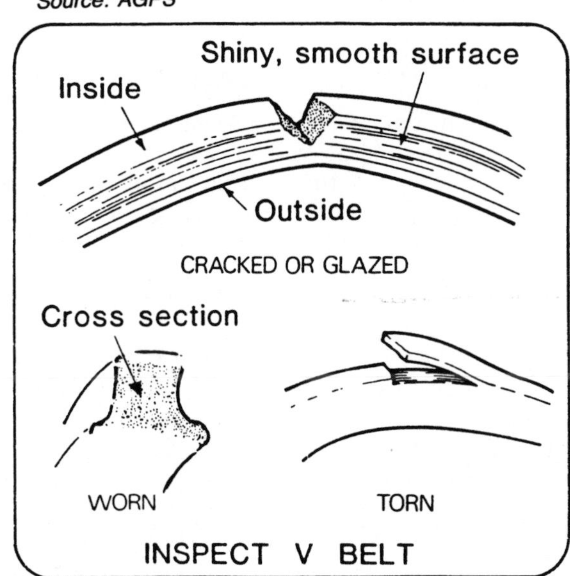

Source: AGPS

4 Inspect internal tooth and internal groove (serpentine) type belts for cracks, separation of rib material and for 'chunking'.

New belt: no cracks or chunks

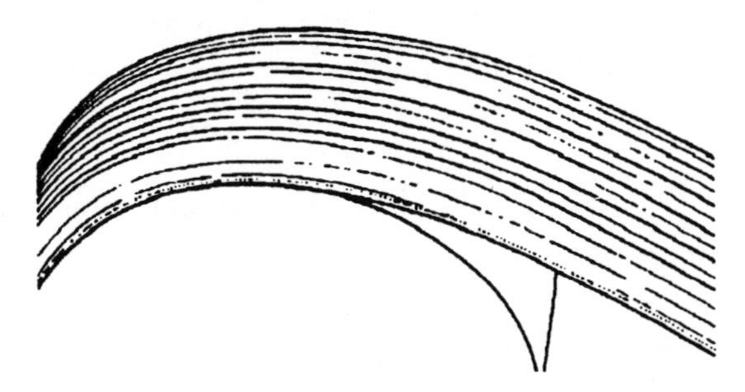

Moderately used belt: few cracks; some wear on ribs and in grooves. Replacement not required.

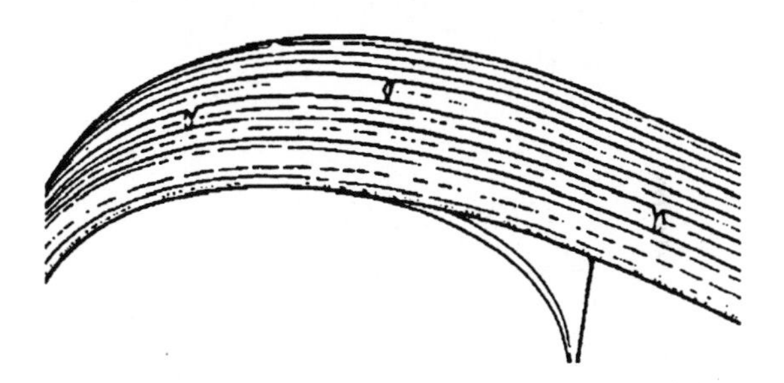

Severely used belt: several cracks per inch. Should be replaced before chunking occurs.

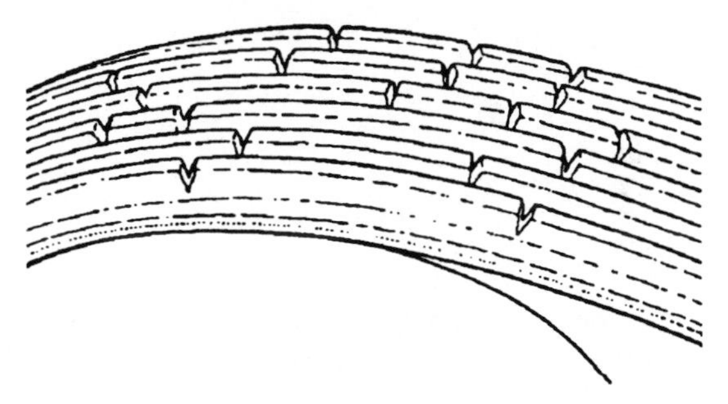

Failed belt: separation of rib material from backing (chunking), replace belt immediately.

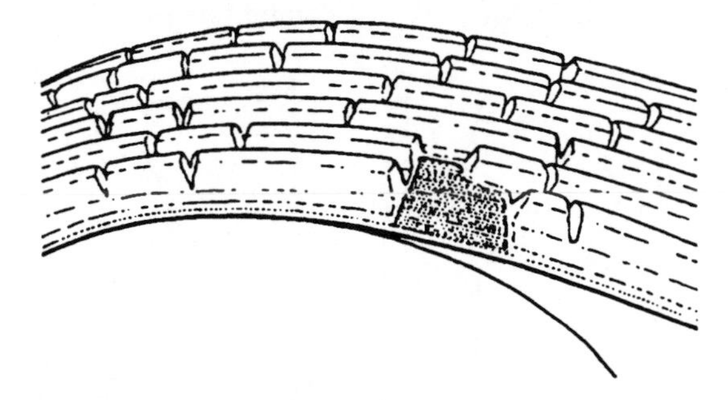

Inspection procedure for internal ribbed serpentine belt.

When a drive belt shows any of these faults it should be replaced, and the cause of the fault be fixed.

3 If pulleys show signs of wear by having burrs, nicks, chips or any sign of damage they should be replaced.

Pulley inspection

1 Ensure all belt pulleys are in correct alignment.

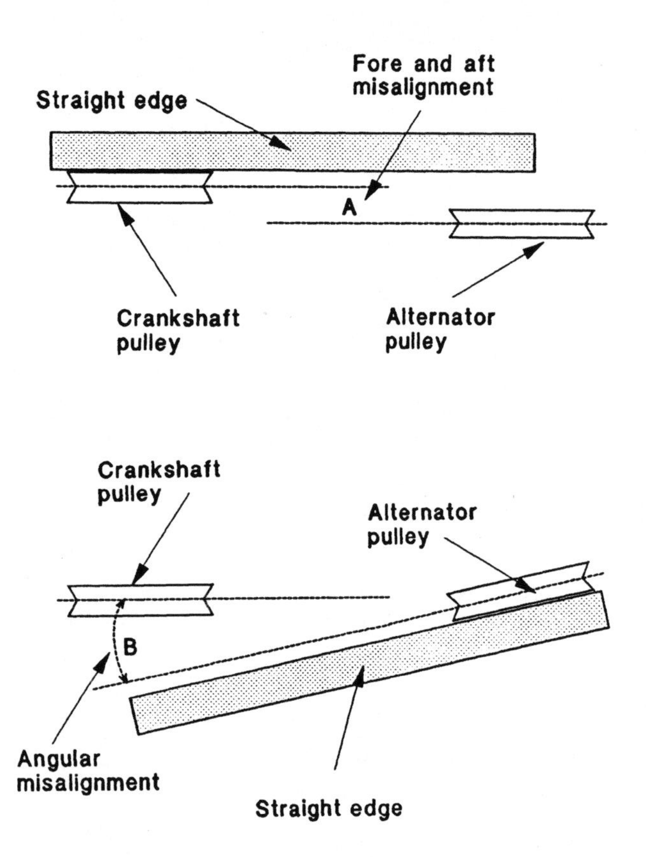

2 Inspect pulleys for wear on the sides and wear in the bottom groove.

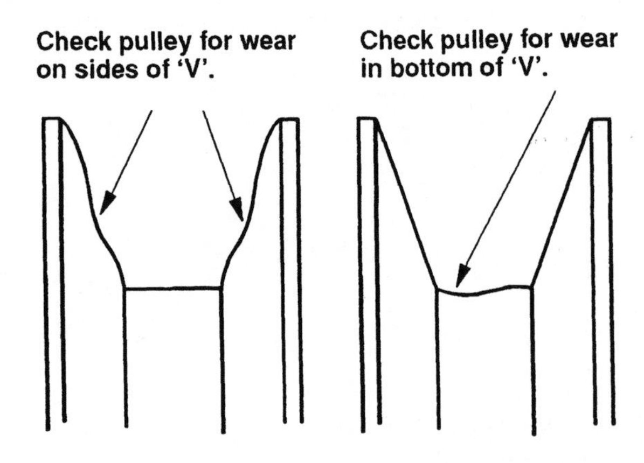

Refit and adjust a fan belt

1 The refitting of the fan belt is the reverse order of removing it.
2 With the alternator pivot bolts and the adjustment clamp bolt loose, a wooden lever is placed between the alternator and the engine block.

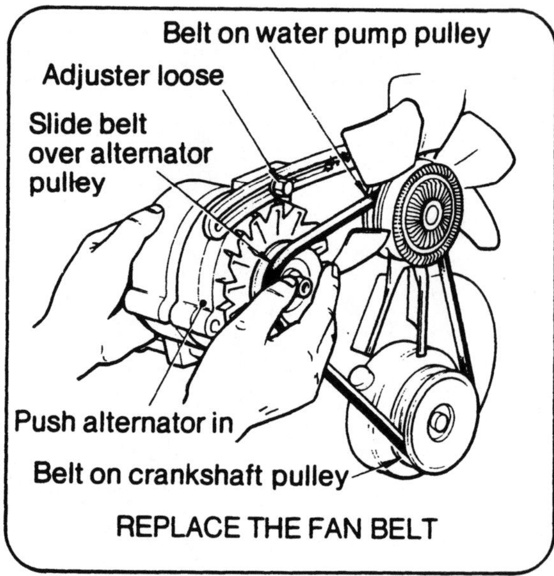

Source: AGPS

3 By applying force to the wooden lever, the fan belt is tensioned.
4 To check for the correct amount of belt tension, special measuring tools are available.
5 On completion of the fan belt adjustment, tighten the alternator pivot bolts and adjustment clamp bolt.

Ensure the fan belt is adjusted correctly, as a loose belt can affect cooling system efficiency, the alternator charging system, and cause premature wear to the fan belt. A fan belt which is too tight can put extra stress onto the water pump, the alternator bearings, and the fan belt, causing premature wear to the components.

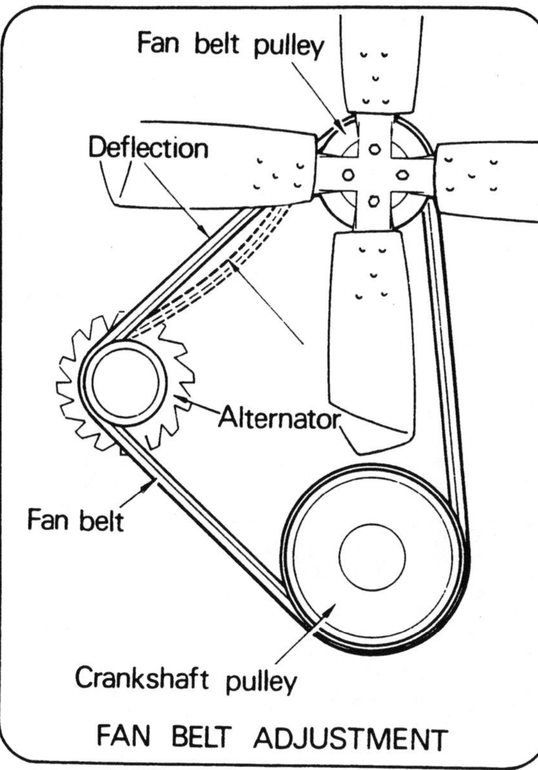

FAN BELT ADJUSTMENT

Source: AGPS

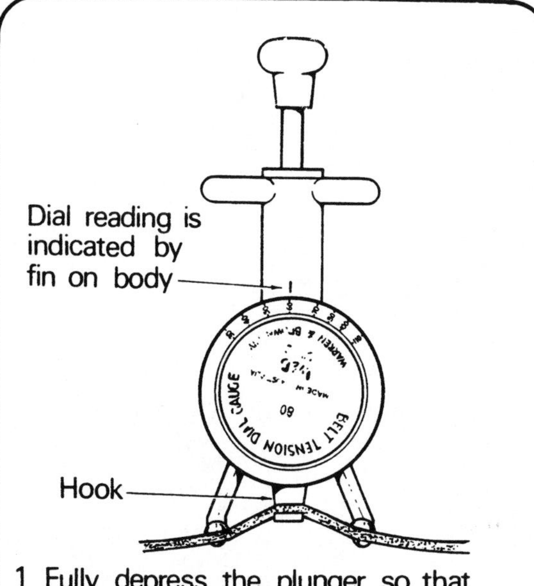

1 Fully depress the plunger so that hook engages the belt.

2 Release the plunger and read the dial.

FAN BELT TENSION

Source: AGPS

AIR CLEANERS

The function of an air cleaner is to:

- filter and remove abrasive dusts from the air going into the engine;
- silence the noise of the air 'rushing' into the inlet manifold;
- act as a spark arrester, in case of a 'backfire' in the inlet manifold.

Types of air cleaners

The four common types of air cleaners are:

1 The paper element type.
2 The oiled metallic gauze type.
3 The oiled foam type.
4 The cyclonic type.

THE PAPER ELEMENT TYPE

This is the most popular type of air cleaner used in today's motor vehicles. It consists of a metal or plastic housing containing a dry paper filter element, a removable top cover and a snorkel inlet tube. The air cleaner filters all incoming air to the engine. Advantages of the paper element air cleaner are:

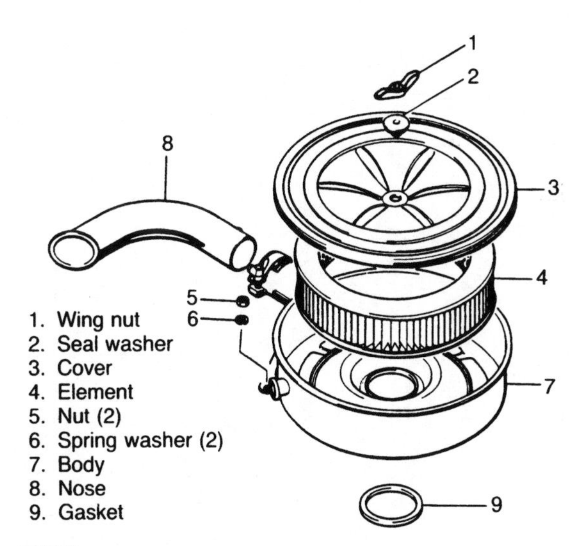

NOTE:
Numbers show order of disassembly.
For reassembly, reverse order of disassembly.

Source: Mitsubishi EA series

1 Having a large filtering area provides maximum filtering efficiency with minimum restriction to air flow.

2 It is lightweight, compact and can be designed to fit into restricted spaces.
3 It is claimed to be able to stop 99 per cent of dust particles down to 1 micron in size.

THE METALLIC GAUZE AND OIL BATH TYPE

This consists of a metal housing containing a measured amount of engine oil and loosely packed metallic gauze, moistened with oil, as the filtering medium. The incoming air is directed down around a baffle then upwards and through the oil soaked gauze. When the air is being directed around the baffle, the heavier dust particles in the air drop into the oil bath. As the air goes through the metallic gauze, the lighter particles of dust are deposited on the oily surfaces of the gauze. This type of filter is now not used as a main air filter.

THE FOAM TYPE

There are various styles of foam air cleaners. Some car sports enthusiasts prefer a dry foam air cleaner fitted to their car, and claim better performance due to less restriction of the incoming air.

Another style consists of a metal or plastic housing filled with oil soaked foam. The air passes through the holes in the housing, the oil soaked foam and into the carburettor. The dust particles in the air are deposited onto the porous oil soaked foam. This type of air cleaner is mainly used on small capacity two-stroke and four-stroke engines.

Component parts in correct sequence

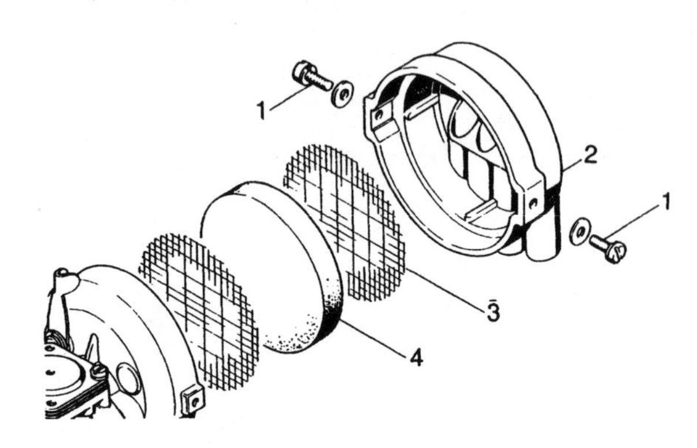

1. ATTACHING SCREWS
2. PLASTIC OUTER COVER
3. WIRE GAUZE
4. OIL SOAKED FOAM

FOAM TYPE

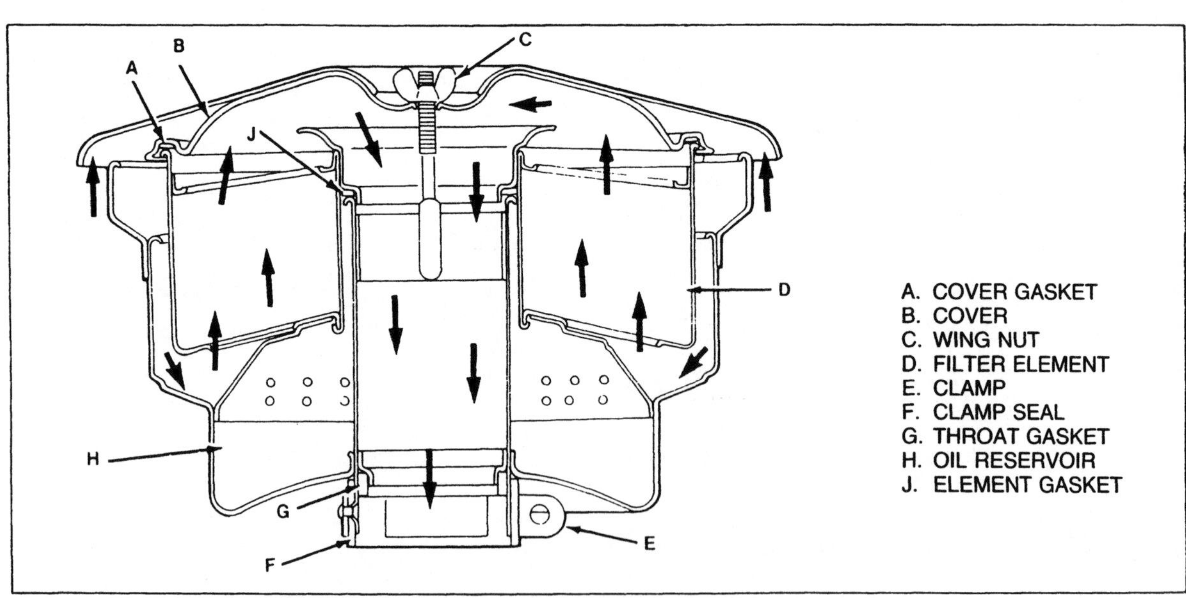

OIL BATH TYPE

THE CYCLONIC TYPE

This type of air filter is mainly used in heavy vehicles and industrial applications. A design feature of this type is that the incoming air is 'swirled' around, causing the heavier dust particles to be thrown to the outside of the filter and the clean air to be drawn through the middle of the filter. This type is sometimes used as a pre-filter in conjunction with another main air filter.

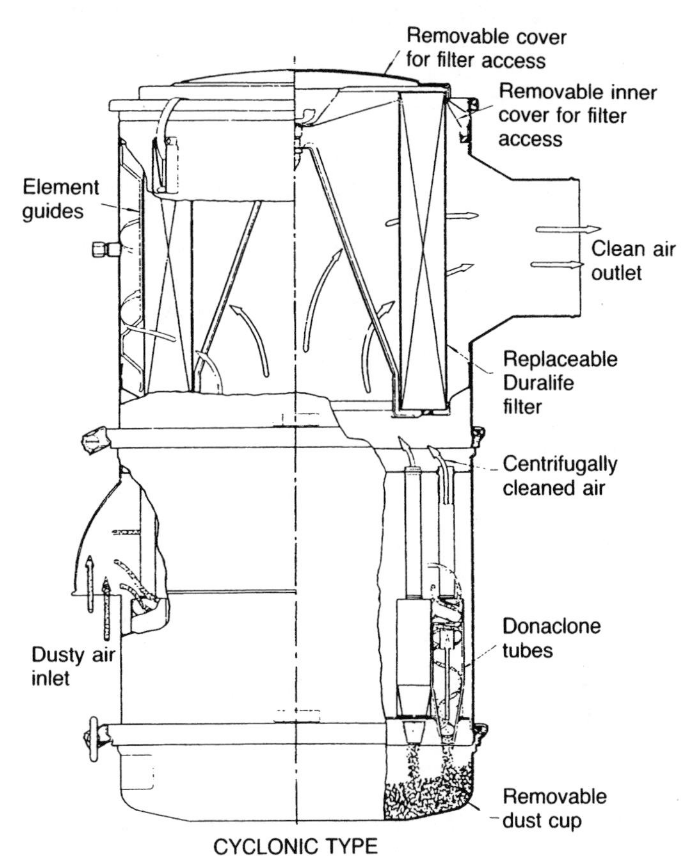

CYCLONIC TYPE

Source: Mitsubishi EA series

Air cleaner servicing (paper element type)

Reasons for servicing air cleaners:

1 A partially blocked air cleaner will restrict the flow of air to the engine. Restricted air flow will decrease the performance of the engine and increase the fuel consumption.
2 A damaged or an incorrectly sealed air cleaner will allow dust particles to be drawn into the engine. These abrasive dust particles will cause premature wear and damage to the engine's components.

The air cleaner element should be checked at scheduled intervals and if the engine is operating in dusty areas, it should be checked more frequently.

SERVICING PROCEDURE

1 Remove the air cleaner from the engine.
2 Remove the top cover from the air cleaner and take out the paper element.
3 Inspect the following.
   - Check that the element has been sealing correctly on both sides of the housing.
   - Check that the housing has been sealing correctly onto the carburettor.
   - Check the condition of the paper filter element. If it is clogged with dust, has oil on it, has signs of not sealing properly or has any holes or small cracks, RENEW it.
4 If the paper filter element is found to be in good condition, it can be cleaned in the following manner.
   - By lightly tapping the paper element on a flat surface, which will dislodge the dust particles.
   - By carefully directing a stream of compressed air through the paper element in a direction opposite to that of normal flow.
5 Thoroughly clean the inside and the outside of the air cleaner housing.
6 Reassemble and refit the air cleaner unit to the engine, according to the manufacturer's service information.

CHAPTER 7 REVISION

1 List the components of a cooling system.
2 Name two types of radiators.
3 Describe the direction of coolant flow through a cooling system.
4 Explain the function of welsh (expansion) plugs.
5 Explain the function of a radiator cap.
6 Explain the function of the valves in a radiator cap.
7 Describe the advantages of a pressurised cooling system.
8 Describe the function of a thermostat.

9 Explain the precautions to be taken when removing a radiator cap
10 Describe the function of additives for a cooling system.
11 Describe a method of checking for internal leaks in a cooling system.
12 Describe a method of determining the serviceability of a radiator cap.
13 List the recommended checks on cooling system components.
14 Describe a method of removing a fan belt.
15 When inspecting a fan belt, list the possible faults.
16 Describe a method of checking the tension on a fan belt.
17 Describe the possible effects of a loose fan belt.
18 Name four types of air cleaners.
19 Describe the function of an air cleaner.
20 Describe the inspection and servicing procedures recommended for the paper element air cleaner.

SEALS AND BEARINGS

Vehicle owners expect their vehicle to operate efficiently and reliably. Generally these requirements are expressed as requiring economical fuel usage and freedom from breakdowns. The vehicle designer seeks a solution to these problems by:

- fitting bearings to rotating parts, and reducing friction. This allows more of the power from the fuel to be used to propel the vehicle;
- using a range of seals, gaskets and sealants to ensure the vehicle operating fluids — coolant, oil, grease, gases, refrigerant, fuels and hydraulic fluids — are effectively contained within their operating circuits. Many mechanical breakdowns are the direct result of a seal or gasket failure, for example transmission bearing failure after loss of oil through a defective seal, or an expensive engine repair after overheating caused by coolant leakage.

This chapter will introduce you to the application and range of bearings, seals and gaskets used within a motor vehicle.

SEALS

I

Function

The function of a seal is to prevent leakage of fluids, primarily between moving and non-moving parts. However, there are a few applications where seals are used yet movement between the faces is not intended. Seals which are used in applications where there is movement between the parts are known as DYNAMIC seals, whereas the name applied to seals used where the parts are not moving against each other is STATIC seals.

Identification and application

O-RING SEAL (STATIC AND DYNAMIC)

The O-ring is a smooth round ring and is generally made of:

- rubber (natural or man-made);
- plastic. It is usually coloured black but it can be made in other colours. This seal is known as an O-ring because of its sectional rather than its circular shape.

Application. O-rings can be fitted in both static and dynamic applications. Typical examples of each (see next page) are:

- between the fittings and pipe connections on an air conditioning system (static);
- between the piston and the cylinder of a hydraulic servo, such as an automatic transmission servo (dynamic).

Fitting. When O-ring seals are fitted there are a number of points which must be considered:

- If the seal is to be fitted into a machined groove in a shaft, then care must be taken not to cut the seal on any sharp edges such as splines or circlip grooves. It is a good working practice to use a seal fitting collar if the correct size is available to suit the shaft.
- Make sure all surfaces are clean, smooth and lubricated with the correct lubricant.
- If the seal is to be rolled into position, then ensure the seal is not twisted when in the

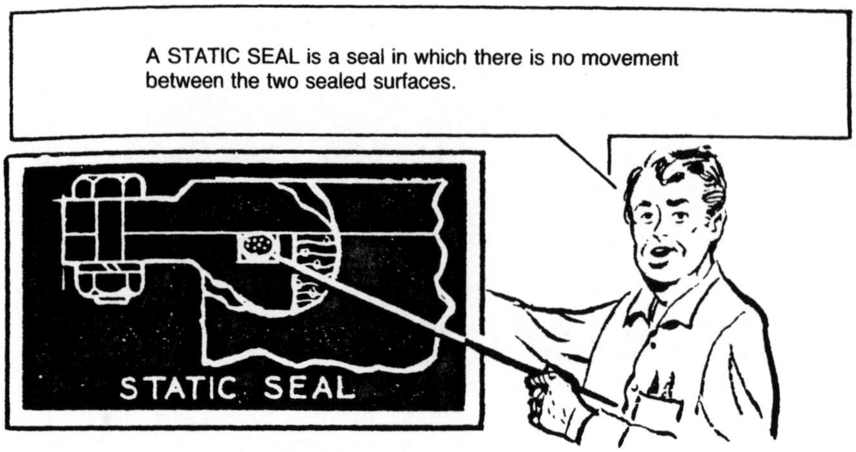

A STATIC SEAL is a seal in which there is no movement between the two sealed surfaces.

STATIC SEAL

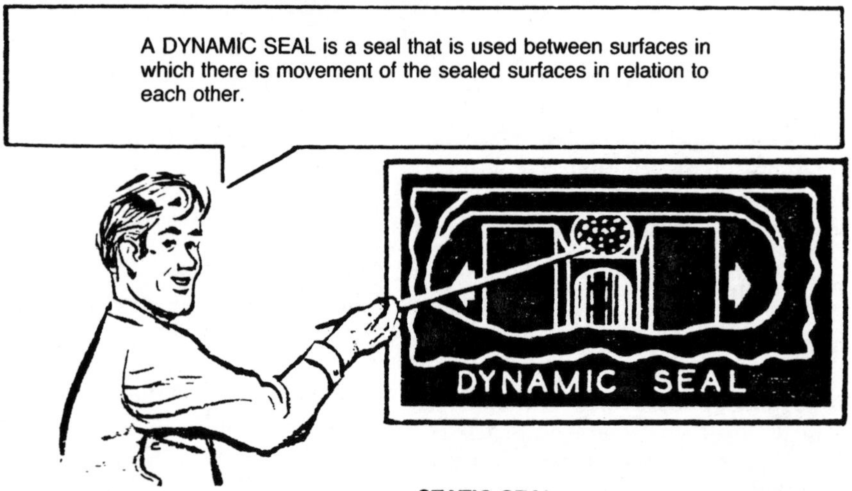

A DYNAMIC SEAL is a seal that is used between surfaces in which there is movement of the sealed surfaces in relation to each other.

DYNAMIC SEAL

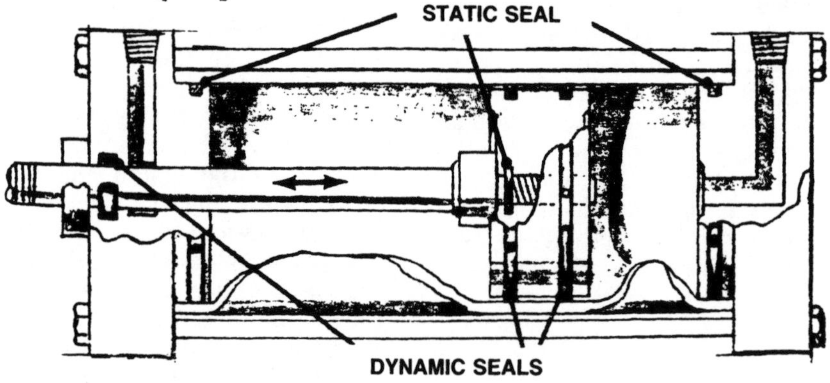

STATIC SEAL

DYNAMIC SEALS

This hydraulic cylinder has both static and dynamic seals. Note the surfaces that move next to the dynamic seals.

Source: Caterpillar Inc.

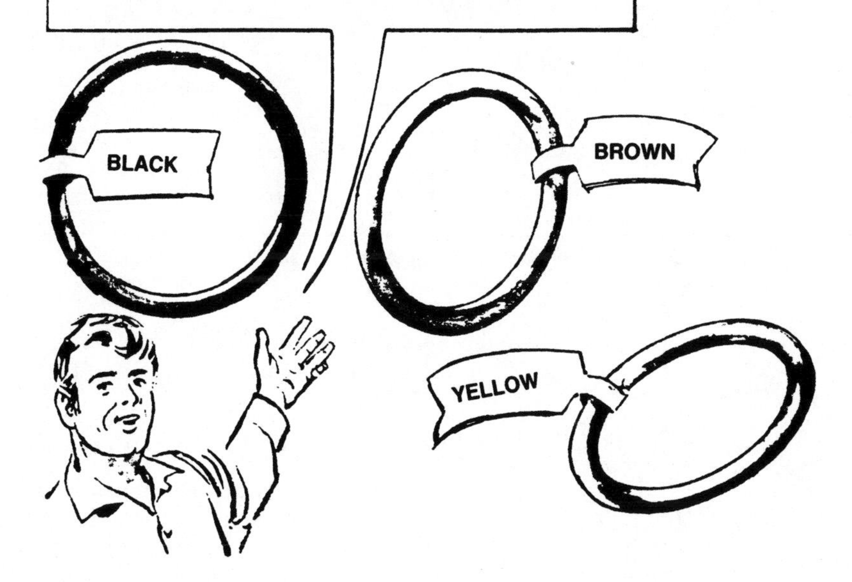

Source: Caterpillar Inc.

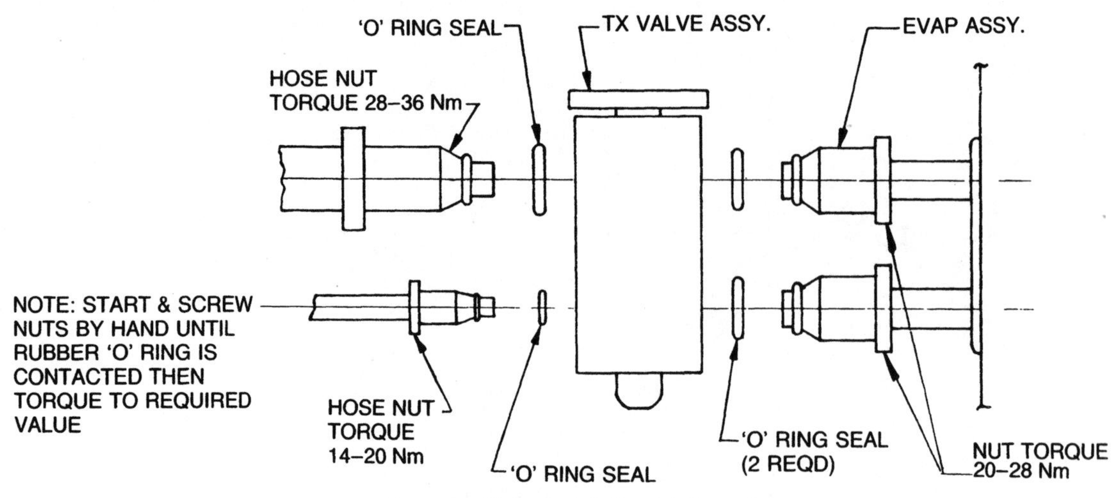

Expansion Valve Installation (static seal)

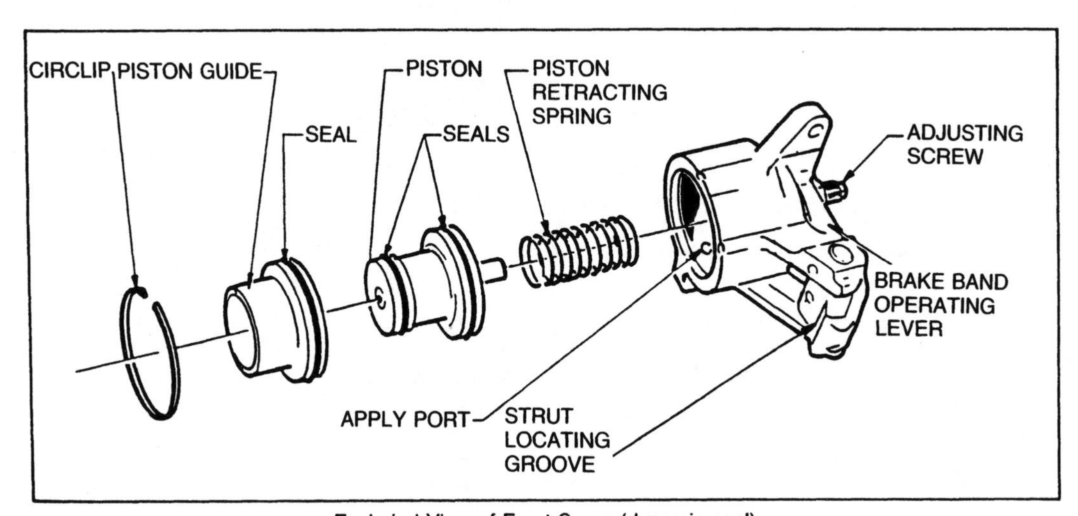

Exploded View of Front Servo (dynamic seal)

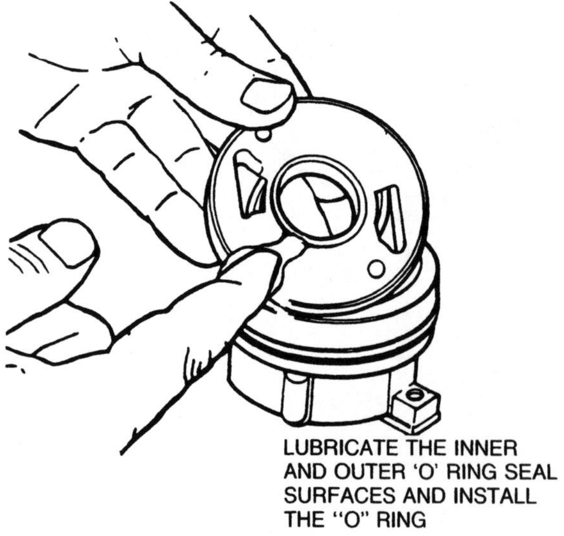

Installing O-Rings

final location. This can be done by checking the seal moulding line is at right angles to the shaft.

LIP OR EDGE SEAL (DYNAMIC)

This seal is possibly the most important type used in the modern motor vehicle. There are many different designs of lip seals, but generally they all use the same principle of operation. The sealing action is achieved by presenting a fine line of contact against a smooth rotating shaft. The seal edge breaks through the oil film almost contacting the shaft, thus preventing oil leakage.

The lip or edge seal consists of a pressed steel frame or case, a natural or man-made rubber sealing element bonded onto the frame or case and a garter spring to increase the load on the seal lip. A number of seals are manufactured where the garter spring is deleted and the rubber sealing element acts as the spring. Alternate materials such as leather have also been used as the sealing element.

Application. The lip or edge seal is a dynamic seal. Therefore, the ideal application is where a rotating shaft passes through a rigid housing which can be used to house the seal. Examples of this type of dynamic seal application are:

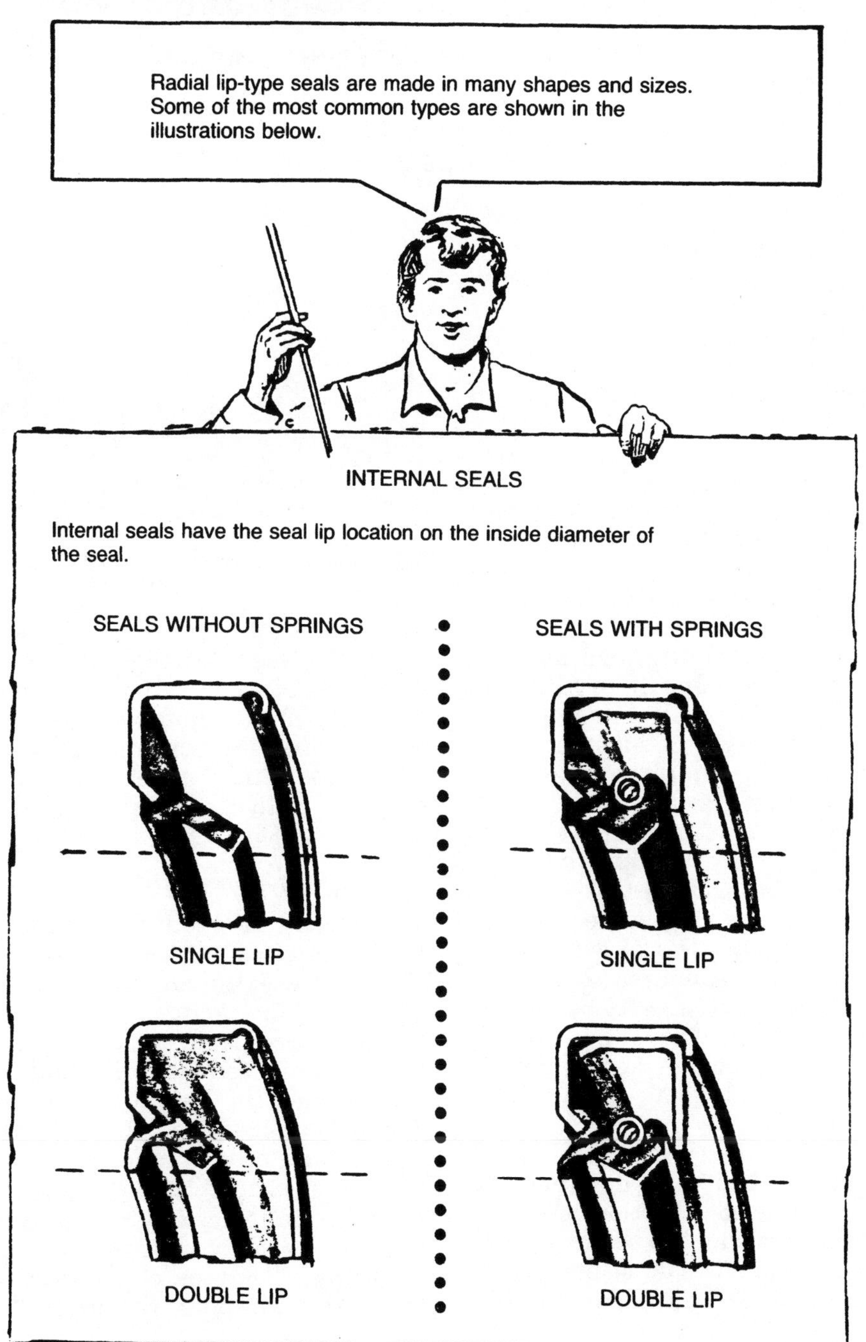

Source: Caterpillar Inc.

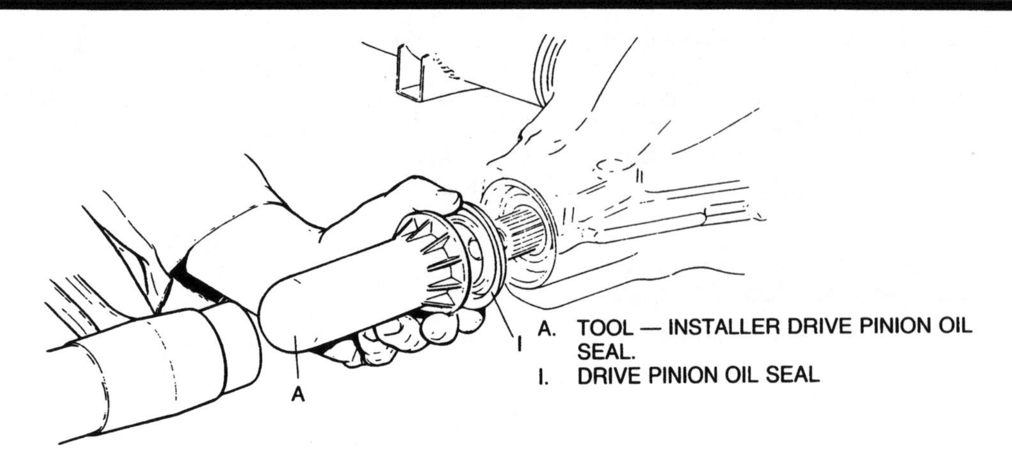

Installing Pinion Seal.

- front and rear crankshaft seals;
- transmission input and output shaft seals;
- steering box shaft seals;
- front wheel bearing seals;
- rear axle bearing seals.

Fitting. This type of seal is robust in construction, however, a few simple fitting precautions are essential to ensure the seal will not suffer premature failure.

1 The seal must be fitted with the open face (garter spring side) of the seal facing the substance to be sealed.
2 If the seal is to be fitted over a shaft that has splines or keyways formed into it, then a protector sleeve or plastic tape must be used to protect the seal during installation.
3 The surface of the seal lip must be lubricated with the type of substance it is going to seal, to prevent the seal from 'burning out' on first start-up.
4 Prevent damage to the seal case, sealing lip and the housing during seal installation by using the correct seal fitting tools.

Hydrodynamic type. Although essentially a lip type, the hydrodynamic seal is becoming more popular with manufacturers because it offers a more positive sealing action, reduced seal 'drag', lower radial load and seal lip temperature. These factors provide greatly reduced seal and shaft wear, thus giving extended service life.

The seal is similar in construction to the standard lip type, but the difference is apparent if the lip or edge section is closely inspected. Current seals have a series of small 'flutes' on the atmospheric side of the sealing face. When the shaft is rotating in the seal these flutes act to prevent oil passing the seal. Only the seal designed for the particular direction of application should be fitted, however, if the shaft were to rotate in the reverse direction (the rear transmission seal when the vehicle is reversing) the seal will operate with the characteristics of a conventional lip seal.

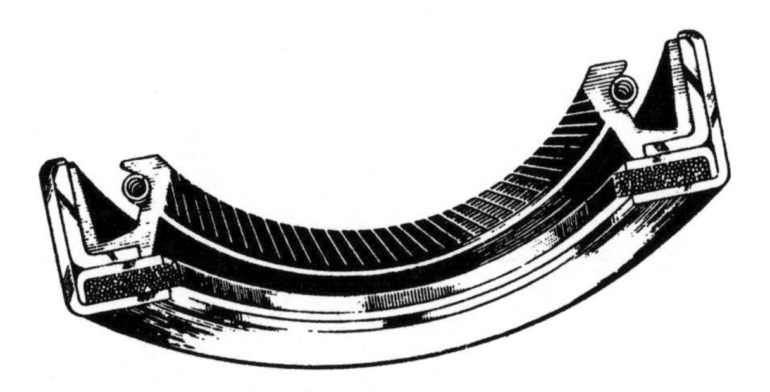

A Sectional view of the Hydrodynamic Oil Seal.

EXTERNAL LIP TYPE

The major automotive application for this type of seal is in vehicle braking systems. They are usually cup shaped with a sealing edge on the outer perimeter, and are manufactured from black synthetic rubber that is unaffected by brake fluid.

Application. This type of seal can be classified as having a dynamic application, but is generally used on sliding plungers instead of rotating shafts. The major applications are:

- brake master cylinder sealing cups;
- brake wheel cylinder sealing cups.

Fitting. This type of seal does not require special tools to press the seal into a housing,

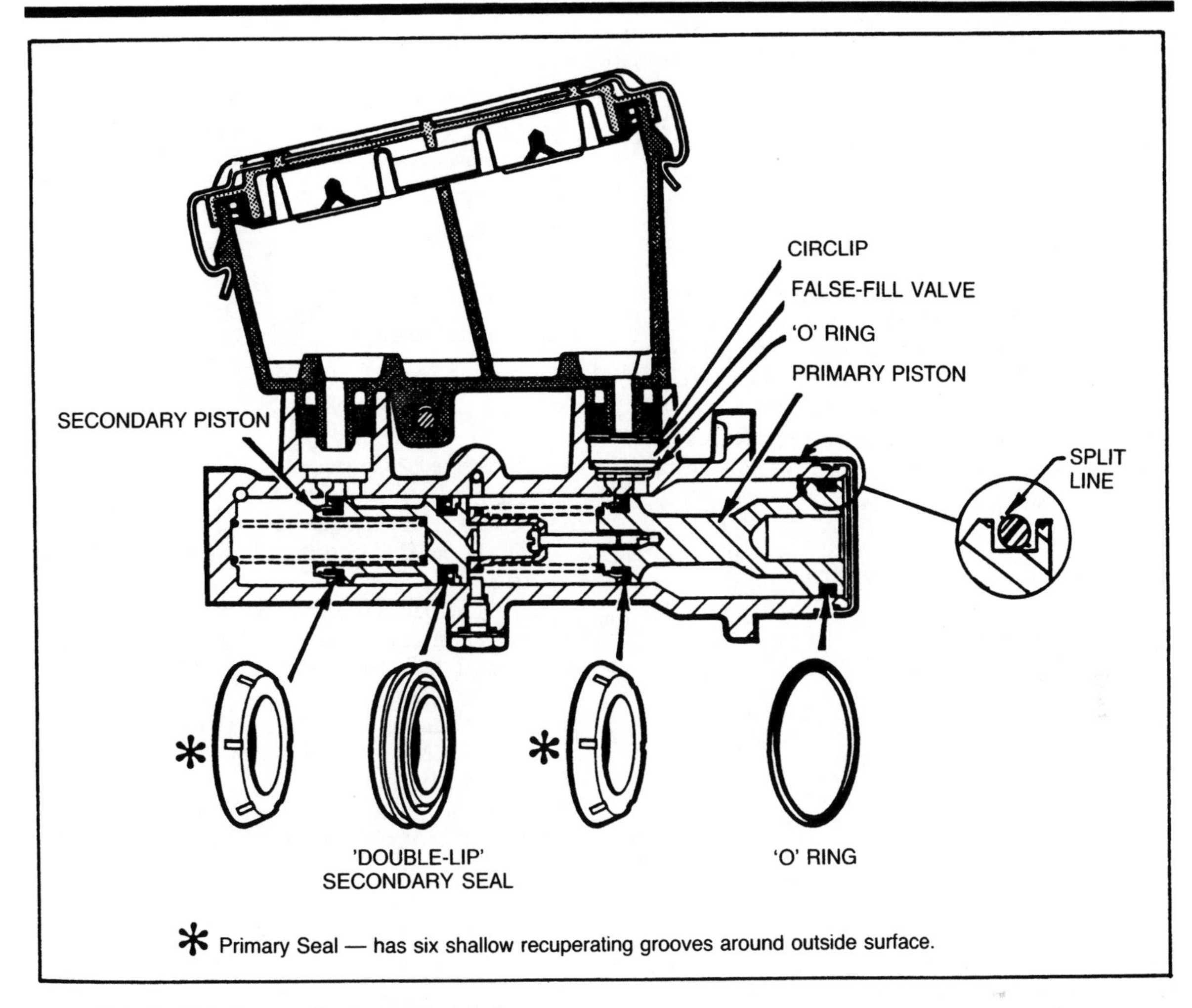

Master Cylinder Main Bore — Component Installed.

nor do the seals have metal frames or garter springs. However, to obtain satisfactory performance from the seal a few simple precautions should be observed.

1 The sealing edge must face the fluid being sealed.
2 Use the type of fluid to be sealed as a lubricant on assembly.
3 Care must be taken, when the seal is to be fitted over a cylindrical section (master cylinder piston) then dropped into a groove, that the seal is not over-stretched and permanently distorted. If you need to use a tool to assist you with this task a plastic knitting needle is ideal, as it will have less chance of damaging either the piston or the seal.
4 When fitting the piston into the cylinder, make sure the sealing cups enter the bore in their correct position, without being rolled over or tucked under, as this will prevent them from sealing.

FELT SEALS (DYNAMIC AND STATIC)

The felt seal is usually of rectangular section and circular shape and is assembled into a pressed steel case. It can be manufactured for both internal and external applications.

Application. Although this type of seal has generally been replaced by spring loaded edge seals, it still has a number of applications as a:

- dust seal in front of a lip seal;
- grease seal in some front wheel bearing applications.

Fitting. This type of seal, although very simple in construction, does have a specific requirement to prevent premature failure.

- The seal must be soaked in engine oil before assembly to prevent damage due to friction burning the felt sealing material.
- The correct seal fitting tools should be used to prevent damage to the seal case, shaft or housing.

SLINGER

The oil slinger is not an oil seal but does reduce or prevent oil and fluid loss where a shaft must pass through a housing. The oil slinger is usually a pressed metal disc attached to the rotating shaft close to the point where the shaft passes through the housing. Oil moving along the shaft reaches the disc and is thrown off by centrifugal force.

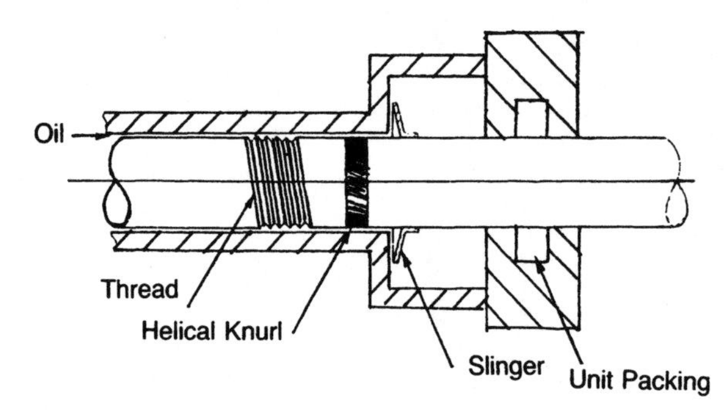

Shaft and housing showing various oil control devices which may be employed.

Application. An oil slinger is often used in conjunction with a lip edge seal and can be an excess oil flow protector (internal) or a shield from dust and water (external). It can be:

- fitted inside the crankshaft front housing (timing case), where it acts as a flood protector for the crankshaft front housing seal;
- fitted inside heavy transmissions between the rear mainshaft bearing and the output yoke lip seal;
- mounted on the outside of the final drive pinion shaft to throw dust and water away from the pinion shaft seal.

SCROLL

The scroll is a screw thread machined or pressed on to a shaft. The screw threaded section of the shaft passes through a close fitting sleeve section of the housing. The thread is either right or left handed, depending on shaft rotation. Any leak is

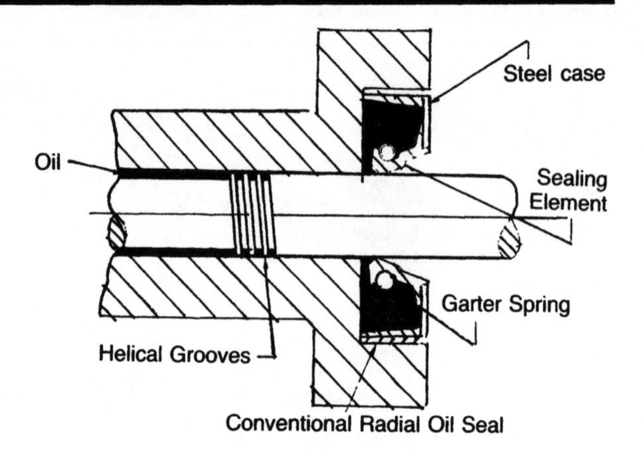

Use of helical groove to supplement lip seal.

reduced as lubricant caught in the threads on the shaft will be wound back in to the housing.

Application. The problem with this type of oil leak control is the necessity to provide shafts and housings with minimum running clearances. The scroll will only function in a dynamic condition (shaft rotating) and unless backed by a lip seal, oil leaks will occur under static conditions.

- Machined onto the input shaft of some manual transmissions. (The front bearing retainer has a thread cut on its outer circumference. This threaded section fits inside a tubular housing.)

WICK SEAL

The wick seal is a section of rope, woven from asbestos material. The woven material is impregnated with either graphite or molybdenum disulphide. Both materials provide excellent lubricating properties. The wick seal is fitted into a machined groove in a housing through which a rotating shaft passes.

Application. This type of seal is generally used as the rear main bearing seal between the engine crankcase and the crankshaft.

Fitting. Unlike most of the other types of seals, wick seals require specific 'tailoring' of the seal when fitting.

1. The seal is formed into its working shape by using a special tool, and working the seal material into the groove machined into the housing.
2. The excess seal material protruding from the groove is trimmed at the ends. (Be

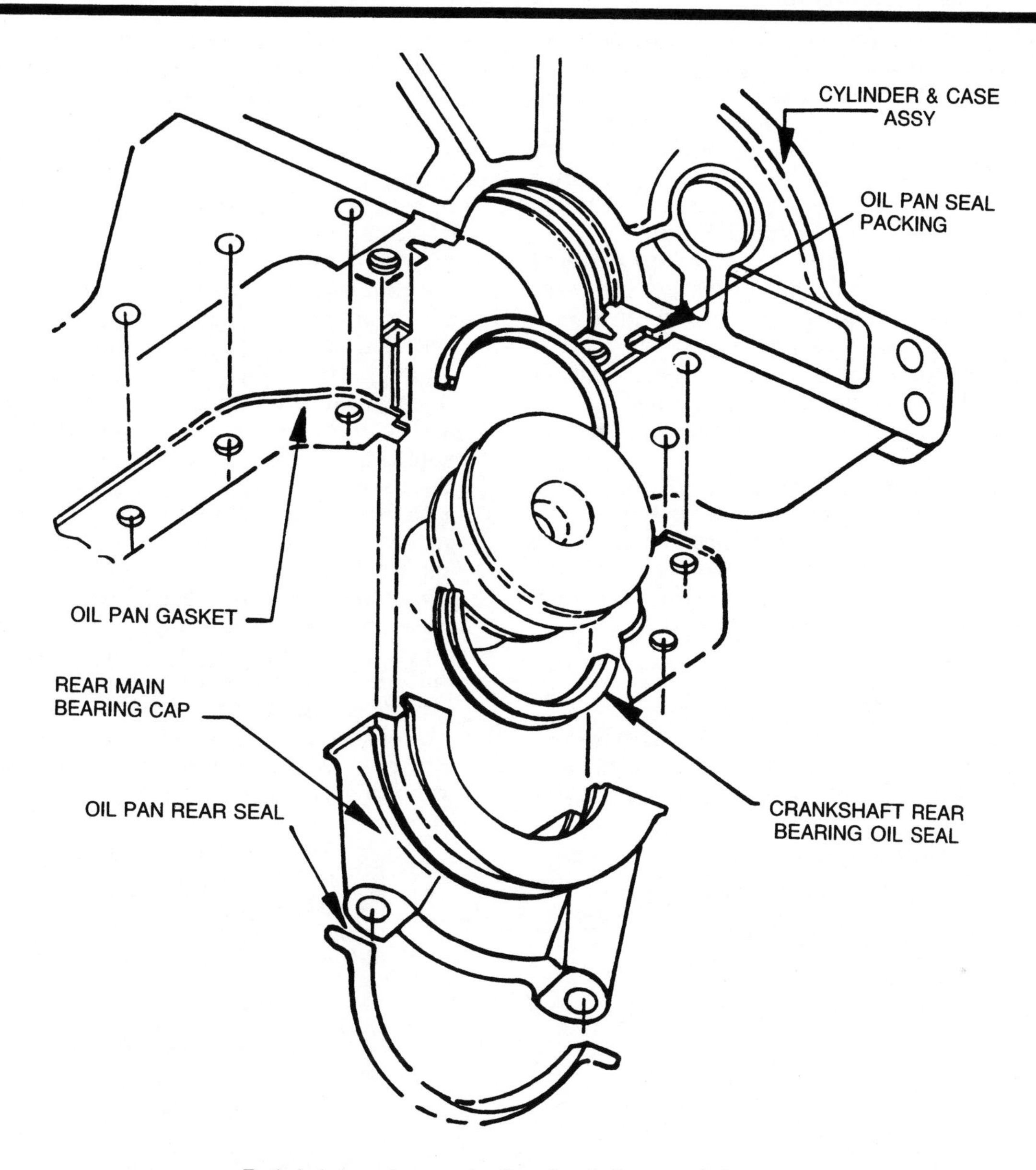

Exploded view of rear main oil seal and oil pan gaskets

careful not to cut the material too short. It should be just above the parting surface of the housing.)

3 Repeat steps 1 and 2 with the lower half of the seal.
4 Lubricate the two pieces of seal material.
5 Install the shaft.
6 Fit the bearing cap and tighten the retaining bolts to the manufacturer's specified torque limit.

GASKETS AND SEALANTS

In simple terms, a gasket is a layer of material fitted between two adjoining parts of an assembly to prevent loss of fluid (gas, coolant etc.). The problem becomes complex when the range of conditions to which the gasket may be subjected is considered.

- When the assembly is bolted together the gasket between the mating surfaces is subjected to pressure and must adapt to the irregularities in the surface finish. The gasket must 'flow' to compensate for the thickness, and create the seal. This quality of a gasket material is referred to as compressibility and extrudability.
- Gaskets must be able to withstand extreme temperatures, contract, expand and still provide an effective seal. This feature of a gasket is known as resilience.
- Permeability is the ability of a gasket material to prevent the contained gas or liquid from passing through the gasket material.
- A critical feature of gaskets, which is often not clearly understood, is the material thickness when installed. Incorrect material or thickness can adversely affect tension of the retaining bolts, critical clearances, part alignment and compression ratios.

Gasket materials

A wide range of materials is used in gasket construction; cork, asbestos, synthetic rubber, paper, steel, copper and aluminium. These materials may be used singly or in combination where the sealing requirement cannot be successfully achieved by a single material. A motor vehicle requires many different types of gasket material and by examining a few of these applications the reason for the selection of certain materials will become evident.

1 **Paper**: generally used where thin gaskets are required. The material requires the use of a sealer when fitting and can become brittle if exposed to high temperatures. By impregnating the paper with resin or rubber the resistance of the material to shrinkage and temperature can be improved. Typical applications for this type of gasket are thermostat housing water outlets, timing case/block seals, final drive assembly gaskets and many transmission applications.

2 **Asbestos**: seldom used as a single material gasket, but when combined with copper or steel, asbestos is an effective material for high pressure and temperature locations. Exhaust and cylinder head gaskets often use this material.

3 **Rubber**: while nominally referred to as 'rubber' gaskets it is probable the material

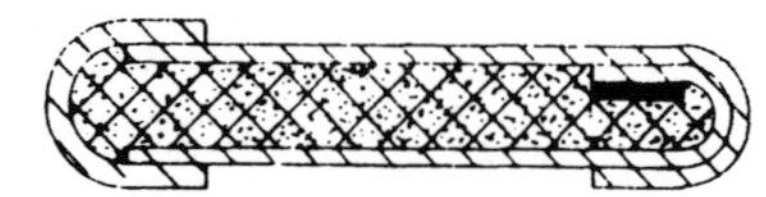

Copper Asbestos Copper Tin Plate Asbestos Tin Plate. (Double Fold Gasket) — the flanges of both metal layers are formed over the pores to give double protection.

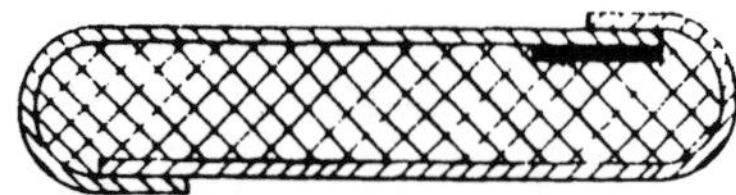

Tin Plate Asbestos Copper. (Reserve push through Gaskets — the flanges of the tin plate are formed over the bores) and the flanges of the copper are formed over the water and oil holes.

CORRUGATED METALS
Tin Plate. Steel. Stainless Steel. Copper. (A single thickness of special hardness metal pre-formed to assure concentration of loading at the points to be sealed.)

Copper Asbestos Copper. Tin Plate Asbestos Tin Plate. (Push through Gaskets — the flanges of one layer of the metal are pushed through and formed over the asbestos for reinforcement and protection of this material.)

COMPOSITE MATERIALS

"Cylgastos". Treated asbestos layers bonded on to a perforated tin plate cote bore tenule in tin plate.

"Cylbestos". Treated asbestos layers bonded on to a plain tin plate core — bore ferrule in tin plate.

Gasket Designs.

used will be synthetic or plastic. This does not mean the material will be inferior — it is more likely that natural rubber would fail in many of the applications where these gaskets are used. To contain oils, fuels and solvents it is necessary to use the synthetic range of materials. However, if air or water only are the fluids in use, natural materials can be used. The synthetic range of gaskets are used as valve cover gaskets, transmission pan gaskets etc.

4 **Cork**: these gaskets normally contain a form of binder to hold the cork granules together and are excellent for applications where temperature extremes are not encountered and significant variation in mating surfaces could be present. This condition is common when a pressed cover is mated to a machined casting. It is possible to obtain satisfactory sealing with low flange pressures and these features make the gaskets suitable for valve cover, engine sump and automatic transmission pan applications.

5 **Steel**: when used as a single material, steel gaskets are usually of the corrugated steel shim type. The forming process used leaves a series of corrugations around the critical sealing areas. This type of gasket is used where extremely fine machining tolerances can be achieved and low loss of bolt torque in service is required. Cylinder head gaskets during the 1950–1960 period were manufactured in this manner. In service replacements for this gasket tended to be a combination (two materials) gasket which possessed greater compressibility and allowed for minor warpage in cylinder head and block.

6 **Composite materials**: this type of gasket has a steel core with a layer of sealing material (asbestos or materials having similar characteristics) bonded to each side. The steel core can be perforated or solid depending on the application for which the gasket is designed. The gasket is suitable for use on exhaust manifolds and exhaust flanges. Variations of this gasket are used for cylinder heads to block sealing, as it is possible to achieve an effective seal with low bolt torque. Fitting procedures should be closely followed as the gaskets are normally fitted dry (no additional sealer applied) and retightening of the retaining bolts or nuts may be required after the engine has reached operating temperature.

7 **Copper/steel and asbestos**: cylinder and exhaust gaskets are subjected to extremes of heat and pressure. By enclosing a layer of asbestos between two thin sheets of metal, the manufacturers have produced a reliable gasket for these applications. A popular version of this gasket has a steel layer uppermost, facing the combustion chambers and cylinder head (particularly aluminium cylinder heads). The lower face is made of copper which is turned through the water openings in the gasket. This effectively lines each water passage with a layer of copper. The cylinder bore holes are usually lined with steel. Copper/steel/asbestos combinations of this gasket type have been in common use for many years.

ADHESIVES AND SEALANTS

Many good quality adhesives and sealants are available, but care must be taken to ensure the correct type is used for each application. Consult the vehicle manufacturer's repair manual for information regarding the correct adhesive to use. Frequently, the recommendation is not located in the specification sections but is included in the overhaul text. The following list of extracts from repair manuals will provide you with an outline of the variety of statements likely to be encountered:

'Apply a 1.5 mm diameter bead of (brand name) or (brand name) to the front face of the extension housing.'

'Apply (brand name) pipe sealant to the water fittings and install.'

'Do not apply sealer to the cylinder head gasket.'

'Coat the thread of the bolt with water resistant sealer.'

'Coat one side of a new gasket with an oil resistant sealer and lay the cemented side of the gasket in place on the cover.'

'Install the extension housing gasket using a light coating of gasket cement.'

TYPICAL INSTALLATION INSTRUCTION

The use of RTV type sealants is tending to displace the use of gaskets and gasket cement. This type of sealant cures in the air and forms an oil and water resistant gasket.

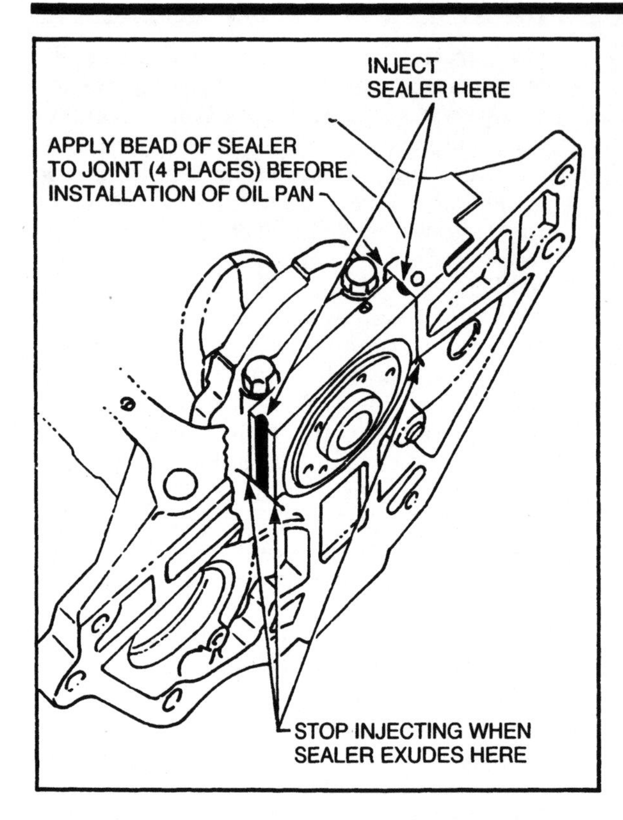

The apparent advantages of this type of sealer explain the rapid expansion of its use, however, use of 'form in place' gaskets should only be as approved by the vehicle manufacturer. Many parts are designed to operate with specific clearances or end floats. These clearances are calculated with a genuine gasket compressed to installed thickness, therefore to delete the gasket and replace with a 'formed in place' gasket of unknown thickness could decrease the operating clearances of the assembly.

The following points provide a general guide to sealant use:

- Use sealants only as directed by the vehicle manufacturer.
- Do not re-use the old gasket by adding gasket cement to assist the gasket sealing action.
- Do not use sealants on rubber gaskets — the gasket tends to be displaced by the use of sealants.
- Do not decrease operating clearances by removing gaskets from assemblies with specified clearances.
- Use gasket cement sparingly — spread cement as a thin film.

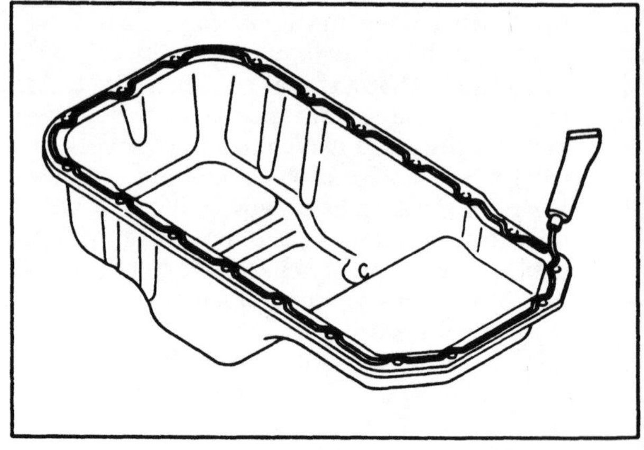

- Formed in place gaskets should be applied as a continuous bead.
- All gaskets, whether bare of sealant, sealant applied, or 'formed in place' should be checked to ensure vital operating holes (oil returns, feeds, water passages etc.) are not blocked by gasket material.
- Do not overtighten the mating components — use torque specifications and tightening sequences where specified.

GASKET STORAGE

Gaskets can be easily damaged in storage by careless handling or poor storage practices. Exposing loose gaskets to direct sunlight will cause 'drying out' and shrinking of the gasket material. Loose gaskets should be stored flat and other objects should not be placed on them. The most suitable form of storage is when the gasket is held in a 'bubble' pack. The packaging of gaskets has been available for many years as a means of convenience when overhauling a major unit, for example engine overhaul gasket sets, but the concept is now extending to individual gaskets with a consequent reduction in storage and handling damage.

BEARINGS

A vehicle is fitted with many different bearings to reduce friction and to provide a relatively cheap method of overhauling assemblies when wear takes place. If replaceable bearings were not present at many locations (engine crankshaft, transmission mainshaft) it would be

Source: Caterpillar Inc.

necessary to replace major components at every overhaul. The wide variety of vehicle bearings can be grouped into major categories:

- where sliding friction is present — the two surfaces are sliding past each other separated only by a thin film of oil. This group is known as *friction bearings*;
- where rolling elements, such as balls or rollers separate the two surfaces, the bearings can be classified as *anti-friction bearings*. This type of bearing minimises frictional losses.

Friction bearings

Bushings, plain bearings, flanged, thrust, insert, shells, precision insert, full round etc., are all names given to bearings in this grouping. These bearings share one feature in common: they are manufactured with contact materials which are softer than the component/housing or shaft to which they will be fitted. By adopting this approach the major wear will be concentrated on the bearing and a cost-effective overhaul to the assembly in which it is fitted is possible. The friction bearing material can be a bronze bushing or can be a very complex combination of layers of steel, tin, copper, tin-lead alloys for use as an engine crankshaft bearing.

Generally, the plain or friction bearing is used to carry 'radial' load (a load applied on the shaft radius). If the bearing has flanges, either attached or fitted to the housing supporting the bearing, then the bearing assembly can also carry 'axial' load (a load applied along the shaft axis or centre).

Friction bearing types

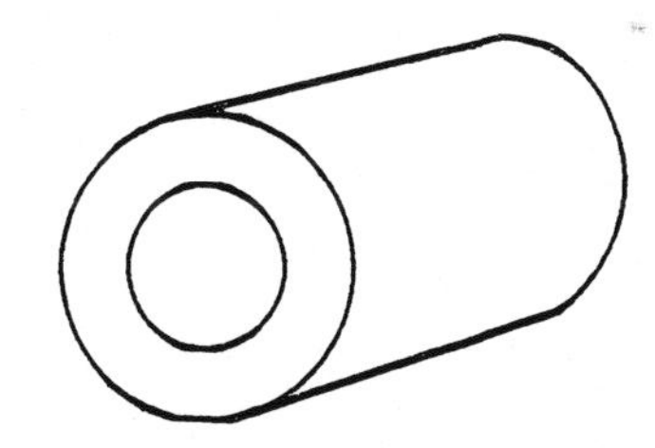

Bushing: a cylindrical, one piece or sintered bronze bush. Bushings have been manufactured with an outer steel casing but the majority are of single piece construction. The bushing is usually pressed into a housing and retained in position by an interference fit on the outer circumference. A shaft rotates, with a preset clearance, in the inner diameter of the bushing. Typical applications for this type of bearing are supporting the armature in a starter motor, or as a 'spigot' bushing (fits over the end of the gearbox input shaft and is pressed into the rear of the crankshaft to support and align the input shaft).

Full round: by description, this bearing sounds similar to the bushing but the construction method is different. This bearing is a precision insert type made from similar materials to crankshaft bearings. The major

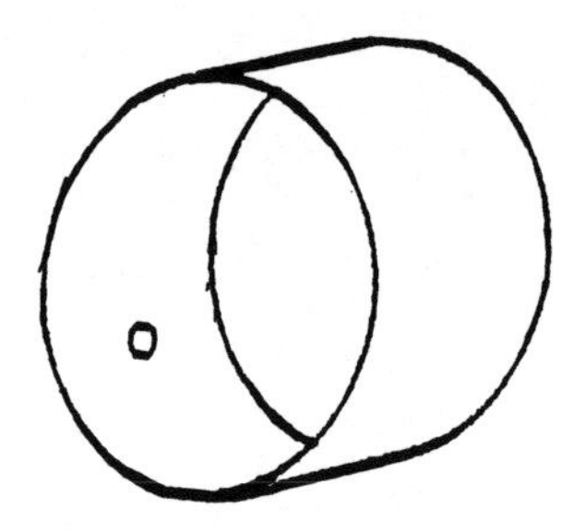

application for these bearings is to support and align engine camshafts. The bearings are pressed into the engine cylinder block and precision bored to size after they are installed. The boring process is to ensure that each bearing along the length of the camshaft will be in perfect alignment and prevent stress being applied to the camshaft when it is rotating.

Precision insert (shell): each bearing consists of two semi-circular halves or shells.

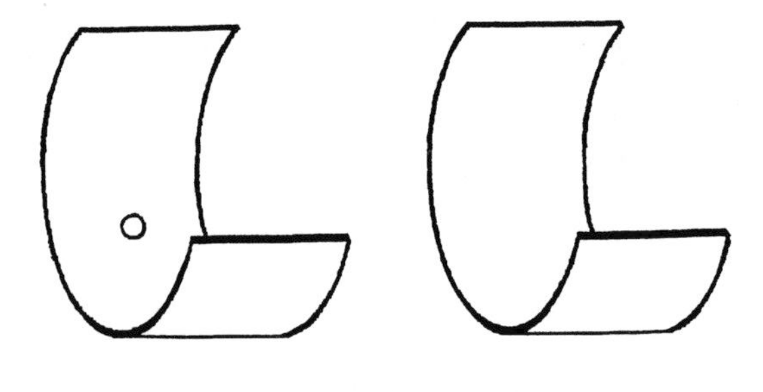

Each half is steel backed and layered with a combination of metals carefully chosen by the engine manufacturer to provide the operational characteristics required. The crankshaft material, engine speed, load, power developed and type will all be factors in deciding the correct mixture of materials for the bearing. Materials that are too hard will cause the crankshaft to wear, however, soft materials may fatigue, wear rapidly and 'pound out' under engine loads. Engine bearings are manufactured in a range of sizes to allow the crankshaft to be reground, and the engine can be overhauled several times before major components require replacement. These bearings are used for both the connecting rod lower end and crankshaft main bearing.

Thrust bearing: resembles a plain steel washer with a bronze (or white metal) layer on one face. The bearing can accept thrust forces in an axial direction only, but prevents the rotating gear or shaft contacting the assembly housing. These thrust bearings (washers) are often available in several thicknesses. This range of sizes allows a mechanic to assemble a transmission with 'as new' end float clearances. The end float is measured and a thickness thrust bearing that will provide the correct clearances is selected. This form of bearing is widely used in manual transmissions to control cluster gear end float and, in automatic transmissions, to control the final end float of the rotating components.

The control of engine crankshaft end float requires the use of slightly different thrust bearings. Two common methods are available:

1 One of the engine crankshaft main bearings has a flange on each side of the bearing. The flange is faced with bearing material

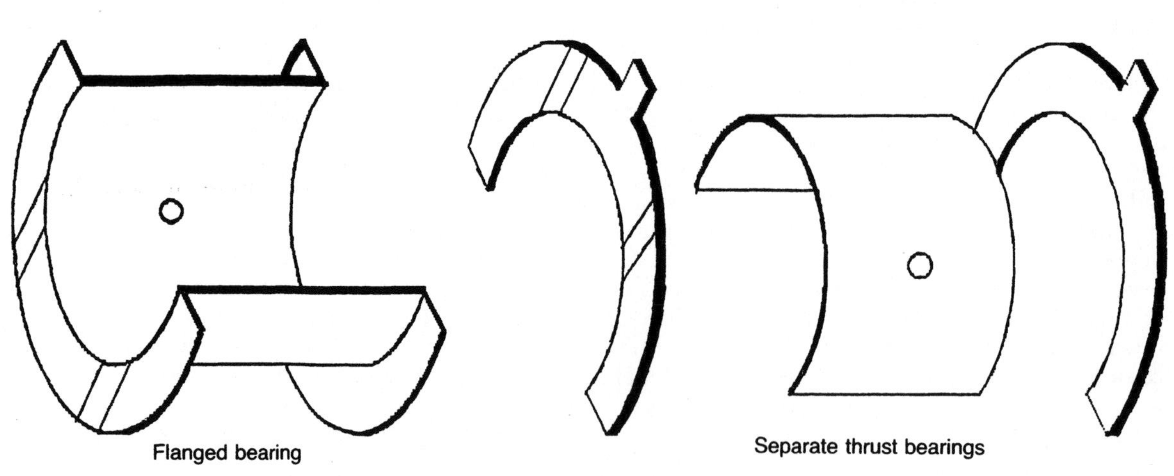

and adjoins machined surfaces on the rotating crankshaft. End thrust on the crankshaft is opposed by the bearing and neither forward or rearward movement of the crankshaft is permitted.

2 The second type functions in the same manner as the flanged bearing, but the flange sections are supplied as separate pieces. These pieces fit in the same location but they are not directly attached to the main bearing and require a projection which locates in the crankcase and prevents the thrust bearing sections rotating with the crankshaft.

This section has introduced you to the types of friction bearings in common use. Although the precision measurement and fitting of these bearings occurs in later stages of your training it is advisable to be aware of the following points and avoid costly mistakes when assembling components.

1 All parts must be clean.
2 Lubricate all bearing working faces before assembly.
3 Worn bearings should not be interchanged between cylinders or housings.
4 Substitute different bearing materials only as directed by the manufacturer.
5 Observe specified bearing clearances. DO NOT START the engine if a component 'locks up' on assembly.

Anti-friction bearings

Anti-friction bearings contain rolling elements to reduce friction between rotating components and the assembly housing. Generally a bearing consists of hardened balls or rollers contained within a hardened inner and outer ring. A separator cage prevents the balls or rollers bunching together.

An extensive range of bearings is available, many of which are designed for special applications. This section will describe the major bearing types commonly used in passenger vehicles. The student should note that different types of bearings may be indicated as having the same applications in the vehicle. This is a normal situation as the designer must consider the space available, loads (forces) applied and the speed at which the bearing will be required to operate, before determining the bearing type to install. Different vehicles may therefore use completely different bearing types, construction and sealing methods on the same application.

Anti-friction bearings can be grouped into three major types: ball, roller and needle.

Ball: the basic bearing has a series of hardened steel balls separated by a metal cage. A groove, formed in a hardened inner and outer ring, provides a track in which the balls run. This type of ball race is suitable for radial loads, but by modifying the basic design other types of forces can be accepted. Deepening the ball track groove will allow the bearing to accept moderate axial forces in addition to radial loads. Further redesign of the tracks will enable the bearing to accept heavier axial forces but there will be reduction in the radial load capacity. The outer track can be reshaped with a buttress enclosing one side of the balls. The inner track also has a buttress but it rises to enclose the opposite side of the balls. This type of bearing is known as an *angular contact bearing*. However, if the axial force is applied in the opposite direction the bearing would be pushed apart. To overcome this problem the angular contact bearings are operated in opposing pairs, providing radial and axial load carrying ability. Ball bearings are used in alternators, transmissions, steering boxes, rear axles, tailshaft centre bearings and, when fitted as a pair, front wheel hubs.

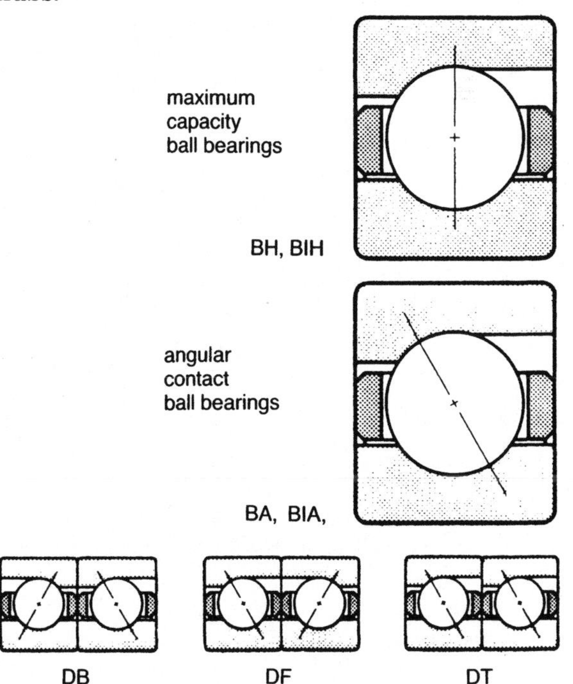

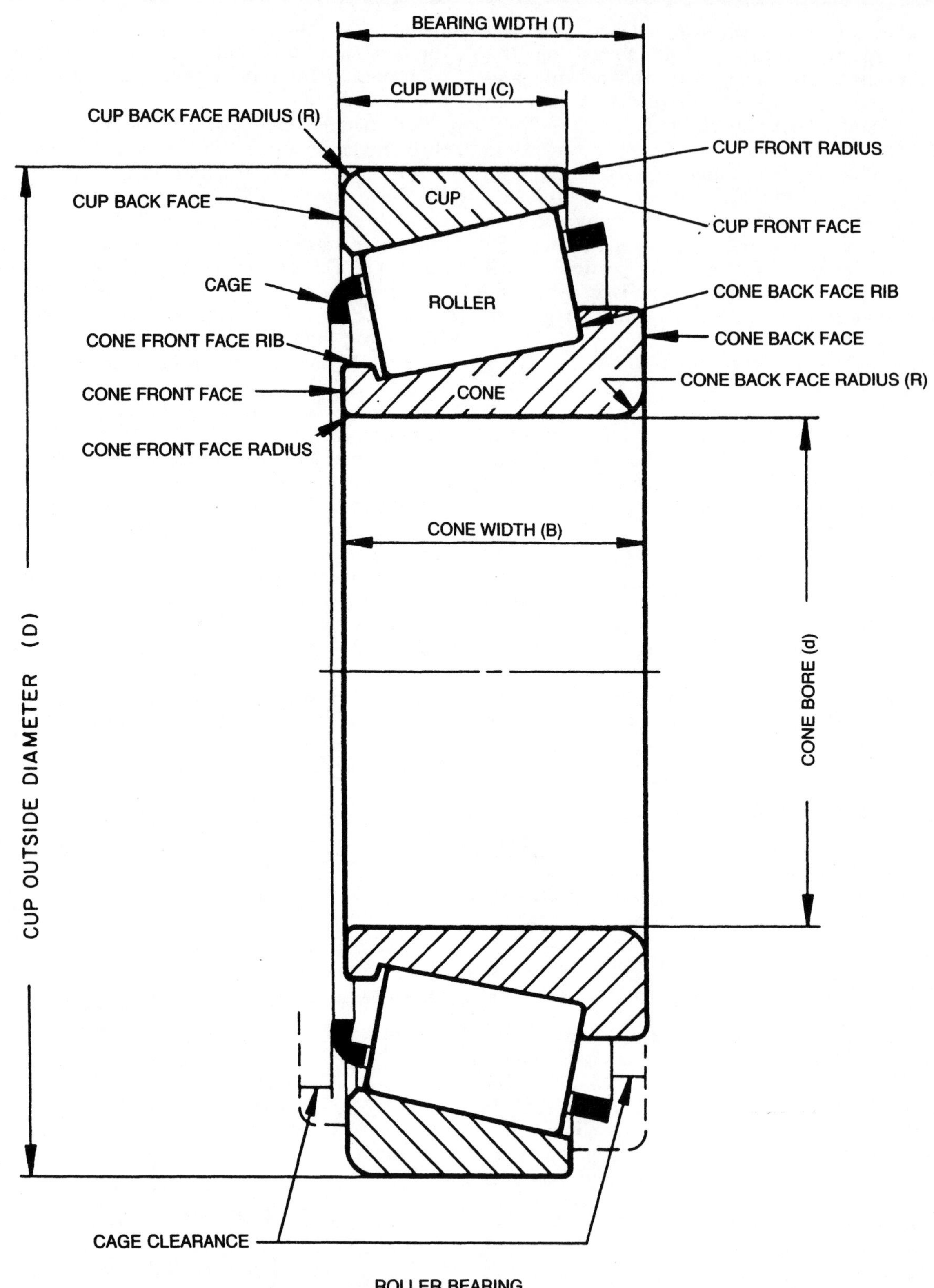

ROLLER BEARING

Roller: a parallel roller bearing will accept heavy radial loads but has no resistance to axial thrust loads. The bearing is available as a single unit but many applications exist where a shaft is hardened to form the inner bearing track and the outer track consists of a metal shield or the hardened internal bore of a gear. The rollers are in direct contact with these hardened surfaces and in many instances are not fitted with a separator cage. This type of bearing is fitted to transmission mainshafts, cluster gear shafts and final drive straddle type pinion gears.

A tapered roller bearing contains rollers which taper over their length. The inner and outer tracks both have angled faces and the rolling elements are separated by a metal cage. This type of bearing can accept heavy, single direction axial forces. When used in opposing pairs the bearings can accept heavy axial and radial loads. This combination is extensively used in front and rear wheel hubs, final drive crown wheel and pinions, transmission mainshafts and cluster gear shafts.

A barrel (spherical) roller is also available. It is similar in external appearance to the more widely used ball or tapered roller bearings.

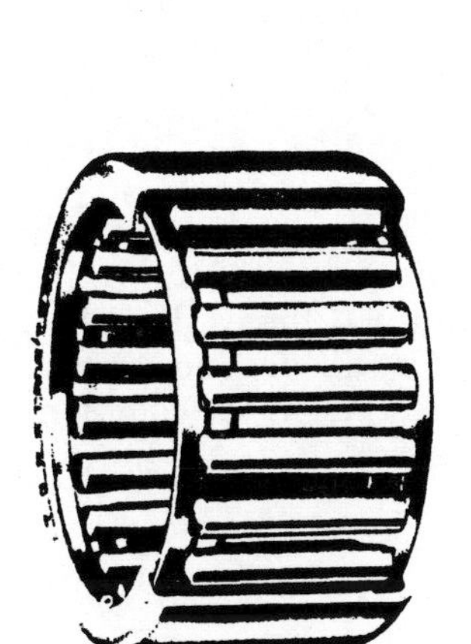

DRAWN CUP
ROLLER BEARINGS

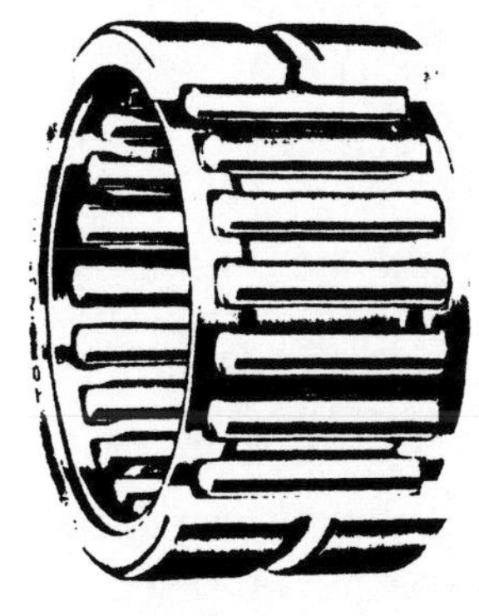

HEAVY DUTY
ROLLER BEARINGS

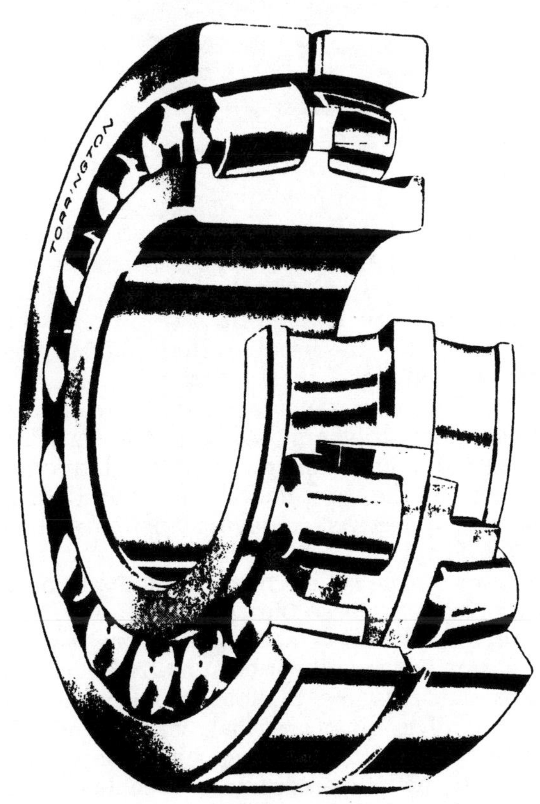

BARREL ROLLER

When used in opposed pairs this bearing is suitable for front wheel hubs. Some applications of this bearing can 'self align' without applying stress to the mounting shaft or housing.

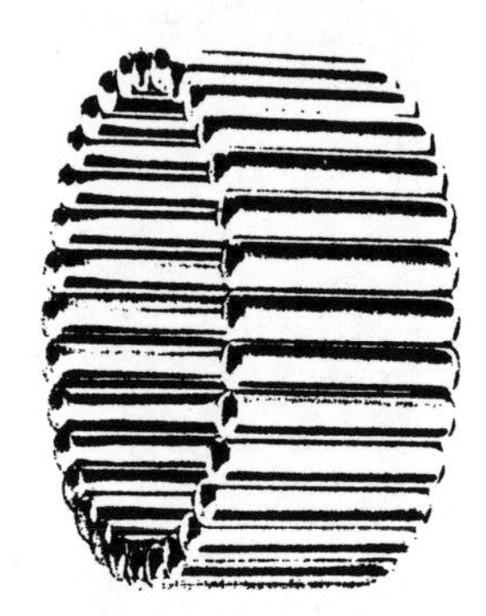

NEEDLE ROLLERS

CAGE AND ROLLER ASSEMBLIES

Needle: the major difference between needle bearings and parallel roller bearings is the relative size of the rollers. Many applications are similar but, because of the compact size of the needle rollers, the bearing can be fitted with a plastic separator cage and inserted into locations, such as between the constant mesh gears and the transmission mainshaft. The rollers contact hardened surfaces on the shaft and the internal bore of the gear, effectively reducing friction at this point. The Hookes type universal joint also contains needle rollers but in this application the rollers are loose (no cage).

Thrust: several vehicle applications require a bearing which will not be subjected to radial loads but must be capable of withstanding high axial loadings. Several bearings of this type are available and they have the rolling elements (ball, tapered roller or needle rollers)

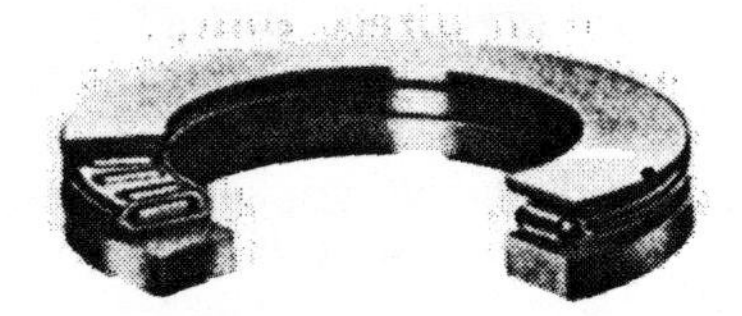

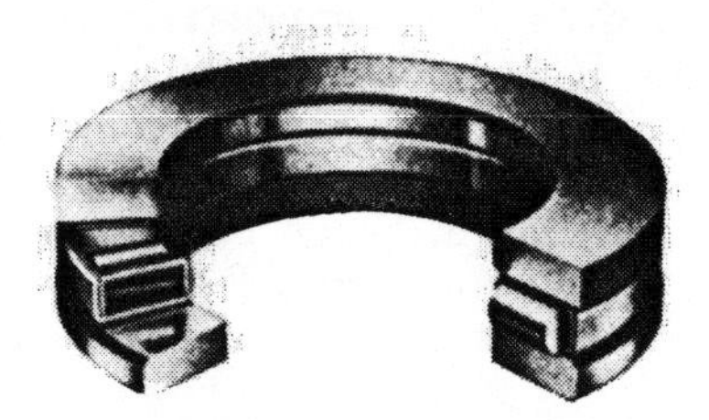

NEEDLE AND ROLLER THRUST BEARINGS

arranged radially like the spokes of a wheel and held in position by a plastic or metal separator. The bearing must have hardened tracks on each side of the rolling elements to accept the axial forces, but the appearance of these tracks will depend on the application of the bearing. The most common use for these bearings are:

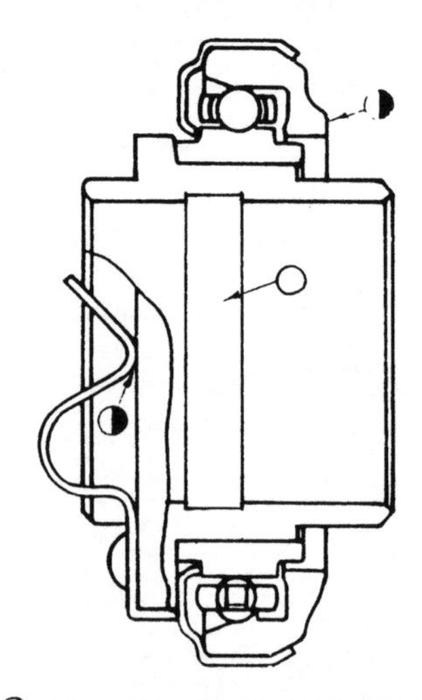

○ FILL GROOVE WITH GREASE

◐ APPLY A LIGHT FILM OF GREASE

RELEASE BEARING

- clutch release bearings — usually ball rolling elements totally enclosed in a casing and lubricated for the life of the bearing when manufactured;
- automatic transmission thrust bearings — usually needle rollers, often operating directly against a hardened face of a gear or drum. This type of bearing, often referred to as Torrington races (a manufacturer), is lubricated from the transmission lubrication system.

Lubrication: terminology

Sealed for life — the bearing is prepacked with grease then sealed on both sides. This type of seal is used for clutch release bearings and rear axle bearings.

Sealed one side only — the seal is usually to protect the bearing from external contamination and the unsealed side always faces the lubricant. The bearing can be grease or oil lubricated. The popular applications are as alternator bearings and rear axle bearings.

Shielded — the shield is not to seal the bearing but is designed to reduce the flow of oil through the bearing and prevent a possible oil seal overload in an adjacent sealing area. This type of shield can be found in transmission bearings.

Open — the bearing is not fitted with seals and may be grease or oil lubricated. Any sealing requirements for the assembly are controlled separately from the bearing.

COMMON SERVICING PROCEDURES

Edge seal: replacement

Replacement of an edge seal is one of the more common tasks a mechanic will encounter. This type of seal is used in many locations and must perform its sealing task under very difficult conditions. The typical transmission extension housing seal has to be capable of withstanding a shaft rotational speed up to 6000 rpm, accept movement of the shaft through the seal (caused by suspension movement), and perform this task in the presence of mud and water.

The replacement method described below is an example which can be modified to suit replacement of most edge seals. The steps of preparing the vehicle and removing the tailshaft have been omitted.

1. Fit an expanding jaw, slide hammer type tool to the seal (refer illustration). A taper chisel may be used to remove the seal, however, the technique requires care as alloy housings can be easily damaged. Well equipped workshops should have a multipurpose tool of the type shown.
2. Repeat blows with the slide hammer until the seal comes free from the housing.
3. Clean and inspect the housing seal recess, remove any burrs before proceeding.
4. Inspect the new seal for storage damage.
5. Metal cased seals may require a thin film of oil resistant sealant on the outer circumference — check the specific manufacturer's repair manual for details.
6. Install the seal into the housing using a tool (illustration on next page) and a hammer. The seal must be correctly aligned to the housing and care must be taken to ensure the sealing lip is not damaged by the shaft. It is possible to install a seal with a suitably sized piece of pipe. Avoid seal distortion by applying an even force around the seal and do not strike the seal directly with the hammer.

Caution: The sealing edge or lip must face the oil or fluid.

7. Coat the sealing lip with lubricant.
8. Inspect the sealing surface (in this example, the front universal yoke) for grooves, nicks

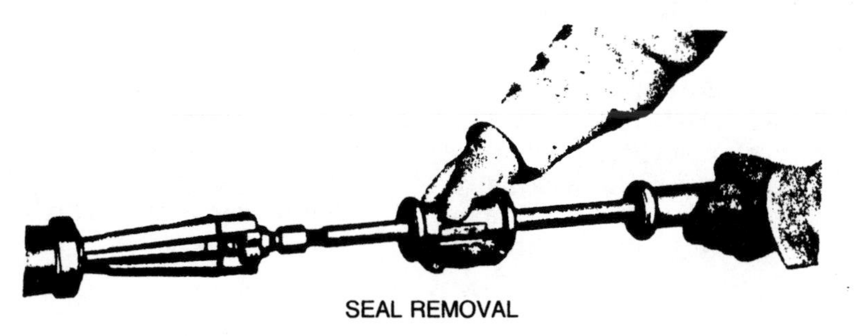

SEAL REMOVAL

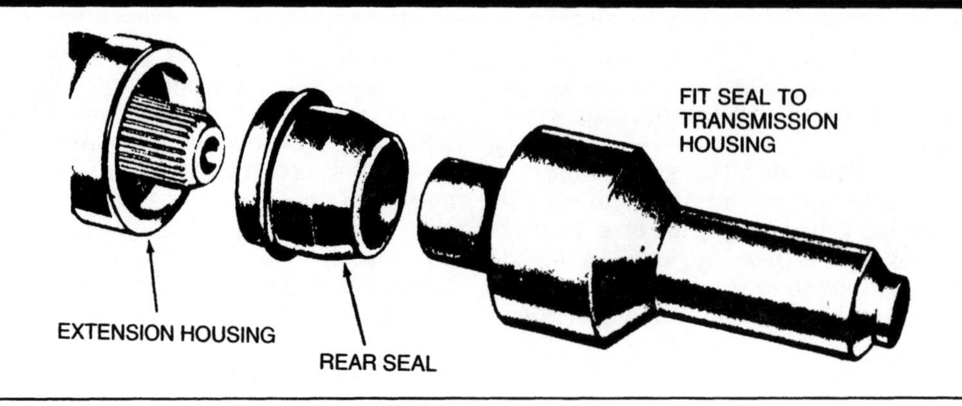

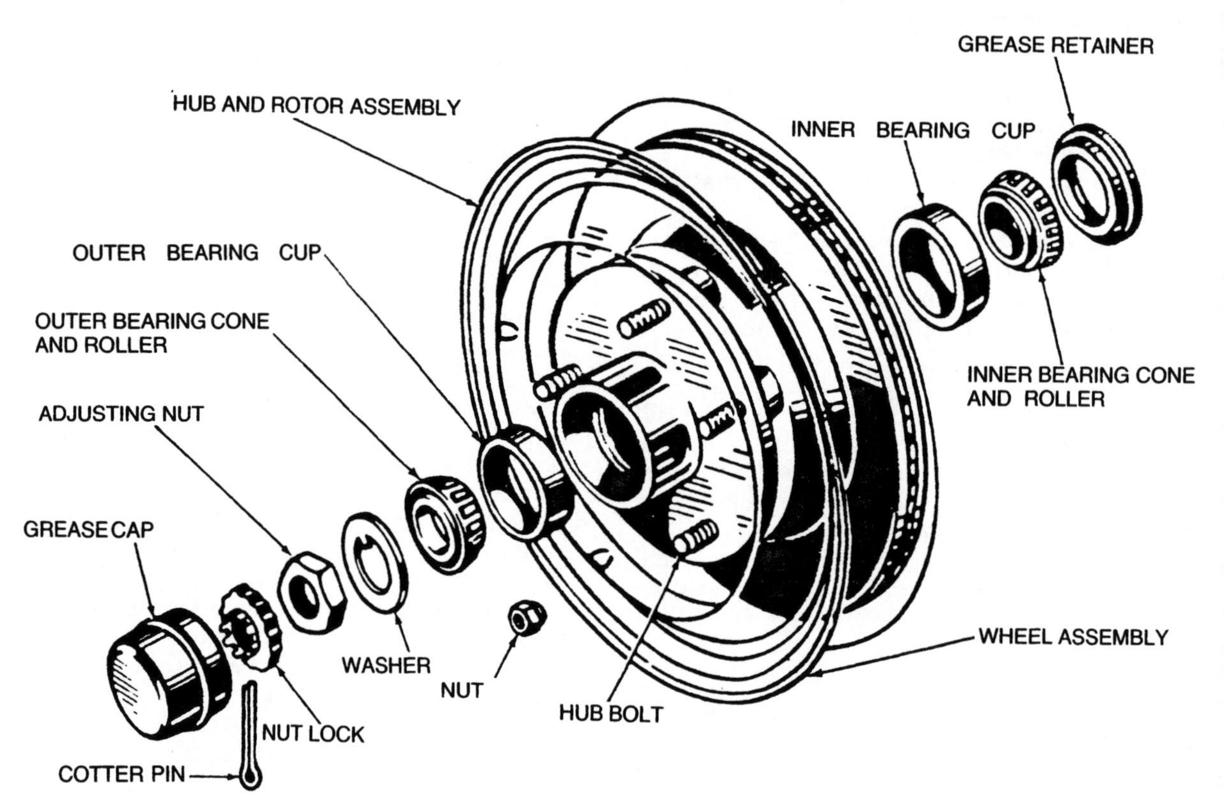

— FRONT HUB, BEARINGS AND GREASE RETAINER —

or burrs. Lubricate the sealing surface before inserting through the seal.

Front hub: service

The majority of vehicles are fitted with taper roller front wheel hubs which require grease repacking at intervals of approximately 40 000 km. This service operation normally requires the removal of the disc brake caliper, therefore servicing costs can be reduced if the hub service is performed when brake replacement work is in progress. It is not the purpose of this chapter to describe brake repair operations, therefore it is assumed the task will commence after the caliper has been removed and will cease prior to caliper replacement.

1 Remove the hub spindle cap (grease cap). The cap may be removed by using two screwdrivers, multigrips or lightly tapping with a soft hammer.

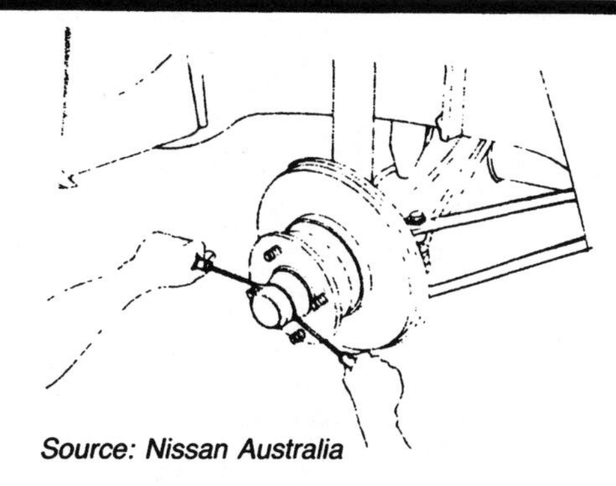
Source: Nissan Australia

2 Remove the cotter pin. Use a pair of side cutting pliers to grip the pin head and lever the pin free.
3 Remove the lock nut and the bearing adjusting nut.

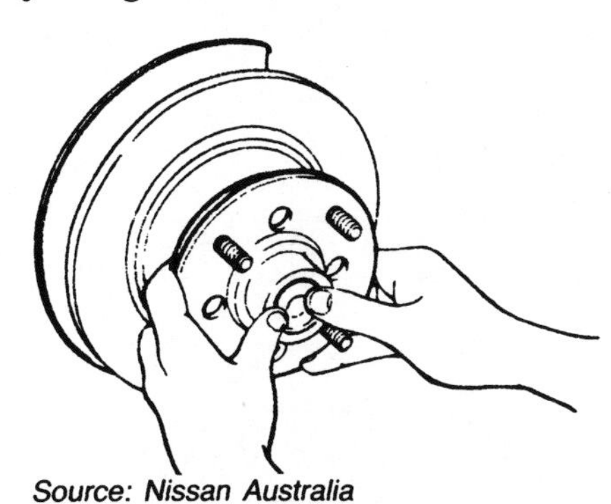
Source: Nissan Australia

4 Remove the wheel hub and disc rotor. Place your thumbs on the edge of the outer bearing washer to prevent the bearing falling on the ground, and pull on the hub.
5 Remove the hub bearings.

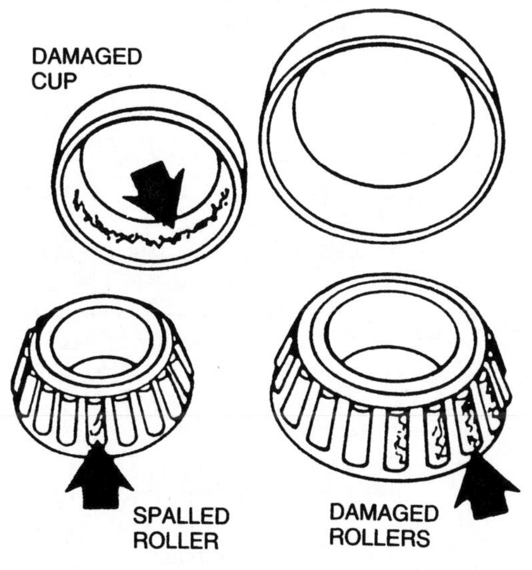

- The outer bearing and washer can be removed and placed in a wash tray.
- Insert a brass drift through the hub until it contacts the inner bearing cone. Strike the drift with a hammer and the inner bearing and oil seal will fall out of the hub.
- Place the inner bearing in a wash tray.

6 Clean and inspect.
- Clean all grease from the spindle, hub and bearings. Do not spin the bearings with compressed air.
- Visually inspect the bearings for fatigue, chips or damage.
- Visually inspect the hub, including the bearing cups, for cracks, wear, heat marks or damage.
- Visually inspect the spindle for thread damage, 'spin' marks where the bearings locate and the seal track area for nicks, cuts or grooves.

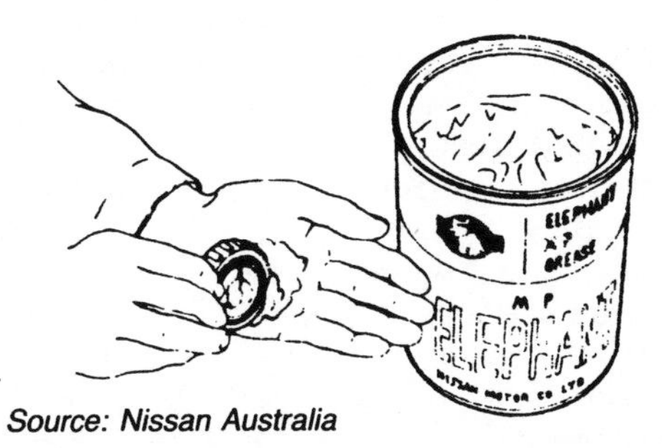
Source: Nissan Australia

7 Repack bearings.
- Work grease, of suitable specification, through each bearing until the roller cavities are completely filled. Use a bearing packer if available.
- Apply a quantity of grease onto the bearing cups and into the hub. Instructions vary on the quantity of grease to be used in the hub. Follow the manufacturer's instructions. The quantity of grease required can vary from a smear to a quarter of the hub cavity, with a few exceptions that specify complete filling of the hub with grease.

8 Place the inner bearing into the hub.
9 Smear grease onto the sealing lip of a NEW hub seal and press or tap the seal evenly into the hub until firmly seated (see next page).
10 Place the outer bearing into the hub.
11 Fit the hub to the spindle.
- Hold the outer bearing in position and place the hub onto the spindle. Keep the

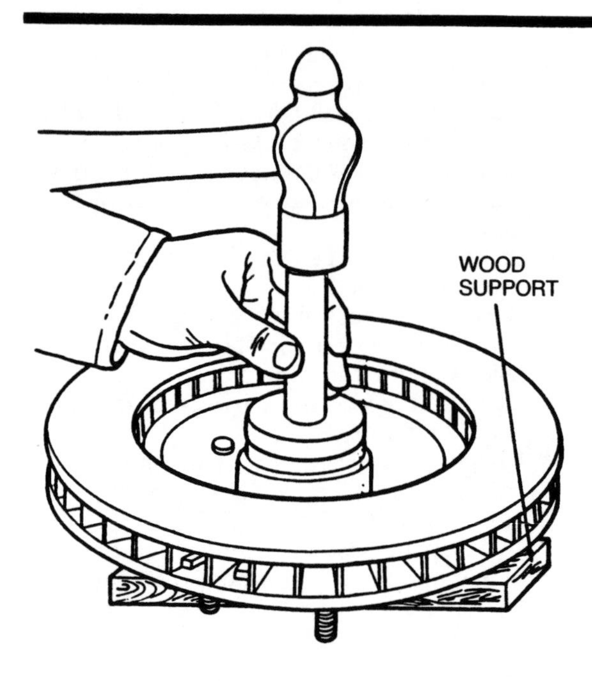

hub aligned during this operation to ensure the seal is not damaged by the spindle thread.

- Fit the retaining washer and the adjusting nut.

12 Adjust the wheel bearing to the manufacturer's specification. The following example is supplied as representative of the methods used by manufacturers. Steps 1 and 3 are similar for most vehicles, but step 2 requires care as different vehicles require different methods for adjustment.

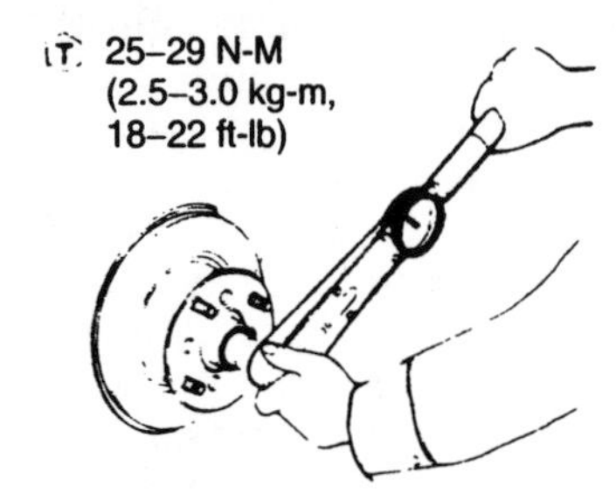

Source: Nissan Australia

Step 1. **Seat bearings:** While rotating the wheel, torque the adjusting nut to specification to seat the bearings.

Step 2. **Set bearing pre-load:** Manufacturers' instructions vary at this step. REMEMBER, all wheel bearings are not adjusted to the same clearances.

Method 1. Release the adjusting nut (up to 1/4 turn) to obtain manufacturer's specified end float.

Alternative preload settings adopted by different manufacturers.

Method 2. Loosen the adjusting nut until the hub can be turned by hand.

With drum/disc and wheel rotating, torque adjusting nut to 17–25 ft lbs.

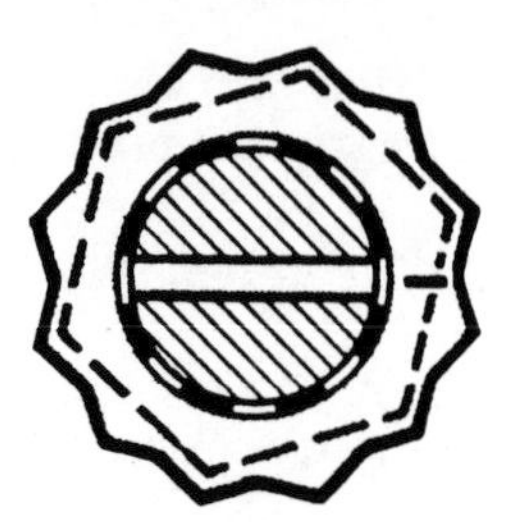

Back off adjusting nut until an end float of .002–0065 for disc brakes and .0005–.0065 for drum brakes is obtained.

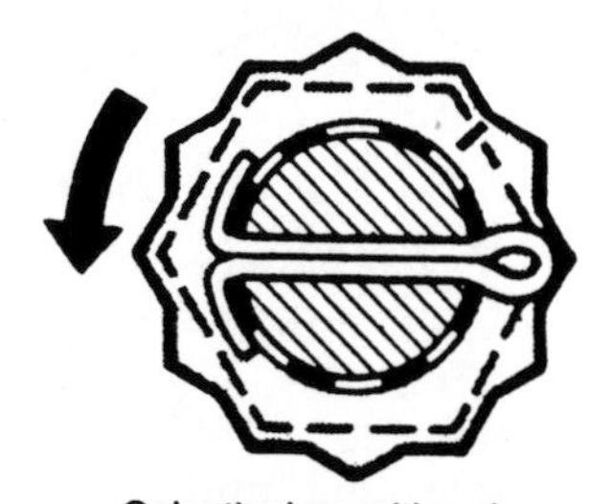

Selectively position the nut lock retainer on the adjusting nut and lock in position with a new cotter pin.

Front Wheel Bearing Adjustment

Attach a spring balance to the wheel hub bolt and measure the drag caused by the hub oil seal. Tighten the adjusting nut until the rotation starting torque (vehicle spec.) plus the hub oil seal drag is recorded on the spring balance.

Method 3. Back off the adjusting nut until just loose. Hand snug the nut but the nut must not be even finger tight.

Step 3. **Install the lock nut:** Loosen the adjusting nut, if necessary, to align the pin hole and insert the cotter pin. Bend the cotter pin legs and install the hub spindle cap.

Front hub-bearing cup: replacement

Many tapered roller bearings are of two piece design — the outer track is separate from the inner track and rolling elements. The outer track (cup) is usually retained in the hub housing and is not removed when routine bearing service — lubricant repack — is being performed. If, during the service inspection, the bearing is detected as unserviceable it will become necessary to remove the bearing cups from the hub assembly. The procedure described is typical of the steps required to remove the bearing cups on the majority of vehicles fitted with this type of bearing.

1 Drive out the inner bearing cup. The inner bearing cup diameter and the counterbore inside the hub may be almost the same diameter, which makes it difficult to position a drift or punch on the bearing cup to drive it from the hub. Many manufacturers provide two notches in the inner hub to provide points where a brass drift can contact the cup. Locate the notches before attempting to strike the bearing.
   - Support the hub on a wooden block to prevent damage to wheel studs or hub faces.
   - Using a brass drift and a hammer, strike the bearing cup on alternate sides until the cup is free from the hub.

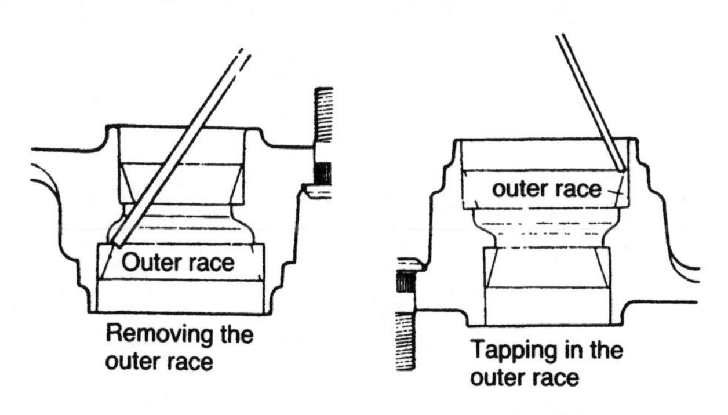

Removing the outer race

Tapping in the outer race

2 Drive out the outer bearing cup. When one bearing has failed, it is essential to replace both inner and outer bearings. Remove the outer bearing cup in the same way as the inner cup.

3 Inspect the hub counterbores. Inspect the counterbores for burrs or damage caused by the bearing spinning in the housing. Repair or replace as required.

4 Install bearing cups.
   - Smear the outer circumference of the cups with grease.
   - Align the cup to the housing and, using a brass punch and hammer, tap the cup on alternate sides until it is firmly seated in the hub counterbore. If a suitable bearing cup installing tool is available it should be used in preference to the brass punch method.

5 Assemble hub. Refer to front hub service for assembly details.

Axle shaft-bearing: replacement

The type of bearing fitted to rear axle shafts may be deep groove ball bearing or possibly a taper roller type. The bearing may be totally sealed or sealed on one face only, however, despite the variations most bearings share a common method of retention to the axle. The bearing is a press fit onto a machined shoulder and retention is ensured by the fitment of an inner retainer ring which is also a press fit on to the axle. The method described below is typical of many repair manuals, but reference should always be made to the specific model repair manual for specialised techniques or clearances.

Caution: avoid eye damage — wear safety glasses. During press operations

place a heavy texture cloth around the bearing.

1 Drill the retaining collar. Drill a 6 mm hole in the outside diameter of the collar. Avoid axle damage by ensuring the hole depth does not exceed 3/4 of the collar thickness.

Removing Rear Wheel Bearing Retainer Ring

2 Split the retaining collar. Support the axle and, using a sharp chisel and hammer, split the collar at the drilled hole.

3 Remove the bearing. Fit a pair of guillotine jaws around the bearing. The jaws must rest against the inner cone so the press force will not be against the outer ring of the bearing. Use the hydraulic press to remove the bearing.

4 Inspect the axle shaft. Check for nicks, burrs, scratches or grooves. Repair or replace if necessary.

5 Assemble the components onto the axle. Check the axle clamp plate for distortion and assembly direction. Check the bearing is facing in the correct direction. Check the retainer ring is NEW. Ensure the axle, bearing and collar are in correct alignment with the press and the guillotine jaws. Misalignment can distort the retaining collar and reduce its holding ability.

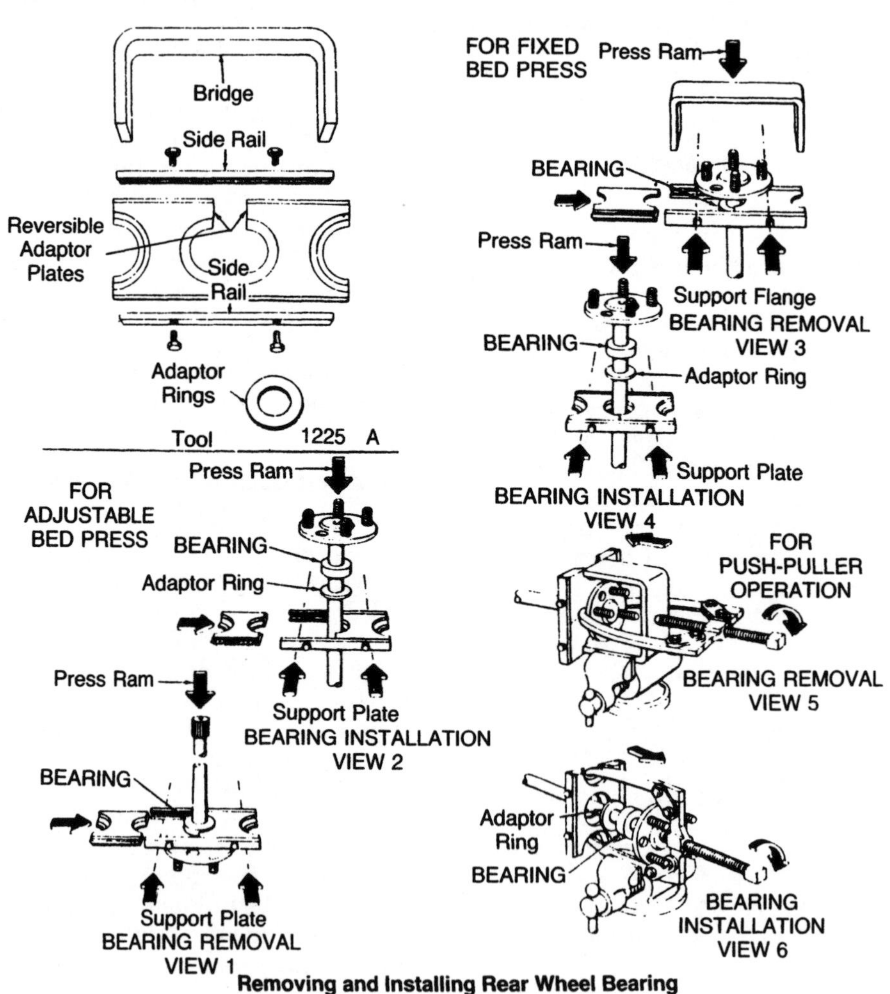

Removing and Installing Rear Wheel Bearing

6 Press fit the bearing. Press the bearing and collar onto the axle in one operation. The press force should apply through the retaining collar and be maintained until the bearing and collar are firmly seated against the axle shaft shoulder.

Bearing: preload

Many components, when assembled, require precise alignment of a gear set to ensure the gear mesh action will be smooth, long-lasting and quiet in operation. Gear alignment must also be maintained after slight wear has occurred and when the gear set is under operating loads. To achieve these objectives it is necessary to apply 'preload' to the bearing set that is supporting the gears.

Involvement in a practical exercise is the most suitable method of understanding what 'bearing preload' is, and how it is applied by the mechanic. Obtain a recirculating ball steering box and mount the unit in a bench vice.

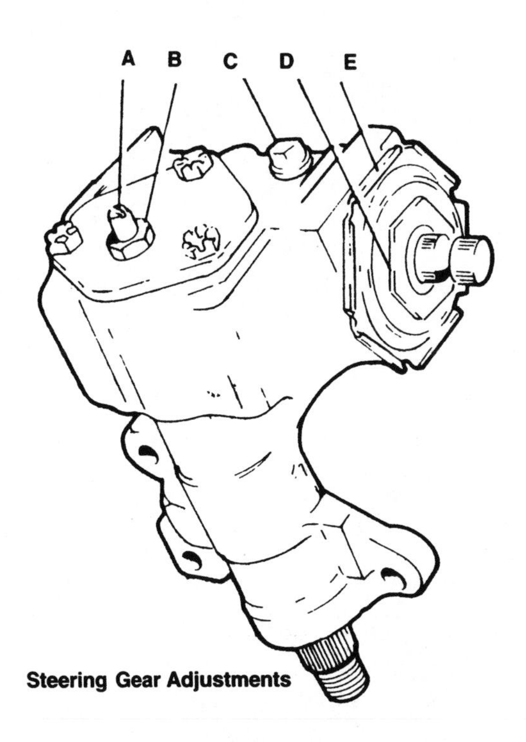

Steering Gear Adjustments

A. Sector Shaft Mesh Preload Adjusting Screw
B. Sector Shaft Mesh Preload Adjusting Screw Locknut.
C. Filler Plug
D. Input Shaft Bearing Preload Adjusting Screw.
E. Input Shaft Bearing Preload Adjusting Screw Locknut.

1 Release the sector shaft adjustment lock nut and turn the adjusting screw counter-clockwise to move the sector shaft out of close mesh with the worm nut.
2 Loosen the input shaft adjustment lock nut.
3 Measure the input shaft bearing 'preload'. This can be done by attaching a small (mNm) torque wrench, or an arm and spring scale, to the upper splines of the input shaft.
4 Tighten the input shaft adjusting nut until the specified (manufacturer spec.) load is indicated on the torque wrench scale. Do not rotate the input shaft more than 1–1½ turns from the centre position or the torque reading will be affected by drag from the sector shaft.

At this point the input shaft preload has been set and it is time to retrace the steps to consider exactly what has been achieved. The input shaft is supported by two bearings, one on each side of a worm scroll on the shaft. If the input shaft has end float or lateral movement, the worm nut which is mounted on the worm scroll will move out of alignment with the sector shaft. This loss of gear alignment will cause increased steering wheel movement before the road wheels and tyres can change direction.

The upper bearing cup is mounted in the shaft adjusting nut and screwing the nut inward reduces the bearing clearance (end float) of the input shaft. Continue rotating the adjusting nut until zero end float of the input shaft is reached. At this point the desired mesh adjustment could be achieved but the gear alignment would not be maintained under load or after slight wear of the bearing set. Any additional rotation of the adjusting nut will place load (known as preload) on the bearing. This loading compensates for bearing wear and operating forces and is the reading obtained on the torque wrench scale.

5 To finalise the adjustment of the steering box it is necessary to rotate the sector shaft mesh adjusting screw clockwise until the preload (as specified by the manufacturer) is obtained as the input shaft is rotated past the centre (high spot) of the steering box movement. Tighten the adjusting screw lock nut.

Bearing preload is applied in transmissions, final drives, steering boxes and axle hubs. In later studies, this simple exercise should be remembered; expensive assemblies can be destroyed by ignoring the manufacturer's preload specifications.

Caution: insufficient preload means loss of gear alignment, too much preload can cause bearing failure. The only acceptable methods of setting preload are by precision measurement or with a torque wrench or similar tool.

Hub bearings: front wheel drive

Front hub construction and the bearing type utilised determine whether the wheel bearings on front wheel drive vehicles require servicing and are adjustable or non-adjustable. Service work on many of these hub assemblies require the use of specialised tools, therefore the manufacturer's repair manual should be consulted prior to undertaking any repairs of this type.

SERVICE CHECKS

Unless the manufacturer has specified a fixed service interval it is good service practice to perform the following two service inspections when the vehicle is in the workshop for periodic maintenance.

End play check: support the vehicle firmly on stands. Grasp the top and bottom of the wheel and attempt to rock the wheel. Any end float present is an indication that hub preload is inadequate and adjustment or repair is required. It is possible the lower ball joint may have movement during this test, therefore

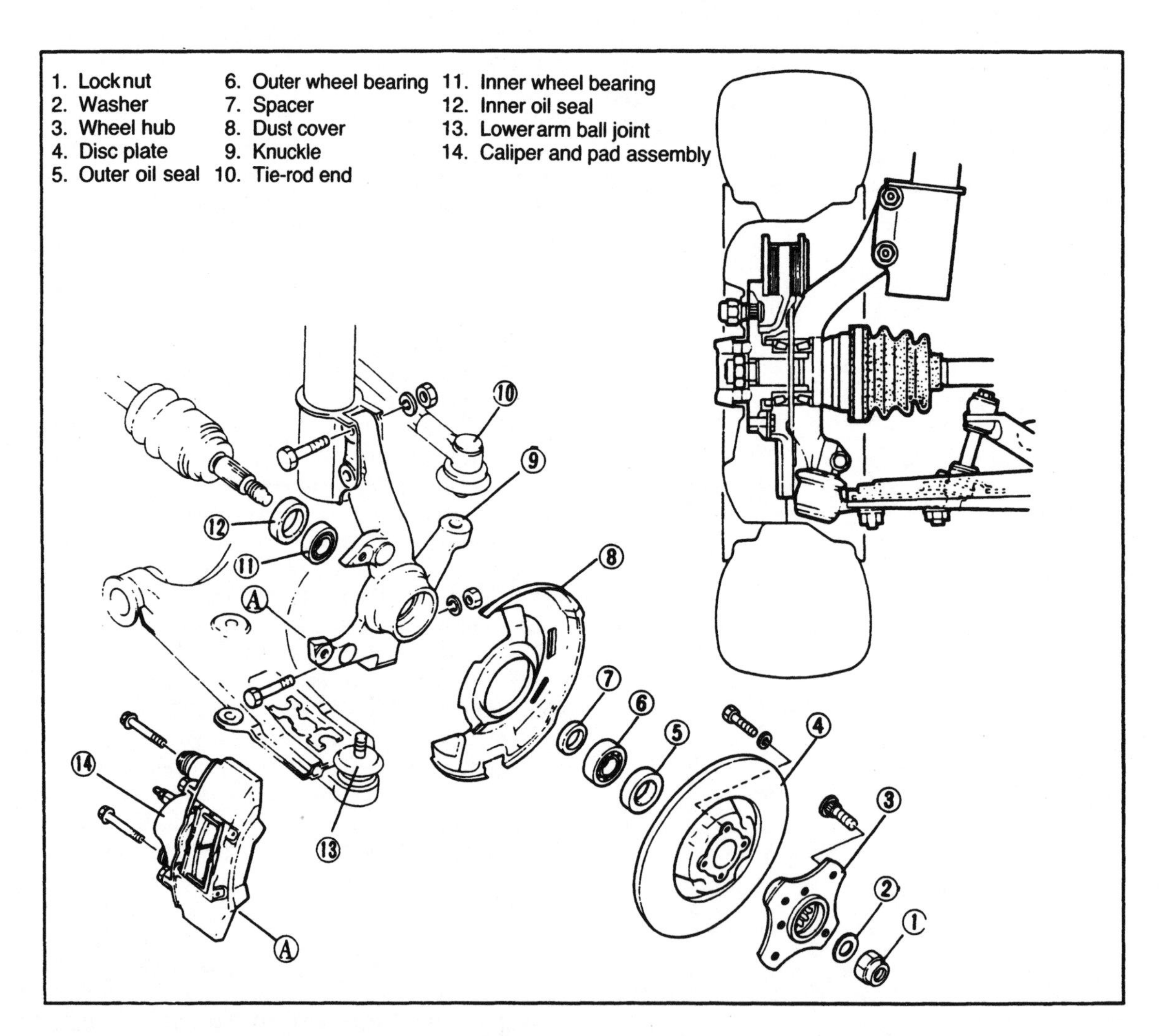

FRONT AXLE (KNUCKLE) AND HUB ASSEMBLY—REMOVE AND INSTALL
Follow the numerical sequence for general disassembly procedures.

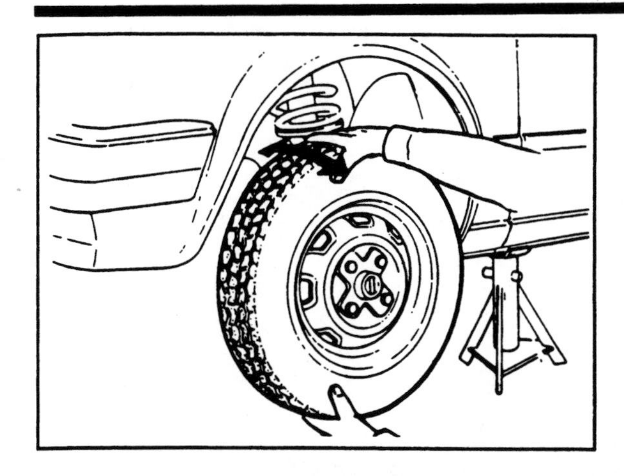

ensure the end float detected is in the hub before proceeding.

Noise check: with the vehicle still supported on stands, rotate the wheel quickly by hand and listen for any abnormal noise from the bearings. If abnormal noise is present the hub must be dismantled and the bearings replaced.

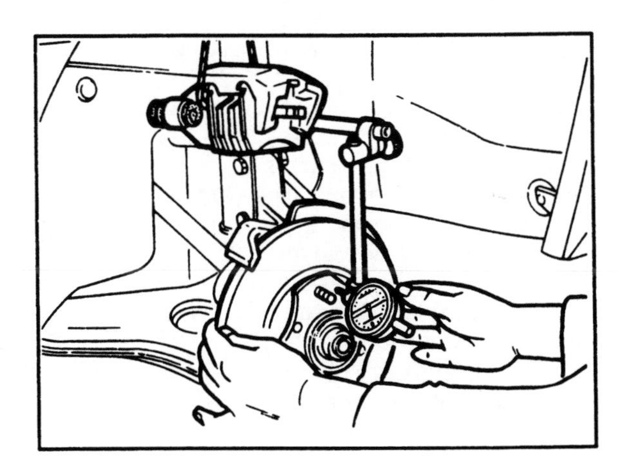

Hub preload: in the majority of units the hub is splined onto the drive shaft and the hub, drive shaft and bearing inner tracks are clamped tightly by the hub nut. In most units the hub nut preload has minimal effect on bearing preload, which is controlled internally in the bearing or by a preload spacer. Therefore if the hub nut is tightened to specification and end float is still present the hub must be dismantled for repair.

SERVICING: ONLY IF REQUIRED

Ball bearing: hubs fitted with twin angular contact ball bearings are normally 'sealed for life', retained in the knuckle support by circlips and are not adjustable for bearing preload. Hub preload is the only adjustment, and end float detected in this type of hub will require replacement of the bearing assembly.

Taper roller bearing: when individual taper roller bearings are fitted it may be possible to service and repack the bearings with grease. A preload spacer is fitted between the two bearings and the thickness of this spacer will determine the bearing preload. After servicing or when new bearings are fitted the bearing preload must be checked and adjusted. The bearings and the old preload spacer are assembled into the knuckle support through which a specialised tool (dummy hub assembly) is inserted and the effort required to rotate the support is recorded on a spring gauge. A reading in excess of specification indicates the bearings are clamped too tightly and a thicker preload spacer must be fitted. When the bearing preload has been established, the hub and bearings are packed with grease, oil seals fitted and the unit is assembled. Final adjustment of hub preload is provided by tightening the hub nut to specification (see diagram on next page).

Caution: use the specified tools and follow the manufacturer's instructions when repairing front wheel drive hubs.

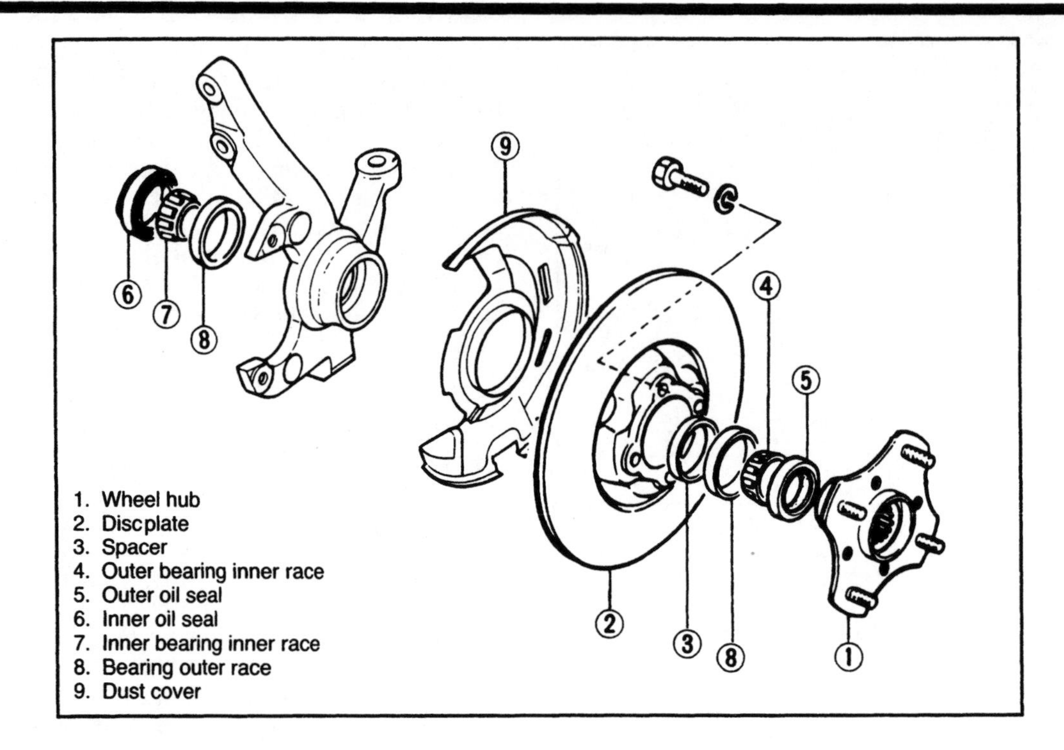

CHAPTER 8 REVISION

1 Why are bearings fitted to vehicles?
2 What is the prime purpose of seals, gaskets and sealants?
3 Define the terms 'static' and 'dynamic' when applied to O-ring seals.
4 Describe the construction, appearance and application of an O-ring seal.
5 List three precautions required when fitting O-ring seals.
6 Describe the construction, appearance and application of a lip or edge oil seal.
7 List three precautions required when fitting lip or edge oil seals.
8 What is a hydrodynamic seal?
9 Describe the construction, appearance and application of an external lip seal.
10 List three precautions required when fitting external lip seals.
11 Describe the method by which an oil slinger helps prevent oil leaks.
12 Describe the method by which a machined scroll helps prevent oil leakage.
13 Describe the fitting procedure required for a wick oil seal.
14 Sketch four housings, each of which is fitted with one of the seals described in this chapter, and clearly indicate the direction in which the sealing edge of the seal must face.
15 Describe one fault which could occur when an original thick gasket is replaced by a non-genuine thin gasket or a thin film of sealant.
16 List six materials used in gaskets.
17 Indicate one vehicle application for each of the gasket materials listed in question 16.
18 List five general usage rules which relate to the use of sealants in vehicles.
19 Describe the difference between friction (plain) bearings and anti-friction bearings.
20 Describe a vehicle application for three different friction bearings and three anti-friction bearings.
21 Describe the difference between radial and axial loads.
22 Describe one application within an engine where axial forces are controlled by a plain or friction type bearing.
23 Sketch and name the forces acting on anti-friction bearings when installed as a:
- front wheel bearing set;
- cluster gear needle roller set;
- rear axle taper roller bearing.

24 Describe the construction of a thrust (Torrington type) bearing of the type fitted in automatic transmissions.

25 Explain the terms 'shielded' and 'sealed for life' when applied to anti-friction bearings.

Suggested practical exercises

All exercises should be performed using the manufacturer's approved repair procedures.

1 Obtain a transmission extension (or similar) housing, remove the seal and install a new seal.
2 Remove a front wheel hub assembly, fit new bearings, repack with lubricant, assemble the hub onto the vehicle and adjust the wheel bearing as required.
3 Obtain a rear axle fitted with a bearing and lock ring. Remove the old bearing and fit a new bearing and lock ring.
4 Obtain a partly dismantled steering box and adjust the input shaft bearing preload to specification.
5 Obtain a knuckle support and hub assembly from a front wheel drive car. Service the hub bearings using the approved tools and the manufacturer's repair procedure.

9

ROADWORTHY AWARENESS

Government or government certified inspectors have the responsibility for determining the roadworthy condition of a vehicle. These inspectors have years of experience and have satisfied a government agency of their capacity to check a vehicle within the limits of carefully designed tests and visual indicators. The contents of this chapter are not intended to prepare you to conduct roadworthy tests, but to alert you that all work you undertake on a vehicle can generate a safety hazard. YOU are charged with a responsibility to ensure any repair, with particular emphasis on modifications, is not only performed in a fast and cost-effective manner but the safety standards of design rules, roadworthy requirements and manufacturers' service wear limits are observed.

The roadworthy regulations, associated interpretation of the regulations and country-of-vehicle use design rules can vary between states and/or countries, therefore at all times you should consult the applicable regulations relative to the location in which you work.

Vehicles are built to stringent design rules and the manufacturers set service wear limits which, if observed by the servicing mechanic, will ensure the vehicle continues to operate in a safe manner. When using testing requirements derived from the basic regulation care must be taken to ensure that the test is totally valid for the specific vehicle under test. You may at times encounter a situation where the result of the test appears to be within the derived testing requirement yet is outside the known service limits as set by the manufacturer. If doubt exists as to the acceptability of the vehicle, refer to the basic regulation. The regulations are drafted in a way to be all-encompassing and provide an answer for the difficult decision (the means of

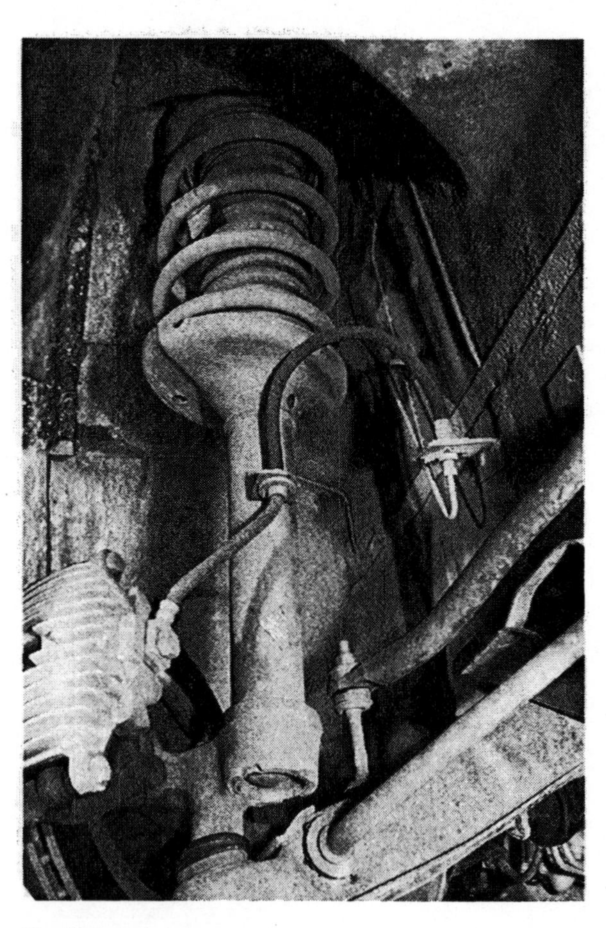

MacPherson type suspension

rejecting a vehicle which while not specifically dangerous is worn to the extent that the vehicle will not operate within the limits set by the manufacturer).

To give an example of the way in which a mechanic may encounter this problem, consider a vehicle fitted with a MacPherson type front suspension.

1 Testing of the lower suspension arm ball joint indicates movement exists in both a vertical and horizontal direction. A preliminary view of the testing requirements indicates an acceptable limit (for the mass of the vehicle under test) of less than 10 mm movement measured at the top of the wheel.

Caution: many footnotes (additional explanations/exclusions) exist in derived testing requirements. Wherever reference is made to a footnote the additional information should be read carefully. In this example the footnote explains that no more than half the freeplay is allowed to be present in any one component and requires a specific test method. The test must be conducted in such a way that the vehicle suspension spring force does not act through the joint under test.

2 For the purpose of this example we will assume the joint under test has 3 mm movement, which appears to meet the testing requirement.

   The vehicle manufacturer has set a service wear limit of no movement in the joint and also states the joint should be under preload. This appears to be a contradiction between the manufacturer and the roadworthy regulations. In many applications the testing requirement applied is correct but the student must appreciate the wide range of products on the market makes it very difficult to have a single *specific* test for all vehicles. The basic regulation makes the problem of applying a further requirement to the vehicle relatively easy.

   The regulation states: The front and rear suspension of any motor vehicle must be of such a nature as to promote safety in the vehicle steering and stability.

3 In the example, the manufacturer indicated that steering instability could be caused by loose or worn lower suspension arm ball joints, therefore the component should be rejected.

This example was provided as a guide to promote the reading and interpreting of the regulations applying to your work. At times it is possible that a consumer action could result from decisions made during a roadworthy inspection. The prime thought which must be remembered is: YOU can, through negligent action, literally kill people. If doubt exists consider if you personally would operate the vehicle in all types of road conditions on the components you are inspecting.

TESTING

As stated earlier in this chapter, roadworthy testing is performed by experienced government approved inspectors. The following tests relate to the serviceability of many of the components checked during a roadworthy inspection. An understanding of the requirements of each area should promote a greater roadworthy awareness in all repairs undertaken. In most cases the tests described are based on specific vehicle manufacturer's service instructions. These procedures are not intended to be prescriptive nor substitute for any test implied within the relevant roadworthy regulation.

BRAKING SYSTEM

Because many of the hydraulic brake parts are hidden from view inside cylinders it is necessary to perform a couple of simple external tests to gauge the condition of the hydraulic system. Also, many vehicles are equipped with power (vacuum assisted) brakes and the internal condition of this part of the system must also be estimated from a series of external tests.

Brake lines/hoses: with the vehicle raised on a hoist, check all brake lines for security, leaks and abrasion or road damage. Flexible brake hoses must also be checked for cracks, fraying or possible fouling with wheel and/or suspension components. Remember to move the front wheels from lock to lock while observing the brake hoses.

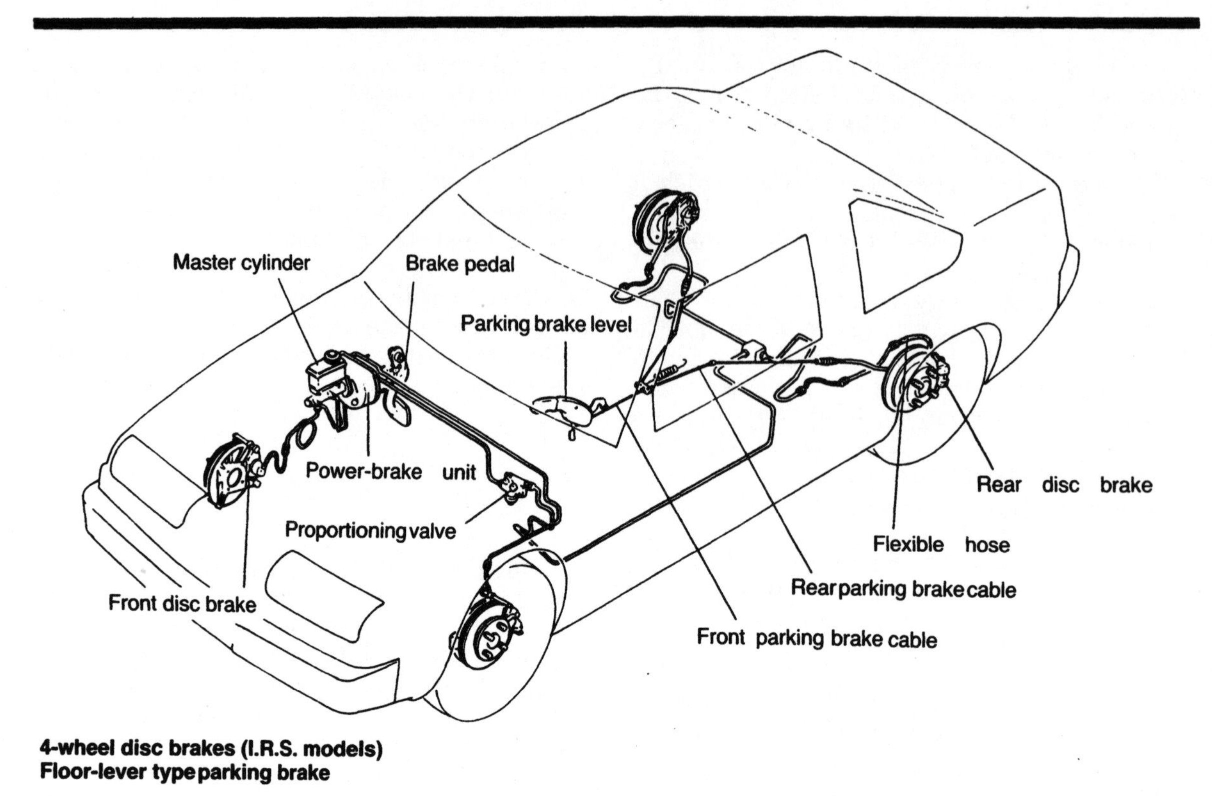

4-wheel disc brakes (I.R.S. models)
Floor-lever type parking brake

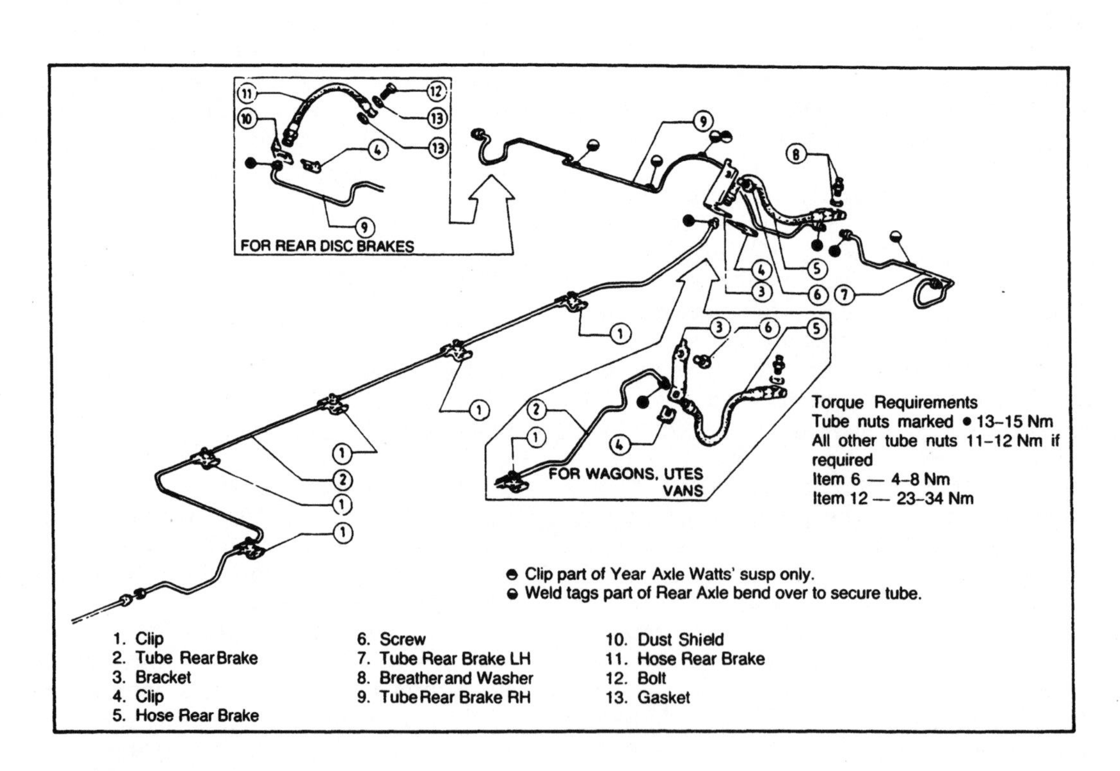

Source: Nissan Australia

Caution: cracks in flexible brake hoses are not always visible on casual inspection. Bend the hose into a tight 'u' shape and inspect using a strong light.

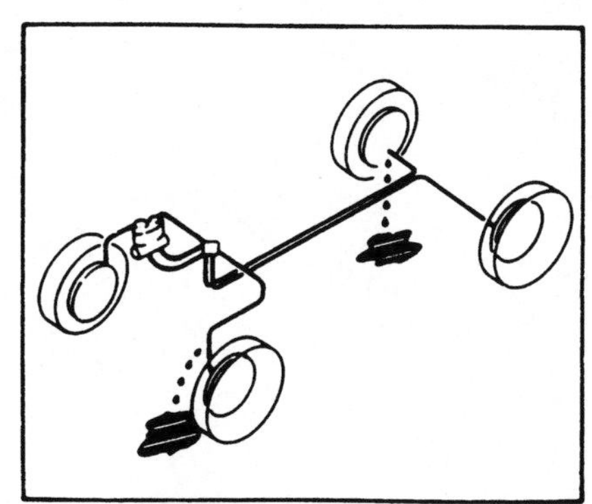

Source: Nissan Australia

Fluid leaks: check the rear of each wheel unit for damp or fluid marked disc shields and backing plates. If oil marks are present the vehicle has need of repair but the source of the fluid must be identified (rear axle oil seal/wheel bearing seal/hydraulic seal) before making a report.

Handbrake: check hand brake cables/links for frayed cables, missing or incorrect pins and worn linkage. Pull on the cable/linkage to ensure movement does occur (seized cables) and if any form of lever or relay is fitted the lever must be at right angles to the applied force when the hand brake is in the on position.

Wheel units: lower vehicle on hoist to a convenient working height. Obviously, to fully check the brake units at each wheel all wheels/drums should be removed. Generally it is sufficient to remove one front and one rear wheel to estimate accurately the lining wear life remaining.

Front wheel: remove one front wheel (the wheel removed should be from the side nearest the kerb as this wheel tends to receive more suspension and brake punishment from the rough road edge).

1 Check for oil, grease or fluid:
- contamination on pad and disc friction surfaces;
- seepage at caliper piston seal areas;
- on the disc inner face and the dust

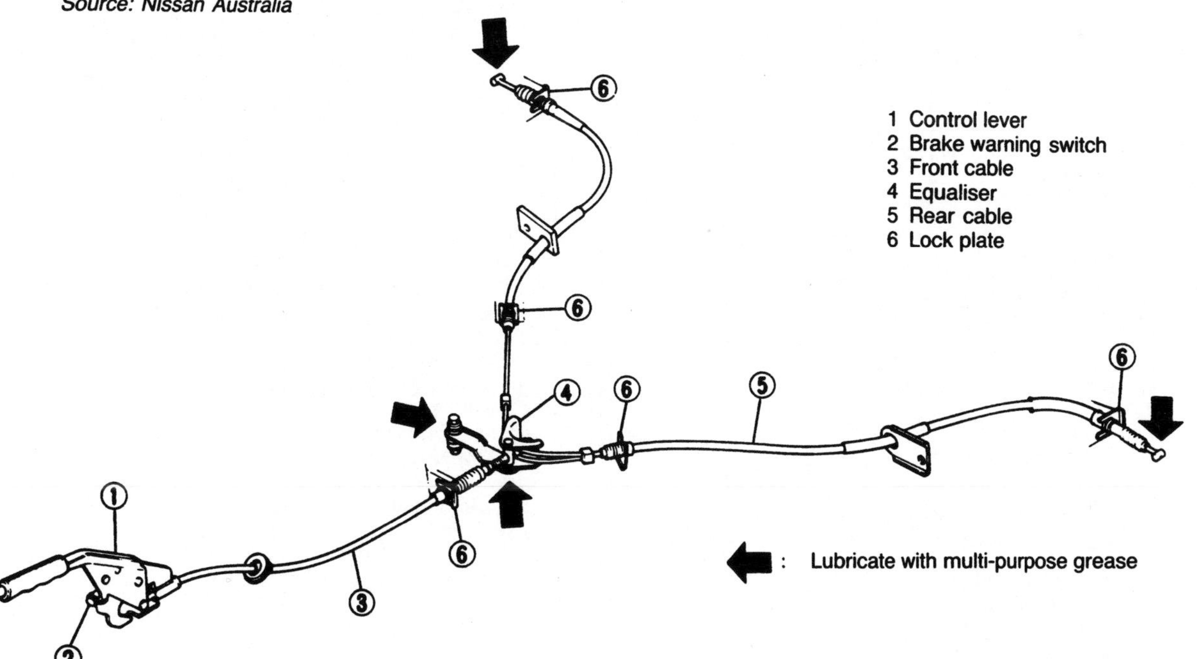

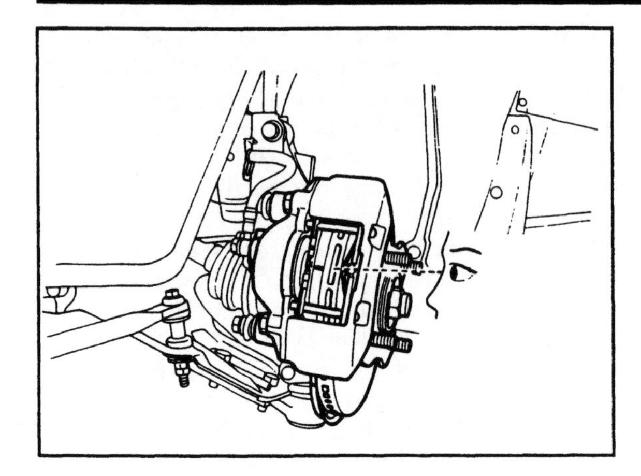

shield, indicating a leaking front hub seal.

2 Compare the disc brake pad lining thickness to the manufacturer's recommended wear limits. The roadworthy requirement is usually less than the manufacturer's wear limit — the manufacturer must estimate wear rate to the next service call, which with extended service periods could be a substantial distance, whereas the roadworthy limit may be as low as thirty days before metal to metal contact occurs at the disc rotor and pad backing.

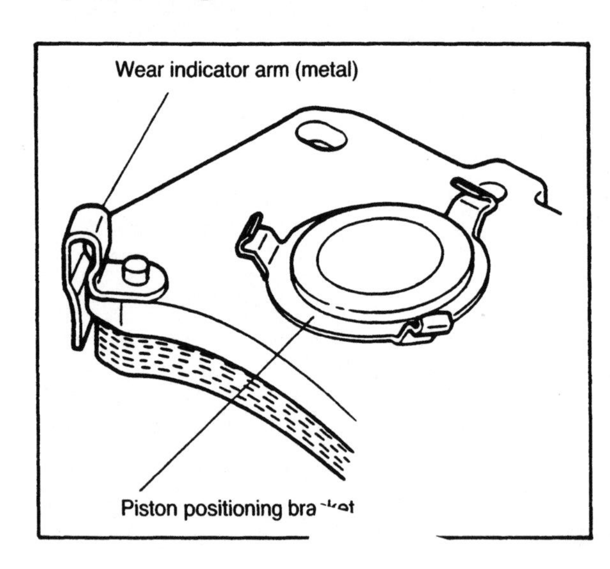

The pad may
indicator which p
as the friction m
its useful service
allow pad thick
thickness set by

impractical to operate the vehicle until the brake pads are replaced.

3 Inspect the disc rotor surface for excessive scoring — light surface scores are acceptable with many disc brakes.

4 Ensure all retaining pins and shims are fitted and undamaged.

5 Refit wheel.

Rear wheel: remove one rear wheel (note comments relating to front wheel for left or right wheel). If the vehicle has rear wheel discs the checking procedure is similar to the front wheel disc check with the added requirement to inspect the handbrake mechanism. Where the vehicle has drum brakes it will be necessary to remove the drum for inspection of the brake unit.

1 Check lining friction surfaces for contamination (oil, grease etc.).

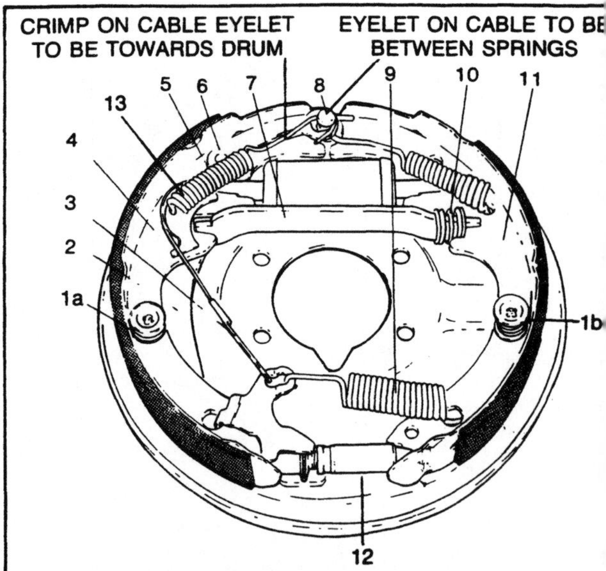

1a. Secondary Shoe Hold Down Shoe (Black)
1b. Primary Shoe Hold Down Shoe (Blue)
NOTE: Brakes without auto adjusters have the same spring on both Primary and Secondary Shoes.
2. Parking Brake Lever
3. Automatic Adjuster Cable
4. Secondary Shoe (Long Lining)
5. Washer
6. Parking Brake Lever Retaining Clip
7. Parking Brake Link
8. Anchor Pin
9. Automatic Adjuster Spring (Blue)
ink Spring
crew
pring (Yellow)

2 Check for visible hydraulic fluid leaks — carefully roll back the edge of the cylinder rubber boot. Slight dampness at this location is normal but drops of fluid are regarded as excessive.
3 Check return springs, retainer clips, self adjusting mechanisms and cables for correct fit and freedom from corrosion. Check handbrake mechanism for movement.

Lining wear limit

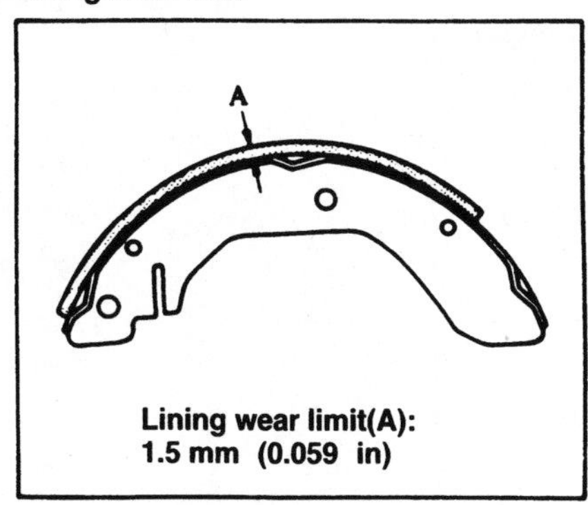

Source: Nissan Australia

4 Compare brake lining thickness to the manufacturer's wear limit. As explained previously, a variation will be evident between manufacturer wear limit and the roadworthy requirement.

Drum repair limit

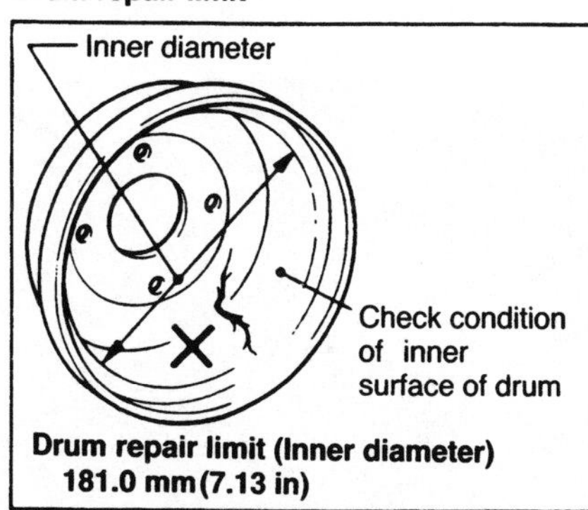

Source: Nissan Australia

5 Before refitting the drum and wheel check brake drum for scoring, bluing, hot spots or signs of distortion.

Under bonnet checks

(lower vehicle to ground level)
1 Visual inspection of all brake lines for leaks, damage and suitable retention clamps.

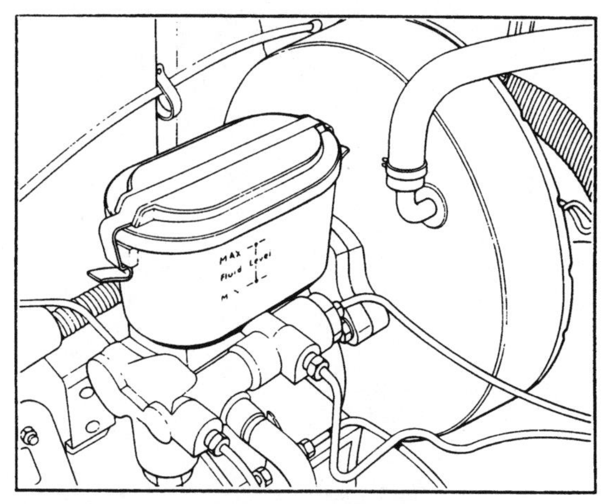

Source: Mitsubishi Magna TM series manual

2 Inspect the hole area where lines pass through the front guards to ensure rubber grommets (where applicable) are fitted and rubbing contact between line and guard is not present.
3 Check the power brake vacuum supply hose for cracks, abrasion and adequate clamping.

Caution: this hose must be an approved type. Substitute materials such as heater hose may collapse, creating a dangerous situation.

4 Ensure power brake unit and master cylinder retaining bolts are secure.
5 Remove master cylinder cover and check fluid level, discoloration of fluid and inspect the diaphragm cover for split and adequate sealing.
6 Check between master cylinder and power brake unit for fluid leakage from rear of cylinder. If vehicle is not fitted with a power assisted brake unit, this check should be made at the point where the cylinder bolts to the firewall of the engine bay. Access to this location may require checking up behind the instrument panel from the interior of the vehicle.

Driver's seat

(check from driver's seat)
1 a Apply handbrake. If adjustment is correct the handle will move less than 2/3 of available travel.
b Test the ratchet by bumping the handle with your hand. The ratchet should not release.

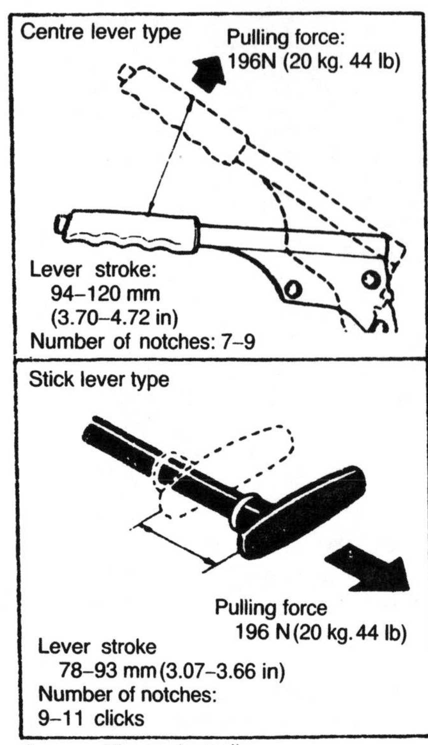

Source: Nissan Australia

2 Check pedal rubber for wear or loose fit.

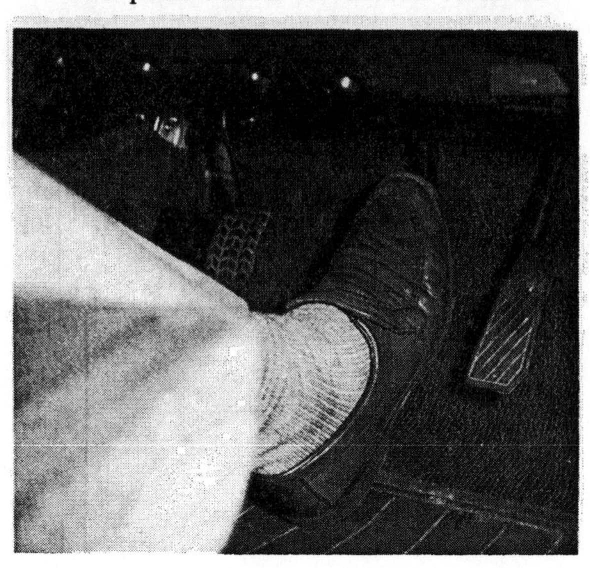

Apply light foot pressure to test master cylinder

3 a Start engine (vehicles with power assisted brakes).
  b Apply foot brake with light foot pressure and maintain pressure for approximately 10–15 seconds. The pedal should not 'creep' downward during this test.
  c Repeat test b with heavy foot pressure. The pedal travel must be less than 2/3 of available movement.

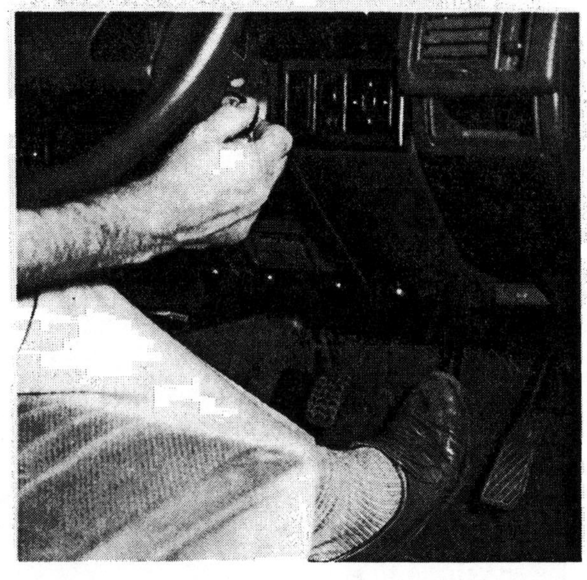

Test power brake function

4 a Switch off engine (steps b, c and d apply to power assisted brakes only).
b After several minutes apply the foot brake: a hissing sound from the power brake unit should be audible. Repeat the application two or three times. The hiss should be evident but diminishing in sound level.
c Apply pedal several times to ensure all vacuum is exhausted from system
d Rest your foot on the pedal and start the engine. The pedal should 'pull down' slightly and require less pressure to hold it down.

Road test

1 When the ignition key is turned to the start position a brake system check light (where fitted) should light up while the engine is cranking then go out when the key is released.
2 Dependent on the test method chosen the vehicle will be fitted with a decelerometer, checked against a measured distance or 'driven' on a brake testing machine. Regardless of the method selected the foot and hand brakes must be checked against an approved set of figures. The normal regulation requirement is not excessive and any modern vehicle with the braking system in good order should pass the test. Road tests refer to testing on even dry surfaces with the transmission in neutral.
3 The handbrake must be capable of holding the vehicle from moving either front or rearward on any normally negotiable slope.
4 The vehicle must not pull or dive under brake application, pedal pulsation should not be present and no abnormal noises should be heard.

EXHAUST SYSTEMS

The requirements for original vehicle equipment are relatively simple. The system (all pipes, joints, mufflers) must be free from leaks, be securely mounted and must not produce excessive noise.

Generally, leaks can be readily detected by the hissing sound at the location where the gas is escaping. Sections of the system which appear to have excessive corrosion can be checked by tapping the suspect area LIGHTLY with the ball peen of a hammer. The system should also be checked to see if poor fitting or mounting sag has allowed the exhaust to come into contact with brake pipes, hoses or suspension components.

Check exhaust for corrosion

Emission control type exhaust (catalytic convertor)

Greater care must be taken when inspecting an exhaust system which has been modified from the original and therefore may not have been approved by the vehicle manufacturer. Exhaust systems are usually modified in an attempt to obtain greater engine power and involve changes in manifolding, internal muffler construction, pipe length, diameter and discharge point of the rear pipe. Where the regulations specifically include the retention of emission control devices a check of the system, to ensure these units are present and functional, is required.

The restrictions on exhaust modification are common-sense and allow substantial freedom in system design, therefore compliance with the following points will usually satisfy the regulations.

1 The system must not produce excessive noise.
2 No part of the system may pass through the passenger compartment.
3 If a side discharge point is fitted the discharge point:
   a must be at least 150 mm in front of the rear wheel;
   b must discharge downward at an angle of 45° or greater;
   c pipe end must not protrude beyond the lower sill when viewed from above;
   d preferably will not be under a passenger opening.

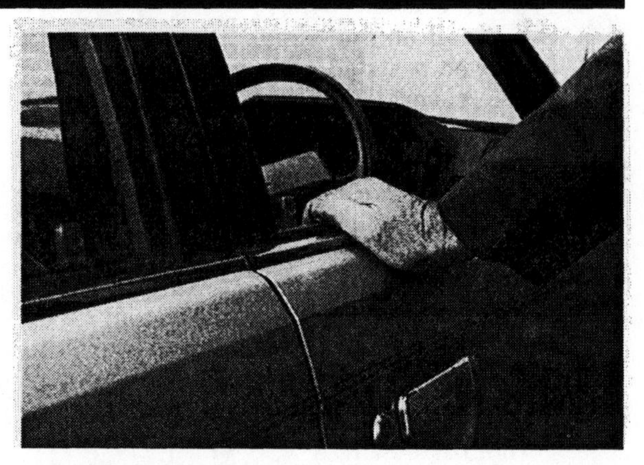

Check door lock is secure.

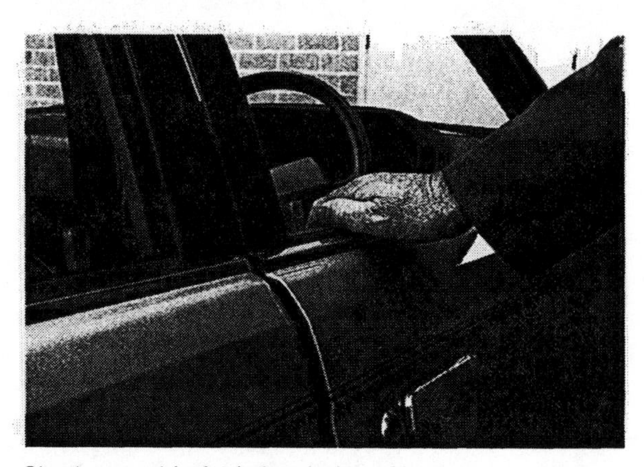

Check second (safety) door lock position.

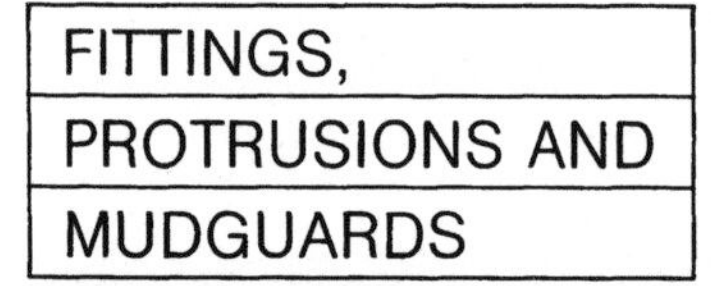

FITTINGS, PROTRUSIONS AND MUDGUARDS

Door catches: do not ignore this easily overlooked item. The danger to other road users and to the vehicle occupants from a door which suddenly opens during a cornering manoeuvre is self-explanatory.

1 Open and close each door several times to test the lock function.
2 Push and pull the door by hand. Excessive movement can indicate worn or poorly adjusted lock striker plates. The lock must 'hold' on both the first (fully closed) and second (safety catch) catch positions.
3 Check the hinges for excessive wear and secure mounting.

Bonnet: the bonnet is fitted with two catches and both must be serviceable.

1 Grasp the opening lip of the bonnet and with a slight 'rocking' motion in an upward direction check the primary catch. The bonnet should not release.
2 Operate the bonnet release and allow the bonnet to 'spring open' onto the safety catch position. Repeat the 'rocking' test. The safety catch should not release under this test.

Fittings: any external fittings must be securely attached. The possibility of sun visors, rear spoilers or mirrors becoming detached and falling into the path of other traffic cannot be ignored. Wind pressure on this type of item can be quite high at speed,

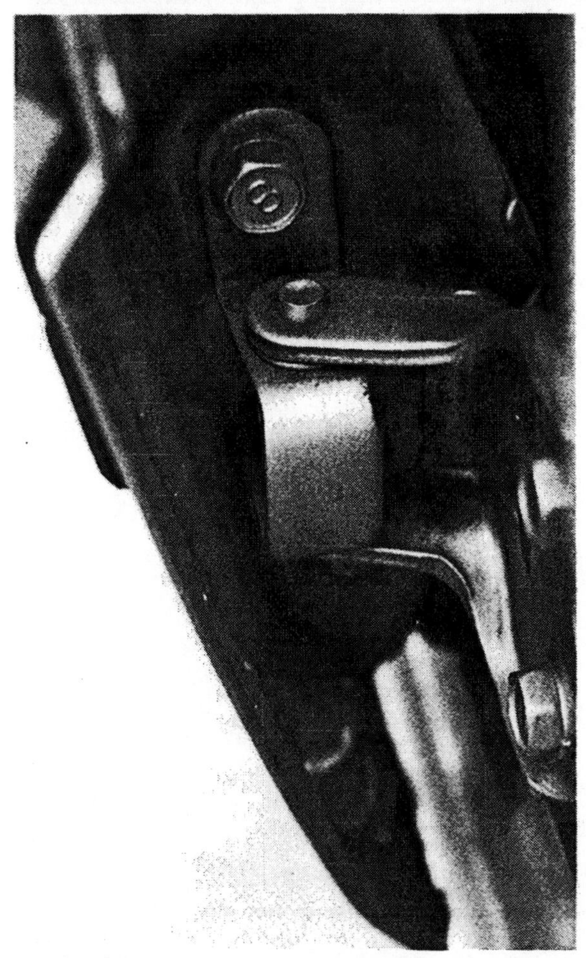
Check door hinges

Check bonnet lock is secure

therefore if the fitting has a large surface area the retaining method must be correspondingly stronger.

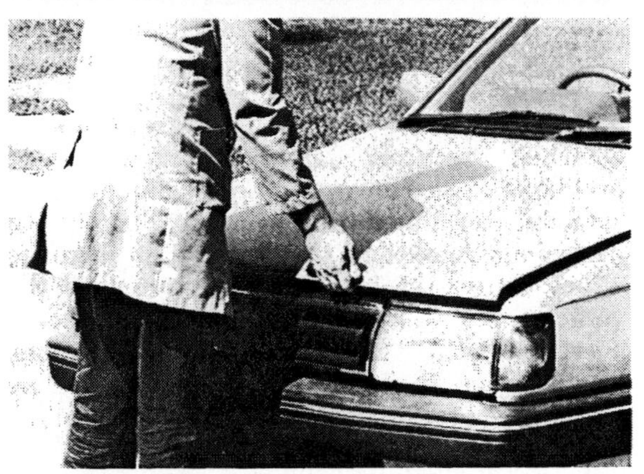
Check bonnet safety (second) lock position

Mudguards: regulations for mudguards can be very complex if considered for all highway traffic (dual wheel trucks). However, for a mass production passenger car it is possible to adequately check the vehicle with a few simple tests.

1 Are the wheels and tyres standard equipment?

   Modified wheels and tyres could protrude beyond the edge of the mudguard. This is not acceptable, as stones could be deflected onto other vehicles. A number of manufacturer-modified vehicles were produced as road use copies of popular racing cars. These vehicles may be fitted with 'bolt-on' flares to cover the wider tyres used: the flares must be fitted and secure if the vehicle still has the wheel and tyre arrangement for which they were designed.

2 Are the mud flaps (if fitted) standard equipment?

   If the mud flap was not designed for the vehicle the lower edge could, under full load conditions, drag on the road surface. Emblems or metal strips on the flap may become detached and be hazardous for other road users. This area requires common-sense decisions but, as a guide, problems will not be encountered if the flap's lower edge ground clearance is comparable to any factory installed mud flap.

3 Are the mudguards damaged in a way that causes sharp edges, protrusions or lack of mounting security?

All these conditions are hazardous to other road users so the vehicle should be rejected.

Bumper bars: if the bars are original, attached securely and undamaged, no problems exist. However, vehicle damage, bar removal or modification may cause a hazardous condition. Check bars for damage which causes the bar to protrude beyond the vehicle or be twisted in a manner that presents a sharp edge to possible contact areas. Removal of the bar may also leave the mounting brackets and/or chassis rails as unacceptable protrusions.

Modification to the bumper bar/frontal section is a controversial area. Additional bar protection is fitted to many vehicles by owners as a means of reducing vehicle damage in an accident and improving occupant safety. Whether these goals are achieved or that the damage is transferred to other vehicles or pedestrians is beyond the scope of this book. Be guided by the general requirements relating to protrusions and fittings not extending beyond the vehicle width or having sharp edges. Keep in close contact with the local regulatory authorities to establish the type of attachment which is acceptable and be aware if specific conditions relating to use become law.

Ornaments: design rules and streamlining of a vehicle have provided the public with a motor car that has a very 'clean' shape. This aspect has obvious advantages if the vehicle strikes a pedestrian. Any fittings added to a vehicle must comply with the requirements of not extending beyond the vehicle width and having no sharp edges.

GLAZING

A vehicle is fitted with glass areas to provide unobscured vision for the vehicle occupants through a material that must not be life threatening in the event of breakage. The type of problem encountered will mainly be related to the windscreen, however, tinting levels

Bull bar (typical)

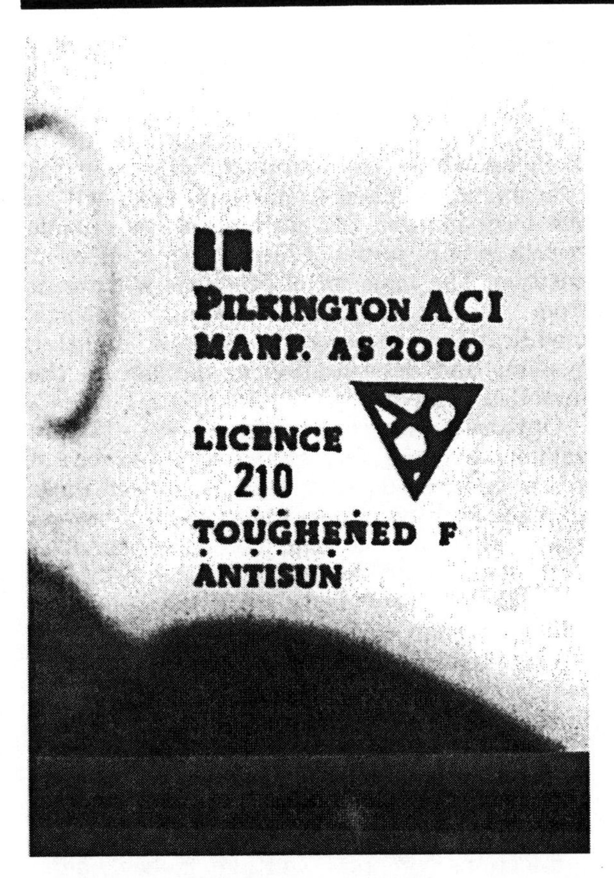

Safety glass marking (typical)

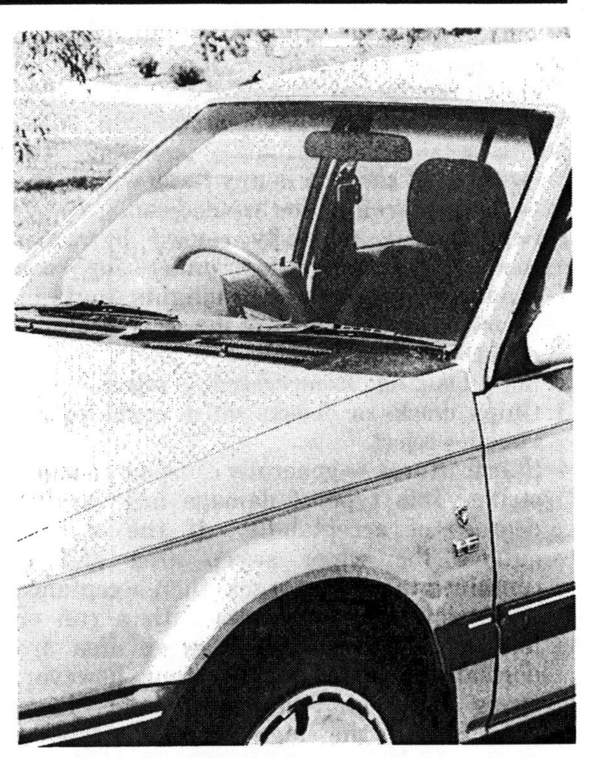
Windscreen band tinting

and material checks of other glazing should be done.

Material: all flat glass areas should be checked for the manufacturer's trademark etching certifying that the glass is 'safety' glass. A difficulty which may be encountered with this inspection is that, unlike tempered glass, laminated safety glass can be cut from larger sheets and the mark may not be present. If the appearance of the glass raises doubts then a discussion with the owner may clarify the origin of the glass. Fortunately the possibility of plain sheet glass being fitted to these areas is remote. However, as the consequences of such a stupid action are so terrible a check should be carried out. It was not illegal to fit glass other than safety glass in vintage vehicles. However, if safety glass replacements are available an owner may be extremely grateful to be advised of a way of reducing the potential danger.

Tinting: band tinting of windscreens and light tinting of all other windows have been available for a number of years as a factory fitted option. The main benefits have been increased comfort to the vehicle occupants and a reduction in load on the air conditioning system during summer.

Problems can arise when modifications to the amount of tinting are made. It is difficult for the mechanic to determine how much tinting is 'too much'. Guides which discuss available light in percentage terms are still difficult to interpret when the sole means of testing is to look through the glass. If the windscreen is clear (upper band tinting allowable) and adequate vision is available to the right and left sides of the vehicle then the vehicle is probably acceptable. If the view to the rear is considered unacceptable then a solution may be reached by fitting an external rear view mirror to the left hand side of the vehicle (this mirror is additional to the compulsory right hand side external mirror).

Windscreen: Early (tempered safety glass) windscreens did not present many problems when determining acceptability. The only damage which the screen could sustain without breaking was scratching. Modern laminated screens can sustain considerable damage, and because of this it is necessary to outline the types of faults on which to base acceptance or rejection.

Most regulatory bodies have published outline drawings of acceptable and non-acceptable conditions, however, the major area of concern is any fault within the wiper sweep area of the windscreen.

1 Scratches — normally caused by wiper blades. If driver vision, under any road condition, rain, sun or headlights, could be impaired by the extent of the scratches then rejection is required.
2 Sand blasting, stone crazed — reject.
3 Chips, cracks or defects which could impair vision — reject.
4 Star fractures — generally caused by a stone strike. This type of damage has varying degrees of acceptability. If the star is outside the wiper sweep area and is contained (no run or crack) then acceptance is possible. Star fractures with a run or located within the wiper sweep area are normally a reason for rejection, however, epoxy injection repair techniques have progressed to the stage where this type of fault can be repaired.

   An acceptable repair is one where the defect has disappeared and the repair will not impair the operation of the wiper blades.
5 Bullseyes — refer to star fractures.
6 Cracks — the location and type of crack provide a guide to acceptability. Unfinished cracks or any crack extending into the wiper sweep area are rejection items. However, if the problem is a 'finished' crack close (within 100 mm) to a screen corner, or a small 'U' shape finished crack outside the wiper sweep, then acceptance is possible.
7 Discoloration — tinting, other than upper band tinting, is unacceptable. If the screen has an area which has a milky-white appearance then moisture has probably entered between the glass laminations. Rejection is possible depending on location, extent and reason for the problem.

Obscured vision: stickers, other than those required by law, posters or decorations must not be placed on any window in a manner which will interfere with a driver's vision.

Window operation: regulations are not, with the exception of the driver's window, specific in requiring all window winders to be in operating condition but it is good practice to check and make a recommendation for repair of any faulty items.

LIGHTING (PASSENGER CAR ONLY)

Vehicles which are manufactured within the country, or supplied from overseas to sell on the local market, are unlikely to be outside regulations in terms of lamp type, number or position. The majority of problems will result from direct private owner imports, modifications to original vehicle lighting systems and deterioration or damage to the installed system.

Obviously, improvements are installed in vehicles by manufacturers over a number of years and regulations tend to reflect these changes by having date cut-off points beyond which an earlier standard is not acceptable. You should read the regulations to establish these dates, particularly when the vehicle is 'different' from current production.

This section will not deal with detailed regulations, but will provide a guide to the faults which a mechanic may encounter during a lighting system inspection.

Headlamps: normal two lamp systems must show WHITE light only and be of equal light intensity (low voltage supply through bad connections, unequal wattage globes and/or discoloured reflectors may lower the light value to an unacceptable standard). The lights must be capable of dipping (low beam).

An item easily overlooked during an inspection is the beam indicator lamp (mounted on the instrument panel). The indicator must be 'on' when high beam is operating.

Caution: vehicles manufactured for use in left hand drive countries have

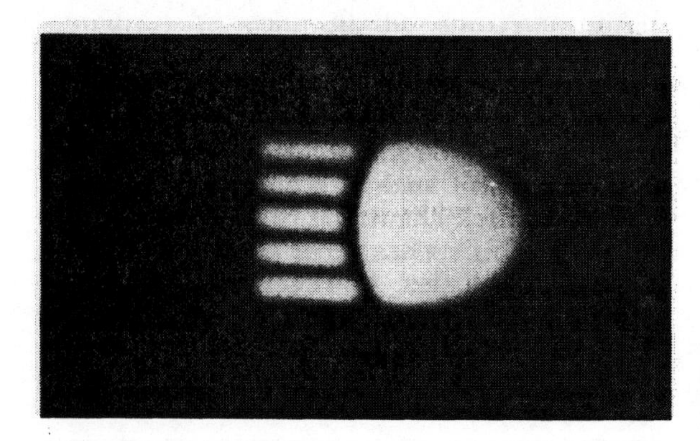

Check high beam indicator lamp operation

Two lamp system (typical)

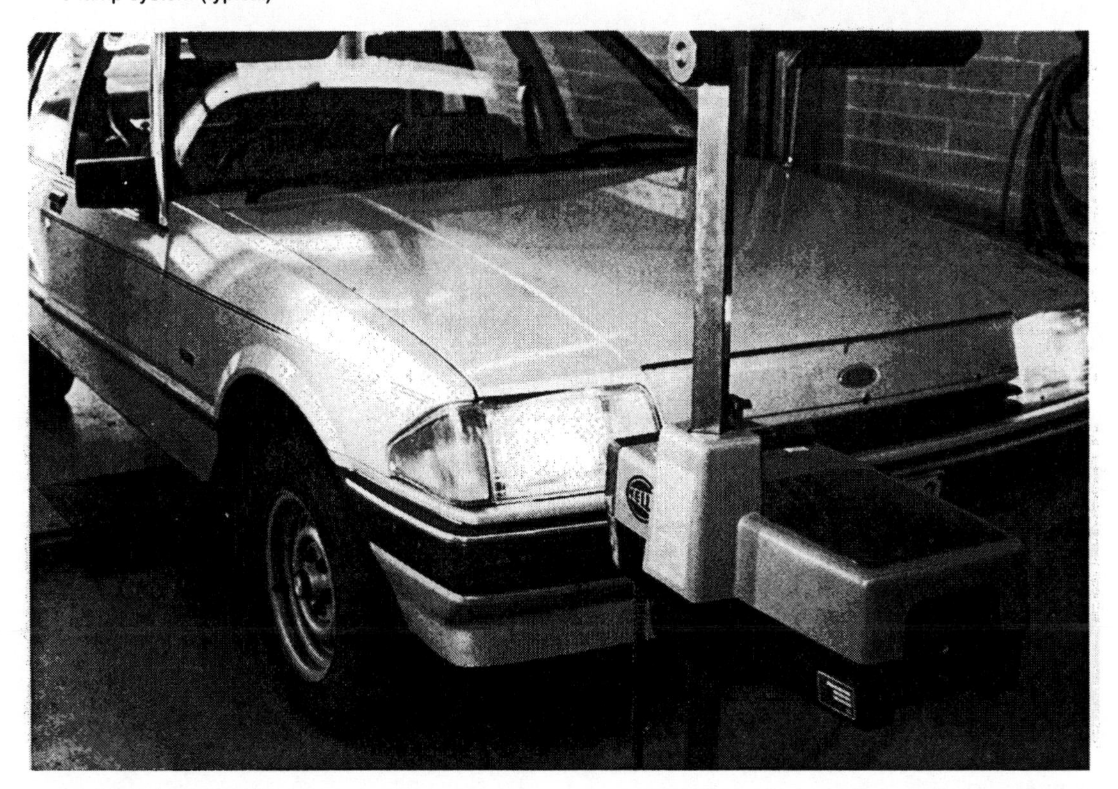

Head lamp aimer (typical)

headlamps which dip to the right. Be careful if the car is a direct import.

Regulations exist for illumination distances and beam heights, however, modern vehicles which are in serviceable condition and aligned to the manufacturer's specification will comply easily with the requirements.

Four lamp (original equipment) systems are subject to the same requirements as two lamp systems, with the further requirement that the system consist of two lamps containing high and low beam which must be mounted in the outer positions and two lamps containing high beam (white light) only are mounted in the inner position. The light switching must be arranged so all four lamps operate on high beam and when low (dipped) beam is required the two outer position lamps only remain illuminated.

Additional lighting: white light driving lamps (two not one) may be added to two-headlamp vehicles (check local regulations) but they must be of approximately the same illumination value, be located the same distance each side of the vehicle centre line, with lamp centres positioned in accordance with local regulations. The driving lamps must not be on when the headlamps are switched to low (dipped) beam.

Additional lighting

Fog lamps (two not one) may be fitted to vehicles in addition to all other lighting. For use only in abnormal atmospheric conditions, the lamps may be amber or white (if white the

Four lamp system (typical)

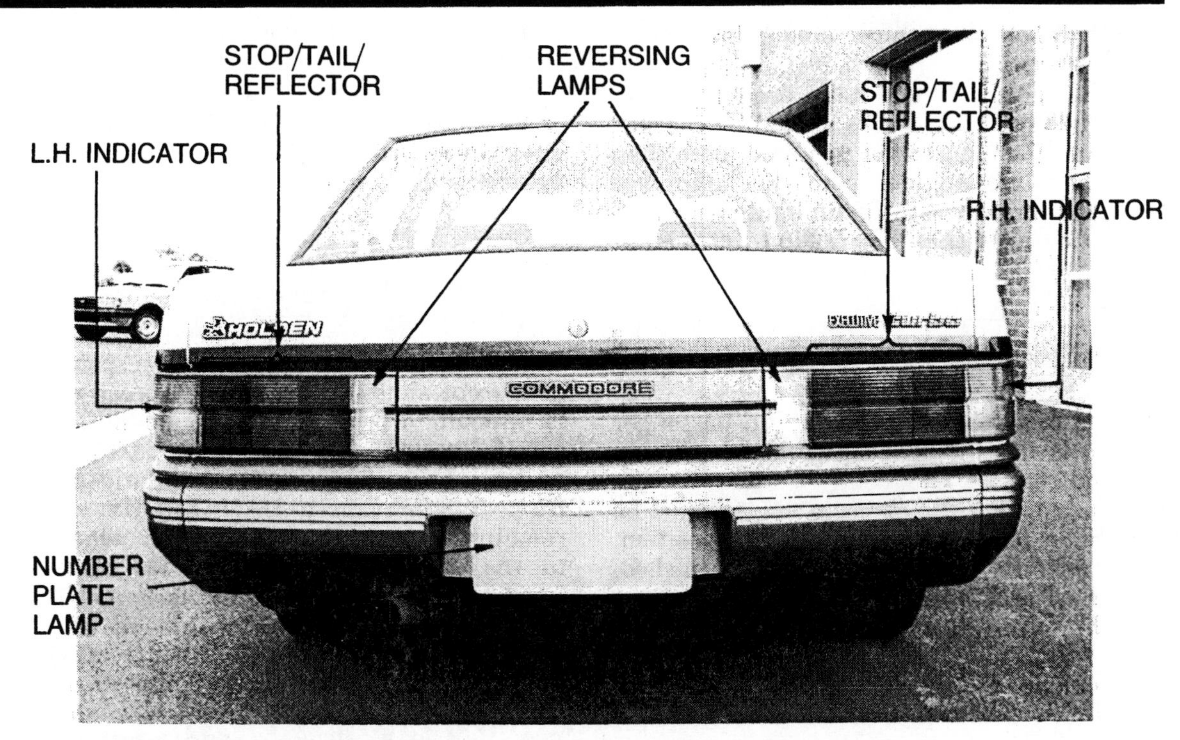

The typical rear lighting group depicted combines the stop lamp function with the tail lamps and the hazard flashers are integral with the indicator units.

lamps can be used as driving lamps). Mounting location, illumination and alignment are similar to driving lamps. When in use no other white light, except park lamps, will be on.

Tail lamps: must be red, of equal intensity and have globes no stronger than 7 W. Lens must be clean, not broken and/or discoloured.

Reflectors (rear): most modern vehicles have the reflector 'built-in' to the rear tail lamp lens. The requirement is that a motor vehicle must have a minimum of two rear reflectors which will provide a red reflection to the driver of a vehicle, with headlamps on dipped beam, approaching from the rear. The colour and positioning of the tail lamps provided manufacturers with the opportunity to include the reflector and the tail lamp in the one unit. The tail lamp lens is of a special type which allows internal light to pass through for the tail light but will reflect external light to provide the reflector. Other vehicles may have part of the lens as reflector and part as tail light.

If the vehicle is not fitted with separate reflectors it is important to check that the rear tail lamps are equipped with original type lens (which are not faded) incorporating the reflector.

Number plate lamps: white light directed on to the number plate (not to the rear except by reflection). The plate must not be obscured and the light must illuminate the number plate so the plate is readable at 20 m.

Flasher lamps: may be white or amber to the front and amber only to the rear. The lens should be clean, not broken or discoloured. The flash rate must be between sixty and 120 flashes per minute. Side repeater lamps (if fitted) and the dash panel indicators should also be working.

Hazard flashers: if front and rear flasher lamps are amber the vehicle can be fitted with an emergency flashing system which causes the flasher lamps to operate as a four way flasher system. The warning lamp on the dash panel and the audible clicker must also be working.

Reversing lights: all modern passenger cars must be fitted with reversing lights. The wiring for the lights must be arranged so that they will light only when the ignition is 'on' and reverse gear is selected. The lamp must show white or amber light only, therefore many manufacturers use the rear flasher lamps as a double purpose unit with both rear amber lamps illuminated when reversing.

Older vehicles which have been upgraded to include reversing lights should be checked to ensure the lamp is not mounted more than 1 m above ground level and the lamps are angled downward so the main light strikes the ground no more than 5 m behind the vehicle.

REAR VISION GROUP

A basic requirement for a modern passenger vehicle is that it must be equipped with an internal rear view mirror and a driver's side external mirror. The external mirror must be adjustable from the driver seating position. The mirror/s must not be tarnished, blemished or loose on the mounting bracket and the mirror surface must NOT be of the convex type. The mounting position must also enable the driver to have adequate vision past any load that may be carried on the vehicle.

If the vehicle is a utility, truck or panel van (station wagons with other than private registration are also in this category) an additional rear view mirror must be fitted to the passenger side of the vehicle.

Further requirements for rear vision mirrors also apply if the vehicle is towing a trailer as the mirrors are usually of the 'hang on' type. Care must be taken that the mirror does not protrude more than 15 cm beyond the vehicle or load width. When towing, the vehicle must have two external mirrors fitted even if the tow vehicle is a passenger car.

STEERING AND SUSPENSION

Many types of steering and suspensions, both front and rear, are in widespread use. It is difficult to provide, in a single statement, specific wear limits that will cater for all designs, therefore most regulations are framed to prevent a DANGEROUS vehicle remaining on the highway. Strict adherence to the maximum limits imposed on wear movement of the components does not mean the vehicle will operate within the design intent of the manufacturer. This is recognised by regulatory bodies by the use of terminology such as 'excessive, unsafe, perceptibly loose, defective and rejection in cases testers consider appropriate'. Testing of suspension components should be carried out in the manner prescribed by regulations, but the informed tester will be aware of the relevant manufacturer's service limits for each component so any acceptance or rejection of components will be based on a

Rear vision mirrors (typical)

comprehensive knowledge of the system under test.

Because of the variation in wear limits this section will concentrate on wear areas and methods of checking for wear. Any practical exercise undertaken should be carried out in conjunction with a workshop manual for the vehicle under test.

Steering tests

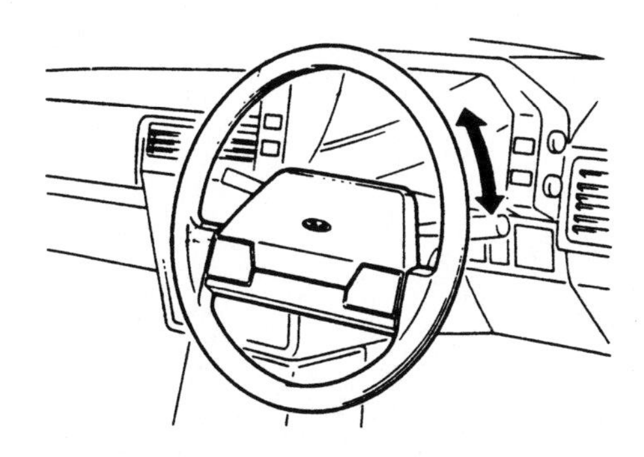

1 *Free movement*
Test the steering wheel for free travel (lash). The free movement is measured at the steering wheel rim and represents the accumulated free play in the total steering linkage. Excessive free movement is a reason for rejection, however, further checks are necessary to determine the points of wear so a recommendation for repair can be made.

Linkage testing requires help from an assistant to move the steering wheel back and forth while an under-body inspection is carried out. Slack in the steering joints will be readily detected if the vehicle weight is on the road wheels to provide resistance to steering movement.

2 *Lock to lock*
Raise the front wheels clear of the ground.

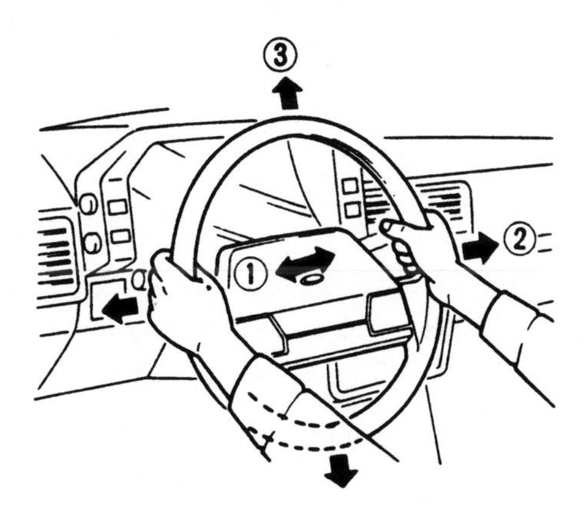

Rotate the steering from lock to lock. The steering wheel should move freely with no tightness, binding, roughness or abnormal noise present. Power assisted steering units should be tested with the engine running. Do not hold the steering hard against a lock for more than a few seconds — the hydraulic valve noise is normal but the system may be damaged by prolonged full lock application.

3 *Steering column and bearing*
Attempt to move the steering wheel up and down. The steering wheel and column must be firmly secured and no movement be apparent in the upper column bearing.

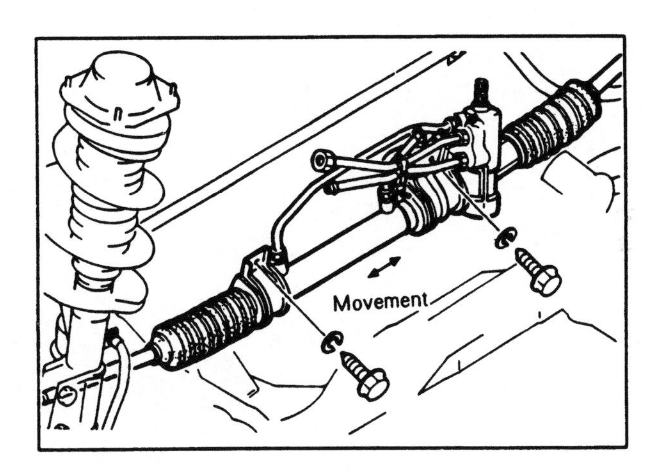

4 *Steering box attachment points*
Check steering box for movement at attaching points by moving the steering from lock to lock while the vehicle load is on the tyres. This check is significant where the unit is rubber mounted as maximum permissible movement may be less than 1.5 mm.

5 *Power steering fluid leakage*

6 *Tie rods*
Check each outer tie rod (and inner if fitted) for perceptible movement. Check seals for cracks and lubricant loss. Check tie rod adjusting threads and retaining nut or clamp for security. Grasp and rotate the tie rod assembly. The assembly should move freely within the limits of the tie rod ball socket.

7 *Ball sockets*
Rack and pinion type only, check ball sockets at rack ends for wear. Check convoluted boots for splits and lubricant loss.

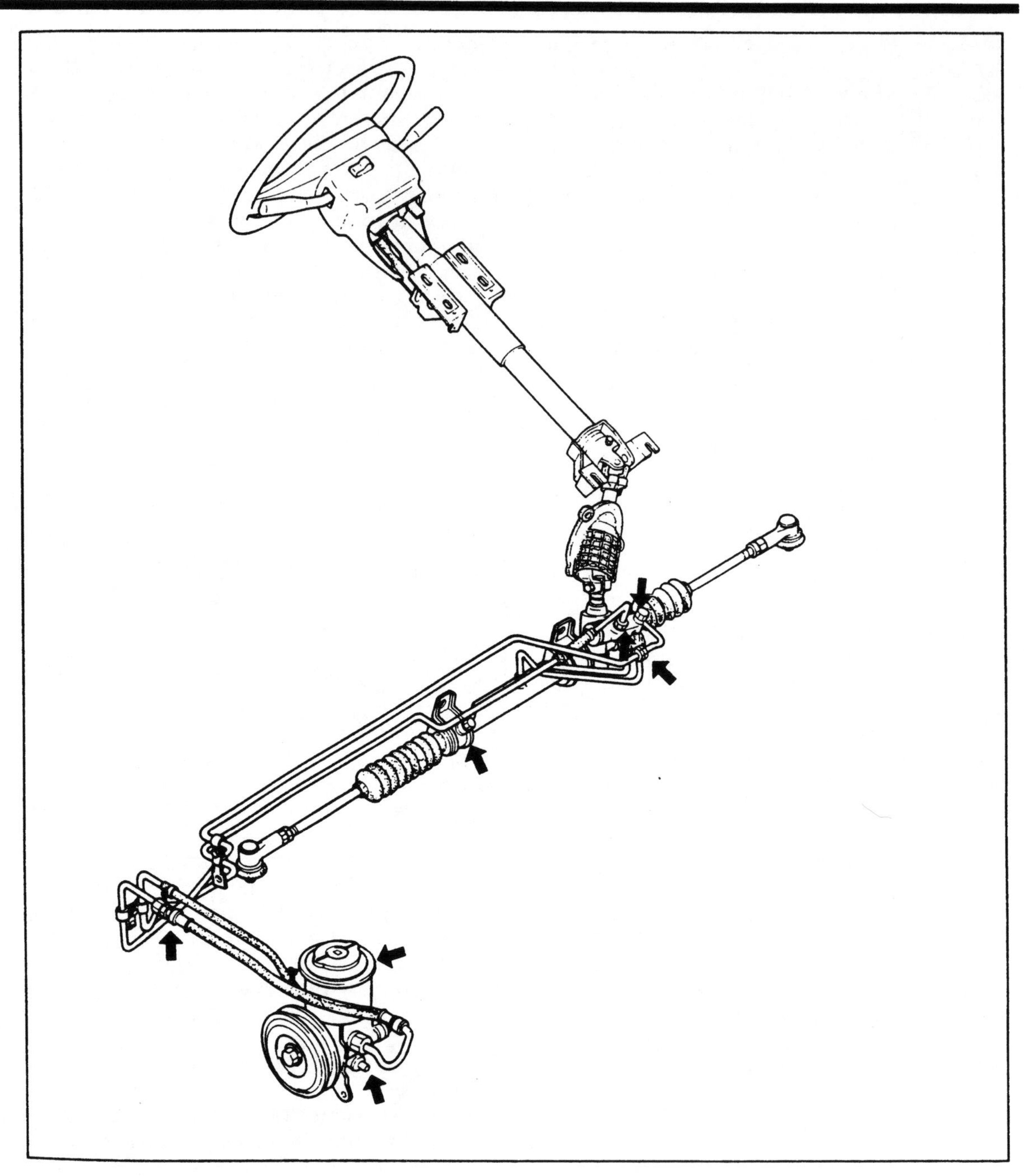

Check for fluid leakage at any of the areas indicated by the arrows.

Note
Start the engine, and check for fluid leakage after turning the steering wheel completely to the left and right in order to apply fluid pressure. Do not, however, keep the steering wheel in the fully turned position for more than 10-15 seconds.

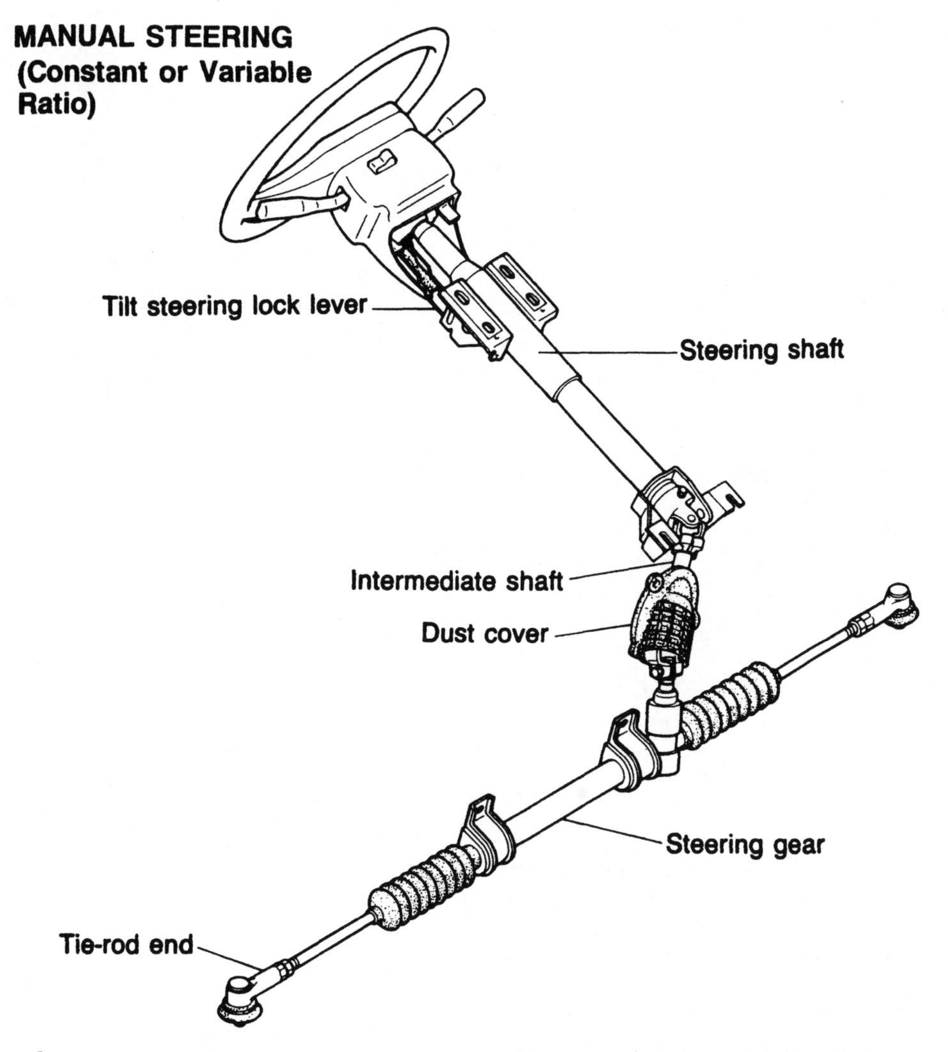

8 *Steering box wear*
Check recirculating ball, or similar types of steering box, for side to side movement at the sector shaft (indicating a worn steering box bush). Rack and pinion boxes can be checked by grasping the rack end and attempting to move it sideways. Repeat at each end of the box.

9 *Idler arm*
Check idler arm (if fitted) for wear by grasping the arm and attempting to obtain movement in a vertical direction. Repeat this test at the pitman arm to check if the pitman arm is secure.

General rule: no steering component should exhibit signs of non-approved repair methods, be loose, bent, foul any other component or have roughness in operation.

Idler arm checking

SUSPENSION

With the majority of passenger vehicles now using ball joint suspension the separation between parts which are steering components and parts which are suspension has become less clear. For this reason the king pin and bushes have been included in suspension testing.

Wheel bearing: this check applies to all vehicles regardless of suspension type.

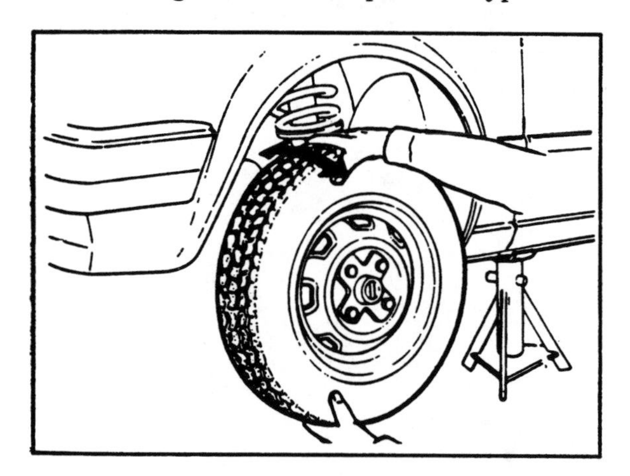

1 Raise the wheel/s clear of the ground.
2 Rock the wheel in and out at the top and bottom of the wheel. If movement is evident, ask an assistant to check for movement between the brake unit and the brake dust shield (or backing plate). If movement is present the wheel bearings should be adjusted prior to checking suspension parts for wear.

King pin and bushes: this arrangement is mainly used on commercial vehicles fitted with a solid front axle. However, a few passenger vehicles using independent front suspension did use king pins.

1 Raise the wheel/s clear of the ground with the vehicle weight being taken through the solid front axle (or lower radius arm if independent suspension).
2 Check the wheel in and out movement at top and bottom. (If the suspension is of the independent type, wear may also be present in the upper and lower pin pivots. A visual check will confirm which points are contributing to the wheel movement.)
3 Compare the movement to the allowable wear limit as stated in regulations or by the manufacturer.

Typical king pin assembly

4 Turn the wheel from lock to lock, check for roughness or binding.
5 Check that the lock pin and welsh plugs are fitted and secure.

Ball joints: ball joints perform a dual function. The road wheel pivots around the ball joints for steering but a pivoting action also takes place in a vertical direction as the suspension moves up and down on the road. Depending on which radius arm the spring forces against, either the upper or lower ball joint will be subjected to the vehicle load

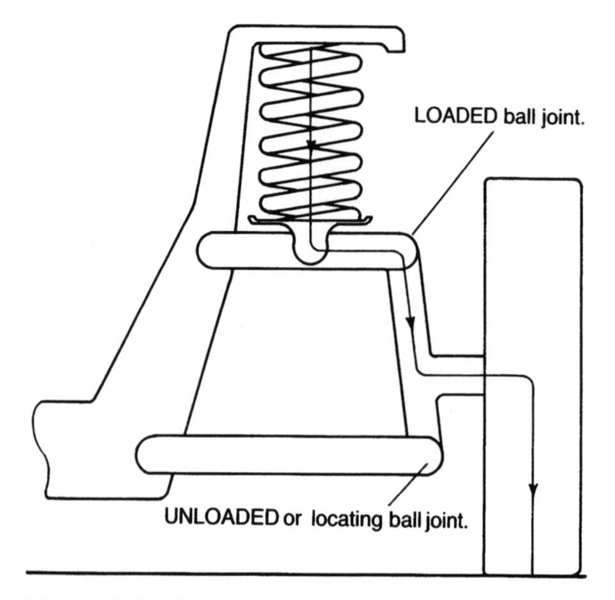

The vehicle mass acts down through the spring, spring seat then through the UPPER BALL JOINT before passing through the axle and tyre to the ground.

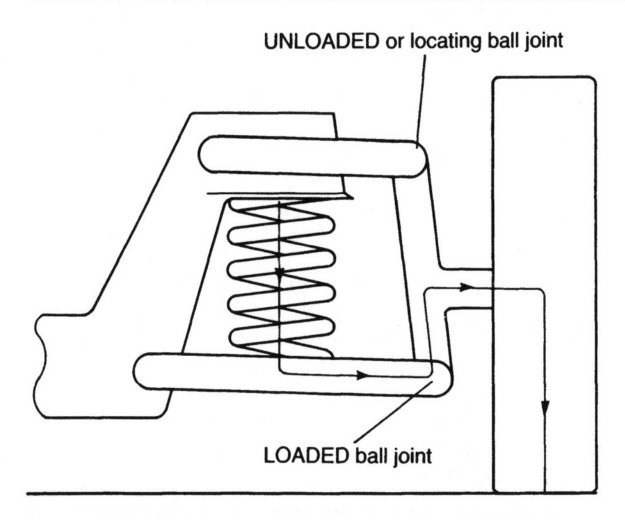

The vehicle mass acts down through the spring and lower arm then through the LOWER BALL JOINT before passing through the axle and tyre to the ground.

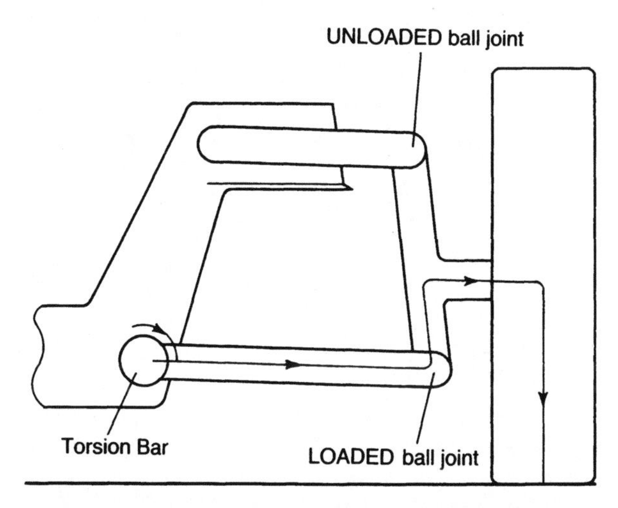

The vehicle mass acts through the torsion bar and lower arm then through the LOWER BALL JOINT before passing through the axle and tyre to the ground.

(loaded joint). The other ball joint is known as the unloaded or locating joint and although not carrying the weight of the vehicle it has an important role in directional control.

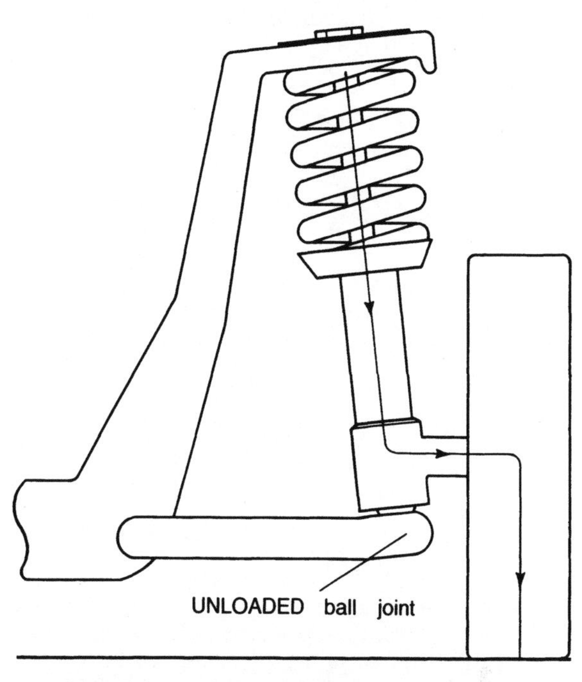

The vehicle mass acts down through the spring, strut, casing then through the axle and tyre to the ground. NOTE: This type of suspension does not have a LOAD CARRYING ball joint.

It is necessary to identify the suspension type and the location of the loaded/unloaded ball joint prior to testing. The identification and location will determine the test technique to be used. Where the spring or torsion bar forces against the upper (radius) control arm the top ball joint will be the loaded joint. On vehicles where the spring or torsion bar forces against the lower arm the lower ball joint will be the loaded joint. MacPherson strut type suspensions do not have an upper radius arm and the spring acts directly between the vehicle body and the strut casing. Load is taken by the upper mounting assembly contained in the strut tower; the single ball joint at the lower end of the strut is of the locating (unloaded) type.

Test procedures

Spring/torsion bar acting on upper control arm (upper joint is loaded, lower joint is of the locating or unloaded type).

1 Raise wheel/s clear of ground. The jack or safety stand MUST be placed under the BODY to allow the spring to extend and the suspension arms to hang down at the limits of their travel. This method will remove the

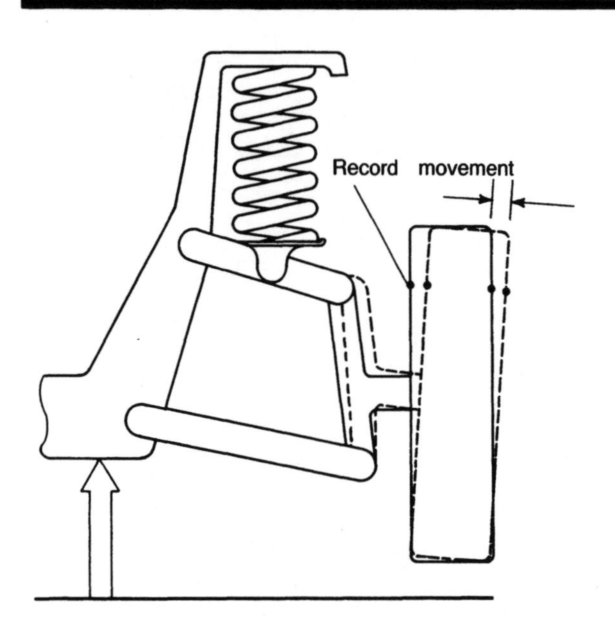

Grasp tyre at top and bottom and slowly move the wheel in and out. Note the radial play at the inner wheel rim adjacent to the upper ball joint. Compare this movement to specification. ANY movement detected at the LOWER ball joint during this check indicates the lower joint is unserviceable.

vehicle/spring forces from the upper joint and allow wear to be detected.

2 Check the lower locating joint (unloaded type) by attempting to move the wheel in/out and up/down. No visible movement is allowable at the joint.

3 Check the upper joint (loaded type) by:
   a moving the wheel in and out and recording the movement noted at the upper edge of the wheel rim;
   b comparing the amount of movement recorded to the regulations/manufacturer's wear limits. Service limits and regulations may vary in determining acceptable wear.

Spring/torsion bar acting on lower control arm (upper joint is of the locating unloaded type, lower joint is loaded).

1 Raise wheel/s clear of ground. The jack or safety stand MUST be placed under the LOWER CONTROL ARM as close to the outer end as possible. This method will support the lower arm and remove the vehicle/spring forces from the lower joint and allow wear to be detected.

2 Check the upper locating ball joint (unloaded type) by attempting to move the wheel in/out and up/down. No visible movement is allowable at the joint.

3 Check the lower joint (loaded type) by:
   a moving the wheel in and out and recording the movement noted at the lower edge of the wheel rim;
   b moving the wheel up and down and recording the movement. It may be necessary to use a lever or pinch bar to assist in making this movement.
   c comparing the amount of movement recorded to the regulations/manufacturer's wear limits. Service limits and

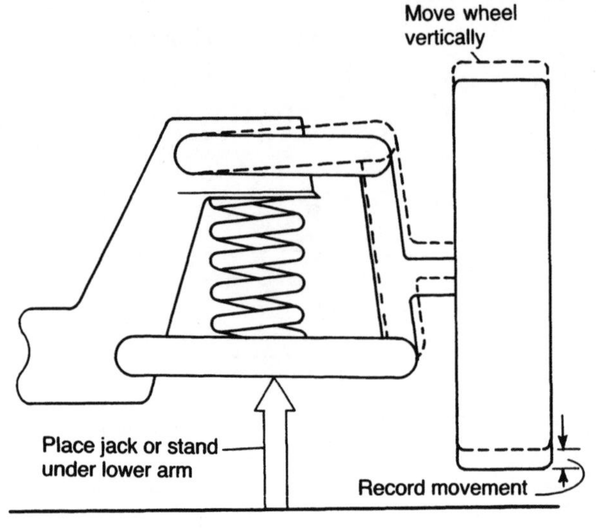

Move wheel in a vertical direction. Compare movement to manufacturer's specification

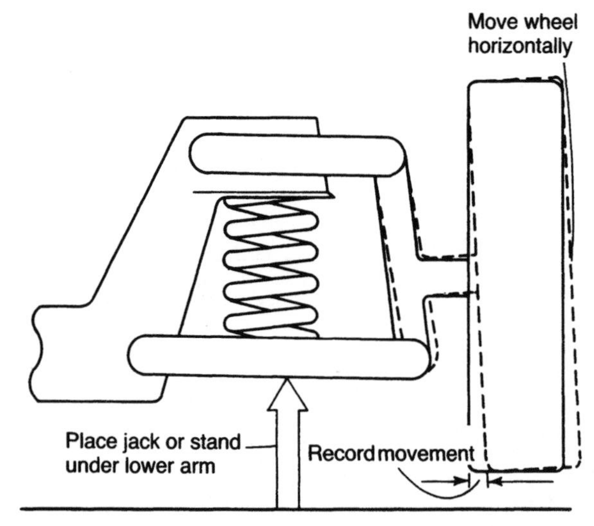

Move wheel in a horizontal direction. Compare movement to manufacturer's specification

regulations may vary in determining acceptable wear.

MacPherson strut type: spring acting through upper mounting assembly (no upper joint, lower joint is a locating unloaded type).

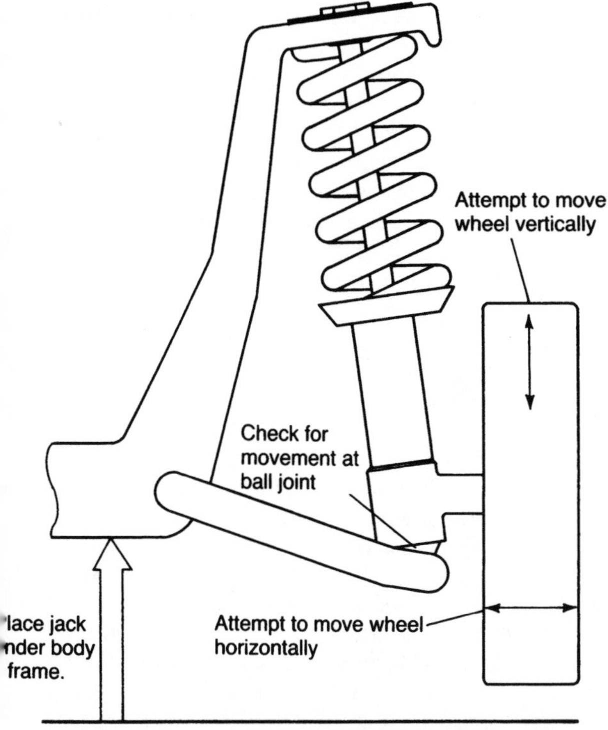

Attempt to move wheel in both horizontal and vertical planes. Movement detected at this joint must be compared to specifications as a wide variation exists between vehicle types i.e. zero up to 3 mm.

1 Raise wheel/s clear of the ground. Follow the manufacturer's instruction relative to the positioning of the jack or safety stands. Generally the stand is placed under the body and the suspension arms are allowed to 'hang' down, effectively removing spring load from the ball joint.
2 Insert a lever between the lower suspension arm and the strut adjacent to the ball joint.
3 Apply force to the joint by moving the lever up and down. Record any movement which occurs at the ball joint.
4 Many joints of this type are not allowed any detectable movement, however, at least two major manufacturers have introduced designs which have a measurable wear limit. These types require a slightly different technique. Prior to measuring the movement a block of wood should be placed under the tyre and the vehicle lowered until approximately half the vehicle weight is on the block. The ball joint movement is then checked in the normal manner.
5 Check the specific model details before classifying the ball joint serviceable or non-serviceable.

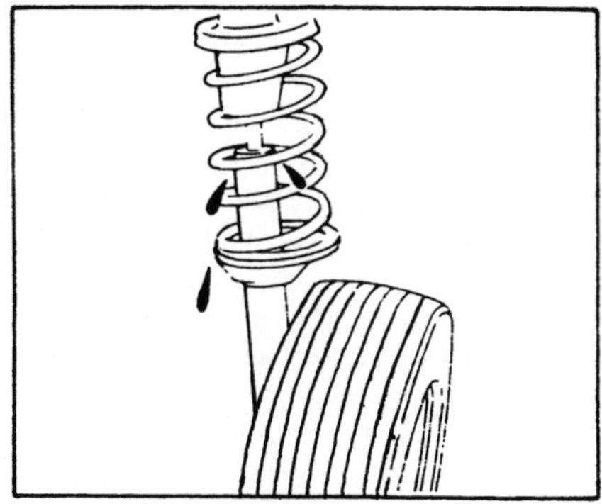

Source: Nissan Australia

6 Check the seal and external casing for signs of fluid leakage.

Source: Nissan Australia

7 The upper mounting assembly consists of a bearing and rubber mounting. The assembly should be checked for wear, cracks, distortion and mounting bolt tightness.

Lower suspension components

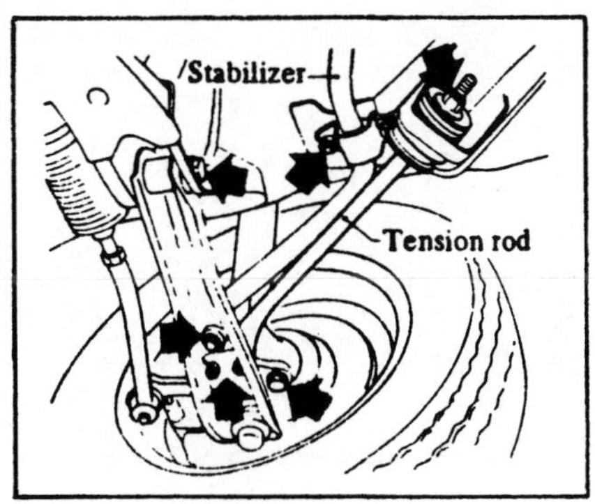

Source: Nissan Australia

Lower arm inner pivot bush: The bushing can be visually inspected for distortion, cracks and movement within the control arm. A pinch bar placed between the control arm and the sub-frame can be used to detect worn components. The pivot must be secure to the sub-frame.

Strut bar/tension rod (if applicable): strut bars are normally bolted to the lower control arm and fitted with a pair of large rubber bushes at the forward subframe attaching point. A number of designs use rubber bushes at both attaching points. Visually inspect all rubber bushes for wear, cracking or distortion. Worn bushes are easily detected by the use of a pinch bar between the strut rod and the sub-frame/lower control arm.

Stabiliser bar and mountings (if applicable): a stabiliser bar will be attached to each of the lower arms and at two points to the sub-frame. The lower arm attachments may consist of a link between the arm and the stabiliser bar. Inspection is relatively simple in that a visual check will readily detect distortion in the bar or worn, distorted or cracked rubber insulators at each of the mounting points.

Upper suspension components: MacPherson strut type suspension upper mounting points are generally checked when ball joint inspections are performed. Inspection of the upper arm inner pivot points of the double arm (wishbone) suspension is more difficult because of their location. On some designs it is possible to place a long lever vertically between the upper and lower arms and, by applying force, move the upper arm so a badly worn component can be detected. Alternatively, a pinch bar inserted between the body and the upper arm adjacent to the pivot/s (if the location allows access) is a useful method of checking wear.

If the pivots are of the rubber bush type a reasonable assessment of condition can be gained by visual inspection. The bushes must not be cracked, distorted or have moved within the control arm. Large retaining washers are often fitted at each end of the pivot and by comparing the clearance between the washer and the control arm at each end it is possible to estimate the extent of bush wear. A large variation in clearance would indicate the control arm has moved in the direction of the smaller clearance and bush wear has occurred. The pivot arm must be secure to the sub-frame.

SPRINGS

The vehicle should be checked for height and sag. It is in the springing/vehicle height area that modifications tend to be made to vehicles when owners are attempting to 'improve' the handling of the vehicle. Regulatory authorities do allow some relaxing of manufacturer's standards, but local requirements should be checked to ensure vehicle compliance is correct. Generally, lowering blocks, traction/torque reaction rods, reset springs (NOT CUT) to engineering standards are acceptable provided the suspension standing height (measured between the bump rubbers and the corresponding suspension contact point is not reduced below 2/3 of manufacturer's specification and the vehicle ground clearance remains greater than 125 mm. Vehicle height reductions must be uniform front and rear.

Coil springs: must not be broken, or have coils cut or sagged in height to the extent that the vehicle attitude is severely affected. Spring mountings must not be distorted or loose.

Leaf springs: check 'U' bolts and mounting rubbers. Check front and rear shackle mounting bushes for movement and/or bush wear. Check leaves for cracks. Check rebound clips to ensure clips are not missing or broken. A combination of a missing clip and a loose 'U' bolt can allow the spring leaves to 'spread'.

Torsion bars: all mounting points, clips and attachments must be present and securely located. Pivot points must not be worn and the

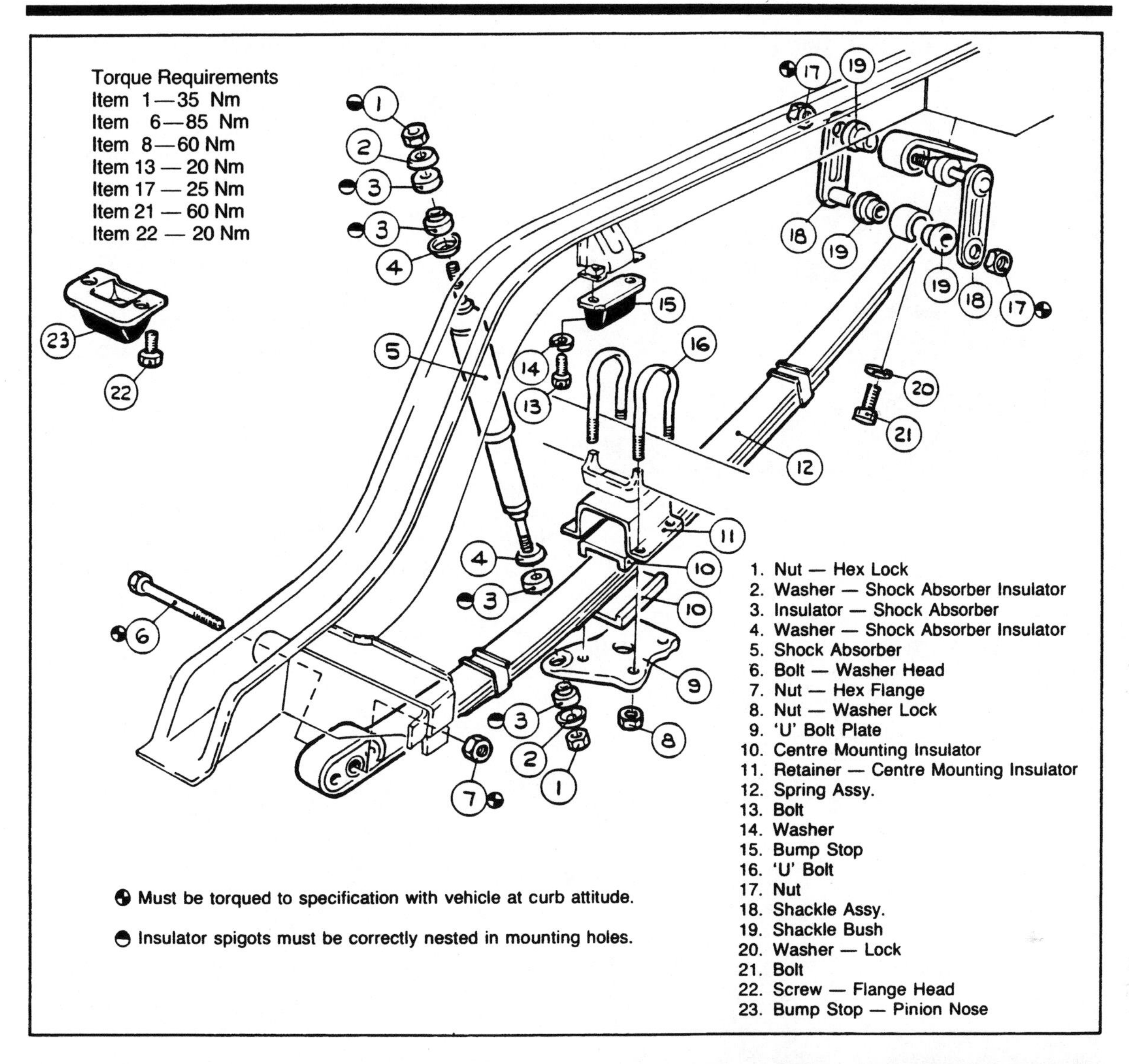

bar adjustment should be within manufacturer's vehicle height requirements.

Fluid: this type of system should be treated the same as any hydraulic unit. It must be free from leaks, the flexible hoses must not be cracked or frayed, and operating units (suspension 'springs') must be securely attached. Vehicle attitude (sag) is an important guide to the operating condition of the system.

Shock absorbers (including the damper unit integral with the MacPherson strut): check all mounting rubbers for wear, distortion or cracking. The attaching points must be secure and nuts and bolts be tight.

The external surfaces should be checked for oil leaks (slight dampness around the seal area is not uncommon and is not an indication of

Raise rubber boot for inspection of seal area if an oil leak is suspected.

shock absorber failure). Road damage should not be evident.

A dynamic test should be made by bouncing the car through the full suspension travel, with the vehicle weight on the road wheels. When the car is released the shock absorbers must prevent the vehicle moving through more than 1 1/2 undulations.

Rear suspension

The rear suspension of a vehicle can be inspected by following the same methods applied to the front suspension. Steering mechanisms are generally not fitted at the rear but wear points, spring problems and shock absorber tests are identical to a system of the same type fitted at the front of the vehicle. Particular care should be taken when inspecting independent suspension designs which are not fitted with longitudinal radius rods. The pivot bushes have a significant role in locating the suspension and excessive wear will allow fore and aft movement with adverse effect on the steering of the vehicle. It may be necessary to place a lever between the lower suspension arm and the sub-frame adjacent to the pivot bushes to adequately test for bush wear.

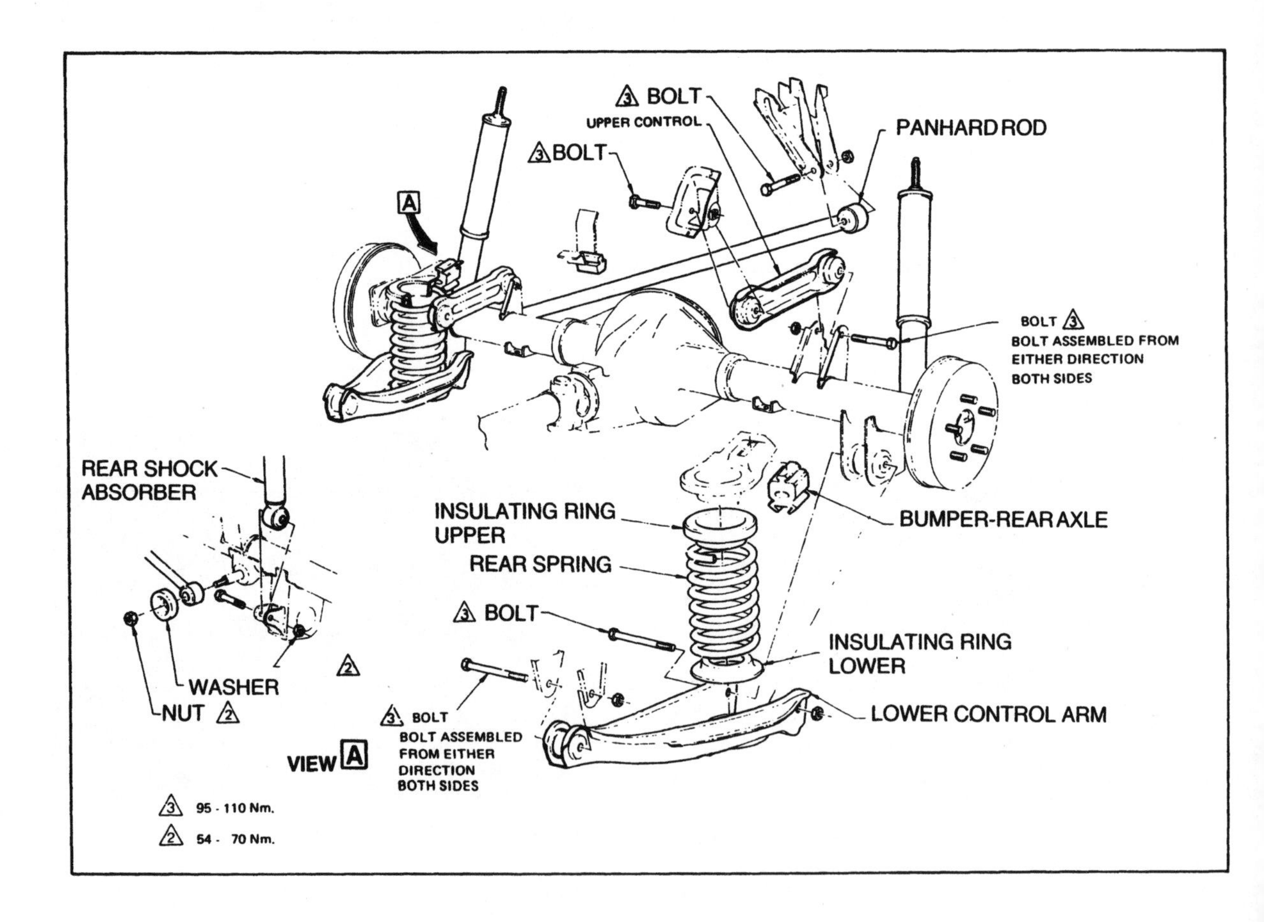

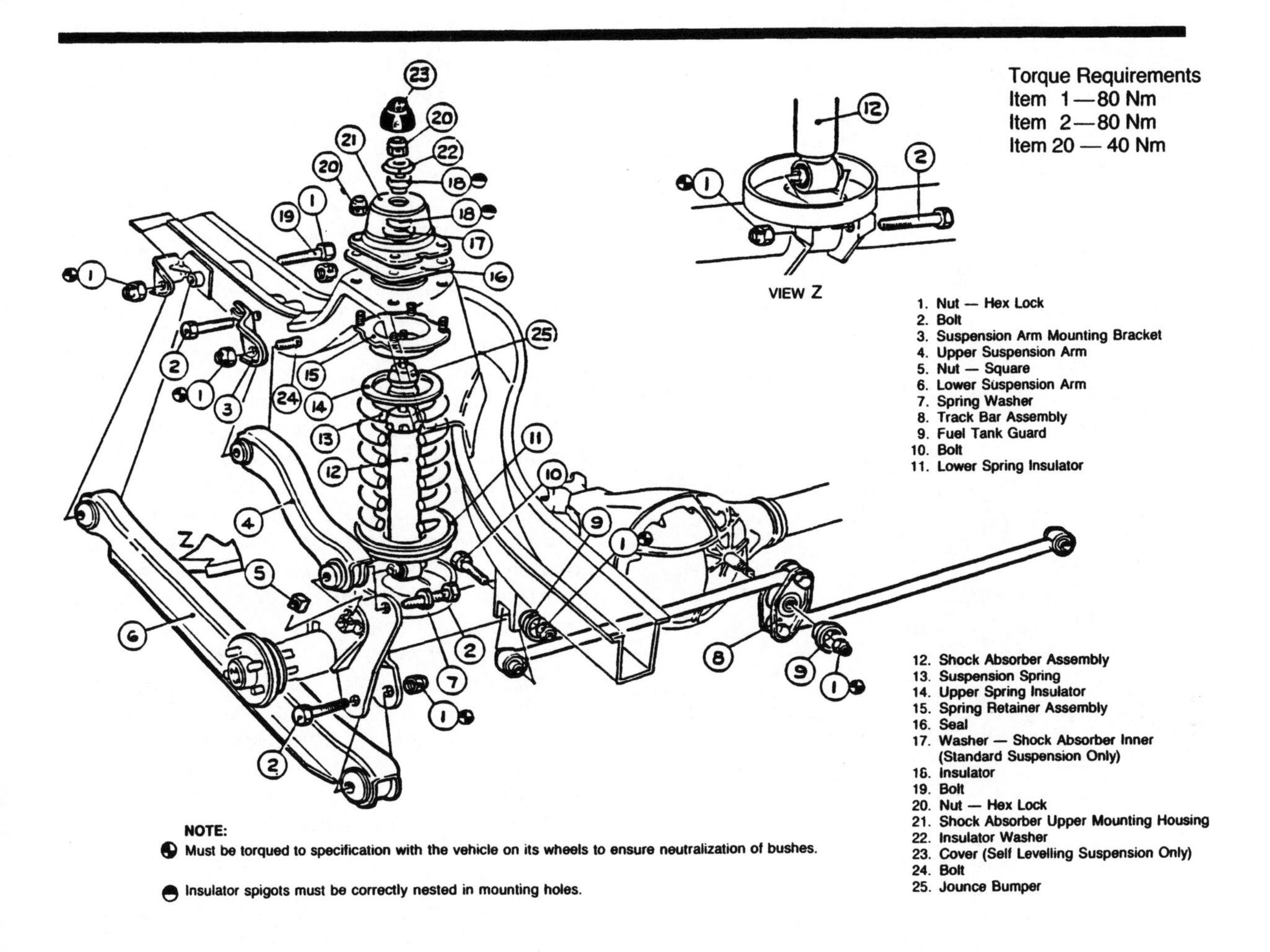

WARNING INSTRUMENT

Press the horn button (which must be mounted in an easily accessible location) and listen to the sound. The sound must be of an acceptable level and if two horns are fitted, both must be working to ensure the sound level and tone is adequate to warn other road users.

Check the horn and wiring for secure mounting and connection. If the horn fitted is not of a conventional (original equipment) type the regulations should be checked, as many authorities ban the use of warning devices which are considered to produce 'objectionable' sound or are of a type reserved for emergency vehicle use. The types of unit generally banned are bells, sirens, repeater horns, exhaust and pressure whistles or any device which produces a similar sound.

HIGH OR LOW TONE IDENTIFICATION

WHEELS AND TYRES

The wheels and tyres of a vehicle are probably the easiest mechanical part of a vehicle for access. Tyres are a normal wear area and every vehicle will have several sets of tyres fitted during its normal service life. This means, in addition to normal wear and road damage, the wheel/tyre combination fitted to a vehicle is also exposed to the possibility of non-standard replacements being fitted several times in the life of the vehicle.

(BH66A)

RIM SIZE	TIRE SIZE	INFLATION PR.	
		FRONT	REAR
5-J	155SR13	200 (29)	180 (26)
	175/70SR13		
5½-JJ	185/60R14 82H		
	P175/65HR14		

THE TYRES FITTED TO THIS CAR SHALL HAVE A MAXIMUM LOAD RATING NOT LESS THAN 425 KG OR A LOAD INDEX OF 78 AND A SPEED CATEGORY NOT LESS THAN 'S'. TYRE PRESSURE IS SUITABLE FOR ALL SPEEDS. WHEN FULLY LOADED, INCREASE REAR TYRE PRESSURE TO 190 KPa(28 PSI). KPa(PSI)

Tyre placard (typical)

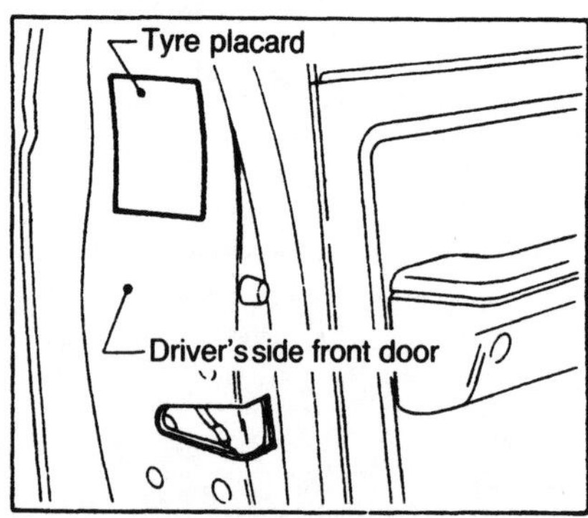

Source: Nissan Australia

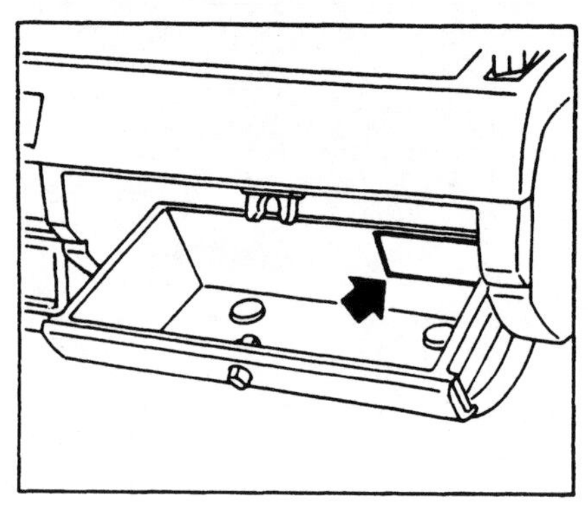

Source: Nissan Australia

Every modern vehicle is fitted with a wheel/tyre application decal (transfer), usually attached to the rear inner edge of the driver's door or inside the glove compartment. Listed on this decal are the manufacturer's recommendations relating to the wheel rim widths, applicable size/rating and inflation pressure of suitable tyres for the vehicle.

Generally, the tread section of a tyre wears out before the sidewall area and in the interests of economy many drivers do not purchase new tyres but look for alternate means of obtaining acceptable tread depth on the tyres. Many tyre companies and independent specialists provide a recapping or retreading process to obtain greater life from the tyre. This is generally an acceptable process provided the driver understands the tyre is now speed limited (original speed rating is not relevant) and the tyre coding is not removed or covered by the repair process.

Caution: regrooving of tyres is not an acceptable method of increasing tread depth. This process can only be used on tyres specifically manufactured for this purpose and labelled accordingly. The possibility of encountering this repair technique on a passenger car tyre is remote, however, be aware: fitting a regrooved standard tyre is illegal.

A further complication to be considered is the modification freedom allowed by many regulatory authorities in the fitting of wider wheels and tyres. This freedom has caused a huge after-market to develop for the supply of non-original wheels and tyres. Owners buying in this market tend to purchase wider wheels and tyres with performance rating superior to the vehicle's original equipment: the problem in inspection is usually the suitability to the vehicle not the product quality.

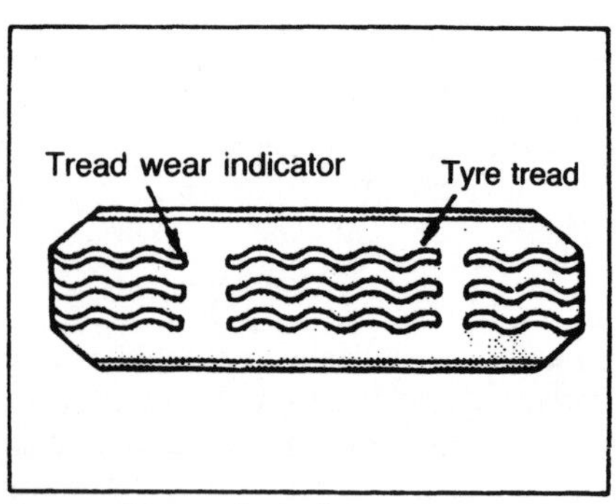

Source: Nissan Australia

Inspection

1 Inspect all tyres for wear: many tyre companies include a tread wear indicator in the tyre pattern. These indicators are sections of rubber moulded into the bottom of the tread grooves and arranged in a line across the tread. A number of these lines are spaced at intervals around the tyre. When the serviceable life of the tyre tread is approached the wear indicators are exposed and are obvious as bands across the tread. A tyre does not always wear evenly and wear in any section to the extent that the tread design is not visible means the tyre is not acceptable.

Check wheel runout.

Toe-in or toe-out wear

Centre wear

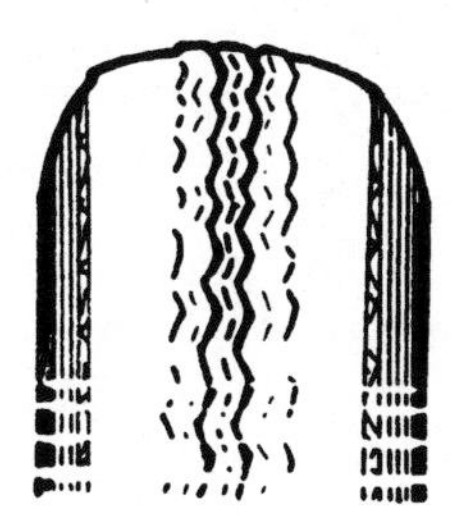

Shoulder wear

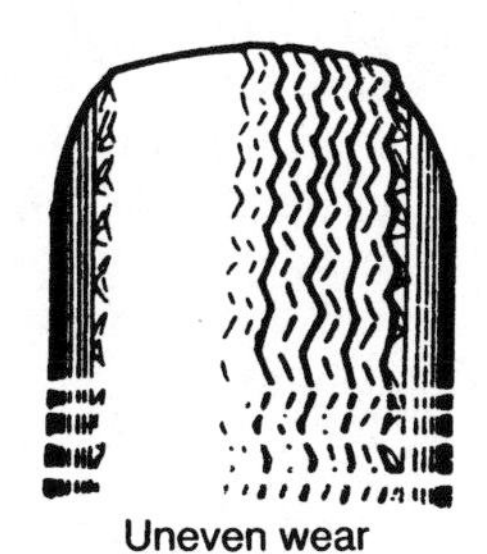

Uneven wear

Source: Nissan Australia

Caution: uneven tread wear is an indicator of possible front suspension problems. Check the suspension thoroughly where this condition is evident.

2 Check each tyre for road damage. Rotate the tyre, checking for objects in the tread, bulges in the side walls or tread areas, lifting of the tread and any cuts or tears which are greater than 25 mm in length or which are deep enough to expose any body cords.

3 Check each wheel and tyre for runout. Excessive runout may be caused by the tyre, its fitting or a buckled wheel rim. Identify the cause of the runout when reporting the fault. The wheel rim must be checked for splits, cracks or rim damage.

4 Remove all hub caps (if fitted) and inspect all wheel retaining bolts/nuts/studs. ANY retainers missing, stripped or loose is unacceptable.

5 Check the tyre construction type: it is not good practice to have tyres of different construction (radial/cross ply) fitted to the same vehicle. The handling characteristics of the tyres are different. Check with the local authority if a problem arises with vehicle compliance but mixing of tyre types is definitely not recommended. Remember the final performance capability of a modern braking system and suspension system depends on road contact through the tyres. The vehicle was designed to perform its task with ALL TYRES OF THE SAME CONSTRUCTION AND TREAD PATTERN.

6 Check the tyre and rim size combination for compliance with tyre and rim manufacturers association standards (Australia — design rule 23). Variations from the recommended tyre use placard are permitted for the wheels and tyres but the combination of tyre and rim size must be checked when non-original wheels or tyres are fitted. Generally, if the tyre and rim size requirements are satisfied and the tyres fitted have equal or greater load-carrying capacity and performance capability, the change is acceptable.

Of course, a limit must exist to the alteration and a number of guidelines are suggested in item 7, 'special applications'.

7 Special applications.

a Knock-off threaded hubs: the retaining

nut must be of the correct thread (left or right hand) in relation to the forward direction of the vehicle.

b Spoke wheels: wire wheels must not have more than six missing, broken or loose spokes and the faults must not be in adjacent locations or any more than two of the faults be in any quarter of the wheel. Any fault should be recommended for repair even though the wheel may comply with the required standard.

c Wider wheels/track: The previous requirements of compliance still apply, with the further conditions that the wheel may have only one peripheral weld in accordance with recognised engineering standards. The wheel and/or tyre must not foul any part of the body or suspension — move the front wheels to full locks and check clearances. Estimate clearances front and rear when the vehicle is loaded to design capacity, for example number of passengers, fuel load, luggage load. Spacers MUST NOT be fitted between the wheels and hubs. The overall increase in track width permitted varies depending on the suspension system in use. Check your local authority but as a guide passenger vehicles are allowed 25 mm increase front and rear where a fully independent system is in use, but the rear track can be increased by 50 mm if the rear system is a solid axle type.

WINDSHIELD WIPERS

Virtually all modern vehicles are equipped with dual speed electric motor operated windshield wipers. Earlier vehicles were fitted with a variety of wiper types, including vacuum operated units. Where these early units are encountered it would be advantageous to check the regulations for cut-off dates concerning acceptable types, but the average mechanic is unlikely to work on these units.

The wipers must operate on at least two speeds (usually >20 cycles/minute and >45 cycles/minute), the blades and arms must be intact and adequately secured and any wear present on the blades must not prevent the wipers maintaining a clear view through the windscreen for the driver. The operating switch must be capable of operation by the driver from the normal driving position and all parts of the system must be serviceable.

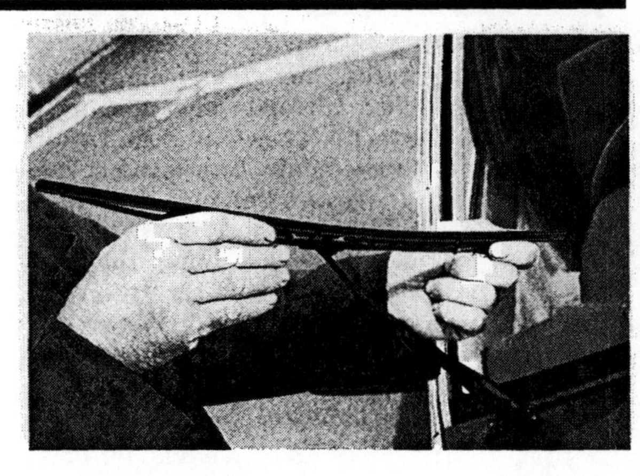

Check wiper blade for wear.

It has been many years since wiper systems were only required to clear the screen in front of the driver and the regulations are drafted in such a way that modern single wiper systems are acceptable because they clear the screen both in front of the driver and a corresponding area in front of the passenger. This means a twin wiper system with the passenger side wiper arm missing would not be acceptable because it could not clear the passenger side screen even though one wiper arm is still functional.

AUTOMATIC TRANSMISSION

The inclusion of the automatic transmission in a roadworthy inspection may cause an owner to question the validity of a safety check for this item. It is not the performance of the transmission or its shift patterns which is of concern but the safety mechanisms which prevent the engine starting in gear, and the selector indicator which informs the driver of the gear position.

Several rules relate to the gear selector layout, but these tend to be outside the area of a mechanic. Unless obvious modifications have been made to the selector position or layout it can be accepted that the gear selection layout will comply with roadworthy requirements.

A requirement does exist for the transmission to include a forward range (other than top) that provides engine braking and, when selected, will not upshift below approximately 40 km/h. This item can be readily checked for correct operation during a road test.

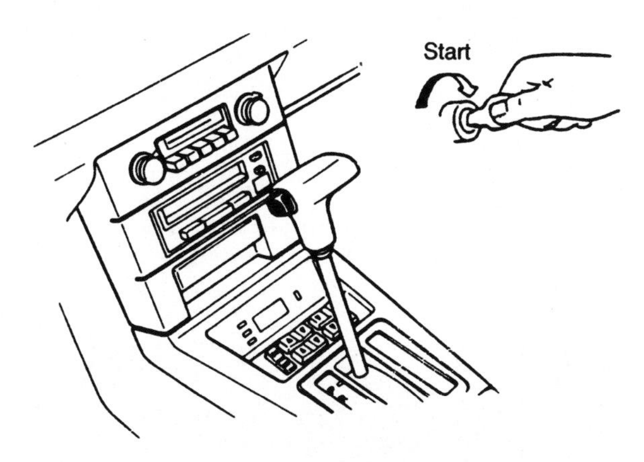

The three adjustable (or subject to wear) items which must be checked for correct operation are:

1 The gear engaged must be correctly shown on the selector lever indicator. Move the lever through each of the selector positions with the engine running and ensure the lever does not 'drop out' of selection position as engine load is applied.
2 The starter motor must be inoperative in all gear positions other than PARK and NEUTRAL. The danger of a vehicle starting while in gear can not be overstated. Under no circumstances is this check to be ignored. Move the lever into each of the selection positions, foot brake applied, and attempt to start the vehicle. If the starter motor cranks in any position other than P or N an urgent repair is required.
3 The vehicle must not be able to roll forward or rearward when PARK is selected. Check this feature by stopping the vehicle on a slight incline, select P, and release the brake. The vehicle should remain stationary. Repeat the test with the vehicle facing in the opposite direction.

Caution: do not select PARK with the vehicle moving. This feature is not a substitute for a parking brake. During normal vehicle use the parking brake should still be applied when the vehicle is unattended.

SEAT ANCHORAGES

The seat runners must be bolted securely to the floor and the front seat slides and locking catches must be serviceable. A further requirement for cars fitted with tilt forward front seats (two door vehicles with a rear seat) is that the tilt latching mechanism must be serviceable AND accessible for operation by the rear seat passenger.

SAFETY BELTS

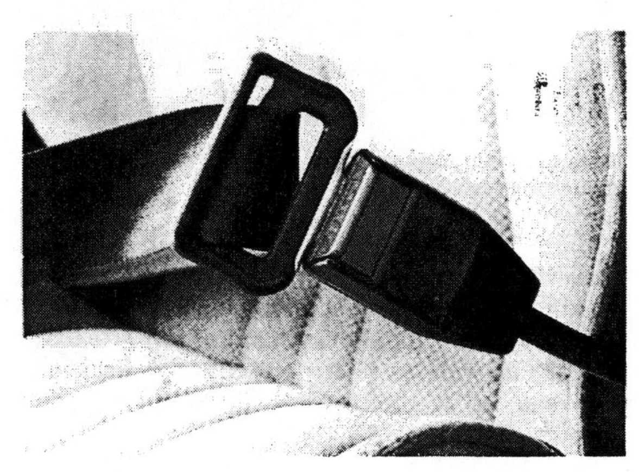

Seat belt buckle (typical)

Safety belts have not always been standard equipment on a passenger car and the regulations were modified several times as belts were accepted by the community. Common requirements for all models, however, are that all anchorages must be tight and located in a reinforced section of the vehicle, the belt material must not be frayed or cut and all buckles must function correctly.

The regulations require many early vehicles (up to thirty-five years old) be retro-fitted with seat belts if a roadworthy certificate is sought. Many of these vehicles do not have suitable mounting points for modern inertia-reel seat belts, so a variety of belt types is acceptable. Provided the belts are of approved type the driving position and the front passenger seat adjacent to the door can be fitted with lap, lap-sash or harness belts.

From the mid 1960s passenger cars were fitted with safety belts in manufacture, and a reasonable guide to the differing belt types which should be fitted to different years of manufacture and seat locations within the vehicle is to check the belt is similar to the original fitment item.

Withdraw belt to check webbing and retractor mechanism

Care must be exercised when testing modern retractor belts (inertia reel). The specific type must be identified as some units will lock if the belt is drawn out of the retractor rapidly as well as under sudden deceleration, whereas other belts will lock only under deceleration forces.

ANTI-THEFT LOCKS

Unhappily cars are attractive to thieves and design rules require the fitting of an anti-theft device in manufacture. The devices can vary from a braking lock or a gear change lock (preventing engagement of forward gears) to a steering lock.

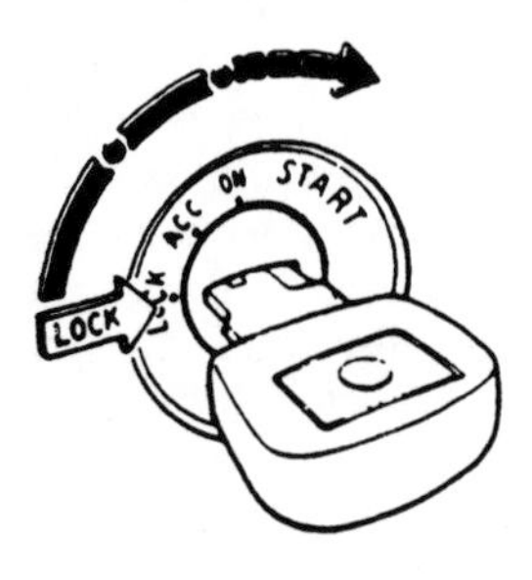

The most popular system is the steering lock, which prevents movement of the steering wheel until the ignition key is inserted and turned to the unlock position. Regardless of anti-theft device type, if the device was fitted in original manufacture it must be in serviceable condition.

WINDSCREEN DEMISTING

Design rules have required the use of demisting systems in passenger cars since 1975. They are integrated into the heating system, therefore the checks are relatively simple.

Check heater operation

The heating system must be connected and operable and the blower fan (where fitted) must be operable.

Turn on the fan with the heater controls in the demist position and the engine at operating temperature. By placing your hand over each of the ducts located at the base of the windshield it should be possible to detect a flow of warm air emerging from the ducts.

WINDSHIELD WASHERS

All modern vehicles must be fitted with a windshield washer system.

The system must be serviceable and capable of directing water to the entire swept area of the windshield. For this requirement to be met each nozzle must be clear and aimed

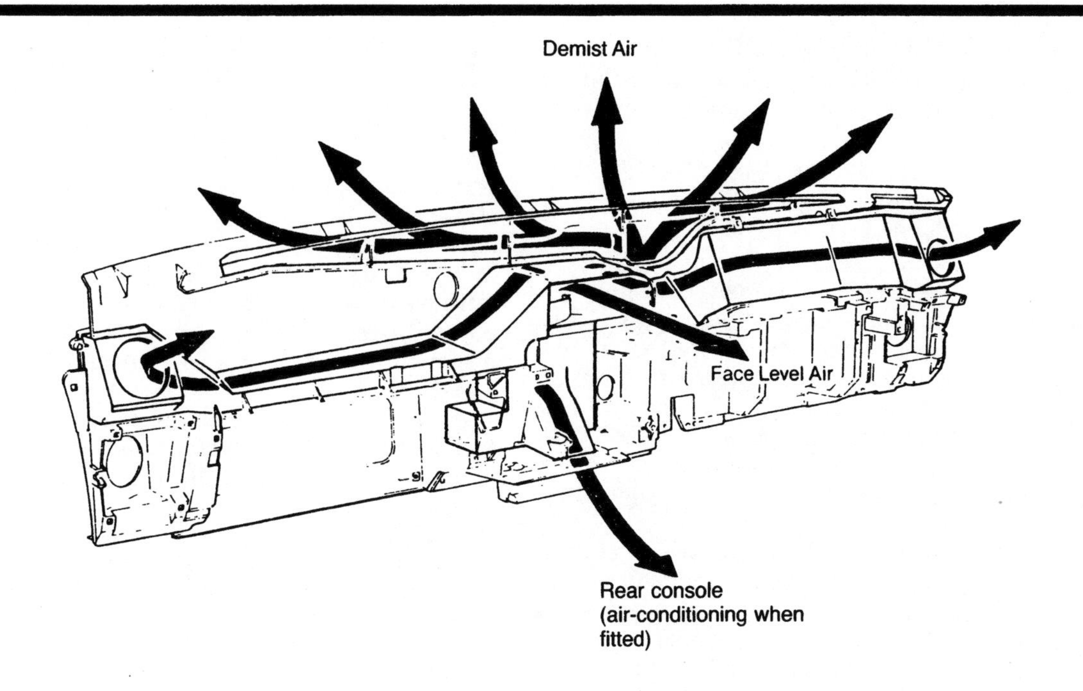
Demist Air
Face Level Air
Rear console
(air-conditioning when
fitted)

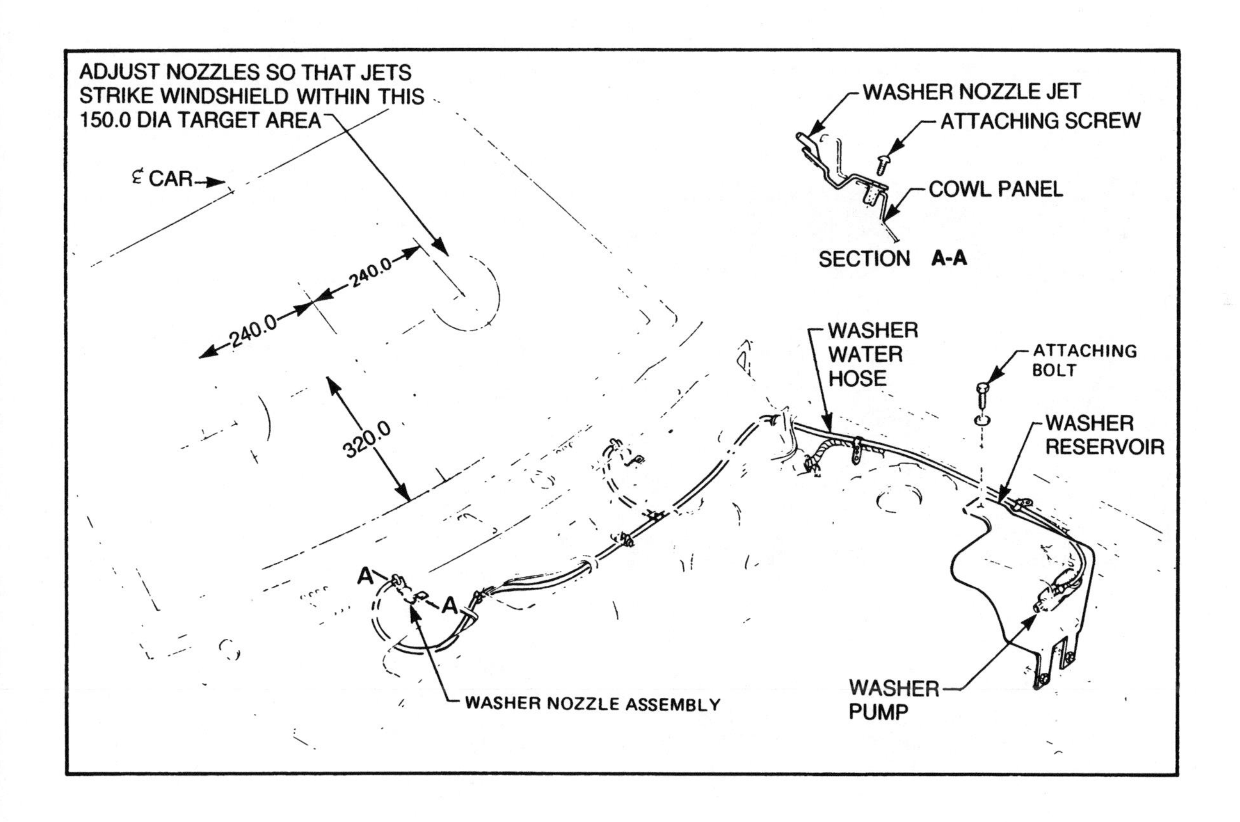
ADJUST NOZZLES SO THAT JETS
STRIKE WINDSHIELD WITHIN THIS
150.0 DIA TARGET AREA
℄ CAR
240.0
240.0
320.0
A
A
WASHER NOZZLE ASSEMBLY
WASHER NOZZLE JET
ATTACHING SCREW
COWL PANEL
SECTION A-A
WASHER
WATER
HOSE
ATTACHING
BOLT
WASHER
RESERVOIR
WASHER
PUMP

at the windshield. The operating switch must be located in a position to permit operation by the driver from the normal driving position.

EMISSIONS

An acceptable level of vehicle emissions is required from the vehicle under test. This level can vary considerably depending on the age of the vehicle (the regulations in force at the time of vehicle manufacture, such as ADR 27 or ADR 37) and the area in which the vehicle is registered (this requirement varies between countries and states for the same model vehicle).

The base requirement of absence of engine fuming and visible exhaust smoke (periods longer than 10 secs) is easily met by modern vehicle in reasonable condition. These vehicles are manufactured to meet design rules with levels of CO far less then the base roadworthy requirements. Visually inspecting the vehicle to ensure the factory-installed emission control devices have not been disconnected, removed or rendered inoperative is a satisfactory method of checking the vehicle for basic roadworthy acceptance.

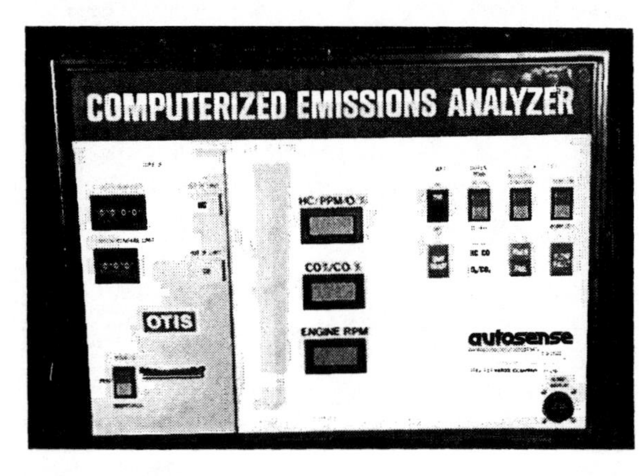

Emission test unit.

Where the regulations have specific instructions relative to modern designs, such as CO levels of 1.5 per cent, a more stringent test will be required.

MISCELLANEOUS ITEMS

An easy statement to describe this category would be: The vehicle must be in a serviceable condition. Immediately, this statement is open to debate about who decides what is serviceable or acceptable. It is virtually unworkable for a regulatory authority to develop legislation to cover all possibilities, therefore the tester must make decisions based on the potential hazard the 'defect' may cause.

A number of examples are listed as a guide to the type of defect which could arise in this category, but the list definitely does not cover all the possibilities.

- Oil leaks of sufficient size which would leave deposits wherever the vehicle parked, creating a visual and possibly a safety hazard.
- Drive or tail shaft universal joints worn and in danger of failing.
- Rust in structural components weakening the strength of the vehicle floor pan, chassis, door pillars or firewall.
- Rust in body sheet metal which could create sharp edges protruding from the vehicle.
- Rust or openings in the body allowing exhaust gases to filter up into the passenger compartment.
- Head restraints removed from vehicle seats (only those vehicles which by year of manufacture are required to have seat head restraints).
- Poor repair of accident damaged components.
- Electrical wiring harnesses in a condition likely to cause a fire hazard.
- Illegal vehicle modifications.

CHAPTER 9 REVISION

1. Describe the method of checking a flexible brake hose.
2. List the items to be checked (visual inspection) on a front wheel disc brake unit.
3. List the steps required when a hydraulic master cylinder and power booster are tested.

4 A brake system can be evaluated for stopping distance by applying the brake from a set speed and checking the distance the vehicle required to stop. List the alternative methods available to conduct this test.
5 List the type of faults which are acceptable in a windscreen.
6 What specific headlamp check should be done on a vehicle which has been converted from LH drive to RH drive?
7 Describe the ball joint checking procedure for a vehicle equipped with a double wishbone suspension with the suspension spring acting on the lower arm.
8 Describe how to test the damper (shock absorber) section of a MacPherson type front suspension system.
9 What is a tyre placard and what information does it provide?
10 List the roadworthy checks applicable to an automatic transmission.

10

ENGINES

INTRODUCTION

The engine of a motor vehicle has captured the interest of motor mechanics more than any other automotive component. Engine designers share this fascination, as engine development has proceeded almost non-stop since the first engines wheezed and clanked into motion. Engines have been designed with one cylinder to as many as thirty-six cylinders. They have also been manufactured in many shapes and sizes. Different types of engines have run on petrol, kerosene, methane, propane, hydrogen or safflower oil. Many of the shapes, designs and fuels have proven impractical for everyday use. The purpose of this chapter is to illustrate the types of engines that became accepted and have found widespread commercial application.

Typical OHV petrol engine

COMPONENT IDENTIFICATION

Modern automotive engine components are manufactured and assembled to exacting specifications. The correct identification of these components is essential for:

- understanding the operation of the engine;
- ordering replacement parts.

ENGINE CONFIGURATIONS

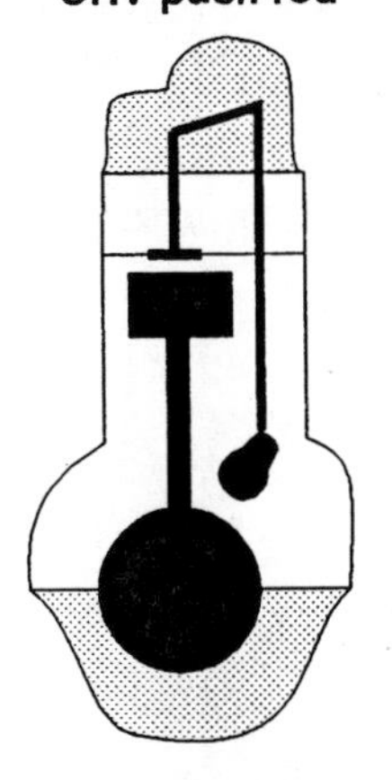

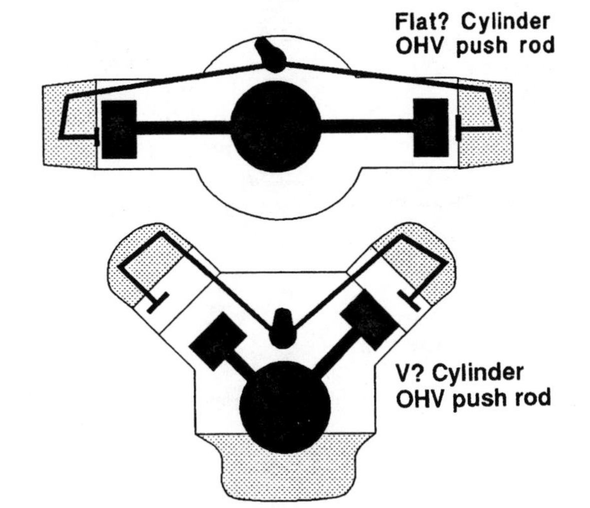

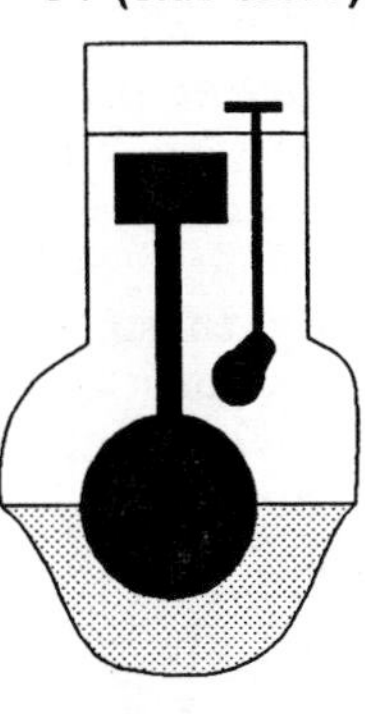

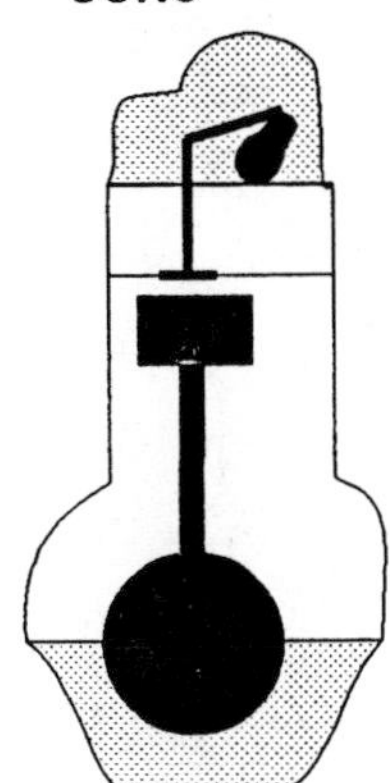

Engine configurations are usually described by the:

1 number of cylinders;
2 arrangement of the cylinders, for example, 'Vee';
3 number of camshafts;
4 position of the camshaft (optional).

These sketches show in-line,
Vee and horizontal (flat) engines. The number of cylinders are not shown but the most popular types are 4 and 6 cylinders for in-line, 6 and 8 cylinders for Vee and 4 cylinders for flat engines. The engine valve action particularly in OHC engines can be direct acting or through rocker arms although this is often not specified in a general description.

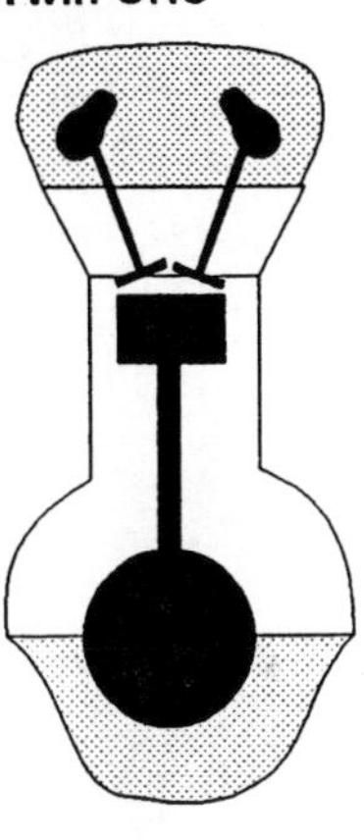

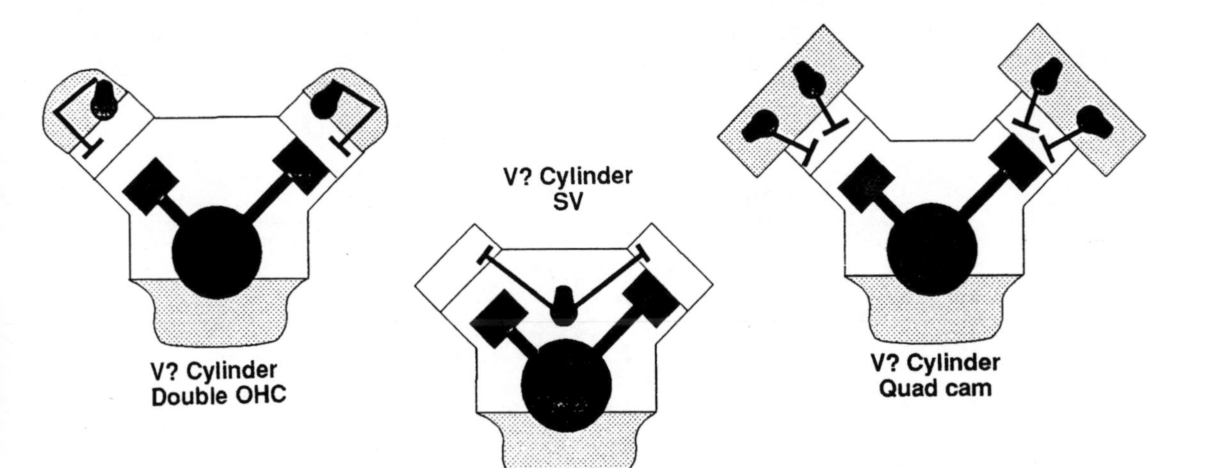

THE FUNCTION OF COMPONENTS

Four-stroke spark ignition engine

The **cylinder block** is usually manufactured from iron or aluminium castings and forms the main body of the engine. A number of circular holes (cylinder bores) are machined vertically through the block and the lower block section (crankcase) is reinforced to provide a rigid structure. All other components are housed in or mounted on this unit.

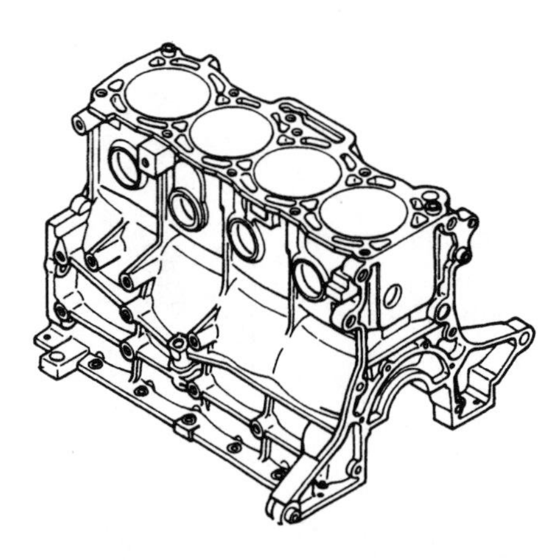

The **crankshaft** can be forged or cast from quality steel or iron. It must be robust to withstand severe shock loadings and, despite its ungainly appearance, it is accurately balanced. The crankshaft converts the reciprocating motion of the piston to rotary motion. It is fitted into the lower section of the block and rotates on a series of oil lubricated plain bearings.

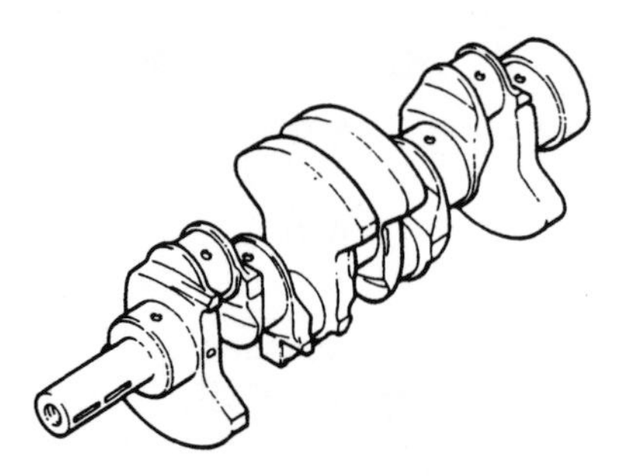

Source: Mitsubishi Engine and Transmission workshop manual

The **pistons**, which are cast from aluminium alloy, form a movable section in their respective cylinder bores. They are forced up and down when the engine is operating. Since it is not practical to manufacture a piston that would completely seal the cylinder bore, rings are fitted to grooves cut into the piston. These rings perform the sealing function of the piston in the cylinder bore. Each piston is attached by a pin to a connecting rod.

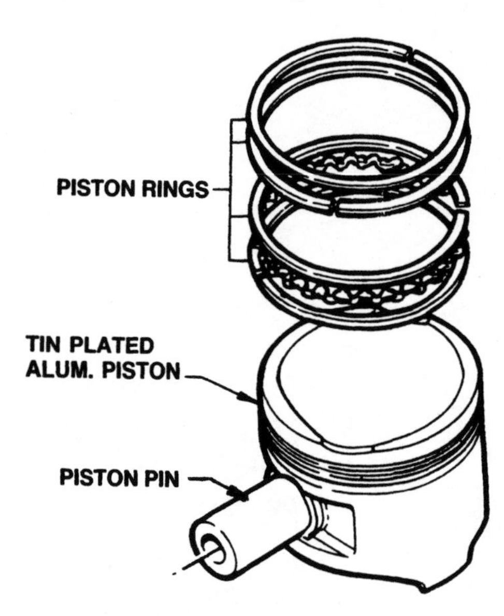

The **connecting rods** are cast or forged from a wide range of metals. The function of a connecting rod is to provide a link between the piston pin and the crankshaft.

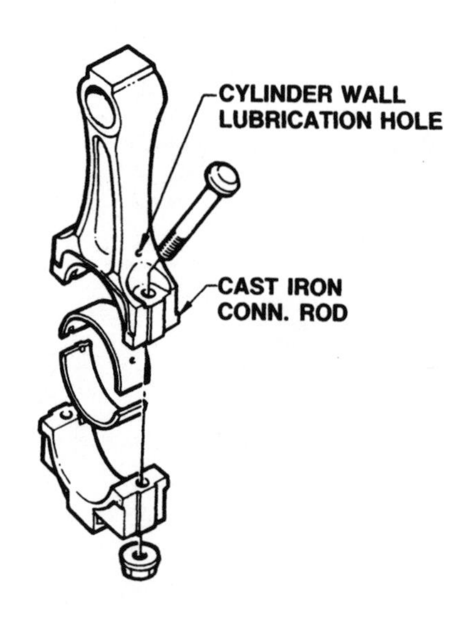

The **cylinder head** is usually an aluminium alloy casting bolted to the top of the cylinder block. Early designs were made of cast iron. In modern designs, the camshaft is placed in the upper sections of the cylinder head (O.H.C), however, other camshaft locations are possible. Recesses (combustion chambers) are cast into the cylinder head in an area directly above each cylinder bore in the block. Each recess contains a threaded hole for the spark plug, an inlet and an outlet hole (ports) for the engine valves.

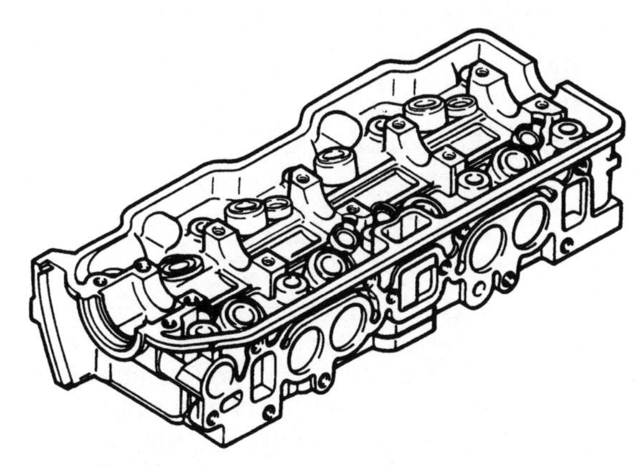

Source: Mitsubishi Engine and Transmission workshop manual

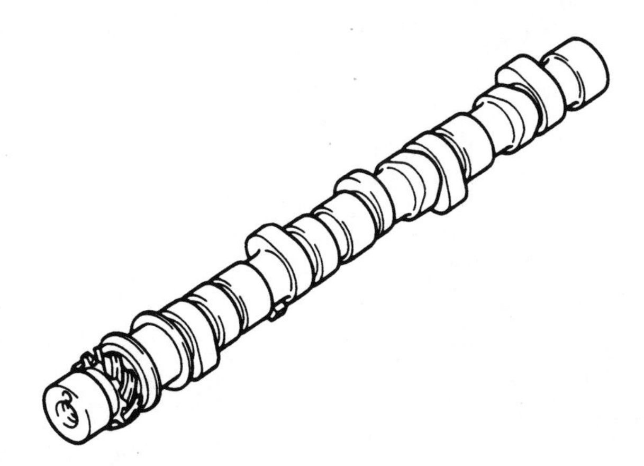

The **camshaft** is usually manufactured from cast iron or cast steel and hardened to provide better wearing properties. It has a series of lobes (bumps) along its length. Each lobe opens and closes a valve at the correct time. The camshaft is driven by the crankshaft through gears, chains or belts and rotates at half the speed of the crankshaft.

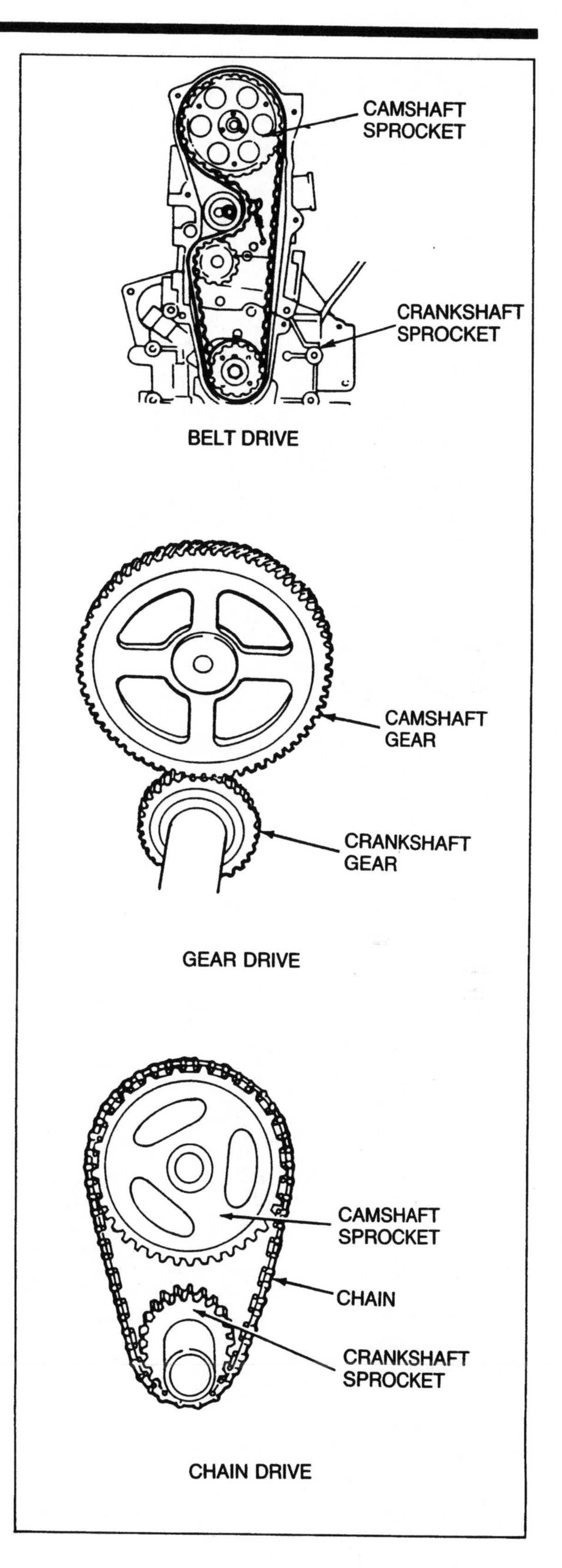

Engine valves are manufactured from high quality alloy steel. Their function is to open and close ports through which the gases enter and leave the cylinders. The valves are supported by valve guides in the cylinder head. Each valve is opened by the camshaft and rocker gear and closed by a valve spring.

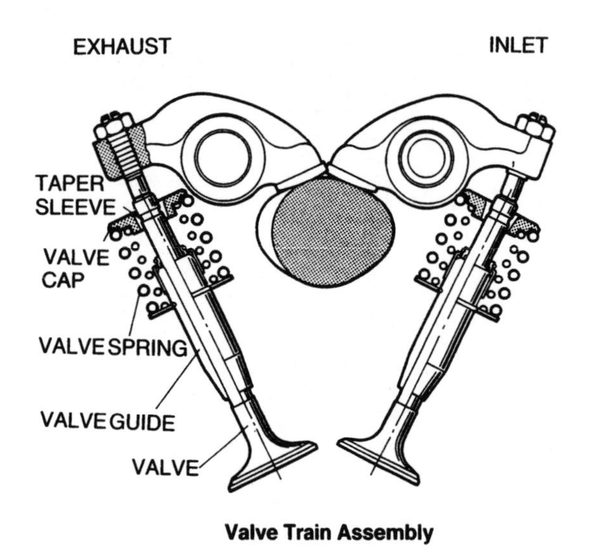

Valve Train Assembly

Engine bearings. An engine contains many sliding or rotating parts. Each of the surfaces at the rotation or contact points must be hardened or provided with a suitable bearing to reduce wear. In all cases, a film of oil on each surface reduces friction and provides long engine life.

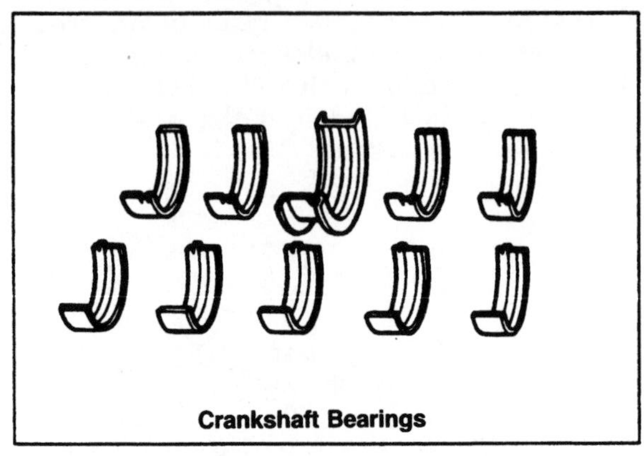

Crankshaft Bearings

The **manifolds**, which are manufactured from cast iron or stainless steel (exhaust) and aluminium alloy (inlet), are bolted onto the cylinder head. The function of the manifolds is to guide the gases into or out of the cylinders. Attached to the inlet manifold is a carburettor or an injection system which provides the fuel to run the engine. The exhaust manifold connects the exhaust (outlet) ports to the exhaust system.

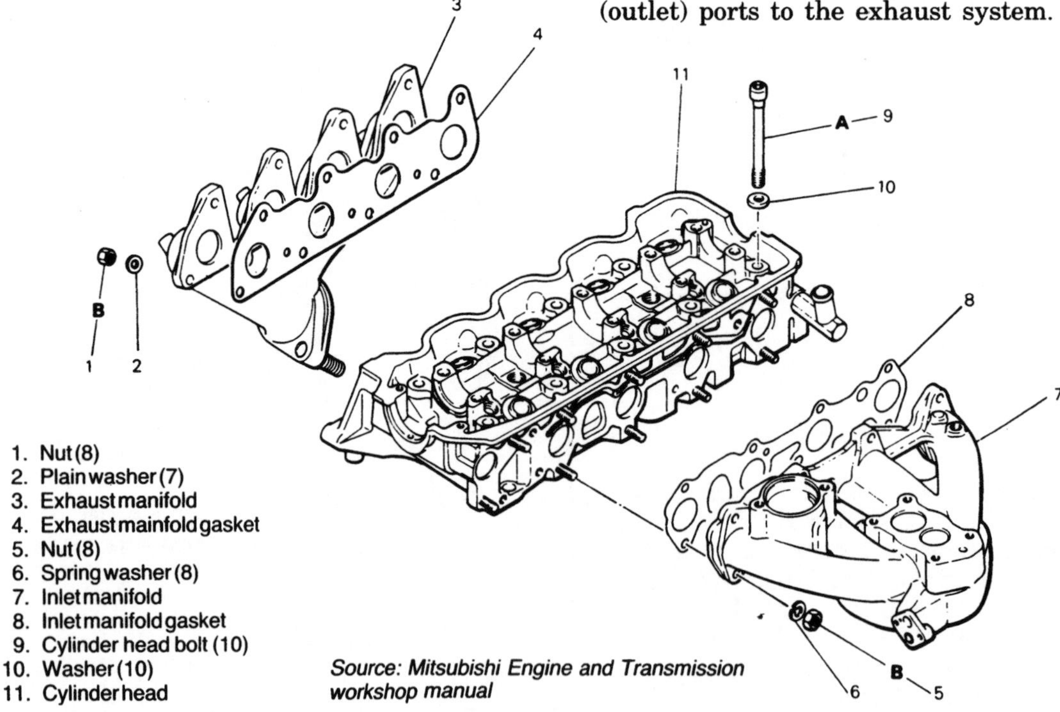

Source: Mitsubishi Engine and Transmission workshop manual

The **flywheel**, which is generally made from a disc-shaped iron casting, is bolted to the rear of the crankshaft to smooth out engine vibrations and to stored energy. It has a clutch assembly attached to the rear face and a toothed ring fitted to its rim. The toothed rim (flywheel ring gear) allows the starter motor to crank the engine.

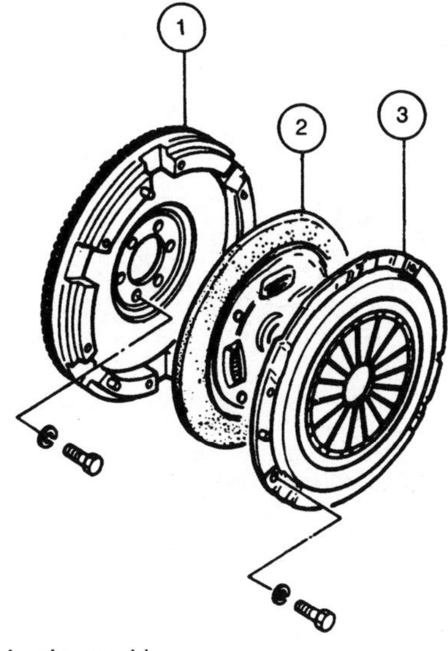

1. Flywheel assembly
2. Clutch
3. Clutch cover assembly

Engine covers are generally made from pressed steel or aluminium alloy castings. They contain the oil within the engine and protect vital working parts from dust while allowing access for maintenance. The types of removable covers fitted to an engine are the:

- tappet cover
- timing cover
- sump, which acts as an engine oil reservoir.

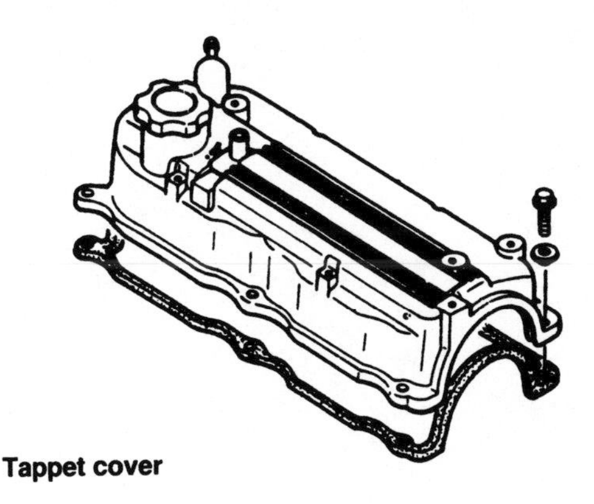

Tappet cover

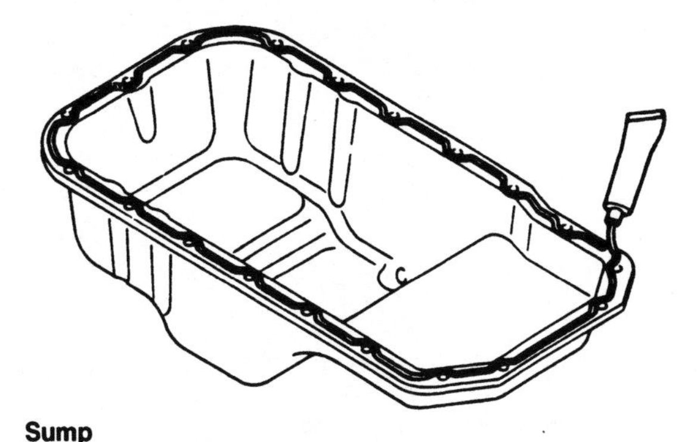

Sump

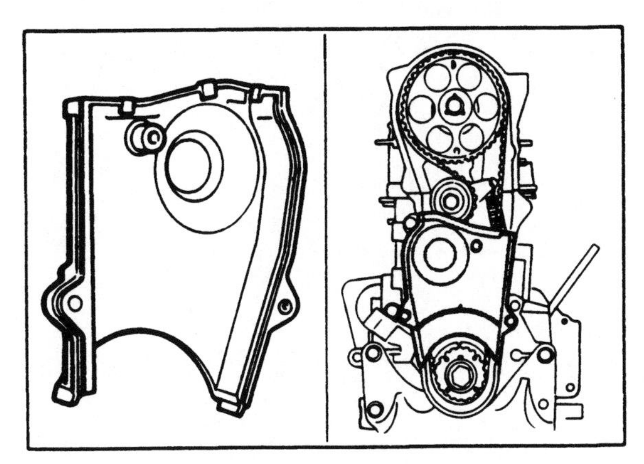

Lower Timing Belt Cover

Ancillary systems are support systems required by the engine to ensure its efficient operation. Each of these systems will be examined separately in your studies, but as they are attached to the engine, their major components must be easily identified. These systems are:

- **Cooling**
- **Electrical**
- **Fuel**
- **Lubrication**

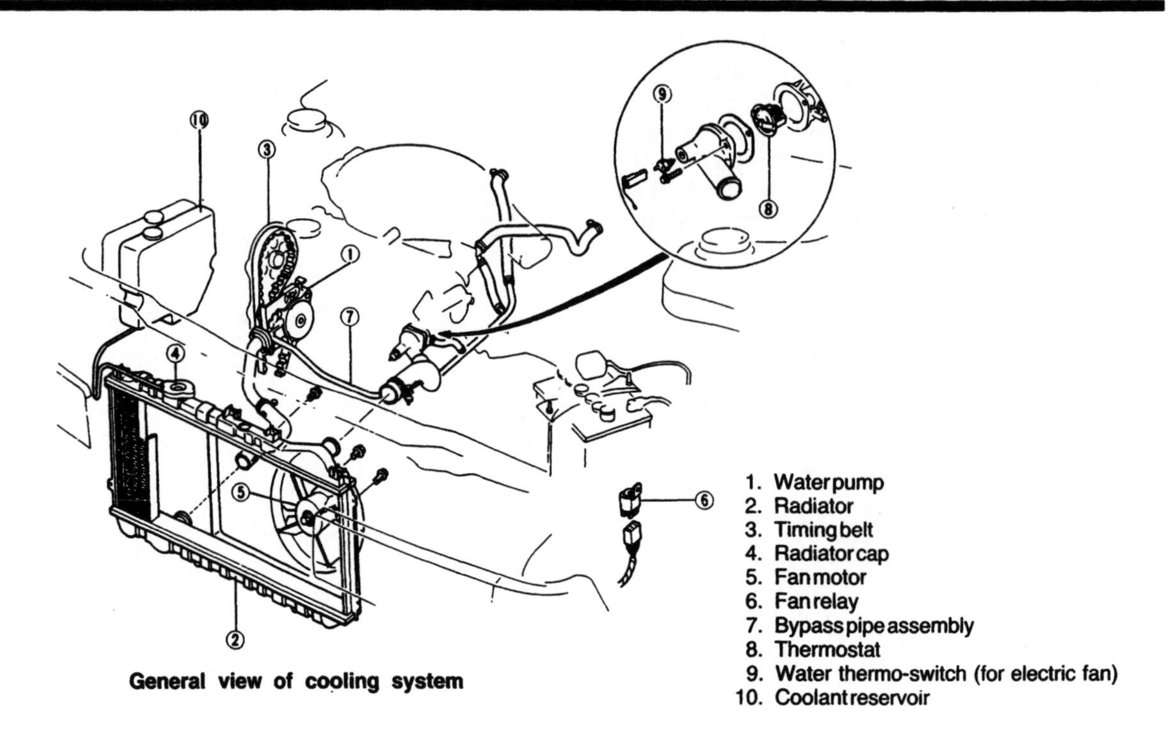

General view of cooling system

1. Water pump
2. Radiator
3. Timing belt
4. Radiator cap
5. Fan motor
6. Fan relay
7. Bypass pipe assembly
8. Thermostat
9. Water thermo-switch (for electric fan)
10. Coolant reservoir

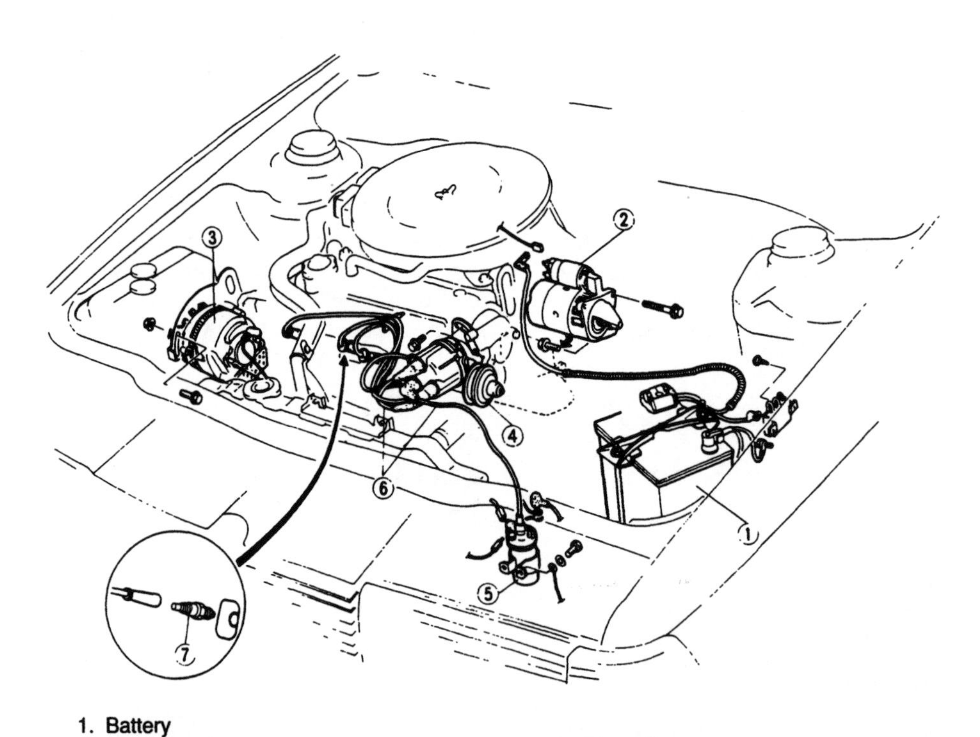

1. Battery
2. Starting motor
3. Alternator
4. Distributor
5. Ignition coil
6. Spark plug cables
7. Spark plug

Engine Electrical

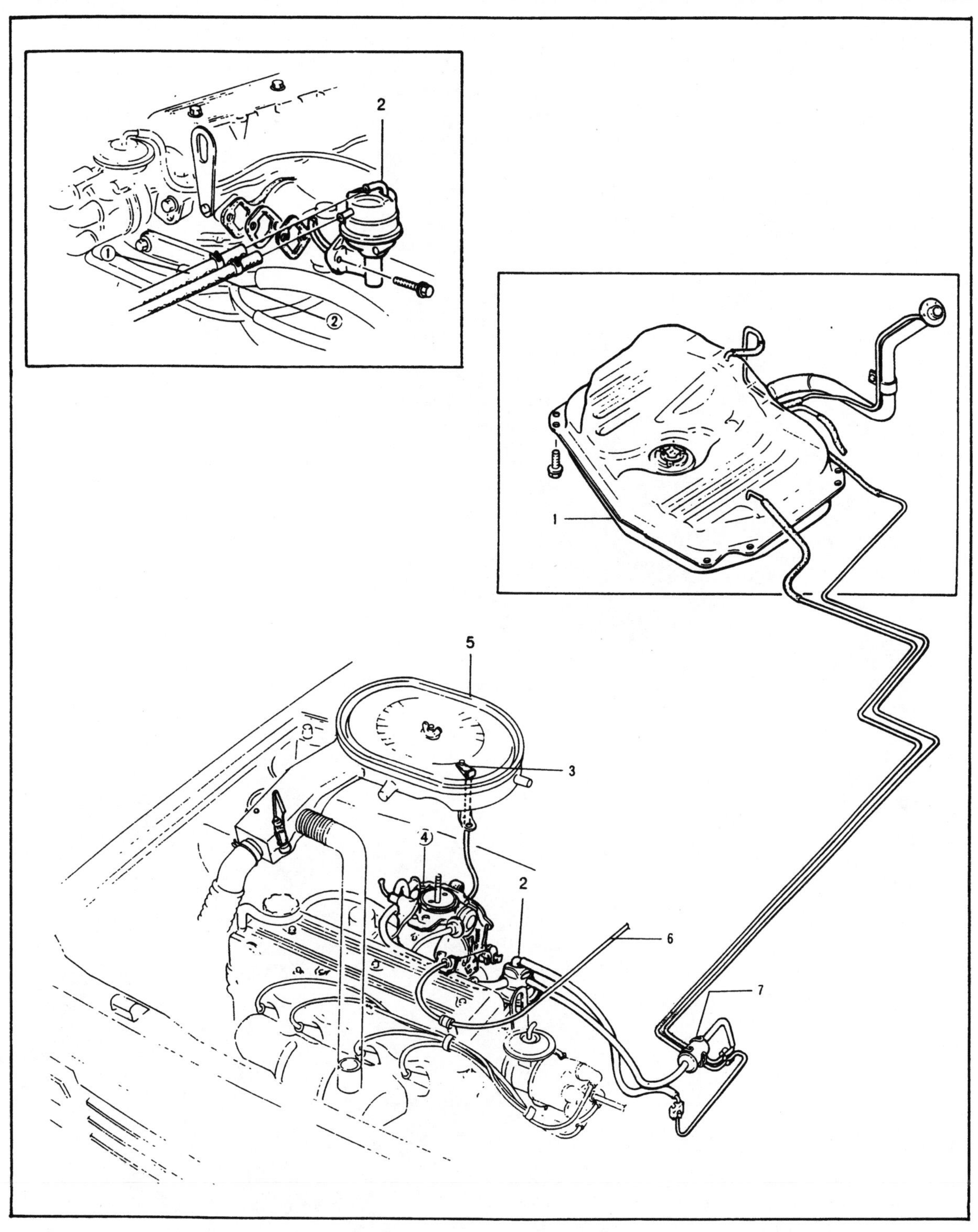

1. Fueltank
2. Fuel pump and lines
3. Idle compensator (if equipped)
4. Carburettor
5. Aircleaner
6. Accel. linkage
7. Fuelfilter

FUEL SYSTEM — HOSE CONNECTIONS AND SCHEMATIC LAYOUT

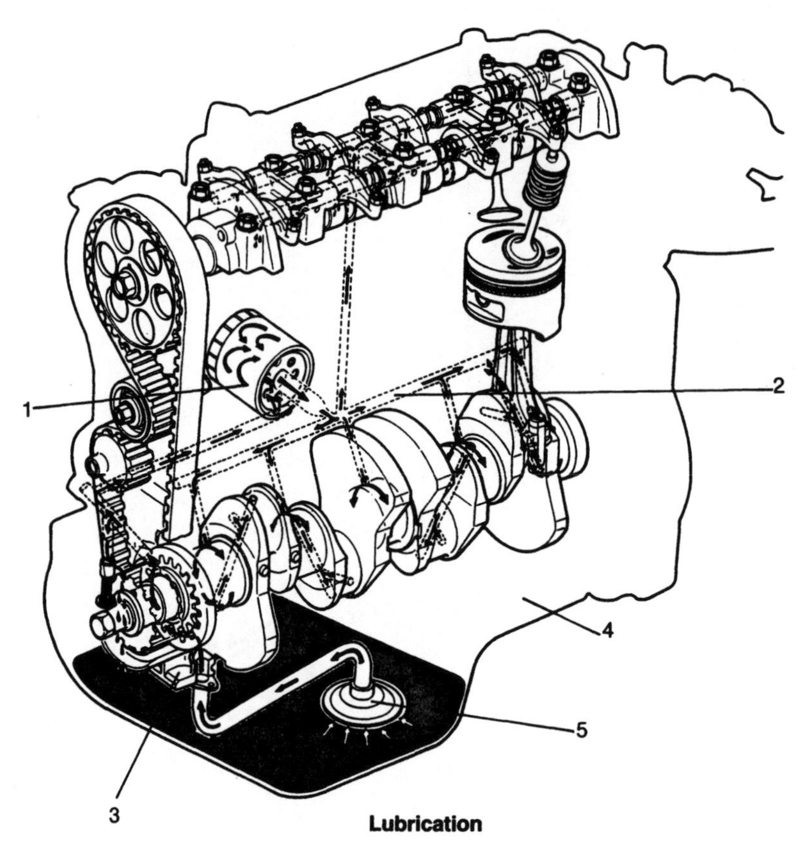

Lubrication

1. Filter
2. Oil gallery
3. Oil pump
4. Sump
5. Strainer and pick-up pipe

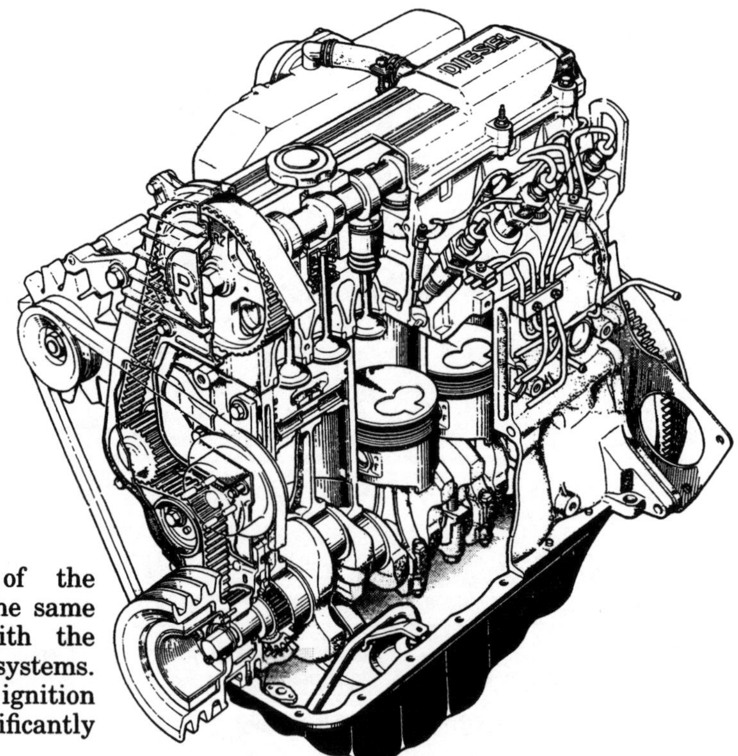

Four-stroke diesel (C.I.) engine

The function and relationship of the components of a diesel engine are the same as the spark ignition engine, with the exception of some of the ancillary systems. These engines do not require a spark ignition system and the fuel system is significantly different.

Two-stroke spark ignition engine

The function and relationship of the cylinder, piston, connecting rod, crankshaft and the cylinder head are similar to the four-stroke spark ignition engine. The piston may have a raised section formed on its crown.

The **reed valve** is a flat metal, flexible plate which acts as a one-way valve between the carburettor and the crankcase.

The **ports**, which are a series of holes in the cylinder bore, perform the same function as the valves fitted to the four-stroke engine. Since the valves have been deleted from this engine then the valve operating mechanisms and the camshaft are not required.

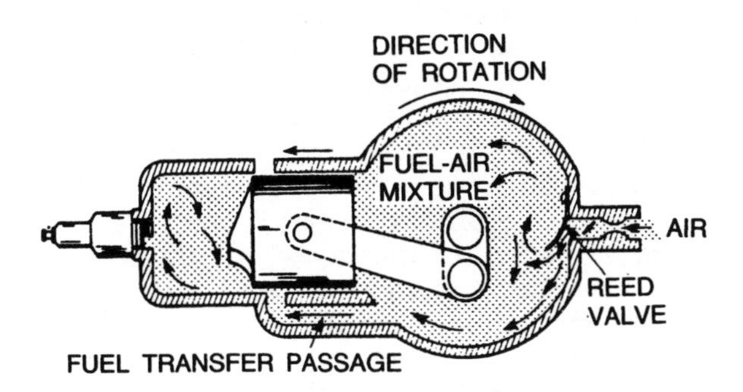

Source: Thiessen/Dales

Two-stroke diesel (C.I.) engine

Although the engine is only suitable for large vehicles, it includes many features of both the two and four-stroke spark ignition engines. The exhaust valves remain in the cylinder head and the inlet ports are located in the cylinder bore. A unique feature of this engine is the large, engine driven blower fitted to the air inlet system.

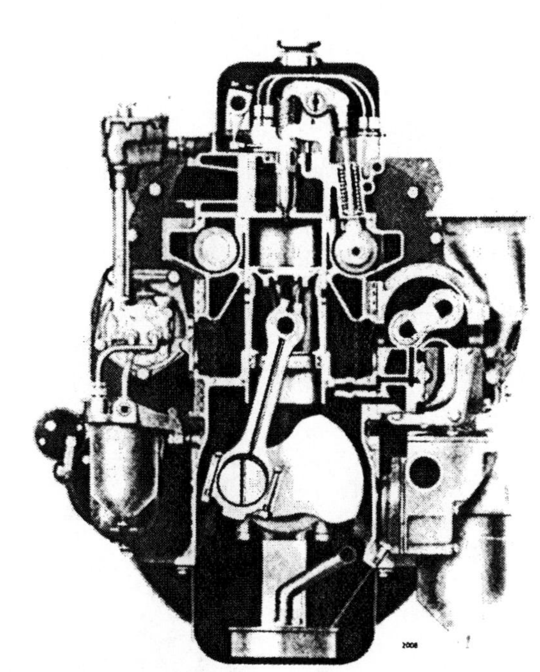

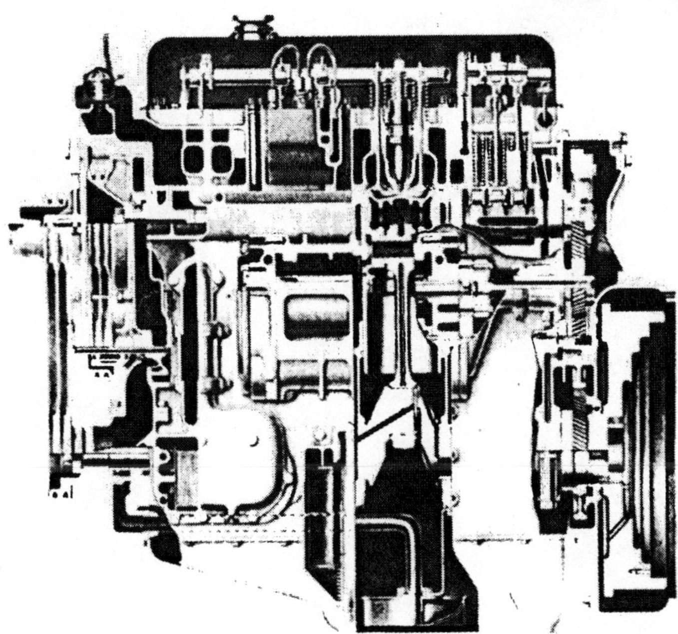

Two-stroke diesel (C.I.) engine

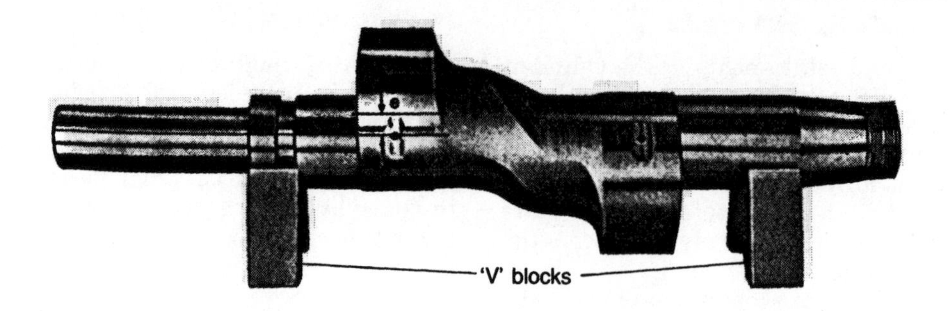

Eccentric shaft (supported on 'V' blocks)

Rotary engine

Basically there are three moving parts in the rotary engine; the eccentric shaft and two rotor assemblies.

The **eccentric shaft**, of one-piece steel construction, has two eccentric rotor journals and two main bearing journals. The function of the eccentric shaft is to transfer the movement of the rotors to the flywheel.

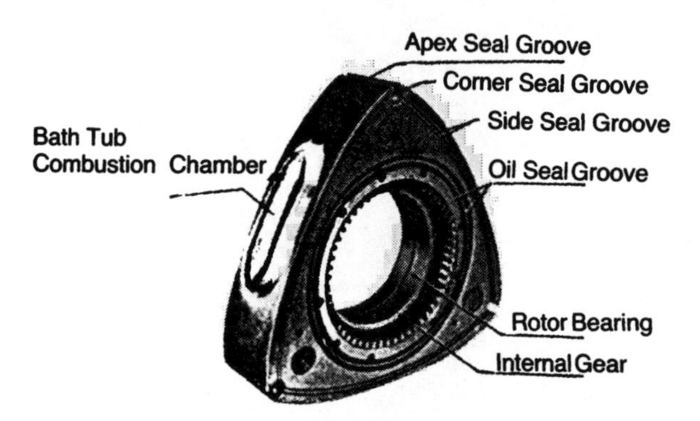

Rotor

The **rotors**, which are triangular shaped, are made from a special cast iron. They have three faces in which combustion chambers are recessed. These rotors have apex seals at each corner and a set of side seals to ensure positive sealing of the chambers. The function of the rotors is to change the space within the housings to create the working areas required for the operating cycles.

The **rotor housings** are aluminium alloy castings containing spark plug holes, exhaust ports and passages for the cooling system. Their inner surfaces, which are plated with a hard-wearing material for long engine life, provide the correct working shapes and volumes for the rotors as they turn inside their respective housings.

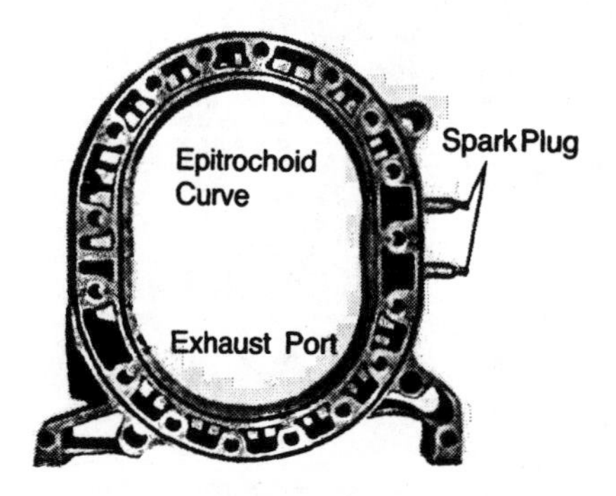

Rotor Housing

Intermediate housing

Rear housing

The three housings which completely seal the rotor housings and support the eccentric shaft are the:

- front housing;
- intermediate housing; and
- rear housing.

Each housing is designed with inlet ports and the front and rear housings are fitted with small timing gears which mesh with the internal gear in the rotors.

The **ancillary systems** are very similar to those of other engines.

ENGINE OPERATION

It has been common practice for many years to describe the operation of the engine in terms of strokes, that is, induction, compression, power and exhaust. This was a convenient arrangement when describing the operation of a four-stroke engine — one phase to one stroke. However, the explanation starts to falter when it is transferred to the real four-stroke engine or engines using only two strokes to complete the operating cycles. There are five phases in the operating process and, regardless of the type of engine, these phases must be completed in a set sequence for each cycle.

The sequence of the phases is:

- induction;
- compression;
- ignition;
- power;
- exhaust.

Note: The method of achieving the ignition phase is an important design feature. The air/fuel mixture may be ignited with an electric spark or by the temperature of the compressed air. These are known as the spark ignition engine and the diesel (compression ignition) engine respectively.

Four-stroke spark ignition engine

Each cylinder of a multi-cylinder, four-stroke, S.I. engine requires the five phases of operation to be completed in two revolutions of the crankshaft.

PHASE 1: INDUCTION

The inlet valve starts to open just before the piston reaches Top Dead Centre (T.D.C). As the piston moves past and away from T.D.C., the volume in the cylinder increases and the pressure decreases below atmospheric pressure. Since the exhaust valve has just closed and the inlet valve has opened, air is forced through the carburettor where it is mixed with fuel. The air/fuel mixture at a ratio of 10–14:1 flows through the inlet manifold and the open inlet valve into the cylinder. The induction phase is complete when the inlet valve closes just after the piston moves away from Bottom Dead Centre (B.D.C.). The crankshaft has rotated more than half a revolution.

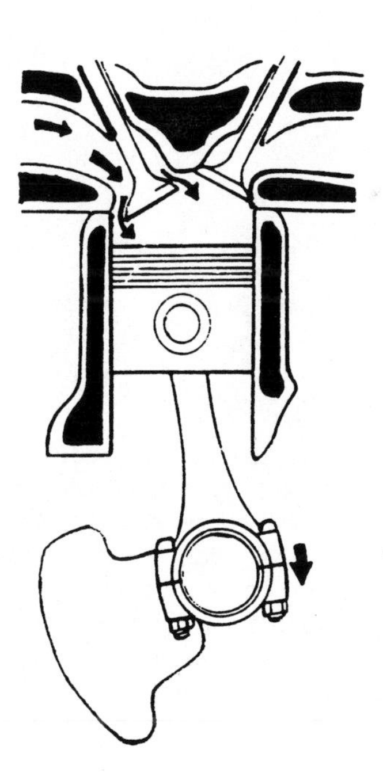

Induction phase

Source: Oklahoma State Dept

PHASE 2: COMPRESSION

As the piston moves away from B.D.C., the volume in the cylinder decreases and the mixture's pressure increases. Since both the inlet and exhaust valves are closed, the mixture is compressed to a pressure eight to ten times greater than atmospheric pressure (800–1000 kPa). The compression phase is complete as the piston reaches T.D.C. and the crankshaft has rotated through an angle less than half a revolution.

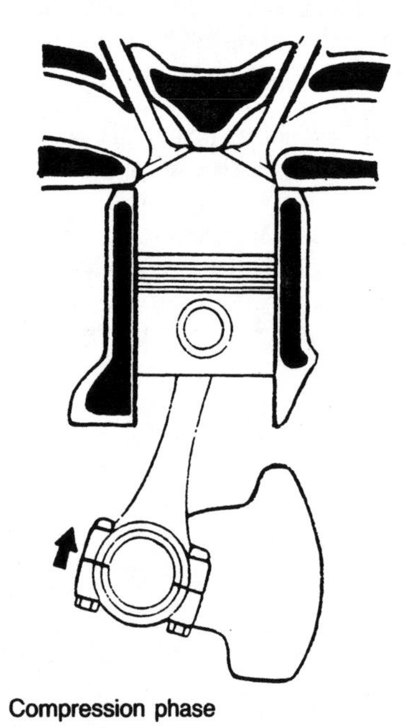

Compression phase

Source: Oklahoma State Dept

PHASE 3: IGNITION

Just before the piston reaches T.D.C., the ignition system produces a very high voltage (8000 V) at the spark plug, which causes a spark in the combustion chamber. The air/fuel mixture is ignited. The heat given off by the burning mixture causes a rapid increase in the temperature and pressure of the remaining mixture.

PHASE 4: POWER

Just after the piston passes T.D.C. the combustion chamber pressure reaches its maximum of two to three times the compression pressure (1600–3000 kPa). This high pressure forces the piston towards the B.D.C. position, causing the crankshaft to rotate. Most of this energy is stored in the flywheel. The power phase is complete when the exhaust valve opens just before the piston reaches B.D.C. The crankshaft has rotated through an angle less than half a revolution.

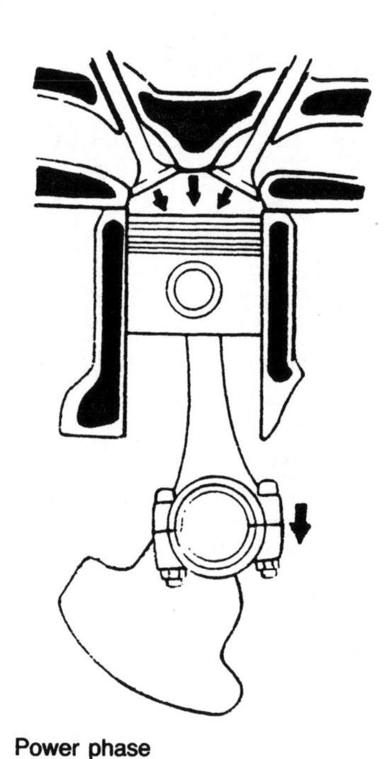

Power phase

Source: Oklahoma State Dept

PHASE 5: EXHAUST

Just before the piston reaches B.D.C., the exhaust valve starts to open. As the piston moves away from B.D.C., the volume in the cylinder decreases and the gas pressure increases to a value slightly higher than atmospheric pressure. This pressure forces the burnt (exhaust) gases through the open exhaust valve, the exhaust manifold and the muffler into the atmosphere. The exhaust phase is complete when the exhaust valve closes just after the piston moves past and away from T.D.C. The crankshaft has rotated through more than half a revolution.

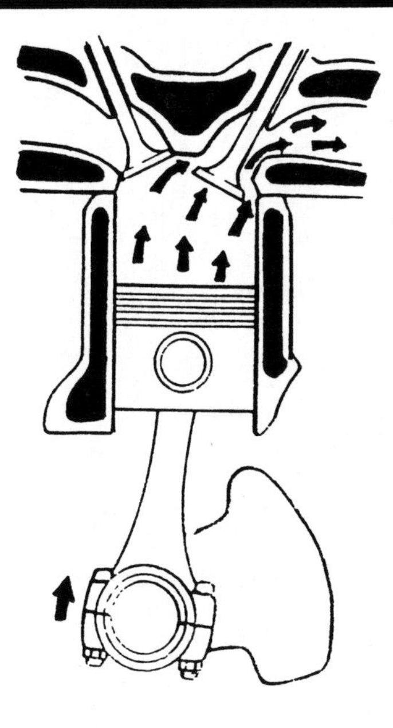

Exhaust phase

Source: Oklahoma State Dept

This cycle (five phases) is completed in a very short time (1/5 second at 600 r.p.m.) and is repeated every two revolutions (720°) of the crankshaft.

The other cylinders of the multi-cylinder, four-stroke engine each produce a power pulse in the same two revolutions of the crankshaft. These power pulses, which are separated by an exact number of crankshaft degrees, occur in a set sequence (firing order). For example, for a six-cylinder, four-stroke engine the power pulses are exactly 120° apart and the firing order may be 1 5 3 6 2 4.

Some of the energy stored in the flywheel is used to rotate the crankshaft through all the phases except the power phase. The remaining flywheel energy is used to propel the vehicle.

The speed of the engine is varied by altering the amount of the air/fuel mixture which enters the cylinders. This is controlled by the throttle butterfly in the carburettor, which is connected to the accelerator pedal.

Four-stroke diesel engine

Each cylinder of a multi-cylinder, four-stroke diesel engine requires the five phases of operation to be completed in two revolutions of the crankshaft.

PHASE 1: INDUCTION

The inlet valve starts to open just before the piston reaches T.D.C. As the piston moves past and away from T.D.C., the volume in the cylinder increases and the pressure decreases below atmospheric pressure. Since the exhaust valve has just closed and the inlet valve has opened, air is forced through the inlet manifold into the combustion chamber. The induction phase is complete when the inlet valve closes just after the piston moves away from B.D.C. The crankshaft has rotated through more than half a revolution.

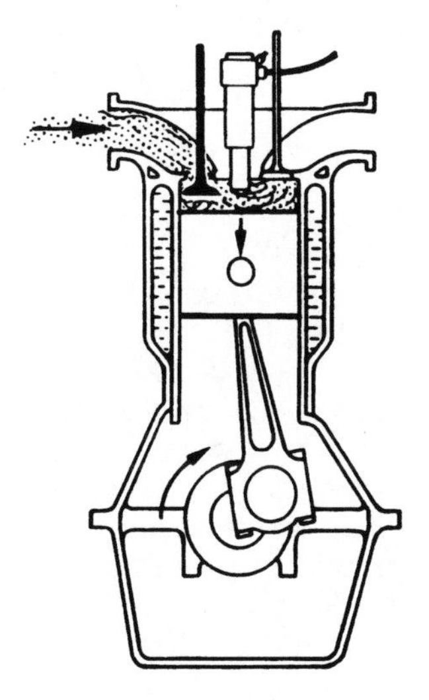

Induction phase

PHASE 2: COMPRESSION

As the piston moves away from B.D.C., the volume in the cylinder decreases and the air pressure increases. Since both the inlet and the exhaust valves are closed, the air is compressed to a pressure sixteen to twenty times greater than atmospheric pressure (1600–2000 kPa). Its temperature is raised to about 550°C. The compression phase is complete as the piston reaches T.D.C. and the crankshaft has rotated through an angle of less than half a revolution.

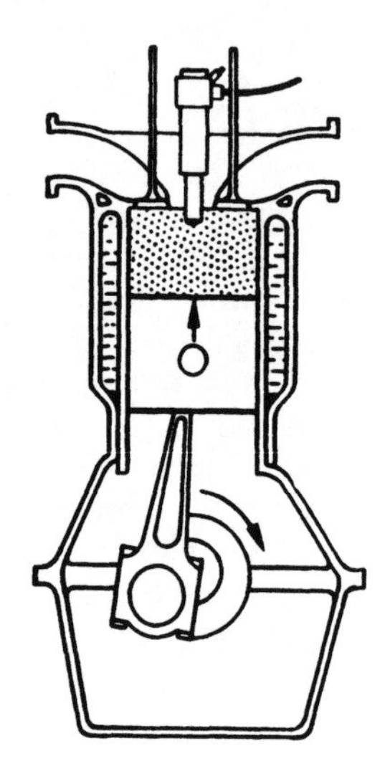

Compression phase

PHASE 3: IGNITION

Just before the piston reaches T.D.C., the fuel is delivered from the fuel system to the injector at 14 000–20 000 kPa. The injector sprays a metered amount of fuel into the combustion chamber where it mixes with, and is ignited by, the hot compressed air. The heat given off by the burning mixture causes a rapid increase in the temperature and pressure of the remaining mixture.

PHASE 4: POWER

Just after the piston passes T.D.C. the combustion chamber pressure reaches its maximum of two to three times the compression pressure (3200–6000 kPa). This high pressure forces the piston towards the B.D.C. position, causing the crankshaft to rotate. Most of this energy is stored in the flywheel. The power phase is complete when the exhaust valve opens just before the piston reaches B.D.C. The crankshaft has rotated through an angle of less than half a revolution.

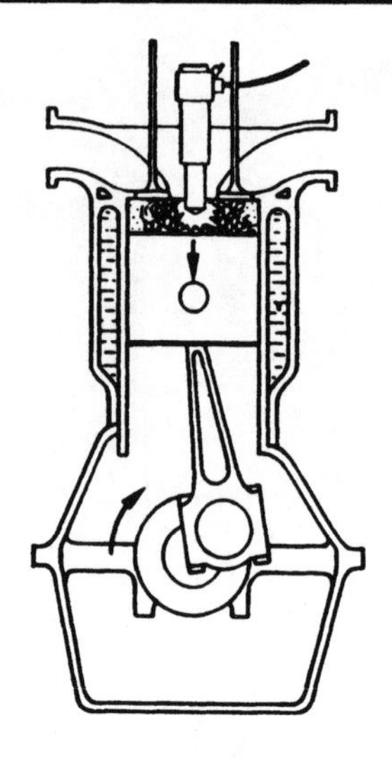

Power phase

PHASE 5: EXHAUST

Just before the piston reaches B.D.C., the exhaust valve starts to open. As the piston moves past and away from B.D.C., the volume in the cylinder decreases and the gas pressure increases to a value slightly higher than atmospheric pressure. This pressure forces the burnt (exhaust) gases through the open exhaust valve, the exhaust manifold and the muffler into the atmosphere. The exhaust phase is complete when the exhaust valve closes just after the piston moves past and away from T.D.C. The crankshaft has rotated through more than half a revolution.

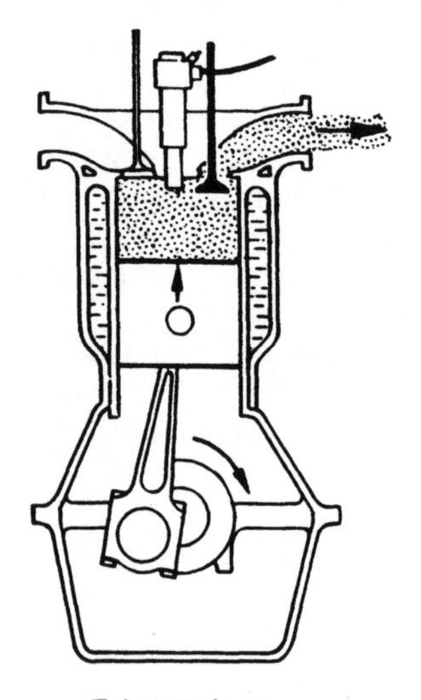

Exhaust phase

This cycle (five phases) is completed in a very short time (1/5 second at 600 r.p.m.) and is repeated every two revolutions (720°) of the crankshaft.

The other cylinders of the multi-cylinder, four-stroke diesel engine each produce a power pulse in the same two revolutions of the crankshaft. These power pulses, which are separated by an exact number of crankshaft degrees, occur in a set sequence (firing order). For example, for a six-cylinder, four-stroke diesel engine the power pulses are exactly 120° apart and the firing order may be 1 5 3 6 2 4.

Some of the energy stored in the flywheel is used to rotate the crankshaft through all the phases except the power phase. The remaining flywheel energy is used to propel the vehicle.

The speed of the engine is varied by altering the amount of fuel injected into the combustion chambers. This is controlled by the pumping elements in the injector pump which are connected to the accelerator pedal.

At the same time, as the piston moves towards T.D.C., the volume in the cylinder decreases and the mixture's pressure increases. Since all the ports are closed, the air/fuel mixture is compressed to a pressure eight to ten times greater than atmospheric pressure (800–1000 kPa). This part of the cycle is complete when the piston reaches T.D.C. and the crankshaft has rotated through half a revolution, or 180°.

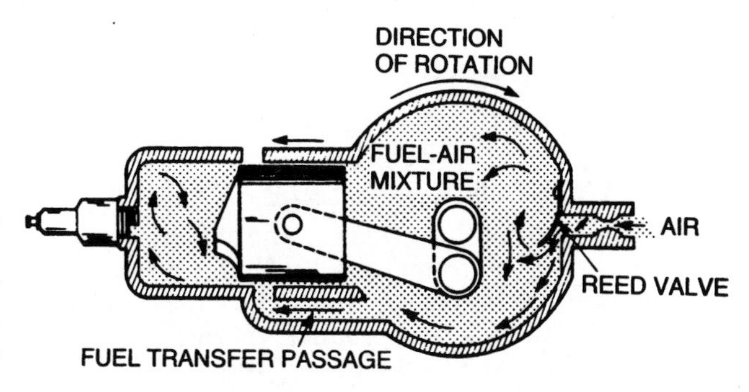

Source: Thiessen/Dales

Two-stroke S.I. engine

Comparatively few engines using a two-stroke cycle have been produced for use in passenger cars. However, this concept found ready acceptance in lightweight applications such as lawnmowers, motor bikes and outboards.

The operation cycle of a two-stroke engine is completed in two piston strokes or one revolution of the crankshaft. For the five phases to be completed in two piston strokes, it it necessary to have more than one phase occurring at the same time.

CRANKCASE INDUCTION AND COMPRESSION PHASE

As the piston moves away from its B.D.C. position, the volume in the crankcase increases and the pressure decreases below atmospheric pressure. This difference in air pressure opens the reed valve.

Air is forced through the carburettor where it is mixed with fuel and oil to obtain a mixture ratio of 10–14:1. This mixture passes through the inlet manifold and the open reed valve into the crankcase.

IGNITION PHASE, POWER PHASE AND CRANKCASE COMPRESSION

Just before the piston reaches T.D.C., the ignition system produces a very high voltage (8000 V) at the spark plug, which causes a spark in the combustion chamber. The air/fuel mixture is ignited. The heat given off by the burning mixture causes a rapid increase in the temperature and pressure of the remaining mixture. Just after the piston passes T.D.C., the combustion chamber pressure reaches its maximum of two to three times the compression pressure (1600–3000 kPa). This high pressure forces the piston to the B.D.C. position, causing the crankshaft to rotate.

Most of this energy is stored in the flywheel. At the same time, as the piston is forced towards B.D.C., the volume in the crankcase decreases and the mixture's pressure increases. This increase in pressure closes the reed valve and the air/fuel mixture in the crankcase is compressed to a pressure greater than atmospheric pressure.

The ignition phase, power phase and crankcase compression are complete when the crown of the piston uncovers the exhaust and transfer ports.

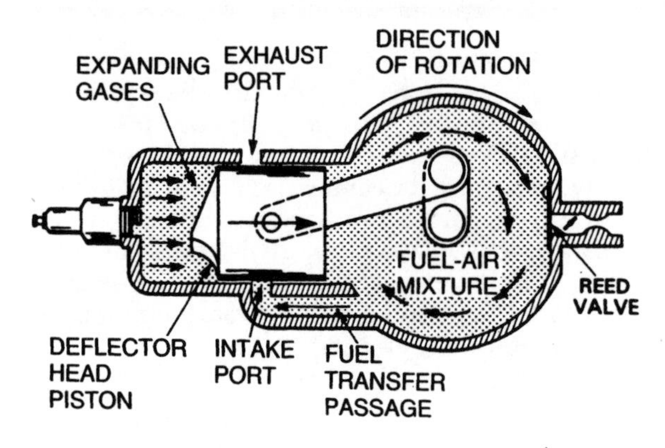

Source: Thiessen/Dales

EXHAUST AND INDUCTION PHASES

As the piston nears B.D.C., the exhaust port is opened. The remaining cylinder pressure from the power phase forces the burnt gases through the port, exhaust manifold and muffler into the atmosphere. Just after the exhaust port is opened, the transfer port opens and the high pressure in the crankcase forces the air/fuel mixture into the cylinder via the transfer passage. The momentum of the air/fuel mixture, as it enters the cylinder, expels the remaining exhaust gas from the cylinder and fills it with a fresh charge. The induction phase and the exhaust phase are complete when the piston has travelled far enough away from B.D.C. to close the transfer and exhaust ports.

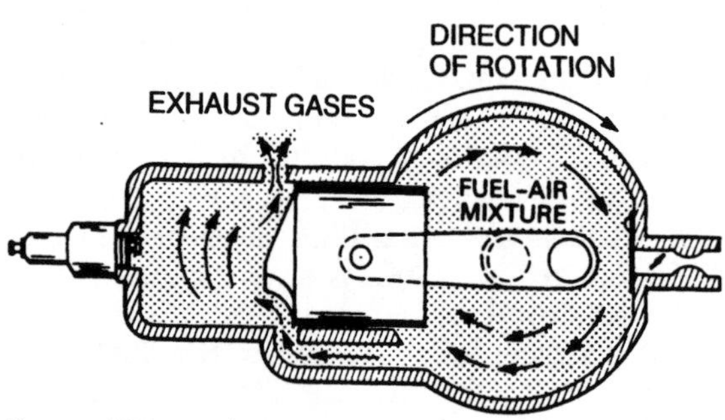

Source: Thiessen/Dales

This cycle (five phases) is completed in a very short time (1/10 second at 600 r.p.m.) and is repeated every revolution (360°) of the crankshaft.

A single-cylinder, two-stroke engine produces one power pulse during each revolution of the crankshaft.

Some of the energy stored in the flywheel is used to rotate the crankshaft during:

- crankcase induction;
- compression phase;
- ignition phase;
- crankcase compression;
- exhaust phase;
- induction phase.

The remaining flywheel energy is used to propel the vehicle.

The speed of the engine is varied by altering the amount of air/fuel mixture entering the cylinder. This is controlled by the throttle butterfly or slide in the carburettor which is connected to the accelerator pedal.

Two-stroke diesel engine

This type of engine uses the two-stroke cycle principle but has an oil lubrication system with the oil contained in its sump. This eliminates the possibility of using the crankcase as a means of storing and transferring the air, as occurs in the two-stroke S.I. engine. The method used to force the air into the cylinder is in the form of an engine driven blower.

The operation cycle of a two-stroke diesel engine must be completed in two piston strokes or one revolution of the crankshaft.

For the five phases to be completed in two piston strokes, it is necessary to have more than one phase occurring at the same time.

INDUCTION AND EXHAUST PHASES

As the piston nears B.D.C., the exhaust valves in the cylinder head are opened by the camshaft and the remaining cylinder pressure starts to force the burnt (exhaust) gases out of the cylinder into the exhaust system. Just after the exhaust valves open, the inlet ports are uncovered by the crown of the piston, enabling the blower to force the air into the cylinder. The fresh air entering the bottom of the cylinder moves towards the open exhaust valves. This action scavenges the cylinder of exhaust gases and fills it with a fresh charge of air.

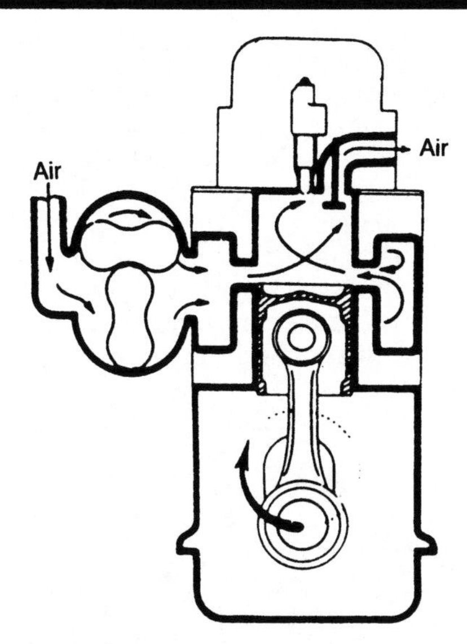

Induction and exhaust phase

COMPRESSION PHASE

Just after the piston moves away from B.D.C., the inlet ports are covered by the crown of the piston and the exhaust valves are closed. Further movement of the piston towards T.D.C. causes a decrease in the cylinder's volume and an increase in the pressure and temperature of the compressed air. The air is compressed to a pressure sixteen to twenty times greater than atmospheric pressure (1600–2000 kPa), causing its temperature to rise to about 550°C. The compression phase is complete when the piston reaches T.D.C.

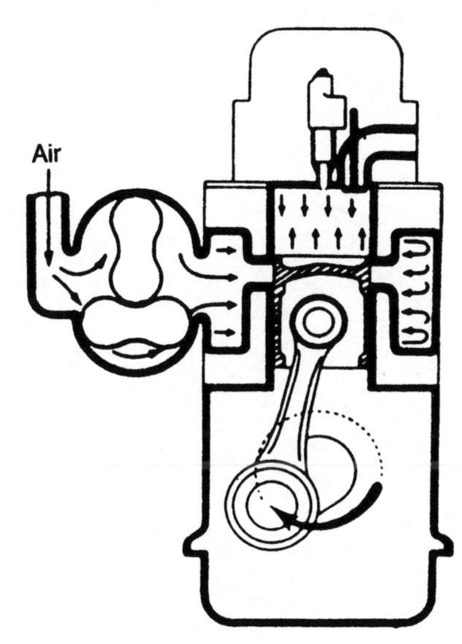

Compression phase

IGNITION PHASE

Just before the piston reaches T.D.C., the fuel pressure in the unit injector is pressurised to an extremely high value (14 000–20 000 kPa). The injector sprays a metered amount of fuel into the combustion chamber, where it mixes with and is ignited by the hot compressed air. The heat given off by the burning mixture causes a rapid increase in the temperature and pressure of the remaining mixture.

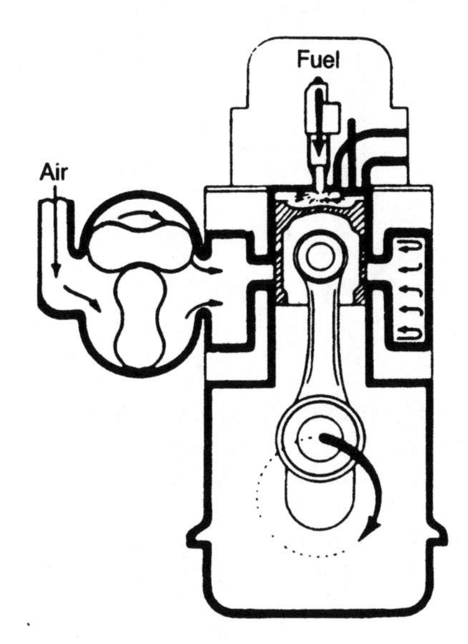

Ignition phase

POWER PHASE

Just after the piston passes T.D.C., the combustion chamber pressure reaches its maximum of two to three times the compression pressure (3200–6000 kPa). This high pressure forces the piston towards the B.D.C. position, causing the crankshaft to rotate. Most of this energy is stored in the flywheel. The power stroke is complete when the exhaust valves open.

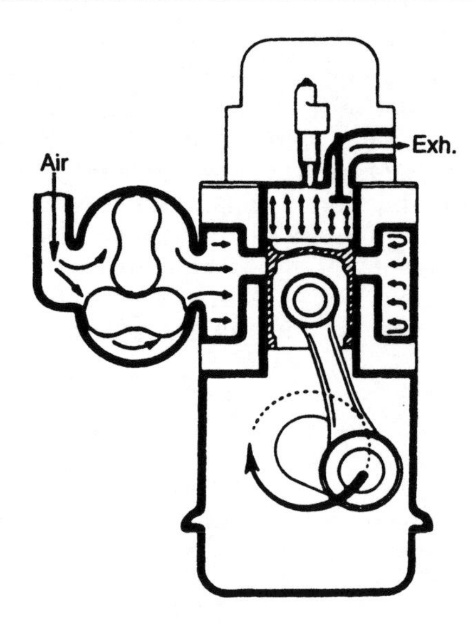

Power phase

This cycle is completed in a very short time (1/10 second at 600 r.p.m.) and is repeated every revolution (360°) of the crankshaft.

A single-cylinder, two-stroke diesel engine produces one power pulse each revolution of the crankshaft. Some of the energy stored in the flywheel is used to rotate the crankshaft during the induction, compression and exhaust phases. The remaining flywheel energy is used to propel the vehicle.

For a multi-cylinder, two-stroke diesel engine with four cylinders, the firing order may be 1 3 4 2. Each of these cylinders would produce a power pulse in the same revolution (360°) of the crankshaft.

The speed of the engine is varied by altering the amount of the fuel injected into the combustion chamber.

Rotary engine

The operation of the rotary engine requires the five phases to be completed when one of the rotors has travelled through one revolution (360°) and the eccentric shaft has turned through three revolutions (1080°).

The following description is for one chamber of a rotor. It must be noted that one of the other two chambers of the same rotor is 120° (rotor) ahead and the remaining chamber is 120° (rotor) behind the one being described.

PHASE 1: INDUCTION

After the turning rotor has uncovered the inlet port, further rotor movement causes the volume in the chamber to increase and the chamber pressure to decrease below atmospheric pressure. This reduced pressure cause the air/fuel mixture to be forced through the open inlet port into the chamber. The induction phase is complete when the turning rotor closes the inlet port.

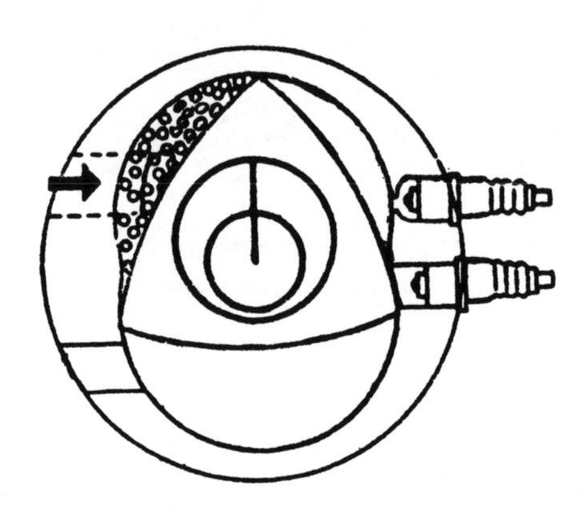

Induction phase

PHASE 2: COMPRESSION

As the turning rotor moves past the inlet port, the volume in the chamber decreases and the mixture's pressure increases. The mixture is compressed eight to ten times greater than atmospheric pressure (800–1000 kPa). The compression phase is complete when the chamber volume is at its minimum and the rotor chamber is facing the spark plugs.

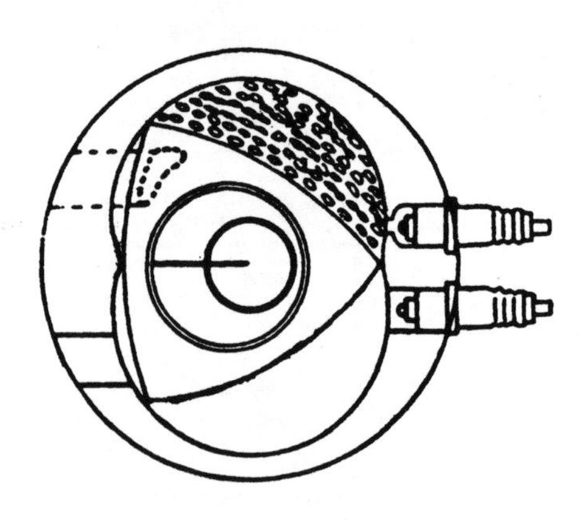

Compression phase

PHASE 3: IGNITION

Just after the rotor reaches a point where the chamber is at a minimum volume, the ignition system produces a very high voltage (8000 V) at the spark plugs, which causes sparks in the combustion chamber. The air/fuel mixture is ignited.

Note: Two spark plugs are used to ensure that all the air/fuel mixture is ignited.

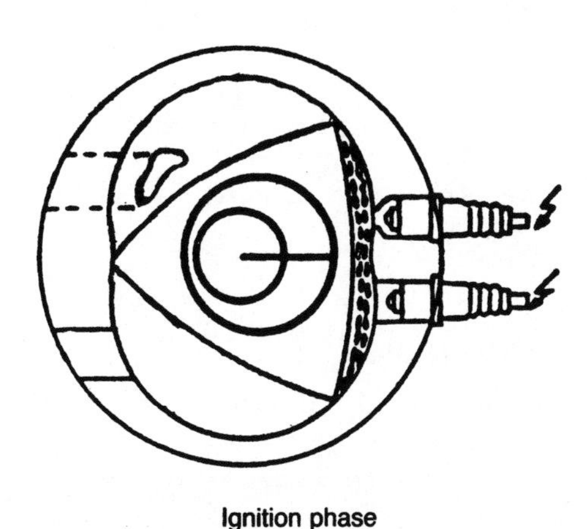

Ignition phase

PHASE 4: POWER

The heat given off by the burning mixture causes a rapid increase in the temperature and pressure of the remaining mixture. This high pressure forces the rotor to turn in the direction of rotation, causing the eccentric shaft to rotate. Most of this energy is stored in the flywheel. The power phase is complete when the leading apex seal opens the exhaust port.

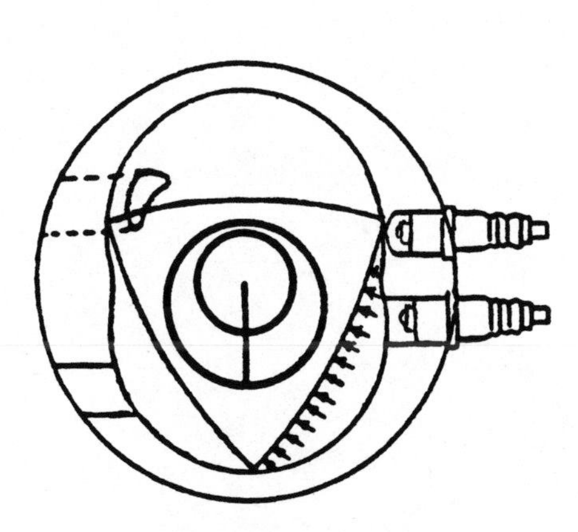

Power phase

PHASE 5: EXHAUST

After the apex seal opens the exhaust port, the volume in the chamber decreases and the chamber pressure increases to a value slightly higher than atmospheric pressure. This pressure forces the burnt (exhaust) gases through the open exhaust port, exhaust manifold and muffler into the atmosphere. The exhaust phase is complete when the trailing apex seal closes the exhaust port.

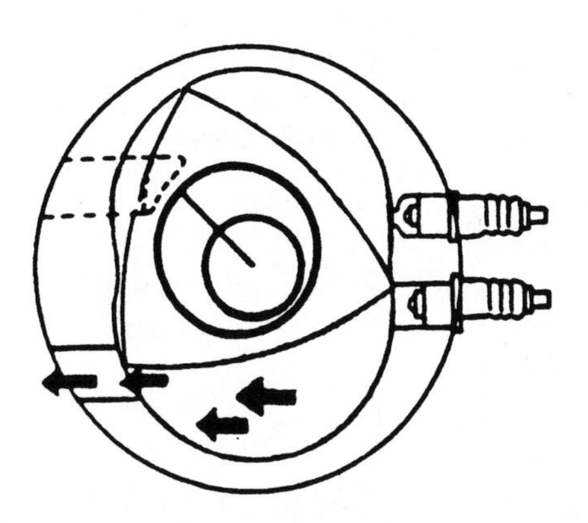

Exhaust phase

This cycle is repeated each revolution of the rotor or every three revolutions of the eccentric shaft. At 600 r.p.m., the cycle takes 3/10th of a second.

For a two-rotor engine, with three chambers on each rotor, there are two power phases every revolution of the eccentric shaft.

Some of the energy in the flywheel is used to rotate the eccentric shaft during the induction, compression and exhaust phases of each chamber. The remaining flywheel energy is used to propel the vehicle.

The speed of the engine is varied by altering the amount of air/fuel mixture entering each chamber. This is controlled by the throttle butterfly in the carburettor, which is connected to the accelerator pedal.

PISTON AND CONNECTING ROD: REMOVAL AND INSTALLATION

This exercise will introduce you to the physical relationship of engine components and provide an insight into the tools and techniques required to dismantle and assemble an engine.

To gain access to the piston and the connecting rod, both the sump and the cylinder head will have to be removed.

Caution: read the engine manufacturer's specifications and procedures.

Sump removal

1 Ensure the engine has been drained of oil.
2 Turn the engine in the stand until the sump is uppermost.
3 Remove the bolts holding the sump to the cylinder block.
- Using the correct sized socket on a speed brace, initially loosen all the bolts one turn. This will prevent sump distortion.
- Unscrew each bolt and place it in a parts container.

4 Remove the sump.
- For a sump that is difficult to dislodge, check all bolts have been removed.
- Strike the side of the sump (close to the rolled bottom edge) with a wooden mallet.

Note: A piece of soft wood placed against the sump and then struck with a hammer can be used instead of the mallet.
- When the sump is loose, lift it directly upwards.

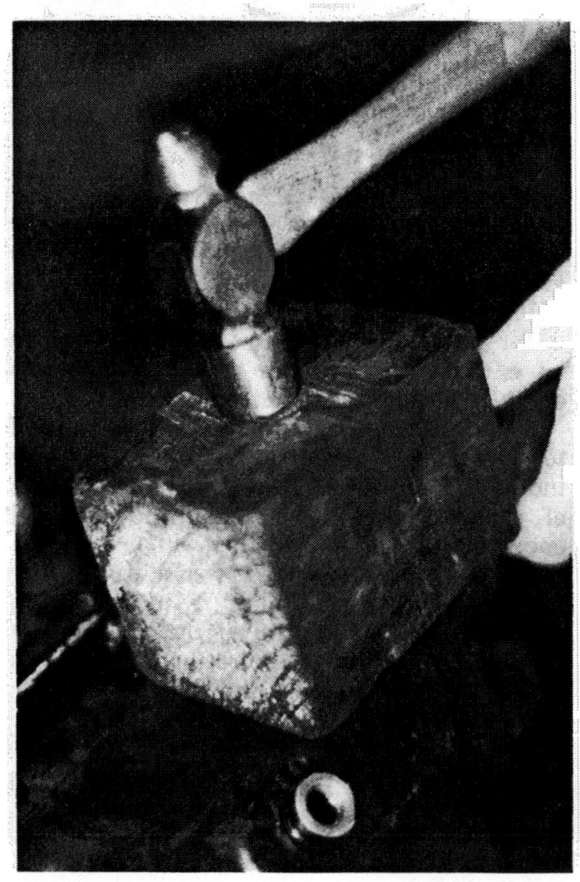
Break hold on sump

Tappet cover removal

1 Turn the engine in the stand until the cylinder head is uppermost.

Remove tappet cover bolts

2 Remove the tappet cover (rocker or valve cover).
- Using the correct sized socket on a speed brace, initially loosen all the bolts or nuts one turn. This will prevent cover distortion.
- Unscrew each bolt or nut and place it in a parts container.

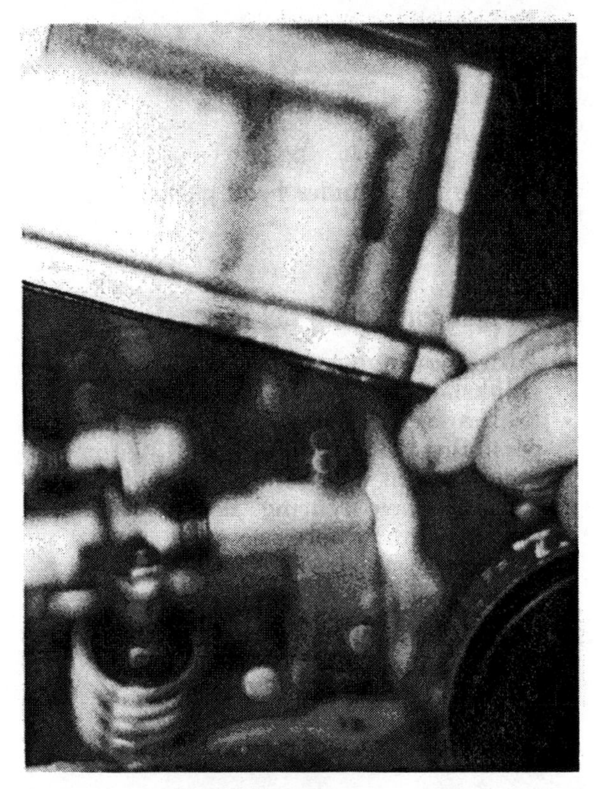

Remove tappet cover

- Lift the cover and gasket away from the cylinder head.

Camshaft drive removal

Caution: Do not rotate the crankshaft until the cylinder head has been removed. On some engines the pistons can strike the valves when the camshaft drive has been disconnected and the crankshaft has been rotated.

TOOTH BELT DRIVE

1 Remove the belt shield.

Remove belt sheild

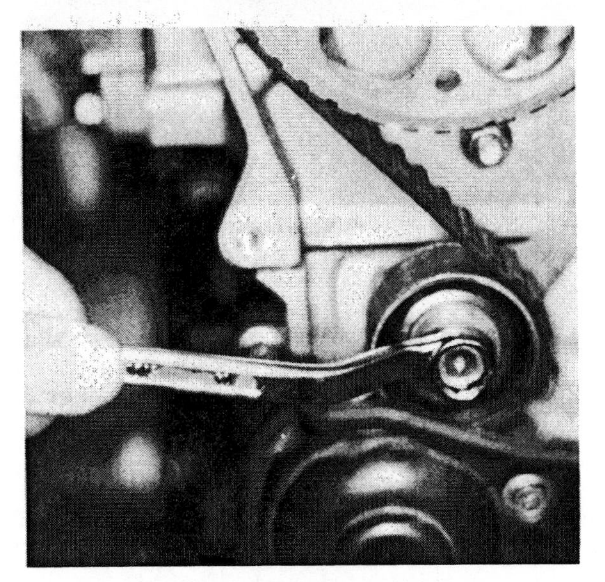

Retract tensioning device

2 Retract the belt tensioning device.
- Locate the device.
- Loosen the belt tensioner clamping bolt or nut.
- Pull or push the tensioner away from the belt.
- Retighten the tensioner clamping bolt or nut.

3 Remove the belt.

Note: Remember, do NOT rotate the engine crankshaft.

CHAIN DRIVE

Chain drives utilise hydraulic or ratchet type tensioning devices to keep the chain taut, so care must be taken with these units to prevent the tensioner extending when the chain is loosened.

Remove belt

1 Refer to the workshop manual for the correct chain securing procedure.
2 Secure the timing chain and/or tensioner.
3 Remove the camshaft drive sprocket retainer (the bolt or nut).
4 Remove the camshaft sprocket.

Note: Remember, do NOT rotate the engine crankshaft.

Cylinder head removal

On some engines it will be necessary to remove the camshaft and/or the rocker mechanisms, prior to loosening the cylinder head bolts.

1 Refer to the workshop manual for the correct procedure.
2 Using the correct sized socket on an extension fitted to a 'breaker' bar, initially loosen the cylinder head retaining bolts or nuts half a turn.
3 Using the correct sized socket on a speed brace, unscrew each bolt or nut and place it in a parts container.

Note: A speed brace is used to reduce the time taken to remove the bolts or nuts.

4 Lift the cylinder head directly away from the cylinder block.

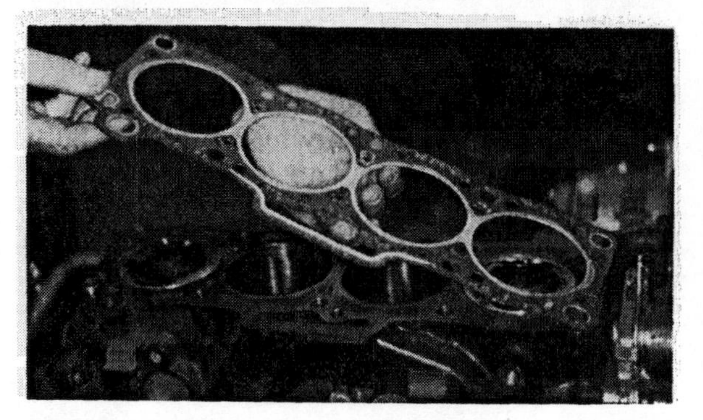

Remove head gasket

5 Remove the cylinder head gasket from the cylinder block.

Piston and connecting rod removal

1 Turn the engine in the stand until the crankshaft is vertical. This will give access to the bearing caps and minimise the possibility of the piston falling on the ground when it is being removed.
2 Remove the securing device from the timing chain (where necessary).

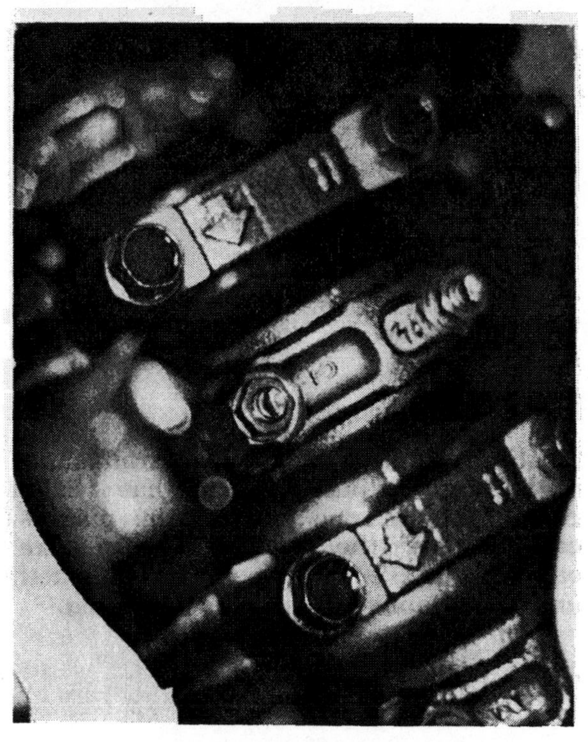

Connecting rod at B.D.C. position

3 Turn the crankshaft until the piston is at B.D.C. in its cylinder.
4 Check the lower section of the connecting rod for cylinder number stampings.

Check number stamping

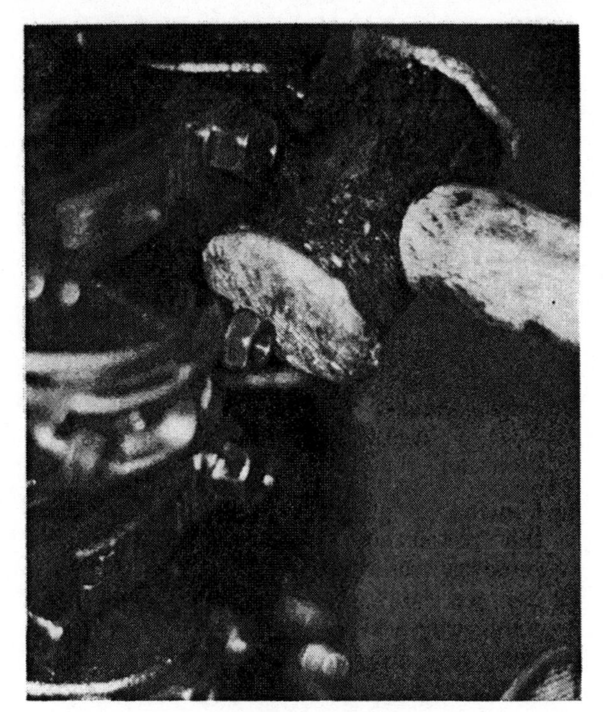

Loosen bearing cap

5 Take particular note to which side of the engine the stampings are facing.
6 Check the top of the piston for an indicating mark or notch which points to the front of the engine.
7 Using the correct sized socket on a breaker bar, loosen the connecting rod bearing cap retaining nuts.
8 Unscrew each retaining nut until it is near the end of its thread.
9 Using a soft faced hammer, gently tap the side of the bearing cap until it is loose.
10 Unscrew the retaining nuts and place them in a parts container.
11 Remove the bearing cap from the connecting rod studs.
12 Using the end of a wooden handle (hammer handle), push on the lower section of the connecting rod so that the piston travels up the bore.
13 Grip the piston as it emerges from its bore.

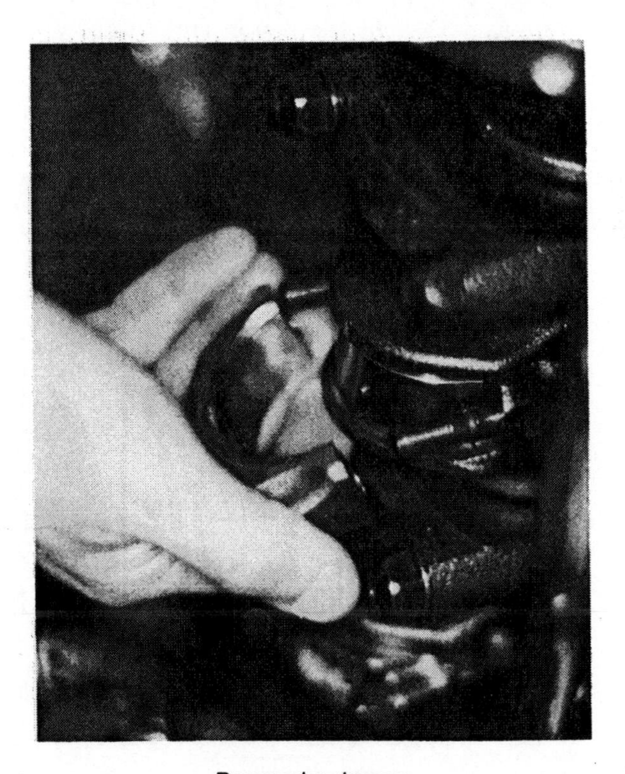

Remove bearing cap

Remove piston from bore

14 Continue to push the connecting rod until the piston and rod assembly can be removed from the bore.
15 Replace the bearing cap onto the connecting rod studs.
16 Screw the nuts onto the studs until they are finger tight.

Piston and connecting rod installation

To fit the piston and connecting rod assembly into the cylinder bore, it is necessary to hold the piston rings tightly in their grooves to prevent breakage as they enter the bore.

1 Select the correct ring compressor.

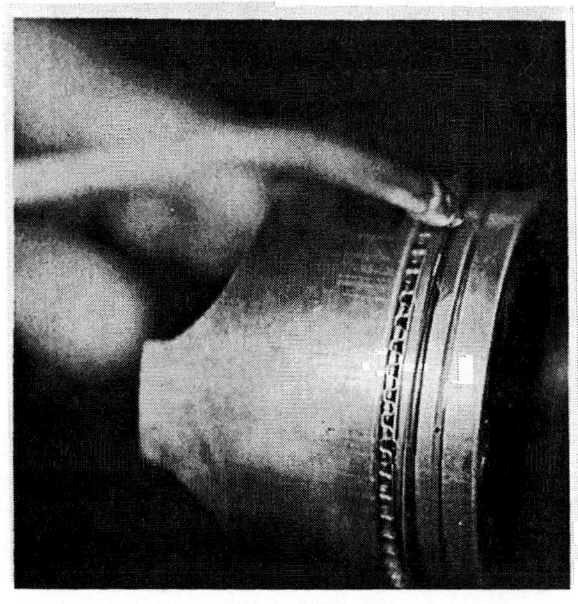

Lightly oil piston rings

2 Lightly oil the piston and the ring area.

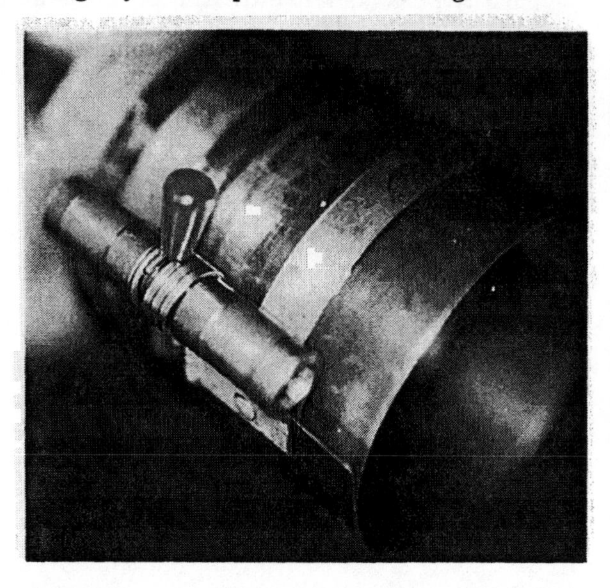

Fit ring compressor

3 Fit the ring compressor.
4 Ensure the crankshaft connecting rod journal is at B.D.C.
5 Check the indicating mark (Front) on the piston is facing to the front of the engine.

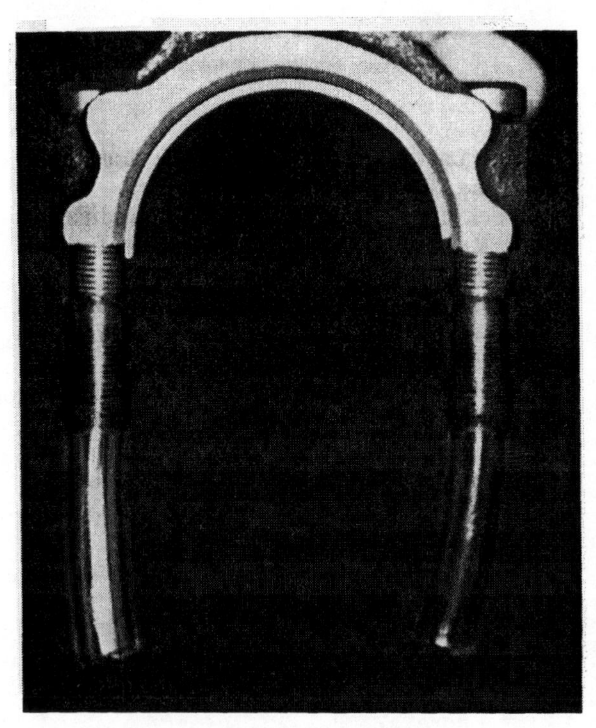

Fit connecting rod stud protectors

Do you remember which side of the engine the rod stamping faced? Does this agree with the indicating mark (Front) on the piston?

6 Remove the bearing cap from the connecting rod.
7 Lightly oil the bearing area.
8 Fit protectors on the connecting rod studs. This will prevent the threads scratching the bearing journal surface.
9 Insert the connecting rod into the cylinder bore until the ring compressor seats firmly against the cylinder block.

Firmly tap piston into bore

10 Using the end of a wooden handle (hammer handle), firmly tap the top of the piston until it has fully entered the bore.

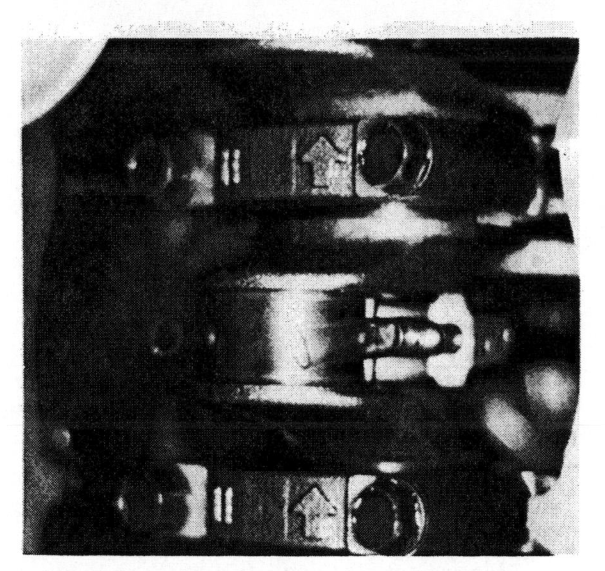

Guide stud protectors past crankshaft journal

11 Continue pushing the piston down the bore while guiding the rod onto the crankshaft journal.
12 Remove the stud protectors from the connecting rod.
13 Lightly oil the bearing area in the bearing cap.

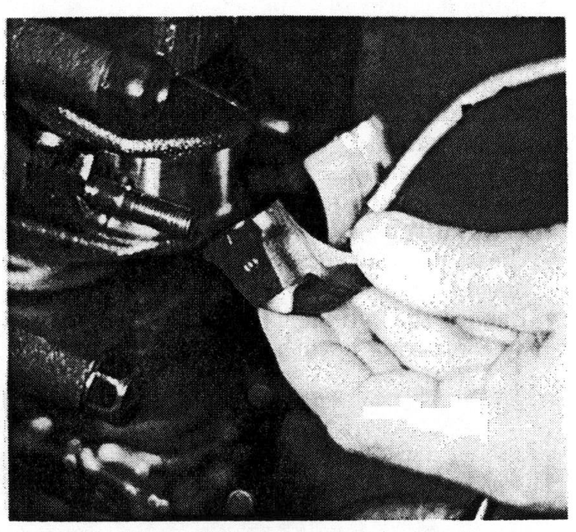

Lightly oil connecting rod bearings

14 Fit the bearing cap onto the connecting rod.
   - Ensure the stampings are on the same side of the engine.

15 Screw the cap retaining nuts onto the studs until they are finger tight.

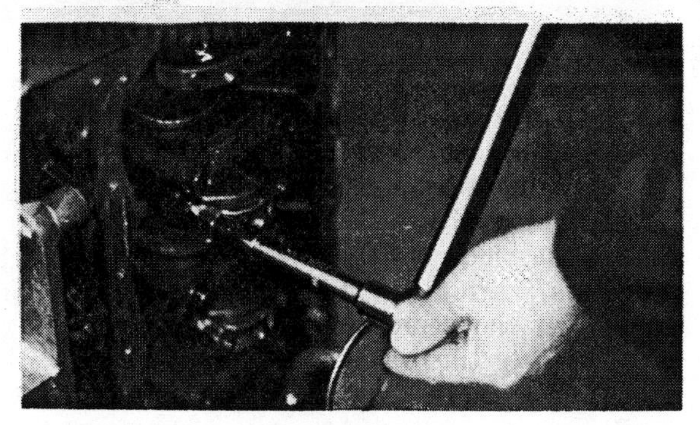

Tension connecting rod

16 Using the correct sized socket on a tension (torque) wrench, progressively tighten the cap nuts to the torque setting specified by the manufacturer.
17 Rotate the crankshaft to ensure it is free to move in the engine block.

Cylinder head installation

Important: consult the workshop manual for specific installation instructions.

1 Turn the engine in the stand until the cylinder head surface of the block is facing upwards.
2 Rotate the crankshaft until the crankshaft timing marks are aligned.
3 Rotate the camshaft in the cylinder head until the timing marks are aligned.

Caution: do not rotate the crankshaft or the camshaft until the timing belt or chain is fitted.

Prior to fitting the cylinder head, the precautions that must be observed are:

- Clean all threads and bolt holes.
- Clean the surfaces of the head and the block.
- Ensure the cylinder head gasket is correct by comparing it with the original gasket.

Place gasket over dowell pin

4 Install the head gasket.
   - Position the head gasket so all the stud or bolts holes and the water jacket holes are aligned.
   - If the block is fitted with studs or dowel pins, carefully slide the head gasket down the studs or dowel pins until it contacts the block.
   - If the block has bolt holes, carefully position the head gasket on the block and screw at least two guide studs into the block.

5 Carefully slide the cylinder head down the studs until it contacts the cylinder head gasket.
6 Fit the cylinder head bolts or nuts.
   - Using the correct sized socket on a speed brace, screw the bolts or nuts down until they are not more than finger tight.
   - When guide studs have been used, install all other bolts before removing the studs.

Fit cylinder head bolts

Caution: cylinder head bolts or nuts must be tightened in a specified pattern to prevent cylinder head distortion (see diagram).

7 Using the correct sized socket on a short extension fitted to a tension (torque) wrench, tighten the cylinder head bolts or nuts to the specified torque in at least two stages.

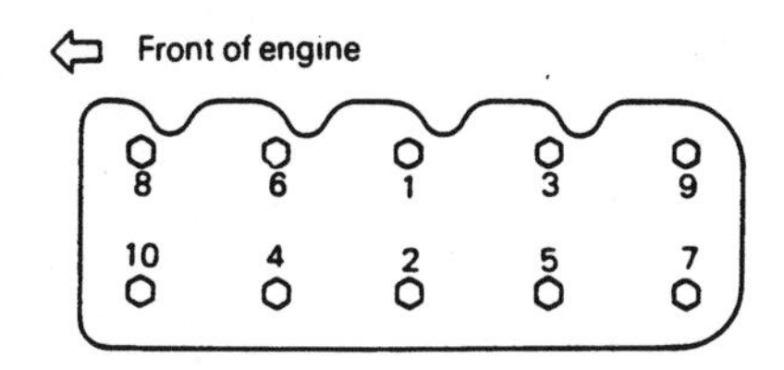

Cylinder head bolt tightening sequence

Source: Mitsubishi Engine and Transmission workshop manual

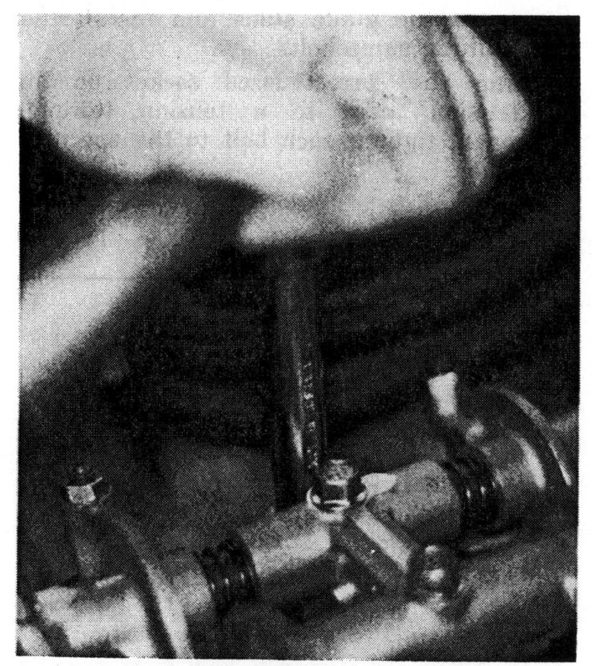

Tension cylinder head bolts

Note: The final bolt or nut tension on some engines is achieved by turning the tension wrench through a specified number of degrees after the initial tightening: for example 100 Nm on each bolt and then turn the tension wrench a further 90° (1/4 of a turn).

Camshaft drive installation

TOOTH BELT DRIVE

1 Confirm the crankshaft and camshaft timing marks are correctly aligned.
2 Fit the timing belt. Ensure the belt is taut on the side opposite the tensioning device.
3 Release the tensioning device clamping bolt or nut to allow tension to be applied to the belt.
4 Tighten the clamping bolt or nut to the manufacturer's specifications.
5 Confirm the crankshaft and camshaft timing marks are still aligned.
6 Rotate the crankshaft through two revolutions.
7 Loosen the tensioning device clamping bolt or nut sufficiently for the tensioner to press freely against the belt.
8 Tighten the clamping bolt or nut to the manufacturer's specifications.
9 Confirm the crankshaft to camshaft timing is correct.
10 Refit the timing belt shields.

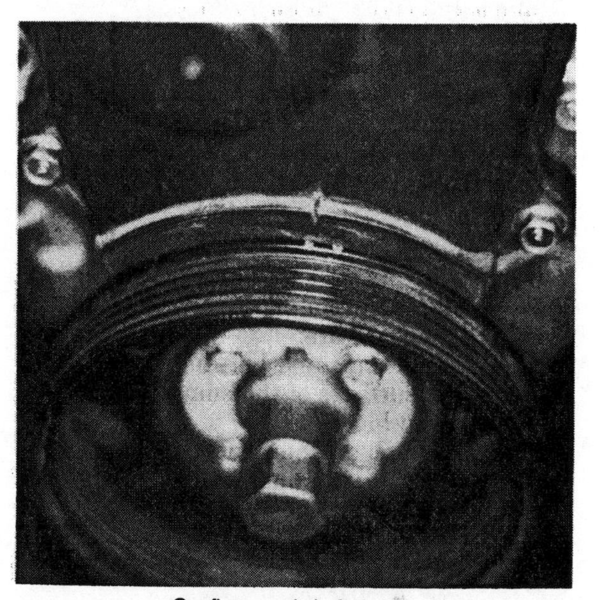

Confirm crankshaft timing

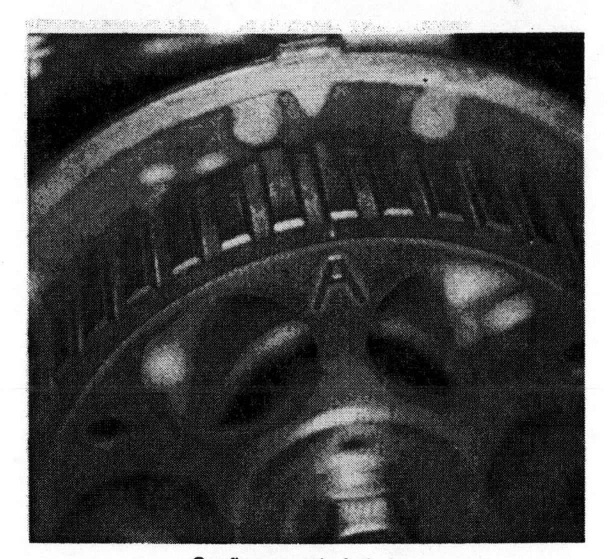

Confirm camshaft timing

CHAIN DRIVE

1 Confirm the crankshaft and camshaft timing marks are correctly aligned.

Note: It may necessary to temporarily fit the camshaft sprocket to the camshaft.

2 Place sprocket in the chain and raise the sprocket into position.
3 Move the sprocket in the chain until the sprocket indexes with the drive pin and/or bolt holes in the camshaft.
4 Fit the sprocket attaching bolts or nut and tighten to the specified torque.
5 Release the tensioning device.
6 Confirm the crankshaft and camshaft timing marks are still aligned.
7 Rotate the crankshaft through two revolutions.
8 Confirm the crankshaft to camshaft timing is correct.

Sump installation

Prior to fitting the sump, the precautions that must be observed are:

- Clean all threads and bolt holes.
- Clean the surfaces of the sump and the pan rails on the block.
- Thoroughly clean the inside of the sump.
- Ensure the oil pump strainer is clear of foreign material.

Position sump gasket onto sump rail

- Ensure the sump gasket is correct by comparing it with the mounting flanges on the sump.

1 Turn the engine in the stand until the cylinder head is downward.
2 Install two guide studs in each rail of the cylinder block.
3 Place the sump gasket in the correct position on the sump rails by sliding it over the guide studs.
4 Carefully lower the sump towards the cylinder block until it is seated on the gasket.
5 Using a small screwdriver or a centre punch, align the other holes in the sump, the gasket and the cylinder block.
6 Fit the bolts to the sump bolts and tighten them finger tight.
7 Check the alignment of the gasket with the sump.
8 Remove the guide studs and install the remaining sump bolts.
9 Using the correct sized socket on an extension fitted to a tension (torque) wrench, tighten each bolt to the specified torque.

Valve clearance (tappet) adjustment

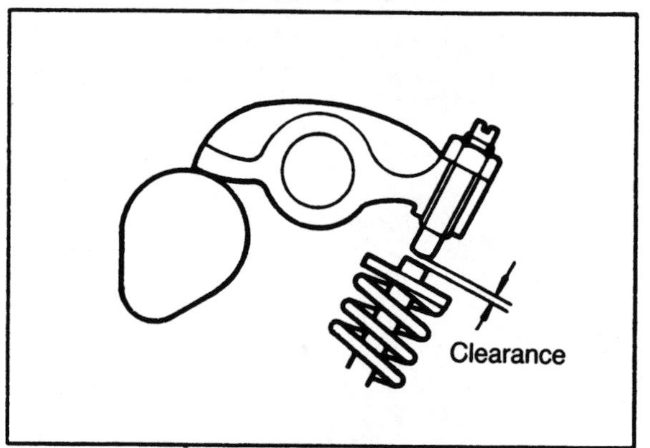

Valve clearance

There are many different methods used to set the position of the camshaft to allow each valve clearance to be adjusted. Consult the engine section of the appropriate workshop manual for the recommended procedure.

The following description is one of the common methods used to adjust the valve (tappet) clearances.

1 Turn the engine in the stand until the cylinder head is upwards.
2 Select the correct sized spanners or screwdriver to suit the tappet adjusting screw and the lock nut.
3 Select the correct sized feeler gauges as specified in the engine section of the appropriate workshop manual, that is, the inlet and the exhaust valve clearances.
4 Turn the crankshaft until one of the camshaft lobes (bumps) is pointing directly away from its rocker arm.
5 Loosen the lock nut on the adjusting screw.

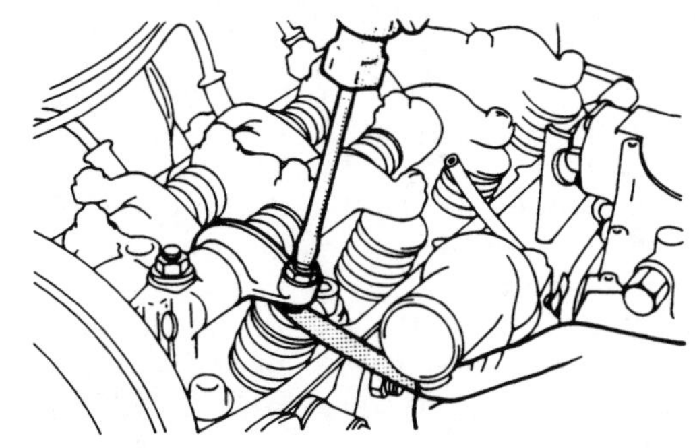

Source: Mitsubishi Engine and Transmission workshop manual

6 Insert the correct feeler gauge into the gap between the bottom of the camshaft lobe and the rocker arm.
   - When the feeler gauge cannot be inserted, turn the adjusting screw in the direction that will increase the clearance and try to insert the feeler gauge.

7 Turn the adjusting screw in the direction that will decrease the clearance until a slight drag is felt on the feeler gauge as it is moved back and forth in the gap.

8 Hold the adjusting screw firmly to prevent movement and tighten the lock nut.

9 Check the feeler gauge still has a slight drag on it as it is moved in the gap.
   - When there is not enough drag felt on the feeler gauge, repeat steps 7 and 8.
   - When there is too much drag felt on the feeler gauge, repeat steps 7 and 8 but turn the adjusting screw in the direction that will increase the clearance.

10 Repeat the above procedure (steps 4 to 9) until all the other valve clearances have been adjusted.

11 Check all the lock nuts are tight.

Tappet cover installation

Prior to fitting the tappet cover, the precautions that must be observed are:

- Clean all threads and bolt holes.
- Clean the surfaces of the tappet cover and the top of the cylinder head.
- Thoroughly clean the inside of the tappet cover.
- Ensure the tappet cover gasket is correct by comparing it with the original gasket.

1 Place the tappet cover gasket in its correct position on the cylinder head.

Note: In some cases, it will be necessary to fit guide studs into the cylinder head to ensure correct alignment of the rocker cover gasket.

2 Carefully lower the tappet cover towards the cylinder head until it is seated on the gasket.

3 Using a small screwdriver or centre punch, carefully align the holes in the cover, gasket and head.

4 Fit the tappet cover bolts or nuts and tighten them finger tight.

5 Check the alignment of the gasket with the cover.

6 Using the correct sized socket on an extension fitted to a tension (torque) wrench, tighten the bolts or nuts to the specified tension.

CHAPTER 10 REVISION

1 Correctly name the following components and briefly describe their function.

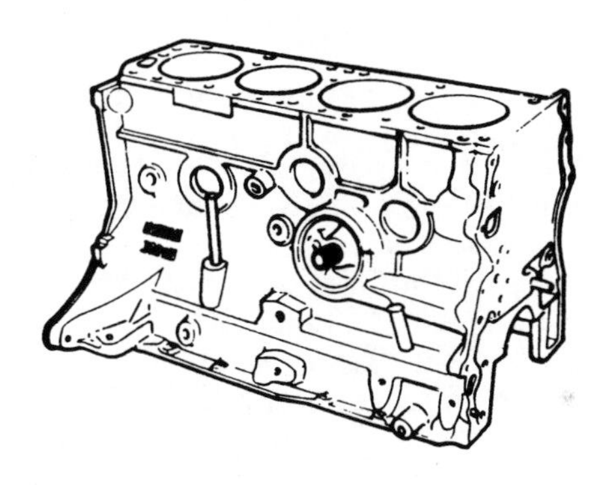

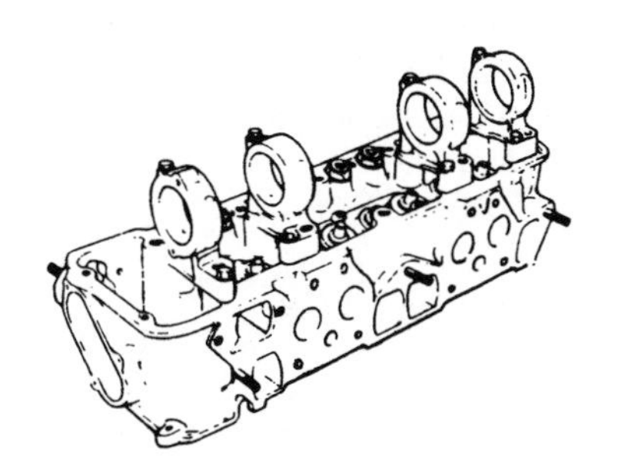

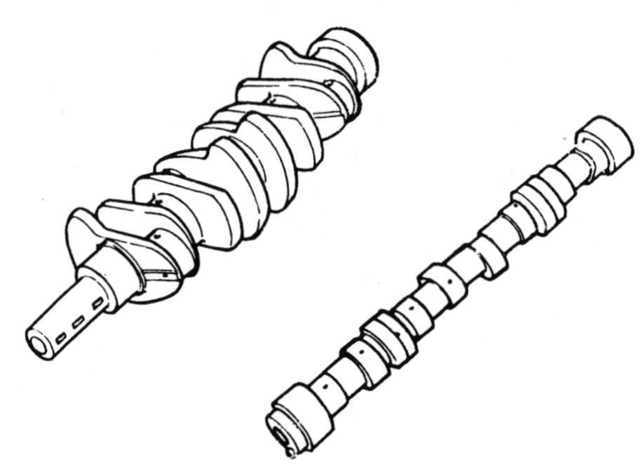

2 What is a combustion chamber?
3 What opens the valves in the cylinder head?
4 How are the cylinder head valves closed?
5 What is done to camshafts to provide better wearing properties?
6 Name the part which connects the exhaust ports to the exhaust system.
7 The purpose of the engine flywheel is to smooth out engine vibration and to store . . .
8 What type of engine would most likely contain a 'reed valve'?
9 What is the function of a reed valve?
10 Name the type of two-stroke engine that uses exhaust valves.
11 Where are the inlet ports located in a two-stroke diesel (C.I.) engine?
12 Which of the following ancillary systems is not required in a diesel (C.I.) engine?
   a Cooling   b Electrical
   c Fuel   d Spark ignition
   e Lubrication
13 Which type of engine would contain an eccentric shaft?
14 Where are combustion chambers located in a rotary engine?
15 Explain the meaning of the following terms.
   a Reciprocating   b T.D.C.
   c B.D.C.   d S.I. engine
   e C.I. engine   f 750 r.p.m.
   g Manufacturer's specifications
16 Name the five phases of engine operation.
17 What is the major difference in the ignition phase between a spark ignition engine and a compression ignition engine?
18 Describe the operating cycle of one of the following engines.
   a Four-stroke S.I. engine
   b Two-stroke S.I. engine
   c Rotary engine
19 What type of tool will 'speed up' the dismantling operation when removing bolts from a sump?
20 Why should all bolts retaining an alloy tappet cover be loosened prior to fully removing any bolt?
21 Why is it inadvisable to rotate the engine crankshaft after the camshaft drive has been removed and the cylinder head is still fitted?
22 What type of tool should be used to initially loosen the cylinder head bolts?
23 What is indicated by a notch, 'F' stamping or an arrow on a piston?
24 What tools are used to fit the piston, rings and connecting rod into the cylinder bore?
25 What should be done to prevent the connecting rod bolts from scratching the crankshaft bearing journal during assembly?
26 Before fitting the cylinder head, what precaution must be taken with the crankshaft and camshaft?
27 What method is used to prevent misalignment of the cylinder head gasket during assembly?
28 Place numbers on the diagram to indicate the order in which the cylinder head bolts should be torqued to specification.

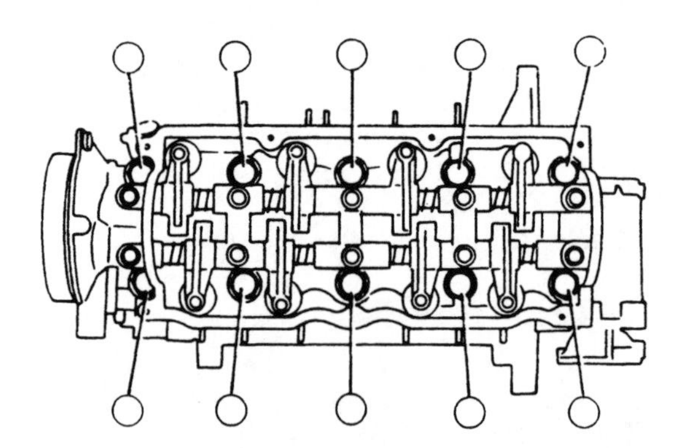

29 Why must the cylinder head bolts be torqued in a specified (by the manufacturer) pattern?
30 Describe a typical method of setting the camshaft valve timing for either a chain or belt drive overhead camshaft.
31 After consulting a service manual for any specified O.H.C. engine, describe the tappet adjusting technique required by the engine manufacturer.

11

TRANSMISSIONS

INTRODUCTION

There have been many variations in transmission systems throughout the years. From gear cogs and chains, belt drive, sliding mesh, constant mesh, with synchromesh, dry clutches, wet clutches, fluid flywheels, torque convertors, automatic transmissions to the latest type of transmission systems, using electronics and microprocessors to control and manage the gear changes. All of these transmissions aim at transferring the maximum amount of torque from the engine to the final drive smoothly and with the minimum loss of power at the drive wheels. In this chapter, some of the more popular types of transmission systems are shown.

Manual transmission

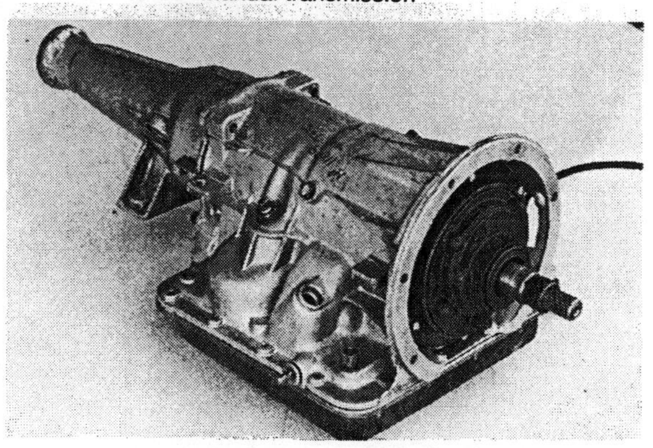
Automatic transmission

CLUTCH

Identification and function of components

It is important to identify and describe the function of the components of a clutch assembly for the following reasons.

1 to understand the operation of the clutch;
2 to assist in diagnosing clutch problems;
3 for the ordering and purchasing of spare parts.

Single plate diaphragm clutch assembly

The design of this type of clutch consists of four main component assemblies:

1 The flywheel;
2 The clutch plate assembly;
3 The clutch pressure plate assembly;
4 The clutch operating mechanism.

Clutch assembly

FLYWHEEL

The flywheel is bolted to the engine's crankshaft. The middle of the flywheel is recessed to accommodate the flywheel to crankshaft mounting bolts and the 'pilot' bearing, which locates and supports the input shaft of the transmission. The rear face of the flywheel is machined to a flat smooth surface. The gear on the outer rim engages with the pinion of the starter motor.

Flywheel

CLUTCH PLATE ASSEMBLY

This is located between the flywheel and the clutch pressure plate. The single disc, dry, friction plate clutch assembly has been developed and improved, and is now a well designed, efficient component.

The linings are the friction surfaces of the clutch assembly. They are manufactured from a mixture of materials including asbestos, resins, brass and other soft metals. Some linings have radial grooves cut into them, to stop the clutch plate from sticking to either the flywheel or the clutch pressure plate during clutch disengagement. The linings are separated by sections of waved spring steel which are called 'cushion springs'. The function of the cushion springs is to ensure a gradual, smooth engagement and disengagement of the clutch assembly and to assist in heat dispersion from the linings. The cushion springs are attached to a central spring retainer plate.

Around the spring retainer plate are elongated slots in which are retained a number of heavy-duty compression springs. The compression springs reduce the torsional vibrations and driving shocks from the engine and transmission.

Stop pins limit the movement between the hub and the spring retaining plate. The internal splines of the hub connect onto the external spline on the input shaft and transfer the drive from the clutch plate to the transmission system.

Clutch plate

CLUTCH PRESSURE PLATE ASSEMBLY

This is bolted onto the engine flywheel and consists of the following components.

The cover, which locates and accommodates the diaphragm spring, the pivot rings and the pressure plate. The diaphragm spring and the pivot rings are part of the cover, and are attached to the cover by special locating rivets. The cover is bolted directly onto the flywheel.

The diaphragm spring applies spring pressure onto the pressure plate. The pivot rings are positioned each side of the diaphragm spring, enabling the spring pressure to be released when the clutch release bearing acts on the fingers of the diaphragm spring.

The pressure plate is usually made from steel or cast iron. One side of the plate is machined to a flat, smooth, surface which presses against the linings of the clutch plate when the clutch is engaged.

Clutch pressure plate

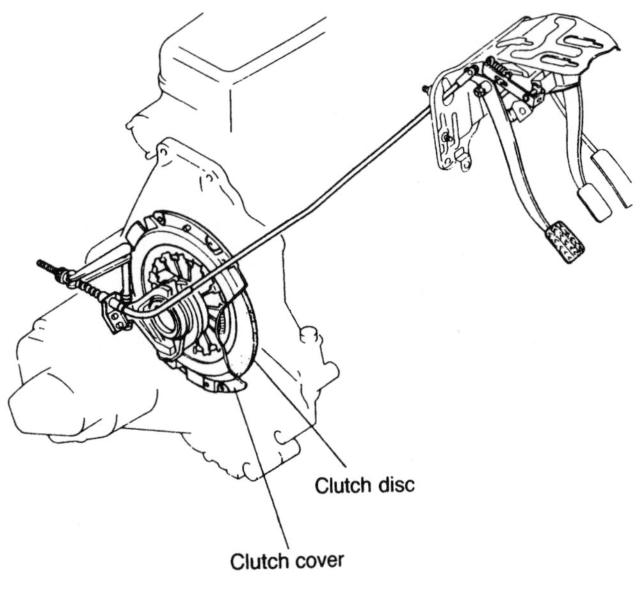

Cable type

Operating mechanism

There are two common types of clutch operating mechanisms:

1. cable type;
2. hydraulic type.

CABLE TYPE

This consists of a flexible inner and outer cable. The outer cable usually fits between the firewall and the clutch housing. The inner cable is connected to the clutch pedal mechanism at one end and to the clutch release bearing fork at the other end. Adjustment of the inner cable is normally provided by a threaded end and an adjusting nut.

HYDRAULIC TYPE

This consists of a clutch pedal and a push rod, acting on a hydraulic master cylinder. A steel pipe and a flexible hose connects the master cylinder to the clutch operating cylinder (slave cylinder). The clutch operating cylinder's push rod is attached to the clutch release bearing fork.

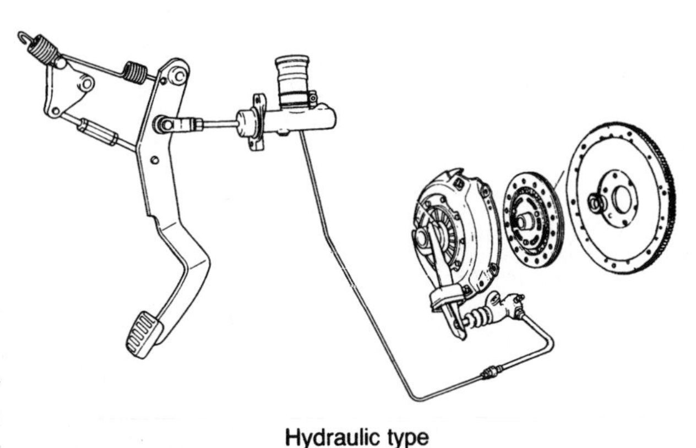

Hydraulic type

Source: Nissan Australia

CLUTCH FUNCTION

The function of the clutch is to provide a gradual, smooth, controlled connection and disconnection of the drive from the engine to the transmission. It must also be capable of transferring all of the driving torque from the engine to the transmission system.

HYDRAULIC CLUTCH CONTROL ASSEMBLY

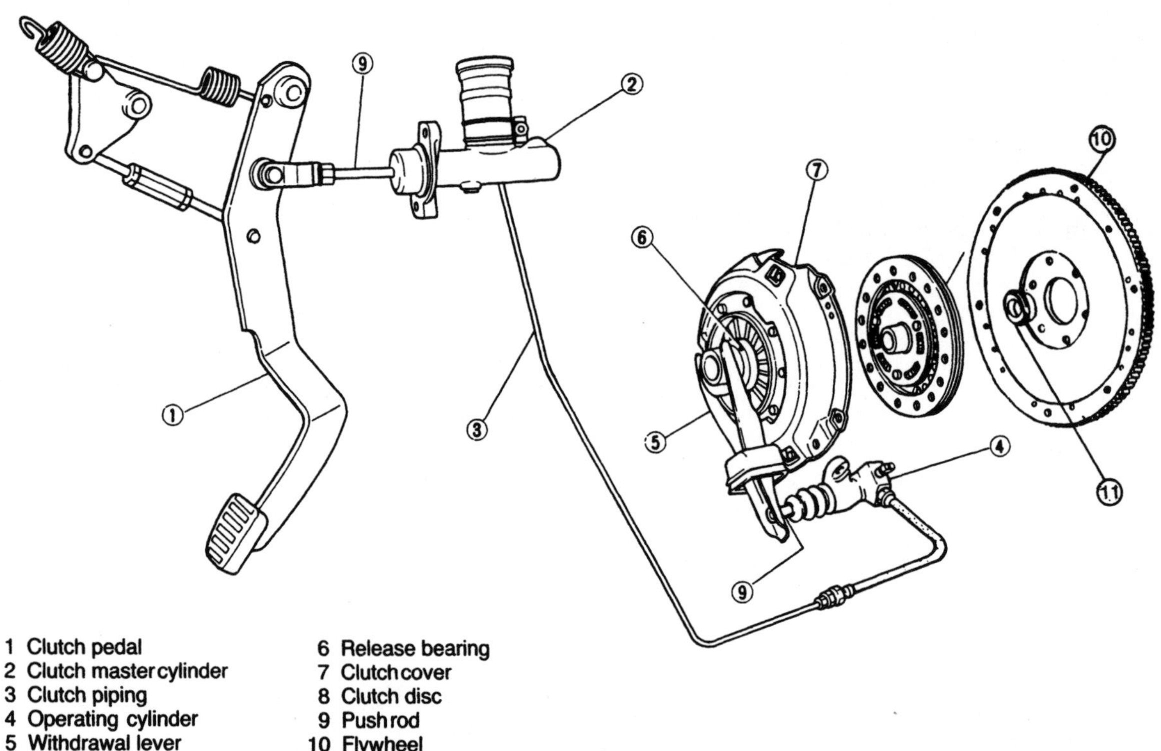

1 Clutch pedal
2 Clutch master cylinder
3 Clutch piping
4 Operating cylinder
5 Withdrawal lever
6 Release bearing
7 Clutch cover
8 Clutch disc
9 Push rod
10 Flywheel
11 Spigot bearing

Source: Nissan Australia

CLUTCH OPERATION

When the clutch is 'engaged' there is a mechanical connection between the engine flywheel and the input shaft of the transmission. This mechanical connection is brought about by the diaphragm spring pressure of the clutch pressure plate acting on the clutch plate, and sandwiching it between the smooth flat surfaces of the flywheel and the clutch pressure plate. Drive from the engine's flywheel and the clutch pressure plate is transferred to the clutch plate. As the clutch plate is splined to the input shaft of the transmission, this mechanical connection accepts the driving torque from the engine and transfers it to the transmission.

When the clutch pedal is depressed, the operating mechanism acts on the clutch release bearing, deflecting the diaphragm spring of the pressure plate. This action releases the pressure on the clutch plate and disconnects the drive from the engine to the transmission, and the clutch is 'disengaged'.

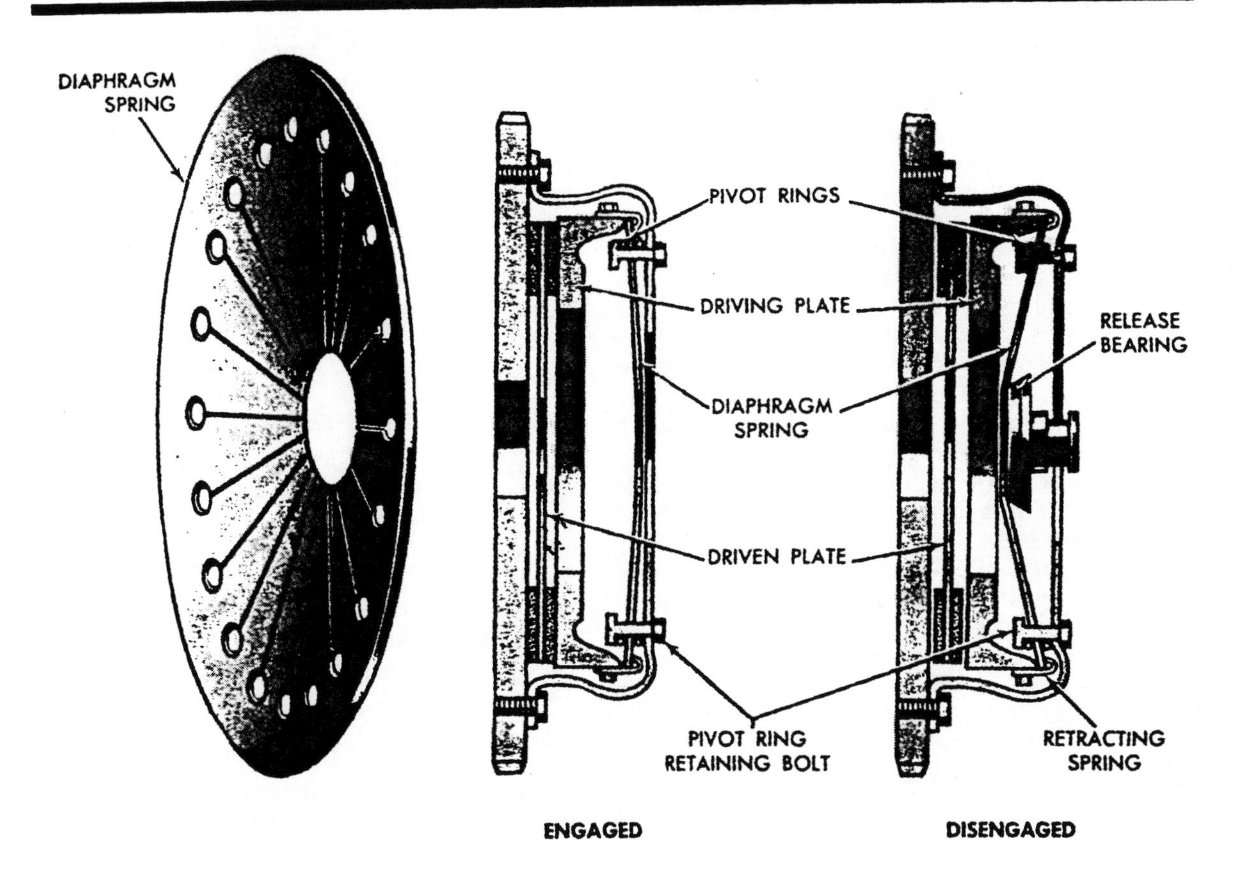

Torque capacity

The clutch assembly must be capable of transferring all of the driving torque from the engine to the transmission. The capacity of the clutch to fully transfer the driving torque from the engine to the transmission depends on the following factors.

1 The applied pressure of the diaphragm spring forcing the pressure plate and the clutch plate against the flywheel.
2 The type and quality of the clutch plate linings, the coefficent of friction (gripping qualities) between the clutch plate linings and the frictional surfaces of the flywheel and the pressure plate.
3 The average or 'mean' radius of the clutch plate's frictional linings.
4 The number of clutch plates in frictional contact.

REMOVING THE CLUTCH ASSEMBLY

Note: Refer to the vehicle manufacturer's service instructions before attempting to remove a clutch assembly. If the manufacturer's service instructions are not available, then the following procedure can be used.

Transmission removal

1 Ensure that the vehicle is supported safely on a hoist or on safety stands.
2 Disconnect the battery to avoid any accidental cranking of the engine.
3 Drain the oil from the transmission into a suitable container.

4 Disconnect the exhaust system (not always necessary)
5 Disconnect the following:
   - gear change linkages;
   - clutch operating mechanism;
   - speedometer cable;
   - drive shaft, and insert a plug to stop any gear oil from leaking out.
6 Support the engine by using a suitable jack and a block of wood under the sump of the engine.
7 Remove the rear support crossmember from the transmission.
8 Remove the transmission to clutch housing, attaching bolts.
9 Support the transmission with a suitable transmission jack and carefully remove the transmission.

Note: Ensure the transmission is supported and does not tilt in relation to the engine until the transmission input shaft is free from the splines of the clutch plate. If the transmission is not supported and it 'hangs' on the clutch plate, it will cause damage to the clutch plate.

10 Before removing the clutch assembly it is a good practice to mark the pressure plate cover and the flywheel to ensure correct reassembly if the clutch has to be used again.
11 Undo the clutch pressure plate attaching bolts progressively, one turn at a time, to ensure that all the pressure from the diaphragm spring is relieved evenly around the pressure plate.
12 Remove the clutch pressure plate attaching bolts and carefully remove the clutch pressure plate and the clutch plate.
13 Inspect the following:
   - the flywheel machined surface for grooves, heat cracks and surface condition, and the pilot bearing for wear and condition;
   - the clutch plate linings for wear and surface condition, and the torsional vibration coil springs for wear and condition, also the splines for wear;
   - the clutch pressure plate surface for grooves, heat cracks and surface condition, and the diaphragm spring for wear and condition;
   - the clutch release mechanism for wear and hydraulic leaks (if applicable) and the clutch release bearing and fork for wear and condition.

On completion of the inspection of clutch assembly, if the assembly is found to be unserviceable then renew the components. It is good practice to renew all the following parts:
- the clutch plate;
- the clutch pressure plate;
- the clutch release bearing;
- the pilot bearing.

Failure to renew all the components could result in faulty clutch operation, premature clutch failure and having to do the job over again.

Refitting clutch assembly

Before refitting the clutch assembly, ensure that all parts are thoroughly cleaned with a non-petroleum based cleaner. The clutch plate linings may be lightly sanded with a medium/fine emery paper. The clutch release bearing should be wiped with a lint free, clean cloth and the bearing mounting sleeve lubricated with a light smear of grease.

Refitting clutch

1 Check that the clutch plate slides easily on the transmission input shaft; smear a slight amount of grease on the splines.
2 Fit the clutch plate and the pressure plate assembly to the flywheel, using an approved clutch plate aligning tool or a 'dummy' input shaft.
3 Fit and progressively tighten the pressure plate holding bolts to the recommended tension wrench setting.
4 Remove the clutch plate aligning tool, visually check that the fingers of the diaphragm spring are even.

Refitting transmission

1 Secure the transmission onto the transmission jack.
2 Enter the transmission input shaft into the clutch housing and the release bearing.
3 Ensure the transmission is correctly aligned with the clutch housing (distance between the front face of the transmission and the clutch housing is even all around).
4 Carefully guide the transmission into the clutch housing until the input shaft enters the clutch plate and the pilot bearing.
5 Enter the transmission to clutch housing attaching bolts and tighten progressively to manufacturer's specifications.

6 Refit:
   a the transmission rear support cross-member;
   b the drive shaft;
   c the speedometer cable;
   d the clutch operating mechanism;
   e the exhaust system;
   f the gear change linkages.
7 Top up the transmission to the recommended level with the specified lubricant.
8 Adjust the clutch pedal free play.
9 Reconnect the battery.
10 Remove the vehicle from the hoist or safety stands.
11 Test the vehicle and ensure that the clutch is smooth in operation, not slipping, and the gear changing is quiet.

Clutch adjustment

CABLE OPERATED

Cable operated clutch release mechanisms have two adjustments.

1 Pedal height adjustment, which is usually measured from the pedal pad to the firewall. Adjustment is by the pedal height adjusting screw.
2 Clutch pedal 'free play', which is usually measured by lightly depressing the clutch pedal by hand until resistance is felt. The 'free play' is the distance the clutch pedal travels from its normal position to when it starts to meet the resistance. Normal 'free play' is between 11 mm and 17 mm.

The 'free play' adjuster may be located at the engine compartment firewall, pedal assembly or near the transmission housing.

14-18 mm Pedal free play
183.5 mm Pedal height
143.5 mm Pedal stroke

Clutch pedal adjustment
Source: Mitsubishi Colt RB/RC series manual

HYDRAULIC OPERATED

Hydraulic operated clutch release mechanisms normally have two adjustments.

1 Pedal height adjustment, usually measured from the clutch pedal pad to the firewall. Adjustment is by the pedal height adjusting screw.

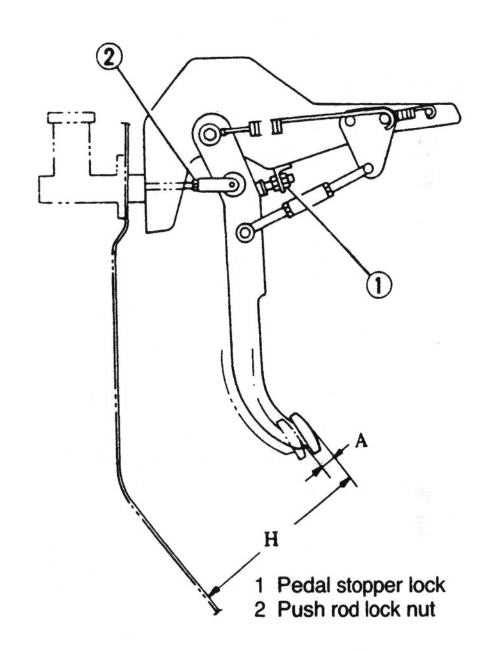

Adjusting clutch pedal height
Source: Nissan Australia

2 Clutch pedal 'free play' is measured by lightly depressing the clutch by hand until resistance is felt. The 'free play' is the distance the clutch pedal travels from its normal position until it meets the resistance. Normal 'free play' is between 2 mm and 6 mm.

The 'free play' adjuster may be on the pedal assembly or, in some cases, at the operating cylinder on the side of the transmission casing.

No free play causes the clutch release bearing to act on the diaphragm spring of the pressure plate, resulting in the 'clutch slipping'.

Excess free play could cause incomplete clutch disengagement, resulting in faulty gear changing.

Note: Refer to the vehicle manufacturer's service instructions for adjustment procedures and specifications. Incorrect adjustment could cause faulty clutch operation and/or premature wear of clutch components parts.

TRANSMISSIONS

Manual transmission

The function of a manual transmission is to provide:

1 A **neutral position** which allows the engine to start and run, without any load being applied to it.
2 A **range of gear ratios** which can easily be selected by the driver. The lower gear ratios allow the vehicle to move slowly and smoothly from a standing position. The higher gear ratios allow the vehicle to increase road speed.
3 A **reverse gear** which allows the vehicle to be driven in a reverse (backwards) direction.
4 **Engine braking** by engaging low gears to slow down the road speed of the vehicle when descending steep gradients and to conserve brakes.

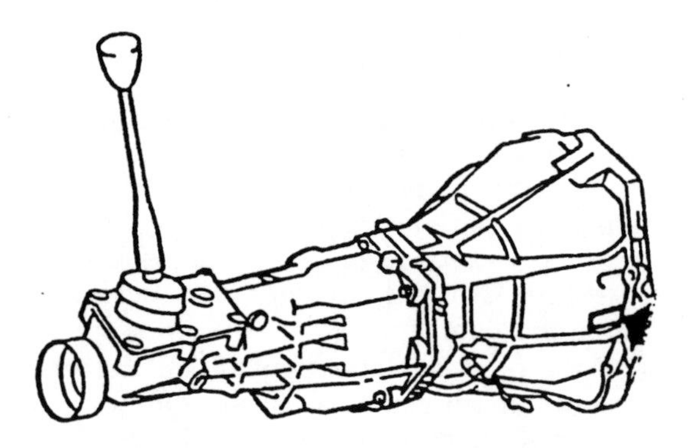

Gear types

There are many different types of gears used in the automotive industry. In this chapter we will be concentrating on three of the more common types.

1 **Spur gears** have gear teeth cut straight across. They are used to connect shafts, which are parallel to each other, as in a sliding mesh gearbox. Spur gears impart a radial load onto their shafts and tend to be noisy in operation.

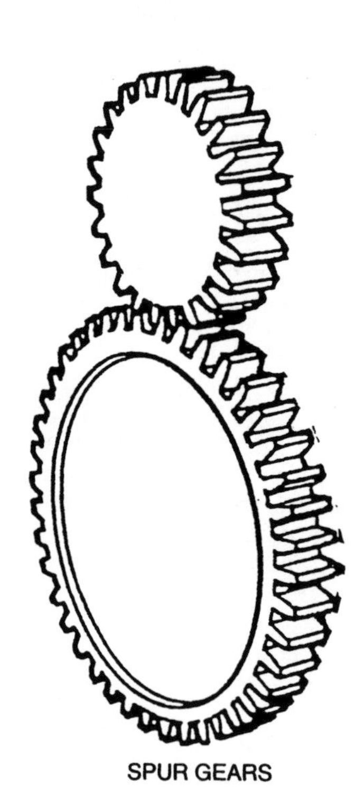

SPUR GEARS

2 **Helical gears** have gear teeth cut at an angle (helix). They are used in constant mesh transmissions. The helical shaped gear teeth provide a greater distribution of the load, as more teeth are in contact. They are quieter than spur gears in operation, however, they exert a side thrust along their shaft.
3 **Epicyclic gears**, or planetary gears, combine internal and external gearing. They are used in automatic transmissions. Epicyclic gears are compact, quiet in operation and can produce a number of gear ratios, including a reverse ratio.

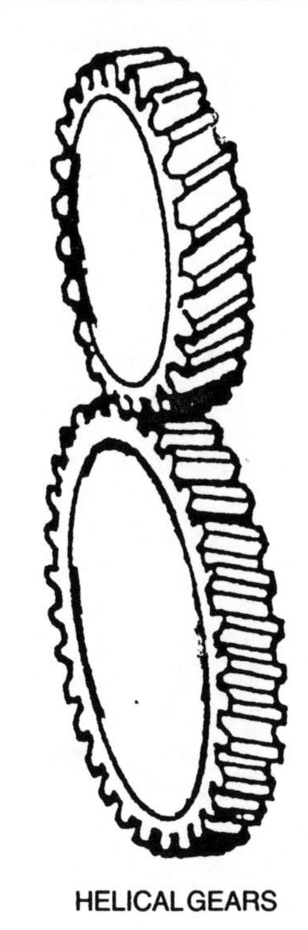

HELICAL GEARS

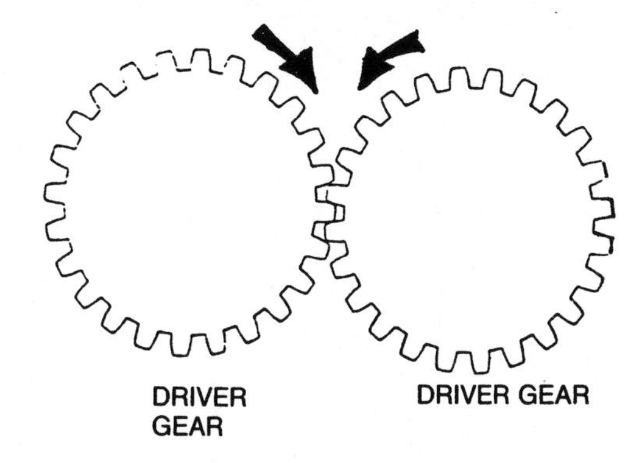

Example

$$\frac{\text{Driven gear} = 24}{\text{Driving gear} = 24} = \frac{1}{1} \text{ Ratio 1:1}$$

Note: Gears revolve in opposite directions.

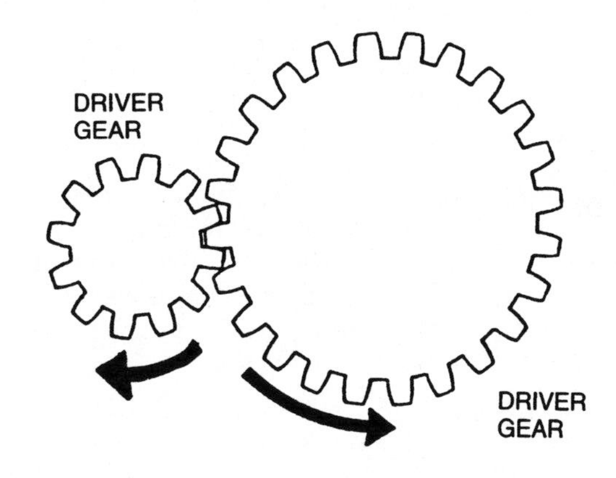

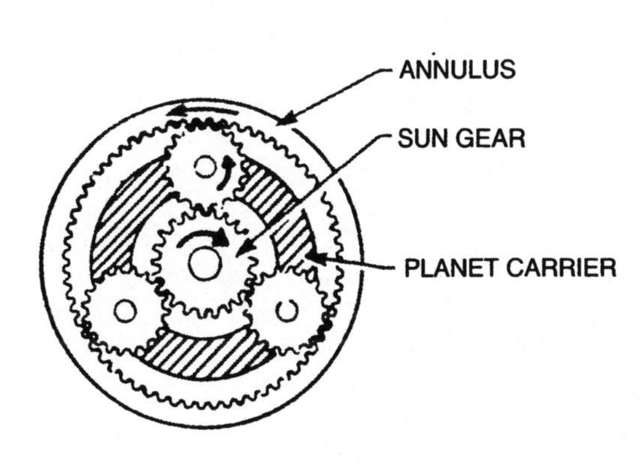

Gear ratios

The speed of rotation between two gears in mesh is determined by the number of teeth in the gears. If two gears in mesh have the same number of teeth, then one revolution of the driving gear will produce one revolution of the driven gear. The gear ratio is calculated by using the following formula:

$$\frac{\text{The number of teeth on the DRIVEN gear}}{\text{The number of teeth on the DRIVING gear}}$$

Example

$$\frac{\text{Driven gear} = 24}{\text{Driving gear} = 12} = \frac{2}{1} \text{ Ratio} = 2:1$$

Note: The driving gear will turn two revolutions for every one revolution of the driven gear

A **simple gear train** is a number of gears, all on separate shafts and in mesh with each other. When calculating the gear ratio of a simple gear train, only the first (driving) gear and the last (driven) gear are considered. All the intermediate gears are known as idler gears and do not affect the gear ratio, however, they do affect the direction of rotation of the driven gear. To calculate the gear ratio of this simple gear train, use the formula

$$\frac{\text{The number of teeth on the DRIVEN gear}}{\text{The number of teeth on the DRIVING gear}}$$

$$\frac{\text{Driven}}{\text{Driving}} \quad \frac{36}{24} = \frac{3}{2} = 1.5:1$$

The gear ratio is 1.5:1

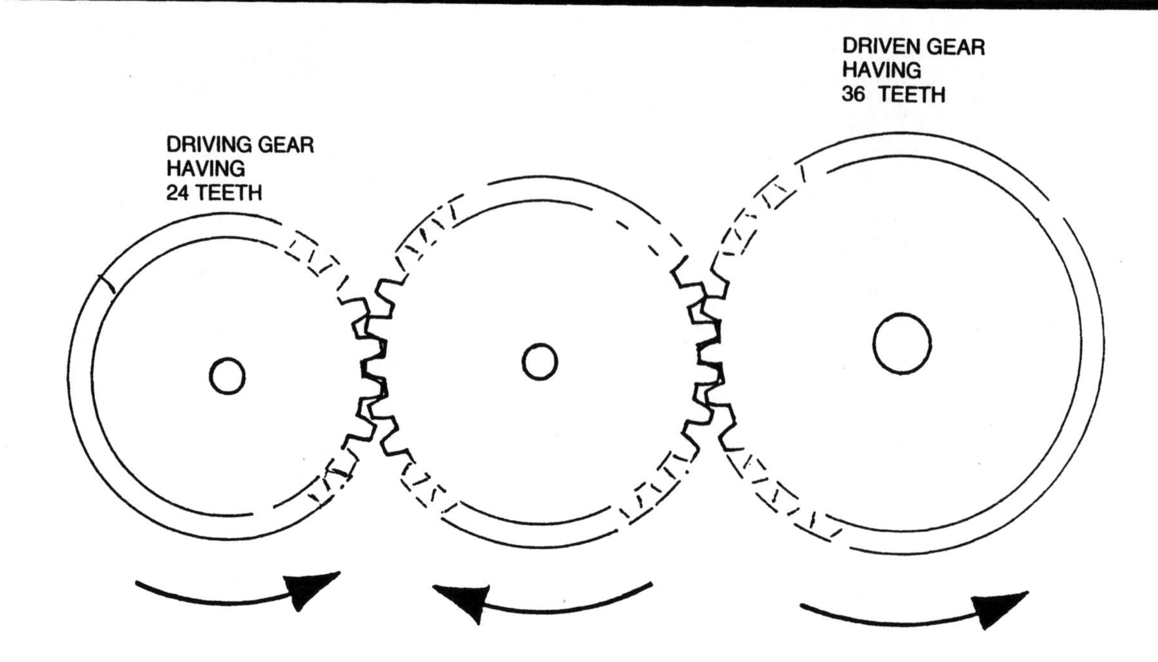

A **compound gear train** is when two gears are on the same shaft and one gear is a driven gear and the other is a driving gear, for example the countershaft, cluster gear in a manual transmission. The gear ratio is calculated by using the formula

$$\frac{\text{The number of teeth on DRIVEN}}{\text{The number of teeth on DRIVING}} \times \frac{\text{The number of teeth on DRIVEN}}{\text{The number of teeth on DRIVING}}$$

$$\frac{B}{A} \times \frac{D}{C}$$

$$\frac{48}{28} \times \frac{56}{24} = \frac{4}{1} = 4{:}1$$

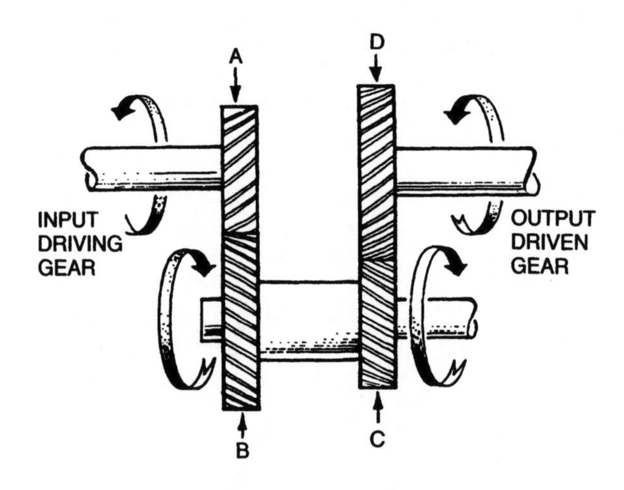

Gear ratio is 4:1
Note: Driving gear (A) will rotate four times, to one revolution of the driven gear (D).

Simple gearbox

The gear ratios in a manual transmission can be likened to various levers of different lengths. These levers (gears) are selected to suit the different load and road conditions. A gear ratio of 4:1 can be likened to a long lever, and this would be suitable for first gear (high torque, low speed). A gear ratio of 1:1 can be likened to a short lever and this would be suitable for top gear (low torque, high speed).
Example
Refer to this illustration. A man tries to push a motor car. He exerts a force of 30 N onto the car and the car does not move. He then uses a lever and exerts the same force of 30 N. The length of the lever is 2 m. The length from B – C is 0.5 m. The length from C–A is 1.5 m. The

ratio of the lever is 3:1, therefore the applied force onto the car would be 90 N. This appears to produce something for nothing. However, that is not so, as the lever has to travel 1.5 m before the car will move 0.5 m.

Refer to the illustration below. Suppose the engine is connected to shaft 'B', and the

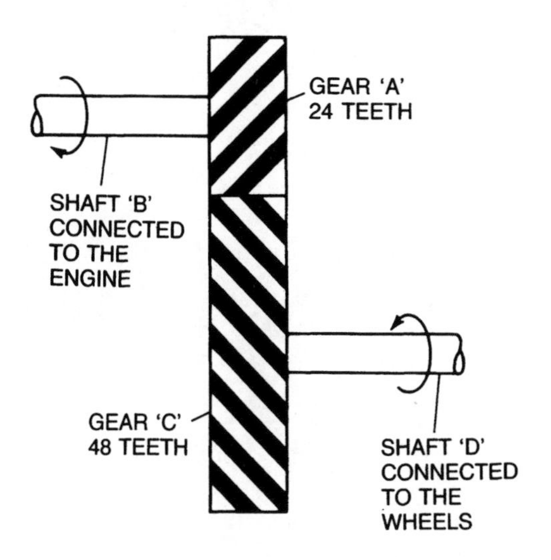

wheels are connected to shaft 'D'. The engine output torque is doubled at 'D' due to the 2:1 ratio. Once again it appears to produce something for nothing. That is not the case, as the engine 'B' will have to complete two revolutions to turn the shaft 'D' one revolution.

Therefore, it can be stated that as torque increases, speed decreases and the power remains the same (assuming the mechanism is 100 per cent efficient).

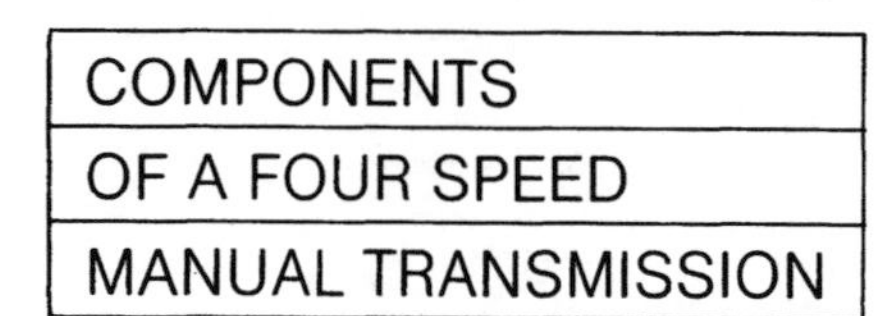

COMPONENTS OF A FOUR SPEED MANUAL TRANSMISSION

Transmission housing is usually made from cast iron or an aluminium alloy. The housing is of a rigid, robust construction to locate and retain the shafts and bearings in their correct position.

Manual transmission

Some transmission housings have the clutch housing as an integral part of their construction, on other types, the clutch housing is separate. The gear change mechanism and the extension housing are attached to the transmission housing. The gear oil is located in the transmission housing.

Input shaft. One end of the input shaft is supported by the pilot bearing in the flywheel. The other end is geared and has a bearing which locates and supports it in the transmission housing. The input shaft is splined, and the clutch plate fits on to these splines.

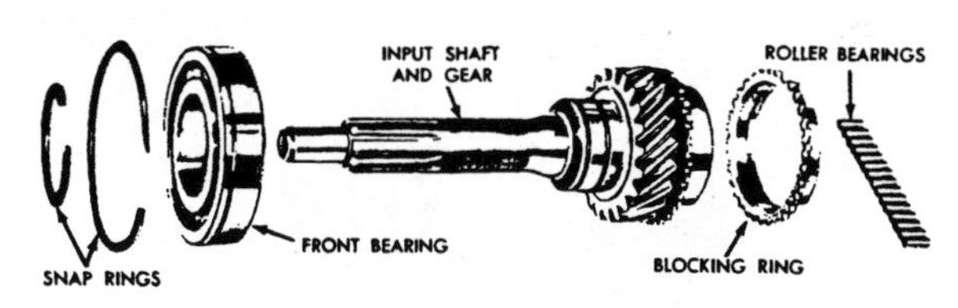

Input shaft

Counter shaft, cluster gear consists of a number of gears cast in the one unit, and rotates on needle roller bearings on the counter shaft. The cluster gear is in constant mesh with the gear on the input shaft.

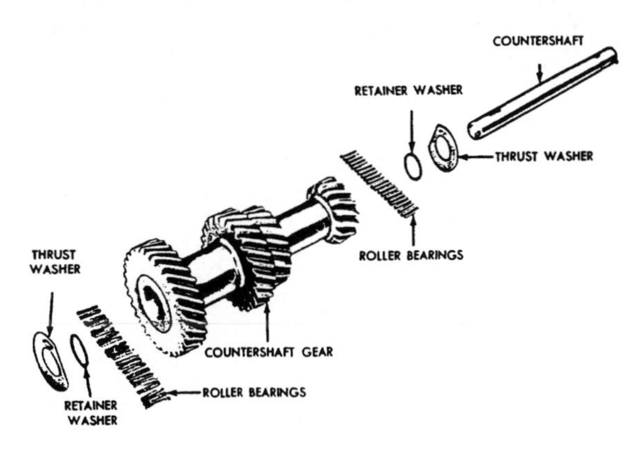

Counter shaft

Main output shaft. One end of the main shaft is supported by roller bearings in the input shaft. The other (output) end is supported in the transmission housing by bearings. First, second and third gears revolve on the main output shaft.
Note: These gears are not attached to the main output shaft and are free to rotate on the shaft. The two synchromesh hub units are pressed on and attached to the main output shaft.

Main shaft assembly

Synchromesh units. There are usually two synchromesh units in a four speed manual transmission. The hub of these units are pressed on and attached to the main output shaft, the synchro sleeves slide on the hubs and connect their respective gears on to the main shaft. One unit is for first and second gears and the other for third and fourth gears.

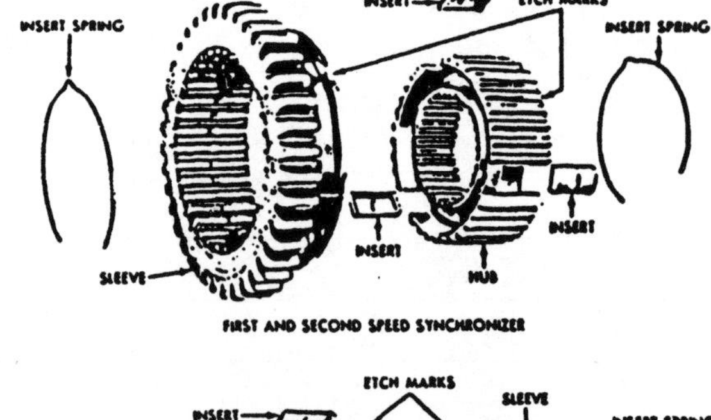

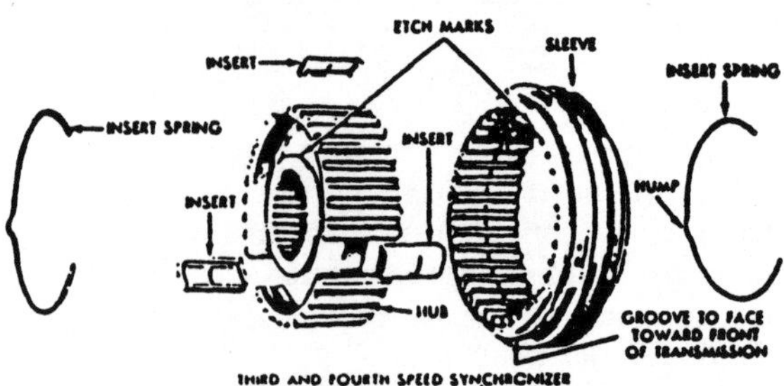

Reverse unit. The reverse unit consists of the reverse 'idler' gear, mounted onto a shaft. There are needle roller bearings between the shaft and the reverse 'idler' gear, and thrust washers located at each end of the shaft. A reverse sliding spur gear is splined to the reverse 'idler' gear.

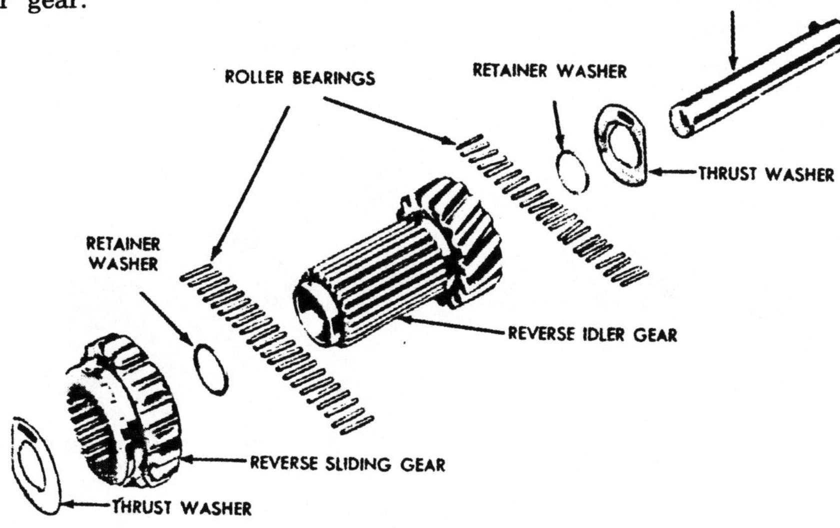

Reverse gear assembly

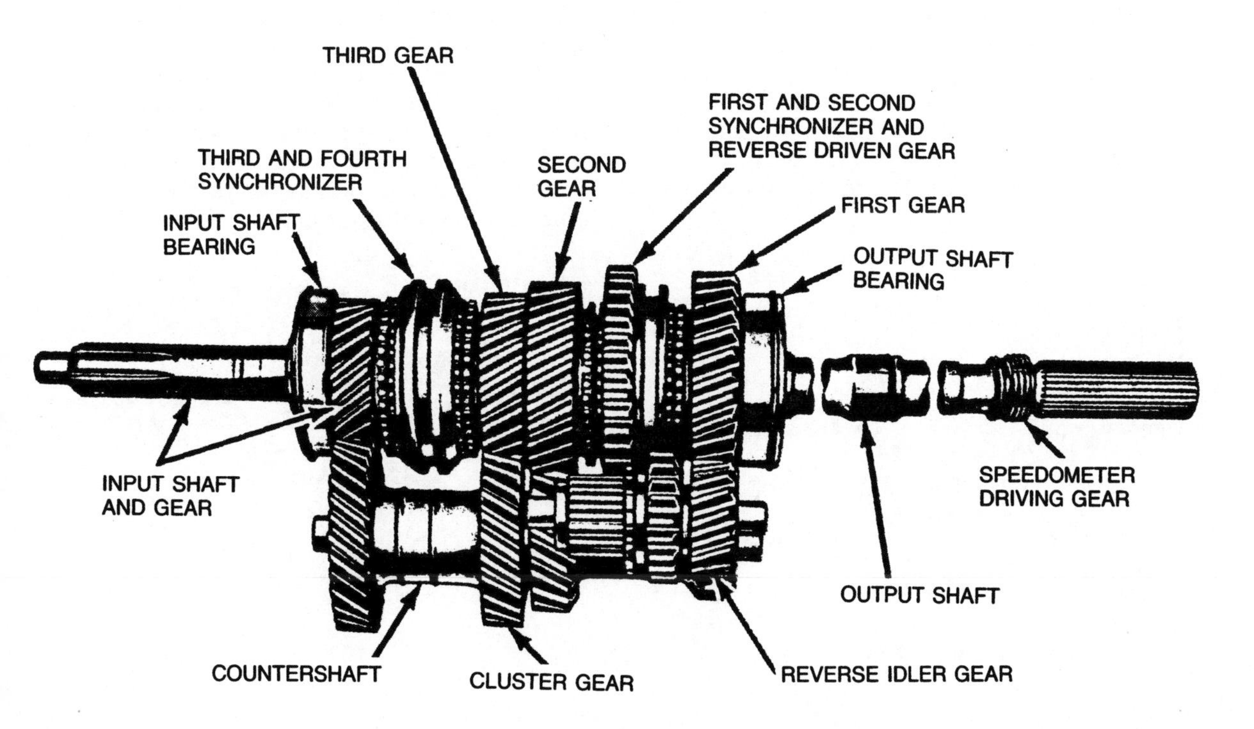

Typical four-speed manual transmission assembly

POWER FLOW

Tracing the power flow through a four speed manual transmission is important for the following reasons.

1 To assist in the understanding of the operation of the transmission.
2 To assist in diagnosing problems in the transmission.

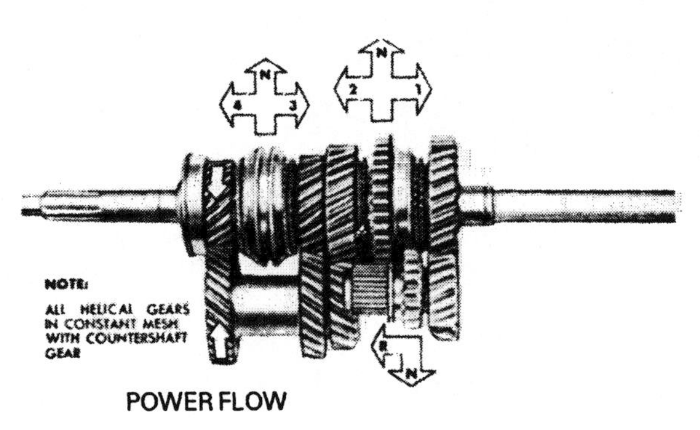

POWER FLOW

Neutral. The driving power from the engine is transferred through the clutch plate to the transmission input shaft. The input shaft transmits the drive to the cluster gear. The cluster gear is in constant mesh with the gears on the main shaft. As there are NO gears 'engaged' on the main output shaft, the gears on the main shaft are free to rotate and NO drive is transmitted through the main output shaft.

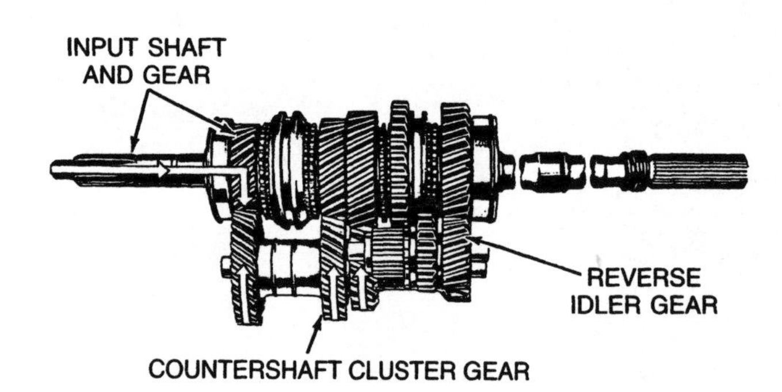

POWER FLOW—NEUTRAL

First gear. When first gear is selected, the synchro sleeve slides over the 'dog teeth' on the first gear, 'engaging' the first gear ratio to the main shaft. The driving power coming from the engine is transferred through the clutch plate to the transmission input shaft. The input shaft transmits the drive to the cluster gear. The cluster gear transfers the drive to the 'engaged' first gear and onto the main output shaft.

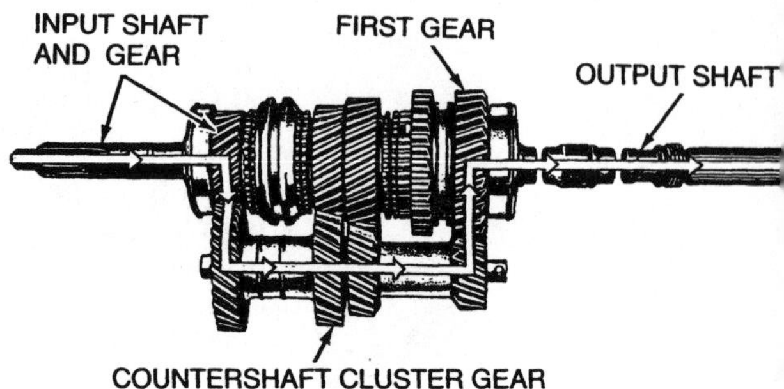

POWER FLOW—FIRST GEAR

Second gear. When second gear is selected, the synchro sleeve slides over the 'dog teeth' on the second gear, 'engaging' the second gear ratio to the main shaft. The driving power coming from the engine is transferred through the clutch plate to the transmission input shaft. The input shaft transmits the drive to the cluster gear. The cluster gear transfers the drive to the 'engaged' second gear and onto the main output shaft.

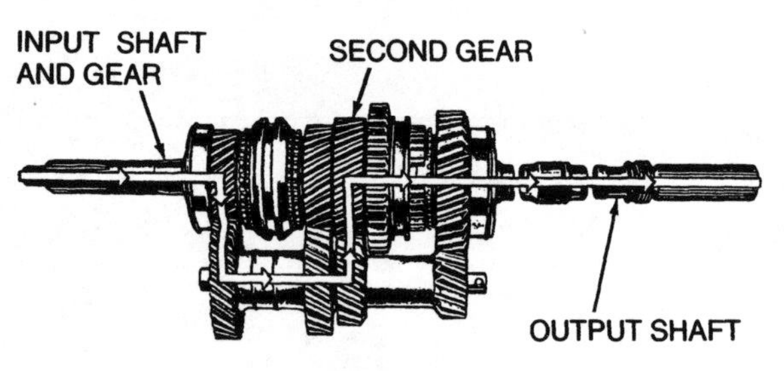

POWER FLOW—SECOND GEAR

Third gear. When third gear is selected, the synchro sleeve slides over the 'dog teeth' on the third gear, 'engaging' the third gear ratio to the main shaft. The driving power coming from the engine is transferred through the clutch plate to the transmission input shaft. The input shaft transmits the drive to the cluster gear. The cluster gear transfers the drive to the 'engaged' third gear and onto the main output shaft.

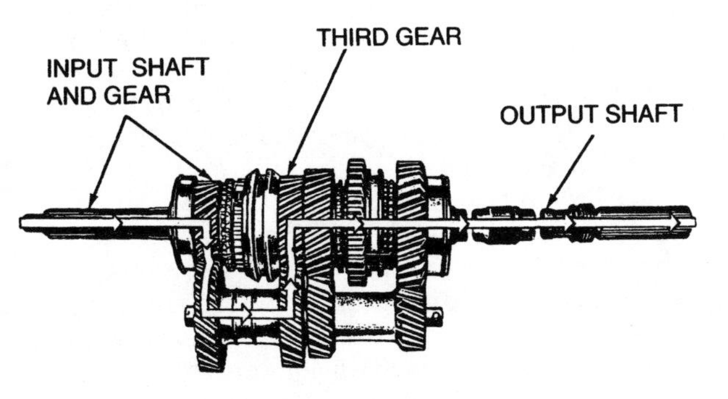

POWER FLOW — THIRD GEAR

Fourth gear. When fourth gear is selected the synchro sleeve slides over the 'dog teeth' on the input shaft gear, 'engaging' the main output shaft directly with the input shaft. The driving power coming from the engine is transferred through the clutch plate to the transmission input shaft. The input shaft transmits the drive directly onto the main output shaft.

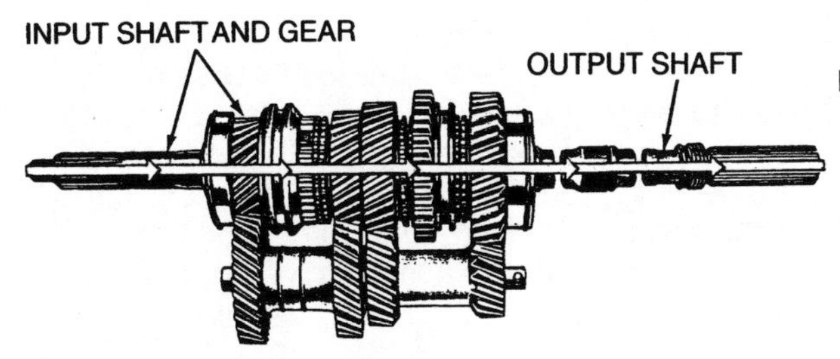

POWER FLOW — HIGH GEAR

Reverse gear. When reverse gear is selected the reverse idler gear is 'engaged' with the cluster gear and the gear on the outside of the first and second synchro sleeve. The driving power coming from the engine is transferred through the clutch to the transmission input shaft. The input shaft transmits the drive to the cluster gear. The cluster gear transfers the drive through the reverse idler gear to the gear on the outside of the first and second synchro sleeve and onto the main output shaft. The reverse direction of rotation is brought about by the following.

If the **D**irection **O**f **R**otation of the input shaft is clockwise, then:

- The **D.O.R.** of the cluster gear would be anti-clockwise.
- The **D.O.R.** of the reverse idler gear would be clockwise.
- The **D.O.R.** of the main output shaft would be anti-clockwise.

Therefore the **D.O.R.** of the main output shaft would be the REVERSE **D.O.R.** to that of the input shaft.

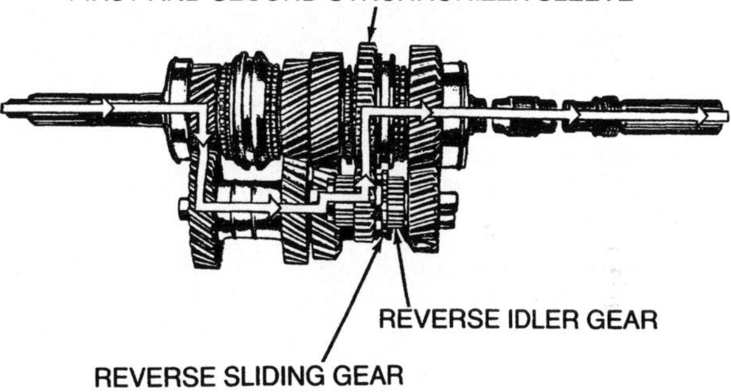

POWER FLOW — REVERSE GEAR

FIVE SPEED MANUAL TRANSMISSION

Most modern vehicles are offered with the option of a five speed manual transmission. The fifth speed gear is usually an overdrive ratio which can improve fuel economy during highway cruising.

The type of components and casings used in five speed manual transmissions are the same as four speed transmissions. Many popular five speed transmissions have been adapted from an earlier design four speed unit and may share many common parts.

The transmission is usually upgraded by the addition of a fifth speed gear to the cluster, a constant mesh fifth gear and a fifth gear synchronizer unit to the output shaft. A favoured location for the additional parts is to mount them to the rear of the main transmission casing. The extension housing may be increased in size to accommodate the additional gear parts and selector. An example of the modifications that could be made to a four speed transmission is shown in the illustration.

Other arrangements of gears and synchronizer units are available to provide a fifth speed overdrive gear. The type illustrated is relatively popular and has been selected as a typical example.

Fifth gear. When fifth gear is selected, the synchro sleeve slides over the 'dog teeth' on the fifth gear, 'engaging' the fifth gear ratio to the output shaft. The driving power coming from the engine is transferred through the clutch plate to the transmission input shaft. The input shaft transmits the drive to the cluster gear. The cluster gear transfers the drive to the 'engaged' fifth gear and onto the main output shaft.

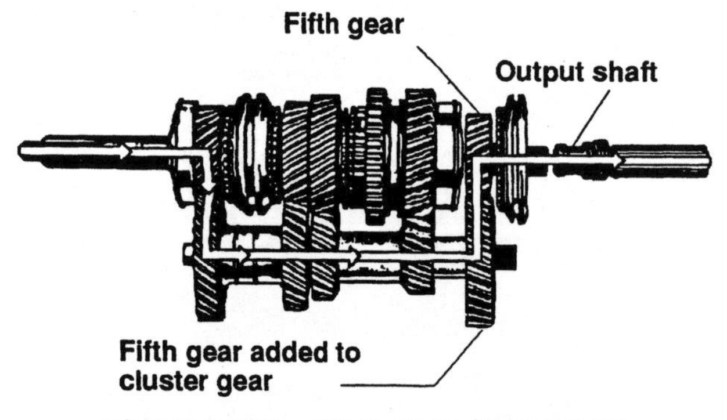

POWER FLOW - FIFTH GEAR (OVERDRIVE)

Transfer case

A transfer case is used to transfer power from the output shaft of a conventional transmission to both the front and rear drive axles of the vehicle. Transfer cases can be fitted to conventional and east-west mounted engines; they also accept input from either manual or automatic transmissions.

The transfer case is a mini-gearbox which has one input shaft and two output shafts (one to the front axle and one to the rear axle). In most applications, the transfer case supplies a 1:1 ratio and reduction ratio of approximately 2:1. The high ratio is suitable for hard surfaces, but the reduction ratio is particularly useful when used off-highway.

The mounting location of the transfer case varies between models. Direct mounting to the gearbox is the popular choice, however there are many applications where the case is mounted away from the vehicle gearbox and driven by a short shaft from the gearbox output.

There are many arrangements of the gears and shafts within transfer cases. The power flow of three different types is explained in the following paragraphs.

POWER FLOW TRANSFER CASE — TYPE A

Two wheel drive — high range. The upper sliding clutch, which is internally splined to the rear axle output shaft, slides over the dog teeth of the input shaft drive gear. The input and rear axle drive shafts are locked together and the drive to the rear axle is available.

Four wheel drive — high range. The upper sliding clutch remains in the position as selected for two wheel drive. The lower sliding clutch, which is internally splined to the front axle output shaft, slides over the dog teeth of the front drive high gear. The front drive output shaft and the front drive high gear are locked together.

The input and rear axle drive shafts are locked together and drive to the rear axle is available. The front axle output is obtained from the input gear, through the idler gear to the front drive high gear and out through the front drive output shaft.

Four wheel drive — low range. The upper sliding clutch slides over the dog teeth of the rear drive low gear locking the gear to the rear drive output shaft. The lower sliding

TRANSFER CASE POWER FLOW (TYPE A)

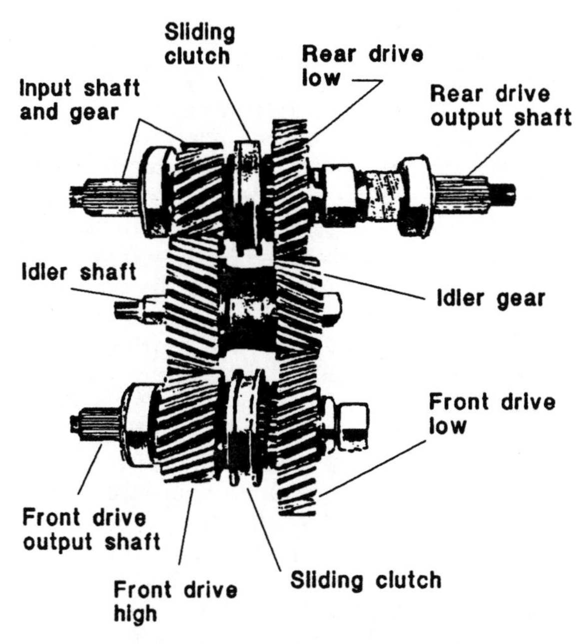

Typical two speed transfer case.

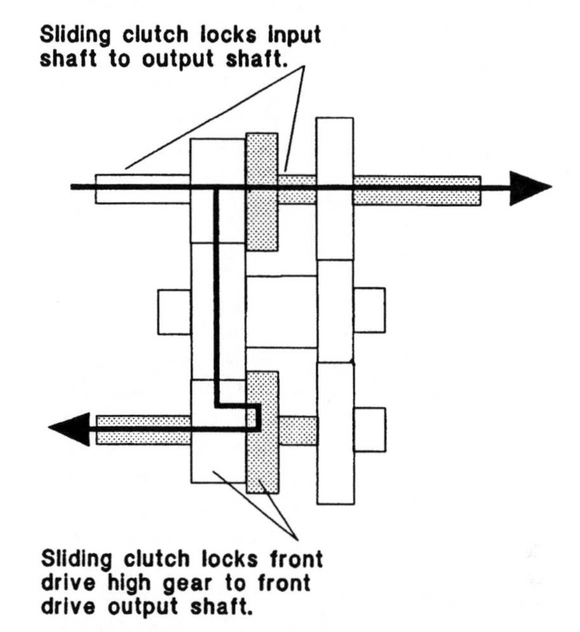

Four wheel drive, high range. Drive is to both the front axle and the rear axle.

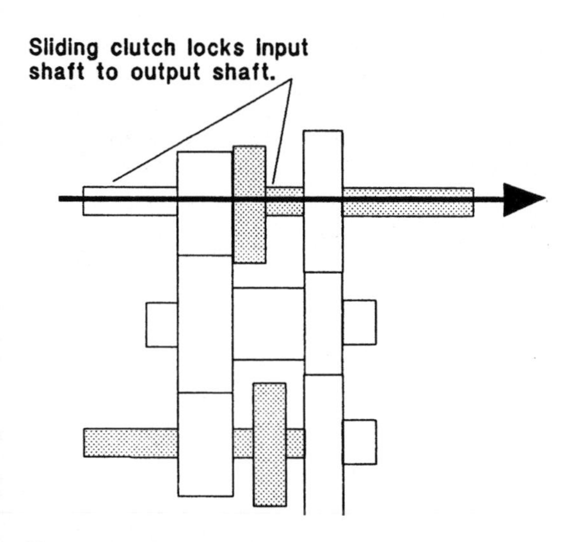

Two wheel drive, high range. Drive is to the rear axle.

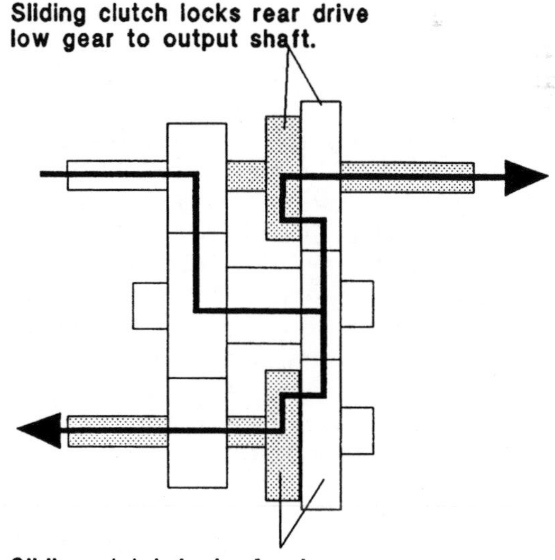

Four wheel drive, low range. Drive is to both the front axle and the rear axle.

clutch slides over the dog teeth of the front drive low gear locking the front drive output shaft. Power flows in the input gear to the large idler gear. Rotation of the large gear causes the small idler gear to also rotate transferring drive to both the rear and front drive low gears. Both rear and front low drive gears are locked to output shafts so drive to the front and rear axles is available. The drive will be at a reduction ratio determined by the number of teeth on the gears.

POWER FLOW TRANSFER CASE — TYPE B

Two wheel drive — high range. The range selection sliding clutch, which is internally splined to the rear drive output shaft, slides over the dog teeth of the high drive gear. Power flows in through the input shaft and gear, through the idler gear to the high drive gear. The high drive gear is locked to the rear axle drive shaft and drive to the rear axle is available.

Four wheel drive — high range. The range selection clutch stays in the position selected for two wheel drive. The front drive selection clutch slides over the dog teeth on an extended part of the rear drive shaft locking the front and rear drive shafts together. Power flows in through the input shaft and gear, through the idler gear to the high drive gear. The high drive gear is locked to the output shaft and drive is available to both rear and front axles.

Four wheel drive — low range. The range selection sliding clutch slides over the dog teeth of the low drive gear. The front drive selection clutch remains in the shafts locked together position. Power flows in through the input shaft and gear and rotates the high side idler gear. The low side of the idler gear also rotates transferring power to the low drive gear. The low drive gear is locked to the output shaft and drive is available to both rear and front axles. The reduction ratio available will depend on the number of teeth on the gears.

POWER FLOW TRANSFER CASE — TYPE C

Two wheel drive — high range. The input shaft sliding clutch, which is internally splined to the input shaft, slides over the dog teeth of the rear output gear. The rear output gear and shaft can be considered as one part. The input and rear output shafts are locked together. Power flows in the input shaft, through the clutch and out the rear output shaft. Drive to the rear axle is available.

Four wheel drive — high range. The upper sliding clutch slides over the dog teeth of the gear and front drive output shaft locking the drive transfer gear to the front output shaft. The input shaft sliding clutch remains in shafts locked together position. Power flows in through the input shaft, the sliding clutch and to the rear output gear and shaft providing drive to the rear axle. The output gear rotates causing the transfer gear to turn. The transfer gear is locked to the front output shaft and drive is available to the front axle.

Four wheel drive — low range. The upper sliding clutch remains in the four wheel high range position. The input shaft sliding clutch slides over the dog teeth of the input drive gear, locking the input shaft and gear together. Power flows in the input shaft through the locked clutch and input gear to the reduction gear. Drive to the front axle is from the reduction gear and front drive shaft. Rotation of the reduction also causes the locked clutch and transfer gear to rotate. The transfer gear rotation forces the rear output gear and output shaft to turn providing drive to the rear axle.

TRANSFER CASE POWER FLOW (TYPE B)

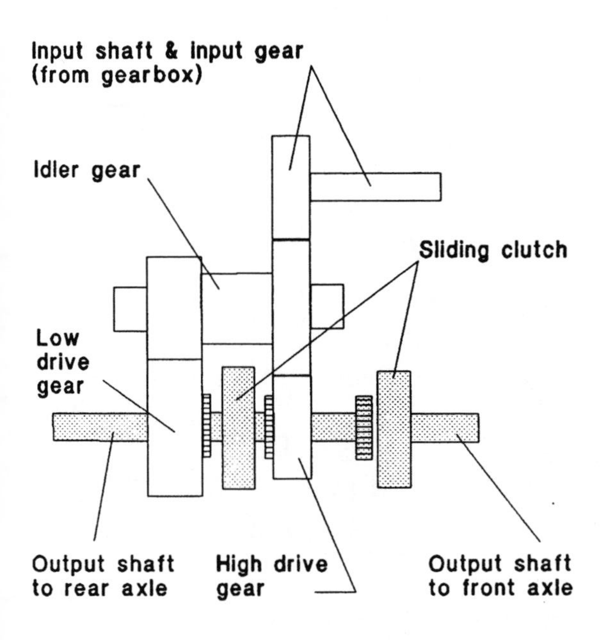

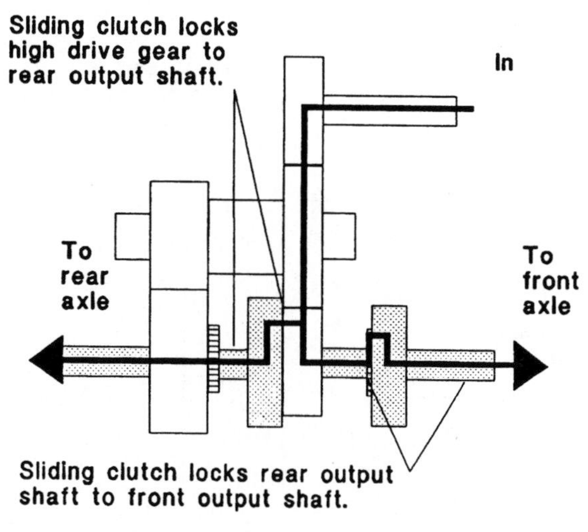

Four wheel drive, high range. Drive is to both the front axle and the rear axle.

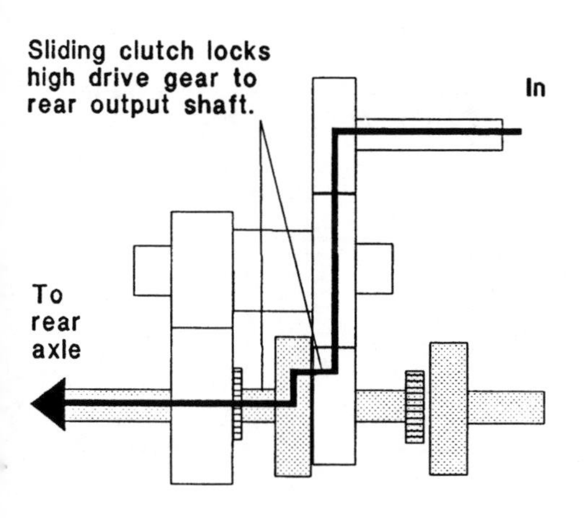

Two wheel drive, high range. Drive is to the rear axle.

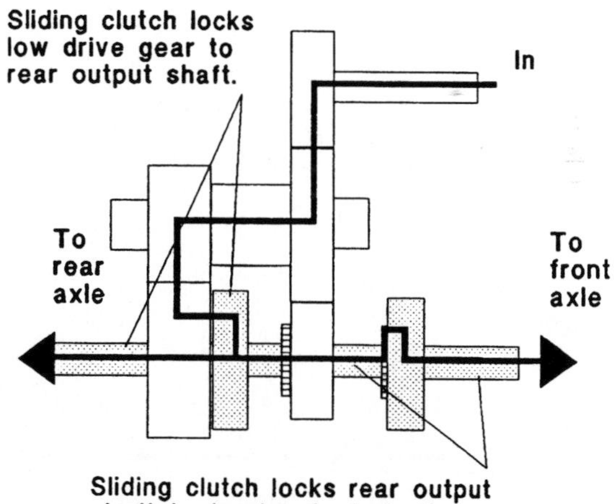

Four wheel drive, low range. Drive is to both the front axle and the rear axle.

TRANSFER CASE POWER FLOW (TYPE C)

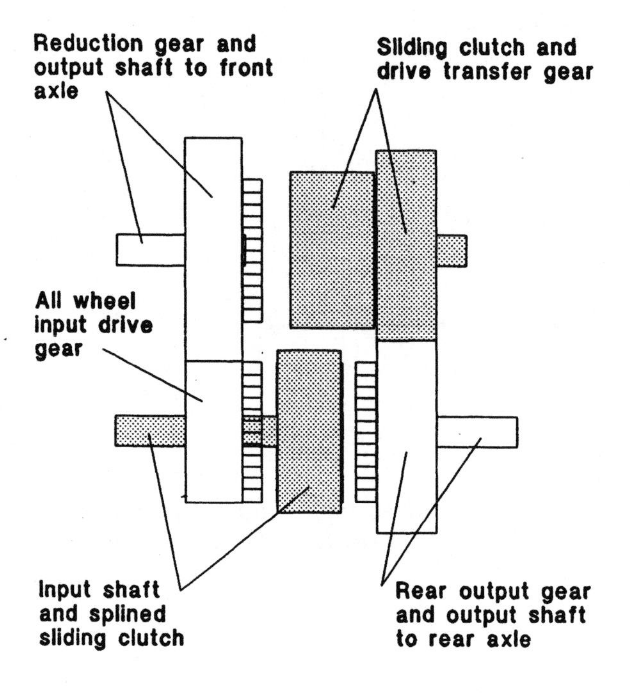

Sliding clutch locks drive transfer gear to front axle output shaft.

To front axle

To rear axle

In

Sliding clutch locks input shaft to rear axle output shaft gear dog teeth.

Four wheel drive, high range. Drive is to both the front axle and the rear axle.

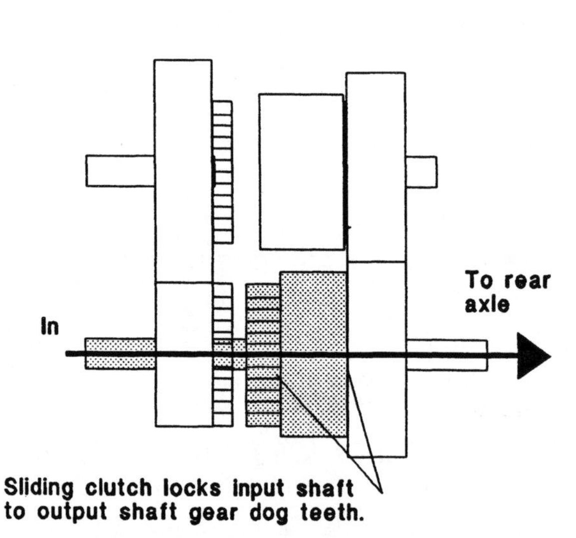

Two wheel drive, high range. Drive is to the rear axle.

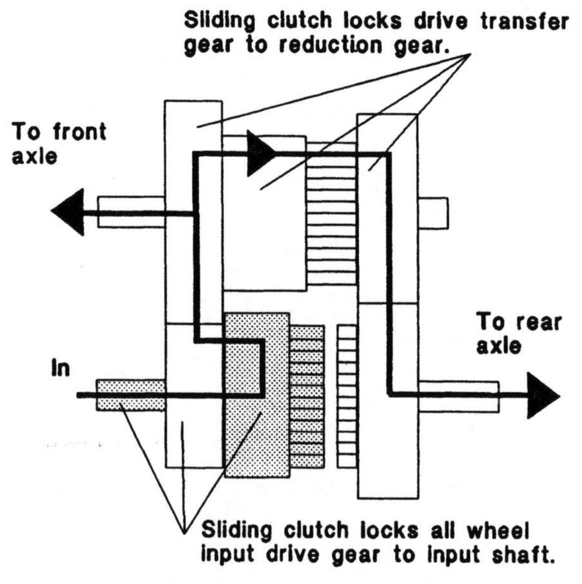

Four wheel drive, low range. Drive is to both the front axle and the rear axle.

TORQUE CONVERTORS

The torque convertor is situated between the engine and the automatic transmission. It is a **fluid coupling** and transmits the drive from the engine to the transmission.

Fluid has been used to transmit power for many years. The hydraulic brake system uses fluid to transmit the pressure applied at the brake pedal. In the torque convertor, fluid is accelerated by the engine. This accelerated fluid assists in transferring the drive from the engine to the transmission.

The function of a torque convertor is:

1 To convert the torque, twisting force of the engine to a higher torque required for smooth take off, low speed and hill climbing.

2 To provide a direct fluid coupling, the torque input from the engine being equal to the torque output to the transmission. This direct coupling is required when the vehicle is in top gear and cruising.

3 To provide a 'No Drive' position when the engine is running at idling speed.

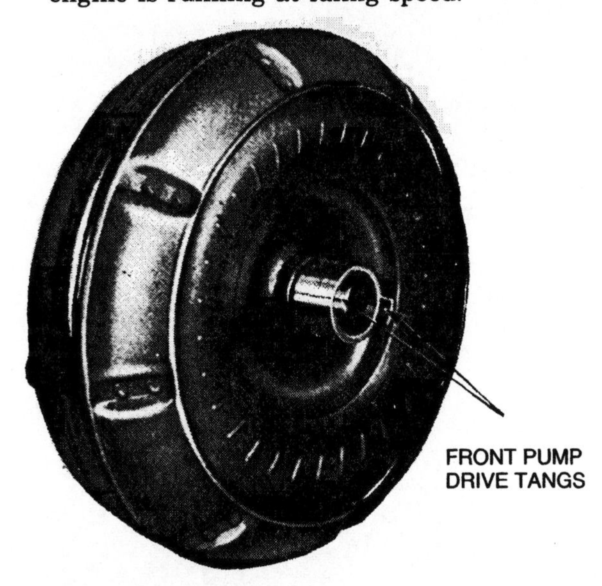

— Typical Torque Convertor

Torque convertor lockup clutch

Torque convertors with an integral lockup clutch are fitted to many modern automatic transmissions. This is not a new idea as a limited number of vehicles in the 1950s were fitted with convertors of this type. The need for greater fuel economy and the use of four speed (fourth gear overdrive) automatic transmissions has seen the re-introduction of the lockup clutch to the torque convertor.

Lockup clutch type torque convertors are similar in appearance and share the same functions as conventional torque convertors. The major difference occurs during the direct fluid coupling function. Conventional torque convertors can maintain direct fluid coupling only under ideal conditions (1:1 top gear ratio and light throttle cruising). Outside these conditions some slippage occurs and fuel economy deteriorates. The slippage would become very large if the top gear was an overdrive ratio and the fuel loss would make overdrives impractical.

The lockup clutch overcomes the slippage problem by forming a direct mechanical coupling between the convertor housing and the transmission input drive shaft. The decision when the clutch should be locked up and when it should be free is made by the transmission operating controls. There are several different lockup clutch units used by vehicle manufacturers with varying methods of application. The illustration is typical of one of the popular types. When lockup is required the damper piston is forced against convertor housing and a direct mechanical coupling between the engine and the transmission input shaft occurs. When the clutch is locked, direct drive third gear can operate through a wide range of operating conditions without slippage. It is also possible for the transmission to operate a fourth overdrive gear without slippage occurring.

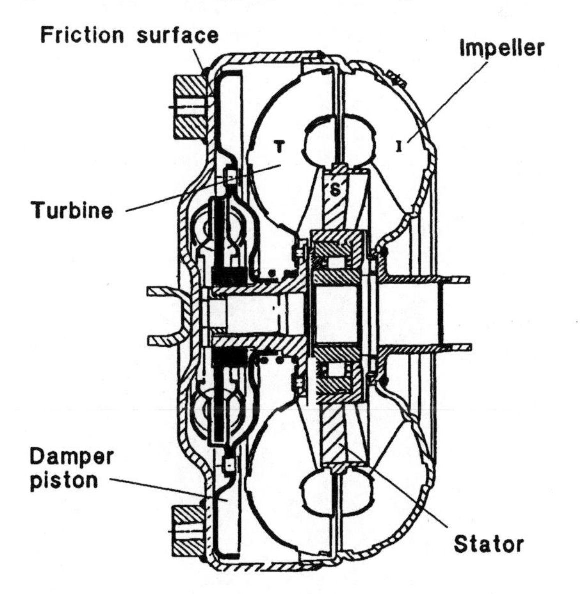

Torque converter with lockup clutch.

AUTOMATIC TRANSMISSION COMPONENTS

Components of a Borg Warner 35 automatic transmission:

Automatic transmission

Transmission housing: is usually made from an aluminium alloy and is of a rigid, robust construction. It retains, locates and supports the many components within the transmission housing.

Transmission housing

Front oil pump: is driven by the engine and is a positive displacement gear oil pump.

Oil pumps

Front clutch: connects the turbine shaft of the torque convertor to the primary sun gear of the epicyclic gear train.

Front clutch assembly

Rear clutch: connects the turbine shaft of the torque convertor to the secondary sun gear of the epicyclic gear train.

Planetary gear assembly

Front band: holds the secondary sun gear stationary.

Rear band: holds the pinion carrier stationary.

Front and rear bands

Front and rear servos: are hydraulically applied to clamp and hold the front and rear bands.

Servo assemblies

The governor: is mounted on the output shaft. Oil pressure from the governor to the shift valves, in the valve body, is controlled by road speed.

Governors

The valve body: is bolted to the bottom of the transmission housing. There are a number of valves and oil passageways in the valve body. Its main function is to direct the oil to apply and release the bands and clutch packs at the correct time.

Valve body

TRANSMISSION SERVICE AND BAND ADJUSTMENT

The Borg Warner 35 automatic transmission has been fitted to many makes and models of vehicles. It is advisable to consult the vehicle manufacturer's service information before proceeding with the servicing of the automatic transmission. Incorrect service procedures could cause premature wear in the transmission assembly, costly breakdowns and repair bills.

Pre-service checks

1 Check the transmission's oil level and top up if required.
2 Road test the vehicle and check for:
   a the correct operation of the neutral safety switch;
   b the correct gear change patterns;
   c the correct throttle linkage adjustment;
   d noises and vibrations.
3 Thoroughly clean underneath the vehicle, especially around the transmission area.
4 Ensure that the vehicle is supported on a hoist or safety stands.
5 Drain the transmission oil into a suitable container.

Caution: the transmission oil may be very hot.

6 Remove the transmission oil pan.
7 Inspect the inside of the oil pan for small particles of metal. This can assist in diagnosing excess wear in transmission components.
8 Remove the oil filter screen, thoroughly clean the filter screen and refit it to the transmission.

Front band adjustment

Special tools required:
1 A measuring gauge block 6.35 mm (1/4″).
2 A tension wrench: milli Newton metres or inch pounds.
3 A special tool.

Note: Incorrect adjustment of the front band can cause:
 a slip in first to second gear change;
 b no first to second gear change;
 c delayed or no second to third change;
 d slip in third to second 'kick down' change.

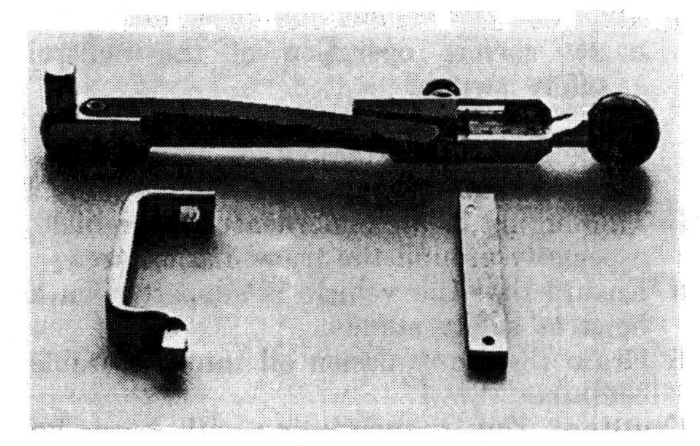

Special tool

Procedure

1 Slacken the locknut on the front band adjusting screw.
2 Move the servo lever outwards and place the measuring gauge (6.35 mm) between the servo pin and the adjusting screw.
3 Using the special tool and the tension wrench tighten the adjusting screw to 1130 mNm (10 inch pounds).
4 Tighten the locknut and remove the gauge block.

Front band adjustment

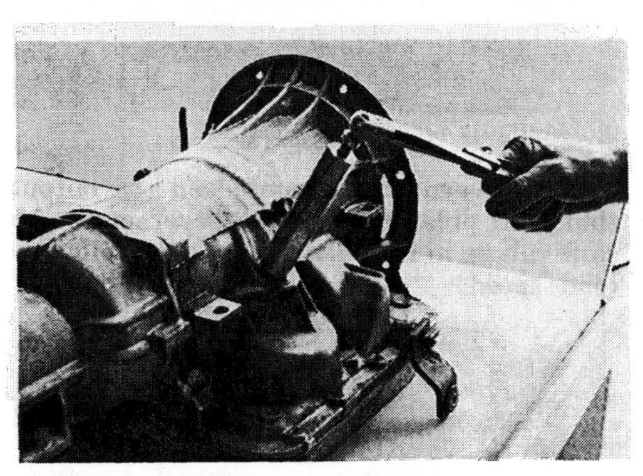

Rear band adjustment

Rear band adjustment

Note: Incorrect adjustment of the rear band can cause:
 a no drive in reverse;
 b no engine braking in low.

Procedure

1 Slacken the adjusting screw locknut, which is located in line with the rear servo on the outside of the transmission housing.
2 Tighten the adjusting screw to 14 Nm (120 inch pounds) by using the tension wrench.
3 Remove the tension wrench and slacken the adjusting screw three-quarters of a turn.
4 Tighten the locknut.

Note: Ensure that the adjusting screw does not turn when the locknut is being tightened.

Reassembly procedure and precautions

1 Ensure that the mating surfaces of the oil pan and the transmission housing are smooth and thoroughly clean.
2 Renew the oil pan gasket and refit the oil pan to the transmission housing. Tighten the bolts to a tension of 11–14 Nm.

Note: Do not apply any gasket cement to the oil pan gasket, as the gasket cement can find its way into the valve body and block oil passages in the valve body.

3 Refill the transmission with the correct amount of the recommended transmission oil.
4 Start the engine and recheck the transmission oil level.
5 Road test the vehicle and check that the transmission is operating correctly and to the manufacturer's specifications.

Electronically controlled automatic transmissions

The external appearance of an electronically controlled automatic transmission is similar to the conventional automatic transmission. The major external difference is the addition of an adaptor plug and one or two electrical wiring harnesses.

Both the electronic and the conventional transmission contain similar mechanical drive components and pumps. Clutches, bands, pistons and servos are still applied by fluid pressure. Planetary gear sets provide the gear ratios. If the wiring, valve bodies and solenoids were removed from the transmission it would be difficult to identify many of the other parts as from conventional or electronically controlled transmissions. Because of this similarity, routine maintenance such as band adjustment can be performed by any competent tradesperson who follows the manufacturer's service manual.

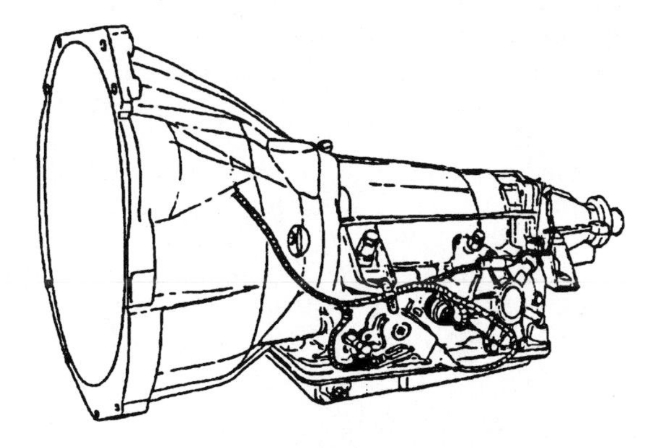

External wiring harness.

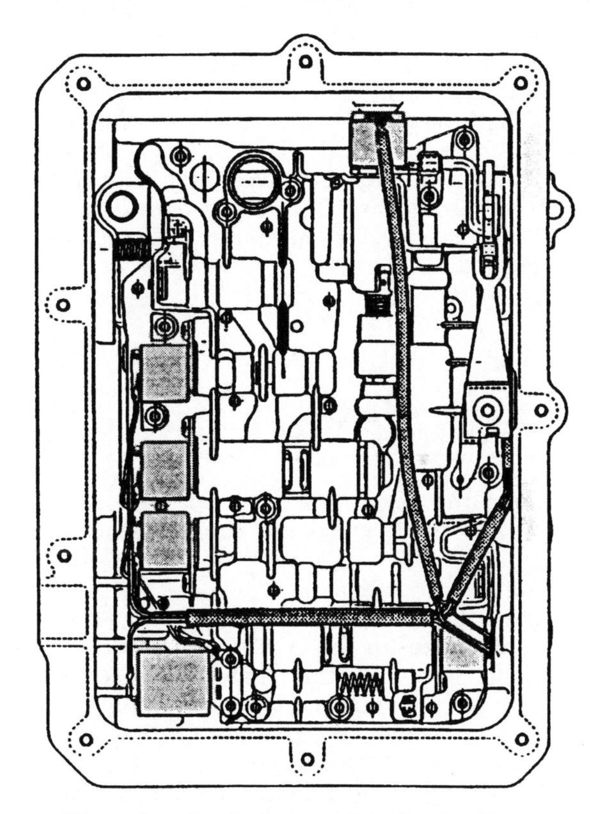

The six shaded areas indicate the positions of the solenoids on the valve body.

The major transmission differences are contained within the valve body. Electrical solenoids, mounted in the valve body and used to initiate gear changes and are controlled by an externally mounted computer. The computer receives information from sensors, references this information to an internal program, and sends output signals to switch the solenoids on or off. The type of information collected by the sensors can be throttle position, rate of throttle opening, battery voltage, engine r.p.m., vehicle road speed, gear selector position, transmission fluid temperature and possibly air conditioner use. A number of units store more than one shift pattern program within the computer and the driver can select the driving mode (power or economy) by pressing a switch.

Diagnostics and trouble shooting this type of transmission should only be performed by service personnel who have received specialist training and have access to the diagnostic units and service information required. As an aid to diagnostics, many transmission computers store fault codes which are released when triggered by the serviceperson. Do not perform these code checks until you have acquired more information at later stages of your studies. Always be aware that solid state electronic units can suffer significant damage from the incorrect use of meters and accidental 'grounding' of wires.

CHAPTER 11 REVISION

1 Name the bearing or bush located in the centre of the flywheel and state its function.
2 State the function of the gear on the outer rim of the flywheel.
3 Name the four main component assemblies which make up the clutch assembly.
4 Name some of the materials used in manufacturing the clutch plate linings.
5 Why do some clutch plate linings have grooves cut in them?
6 What is the function of cushion springs, and where are they located on the clutch plate?
7 Describe the function of the coil springs located around the clutch plate.
8 Name the large spring steel disc in the clutch pressure plate.
9 Correctly name and describe the function of the rings located at each side of the spring steel disc.
10 How is the large steel spring disc attached to the cover?
11 Why is the correct clutch adjustment essential? Describe the possible results if the clutch clearance is incorrect.
12 Describe a method of checking the clutch pedal free play on a cable operated clutch.
13 Describe a method of checking the clutch pedal height on a hydraulic operated clutch.
14 Describe the function of the clutch.
15 Describe the operation of the clutch in the engaged and disengaged positions.
16 Explain the meaning of torque capacity in reference to the clutch assembly.
17 Describe the factors affecting the torque capacity of a clutch assembly.
18 List the inspection checks recommended on clutch assembly components.
19 Describe the method of removing and replacing a clutch plate and pressure plate assembly.
20 Describe the function of a manual transmission.
21 Briefly describe the procedure for removing and refitting a manual transmission. Include all safety precautions.
22 Explain the meaning of the term 'gear ratio'.
23 Calculate the following gear ratio. The driver gear has seven gear teeth and the driven gear has twenty-one gear teeth.
24 Name ten components of a manual transmission.
25 Describe the power flow through third gear and reverse gears.
26 Describe the function of a torque convertor.
27 Correctly name ten components of an automatic transmission.
28 Describe the pre-tests recommended before servicing an automatic transmission.
29 Describe the method of adjusting the front band of a Borg Warner 35 automatic transmission.
30 Describe the method of adjusting the rear band of a Borg Warner 35 automatic transmission.

12

ELECTRICAL

The electrical/electronics system of a motor vehicle has increased in complexity at a rate which is unmatched by any other vehicle system. Early vehicles had very simple systems with relatively simple maintenance requirements. Today's vehicles have extremely complex systems but still only have simple maintenance needs. By performing these maintenance tasks correctly the motor mechanic can prevent many problems being introduced into complex electrical/electronics systems.

Modern vehicle systems are proving reliable, and for the majority of vehicle owners this means increased confidence in the product and less vehicle down-time for repair. However, as with all devices, a number will fail either by introduced fault (poor servicing procedure) or by product fault. Trouble-shooting this type of electrical

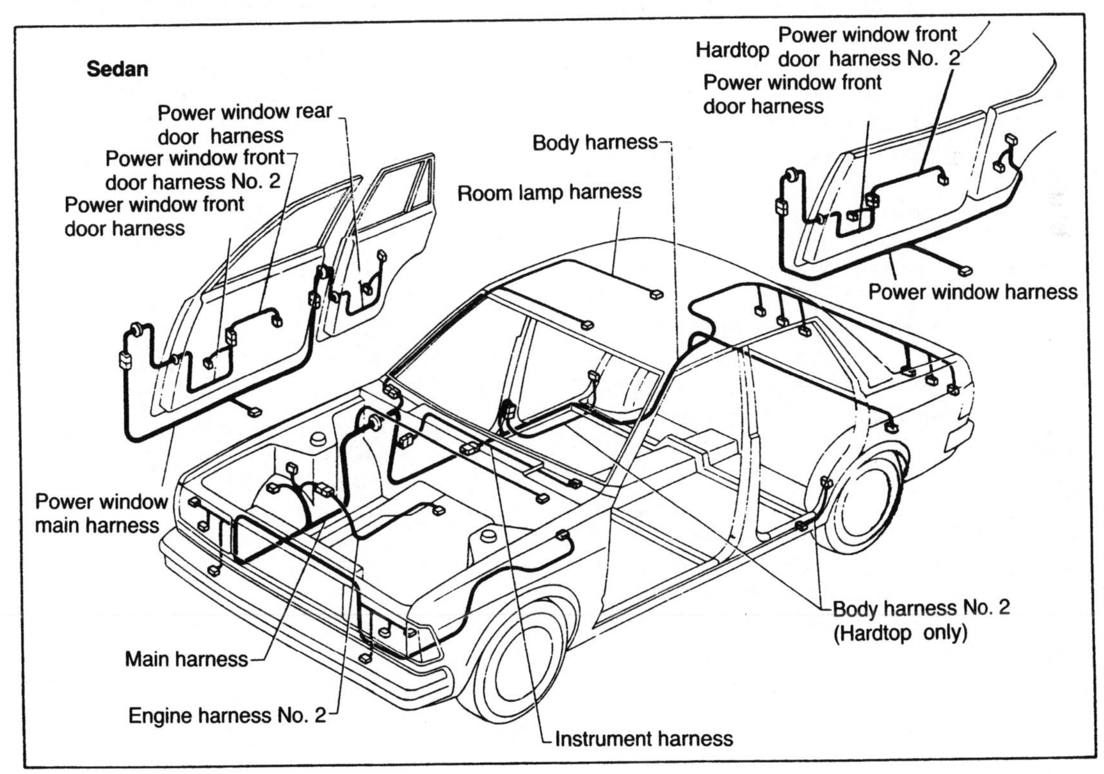

Source: Nissan Australia

problem usually requires the services of an extremely competent automotive technician and/or the use of specialised test equipment. To become this person you will require many years of experience and study. This chapter will introduce you to the first stage in understanding and servicing vehicle electronics.

COMPONENT IDENTIFICATION AND FUNCTION

The electrical components of the modern vehicle are designed and manufactured to the highest standards. The correct identification of these components and their function is essential for:

- understanding the operation of the electrical system;
- ordering replacement parts;
- servicing the system in a safe and efficient manner.

The vehicle's electrical system comprises of several systems which operate independently. They are:

- **Starting**: provides a method of cranking the engine prior to starting.
- **Ignition**: produces a high voltage spark at a spark plug located in the engine's cylinder head.
- **Lighting**: illuminates the roadway and marks the vehicle's width for night driving.
- **Charging**: ensures the battery is kept at a voltage level that will sustain the operation of all the electrical systems for a reasonable period.
- **Warning**: allows the driver to alert other traffic and/or pedestrians of the vehicle's presence and directional movements in the roadway.
- **Safety**: ensures the driver has a clear vision at all times and that the vehicle is protected from electrical fires.
- **Instruments**: provide the driver with vital operational information from all sections of the vehicle.

MAJOR COMPONENTS OF THE SYSTEMS

Each electrical circuit in a system comprises of a:

- power source;
- protection device/s;
- control unit/s;
- wiring harness;
- work unit/s.

POWER SOURCE

The source of electrical power in motor vehicles is the battery. The battery stores electricity in a chemical form until it is needed by a system.

Generally, the battery is located under the bonnet near the front of the vehicle and is attached to the electrical system by two cables. The 12 V direct current supply required for the ignition system is supplied from the battery.

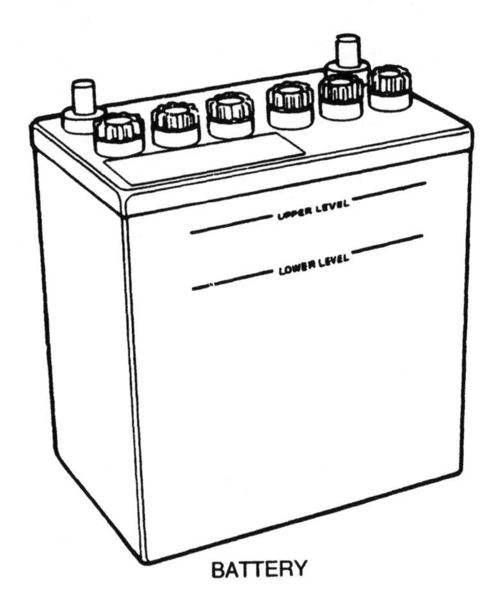

BATTERY

PROTECTION DEVICE

A protection device is included in an electrical circuit to minimise the chance of fire being started by a faulty electrical component. Three main types are:

1 fusible links — which are located near the battery;
2 fuses — located in a fusebox or fuse holder;
3 circuit breakers — located in or near the fusebox.

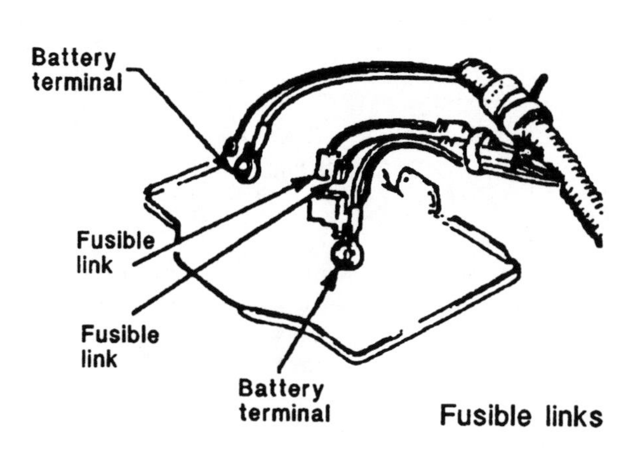

Fusible links

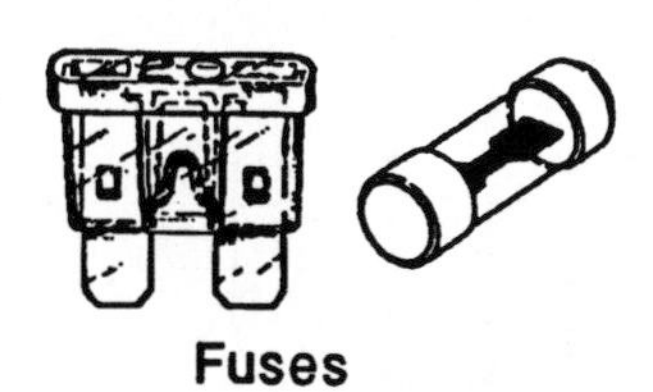
Fuses

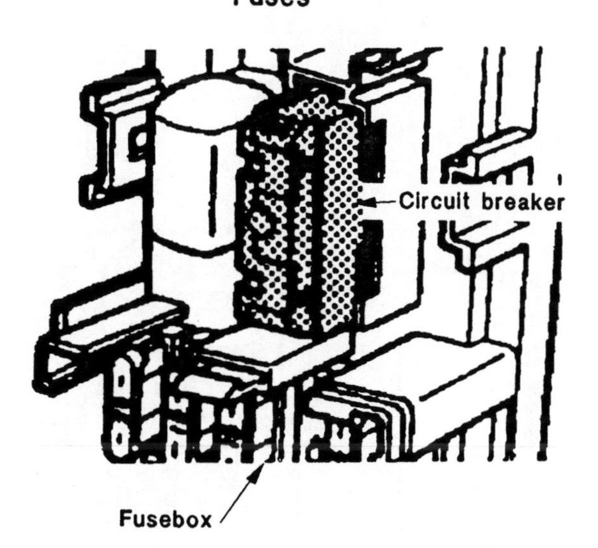

CONTROL UNIT

The main types of control units are switches and relays. Their purpose is to connect and disconnect the power source to and from the work unit, respectively. Generally, the switches are located in reach of the driver and the relays may be included in the fusebox or near the work unit.

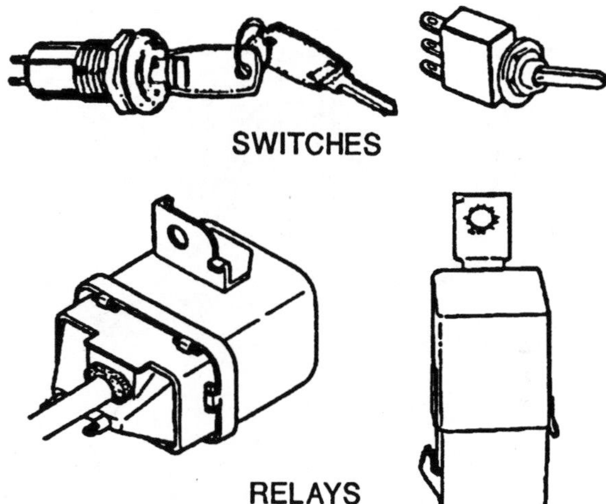

WIRING HARNESS

Each work unit is connected to the battery by two wires — the supply wire and the return wire. The return wire is the shorter because the vehicle's chassis is a part of the return to the battery. All the wires from the various systems are neatly bundled together and wrapped with tape to form a wiring harness. The wiring harness can be observed at various locations throughout the body of the vehicle.

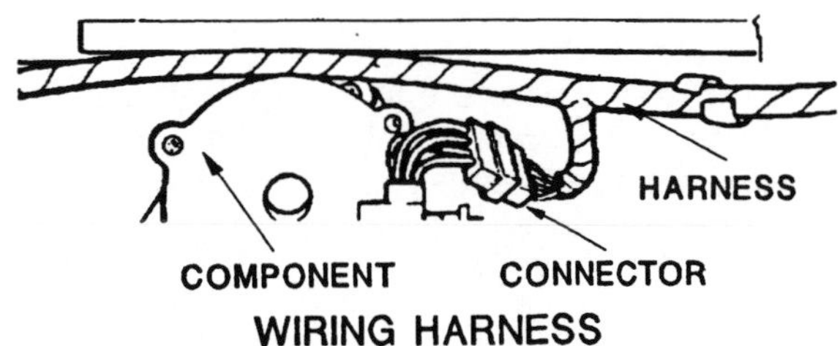

WIRING HARNESS

WORK UNIT

A work unit converts electrical energy into another form of energy that can be used to produce the required effect. The basic work units are:

- Motors, which convert electrical energy into mechanical energy. They include starter motors and windscreen wiper motors.

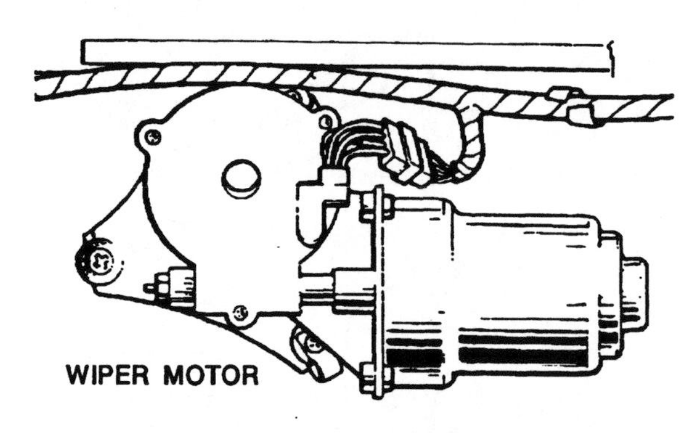

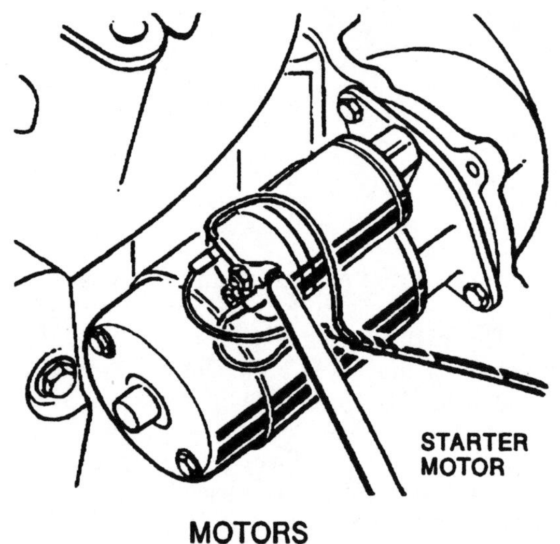

MOTORS

- Lights, which convert electrical energy into light rays. There are many applications for lights in the motor vehicle, but the main ones are head lights, tail lights, stop lights and trafficators (turn indicators).

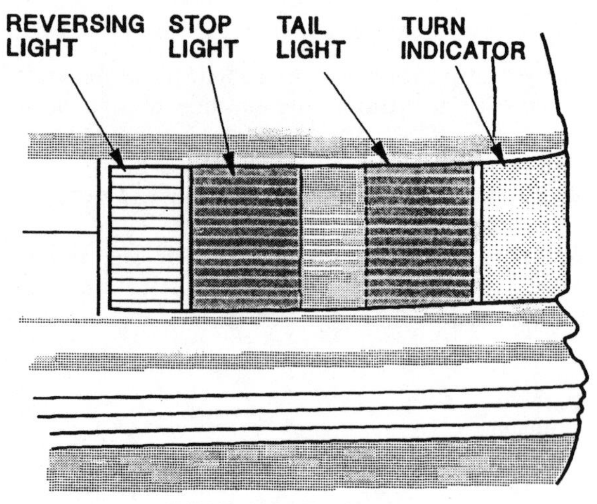

REAR LIGHTING

- Electro-magnets, which convert electrical energy into magnetic fields. Some of the devices which are operated by an electro-magnet are horns, relays and fuel injectors.

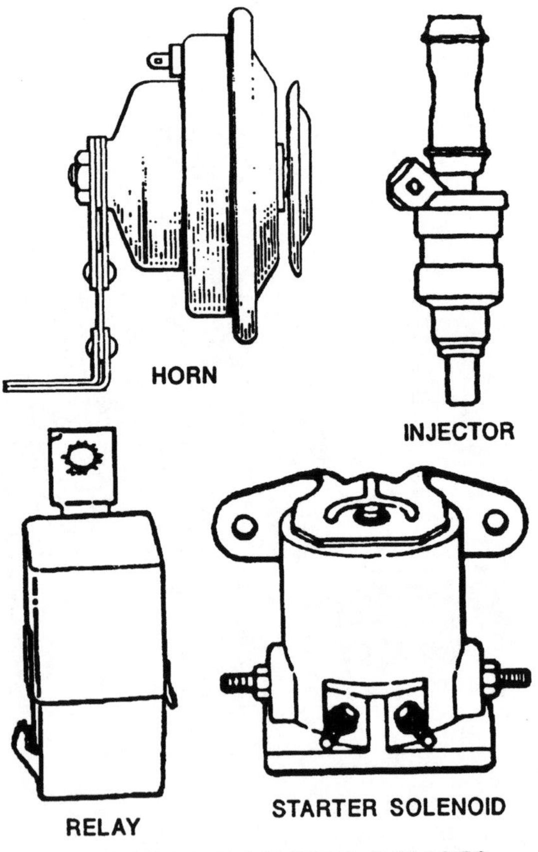

ELECTRO-MAGNETIC DEVICES

- Heaters, which convert electrical energy into heat. The main applications for heaters are in rear window demisters, thermal type instruments and protection devices.

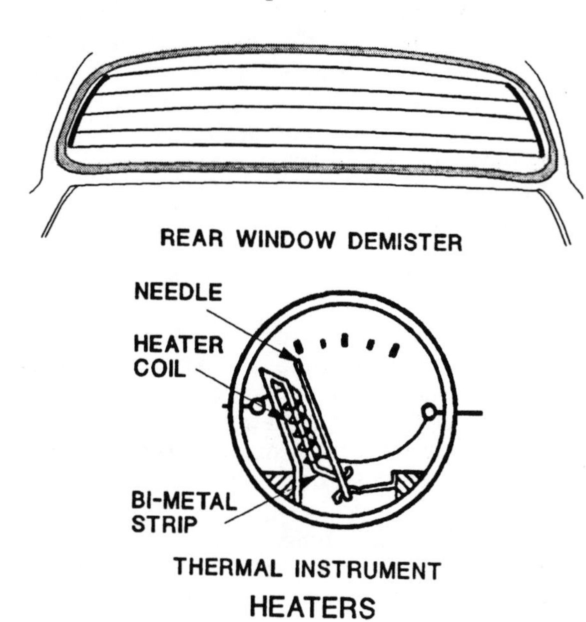

HEATERS

IGNITION SYSTEM

Mechanically triggered ignition

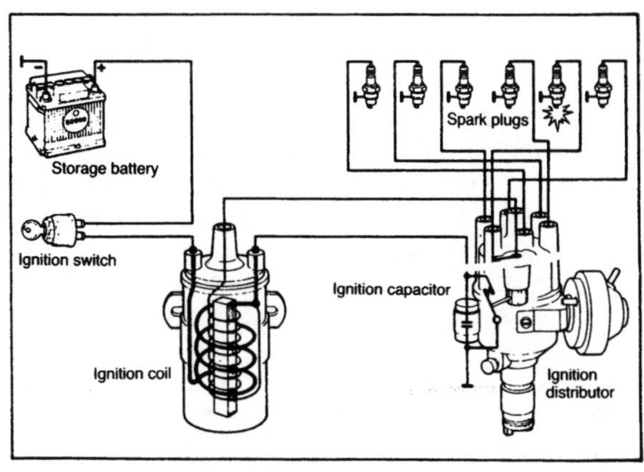

—Conventional coil ignition system (six cylinder engine).

A typical ignition system requires a voltage supply from the vehicle battery via the ignition switch. The major parts of the system are the:

- battery;
- ignition switch;
- ignition coil;
- distributor;
- high tension leads;
- spark plugs;
- low tension wiring;
- ballast resistor (if fitted).

The two functions of an ignition system are:

1 to produce, from a 12 V battery source, a voltage (8000 to 40 000 V) of sufficient intensity to 'jump' a spark plug gap and ignite the air/fuel mixture in the combustion chamber of the engine.

2 to ensure the high intensity voltage impulse is delivered to the correct spark plug at the precise instant for full power development.

IGNITION SWITCH

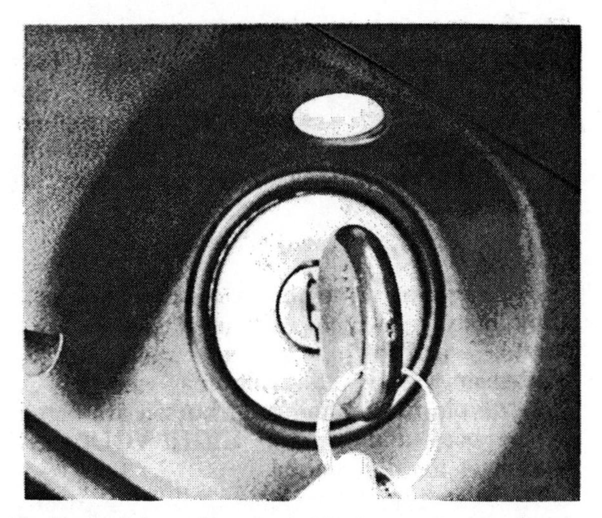

Ignition switch combines Lock, Off, Accessories, On (Ign) and crank functions

The ignition switch, which is operated by the driver, is usually part of a combined key switch located on or near the steering column. The purpose of the switch is to turn the ignition on and off, enabling the engine to be started and stopped.

IGNITION COIL

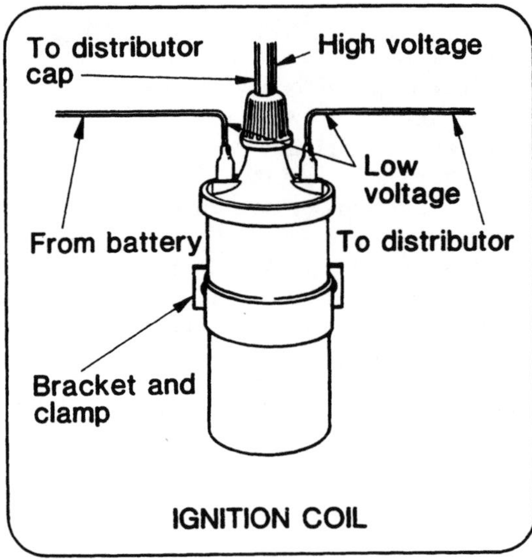

Source: AGPS

The ignition coil is securely mounted to the engine or inner guard by a bracket. Two internal windings form a 'step up' transformer which can raise the 12 V supplied from the battery to the voltage (8000–40 000 V) required to ignite the air/fuel mixture in the engine cylinder.

DISTRIBUTOR

The distributor assembly consists of a cast alloy body in which is housed the contact points, ignition advance mechanisms, condenser and distributor shaft. A bakelite/plastic cap and rotor button are fitted to the body and are the HIGH VOLTAGE section of the distributor.

The distributor is usually driven through gears from the camshaft and has been fitted in many different locations around the engine. The function of a distributor is to:

- switch on and off the current flow through the coil;
- receive the high voltage impulse from the ignition coil and 'distribute' the voltage to the correct cylinder;
- precisely 'time' the voltage delivery to the required cylinder taking into consideration the engine speed (r.p.m.) and load conditions.

The major parts of the distributor are:

Distributor cap assembly. The distributor cap (moulded from bakelite or similar material) must have excellent insulation properties. Each H.T. tower has a metal insert through the cap in a position that will enable the rotor button to pass with a small clearance (1 mm). The coil lead insert is fitted with a carbon brush and spring which allows a rubbing contact to be maintained with the rotor button.

The distributor cap and the rotor button form a rotary switch that applies high voltage impulses from the coil to the spark plugs in the correct sequence.

Rotor button (rotor head assembly). The rotor button is formed of similar insulation material to the distributor cap. It is moulded to fit closely over the distributor shaft and a lug on the rotor button ensures the shaft and button rotate together. The purpose of the rotor button is to accept the high voltage impulse from the ignition coil after it has travelled through the high tension coil lead and the carbon insert in the distributor cap. The voltage is conducted along the metal strip in the rotor button and directed to a distributor cap spark plug insert at a precise time.

Contact set. The contact set, usually referred to as distributor points, consists of two points:

1 a fixed or stationary contact point secured to the distributor plate assembly;
2 a movable breaker arm point fitted with a fibre rubbing block and a spring strip.

The contact point set opens and closes the ignition coil circuit.

Vacuum advance controller. The assembly is connected by a small rubberised hose to the carburettor or engine inlet manifold. The internal diaphragm is linked to the moveable contact breaker plate. Ignition timing can be altered by the vacuum advance unit to suit engine load conditions.

Distributor body (housing). The distributor body, which is the lower section of the distributor, is a die cast housing containing the shaft bushings, cap clips and a mounting flange. It provides a rigid mounting structure for the other distributor parts. A clamping bolt locks the body to the engine. The body is

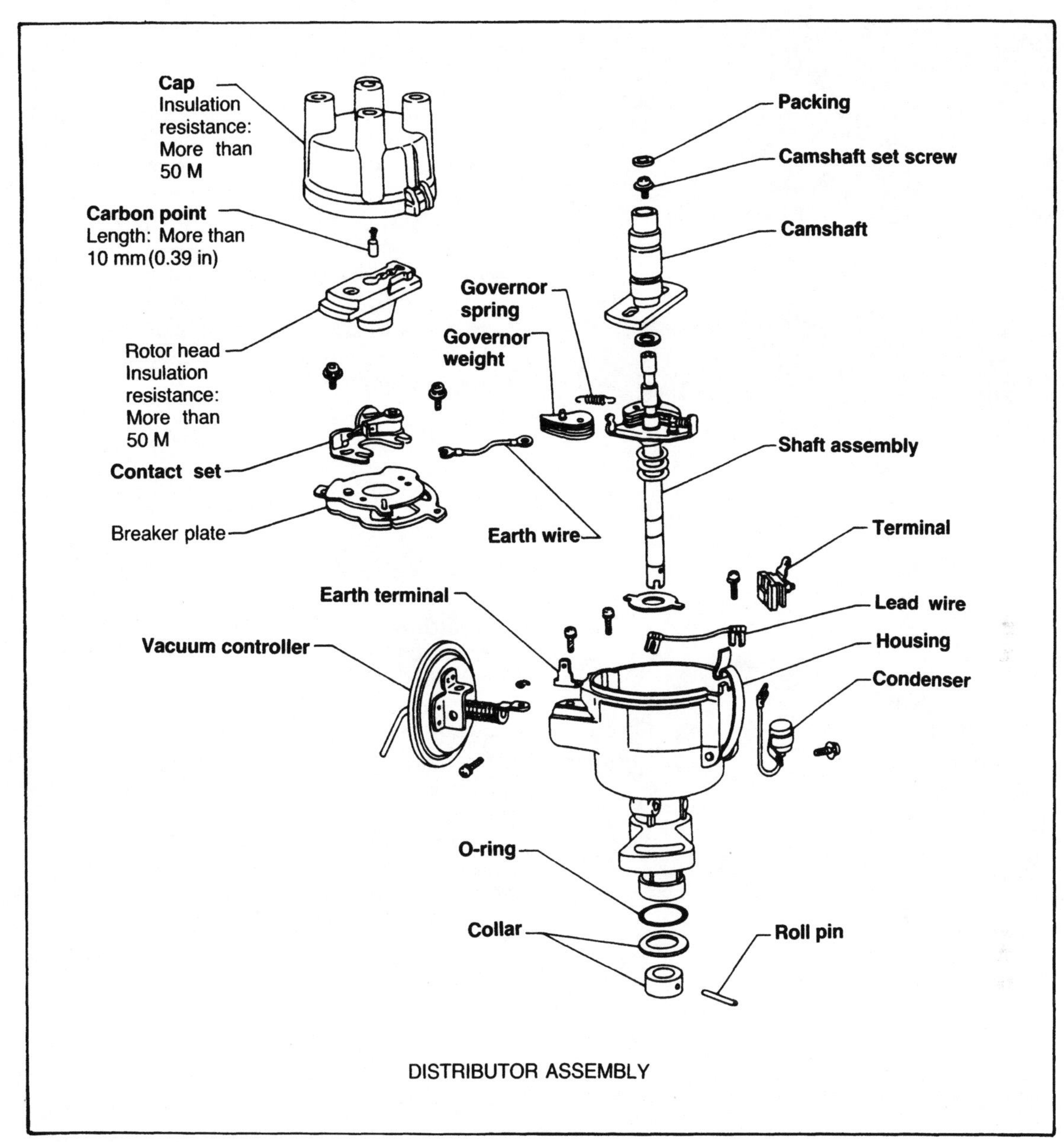

DISTRIBUTOR ASSEMBLY

Source: Nissan Australia

usually fitted into a circular hole in the engine to allow the initial ignition timing to be set to manufacturer's specifications.

Condenser assembly. The condenser is a small metal canister which is securely clamped to the distributor. The condenser may be mounted externally or internally within the distributor body. An insulated 'pigtail' wire protrudes from the condenser and is connected to the movable contact breaker arm terminal. The condenser is fitted to speed up the collapsing action of the magnetic field in the ignition coil and to prevent arcing across the distributor contact points.

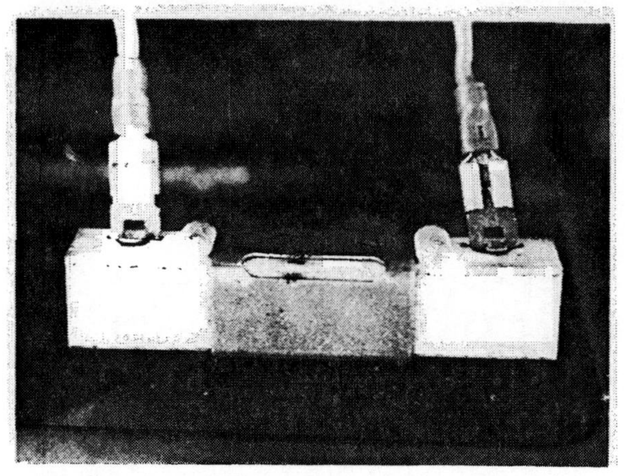

Ceramic block type ballast resistor

BALLAST RESISTOR

The ballast resistor, if any is fitted, may consist of a specified length of special resistance wire or may take the form of a resistance wire embedded in a ceramic block. The resistor is connected into the low tension wire between the IGN. terminal of the ignition switch and the positive (+) terminal of the ignition coil. The function of the ballast resistor is to control low tension current flow through the ignition coil, particularly at low and high engine speeds.

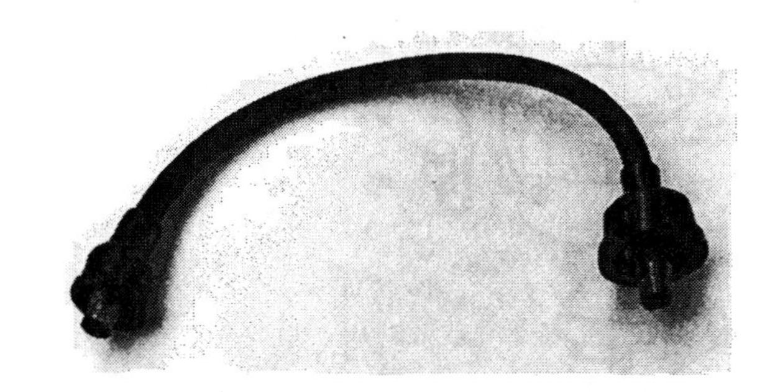

High tension lead

HIGH TENSION LEAD

The high tension leads are manufactured with a conductive centre which is covered by a thick insulating material. All current vehicles have leads with a silicone or carbon impregnated centre to prevent interference to radio and television reception. A metal H.T. terminal is fitted to each end of the lead and covered by a rubberised insulator 'boot'. The primary function of the H.T. leads is to conduct the high voltage produced by the ignition coil to the distributor and from the distributor to the spark plugs.

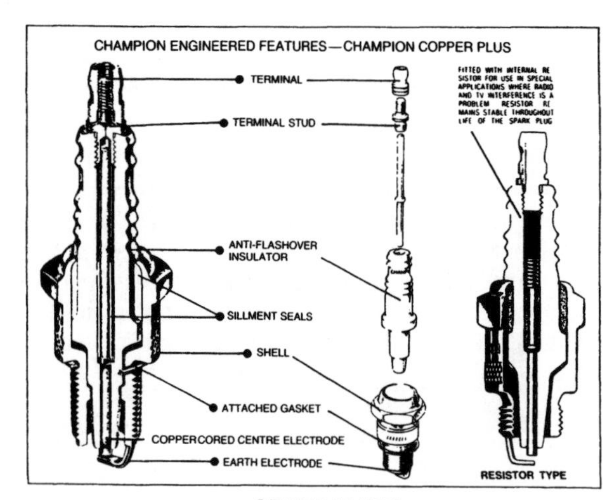

SPARK PLUGS

SPARK PLUGS

The spark plug screws into the cylinder head in a position that allows its centre and earth electrodes to protrude into the combustion chamber.

Generally, spark ignition engines have one spark plug for each cylinder (six cylinders — six spark plugs), but there are a number of special applications which use two spark plugs in each cylinder.

The spark plug provides a spark gap inside the combustion chamber of the engine. When high voltage 'jumps' the plug gap a spark is created, which ignites the air/fuel mixture.

Electronically triggered ignition system

The electronic ignition system contains many components which are similar in external appearance to those in the mechanically triggered system. There are several different types of electronic ignition available. These include systems based on magnetic pulse generators, Hall effect generator, infra-red pulses or flywheel induction to trigger the ignition module or computer.

A typical electronic ignition system consists of a:

- battery;
- ignition switch;
- ignition coil;
- distributor;
- control module (trigger box);
- H.T. leads;
- spark plugs;
- low tension wiring;
- ballast resistor (if fitted).

The three components, in a basic electronic ignition system, which vary significantly from the mechanically triggered point contact system are the:

- ignition coil;
- distributor;
- control module.

You will be required, in later studies, to develop an understanding of several of the more advanced electronic ignition control systems. The system described below is relatively simple by modern standards.

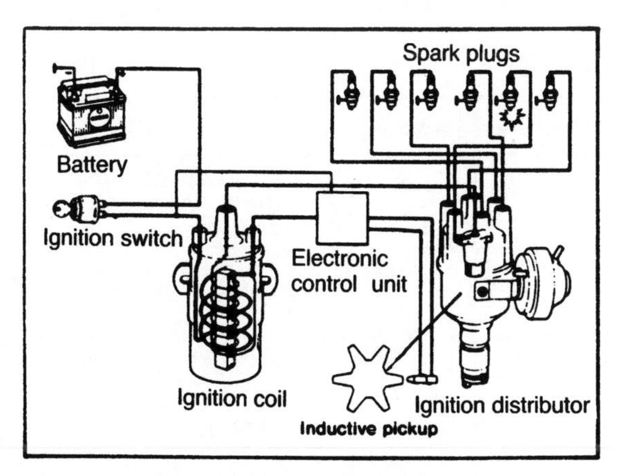

Breakerless high energy ignition system with miniature electronic control unit.

IGNITION COIL

Ignition coil suitable for electronic ignition

The ignition coil appears similar in appearance to the conventional unit but the internal construction allows higher current flow and increased voltage output.

ELECTRONIC DISTRIBUTOR

The distributor has a magnetic pulse generator in place of the contact point set. A typical pulse generator consists of a:

- pick up coil — a circular coil of wire from which two wires protrude to provide a connection to the control module (trigger box);
- permanent magnet — circular in shape and clamped to the generator coil;
- reluctor (rotor) — replaces the upper cam section of the conventional distributor shaft;
- stator — provides a series of protrusions, usually one for each cylinder, which momentarily align with the rotating reluctor and produce an electrical signal from the pulse generator.

CONTROL MODULE (TRIGGER BOX/ IGNITER)

The control module receives the signal from the pulse generator and turns the ignition coil on and off in a similar manner to the mechanical contact point set. Modern versions of the control module limit coil primary current flow and can automatically increase the time current flows to the ignition coil as engine speed increases (dwell period).

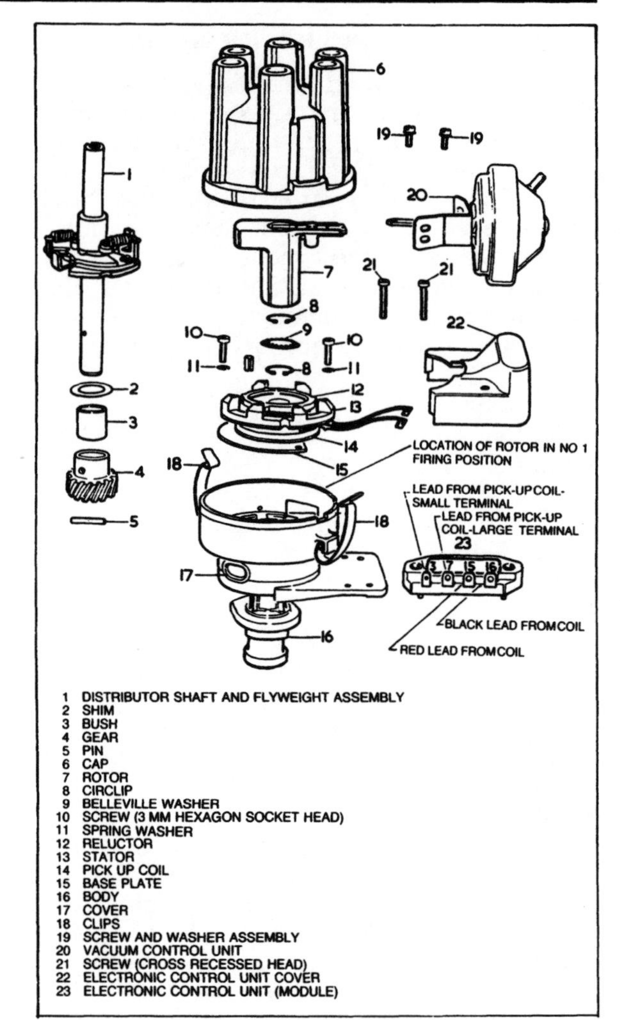

HIGH ENERGY IGNITION SYSTEM

Connections to the electronic control unit

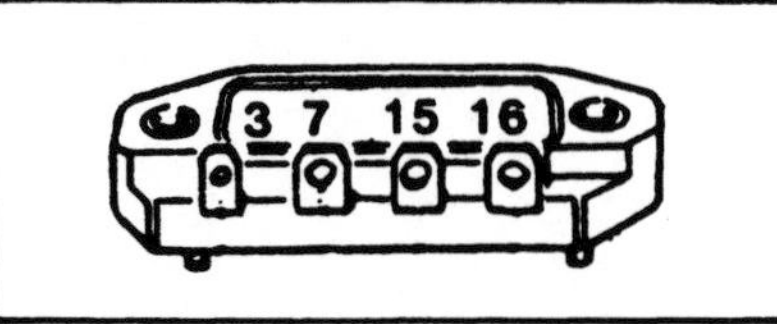

IGNITION CONTROL MODULE

TERMINAL	WIRE COLOUR	ROUTING
3	BLACK	FROM PULSE GENERATOR SMALL TERMINAL
7	BLACK	FROM PULSE GENERATOR STD TERMINAL
15	RED	POSITIVE FEED FROM IGNITION COIL POSITIVE TERMINAL
16	BLACK	FROM IGNITION COIL NEGATIVE TERMINAL

ELECTRONIC CONTROL UNIT WIRING CONNECTION

IGNITION SYSTEM

TUNE-UP

Ignition systems deteriorate in use and require servicing at specified intervals to maintain engine performance. Most potential ignition faults can be detected and/or eliminated during a normal tune-up.

Since the procedure for tuning a mechanically triggered ignition system and a basic electronic ignition system are similar, the following description will be of a general nature except where specific detail needs to be given.

Preliminary visual inspection: ignition components

1 Check H.T. leads for cracked or burnt insulation.
2 Check low tension wiring for frayed, burnt or worn insulation and/or loose terminals.
3 Check ignition coil for case damage, oil leakage or damaged insulation.

Removal, inspection and installation: ignition components

DISTRIBUTOR CAP AND H.T. LEADS

1 Check spark plug leads for cylinder numbering. If marks are not present it may be necessary to mark leads with masking tape to assist reassembly.

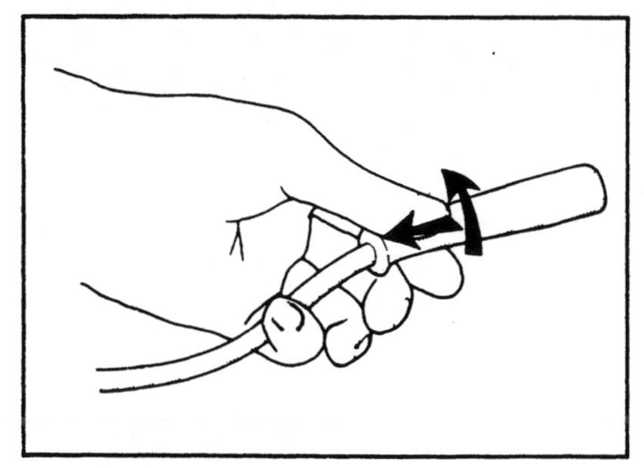

Removing spark plug cable

Source: Mitsubishi Magna TM series manual

Check spark plug leads for cylinder numbering

2 Remove the H.T. leads. Avoid damage to the inner core of the H.T. lead by twisting the insulator 'boot' slightly as the lead is removed. Never pull on the lead.

3 Remove the coil H.T. lead at the ignition coil tower and inspect the socket for corrosion.

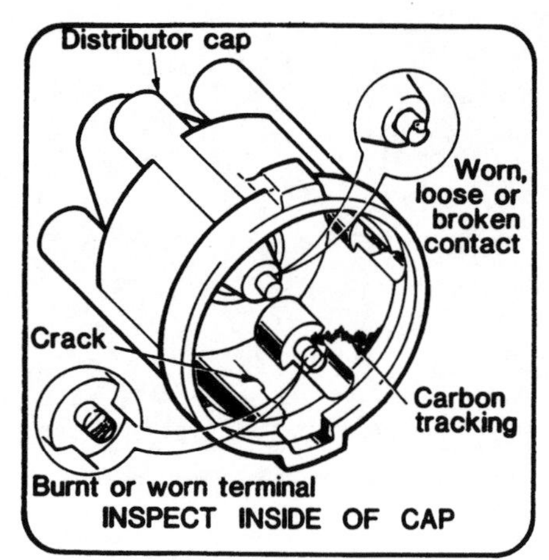

Source: AGPS

4 Release the distributor cap clips, remove and inspect the cap for tracking, cracks, corrosion or burnt terminals.

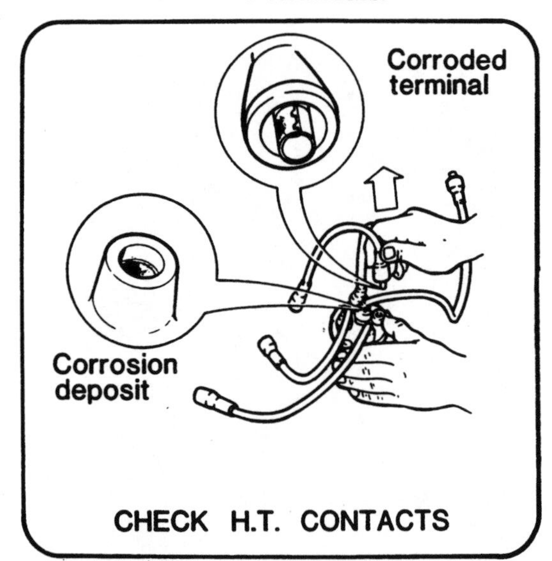

Source: AGPS

5 Remove the H.T. leads one at a time from the distributor cap and check each tower socket and lead terminal for corrosion.
6 Check the H.T. lead for breaks (continuity) with an ohmmeter switched to the 1000 scale. Refer to the section on test instruments for the technique of using an ohmmeter.
7 Clean the distributor cap and leads.
8 Attach the H.T. lead to its respective tower socket.

ROTOR BUTTON

Remove, clean and inspect rotor button.
Caution: the removal of metal from the blade tip will increase the air gap and cause a loss in voltage at the spark plugs.

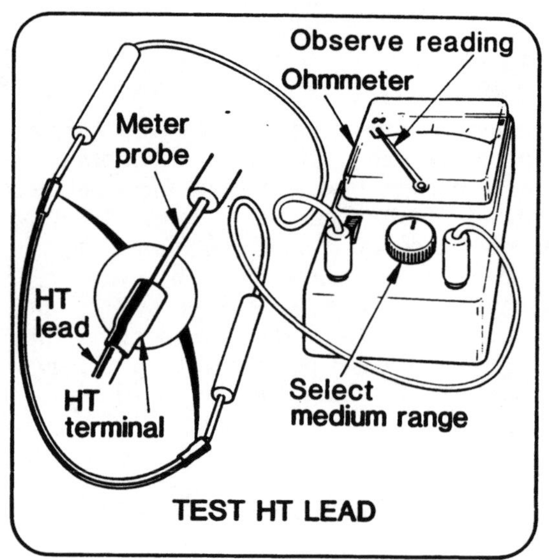

Source: AGPS

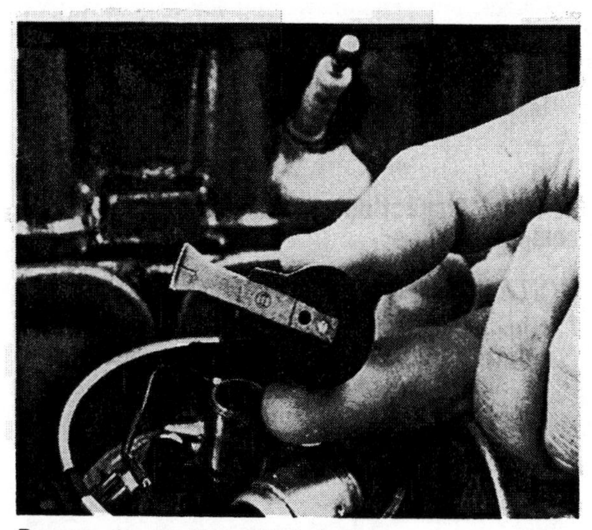

Remove, clean and inspect rotorbutton.

SPARK PLUGS

1 Select the correct spark plug socket.
2 Loosen each spark plug a half to one full turn.
3 Remove loose dirt from spark plug recess with an air nozzle (wear eye protection).
4 Unscrew each spark plug and place them on a bench in cylinder order.

CLEANING SPARK PLUG AREA WITH COMPRESSED AIR

SPARK PLUGS

Note: A skilled mechanic can determine the condition of an engine cylinder by inspecting the spark plug.

Source: AGPS

5 Clean spark plug threads.
6 Set spark gap to manufacturer's specification with a gap adjusting tool.
7 Install spark plugs.

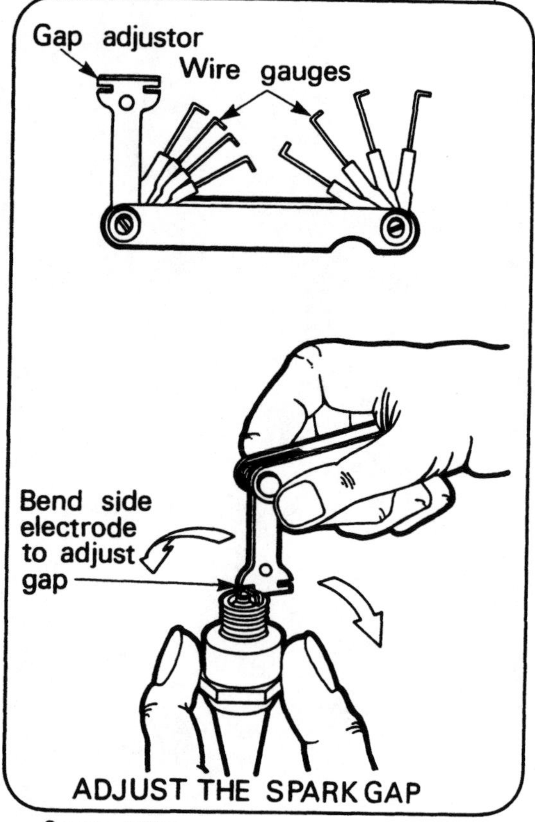

Source: AGPS

SPARK PLUG INSTALLATION

Always make sure that the area around the spark plug is clean before removing the spark plug. If necessary, use compressed air.
Always use a plug spanner, making sure the socket is properly seated on the spark plug hexagon.
Champion's attached steel plug gaskets do not fall off during installation — and they can be reused.
Tightening spark plugs is important. Following are recommendations for final tightening, after the plug has been seated by hand::

Spark Plug with Gasket

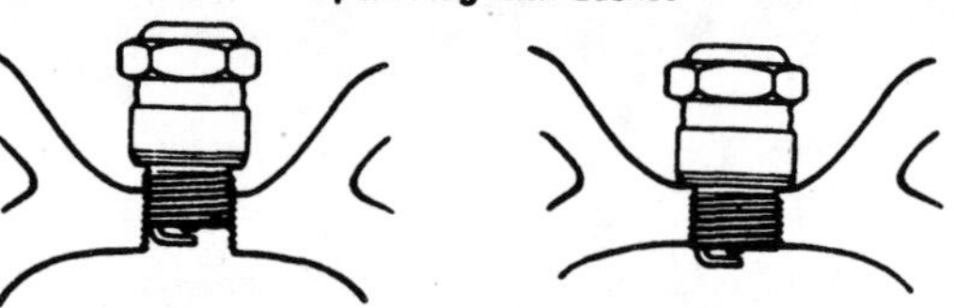

Spark plugs with gaskets should be tightened about 1/4 turn past finger tight to effect a gas-tight seal. (See illustration «A»).

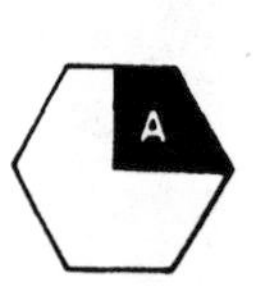

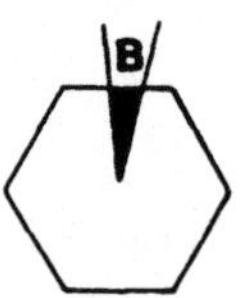

Taper Seat Spark Plug

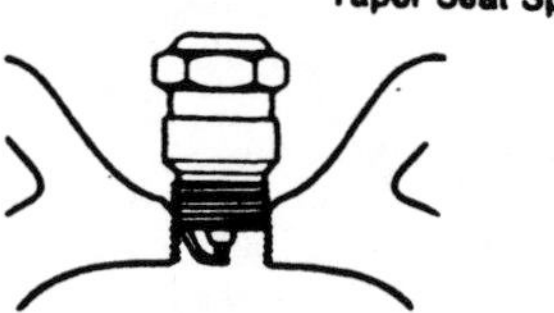

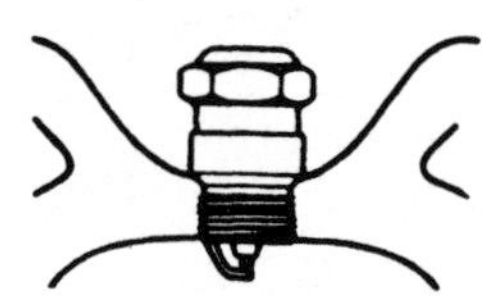

Spark plugs without gaskets (taper seat design) should be tightened only firmly enough to assure a gas-tight seal.
(See illustration «B»).

Care should be taken not to overtighten spark plugs, particularly . . . taper seat types . . . types fitted to Mazda rotary engines . . . types fitted to Aluminium cylinder heads.

Distributor service: contact breaker points

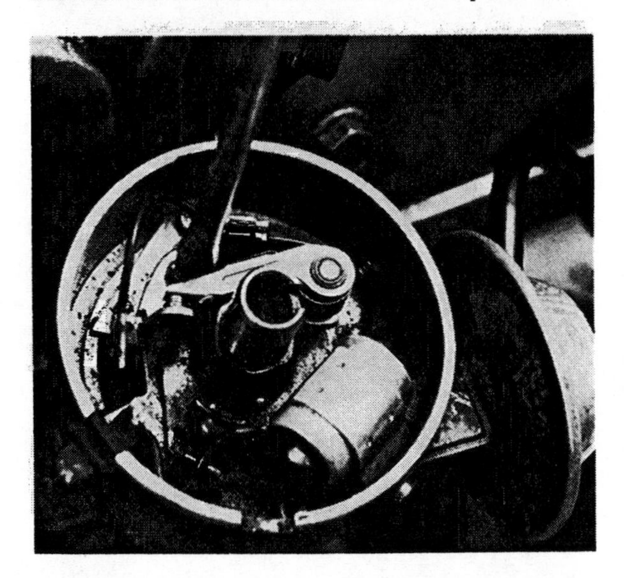

1 Remove contact breaker points.
- Disconnect the L.T. lead at the breaker arm. It usually has a slide or bolt-on L.T. terminal.
- Unscrew the retaining screws holding the fixed contact point. Fold the earth wire (if fitted) clear of the point.

Inspect contact points

2 Inspect the contact point set for:
- frayed L.T. lead;
- wear of the fibre rubbing block;
- wear, pitting, burning or bluing of the point faces. Clean and reface contact points only when new parts are not available.

An enlarged view of the contact set points

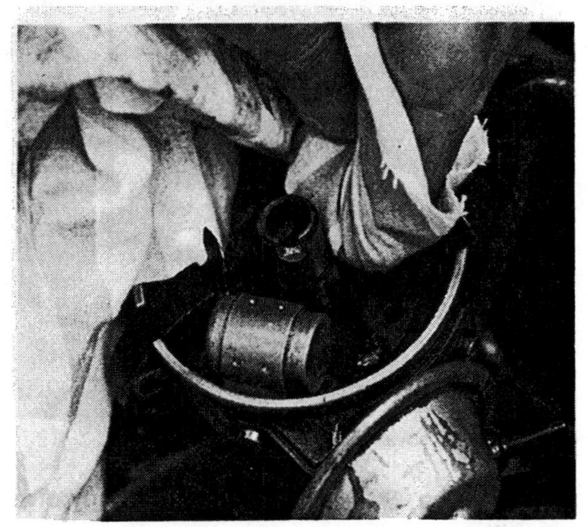

3 Wipe the cam, movable plate and internal parts with cloth soaked with solvent, then

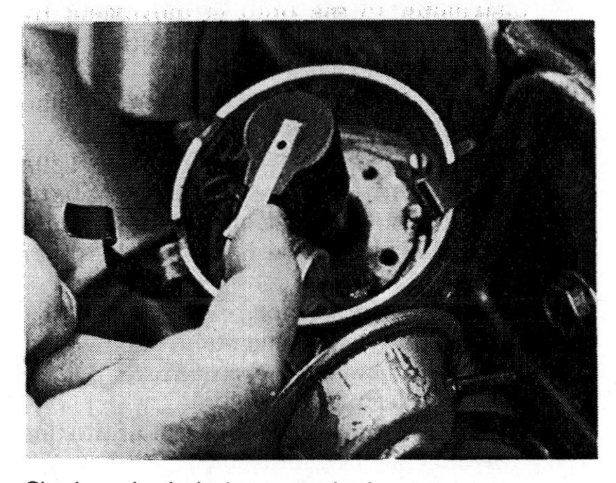

Check mechanical advance mechanism

wipe dry. Inspect all parts for wear or damage.

4 Check the operation of the mechanical advance mechanism.
 - Temporarily refit the rotor button.
 - Press against the button in the direction of rotation.
 - Note that movement should occur against spring tension in one direction only.
 - Release the cam and it must spring back to its original position.

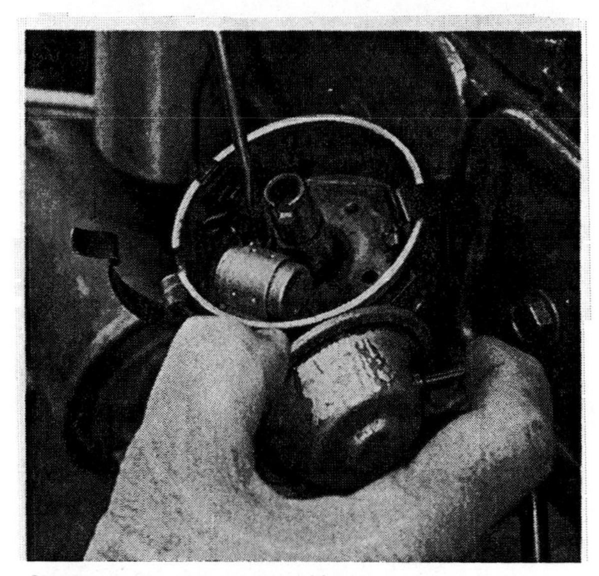

Check vacuum advance assembly

5 Check the operation of the vacuum advance unit.
 - Remove the vacuum pipe to the advance unit.
 - Rotate the movable plate of the distributor to the limit of movement by applying force to the connecting link with a screwdriver.
 - Place your finger over the unit pipe fitting, then release the movable plate.
 - Wait two seconds then remove your finger from the pipe fitting. Slight suction should be felt and in many units a slight click will be heard as the plate seats.
 - Refit the vacuum pipe.

6 Install the new contact points in the reverse order to the dismantling procedure.

7 Set the contact point gap.
 - Turn the engine crankshaft until the breaker arm rubbing block is resting on the high point of one of the cam lobes.

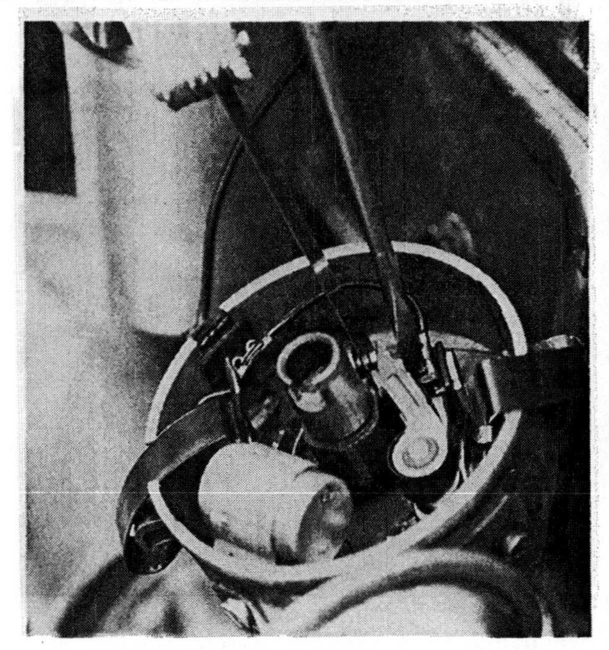

 - Adjust the gap until a feeler strip (as specified by manufacturer) is a sliding fit between the contact points.
 - Tighten the retaining screws.

8 Lubricate the distributor.
 - Place one or two drops of light oil at the points indicated in the diagram.
 - Smear a light film of grease onto the cam on the rubbing block areas.

To install the rotor button, distributor cap and leads:

1 Refit the rotor button, ensuring the button lug seats on the distributor shaft.

Install rotor button, disk cap and leads

Note: The rotor button should be a neat fit on the distributor shaft.

2 Refit the distributor cap.
  - Ensure that the lug is engaged in the slot.
  - Hold the cap firmly in place with one hand and push the clips on with your other hand.
  - Ensure that the clips are secure in their slots.

3 Refit the H.T. leads, ensuring the correct lead is placed on each spark plug (firing order).

IGNITION SYSTEM ADJUSTMENTS

Two adjustments are required after the contact points have been serviced:
1 dwell angle;
2 ignition timing.

Dwell angle adjustment

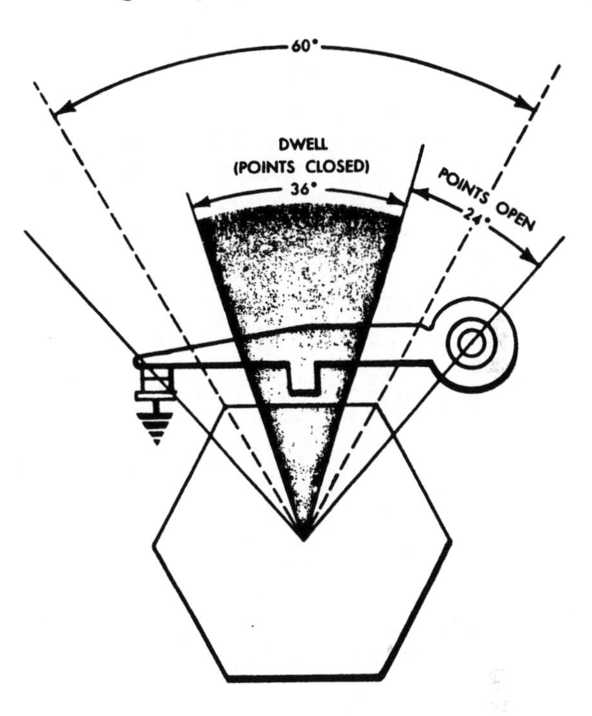

Dwell angle is the number of degrees the distributor shaft turns from where the contact points close, to where the rubbing block and cam open the points.

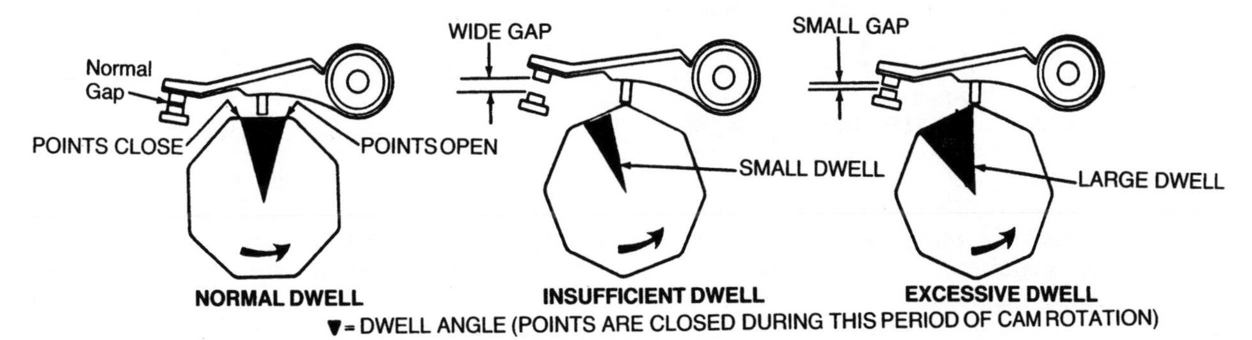

— Breaker Point Dwell

1 Connect a dwell meter to the engine.
Caution: carefully follow the manufacturer's instructions for the meter connection and operation.
2 Start the engine.
3 Compare meter reading to manufacturer's specification.
- When the dwell angle is outside specifications, stop the engine and re-adjust the distributor point gap.
- Repeat the dwell angle check.

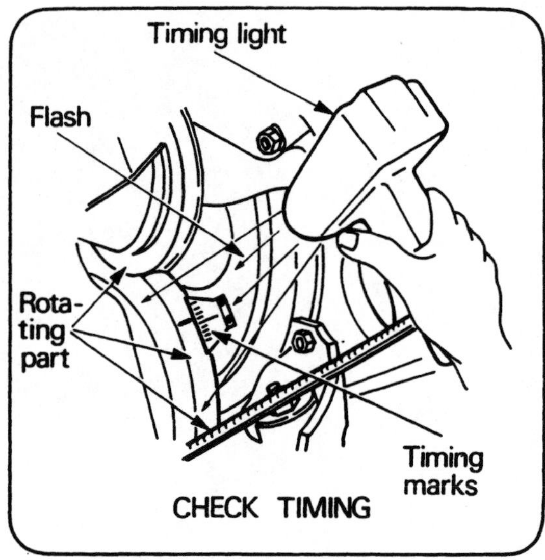

Source: AGPS

Ignition timing adjustment

The engine must be at normal operating temperature, idle speed within specification and timing marks located and wiped clean.
To check the ignition timing:
1 Disconnect and plug the vacuum advance pipe.

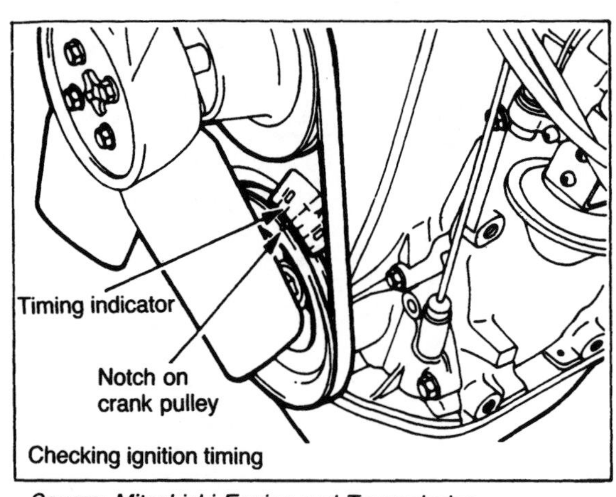

Checking ignition timing

Source: Mitsubishi Engine and Transmission workshop manual

2 Locate and clean the timing marks.
3 Connect a stroboscope (flashing) timing light to the engine as recommended by its manufacturer.
4 Start the engine and allow to idle.
5 Aim the timing light at the engine timing marks and check the alignment of the single mark against the degree scale each time the light flashes.
6 Compare reading to specification.

To adjust the ignition timing while the engine is running:
1 Loosen the distributor clamp bolt.
2 Rotate the distributor body until the timing marks are correctly aligned.
3 Tighten the distributor clamping bolt.

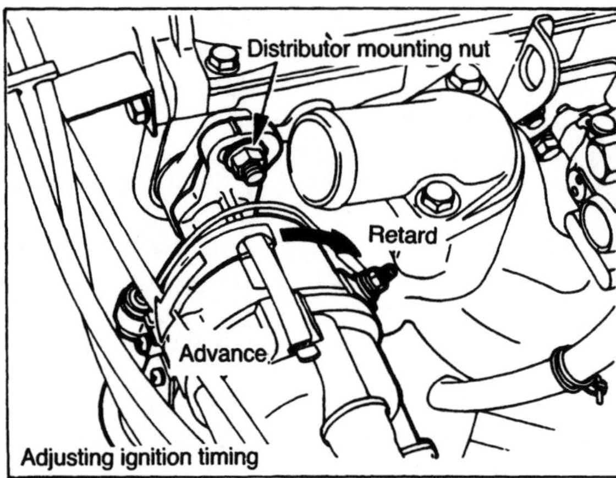

Adjusting ignition timing

Source: Mitsubishi Engine and Transmission workshop manual

4 Stop the engine, refit the vacuum advance unit pipe, remove all tools and test equipment.
5 Restart engine, adjust engine idle if required.

TUNING: ELECTRONIC IGNITION

Electronic ignition systems share many common components with the earlier design mechanical contact points systems. The servicing procedure for spark plugs, H.T. leads, distributor caps, rotor buttons and ignition timing is identical to the contact point systems. The distributor does not have contact points but will require mechanical and vacuum advance unit checks. (Current models fitted with engine management systems or electronic spark timing require special techniques to check ignition timing and may have computer-controlled spark advance curves.)

For your safety, learn the danger points on electronic high energy ignition systems.

Caution: when servicing an electronic ignition system, care must be taken to ensure both personal and equipment safety. The voltages present are potentially fatal. Be careful — learn the danger points and follow the rules.

The rules are:

1 DO NOT CONTACT THE IGNITION SYSTEM AT ANY OF THE TERMINAL POINTS SHOWN WITH A JAGGED ARROW.
2 DO NOT ROTATE THE ENGINE ON THE STARTER MOTOR WITH ANY H.T. LEAD DISCONNECTED.
3 DO NOT CONNECT OR DISCONNECT TEST EQUIPMENT OF ANY TYPE WITH THE ENGINE RUNNING OR THE IGNITION SWITCH IN THE 'ON' POSITION.
4 DO NOT CONNECT OR DISCONNECT THE BATTERY WITH THE IGNITION OR ANY SYSTEMS SWITCH IN THE 'ON' POSITION.
5 DO NOT CONNECT NON-APPROVED ACCESSORIES TO THE IGNITION SYSTEM.

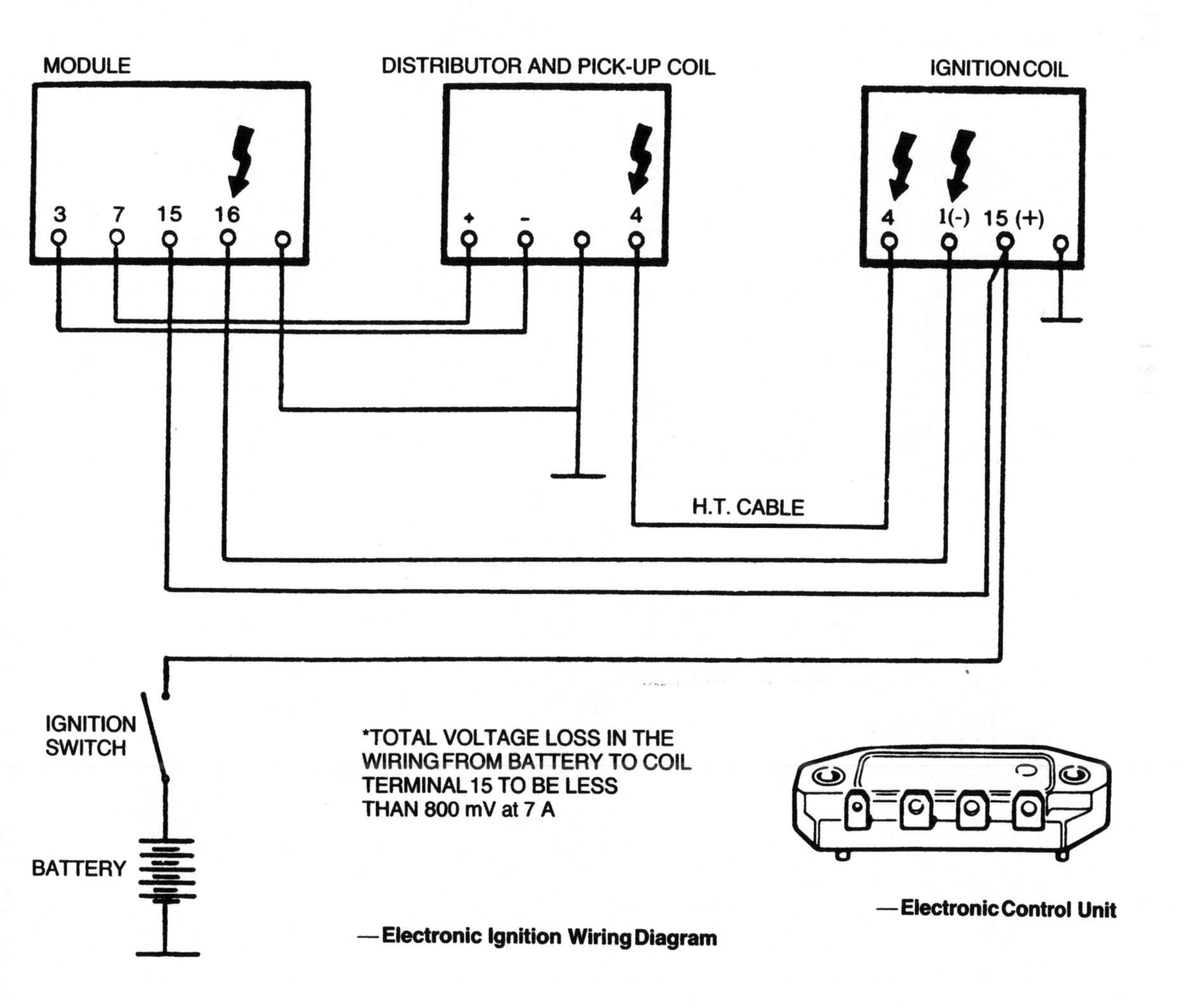

—Electronic Ignition Wiring Diagram

—Electronic Control Unit

SPARK PLUG IDENTIFICATION

There are many different types of engines and each engine has defined operating requirements for the spark plugs. To ensure the correct spark plug is fitted to an engine, the spark plugs are identified by a number code either on the insulator or stamped into the body. This coding describes, in conjunction with a chart, the following features of a spark plug.

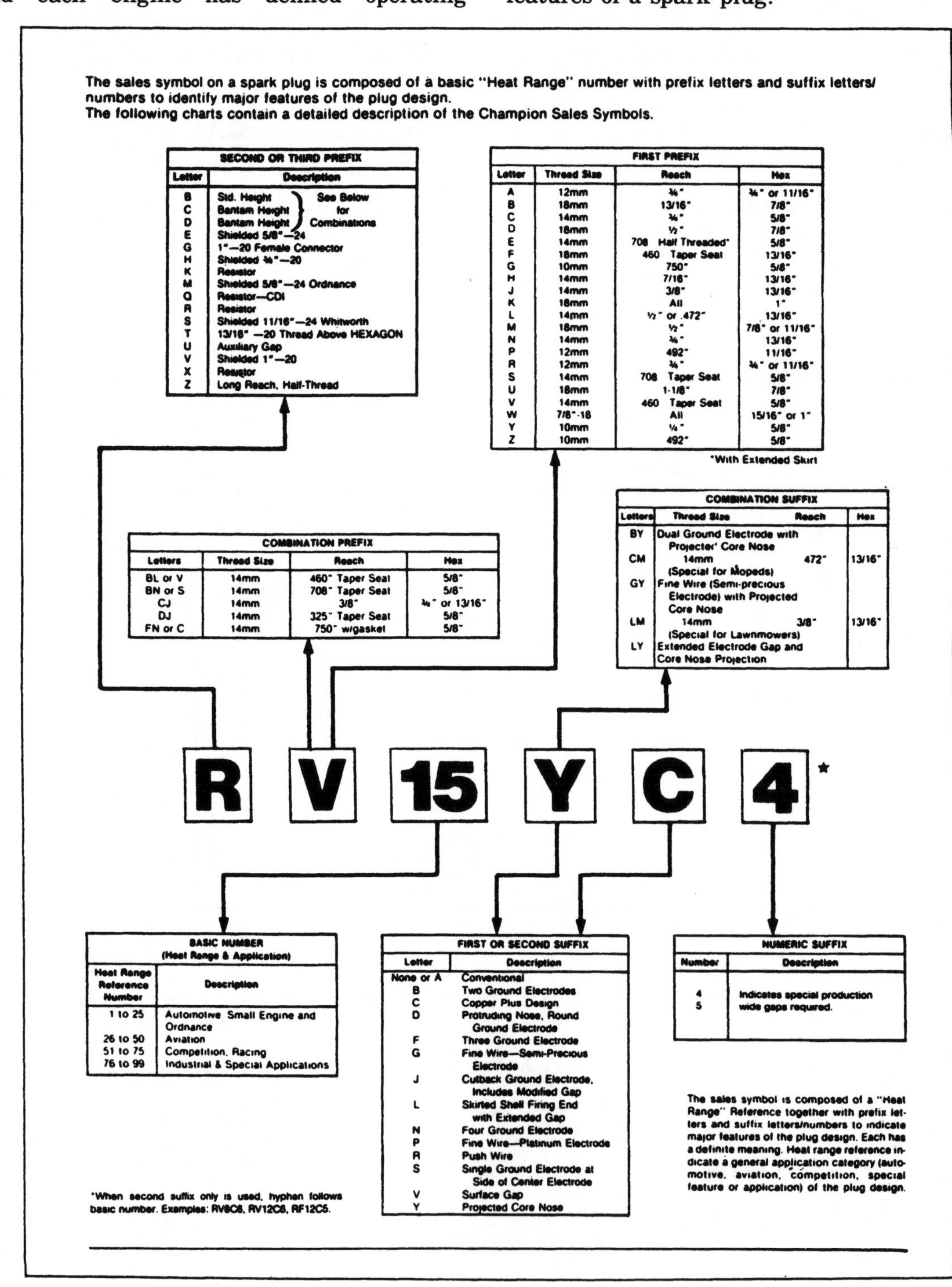

The sales symbol on a spark plug is composed of a basic "Heat Range" number with prefix letters and suffix letters/numbers to identify major features of the plug design.
The following charts contain a detailed description of the Champion Sales Symbols.

SECOND OR THIRD PREFIX

Letter	Description
B	Std. Height } See Below for Combinations
C	Bantam Height } See Below for Combinations
D	Bantam Height } See Below for Combinations
E	Shielded 5/8"—24
G	1"—20 Female Connector
H	Shielded ¾"—20
K	Resistor
M	Shielded 5/8"—24 Ordnance
Q	Resistor—CDI
R	Resistor
S	Shielded 11/16"—24 Whitworth
T	13/16" —20 Thread Above HEXAGON
U	Auxiliary Gap
V	Shielded 1"—20
X	Resistor
Z	Long Reach, Half-Thread

FIRST PREFIX

Letter	Thread Size	Reach	Hex
A	12mm	¾"	¾" or 11/16"
B	18mm	13/16"	7/8"
C	14mm	¾"	5/8"
D	18mm	½"	7/8"
E	14mm	708 Half Threaded*	5/8"
F	18mm	460 Taper Seat	13/16"
G	10mm	750"	5/8"
H	14mm	7/16"	13/16"
J	14mm	3/8"	13/16"
K	18mm	All	1"
L	14mm	½" or .472"	13/16"
M	18mm	½"	7/8" or 11/16"
N	14mm	¾"	13/16"
P	12mm	492"	11/16"
R	12mm	¾"	¾" or 11/16"
S	14mm	708 Taper Seat	5/8"
U	18mm	1-1/8"	7/8"
V	14mm	460 Taper Seat	5/8"
W	7/8"-18	All	15/16" or 1"
Y	10mm	¼"	5/8"
Z	10mm	492"	5/8"

*With Extended Skirt

COMBINATION PREFIX

Letters	Thread Size	Reach	Hex
BL or V	14mm	460" Taper Seat	5/8"
BN or S	14mm	708" Taper Seat	5/8"
CJ	14mm	3/8"	¾" or 13/16"
DJ	14mm	325" Taper Seat	5/8"
FN or C	14mm	750" w/gasket	5/8"

COMBINATION SUFFIX

Letters	Thread Size	Reach	Hex
BY	Dual Ground Electrode with Projecter' Core Nose		
CM	14mm (Special for Mopeds)	472"	13/16"
GY	Fine Wire (Semi-precious Electrode) with Projected Core Nose		
LM	14mm (Special for Lawnmowers)	3/8"	13/16"
LY	Extended Electrode Gap and Core Nose Projection		

BASIC NUMBER (Heat Range & Application)

Heat Range Reference Number	Description
1 to 25	Automotive Small Engine and Ordnance
26 to 50	Aviation
51 to 75	Competition, Racing
76 to 99	Industrial & Special Applications

*When second suffix only is used, hyphen follows basic number. Examples: RV8C8, RV12C8, RF12C5.

FIRST OR SECOND SUFFIX

Letter	Description
None or A	Conventional
B	Two Ground Electrodes
C	Copper Plus Design
D	Protruding Nose, Round Ground Electrode
F	Three Ground Electrode
G	Fine Wire—Semi-Precious Electrode
J	Cutback Ground Electrode, Includes Modified Gap
L	Skirted Shell Firing End with Extended Gap
N	Four Ground Electrode
P	Fine Wire—Platinum Electrode
R	Push Wire
S	Single Ground Electrode at Side of Center Electrode
V	Surface Gap
Y	Projected Core Nose

NUMERIC SUFFIX

Number	Description
4 5	Indicates special production wide gaps required.

The sales symbol is composed of a "Heat Range" Reference together with prefix letters and suffix letters/numbers to indicate major features of the plug design. Each has a definite meaning. Heat range reference indicate a general application category (automotive, aviation, competition, special feature or application) of the plug design.

a Reach: length of the threaded section
b Thread size: diameter of the tapped thread in the cylinder head.
c Body hexagonal: socket size required for the steel body.
d Resistor: centre electrode has inbuilt resistor for use in special application where radio or television interference is a problem.
e Seat type: tapered seat or sealing washer used as a sealing face to the cylinder head.
f Electrode: recessed, normal or projected core, positioning, number of electrodes, special features.
g Heat range: spark plugs are designed to operate within a specified temperature range. Temperature is controlled by the distance the heat must travel from the centre electrode to reach the metal body and enter the cylinder head. By increasing the length of the insulator foot the heat dissipation rate is slowed down and the plug runs HOTTER.

Caution: never use spark plugs with a heat range above the specification as severe engine damage may result.

The earlier chart is a typical example of the type of coding used by spark plug manufacturers. In addition to this type of information manufacturers also supply a chart so equivalent codings from other spark plug manufacturers can be cross-referenced.

LIGHTING

IDENTIFICATION

A motor vehicle uses globes in many locations. For each application it is necessary to consider the following globe specifications:

- socket type;
- space available for mounting;
- shape of globe;
- single or double filament;
- amount of illumination;
- wattage (power consumption);
- voltage.

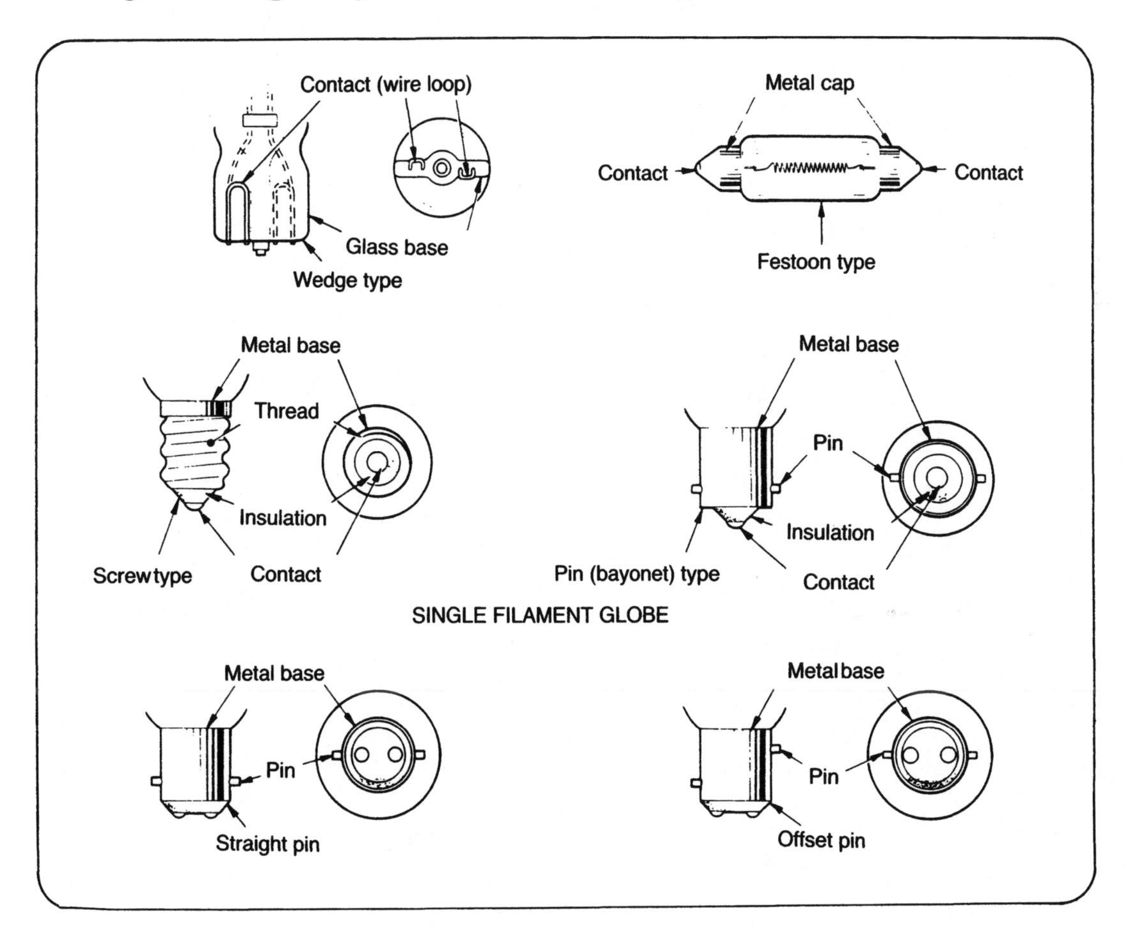

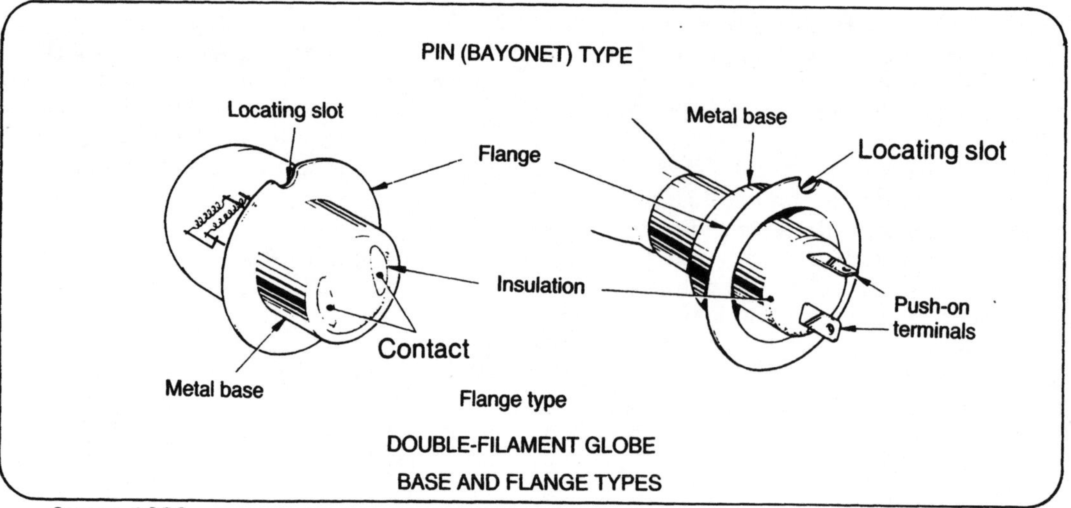

DOUBLE-FILAMENT GLOBE
BASE AND FLANGE TYPES

Source: AGPS

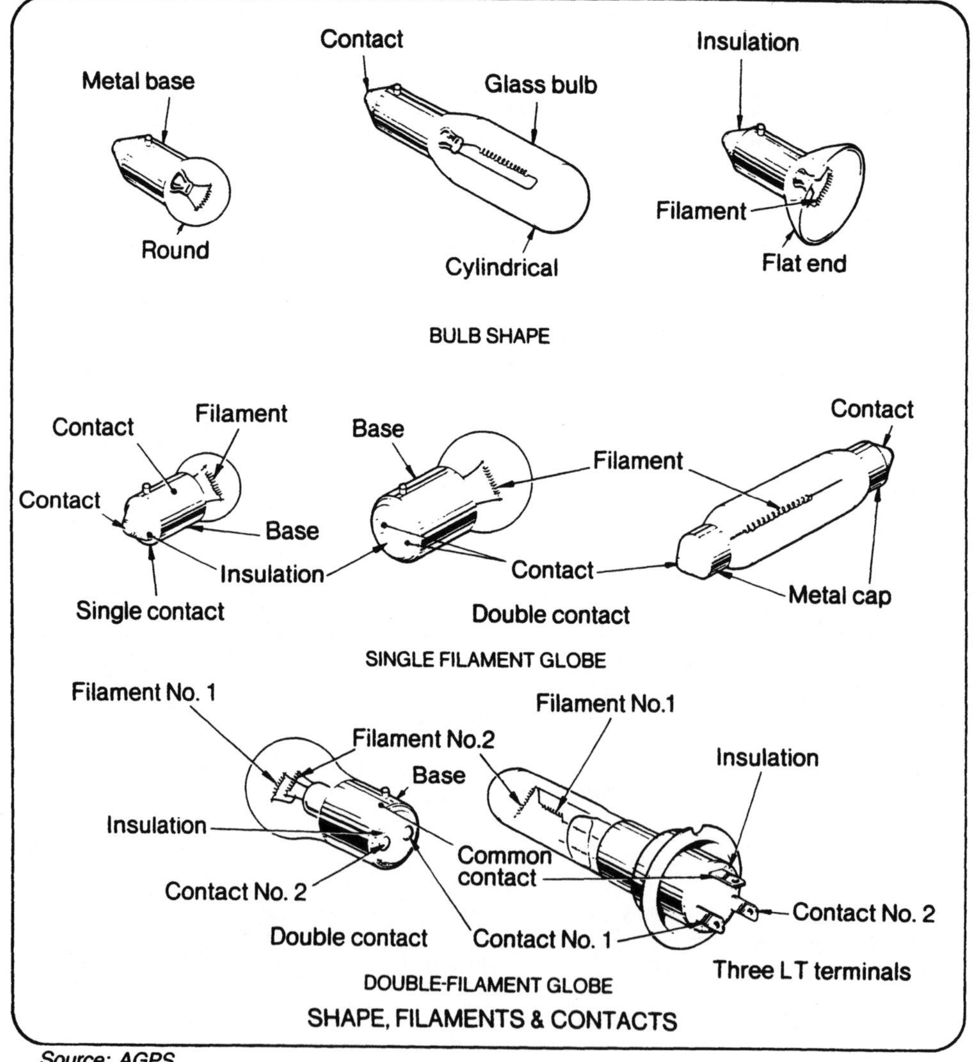

SHAPE, FILAMENTS & CONTACTS

Source: AGPS

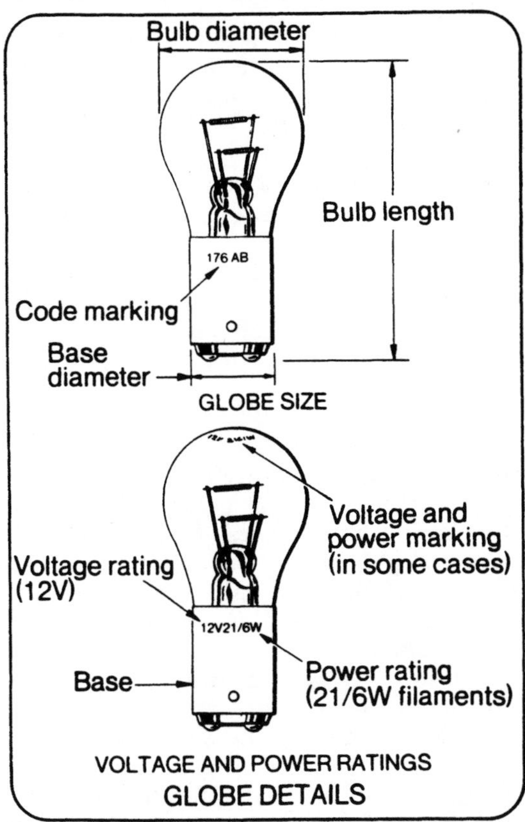

GLOBE DETAILS

Source: AGPS

Generally the higher the globe wattage value (for a given voltage) the higher the light output. This rule only applies when the globes being compared are based on the same manufacturing process. The best example of this variation is the headlamp globe. Modern headlamp globes contain a halogen vapour and will produce a light output far greater than a conventional globe of the same wattage.

Problems which could arise from fitting a globe with the incorrect wattage are:

- One headlamp may be brighter than the other. This is a roadworthy offence.
- The lens compartment may not be large enough to remove the heat supplied by the globe, leading to a distorted interior lamp lens.
- A light 'out' checking system may malfunction.
- Trafficator flasher rates may vary.

Caution: a replacement globe must match the original globe specification.
The major types of globes in common use are described in the following paragraphs. This information will be useful in identifying globe types and understanding the terminology applied to globe descriptions.

HEADLAMPS

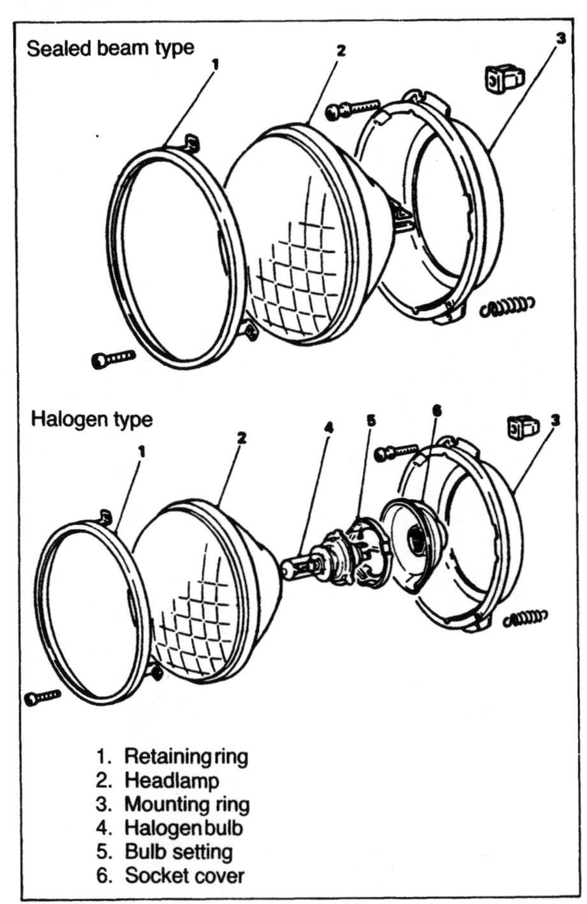

Source: Mitsubishi Pajero manual 1983

Headlamps may be sealed beam, conventional globe or halogen types.

Sealed beam headlamps have the reflector, lens and filament sealed into one unit. The units are available in circular shape (178 mm or 146 mm) or oblong (usually 200 by 142 mm or 165 by 100 mm). Single or double filament types can be obtained in either shape in a range from 40 to 75 W.

Semi-sealed beam (conventional globe) headlamps have the reflector and lens sealed into one unit, but a socket opening is left in the rear to accept a conventional type globe. The globe base is indexed (notched) so the filament positioning within the semi-sealed unit is correct for focus of the light beam. This type of globe is sometimes referred to as a pre-focus globe, and available in a range from 40 to 75 W.

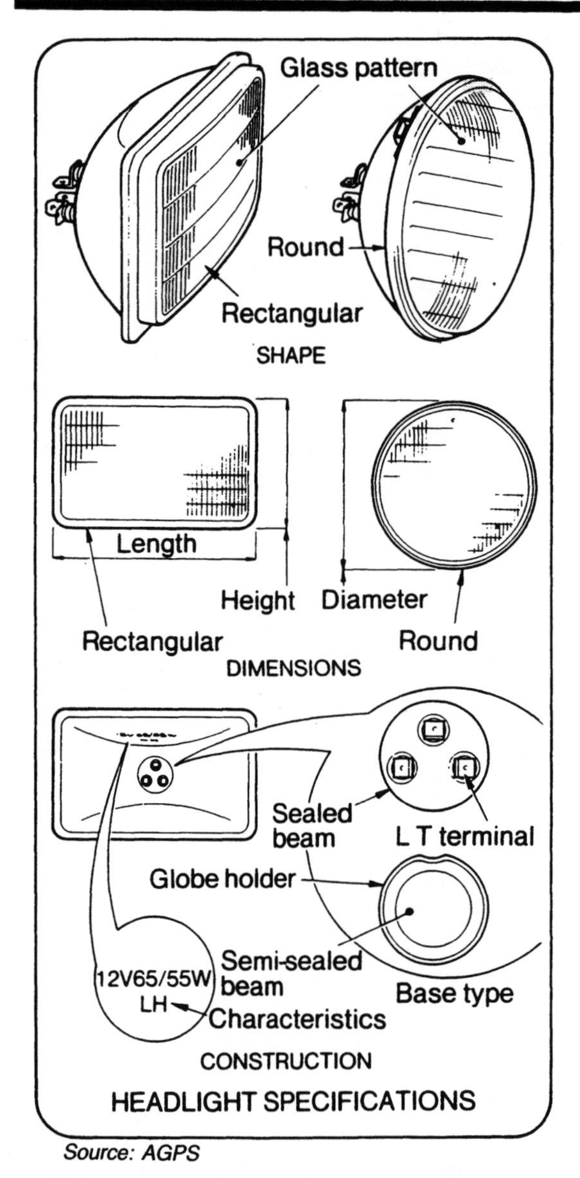

Source: AGPS

Halogen type refers to the globe construction. The globe contains one or two filaments and is filled with halogen vapour. The reflector and lens are similar to the semi-sealed units but are usually available as separate items. These units have a wattage range from 55 to 100 W, but produce a greater light output than conventional globes for the same power consumption.

LIGHTING SYSTEMS

A vehicle may be fitted with two or four headlamps.

Two headlamp systems require each lamp to be dual purpose, meaning that each lamp must contain a low and high beam filament. The two filaments are contained within the same envelope to form a 'double filament' globe. Double filament headlamp globes have three connection points: high beam, low beam and earth.

Four headlamp systems use the same type of lamp as the two headlamp system in the outer mounting positions on the vehicle. The inner mounted lamps are single filament and are designed to operate on high beam only as a driving or long range beam. Single filament headlamp globes have two connection points: supply and earth.

Trafficator (flasher) globe has a single base contact with the earth return through the socket and its retaining/locating pins are at equal heights. Their wattage values range from 18 to 24 W, which makes them also suitable for stop lights when the mounting location requires one filament only. This globe can be described as single filament, single contact, straight pin.

Stop/tail lamp globe contains two filaments in a wattage combination such as 18/6, 21/6 or 24/6 W. The base has two supply contacts and the retaining pins are offset to ensure the globe will mount in one position only. This globe is described as a double filament, double contact, offset pin.

Park or tail lamp globe. A single filament, single contact, straight pin globe is used in many locations. The wattage of this type of globe is in the range from 5 to 10 W.

Dash panel globe. Instrument lighting, warning displays and indicator lights require very small globes of about 1.2 to 3 W. Many of the applications require special sockets such as wedge, screw or straight pin. The physical size of the globe can be critical to correct installation, as mounting locations are restricted.

Interior and number plate globe. The conventional single filament, single contact, straight pin globes of 3 to 5 W are often used in these applications but many manufacturers favour the tubular festoon. This type of globe fits between two spring clips (supply and earth) and has the advantage of a very flat mounting space.

HEADLAMP AIMING

It is essential for headlamps to be aimed correctly. Headlamps aimed too low may not illuminate a pedestrian standing by the kerb, or an object on the road. Alternatively, headlamps aimed too high or too far to the right may impair an oncoming driver's vision and a serious accident could result.

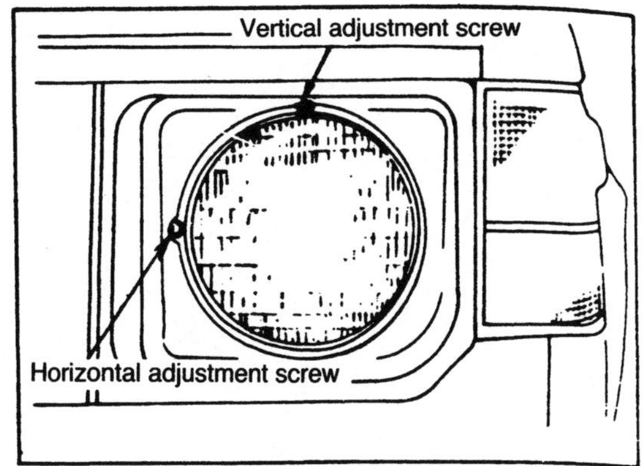

HEADLAMP ADJUSTMENT

Headlamps are pivoted in their mounting arrangement with the amount of movement being controlled by spring loaded adjusting screws. Different techniques are available for aiming headlamps, using specialised test instruments. Where these are available the equipment manufacturer's instructions should be followed. However, one of the simplest methods uses a flat test screen (wall) and a level floor. Vehicle manufacturers provide the information required to use this method, and illustrated below is an aiming chart suitable for both two and four headlamp systems.

PREPARATION

The preliminary requirements for aiming headlamps are:

1 Locate a flat vertical test screen (wall).
2 Park vehicle on level ground with headlamp centre 5 m from test screen.
3 Inflate the tyres to the correct pressure.
4 Check vehicle loading (as required by manufacturer): amount of fuel in tank, number of passengers.
5 Ensure the hand (park) brake is in the off position.
6 Bounce the vehicle gently to 'normalise' suspension.

LOW BEAM ADJUSTMENT

1 Refer to the illustration and measure distance (H) (the height of the lamp bulb centre above ground). Vehicle heights will vary due to vehicle production tolerances.
2 Set the horizontal break line on the test screen to a height (H = H–e) where (e) is the aim setting dimension.
3 Measure (A1) the distance between the bulb centres. Using this dimension set up the low beam aim centre (C1) and construct the 15° inclined breakdown from this point on the test screen.
4 Switch on the lights and cover all lamps except the one that is being aimed.
5 Using the horizontal and vertical aim adjusting screws, shift the light pattern of the headlamp on the test screen until the light/dark boundaries of the light pattern coincides with the breaklines on the test screen. The low beam pattern should now be centred on point (C1).
6 Repeat the above procedure to adjust the other headlamp.

HIGH BEAM ADJUSTMENT

1 The high beam is fixed internally within the headlamp relative to the low beam. Whenever the low beam is adjusted the high beam aim is automatically adjusted.
2 The high beam aim should be centred on point (C2). As a check on the low beam setting the high beam aim centre should lie within the specified tolerance area.

DRIVING BEAM ADJUSTMENT

For four headlamp systems follow this method:

1 Vehicle and test screen requirements are as for low beam adjustment.
2 Measure (H) the height of the lamp bulb centre above the ground and (A2) the distance between the lamp bulb centres. Using these dimensions, set the driving beam aim centre (C3) on the test screen.
3 Switch on the lights and cover all lamps except the one that is being aimed.
4 Using the horizontal and vertical aim adjusting screws, shift the light pattern of the driving lamp on the test screen until the bright spot of the light pattern is centred on point C3.
5 Repeat the above procedure to adjust the remaining driving lamp.

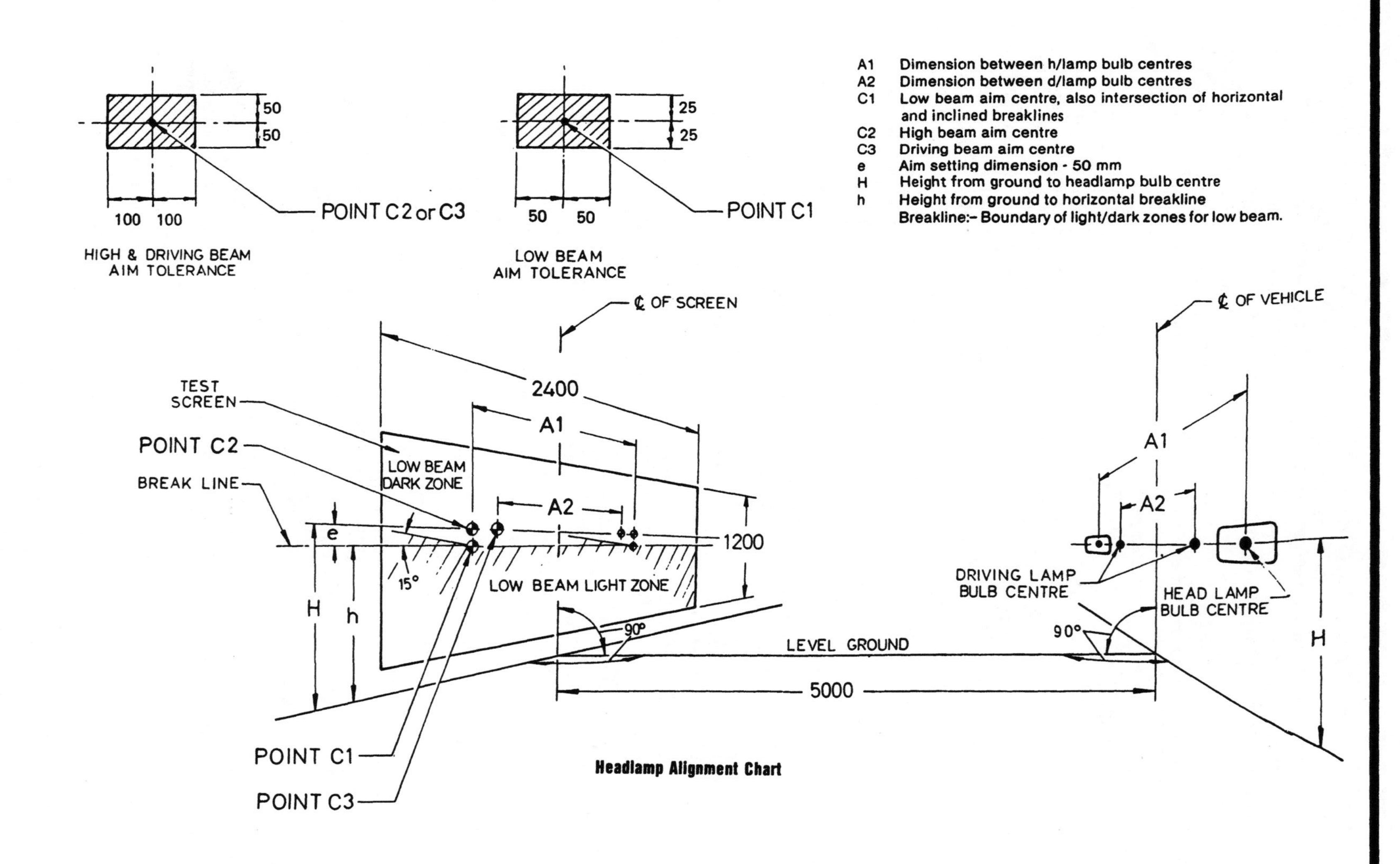

Headlamp Alignment Chart

WIRING AND CONNECTIONS

The most common tasks a mechanic will encounter in vehicle electrical service work are replacement of wiring terminals and the addition of a trailer/caravan socket to the wiring loom.

Wire type and size

As already stated, the work unit is connected to the battery by wires. The type of wire used for these connections is very important. It must be insulated, multi-strand and preferably have a copper core — copper is easy to solder. Generally, the insulation is formed from plastic like PVC.

The size must also be carefully considered. A wire with a small diameter core is used for low current circuits such as warning lights. A wire with a large diameter core — called a cable — is used for high current circuits such as the starter motor. The chart gives an indication of the wire size for various applications. The main point to remember is that when a wire is too small it will cause excessive voltage drop — loss of voltage — and it could 'burn out' due to overheating.

APPLICATION	SIZE (mm)	APPLICATION	SIZE (mm)
Alternator B+	6	Ignition coil	3
Alternator D+	3	Number plate light	2
Battery (cables)	8	Relays (85 & 86)	2
Battery (fusebox)	6	Relays (87, 87a & 88a)	3-4
Cigarette lighter	3	Stop lights	3
Dash lights	2	Temperature sender	2
Dome lights	2	Turning indicators	3
Fuel pump	3	Windscreen washer	2
Headlights	3-4	Windscreen wiper	3
Horn	3		

Terminals

The two groups of terminals, used for automotive electrical circuits, indicate the method used to secure them to the wire's core. One group is soldered onto the wire's core and the other group is designed to be crimped onto the wire's core. Each group of terminals have a similar range of ends that attach to the work unit. The following chart shows some of the terminals and their application.

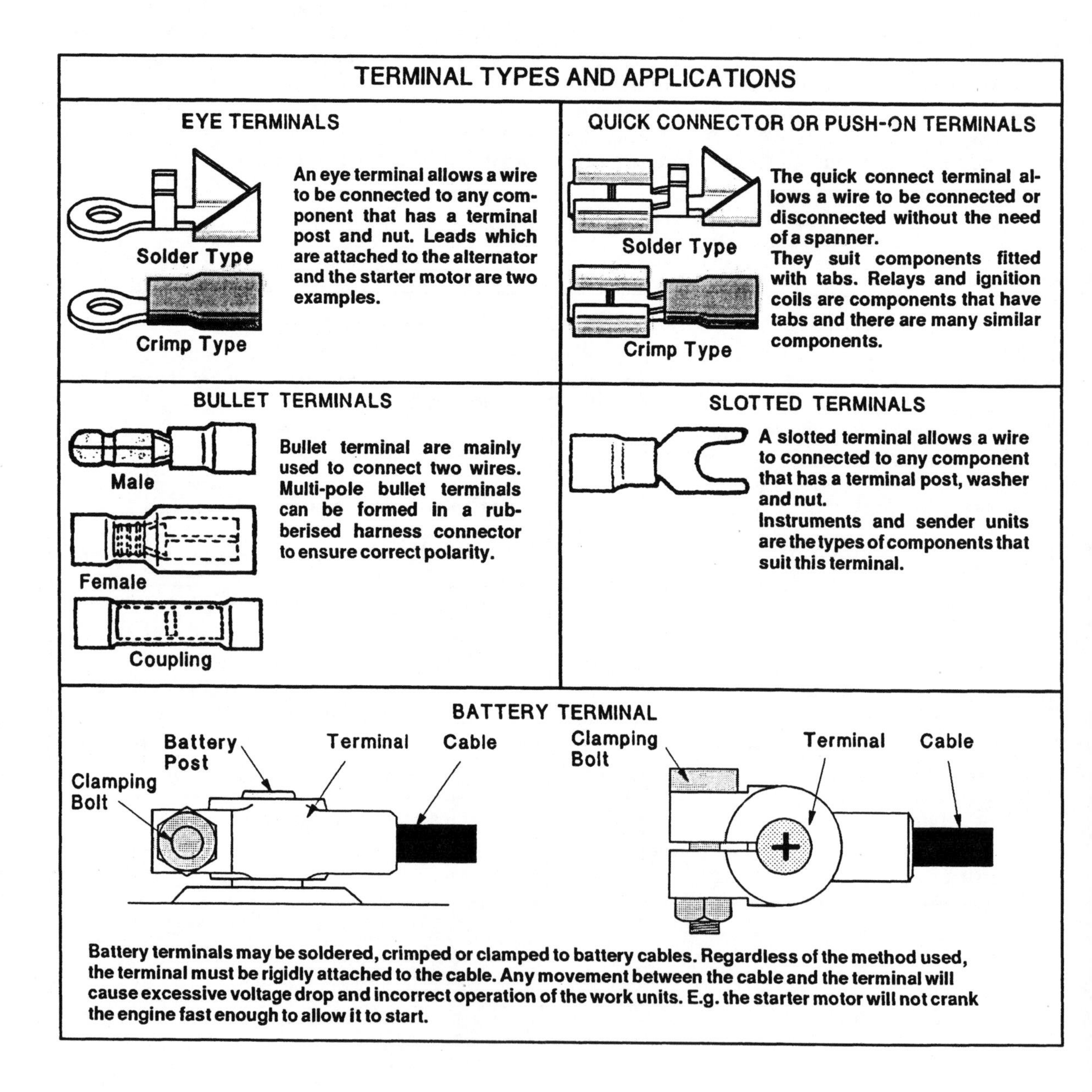

Terminal fitting

The selection of terminal is important, as the type of terminal selected must match the unit connecting post or slide-on prong. Full ring terminals are probably the most suitable, particularly in heavy current applications, as the terminal will not fall off if the connection nut becomes loose. When placed on the connecting post the terminal should be a neat fit and have sufficient material for the retaining nut to butt against. Material thickness should be sufficient to prevent breakage by wire movement in service. The terminal must also be suitable for the wire to which it is to be fitted. The wire core and the insulation must fit neatly into the terminal barrel or tangs.

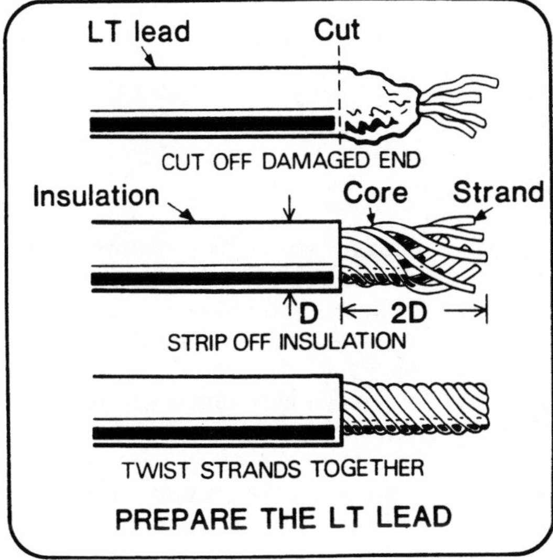

Source: AGPS

Similarly, the wire preparation is just as important as the terminal selection. The wire must be squared off on the damaged end and the insulation stripped back for a distance equivalent to the terminal barrel or folded wing area. It is possible to use a knife or cutting pliers to strip the insulation, but a specialised crimping or stripping tool is preferred.

The two methods of attaching a terminal to a wire are by soldering and crimping.

Soldering a terminal to a wire:

1 'Tin' the wire core.
- When solder starts to flow remove the soldering iron.

Caution: use only rosin core solder as other flux contain acid which will corrode the wire.

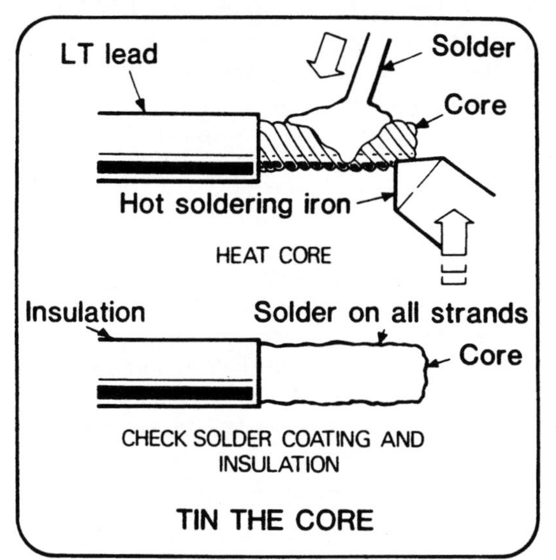

Source: AGPS

2 Fit the terminal to the wire as depicted in the illustration.

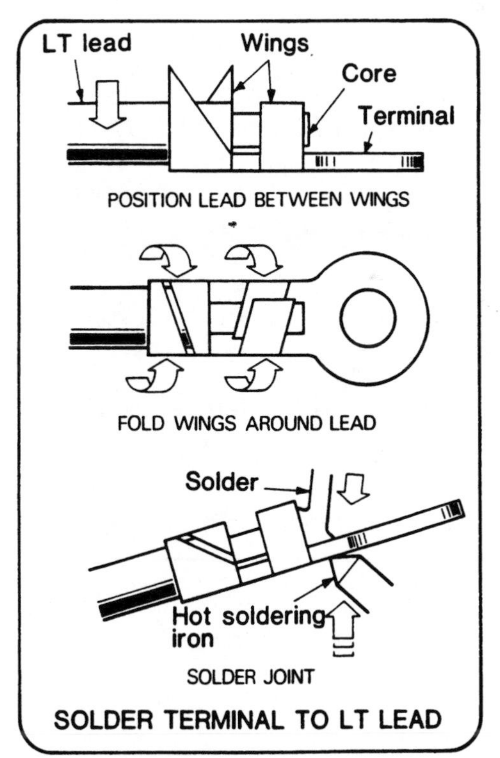

Source: AGPS

Caution: the wire must be hot enough to allow the solder to flow and prevent a dry joint, but the heat must not damage the insulation on the lead.

Barrel terminals are prepared in the same way for soldering but require a different soldering technique for satisfactory results.

Crimping a terminal to a wire:

1 Strip the wire and insert it into the terminal barrel.
2 Trim wire to length if necessary.
3 Using a special crimping tool, crimp the barrel in the location shown.
4 Check that the wire core is firmly fixed within the terminal barrel and no movement between barrel and wire is evident.

Note: This type of terminal does not require soldering, making it particularly suitable for aluminium wire as well as the more common copper wire application.

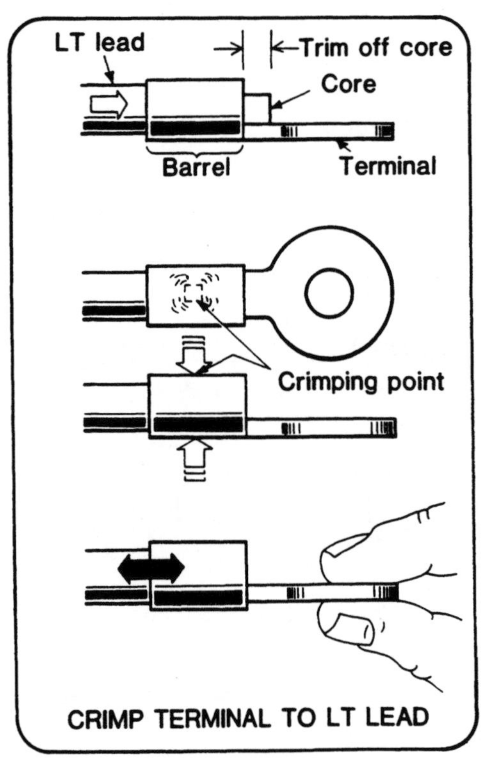

Source: AGPS

Wiring trailer/caravan socket

One of the most common vehicle accessories is a tow bar. Towed units must comply with vehicle lighting regulations, therefore it is necessary to provide a coupling point to connect the vehicle system to the trailer/caravan.

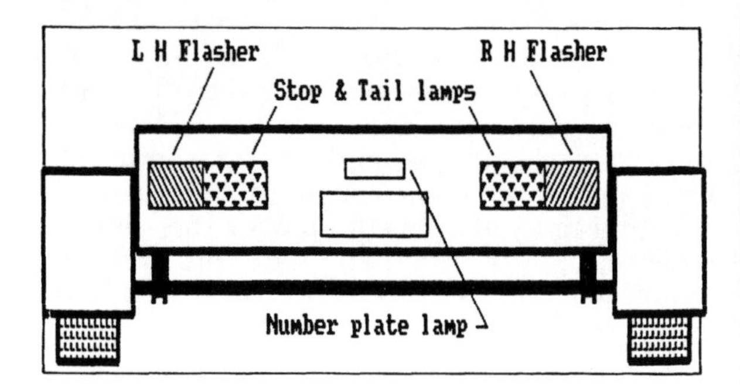

The trailer must be fitted with:

- tail lamp/s;
- number plate lamp;
- stop lamp/s;
- flasher lamps;
- earth return to battery.

In addition to these basic requirements reversing lights will be automatically provided if the vehicle uses the flasher lamps as reversing lamps. If the towed unit is to be a caravan the owner will also require battery (12 V) power for internal lights and refrigerator when the caravan is not on 240 V A.C. site power.

The two common coupling arrangements used are the five (5) pole plug/socket for the basic trailer link and the seven (7) pole plug/socket for caravans.

The standard colour wiring code should be used on either application but the connection to the socket will vary between types. The recommended colour code is:

Circuit	**Wire colour**
Left flasher	Yellow
Right flasher	Green
Stop lamp	Red
Tail/aux. lamps	Brown
12 V power No. 1	Blue
12 V power No. 2	Black
Earth	White

To wire a coupling socket to a vehicle:

1 Disconnect vehicle battery.
2 Select a secure mounting location for the coupling socket. The selected location must be:

- accessible from the rear for the wiring to be attached;
- close to the draw bar yet not interfere with the draw bar coupling or obscure the vehicle number plate.

3 Drill holes for the socket mounting screws and, if necessary, the wiring loom.

4 Locate the vehicle rear lighting loom and unwrap approximately 12 cm of loom tape.

- The wiring area to be uncovered should be located away from the luggage area and preferably at a location behind a removable panel.

5 Obtain a roll of wire to match each of the required colours, or a roll of multi-core flex.

6 Consult the vehicle wiring diagram and locate each of the supply wires required.

7 Each trailer wire must be connected securely into the vehicle loom. Two widely used methods of wire connection are:

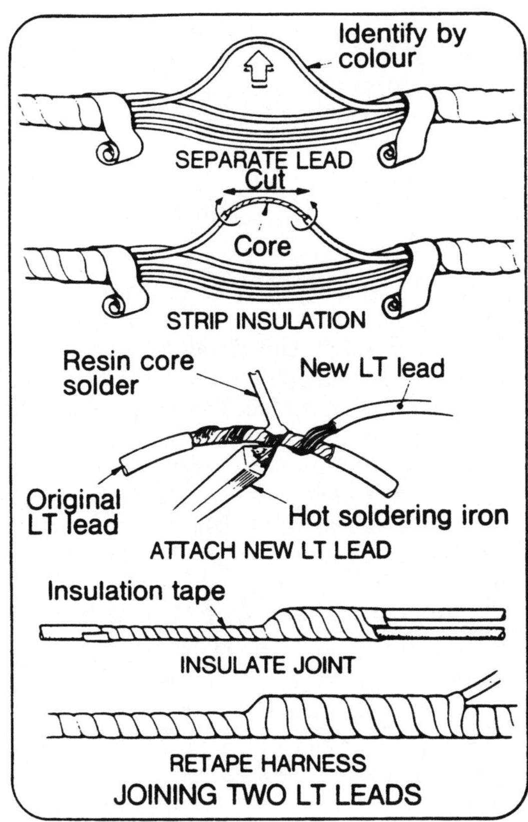

Source: AGPS

a **Soldered**

- The loom wire is stripped of insulation for approximately 1 cm and the add-on wire bared end is wound onto the loom wire.
- The two wires are soldered together and the joint is covered by insulation tape.

b **'Scotchlok'**

- An effective alternative to soldering of each wire is to use a plastic coated connector (similar to 'Scotchlok'). The connector uses small metal knife edges to perforate the loom wire and the add-on wire simultaneously and acts as a bridge for current supply.

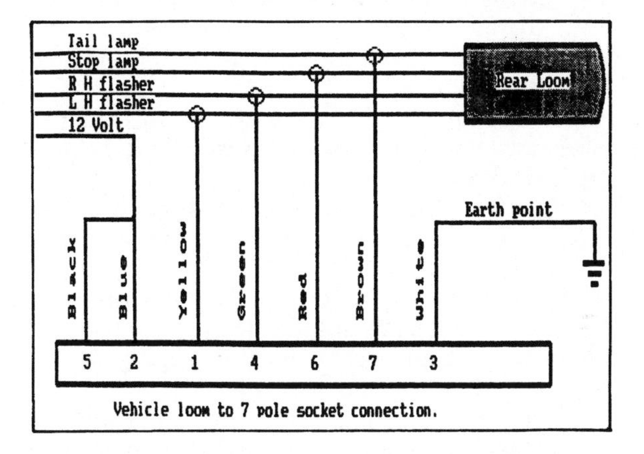

Vehicle loom to 7 pole socket connection.

8 Connect the yellow wire to the L.H. flasher wire.

9 Connect the green wire to the R.H. flasher wire.

Caution: turn signal flash rates are part of the roadworthy regulations. When the flash rate is outside the accepted range, replace the vehicle standard unit with a flasher unit suitable for trailer applications.

10 Connect the red wire to the 'stop' lamp wire.

Caution: vehicles fitted with a 'brake light out' checking system may require direct connection of the stop lamp wire to the stop lamp switch to prevent damage from current overload. Check manufacturer's instructions.

11 Connect the brown wire to the tail lamp/auxiliary lamps wire.

Caution: vehicles fitted with a 'corner light out' checking system may require the tail lamp wire to be connected directly after the main switch to prevent damage from current overload. Check manufacturer's instructions.

12 Connect the white wire to the earth wire in the loom or earth point near the rear of the vehicle.

Note: When a standard point is not available, drill a hole in a metal area and clean paint from the area adjacent to the hole. Fit a full circle terminal to the white wire and fasten securely to the earth point.

13 Adding one or two 12 V power supplies is achieved by running a single (5 mm) wire through the vehicle to a 12 V supply source.
   - The source point should be fused and not pass through the ignition switch.
   - When a suitable fused point is not available, an in-line fuse holder must be fitted close to the connection point (usually at the starter or battery terminals).

Note: The single 12 V wire can be split into a blue wire and a black wire adjacent to the trailer socket.

14 Position all wires along the vehicle harness between their connection points and the trailer socket.

15 Cut all wires to length. Ensure there is sufficient length to allow for the connection to the trailer socket and the taping to the harness.

16 Using insulation tape and starting from the connection points, wrap the connections and add-on wires into a loom to provide a neat and professional appearance to the task.

17 Strip the insulation from the end of each wire for a distance of 6 mm.

18 Connect each wire into the socket. Each socket pin is numbered and the pin/wire combinations are listed below.

19 Attach the socket to the mounting point.

20 Tape the trailer wiring loom to the main vehicle harness.

21 Refit any panels removed to attach or route the trailer wiring loom.

22 Connect vehicle battery.

23 Test each pin connection using a 12 V test lamp to ensure the correct pin is 'live' when the appropriate switch is depressed or turned on.

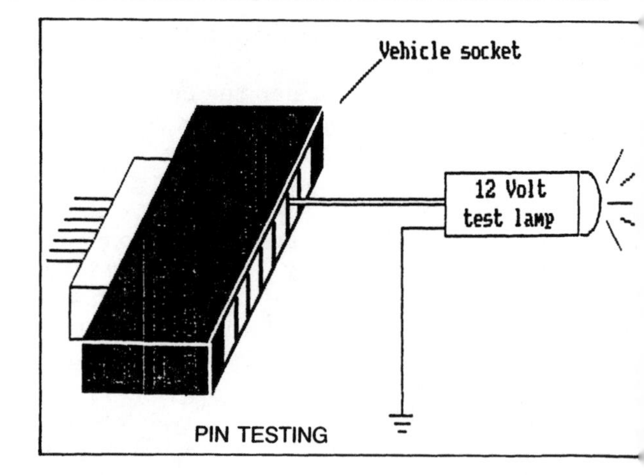

PIN TESTING

24 Connect the trailer or caravan to the vehicle and visually check the operation of all lamps.

Note: When wiring a plug to a trailer or caravan, use the same wiring code as used when installing the socket.

BATTERY

Identification

Before selecting a replacement battery for a vehicle it is necessary to ensure the new battery will meet the demands likely to be placed on it. This is usually a simple task of consulting the manufacturer's selection chart for the correct battery, however, the vehicle owner may have modified the original vehicle (power winch, spot lights, or high current draw accessories) and placed demands on the battery not anticipated by the manufacturer.

An understanding of the following terms will enable you to compare the relative

Wiring code

Pin no.	5 pole circuit connection	Colour	7 pole circuit connection	Colour
1	Not used		Left flasher	Yellow
2	Left flasher	Yellow	12 V supply no. 1	Blue
3	Earth	White	Earth	White
4	Not used		Right flasher	Green
5	Right flasher	Green	12 V supply no. 2	Black
6	Stop lamp	Red	Stop lamp	Red
7	Tail/aux.	Brown	Tail/aux.	Brown

performance of various batteries. Obviously physical size, terminal placement and retention method in the vehicle must be satisfied before any further consideration of a replacement battery.

VOLTAGE

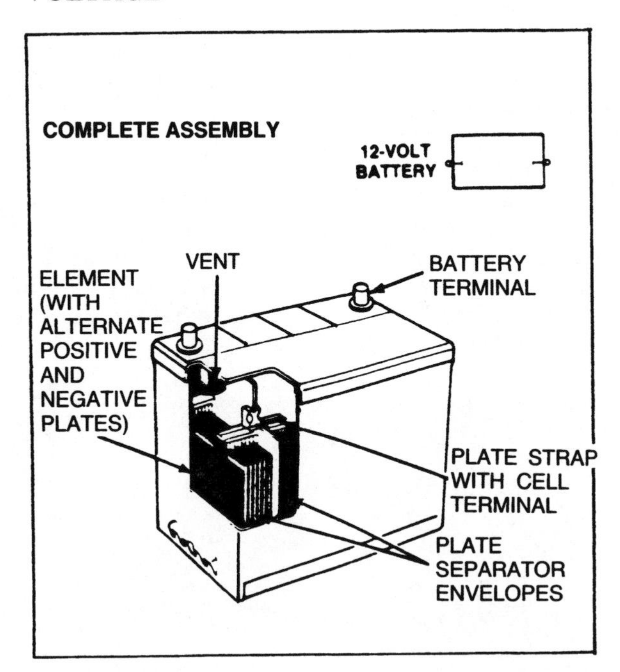

BATTERY AND SYMBOL

Most car batteries are rated at a nominal 12 V. A small number of 6 V applications are available, but the 12 V battery is almost universal in passenger car installations. **Caution: severe electrical system damage will result from fitting a higher voltage battery.**

BATTERY TYPE

A choice is usually available between conventional, low maintenance or maintenance-free batteries. The appearance of conventional and low maintenance batteries is almost identical, therefore it may be necessary to check the decal on the battery or refer to the manufacturer's chart for verification of battery type.

Maintenance-free batteries do not require topping up and are identifiable in most applications by the absence of filler caps.

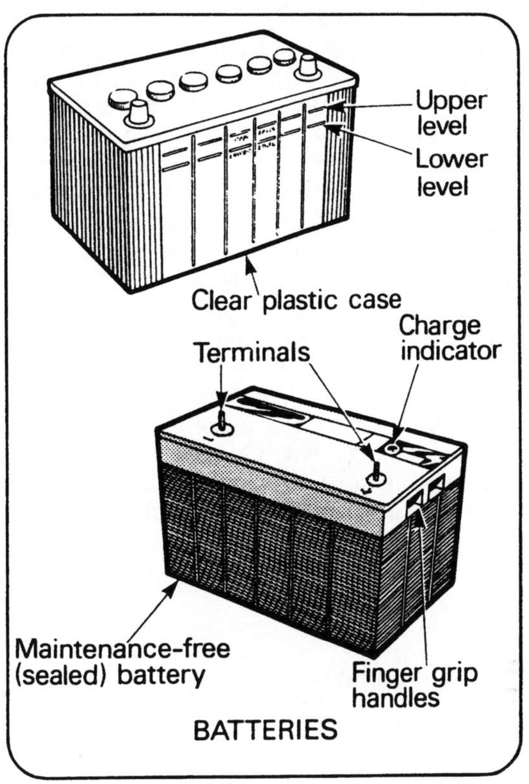

BATTERIES

Source: AGPS

Battery performance

COLD CRANKING PERFORMANCE

This relates to the discharge rate a battery can provide for a period of thirty seconds under specified test conditions. This information is usually provided on a decal on the battery, and a typical example would be 380 AMP.

RESERVE CAPACITY

The length of time in minutes that a battery can sustain a set electrical load (assuming the alternator is not charging) before the cell voltage drops below 1.75 V per cell. For example, loading a 12 V battery with a 25 A load at 26.6°C, and the battery voltage does not drop below 9.5 V in 120 minutes rates the reserve capacity of this battery at 120 minutes.

AMPERE HOUR CAPACITY

The ampere hour capacity is calculated at a twenty-hour rate. For example, when 2.5 A is supplied for twenty hours from the battery, its ampere hour capacity is fifty (2.5 multiplied by 20 = 50 ampere hours).

Attempts to compare reserve capacity and ampere hour capacity must be treated with caution. In theory, a battery rated at two hours (120 minutes) on a 25 A load by the reserve capacity method would appear to be equal to a battery rated at twenty hours on a 2.5 A load by the ampere hour capacity method. Both batteries calculate to fifty ampere hours (time × load), but actual performance may be different. Because of chemical action within the battery under heavy current draw conditions, the ampere hour rated battery is unlikely to match the performance of the reserve capacity rated battery.

Caution: the replacement of a modern lightweight maintenance-free battery with a conventional battery of the same dimensions will result in poor starting system performance.

Battery removal

Batteries are heavy, fragile and contain corrosive acid. Frequently they are located in difficult access areas: recessed under guards or surrounded by washer bottles and under bonnet components. The removal of the battery requires the terminals to be disconnected from the battery posts.

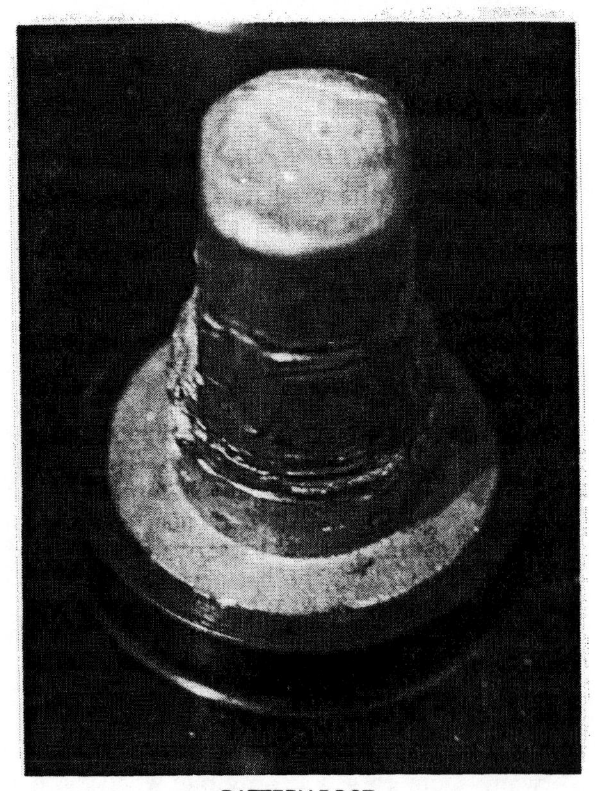

BATTERY POST

TERMINAL IDENTIFICATION AND REMOVAL

It is essential to identify the terminal polarity of a battery before commencing work on the battery or electrical system.

Modern vehicles tend to use a 'negative' earth system. The negative (black) terminal will be marked '–', 'NEG.' or 'negative' on the casing adjacent to the terminal. It is slightly smaller in size than the positive terminal. The positive (red) terminal will be marked '+', 'POS.' or 'positive' on the casing adjacent to the terminal.

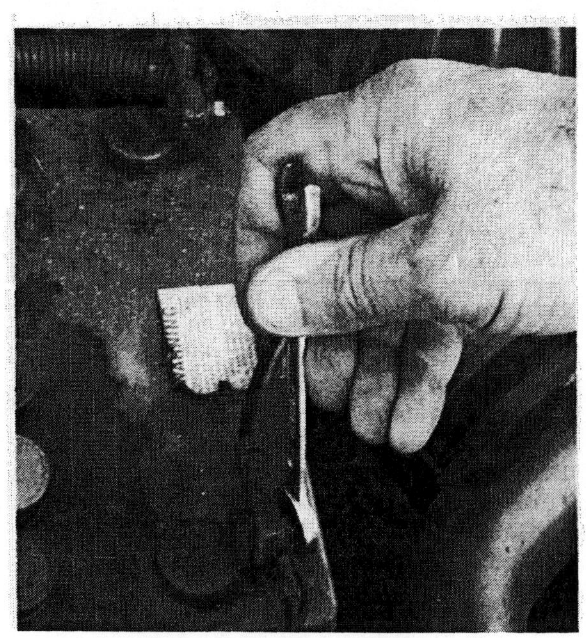

BATTERY TERMINAL REMOVAL

Several methods have been used to secure the terminals, but the two major types are the:

1 flat terminal to battery lug type. The battery terminal post (lug) is drilled and a flanged cable (flat) terminal is attached by a bolt and nut to the battery post.
2 clamp type. A cast lead or brass clamp fits over the battery post and is held tightly by either a bolt and nut or spring clip.

To remove the battery:

1 Ensure all electrical system switches and the ignition switch are in the 'off' position.

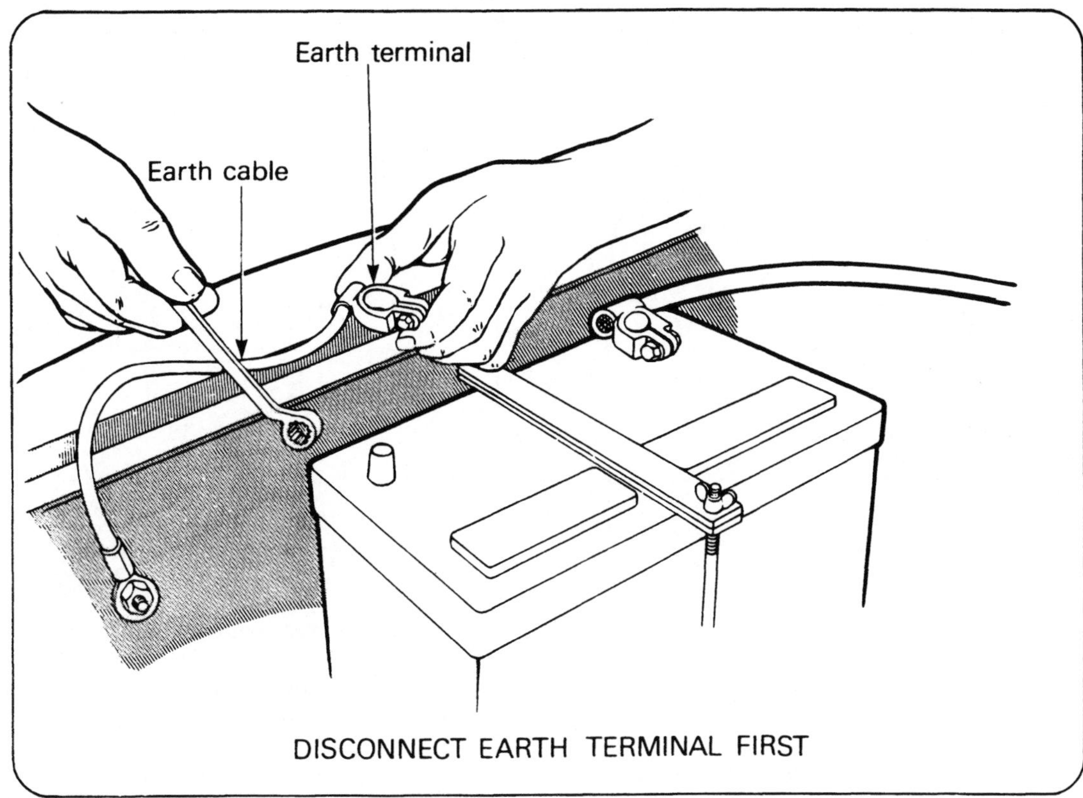

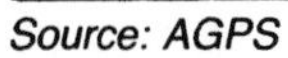
Source: AGPS

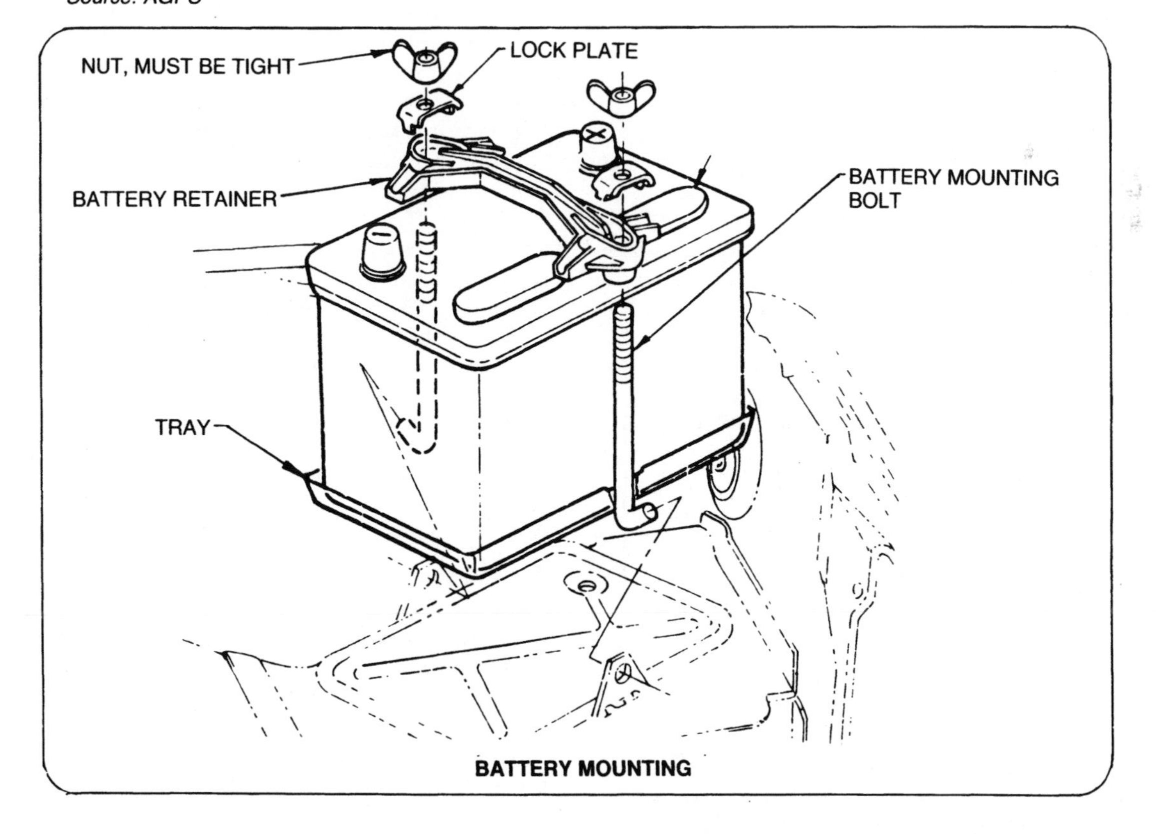

BATTERY MOUNTING

2 Disconnect the negative battery terminal.
3 Disconnect the positive battery terminal.
4 Unbolt the battery retaining clamps.
5 Lift the retaining clamp clear of the battery.

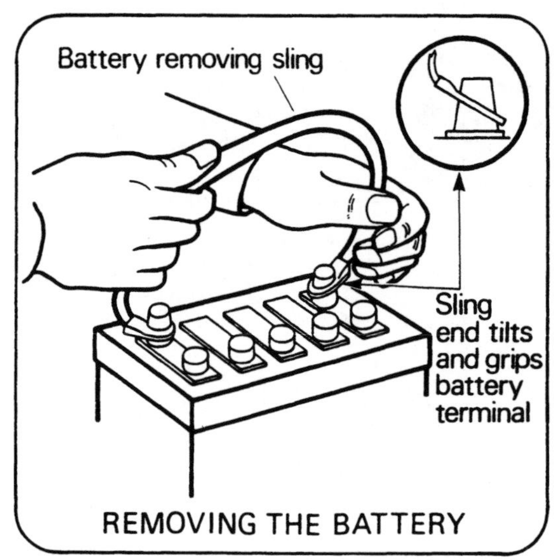

Source: AGPS

6 Lift the battery from vehicle (use a battery sling if necessary).

Caution: batteries are very heavy for their size so ensure that a firm grip is obtained before starting to lift.

CLEANING

Corrosion around a battery may be an indication of future electrical problems. If corrosion appears excessive, carefully examine the battery for cracks or bulged case and arrange for the charging system to be checked for excessive charge rate. The cleaning procedure is:

1 Clean battery posts and battery case.

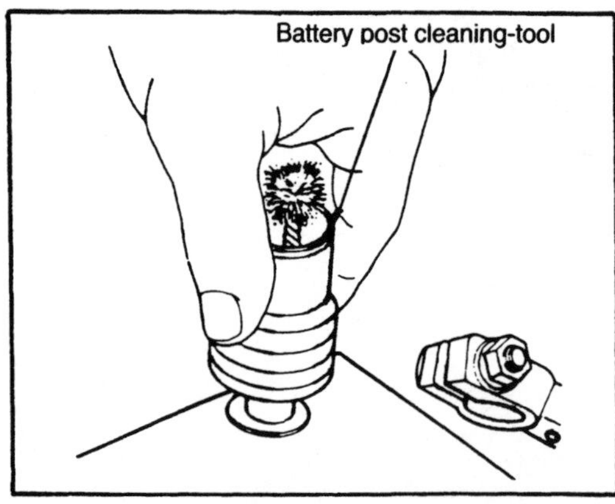

Source: Mitsubishi Sigma GJ series manual

- Rub the posts with emery tape or a wire brush (terminal post tool).
- Wipe the battery case with a water-soaked rag and flush any residue away with water.

2 Clean battery cable terminals.

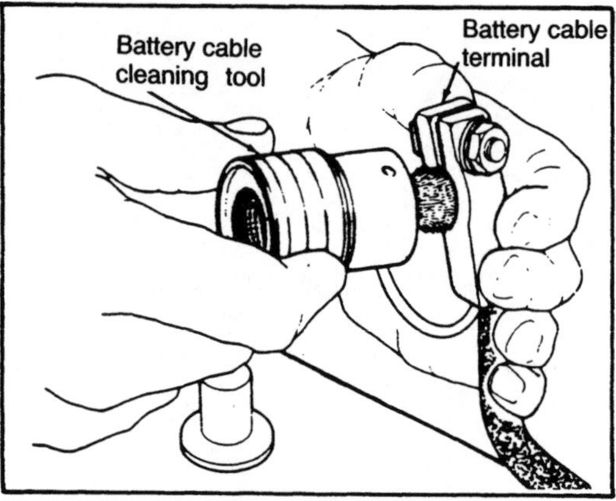

Source: Mitsubishi Sigma GJ series manual

- Soak cable terminals in hot water until all corrosion has been removed.
- Baking soda may be added to the hot water to assist with corrosion removal.
- Clean the terminal with a battery terminal cleaning tool until bright metal is visible.

Caution: never allow water/baking soda solution to spill over the battery near the filler caps as the solution will dilute the acid inside the battery.

3 Dry battery, posts and terminals with a cloth.
4 Replace corroded terminals or clamp bolts.
5 Inspect the battery tray, mounting brackets and retaining clamp for corrosion.
- A solution of hot water and baking soda will effectively remove corrosion and acid.
- Dry the area and repaint any affected clamps or body parts.

Battery replacement

1 Refit battery into vehicle. Ensure battery is not reversed in tray.
2 Secure battery with retaining clamp/bolts and nuts.
3 Connect positive terminal (negative earth vehicles).
4 Connect negative terminal.
5 Coat terminals and posts with petroleum jelly.

6 Check the level of the electrolyte (battery acid).

Note: Maintenance-free (sealed) batteries do not require topping up of the electrolyte with distilled water.

Caution: to prolong battery life use only distilled water.

To check the electrolyte level:

1 Brush away any loose dirt from the filler cap area.
2 Remove the caps and observe the electrolyte level. The top of the separators must be covered (approx. 3 mm) by the electrolyte.

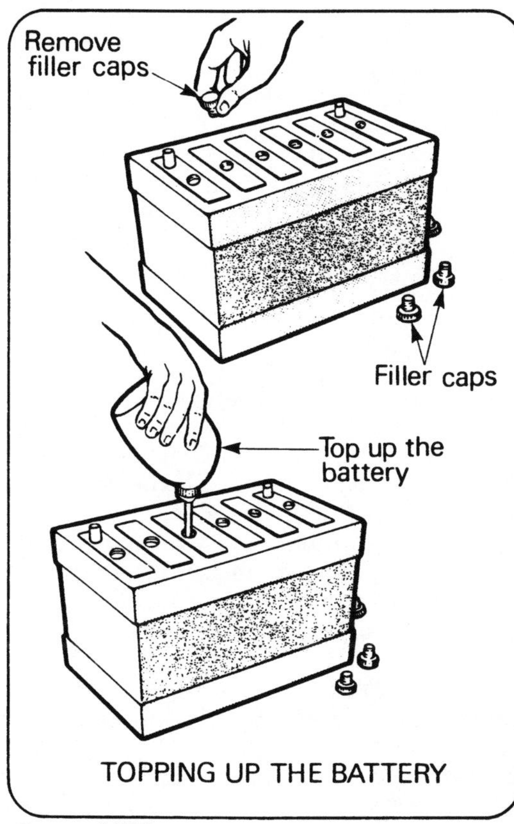

Source: AGPS

3 Top up the battery with distilled water. Use a dispenser to minimise spillage. Add water until the electrolyte level is approx. 3 mm above the top of the plate separators.
4 Wipe battery dry of any spillage.

Caution: never overfill the battery with distilled water as it will dilute the electrolyte and the spillage will cause corrosion problems.

5 Remove excess electrolyte (if applicable). Use a battery hydrometer to lower the electrolyte to the correct operating level.

Note: Modern batteries have translucent cases with level markings. The level is visible to the mechanic and filler cap removal is not necessary unless top-up is required.

Battery testing

Battery testing can be divided into two groupings.

STATE OF CHARGE

A fully charged battery will contain an electrolyte with a high acid concentration. This will result in a high specific gravity reading (1.280 s.g.) when checked with a hydrometer. As the battery discharges (becomes 'flat') the acid concentration in the electrolyte decreases and the hydrometer reading becomes less (1.150 s.g.).

BATTERY CONDITION

Battery condition testing may be referred to as a battery capacity test, and is performed using a high rate discharge tester. Follow the equipment manufacturer's instructions when using the tester. The test method will depend on the equipment used, but generally a form of load control is adjusted until an ammeter reading of three times battery ampere hour rating is obtained. For example, a 61 A/hr battery would require a setting of 180 A.

Note: Before this test is carried out, the battery must be disconnected from the vehicle's electrical system.

Caution: to reduce the possibility of an explosion due to the ignition of the hydrogen and oxygen gases given off during normal battery operation, cover the vent openings with a damp cloth. Never probe or pierce the cover of a modern battery in an attempt to perform a single cell test.

Check state of charge

This test is not suitable for maintenance-free batteries.

The following points must be observed when using a hydrometer.

1 Withdraw sufficient electrolyte from the cell to float the graduated float.
2 Record the reading (take readings at eye level).
3 Obtain a reading from each cell. Variation in reading should not exceed 0.025 s.g.
4 Recharge the battery when the average reading is below 1.200 s.g.

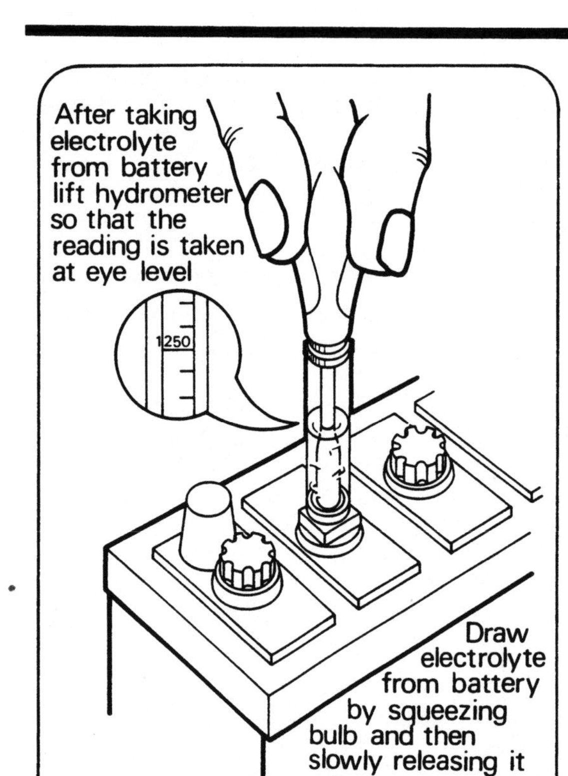

Source: AGPS

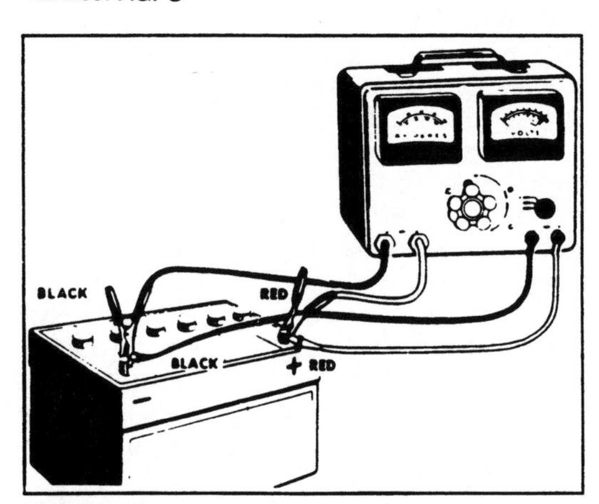

High rate discharge test
Source: Mitsubishi Magna TM series manual

Check battery condition

1 Check specific gravity of the battery using a hydrometer (see state of charge).
   - When the s.g. is below 1.210 do not proceed with the high rate discharge test.
   - Place the battery on charge until fully charged.

2 Connect the tester to the battery (VEHICLE BATTERY LEADS MUST BE DISCONNECTED TO SAFEGUARD VEHICLE ELECTRONIC SYSTEMS).
   - Ensure the load control is fully 'off'.
   - Red lead to battery positive post.
   - Black lead to battery negative post.
   - Ensure leads are firmly clamped to posts.

3 Turn 'on' the load and adjust ammeter reading.

Caution: the load must not be 'on' for more than fifteen seconds.

4 Read the value shown on the voltmeter.
   - When the reading is 9.6 V or greater the battery is in good order.
   - When the reading is less than 9.6 V the specific gravity should be rechecked for variation in readings between cells.
   - When variation does not exceed 0.050 s.g., recharge the battery and repeat all tests.
   - Replace the battery where the s.g. variation is greater than 0.050.

5 Refit vent caps.

6 Install battery in vehicle and connect terminals.

Battery charging

Battery chargers are available in a wide range of capacities. Generally they are switchable between 6 V and 12 V and have variable current (charge rate) outputs. If the charger is for home use the number of features on the unit will be minimal and variable output controls may not be included. Industrial battery chargers, as used in workshops, can have varying outputs ranging from 2 A to 60 A.

Caution: never use high ampere settings as they will damage the battery.

When time permits, it is advisable to restrict the charge rate to 7–10 per cent of the ampere hour capacity of the battery, for example approximately 6 A charge rate for a 61 A/hr battery. Charge rates in excess of 10 per cent of battery rating will not harm a battery in sound mechanical condition, but temperature rise within the battery must be monitored and the charge rate reduced immediately temperatures in excess of 50°C are detected.

A battery may be charged as a single unit or as a group of batteries. The number of batteries being charged as a group will depend on the capacity of the battery charger.

To charge a single battery:

1 Remove the battery from the vehicle.

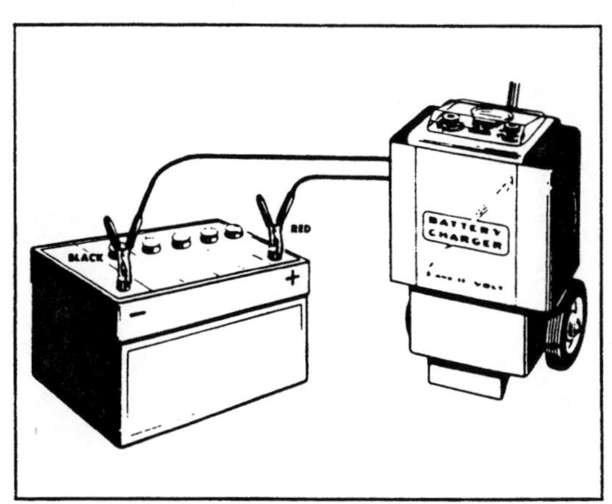

Source: Mitsubishi Magna TM series manual

Caution: It is possible to charge a battery while it is in the vehicle, but the battery leads must be disconnected as a precaution to prevent damage to the electrical system.

2 Remove filler caps, check electrolyte level and replace caps loosely in battery filler holes.

3 Set battery charger voltage to suit battery on charge (6 or 12 V).

4 Ensure charge rate control is on zero.

Caution: ensure the battery charger switch is off.

5 Connect battery charger to battery. Red lead to positive battery post, black lead to negative battery post.

6 Switch on battery charger.

7 Adjust charge rate control until the ammeter indicates an acceptable charge rate. The rate will vary depending on battery capacity and time available for charging. **Remember** a low charge rate is preferred.

8 Check electrolyte level at intervals while battery is on charge.

9 Switch off charger when battery is fully charged. This condition can be detected by:

- a charge rate on ammeter reducing to low scale reading;
- b electrolyte gassing freely. This visual test is an approximation only and is not an accurate method with low maintenance batteries;
- c using a hydrometer to check the specific gravity of the battery.

10 Disconnect battery charger leads.

Caution: ensure the battery charger is switched off before disconnecting the leads from the battery.

11 Check electrolyte level (top up if necessary), refit filler caps and clean away any electrolyte spillage.

ALTERNATOR

Removal

1 Disconnect the negative battery terminal.

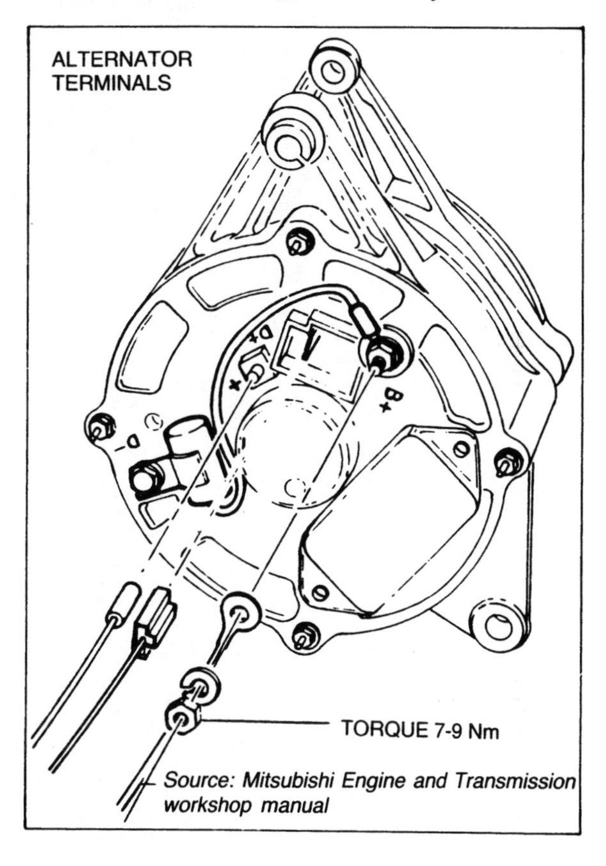

Source: Mitsubishi Engine and Transmission workshop manual

2 Remove the terminals from the rear of the alternator.

Note: The type and number of terminals will vary, however, there is normally a relatively large terminal on the main B+ lead, retained by a nut. The DF (field) and D+ (ind.) leads may be separate or in a single connection. Slide-on or bolted terminals may be used.

Caution: when the alternator is mounted on rubber bushes an earthing wire will be fitted between the alternator body and the engine.

3 Remove the adjusting link clamping bolt.

4 Loosen the bolt securing the adjusting link to the engine.

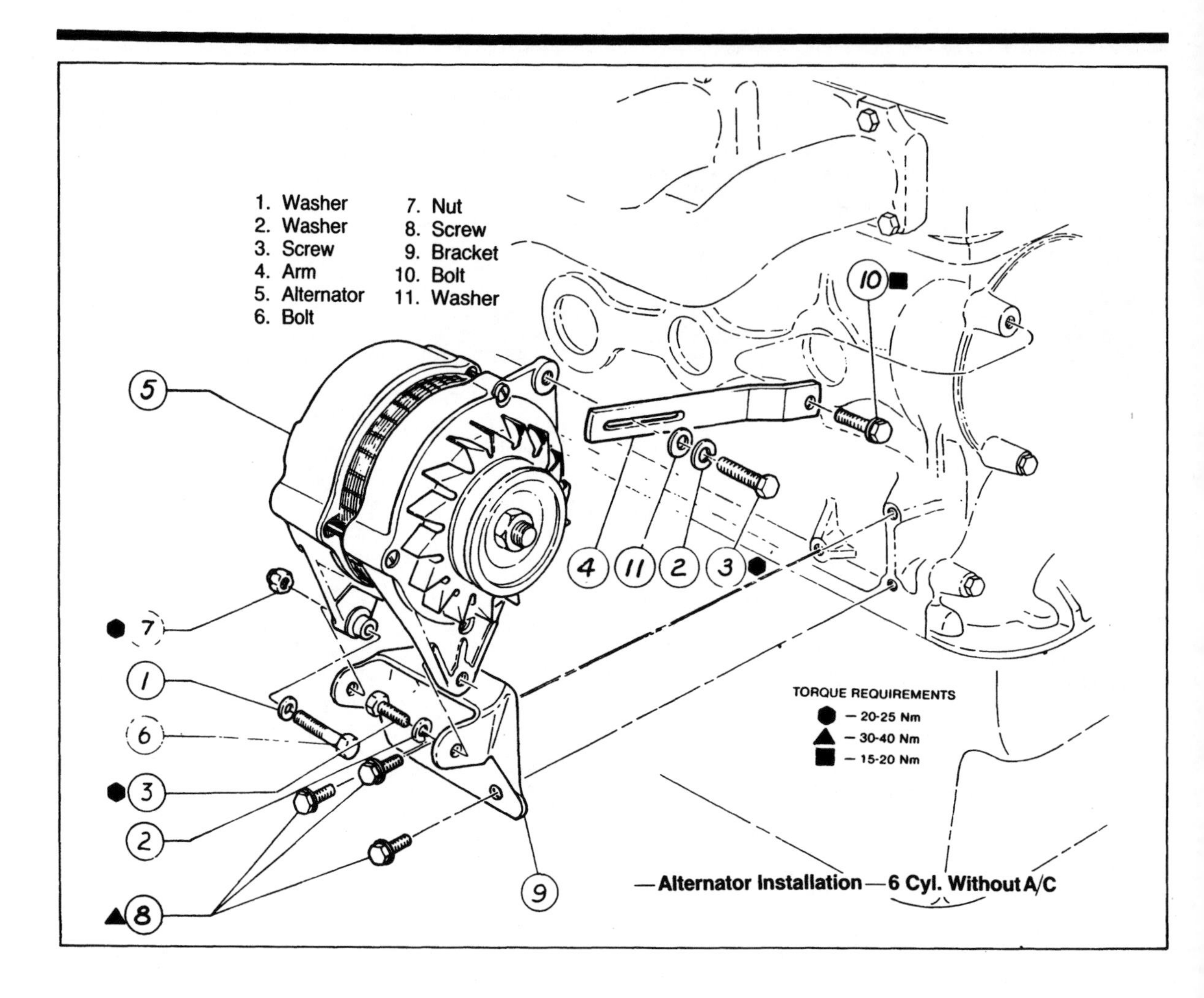

5 Remove the mounting pivot bolt.
6 Push the alternator towards the engine and slip the drive belt over the pulley flange.
7 Remove the mounting pivot bolt and lift the alternator clear of the engine.
8 Slide the drive belt past the fan (when fitted in this application) and remove the belt for inspection.

Refit

1 Fit drive belt over pulleys (fan/crankshaft).
2 Hold the alternator in position and insert mounting pivot bolt and nut.
3 Push the alternator toward engine and loop the drive belt over the pulley flange.
4 Pull the alternator away from the engine, ensuring drive belt is seated in all pulleys, and insert the adjusting link bolt.

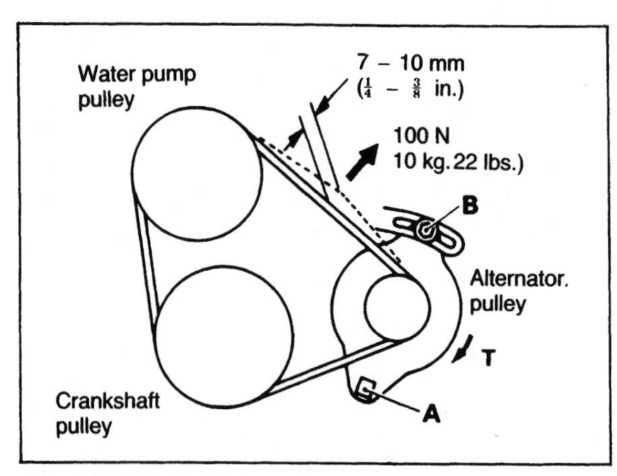

Adjustment of drive belt tension

5 Apply hand force to the alternator until belt tension is correct (deflection or gauge

method) then tighten clamp bolt to specification.

6 Torque the mounting pivot bolt/nut and adjusting link to engine bolt to the correct specifications.
7 Refit leads to rear of alternator.
8 Connect negative terminal to battery.
9 Start engine and check for correct alternator operation.
10 Stop engine, recheck belt tension, and adjust if necessary.

STARTER MOTOR (INTEGRAL SOLENOID TYPE)

Removal

1 Disconnect the negative battery terminal.

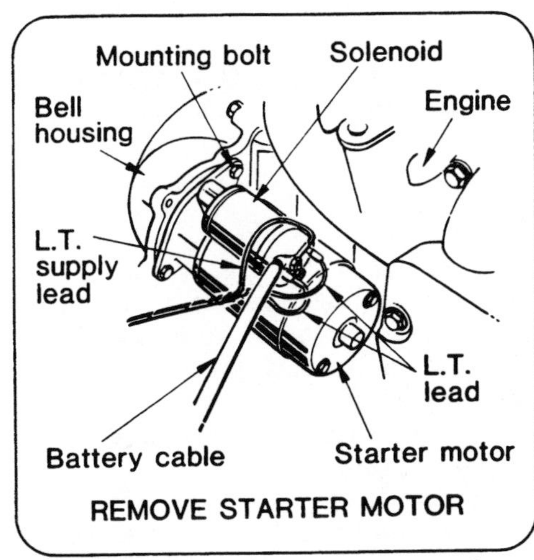

Source: AGPS

2 Loosen and remove the nut retaining the battery cable to the starter motor solenoid.
3 Remove the battery cable.
4 Remove the actuating and/or bypass wires from the solenoid by pulling on the terminal block of the slide-on type connectors.

Caution: when more than one wire is fitted to the solenoid, tag each wire to ensure correct fitment during assembly.

5 Loosen each starter bolt then remove all bolts.
6 Lift the starter motor clear of the engine.

STARTING SYSTEM INSPECTION

1 Check the flywheel ring teeth for burrs and damage. Turn the engine a few centimetres at a time and 'sight' the ring gear teeth through the starter motor mounting aperture. Continue until all teeth have been checked.
2 Clean the mounting flange of the starter motor and the housing.

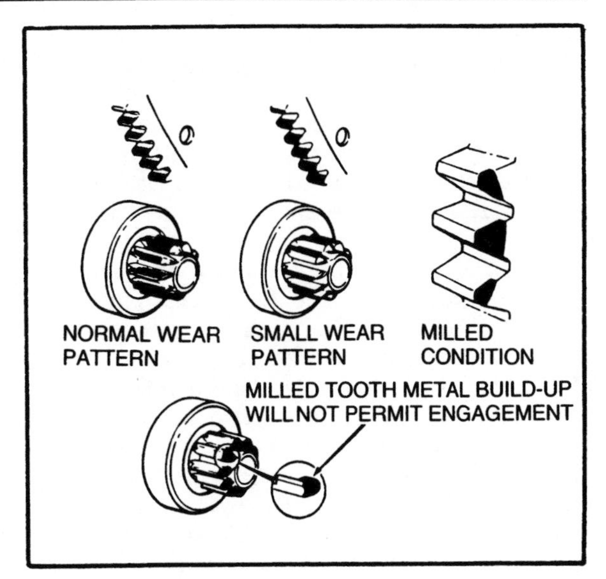

TOOTH WEAR PATTERNS

Refit

1 Insert the starter motor and 'start' all mounting bolts by hand.
2 Rock the starter motor slightly by hand and continue tightening the mounting bolts until spring washer tension is felt.

Caution: to prevent damage to the flange and housing, align the mating surfaces before applying spanner force to the mounting bolts.

3 Torque the mounting bolts evenly until they are correctly tensioned.
4 Fit the battery cable and nut to the solenoid and torque to specification.
5 Slide the actuating and bypass connectors onto their correct terminals.
6 Connect the negative terminal to the battery.
7 Check the starter motor operation by cranking engine. Listen for abnormal noise or harshness in operation.

TEST INSTRUMENTS

Many of the component operations on a vehicle are physical, the method of operation is visible and can be readily understood. When considering electrical and electronic systems the problem becomes more difficult. The effect of electrical current is observable — a starter motor turns or a lamp is illuminated — but electricity cannot be 'seen' in the same way as it is possible to observe oil or fuel flow.

To test electrical circuits and components a means of measuring the flow rate (current), the flow pressure (voltage) and the resistance to flow (resistance) of electricity is required. The majority of electrical system testing can be achieved by the use of a voltmeter, ammeter and an ohmmeter. The instruments can be analogue (moving pointer on a printed scale) or digital (numbers appearing on a liquid crystal display). Selection of a test instrument should be based on the type of work it will be used to test. Relatively cheap moving coil meters are adequate when high current flow is present and/or voltage reading accuracy is not critical. Extremely accurate moving coil meters are available but these tend to be expensive. The type of meter gaining popularity amongst mechanics and technicians is the digital multi-meter. These units are robust and many are auto-ranging (not necessary to change scales for accuracy), overload protected in some ranges and can be switched to read amps, volts and ohms. The accuracy of these units on a cost basis is superior to analogue meters.

The three forms of electrical circuits are the:
1 series;
2 parallel;
3 and a combination of both known as series/parallel.

A working knowledge of these circuits, coupled with an appreciation of the volt, ampere and the ohm will be of invaluable assistance when using instruments to test.

Ampere

Current is the flow rate of electrons/sec. in a conductor. An electron is very small and the number passing a point in the conductor is huge (6.8 billion billion electrons per second), therefore to describe current flow in electrons/sec. is not practical for everyday use. For convenience the term 'ampere' is used (1 A = 6.8 billion billion electrons/sec.). A general statement to describe current would be:

The flow of electrons through a wire (conductor) is called current and is measured in amperes.

Voltage

Electrical current will not flow unless a force is available to cause the electrons to move along a wire. To supply this force it is necessary to have a positive charge at one end and a negative charge at the other end of a piece of wire. As long as a charge difference exists between the ends of the wire, electricity will flow through the wire. By increasing the difference in charge values at the ends of the wire a greater force can be produced. This force can be described as a potential difference or by the more commonly used term, voltage. The unit of measurement is the volt.

Voltage is produced between two points when a positive charge exists at one point and a negative charge exists at the other point.

Resistance

All materials have some resistance to current flow. By careful selection of materials it is possible to have virtually no flow of electrons (this type of material would be an insulator such as plastic, glass etc.) or extremely high electron flow rate (a conductor such as copper, aluminium or gold).

When a small section of wire, much smaller in diameter than the main wire, is placed in the circuit it acts as a restrictor. This feature, coupled with selection of materials, can be used to create a heating effect such as is in a headlamp globe. The special wire in the globe is heated by the flow of electricity and glows brightly, producing light for the headlamp. Unfortunately, resistance can be a problem as well as a benefit. Wiring which is too small in cross-section or a corroded connection can also act as a resistance, slowing the flow rate of electricity and preventing the component at the end of the wire from operating correctly. Resistance values are measured in ohms.

The basic unit of resistance is the ohm. It is defined as the resistance that will allow one ampere to flow when the potential difference is one volt.

CIRCUITS

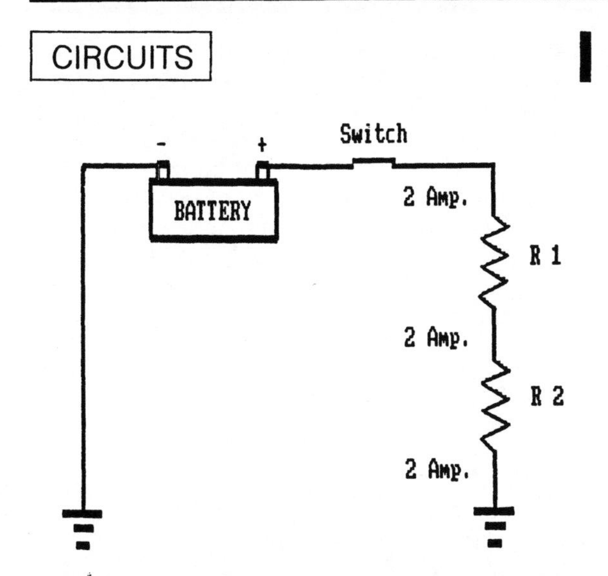

SERIES CIRCUIT — Any current (e.g. 2 Amperes) flowing will pass through both R1 and R2 and will have the same value at any test point.

Series circuit

A series circuit is a single path where all the current that flows through one resistor also flows through all other resistances in the circuit.

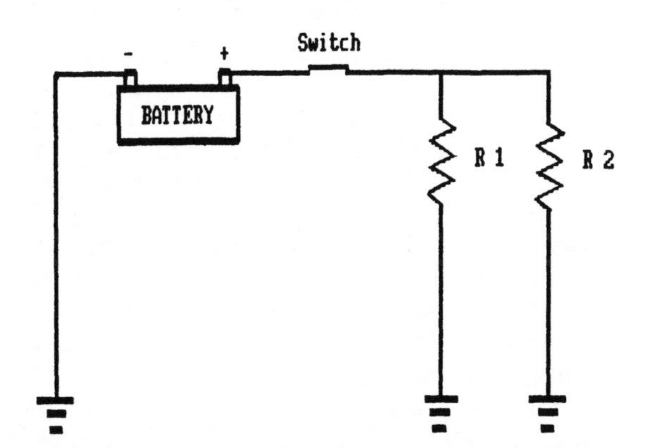

PARALLEL CIRCUIT: R1 and R2 operate independently of each other and the current flow in the alternate paths to earth will be determined by the resistance value in the path.

Parallel circuit

A parallel circuit allows the current to follow alternative paths to complete the circuit. The amount of current flow through each path will depend on the resistance value in each path. As the resistance value is increased the current flow will be decreased.

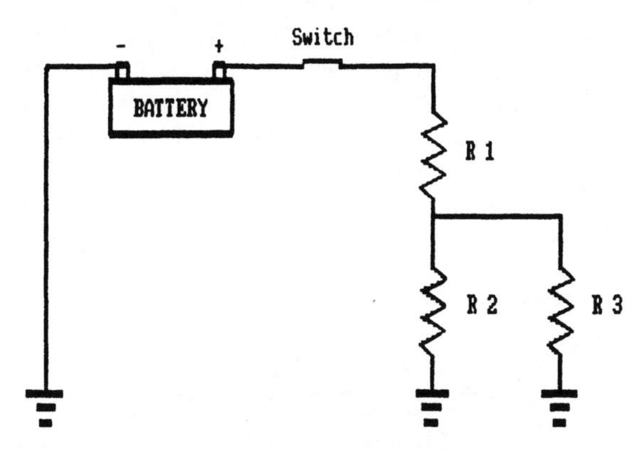

SERIES PARALLEL CIRCUIT: All current in the circuit flows through R1 but is then split through R2 and R3. The combined current flow through R2 and R3 will be equal to the current flow at R1.

Series/parallel circuit

As the name suggests, this circuit is a combination of a series circuit and a parallel circuit. In the example shown ALL current must pass through resistor 1, but alternative paths are available through resistor 2 or resistor 3.

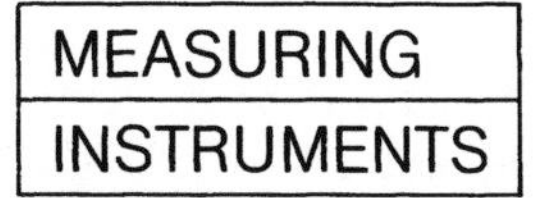

MEASURING INSTRUMENTS

Voltmeter

A voltmeter is used to measure the voltage (potential difference) present between two

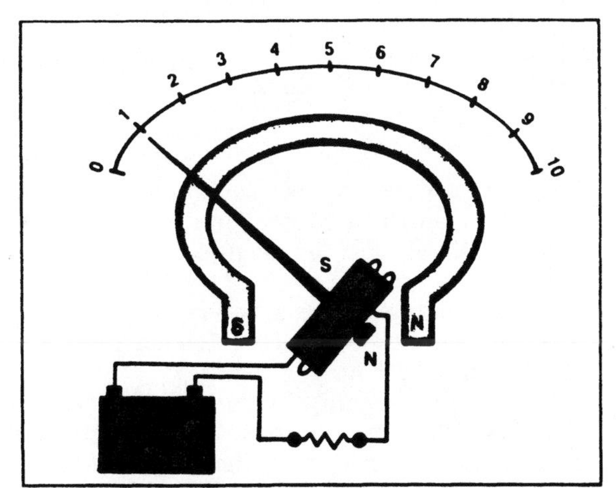

— BASIC VOLTMETER

points in a circuit. The connection of a voltmeter into a circuit is in parallel. This means it is not necessary to disconnect any wires to connect a voltmeter into a circuit.
Caution: when the voltmeter is not auto-ranging, ensure the voltage scale selected is in excess of the possible voltage in the circuit. Also ensure the control is switched to D.C. volts.

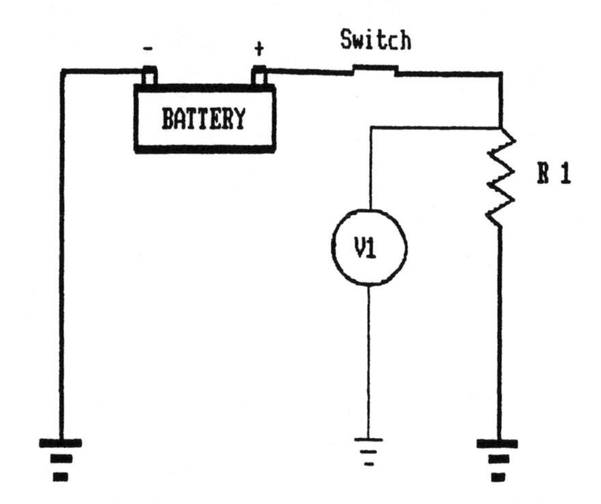

In this example the voltmeter is connected across a single resistance. The reading obtained will be battery voltage.

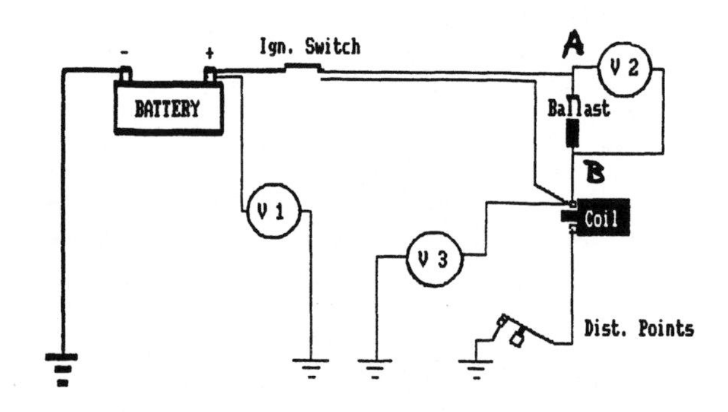

Three voltmeters, V1, V2 and V3 are connected into the simple ignition circuit as shown. Note all the connections are in parallel.

V1 is reading between the positive battery terminal and earth (negative). The meter would display battery voltage.

V2 is fitted across the ballast resistor. REMEMBER the meter reads potential difference between the points of measurement. V2 will read the difference in voltage present between points A and B. With the distributor points closed a typical voltage value would be 5 V. This means there is a voltage drop of 5 V across the ballast resistor.

V3 is connected across the ignition coil terminal and earth (negative). Because of the ballast resistor 12 V will not be present at the terminal (distributor points closed).

12 V (battery) – 5 V (drop across resistor) = 7 V. The reading at the IGN. SW. terminal of the coil will be 7 V. Note the sum of V2 and V3 equals V1.

Ammeter

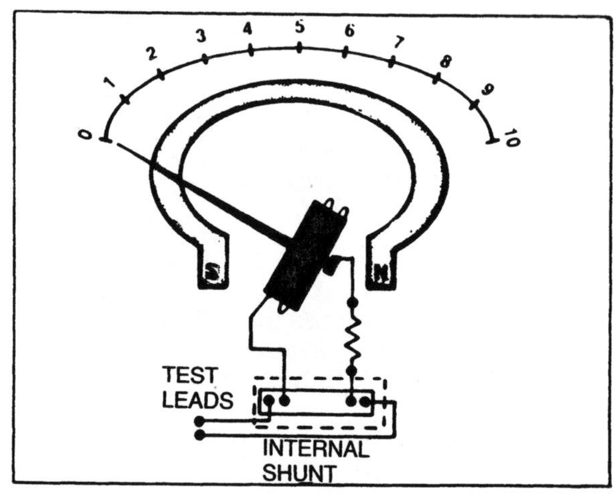

AMMETER

An ammeter is used to measure the current flow (amperes) in a circuit. The ammeter is connected in series. To measure currents greater than the meter full scale deflection a shunt must be used to divert current around the meter internal movement. A switch and additional shunts may be used to provide more than one range. The shunt is fitted either internally in the meter or may be external if large current values are to be measured. Follow the meter manufacturer's instructions for shunt application.
Caution: always connect the ammeter in series with the circuit, as an accidental parallel connection will burn out the meter. Ensure the shunt and/or scale selected is greater than the maximum flow in the circuit. Switch multi-meters to amps.
The ammeter will record all current flow in the circuit. The value will depend on the value of the resistor.
A1, A2 and A3 will record identical values. Current flow in a series circuits is the same

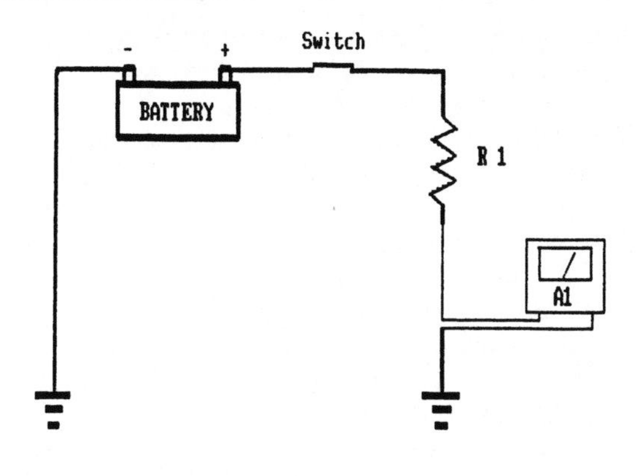

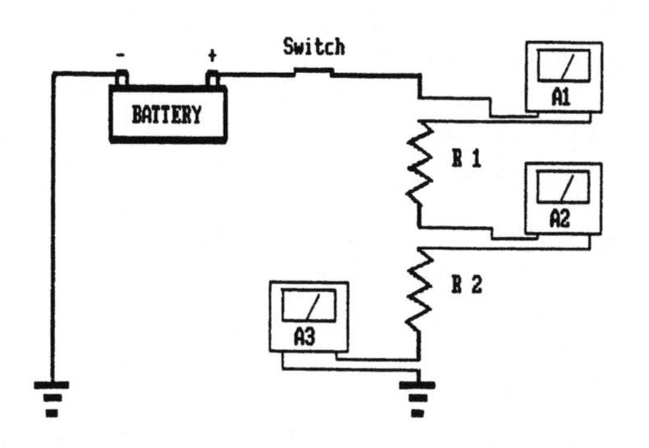

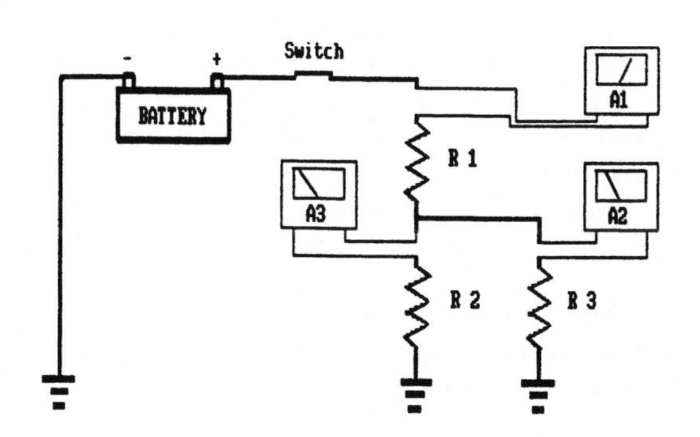

at any point within the circuit.
A1 will record the total current flow in the circuit. A2 and A3 will record the current flow through R2 and R3 respectively. The sum of A2 and A3 will equal A1.

Ohmmeter

An ohmmeter is used to measure the resistance present in a circuit. The meter is

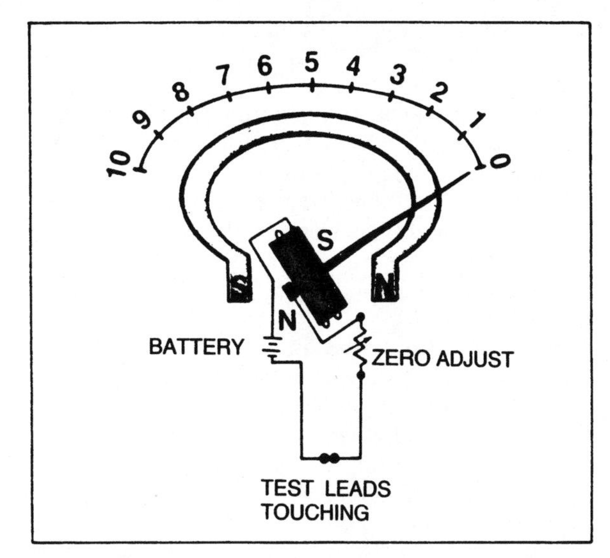

OHMMETER

connected in parallel across the resistance to be measured.
Caution: to safeguard the meter, the following steps must be carefully followed:

1 Power must be off in the circuit before the meter is connected.
2 Select the correct scale
3 Multi-meters should be switched to ohms.
4 Ohmmeters are self powered and contain a small battery. Never use the ohmmeter on an electronic circuit unless the test is approved by the vehicle manufacturer. Mistakes in this area can be extremely expensive.

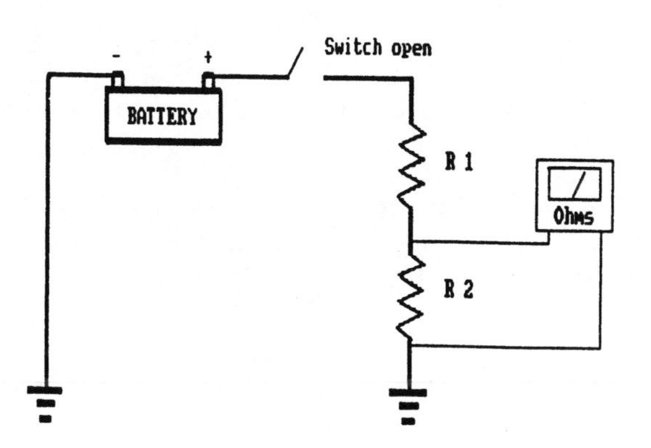

The ohmmeter will record the resistance value of R2.

CHAPTER 12 REVISION

1 List the general safety rules which should be observed when working on or near vehicle electrical systems.
2 List the five safety rules which must be observed when servicing vehicle electronic systems.
3 Describe the function of an ignition system.
4 What is the name of the device which alters the engine ignition timing in response to engine load conditions?
5 Name the component which reduces arcing at the distributor contact points.
6 What is the function of the ballast resistor in an ignition system?
7 What are the major differences between a mechanically triggered ignition system and an electronic ignition system?
8 Describe the correct method of removing an ignition system spark plug high tension lead from a spark plug.
9 What may happen if metal is removed from the blade tip of the rotor button during cleaning?
10 Describe the method normally used to prevent dirt entering the engine cylinders when spark plugs are being removed for cleaning.
11 List the steps required when changing the contact points in a mechanically triggered distributor.
12 What is 'dwell angle' and how is it checked?
13 Describe the method of checking an engine for correct ignition timing.
14 Obtain the manufacturer's coding from the spark plugs in three different engines. Use a spark plug chart and change the code into a description (heat range, reach, seat type, electrode etc.).
15 What is the major difference between a semi-sealed lamp and a sealed beam lamp?
16 In a four headlamp system, do all the lamps have two filaments and is there a requirement as to which lamps are fitted to the inner or outer positions?
17 Describe a globe suitable for use as (a) a trafficator globe and (b) a stop/tail lamp globe.
18 List the preliminary steps prior to aiming the headlamps of a vehicle.
19 What is the purpose of the 'cover' used during headlamp aiming?
20 What are the two methods of attaching terminals to wires?
21 What problem may occur if side cutting pliers are used to 'crimp' terminals?
22 List the CAUTION steps to be observed when installing a trailer or caravan socket to a late model vehicle.
23 An owner complains that each time the vehicle is connected to a trailer the flash rate of the trafficator is faster than the law allows. What can be done to rectify the problem?
24 Do the wiring standards for five-pin and seven-pin sockets allow interchanging of plugs between the two types when towing different trailers?
25 What rating or specification of a passenger car battery would the salesman be referring to if he described a battery as 420 AMP?
26 Why should two spanners be used when loosening or tightening battery terminals?
27 What type of solution is widely used to remove corrosion from battery terminals and trays?
28 List the steps required to perform (a) a battery test using a hydrometer and (b) a battery test using a high rate discharge tester.
29 Write down a brief description for 'current', 'voltage' and 'resistance'.
30 Name the three types of electrical circuit in common use.
31 List the precautions which should be observed when testing vehicle circuits with a voltmeter, ammeter and ohmmeter.

13

FUEL SYSTEMS

INTRODUCTION

Of the ancillary systems fitted to the motor car engine, the fuel system has always been close to the centre of technological advance. The purpose of the fuel system is to supply the engine with the correct air/fuel mixture for all speed and load conditions.

More accurate control of the fuel mixing system has been made possible with the application of electronic microcomputers. These computers (via sensors) can control the supply of fuel into the engine cylinder, then monitor the make-up of the exhaust gases leaving the engine and provide a correction if too little or too much fuel is being burnt. Because of these control systems it has become possible to develop greater engine power (kW) and improve fuel economy while complying with vehicle exhaust emission laws.

The purpose of this chapter is to describe and highlight the different types of fuel systems fitted to spark ignition and compression ignition engines.

COMPONENT IDENTIFICATION AND FUNCTION

Fuel systems, as fitted to modern vehicles, can be divided into three general groups:

1 spark ignition—carburettor
2 spark ignition — fuel injection
3 compression ignition — fuel injection

The correct identification of these components and their function is essential for:

- locating components on individual vehicles;
- ordering correct replacement parts;
- servicing the systems in a safe and correct manner.

Spark ignition — carburettor

A typical fuel system consists of a:

- fuel tank;
- fuel lines (pipes and hoses);
- fuel pump;
- fuel filter;
- carburettor;
- inlet manifold.

The functions of a fuel system are to:

- store sufficient fuel for long journeys;
- supply the carburettor with clean fuel;
- distribute the correct air/fuel mixture to each cylinder.

The **fuel tank** is usually located in the rear of the vehicle and is constructed from either zinc coated steel or reinforced plastic. The main components of the fuel tank are the:

- filler tube;
- vent tube;
- fuel pickup and filter;
- fuel gauge electrical sender;
- fuel tank.

Fuel tanks are able to:

- receive fuel from a petrol bowser;
- store 40–80 L of fuel;
- deliver fuel in as safe a manner as possible.

Fuel lines can be manufactured from either zinc coated steel or petrol resistant

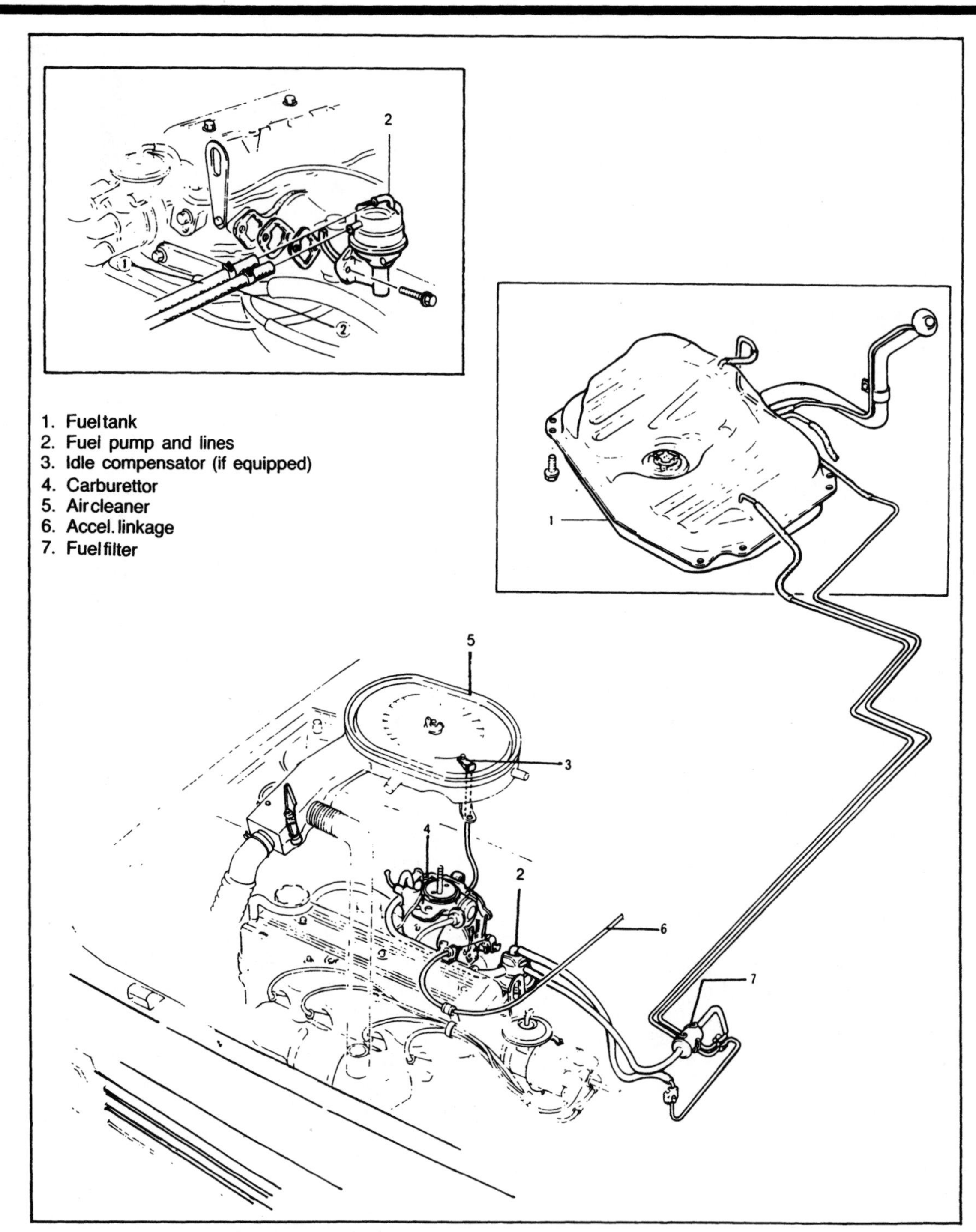

FUEL SYSTEM – SPARK IGNITION – CARBURETTOR

tubing. The latter is becoming more popular because it does not require a separate flexible hose between relative moving parts. The fuel lines are usually attached to the underside of the vehicle floor pan with steel or plastic clips. They allow liquid fuel to be transported from the fuel tank to the fuel mixing device on the engine.

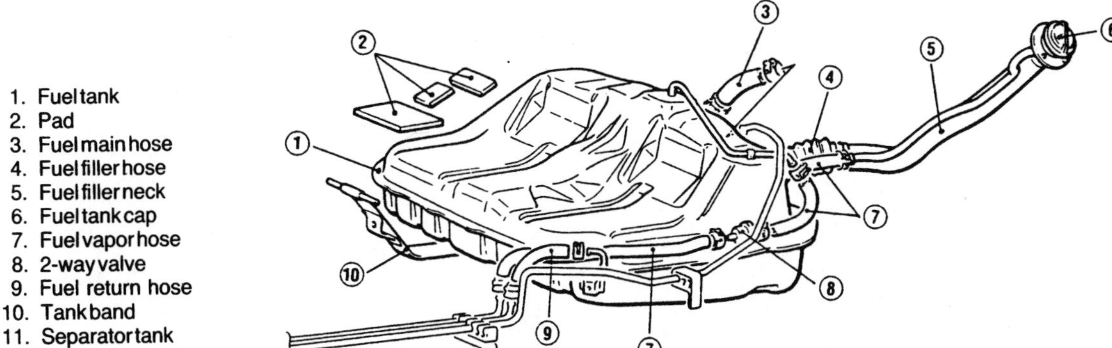

Fuel tank components

Source: Mitsubishi Magna TM series manual

The **fuel pump** can be either a mechanical type, driven from the engine, or an electrical type driven by electricity when the ignition key is turned on.

A mechanical fuel pump is generally located in one of four places on the engine. These are the:

1 timing cover;
2 cam/rocker cover (O.H.C. engine);
3 camshaft side of crankcase;
4 cylinder head.

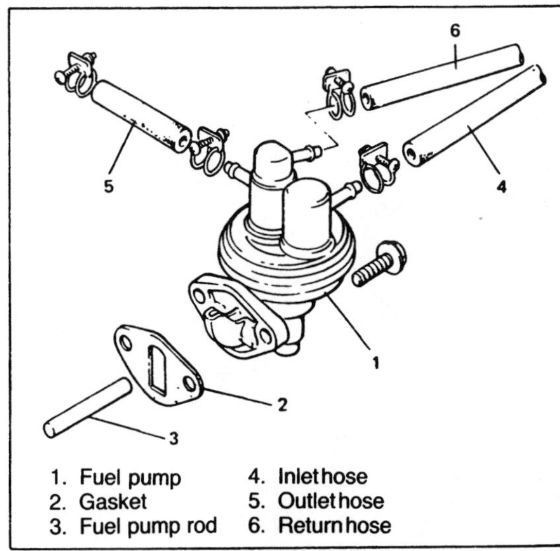

Mechanical Fuel pump

An electric fuel pump can be located either at the front of the vehicle, in the engine compartment, or at the rear near the fuel tank. It is common to find an electric fuel pump at the rear of the vehicle for reasons of quietness.

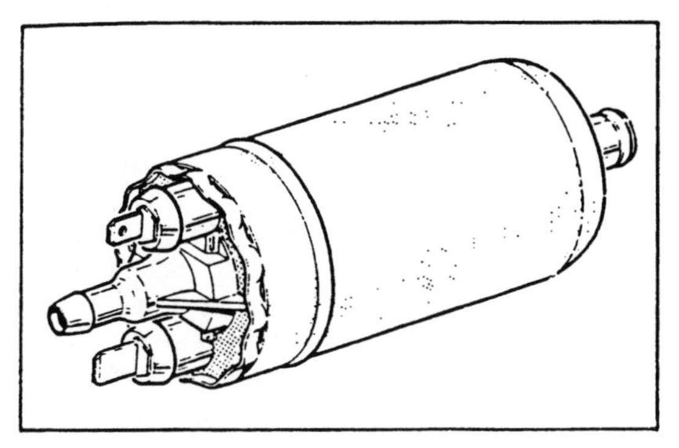

Electric Fuel Pump

The fuel pump, either mechanical or electrical, causes the fuel to be moved from the tank through the lines to the fuel/air mixing device.

Fuel filters on most vehicle engines fitted with a carburettor have a paper element housed in a plastic cylinder with fittings at either end to connect it into the fuel line between the pump and the fuel/air mixing device (carburettor). This unit is of a 'throw-away' design and should be replaced at manufacturer's recommended service intervals. When a paper element filter is not fitted, a brass wire gauze or a ceramic filter element will be installed in either the carburettor or the fuel pump bowl.

The fuel filter removes from the fuel small foreign particles that could block or damage the drillings and passages inside the fuel/air mixing device (carburettor).

Note: An arrow or marker on the body of the fuel filter indicates the direction of fuel flow.

Caution: filters must always be fitted in the direction of fuel flow.

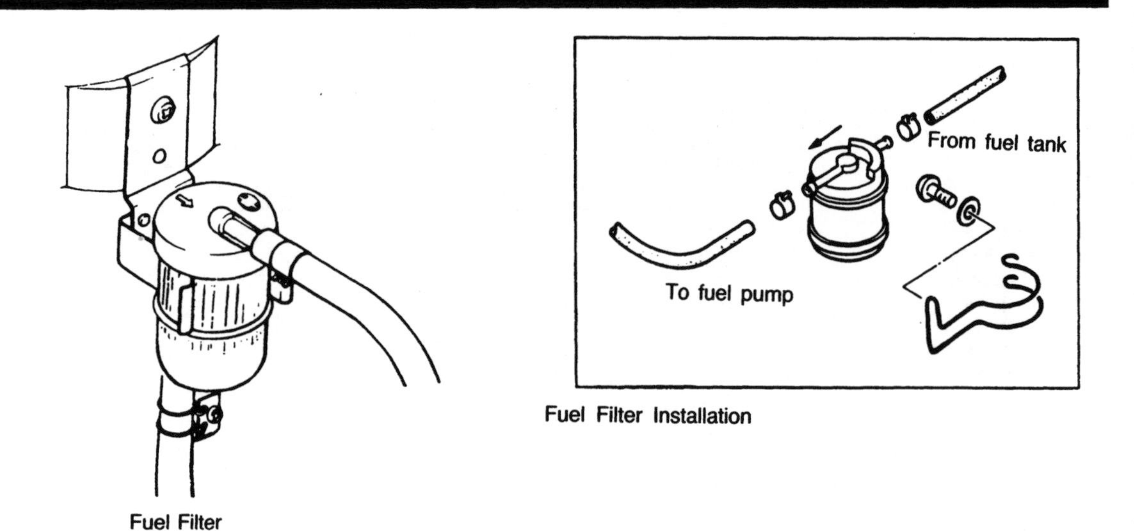

Fuel Filter

Fuel Filter Installation

The **carburettor** (fuel/air mixing device) is generally manufactured by the die casting process, but moulded plastic has been used in some designs. The carburettor consists of a number of fuel and air passages and drillings, a fuel reservoir and a throttle butterfly. A linkage connects the accelerator pedal (inside the vehicle) to the carburettor throttle butterfly, thus giving control of engine speed to the driver. Carburettors:

- mix the fuel and air in the correct proportions for all speed (engine) and load conditions;
- atomise and start vaporising the fuel;
- allow the driver to control the quantity of vaporised mixed fuel/air being delivered to the inlet manifold.

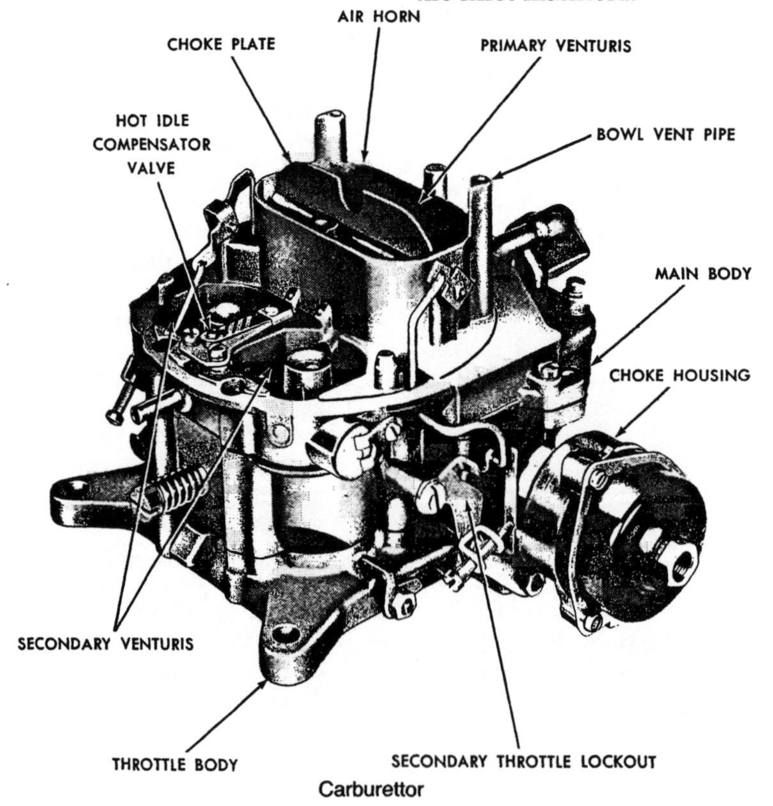

Carburettor

Inlet manifolds can be manufactured from cast iron, cast aluminium or, in some cases, steel tube. A heating passage is provided on all carburettor manifolds. Heat is provided by passing either engine coolant or exhaust gases through the heating passage. The inlet manifold is usually a detachable component bolted to the cylinder head. Some manufacturers have produced engines with the inlet manifold cast into the cylinder head.

The inlet manifold continues to vaporise the fuel (with the addition of heat from the heater passage) and distributes the fuel/air charge evenly to all engine cylinders.

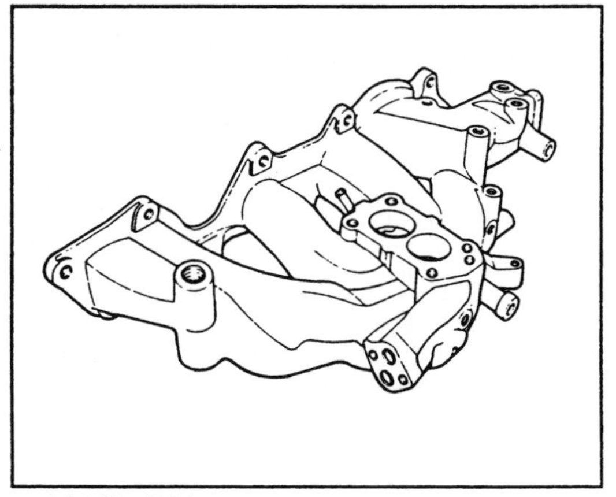

— Inlet Manifold

Source: Mitsubishi Magna TM series manual

Spark ignition — fuel injection

Fuel tanks are constructed of either zinc coated steel or of a formed plastic material. They are usually located at the rear of the vehicle. A fuel tank consists of:

- filler tube;
- vent hose;
- fuel pickup pipe and strainer;
- fuel return pipe;
- fuel gauge sender unit;
- baffles to reduce fuel surge;
- tank body.

Note: Some fuel injection systems also have a fuel pump in the fuel tank.

The fuel tank assembly stores and carries fuel in as safe a manner as possible.

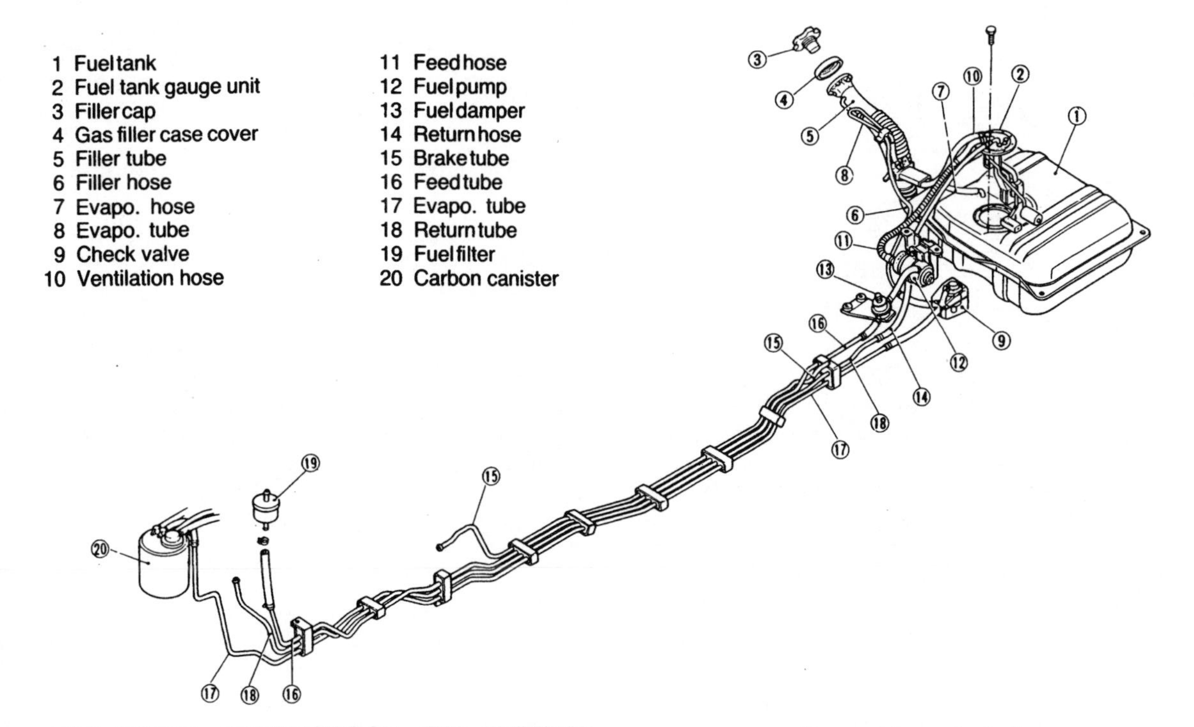

FUEL SYSTEM - SPARK IGNITION - FUEL INJECTION

Source: Nissan Australia

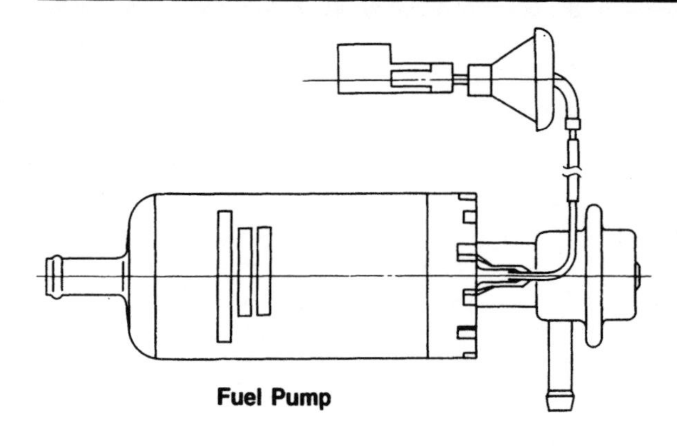
Fuel Pump

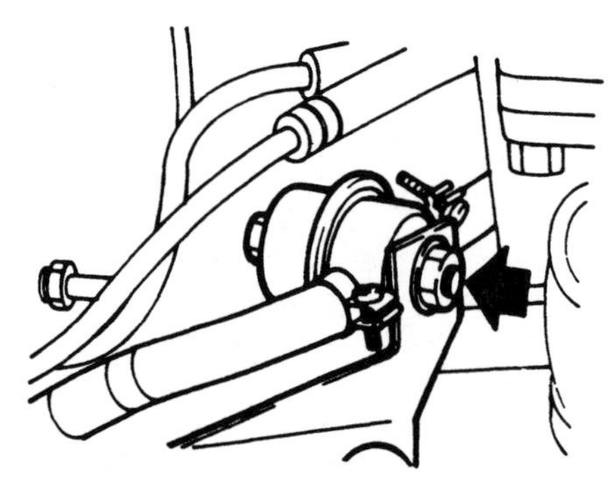
Fuel Damper
Source: Nissan Australia

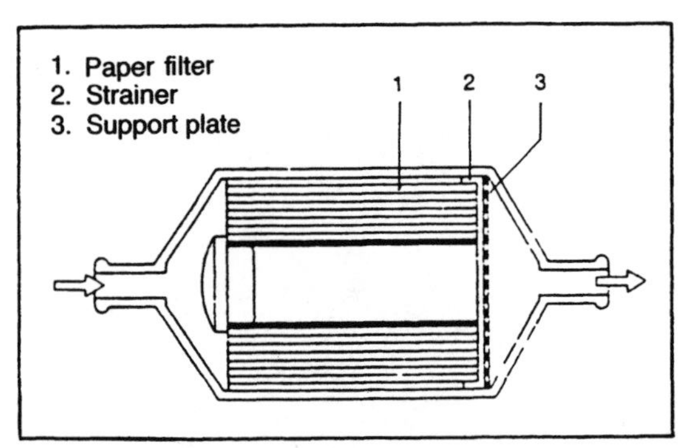

Fuel Filter

The **fuel filter** body is generally constructed of aluminium, although some designs use a steel pressing. An element inside the filter body is made of very fine filter paper. This unit can be located in the fuel supply line, either at the rear or the front of the vehicle. The fuel filter removes from the fuel dirt and foreign particles that could block and/or damage the fuel injection equipment.
Note: An arrow on the filter body indicates the direction of the fuel flow.
Caution: filters must always be fitted in the direction of fuel flow.

The **fuel pump** can be located either inside or outside the fuel tank. On some vehicles a small pump inside the tank supplies fuel to the externally mounted output pump. These fuel pumps are operated electrically and will produce an output pressure of 100–400 kPa. Their internal parts are not serviced separately, so they must be replaced as a unit. **Caution: never attempt to test an E.F.I. fuel pump by connecting it to a battery. An explosion could occur due to lack of fuel to cool and lubricate the unit.**
The fuel pumps move the fuel from the tank to the fuel injector rail and develop the fuel system pressure of 100–400 kPa.

A **fuel damper** is located between the fuel pump and the pressure regulator. It is a spring loaded diaphragm in a cylindrically shaped pressed steel container. Two pipe fittings are provided at one end of the container to allow the fuel damper to be connected into the supply line. A fuel damper is used to reduce pressure pulsations in the fuel supply system.

Fuel lines or pipes, in a fuel injected system, are usually constructed of zinc coated steel. The reason for using steel is because of the high pressures (100–400 kPa) used in these types of fuel systems. At all points in the fuel lines where flexibility is required, high pressure rubber hoses are used. Two lines are normally used in these systems. They are:

1 Fuel supply line from tank to pressure regulator.
2 Fuel return line from pressure regulator to tank.

The fuel lines deliver fuel from the fuel tank to the injection system and return excess fuel to the fuel tank.

The **fuel supply rail** is a tube of either aluminium or steel construction, which connects the inlets of all fuel injectors to the fuel supply line. The fuel supply rail is designed to keep a quantity of fuel under pressure available to all injectors for all engine load and speed conditions.
See diagram on next page

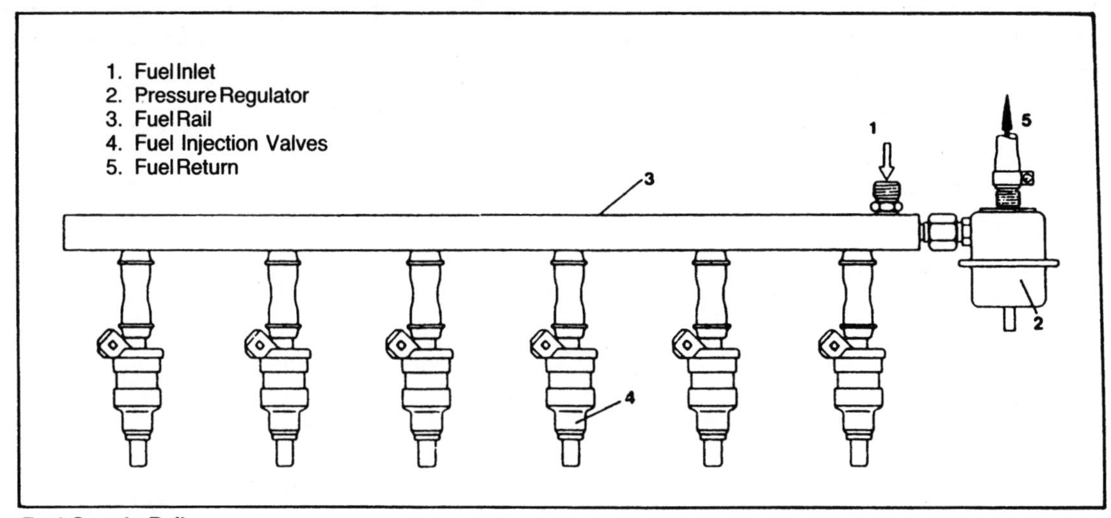

Fuel Supply Rail

The **fuel pressure regulator** is of pressed steel construction and is a cylinder-like shape. Three pipe connections are fitted to the regulator. They are the:

1 fuel supply from the fuel supply rail;
2 fuel return to the fuel tank;
3 engine vacuum connected to engine inlet manifold.

A fuel pressure regulator, which is usually located on or very close to the fuel supply rail, controls the fuel system pressure at all times for all engine speed and load conditions.

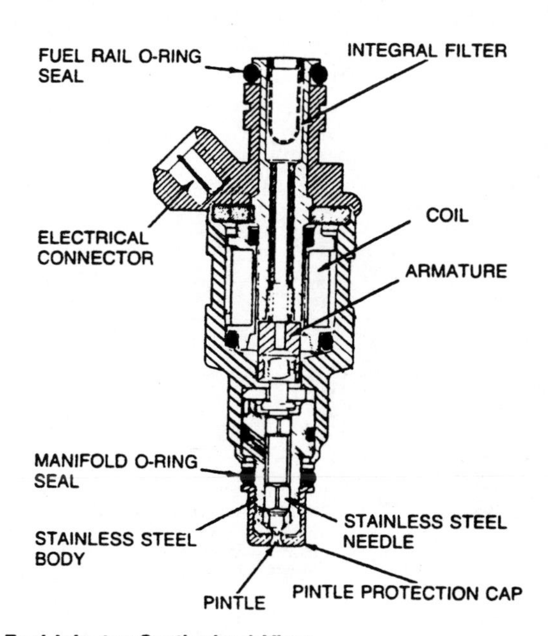

Fuel Injector Sectionized View

Fuel Pressure Regulator

Fuel injectors are electrically operated valves. They are usually fitted into the inlet manifold, just in front of the engine inlet valves. The other end of each injector (opposite its mounting) is connected to the fuel supply rail, where it receives fuel under pressure for injection into the engine inlet manifold. An electrical connection is made by a two-pin plug from the engine loom to the two-pin socket on the end of each injector. The fuel injector acts as a tap and as a spray valve. When it is activated by a signal from the 'electronic control unit' it must inject fuel, in the right spray pattern, into the inlet port of the engine cylinder.

An **inlet manifold** can be constructed of steel but is usually made from an aluminium casting. When a fuel injector is used for each engine cylinder (multi-point injection), then the inlet manifold (sometimes called a ram tuned manifold) will generally have long sweeping tubes that connect between the inlet ports and the throttle body. For throttle body injection (T.B.I.), or centre point injection (C.P.I.), the manifold is similar to that of a carburettor engine. The function of an inlet manifold is to:

- provide an even distribution of air or air/fuel to all engine cylinders;
- provide mounting points for injectors and a throttle body;
- provide vacuum ports for all vacuum operated devices.

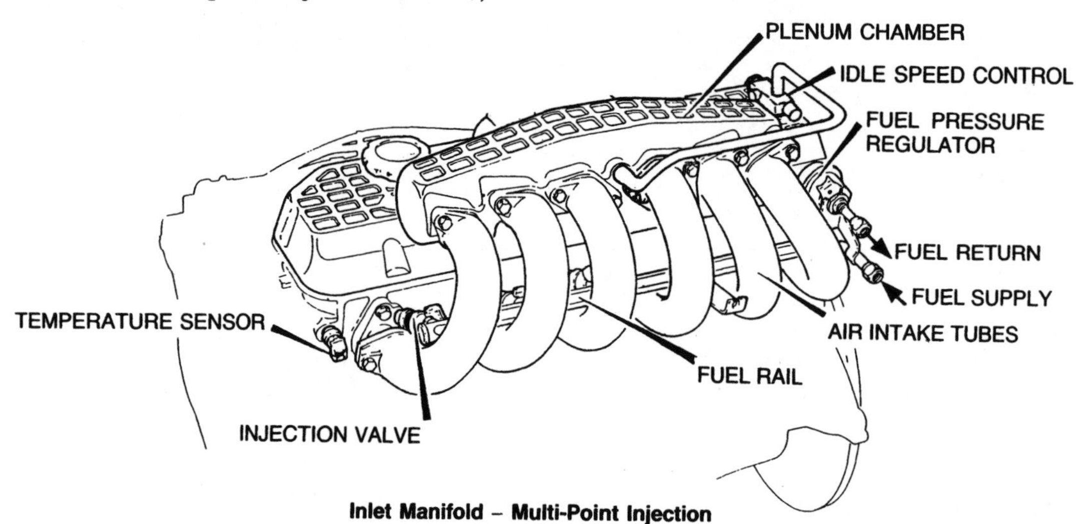

Inlet Manifold – Multi-Point Injection

The **engine air filter** will most likely be of a 'throw away' paper element type, fitted into a reinforced detachable plastic housing. The air filter is usually located behind the right or left hand headlight assembly, in the engine compartment. An engine air cleaner removes particles of dust and foreign materials from the air before it enters the engine. A failure to do this would dramatically shorten the life of the engine.

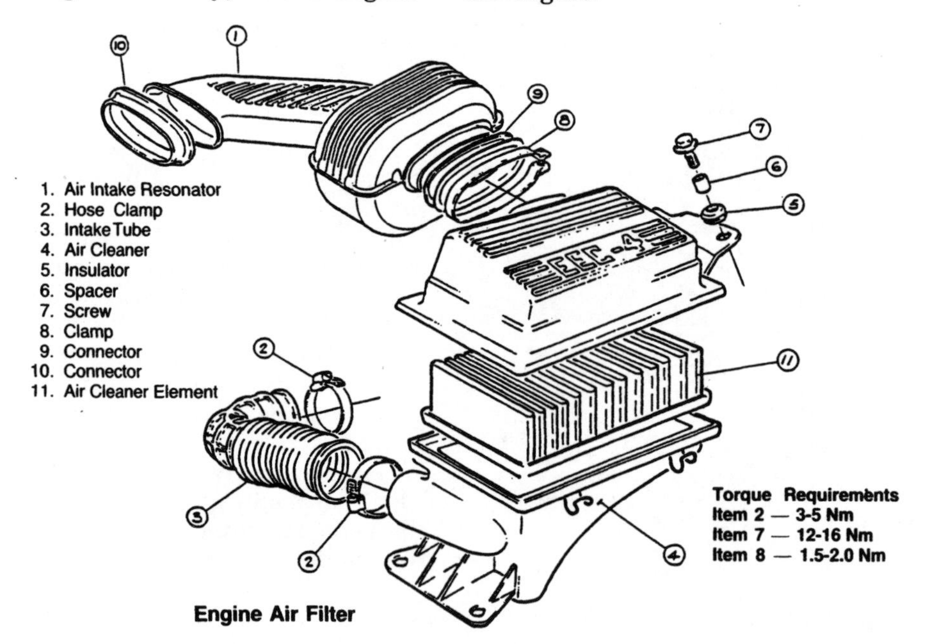

Engine Air Filter

The **air flow meter** is a unit usually located between the air filter and the throttle body and, in most cases, is attached to the air filter housing. Different types of air flow meters can be used on different E.F.I. systems. They include:
- swinging flap;
- hot wire;
- ultra sonic sound (Karmman vortex type).

Each type of air flow meter is connected to the electronic control unit by a loom (wiring) and plugs. The air flow meter tells the electronic control unit how much air is entering the engine at any time.

A number of manufacturers have manifold absolute pressure (M.A.P.) devices instead of air flow meters.

An **air duct**, made from a rubberised or plastic material, is connected between the air flow meter and the throttle body assembly. The function of the air duct is to deliver the air passing through the air flow meter to the throttle body assembly.
Note: This duct must be airtight to prevent air bypassing the air flow meter, as this will upset the engine fuel/air mixtures.

The **throttle body assembly** is generally of cast aluminium construction and consists of a:
- body;
- throttle butterfly valve;
- throttle position sensor;
- vacuum ports (ignition advance port; exhaust gas recirculation control valve port);
- idle speed adjustment.

The throttle valve assembly is connected between the air duct and the inlet manifold. The function of the throttle valve assembly is to:
- allow driver control (via linkage from the accelerator pedal to throttle butterfly) of air entering the engine and therefore of engine speed;
- provide vacuum to operate devices at engine speeds above idle throttle position;
- allow the engine idle speed to be adjusted.

A **throttle position sensor**, located on the throttle body, is attached to the end of the throttle butterfly shaft. It is connected electrically, by the wiring loom and plugs, to the electronic control unit. The throttle position sensor sends information about the throttle valve position to the electronic control unit.

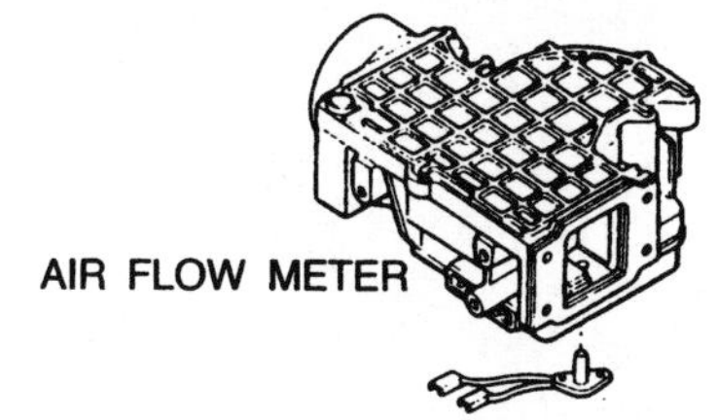

Source: Nissan Australia

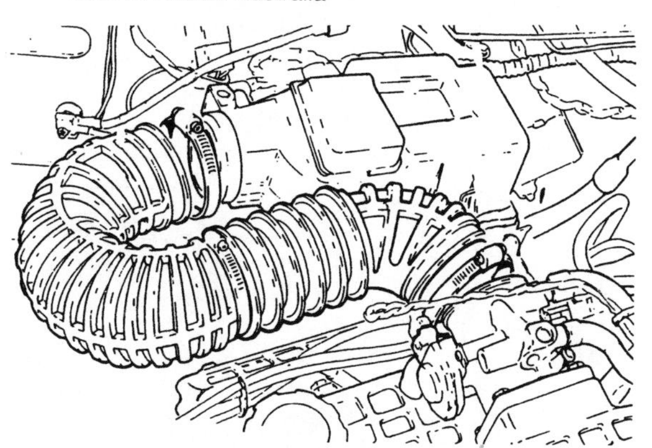
AIR DUCT

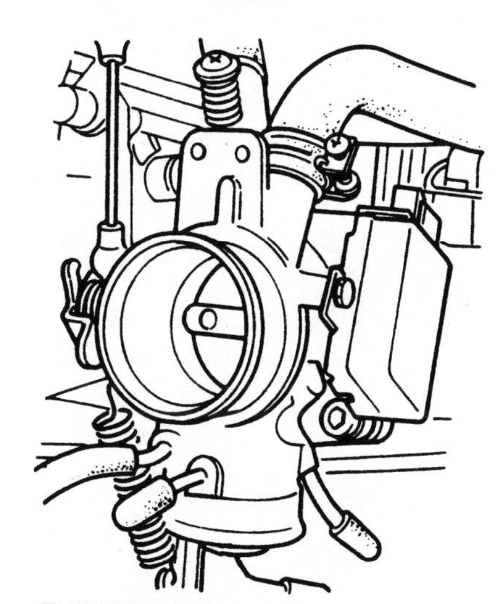
THROTTLE BODY ASSEMBLY
Source: Nissan Australia

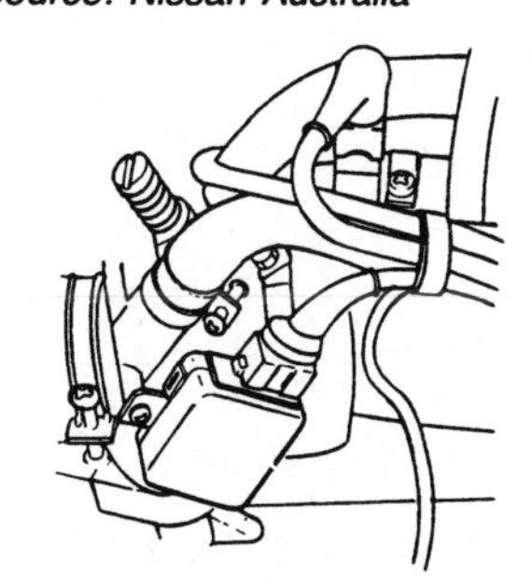
THROTTLE POSITION SENSOR

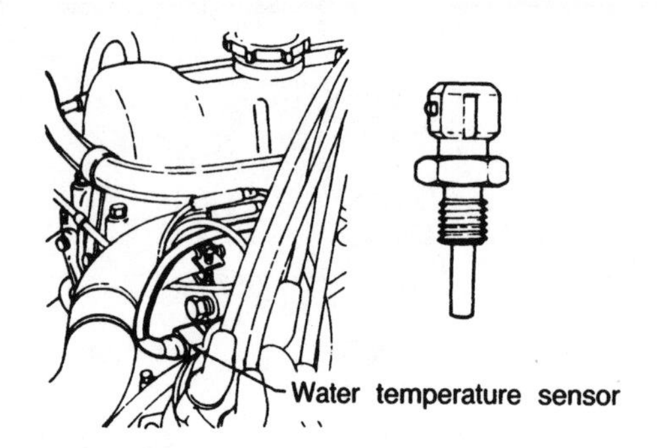

Source: Nissan Australia

A **coolant temperature sensor** is fitted into the engine cooling system. It is an electrical device connected by the wiring loom and plugs to the electronic control unit. The coolant temperature sensor sends information about the engine's temperature to the electronic control unit.

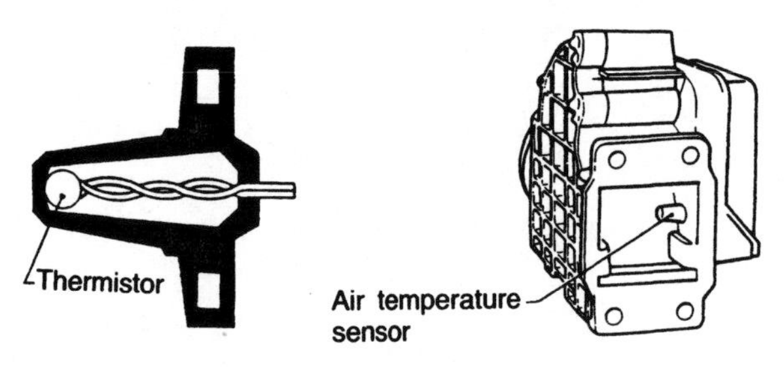

Source: Nissan Australia

An **air temperature sensor** is fitted into the air flow meter. It is an electrical device connected by the wiring loom and plugs to the electronic control unit. The function of the air temperature sensor is to send information to the electronic control unit about the temperature of the air entering the engine.

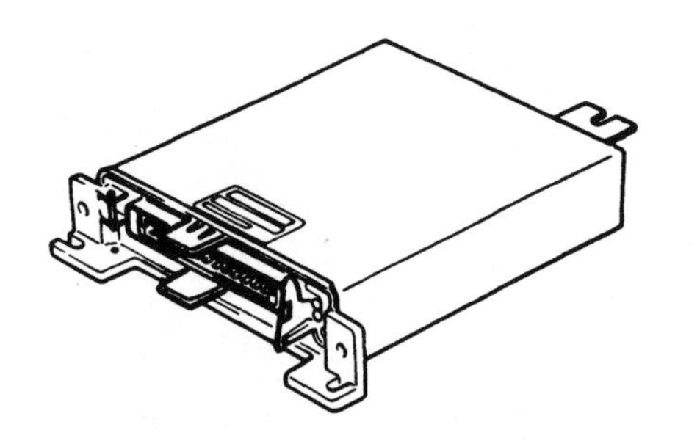

Electronic Control Unit
Source: Nissan Australia

The **electronic control unit** is an electronic microcomputer, housed in a pressed steel case. It is usually located inside the passenger compartment rather than the engine area, as it can be affected by heat. This unit is connected to all sensors, injectors and the vehicle electrical supply by the wiring loom and plugs. The function of the electronic control unit is to receive and process the information sent from the sensors, calculate the fuel quantity for the engine speed and load condition and then activate the fuel injectors to allow them to deliver this fuel quantity.

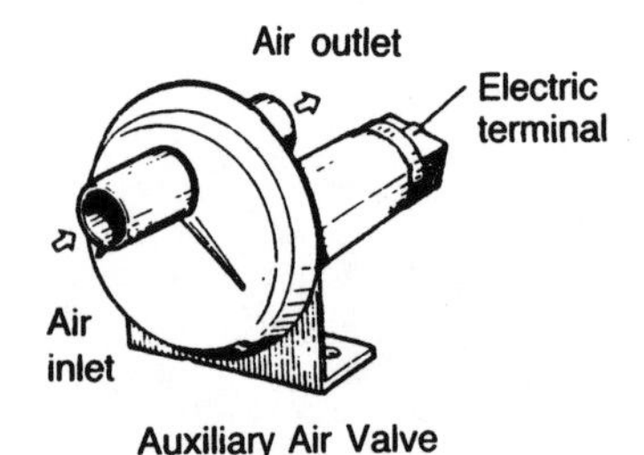

Auxiliary Air Valve

Source: Nissan Australia

An **auxiliary air valve** is usually located on or near the engine cylinder head. A hose attaches one end of the air valve to the air duct and another hose attaches the remaining end of the air valve to the inlet manifold. It is supplied with power from the vehicle electrical system via the E.F.I. relay.
Note: This unit is not connected to the electronic control unit.

The auxiliary air valve, which bypasses the throttle butterfly, controls the engine idle speed until the engine reaches its normal operating temperature.

A **cold start injector** (if fitted) is located centrally on the inlet manifold and is supplied with fuel from the fuel supply rail. It is connected electrically by the wiring loom and plugs to the vehicle power supply system via the E.F.I. relay and thermotime switch.
Note: This unit is not connected to the electronic control unit.

The function of the cold start injector is to supply extra fuel for cold start conditions.
Note: Cold start injectors are not fitted to late model E.F.I. systems.

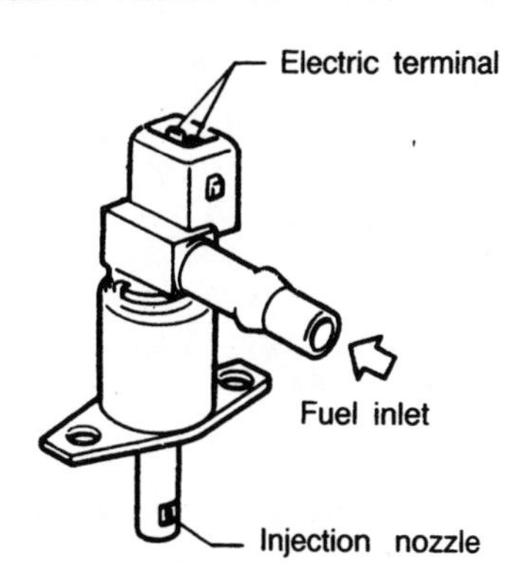

Cold Start Injector
Source: Nissan Australia

A **thermotime switch** (if fitted) is located in the engine cooling system on the cylinder head or block and is connected electrically between the E.F.I. relay (vehicle power supply) and the cold start injector by a wiring loom and plugs.
Note: This unit is not connected to the electronic control unit. Also, it will not be found on late model vehicles.

The function of the thermotime switch is to control the cold start injector during cold engine starts.

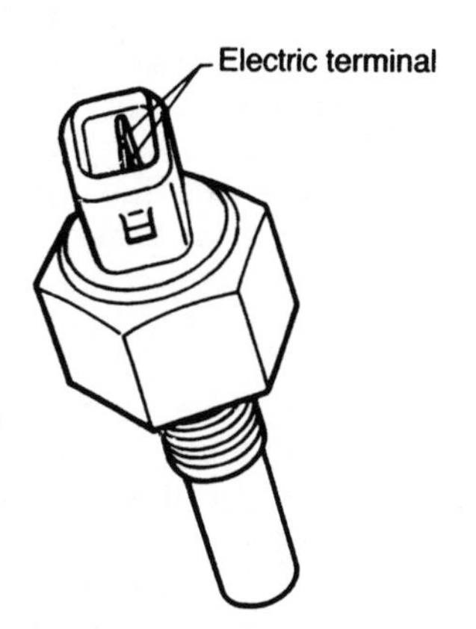

Thermotime Switch
Source: Nissan Australia

The **E.F.I. relay** is usually located in the fuse or relay box in the engine compartment. It is connected by the wiring loom and plugs to the:

- electronic control unit;
- injectors;
- fuel pump;
- auxiliary air valve;
- cold start injector and thermotime switch;
- engine ignition coil negative terminal;
- vehicle power supply.

The E.F.I. relay controls power supply to all E.F.I. system components during engine cranking and running conditions.

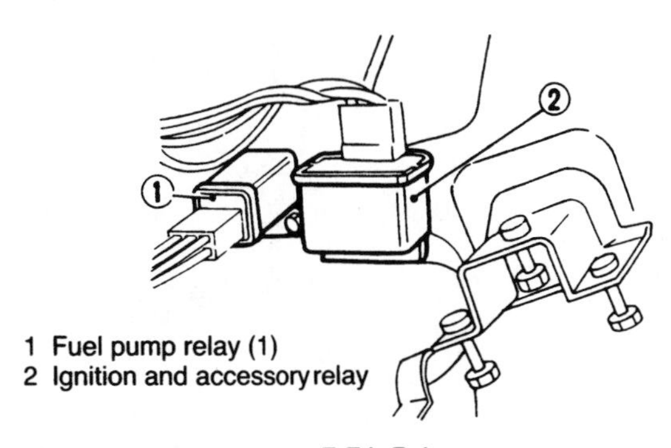

E.F.I. Relay
Source: Nissan Australia

Compression ignition — fuel injection

The **fuel tank** for the diesel or compression ignition engine is usually constructed of zinc coated steel similar to that of the spark ignition engine fuel tank, but it can also be constructed in a cylindrical shape of thicker steel or aluminium on larger vehicles. In either case, the tank consists of:

- fill tube and cap;
- vent;
- fuel supply pickup and strainer (filter);
- fuel return line connection;
- fuel gauge sender unit;
- anti-surge baffles (when fitted);
- fuel tank body.

The function of the fuel tank is to store and supply fuel in as safe a manner as possible.

Fuel lines are usually constructed of zinc coated steel tubing, but diesel fuel resistant plastic piping may be used. The fuel lines are attached to the vehicle floor pan or frame rail with either steel or plastic clips. Both the supply and return lines have a low pressure rating.

Note: All connections in the fuel lines must seal correctly to prevent the entry of air into the fuel system, which will stop the engine from starting and running.

The fuel lines allow the fuel to be transported from the fuel tank to the engine fuel system, and return the excess fuel to the fuel tank.

The **fuel filters** are located in the fuel supply line and their position will determine whether they are a primary (sedimentor) or secondary filter. A filter may be placed:

1 before the fuel lift pump.
   - Generally known as a sedimentor (primary type), and may or may not have a filter element.
   - Usually located on the frame rail on truck type vehicles and in the engine compartment on car type vehicles.
   - Always include a drain plug or tap at the base of the glass or metal filter bowl.

Note: On some systems a strainer is fitted to the fuel pickup pipe in the tank.

2 after the fuel lift pump.
   - Generally known as a secondary filter and has a fine filter paper supported in a cylindrical pressed metal body.
   - Located either on or close to the engine.
   - It is a 'throw away' filter and must be changed at manufacturer's service intervals.

Sedimentors remove water and large heavy particles, as the first step in fuel filtering.

Secondary filters remove the last traces of water and the very fine particles that will cause damage to the precision fuel injection equipment.

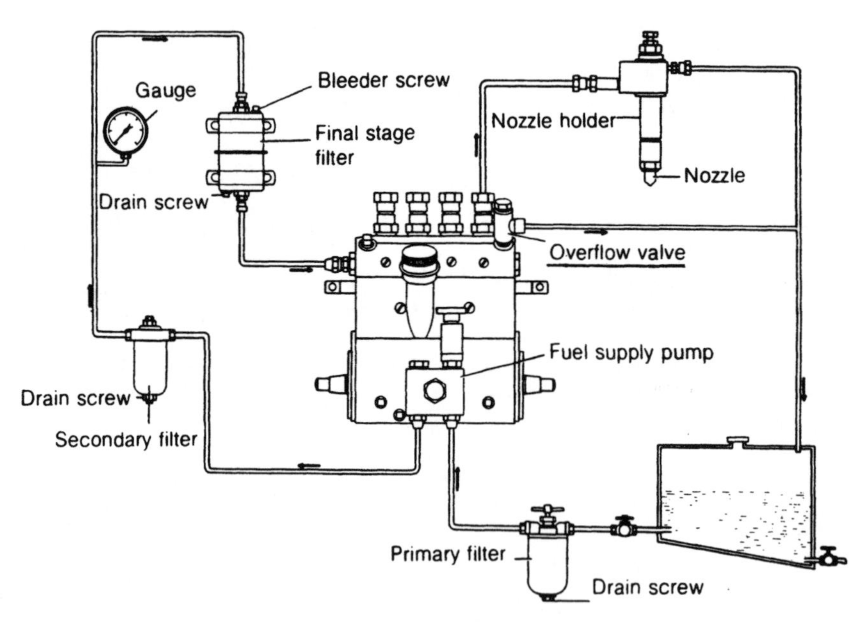

C.I. Fuel Injection System

The **fuel lift pump** will generally be one of two mechanical types, which are the:
1 plunger type.

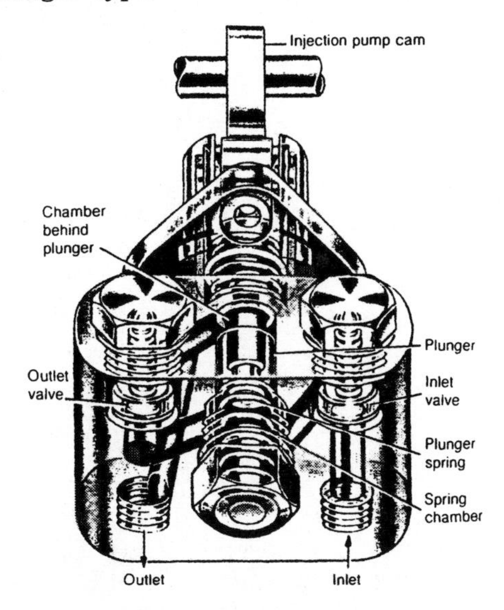

- Constructed of steel and aluminium.
- Usually attached to the side of an in-line fuel injection pump.
- Fitted with a priming device to allow fuel to be pumped during the 'bleeding' process, without the engine running.

2 diaphragm type.

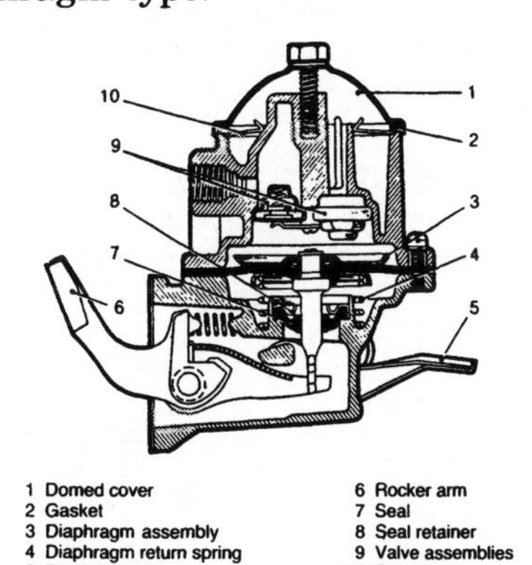

- Manufactured by the die casting process.
- Can be attached to either the side of an in-line type fuel injection pump or to the side of the engine.
- This pump is similar in appearance to the one fitted to most S.I. engines, but has the addition of a lever at its base.
- Its lever is used to pump fuel during the 'bleeding' process, without the engine running.

Note: Although they are not common, electric pumps similar to those used on some S.I. engines have been used on C.I. engine fuel systems. With this type of pump a primer is not required as the fuel lift pump will operate when the engine key is turned on.

The function of a fuel lift pump is to cause the fuel in the tank to be lifted or moved to the fuel injection equipment, where it will be either injected or returned to the tank.

Note: These types of pumps supply fuel at a pressure between 25 and 60 kPa.

The **fuel injection pump** can be either of the rotary or in-line design. Both pumps are generally constructed of aluminium for the body, and high quality steel for the precision, highly stressed components. In-line pumps are generally bolted to the timing cover area, parallel to and driven by the crankshaft. The method of driving the injection pump may be either gears, chain or toothed rubber belt.

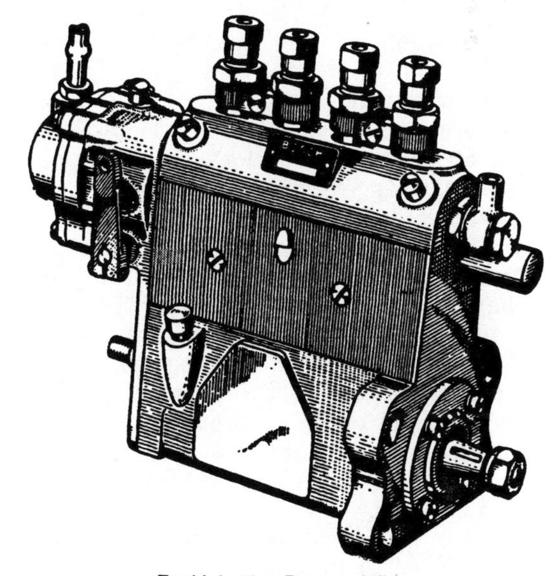
Fuel Injection Pump – Inline

Rotary pumps may be either fitted like the in-line version or bolted on to the engine cylinder block in a way that is similar to a S.I. engine ignition distributor. It is easy to identify the difference between the two pumps. On the in-line injection pump, as its name implies, all the injection pipes are in a line along the top of the pump unit. On the rotary injection pump, all the injection pipes come out of the pump in a circular pattern, at one end.

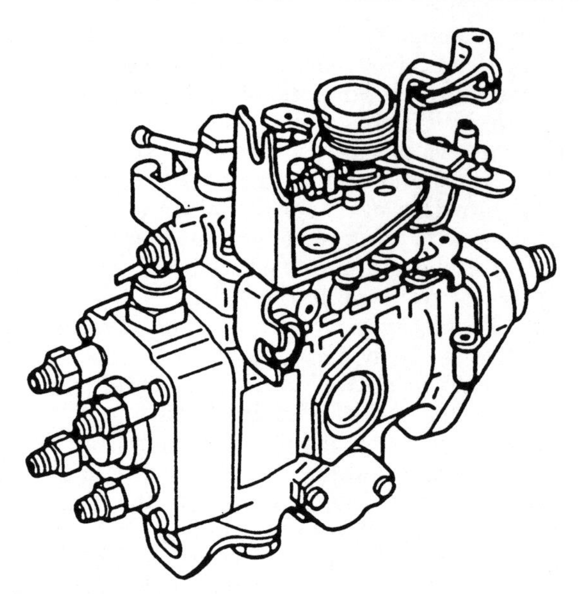

Fuel Injection Pump – Rotary
Source: Nissan Australia

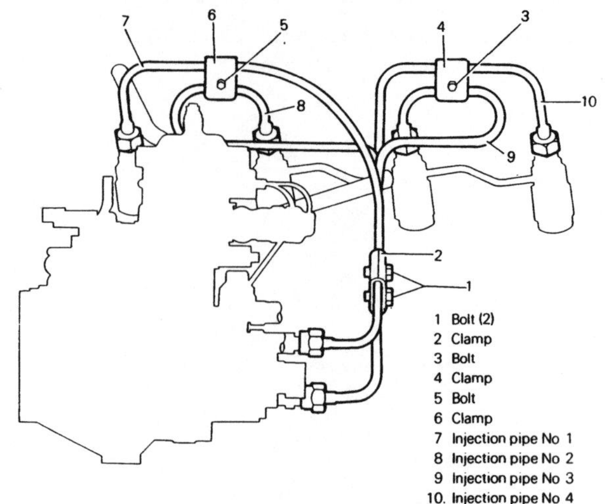

High Pressure Fuel Injection Pipes

The function of the injection pump is threefold.

1 It raises the fuel pressure from lift pump pressure to injection pressure (for example from 60 kPa to 18 000 kPa).
2 It accurately calibrates or meters the quantity of injected fuel.
3 It phases or times the fuel injection point.

The **high pressure fuel injection pipes** are manufactured from high quality thick wall steel piping. There is one pipe for each engine cylinder connected between its injector and the injection pump. To prevent possible breakages from engine vibrations, the pipes are clamped together at a bracket which is rigidly attached to an engine component.

Note: Pipes between injectors close to the injection pump MUST be of the same length as those fitted to injectors which are further from the pump. This ensures that the pressure remains constant at each injector.

The high pressure fuel injection pipes carry the very high pressure fuel to the fuel injector without leaks or pressure loss.

The **injector** is generally located at the top of the cylinder head, where it will be secured by one of the following methods:

- bolted through a flange formed on the injector body;
- tubular threaded nut;
- clamp plate and bolt.

The spray tip or nozzle protrudes into the engine's combustion chamber, where it injects the fuel for combustion. The function of the fuel injector is to inject fuel into the engine's cylinder in the correct form or spray pattern.

Note: The correct spray pattern depends on a combination of direction of spray, penetration distance and atomisation quality.

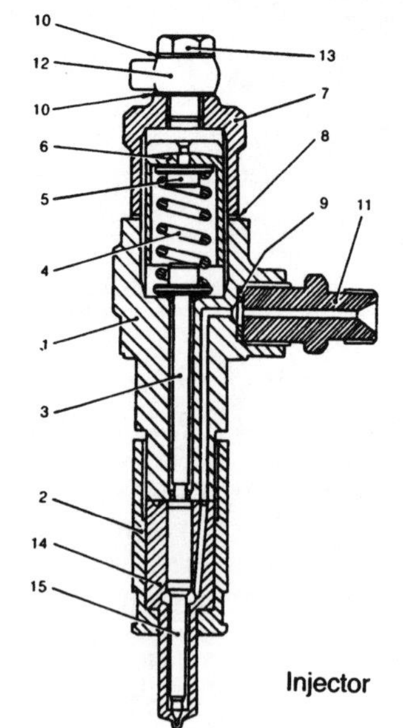

1 Nozzle holder
2 Nozzle nut
3 Spindle
4 Spring
5 Upper spring plate
6 Spring cap nut
7 Cap nut
8 Joint washer
9 Joint washer
10 Joint washer
11 Inlet adaptor
12 Leak-off connection
13 Banjo screw
14 Nozzle
15 Needle valve

Injector

The **fuel injector leak-off line** is a small diameter steel or plastic tube, connected between the 'leak-off' connection on each injector and the secondary filter or the fuel tank. Its function is to collect the leak-off fuel at each injector and return it to the low pressure side of the fuel system (before the inlet side of the lift pump).

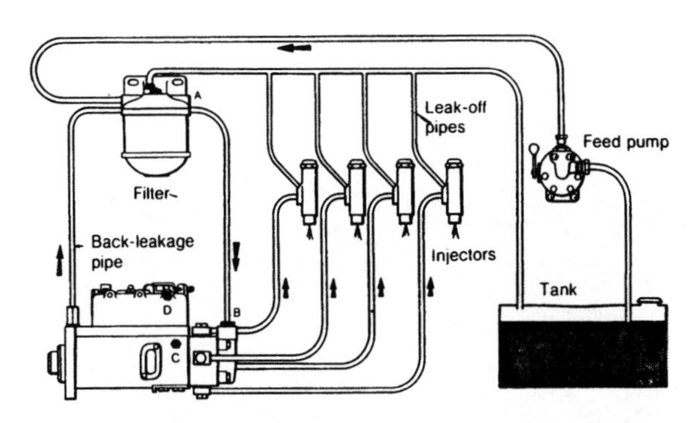

Fuel Injection Leak-Off Lines

LIQUEFIED PETROLEUM GAS

Liquefied petroleum gas (L.P.G.) is a hydro-carbon fuel obtained from either naturally occurring field gases or from the refining of crude oil. At normal atmospheric temperature and pressure L.P.G. is a vapour (gas) but can be liquefied at moderate pressure (600 kPa at 20°C). L.P.G. is colourless, odourless and non-toxic but an odour is added to make it possible to detect leakage into the atmosphere. Automotive L.P.G. is generally 95 per cent propane and 5 per cent butane.

SOME SPECIFIC FACTS ABOUT L.P.G.

1 The vaporisation ratio of L.P.G. is generally in the order of 270:1. ie. When 1 litre of liquid L.P.G. is released into the atmosphere, then it will expand to 270 litres of L.P.G. vapour.
2 L.P.G. vapour, like petrol, is heavier than air. So when L.P.G. has been released or leaked into the atmosphere then it will tend to drift to the lowest spots ie. workpits, dynamometers pits, drainage pits or any other holes.

Repairing L.P.G. fuelled vehicles

Work can be performed on L.P.G. fuelled vehicles within the confines of a workshop provided basic precautions are observed. The following list of precautions is basic and is not intended to supersede or substitute for any safety ordinance required by state or local authorities. The prime considerations are to avoid the discharge of fuel (gas) and keep the vehicle away from ignition sources (flame or cutting operations).

- Do not park the vehicle near a service pit. L.P.G. fumes will collect in a pit.
- Ensure the work area has adequate ventilation.
- Check the L.P.G. system is not leaking.
- Check the fuel level does not exceed 80 per cent. Subsequent warming of the tank on a hot day could cause fuel discharge inside the workshop.
- Close the tank service valve if the vehicle will be parked for a prolonged period.

COMPONENT REPAIR

The removal of repair of L.P.G. components should be performed by suitably trained and authorised service personnel. Prior to commencing work, the service valve should be closed and the engine run until fuel in the lines is exhausted.

ACCIDENT DAMAGED VEHICLES

The service valve must be closed prior to the vehicle entering the workshop. If welding is required on the vehicle (within 2 m of the tank) the supply tank should be removed from the vehicle.

EMERGENCY SITUATIONS

To attempt fire fighting or emergency leak procedures at this stage of your training could be very hazardous. If a situation arises where you are involved in an L.P.G. fuelled vehicle emergency, immediately close the service valve (if it is safe to do so), clear the area and obtain assistance. On the job training is essential to identify correctly the type of hazard and apply the correct emergency technique. For example, under certain circumstances extinguishing the flame can change a controlled fire into a potentially explosive situation.

L.P.G. fuel system

The **L.P.G. fuel tank** is a pressure vessel (test pressure 3.3 MPa) and may be constructed from high tensile steel or stainless steel. In either case, the tank will be cylindrical in shape and have domed ends. The most likely location for the tank in a sedan vehicle is in the boot space above the rear wheels. For wagon and utility vehicles the tank is generally located in the spare wheel area. The function of the L.P.G. fuel tank is to store and carry fuel (L.P.G. in liquid form) in as safe a manner as possible.

The **L.P.G. lines** must be manufactured from matèrials as indicated in current Australian Standard 1425-●●, *The Use of L.P. Gas in Internal Combustion Engines.* The L.P.G. lines fitted are the:

- fill line. Connected between remote fill point and A.F.L. valve, it allows the L.P.G. tank to be filled without opening boot or rear of wagon.
- service line. Connected between the tank service valve and gas filter lock off valve, in engine compartment, it allows the fuel to be delivered to the gas filter lock off from the tank service valve.
- tank pressure relief line. Connects the tank pressure relief valve to atmosphere outside the vehicle. Must have a melting point higher than aluminium. Allows vaporised fuel to be safely vented to outside of vehicle, even when the tank area is on fire. Must have a sealing plug fitted to the open end of line so that it will seal after venting has occurred. Prevents the entry of dirt and water that could block line, thus making it impossible for tank pressure to be relieved.

Note: A blocked pressure relief line can pose a dangerous situation.

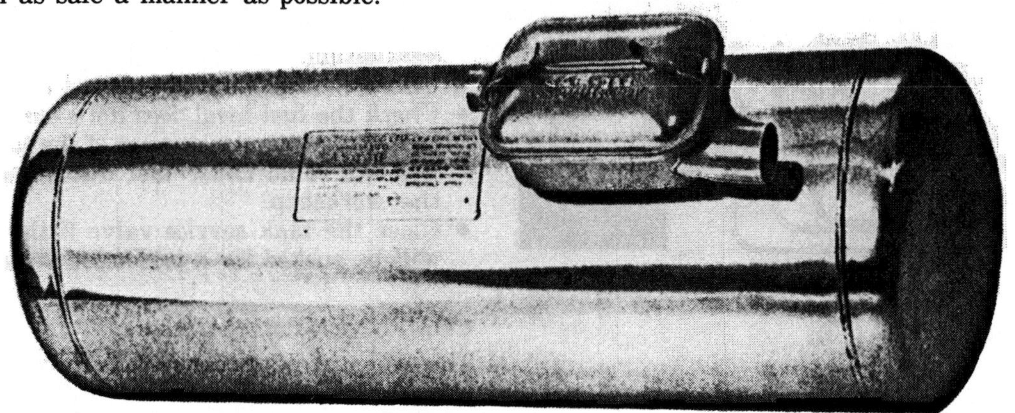

L.P.G. Fuel Tank

The **valves** and **fittings** that must be installed on an automotive L.P.G. fuel tank are the:

- service valve — turns the fuel supply on and off.
- excess flow control valve — fitted in base of service valve, turns off the fuel supply when a break occurs in service line.
- tank pressure relief valve — vents L.P.G. vapour when the tank pressure exceeds 2.55 MPa.
- automatic fill limiter (A.F.L.) – limits entry of fuel (during refuelling) to 80 per cent of the tank's volume.
- fixed liquid level gauge (ullage gauge) is a manual method of testing the 80 per cent fill limiting.
- contents gauge — gives an indication of fuel level in tank.
- ventilation box — covers valves and fittings in case of gas leakage.
- vent conduits (hoses) — provide connection between ventilation box and outside of vehicle, in case of L.P.G. leakage.

The **L.P.G. filter lock off valve** is, as its name implies, a filter and a lock off valve (tap). It is manufactured from aluminium and pressed steel and may be either vacuum or electrically operated. It is usually located in the engine compartment as close as possible to the vaporiser and regulator. The L.P.G. filter lock off valve provides a means by which the liquid L.P.G. supply may be closed off when not required for engine operation.

DESCRIPTION

1. Filler valve with 80% liquid level indicator
2. L.P.G. tank
3. L.P.G. lock off device
3. L.P.G. lock off device
4. Petrol lock off device
5. Converter
6. Carburettor adaptor gas ring
7. Switch with choke
8. Cooling system
9. Carburettor
10. Inlet manifold
11. Petrol-pump
12. Petrol-tank
13. Starter motor
14. Ignition key

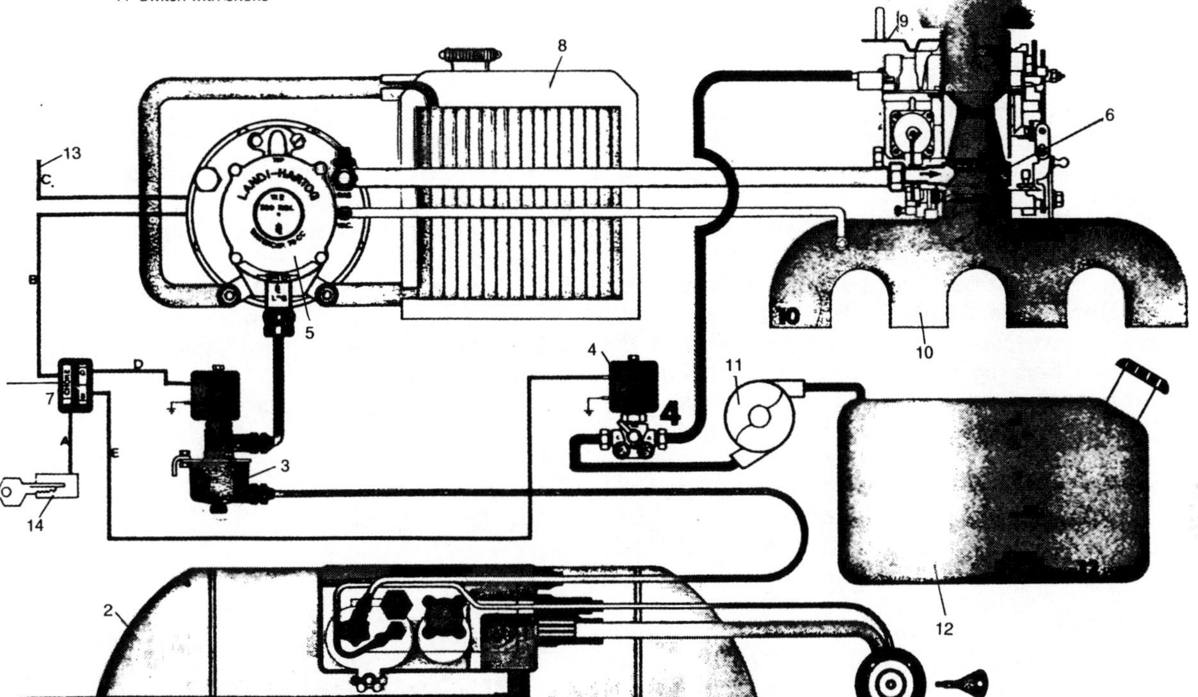

L.P.G. Fuel System

The **safety controller** (filter lock off valve) is an electronic device connected to the:

- fuel selection switch;
- ignition coil;
- vehicle earth;
- electric filter lock off device.

The safety controller automatically turns off the gas filter lock off valve, when the engine is not turning, even when the ignition system is left on. It is a safety override in case of an accident.

The **petrol lock off valve** is an electrically operated valve fitted into the petrol supply line between the fuel pump and the carburettor. It is connected electrically to the fuel selector switch. The function of the petrol lock off valve is to turn off the petrol supply when the engine is to run on L.P.G. and to turn the supply on when it is to run on petrol.

The **fuel selector switch** is located in the driver's area of the vehicle and is simply a switch with three positions:

1 L.P.G. ON – PETROL OFF.
2 L.P.G. OFF – PETROL OFF.
3 L.P.G. OFF – PETROL ON.

Note: The switch is designed to allow only one fuel to be supplied to the engine at any one time.

The function of the fuel selector switch is to provide a convenient and simple means by which the driver can change from one fuel to another.

The **L.P.G. vaporiser/regulator/converter** is usually manufactured from cast aluminium but plastic housings have been used on some models. It is generally located in the engine compartment, as close as possible to the gas filter lock off valve, to which it is connected by a union or a pipe. Two hoses connect the vaporiser/regulator to the engine's cooling system. A single line connects the L.P.G. vapour outlet on the vaporiser/regulator to the mixer (carburettor). The function of the vaporiser/regulator is to:

- regulate L.P.G. pressure. Reduce tank pressure to a constant pressure just less than atmospheric pressure, irrespective of engine speed or load.
- vaporise the L.P.G. Add heat during the time the liquid fuel is changing into a vapour.

Note: Heat is supplied from engine cooling system.

The **L.P.G. mixer (carburettor)** is a simple device which mixes the air and fuel (L.P.G.). The main materials used in its construction are aluminium, plastics and steel. As most vehicles are converted to run dual fuel it is usual to find the L.P.G. mixer fitted between the original carburettor and the air cleaner. An alternative arrangement is to insert a gas ring in the base of the carburettor above the throttle plates. The function of the L.P.G. mixer is to mix air and L.P.G. vapour in the right proportions to give the correct air/fuel mixture for all engine speed and load conditions.

Refuelling

Refuelling a vehicle powered by L.P.G. is a relatively simple task as long as the following conditions are adhered to:

1 Refuelling must be carried out by a trained person.
2 Refuelling must only be done in a well ventilated outdoor area.

Note: The refuelling area must comply with current L.P.G. fuelled vehicle regulations.

3 Occupants must not remain in the vehicle during the refuelling process.
4 Apply the vehicle handbrake and make sure the vehicle engine's ignition is turned off.
5 Do not overfill the L.P.G. cylinder. It MUST NOT be filled beyond 80 per cent of capacity.

Note: The problem of overfilling an L.P.G. cylinder has been reduced because all new installations MUST be fitted with an A.F.L. (automatic fill limiter) valve.

6 Smoking must not be allowed within 3 m of the fill connector.
7 Ignition sources, of any type, must not be allowed within 3 m of the fill connection.
8 Wear gloves to protect your hands in case of an accidental spill of L.P.G.

Caution: L.P.G., when released into the atmosphere, will freeze whatever it contacts, including human tissue. This freezing can cause frost bite burns or

FUEL SYSTEM MAINTENANCE

Carburettor

CARBURETTOR REMOVAL

1 Disconnect all vacuum and air hoses from the air filter assembly.
2 Remove air filter assembly and store in a safe place.
3 Disconnect choke cable or auto choke wires/pipes.
4 Disconnect idle control solenoid wires, when fitted.
5 Disconnect throttle cable or linkage from carburettor.
6 Label, then disconnect, all vacuum lines.
7 Disconnect fuel line from carburettor.

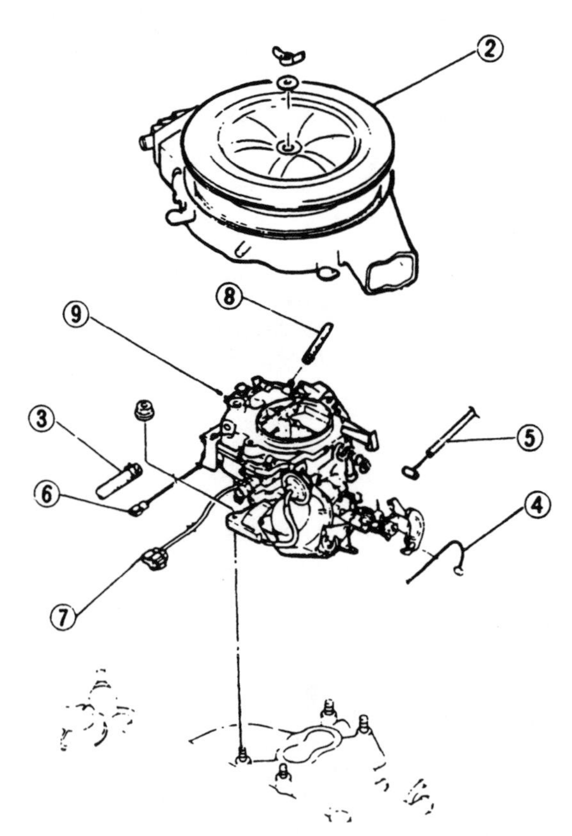

Carburettor Removal

Caution: Ensure fuel cannot spill onto hot engine manifolds.

8 Loosen and remove nuts or setscrews securing carburettor base to inlet manifold.
9 Remove carburettor, and plug inlet manifold opening with clean rag.
10 Clean old gasket material from the inlet manifold flange and carburettor base.

CARBURETTOR REPLACEMENT

1 Remove the plugging cloth and fit a new gasket to the manifold flange.
2 Place the carburettor on the manifold flange and loosely fit the attaching nuts or setscrews.
3 Fit the fuel line into the carburettor union and carefully screw the fuel line flare nut into its union, ensuring the threads are not **cross threaded**.
4 Tighten the carburettor base flange nuts or setscrews evenly to prevent flange distortion.
5 Tighten the fuel line flare nut.
6 Refit all vacuum lines (check each line is connected to the correct carburettor port).
7 Connect and adjust throttle cable or linkage to the carburettor.

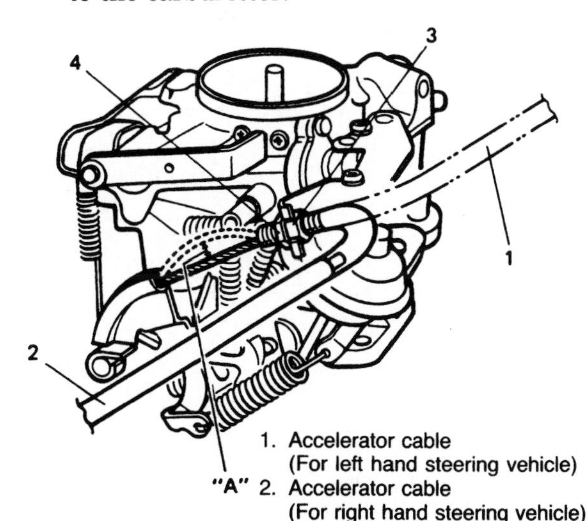

1. Accelerator cable (For left hand steering vehicle)
2. Accelerator cable (For right hand steering vehicle)
3. Adjusting nut
4. Lock nut

Note: Check that full throttle is available by asking an assistant to depress the accelerator pedal and observe the throttle position.

8 Refit and adjust the choke cable (manual choke). Ensure that the dash knob is fully in, the choke valve is fully open, and operate the choke cable to ensure that the cable is not fouling. Refit automatic choke wires or pipes (when fitted).
9 Connect the idle control solenoid wires (when fitted).
10 Fit the air filter assembly.
11 Check all hoses (vacuum and air) for correct location and tightness.
12 Start the engine and ensure there are no vacuum, air or fuel leaks.
13 Adjust the idle speed and fuel mixture.

Idle speed and mixture adjustments

All manufacturers use different methods to check and adjust the idle speed and mixture, so it is necessary to read their service information before starting the job. Most vehicles display the adjustment specifications on a decal stuck on the underside of their bonnets. A common method of adjusting the idle speed and mixture is:

1 Start engine and run it until operating temperature is reached.
2 Stop the engine and connect a tachometer to its ignition system.
3 Connect a non-dispersive infra-red exhaust gas analyser to the vehicle exhaust outlet.
4 Start the engine and adjust its idle speed to the specified value (see decal for value).

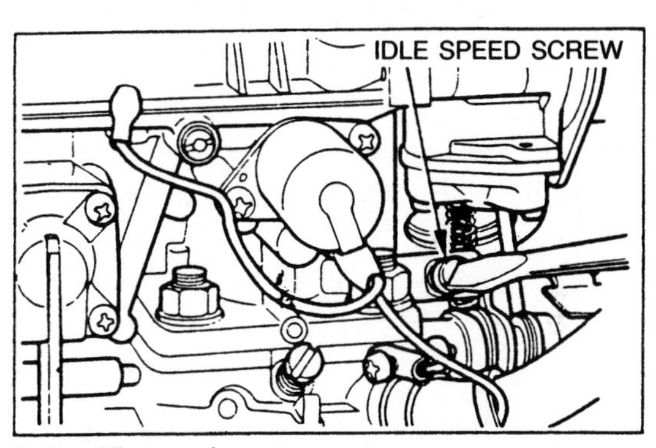

Setting idle speed

5 Turn the mixture screw to the right (clockwise) until the engine runs roughly.

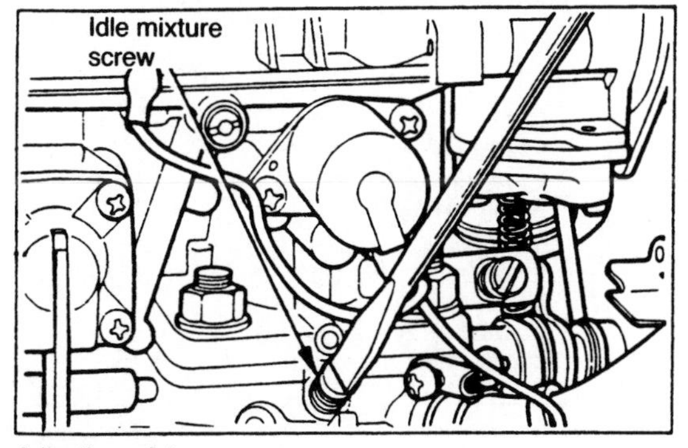

Adjusting mixture screw

6 Turn the mixture screw to the left until the engine runs smoothly.
7 Re-adjust the idle speed when necessary.
8 Run the engine for several minutes to allow the gas analyser to settle.
9 Observe and record the gas analyser readings.
10 When these readings are not in the range stated on the decal, slowly turn the mixture screw in a direction that will bring these values into the correct range.
11 Stop the engine and disconnect the gas analyser and tachometer.

MECHANICAL FUEL PUMP

Fuel pump removal

1 Disconnect the fuel lines from the fuel pump. Use two spanners for this task, one to hold the fitting on the fuel pump and the other to turn the fitting on the pipe.
2 Loosen and remove the securing setscrews on the pump base.
3 When necessary, tap the base of the fuel pump with a soft faced hammer to loosen it from the engine.
4 Slowly pull the pump away from the engine. A slight twist may be needed to withdraw the arm from the hole.
5 Clean old gasket material from the engine block and pump base flange.

Fuel pump replacement

1 Apply a thin cover of gasket cement or silicon compound to a new gasket.
2 Fit the new gasket to the engine mounting flange.
3 Insert the pump into the opening in the engine.
4 Place both setscrews through the base holes and start them into the engine mounting flange threads.
5 Tighten both setscrews evenly, to manufacturer's torque specifications.
6 Fit the fuel lines into the fuel pump unions and carefully screw each fuel line flare nut into its union, ensuring the threads are not **cross threaded**.
7 Tighten both pipe fittings.

Note: Remember to hold the pump union with one spanner and to tighten the flare nut with another.

8 Start the engine and check for fuel, air and oil leaks.

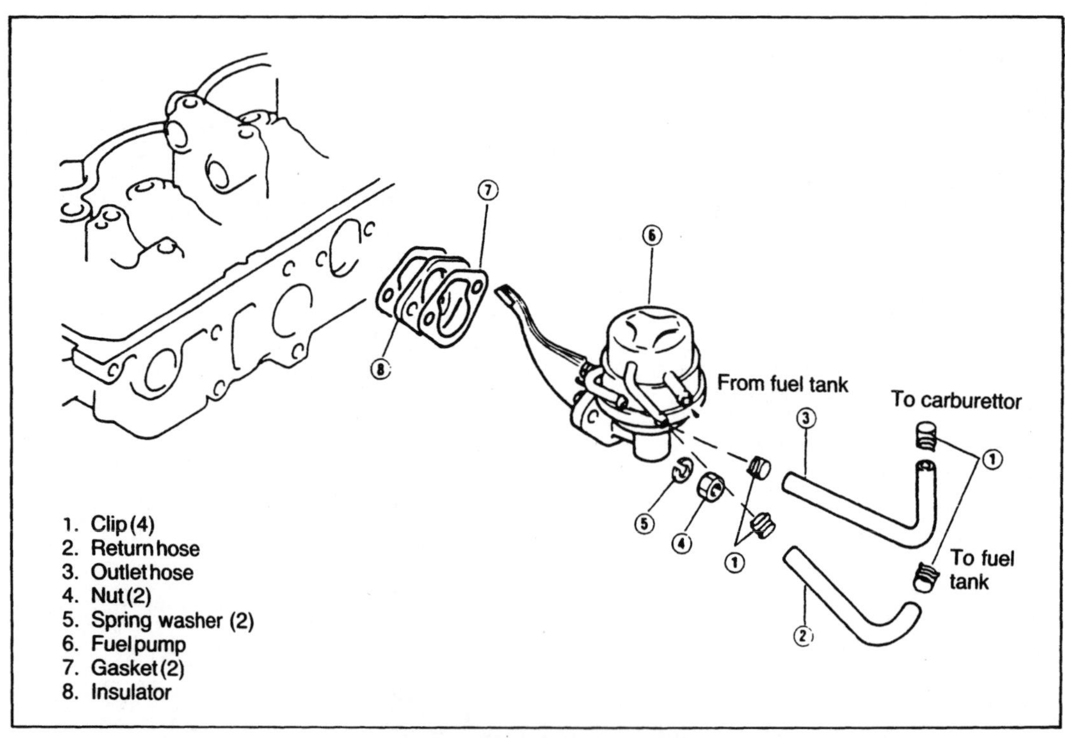

Mechanical Fuel pump removal

COMPRESSION
IGNITION ENGINE FUEL
FILTER SERVICE

Fuel filter removal

1 Remove the central securing bolt from the filter head.
2 Remove the base and the filter element.
3 Discard all sealing rings.
4 Wash all parts, except the filter element, in cleaning solvent and dry them with compressed air.
5 Check the condition of all parts, especially seal surface areas.

Fuel filter replacement

1 Fit new O-ring to centre of filter head.
2 Fit new sealing rings to the filter head and base.
3 Position a new filter element in the filter head.
4 While holding the filter element, fit the filter base and hold it against the filter head.
5 Fit the central clamp bolt through the filter head and screw it into base thread.
6 Tighten bolt to the filter manufacturer's specifications.
7 Check that all filter seals, fittings and connections are correctly installed and are tight.

Air removal or system 'bleeding'.

To remove the air from a filter fitted with an automatic bleeding device, operate the manual prime pump until resistance is felt.

To remove the air from a filter fitted with a manual bleeding device:

1 Open the bleeder vent on the filter head.

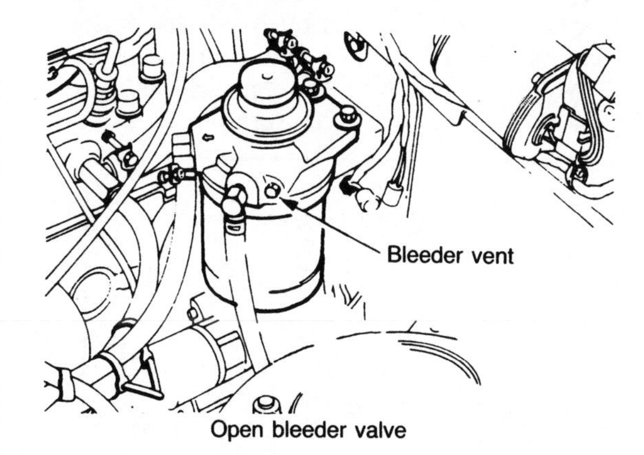

Open bleeder valve

2 Pump the prime or lift pump until fuel flows from the bleeder vent.

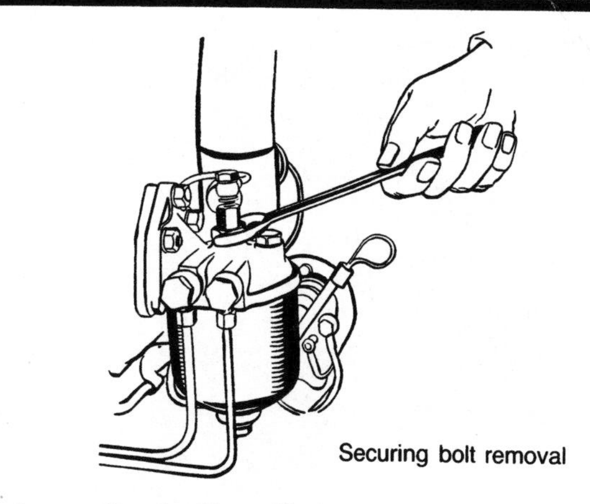

Securing bolt removal

Source: Chrysler Sigma Bk 4

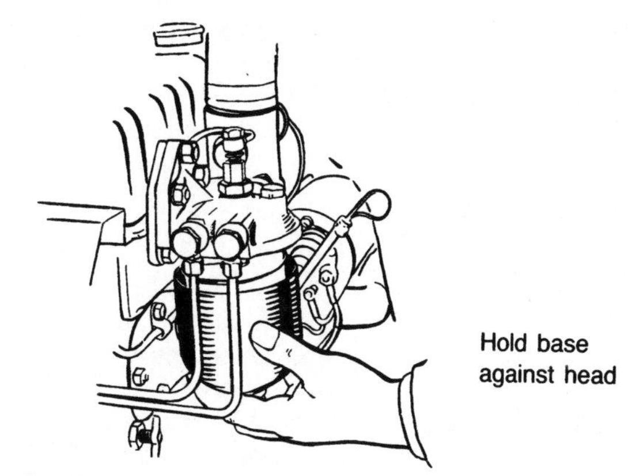

Source: Mitsubishi Magna TM series manual

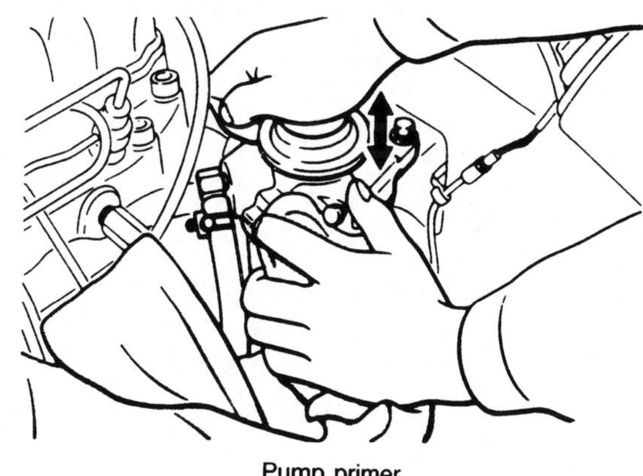

Pump primer

3 Continue pumping until the fuel flow is free of air bubbles.
4 Close the bleeder vent.
5 Open the bleeder vent on the fuel injection pump.
6 Pump the prime or lift pump until fuel flows from the bleeder vent.
7 Continue pumping until the fuel flow is free of air bubbles.

8 Close the bleeder vent.
9 Loosen all injector pipes at the injectors.

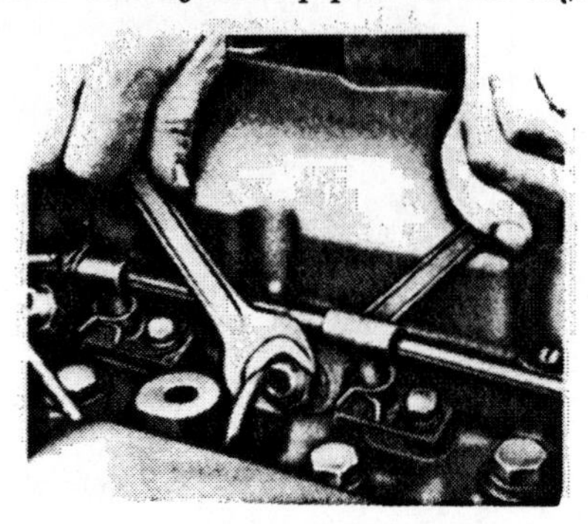

10 Using the starter motor, crank the engine until fuel is visible at each injector.
11 Tighten all injector pipes to the manufacturer's specifications.
12 Start the engine and inspect for fuel and air leaks.

EMISSION CONTROL

Vapours and gases that either escape or are forced into the atmosphere by motor vehicles contribute to the pollution of the air. Air pollution has become such a problem in Australia that the government clean air agencies regularly monitor the air for the type and quantity of toxic material that is present.

Motor vehicle emissions are:
- unburnt fuel (hydrocarbons — HC);
- incomplete combustion of fuel (carbon monoxide — CO).
- bonding oxygen with nitrogen (oxides of nitrogen — NO_x).

Their sources are:
- 5 per cent from the fuel tank (HC);
- 5 per cent from the carburettor (HC);
- 30 per cent from the crankcase vent (HC, CO and NO_x);
- 60 per cent from the exhaust system (HC, CO and NO_x).

Hydrocarbons have been described as carcinogenic agents (cancer causing) and, when combined with oxides of nitrogen and sunlight, cause photo-chemical smog. This smog has three effects.
1 It produces a 'greenhouse' effect by forming an envelope over an area, thus raising its temperature.
2 When it forms close to the ground, it acts as a partial filter and screens out some of the sun's rays that are necessary for plant growth.
3 It irritates the eye, nose, throat and lung tissues, causing severe discomfort for people who suffer from respiratory diseases, such as asthma, or heart diseases. It can be fatal.

Carbon monoxide is a poisonous substance, that, when inhaled in large enough quantities, can cause brain and nerve damage leading to death.

Because of the nature of the toxic vapours and gases released by early model vehicles, the government, through the Australian Design Rules (A.D.R.), has compelled the motor vehicle manufacturers to incorporate systems on their vehicles that will reduce these harmful emissions.

The first system introduced was positive crankcase ventilation (P.C.V.). Several changes have led to the closed system which draws the gases and fumes from the crankcase into the combustion chambers of the engine where they are burnt instead of being released directly into the atmosphere. These gases and fumes are the result of small quantities of combustion material escaping past the piston rings, into the crankcase. Some of them mix with water and oil mist to form acids and oil sludge which drop into the sump.

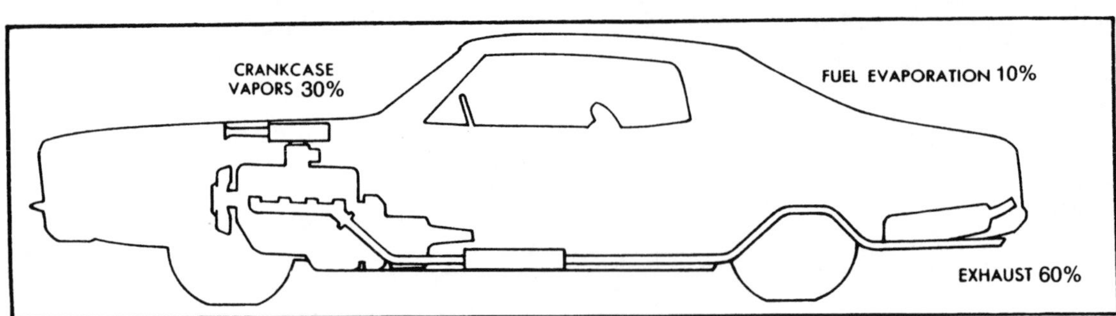

Automotive Air Pollutants

Positive crankcase ventilation

A positive crankcase ventilation system consists of:

1 a P.C.V. valve.
- Located in a grommet in the rocker cover or oil filler cap.
- Controls the flow of gas and fumes from the crankcase to the inlet manifold.

2 several hoses.
- One rubberised hose connects the P.C.V. valve to the inlet manifold.
- Another rubberised hose connects the air cleaner to the rocker cover or oil filler cap.
- They ensure the correct circulation of air or gas.

3 a sealed oil filler cap.
- Located on the rocker cover.
- Prevents gases and fumes from escaping into the atmosphere.

4 several rubber grommet seals.
- Located in the rocker cover, oil filler cap or the air cleaner.
- Provides mounting points and seals for the P.C.V. valve and hoses.

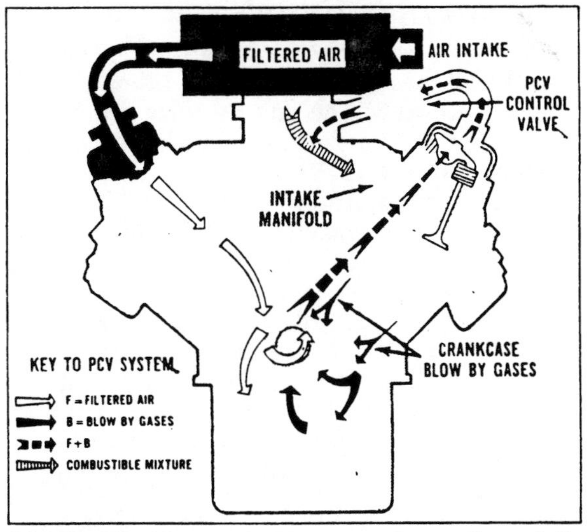

SCHEMATIC OF PCV SYSTEM *Source: Deere & Co.*

5 air filter pad (when fitted).
- Located inside the engine air cleaner housing.
- Removes dust and dirt from the air entering the engine through the P.C.V. system.

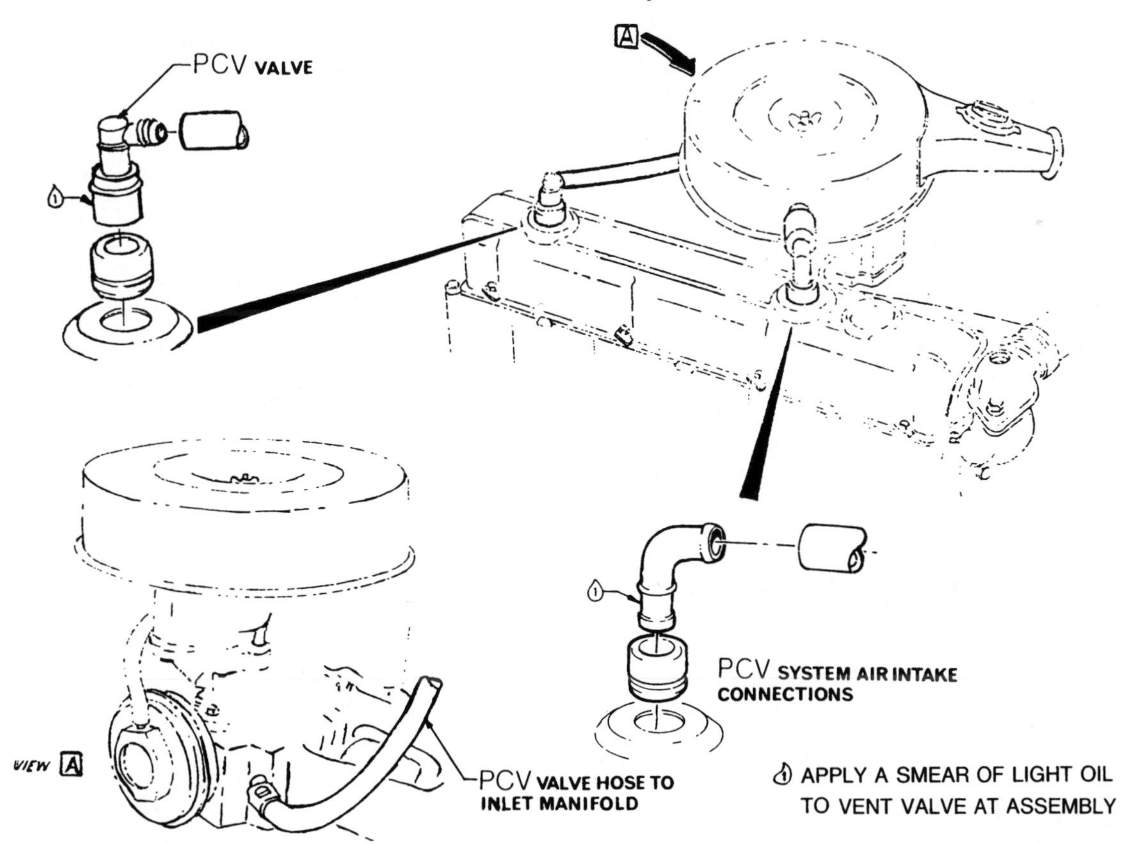

POSITIVE CRANKCASE VENTILATOR SYSTEM

TESTING THE P.C.V. SYSTEM

The positive crankcase ventilation system should be checked at a regular interval of 30 000 km, or when the engine idles roughly or stalls. There are several simple methods of testing the system.

R.p.m. drop method

1 Connect a tachometer to the engine.
2 Start the engine and allow it to reach normal operating temperature.
3 Observe and record the engine speed.
4 Either 'pinch' the hose connected to the P.C.V. valve or remove valve (with hose still attached to inlet manifold) and put your thumb over its end.
5 Observe and record the drop in the engine's speed.

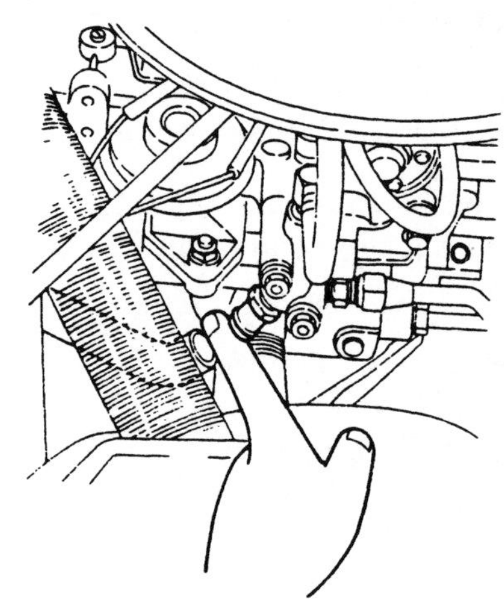

Checking P.C.V. valve

- When the engine speed drops by 50 r.p.m. or more, the system is working correctly.
- When the engine speed drop is less than 50 r.p.m., the valve needs to be cleaned or replaced and the hoses checked for blockages.

6 Refit the valve.
7 Stop the engine and remove the tachometer.

Vacuum test method

1 Remove the oil filler cap. This test will not work when the P.C.V. valve is fitted to the oil filler cap.
2 Start the engine and allow it to reach normal operating temperature.
3 Disconnect the inlet hose at the air cleaner and plug the hose.
4 Put a piece of stiff paper or a P.C.V. tester over the oil filler cap hole.
5 Allow about a minute for the vacuum in the crankcase to start pulling the card into the hole.
6 Check the vacuum in the crankcase.
   - When the card is difficult to remove, the P.C.V. system passes this test.
   - When the card is not held to the hole, the P.C.V. does not pass this test. The fault must be located and repaired.
7 Stop the engine and refit the oil filler cap.

Rattle test method

1 Remove the P.C.V. valve from the system.
2 Shake the valve.
3 Listen for a noise.
   - When the valve does not make a clicking or a rattling sound, it is faulty. Clean or replace the P.C.V.
4 Refit the valve to the system.

SERVICING THE P.C.V. SYSTEM

To inspect the hose of a P.C.V. system:

1 Remove the hose from the system.
2 Inspect hose outside surface. If cracked, burnt or very soft, replace the hose.
3 Inspect hose inside surface. When a flaking condition is present, replace hose. If the hose is blocked, clean it.
4 Refit the hose to the system.

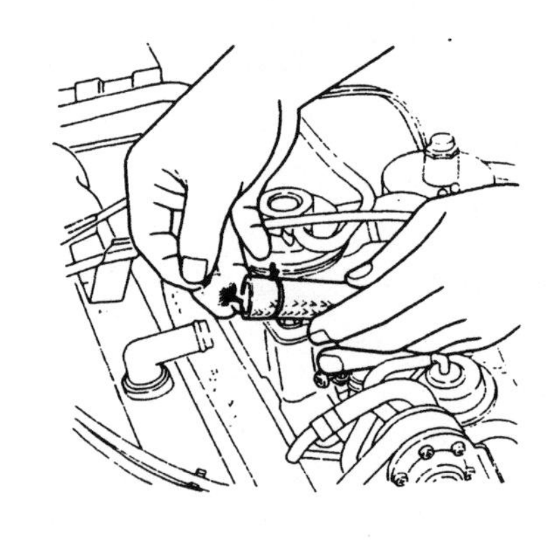

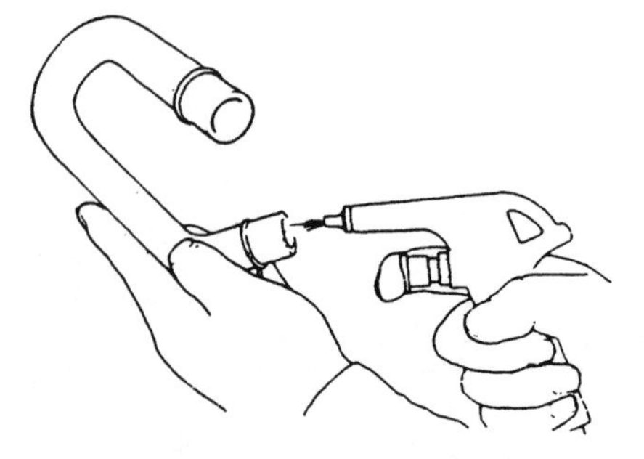

Cleaning ventilation hose *Source: Nissan Australia*

5 Ensure that the hose is not kinked during assembly.

To clean the hose of a P.C.V. system:
1 Remove the hose from the system.
2 Flush it with water.
3 Dry the inner surface with compressed air.
4 Inspect the inner surface for scaling, cracking or restrictions.
5 Refit the hose to the system.
6 Ensure that the hose is not kinked.

To service a P.C.V. valve:
1 Remove the valve from the system.
2 Soak it in 'degreasing' solution.
3 Flush the valve throughly with water.
4 Dry it with compressed air.
5 Refit the valve to the system.

Note: Most modern P.C.V. valves are not serviceable (cannot be dismantled for cleaning) and must be renewed as an assembly.

To service the air filter pad:
1 Remove the top of the engine air cleaner housing.

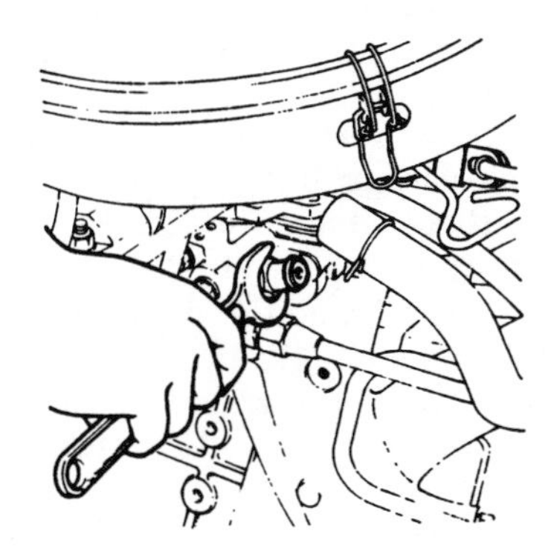

Removing P.C.V. valve

2 Carefully pull the air filter pad from its clip or bracket.
3 Wash the pad in the cleaning solution.
4 Dry it with compressed air.
5 Apply a few drops of light oil to one surface.
6 Squeeze the pad to disperse the oil even through the foam (foam type only).
7 Install the filter pad in the retaining bracket or clip.
8 Refit the top of the engine air cleaner housing.

Note: The air filter pad should be serviced when the engine air filter is being serviced.

Clean P.C.V. filter

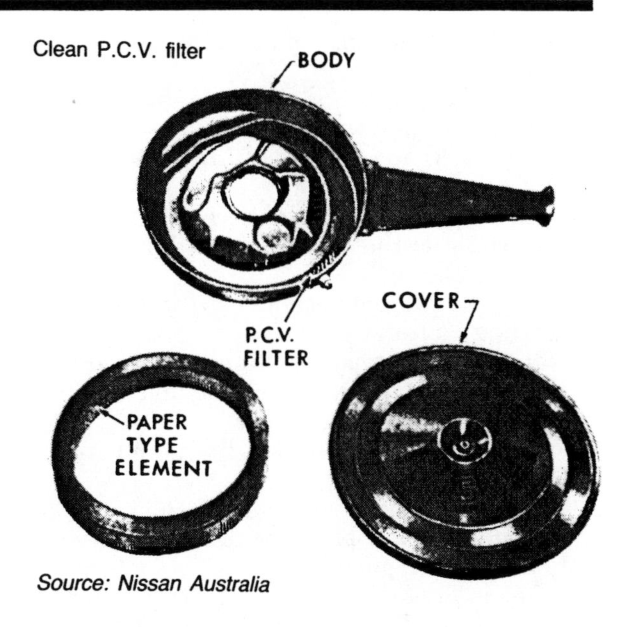

Source: Nissan Australia

CHAPTER 13 REVISION

1 What is the purpose of a fuel tank?
2 Name the material used in the manufacture of the modern fuel line.
3 Name two types of fuel pumps.
4 What is the purpose of a fuel filter?
5 What is a carburettor?
6 Name the component that distributes the air/fuel mixture evenly to all engine cylinders.
7 Where is an E.F.I. fuel pump located?
8 What could happen when an E.F.I. fuel pump is run without fuel?
9 Name the component that reduces the fuel pump pulsations.
10 Explain the meaning of an arrow on the body of a fuel filter.
11 Why are two fuel lines connected to the fuel tank?
12 What is the purpose of a fuel rail?
13 Name the component that controls the fuel pressure in the fuel rail, and give its location.
14 Name the component that sprays fuel into the inlet port of a cylinder.
15 List the functions of an E.F.I. inlet manifold.
16 What is the purpose of an engine air cleaner?
17 Name the component that tells the electronic control unit how much air is entering the engine.

18 List the functions of a throttle body assembly.
19 Name three sensors fitted to an E.F.I. engine and give their locations.
20 What is the function of an electronic control unit?
21 Name the component in an E.F.I. system that controls the engine speed during the warm-up period.
22 Name the two components used in the cold starting circuit.
23 Name the two types of fuel filters used in the fuel system of a C.I. engine and briefly describe the function of each.
24 What is the function of a lift pump in a C.I. engine fuel system?
25 How is an injector pump driven?
26 How many high pressure pipes are fitted to an injector pump?
27 What is the function of an injector pump?
28 Why must all the high pressure injector pipes be the same length?
29 To what point is the leak-off fuel delivered?
30 When a carburettor is being removed, why is it necessary to label the vacuum hoses before they are disconnected?
31 What should be used to plug an inlet manifold opening?
32 Name the adjustments that should be carried out on a carburettor after it has been fitted to an engine.
33 How is a fuel pump loosened on the engine after the securing setscrews have been removed?
34 What method should be used to tighten the flare nuts on a fuel pump?
35 Name the important area that must be inspected when a fuel filter on a C.I. engine has been dismantled.
36 Explain the term 'system bleeding' as related to the fuel system of a C.I. engine.
37 What is the meaning of the abbreviation 'L.P.G.'?
38 List the properties of L.P.G.
39 Name the material that is used in the construction of an L.P.G. fuel tank.
40 List the three valves that are located on an L.P.G. fuel tank and briefly describe the purpose of each.
41 List the three lines that are fitted to an L.P.G. system and briefly describe the purpose of each.
42 What is the function of an L.P.G. filter lock off valve?
43 Name the component that turns off the filter lock off valve when the ignition is on but the engine is not running.
44 What is the name of the valve, fitted to a dual fuel system, that controls the petrol system?
45 Which component allows the driver to change from one fuel to the other?
46 Describe an L.P.G. vaporiser/regulator.
47 What is the function of an L.P.G. mixer?
48 Why is escaping L.P.G. dangerous?
49 Name three harmful gases that are emitted by a motor vehicle.
50 What is the function of a P.C.V. system?
51 List the components of a P.C.V. system and state their locations.
52 Name three methods of testing a P.C.V. system.
53 Name the component of a P.C.V. system that should be attended to during the servicing of the engine air cleaner.

14

BRAKES

INTRODUCTION

The wheel was probably humanity's greatest invention but, like most great inventions, there are some disadvantages. The two main problems with the wheel is how to hold it stationary and how to stop it when it is turning. In past years, many different methods have been used to overcome these problems. Most of the methods use friction to gain the necessary results, and they are called brakes.

The braking system has been developed into an extremely sophisticated unit of the motor vehicle. It has two sections which work independently of one another but, in some cases, share major components. These sections are:

1 the hand (park) brake, which is operated by a mechanical device;
2 the foot brake, which is operated by a hydraulic system.

These sections are included in several layouts which are designed to suit the various types of motor vehicles currently being produced. The common layout has a dual hydraulic system which allows two of the four brake assemblies to operate independently of the other two.

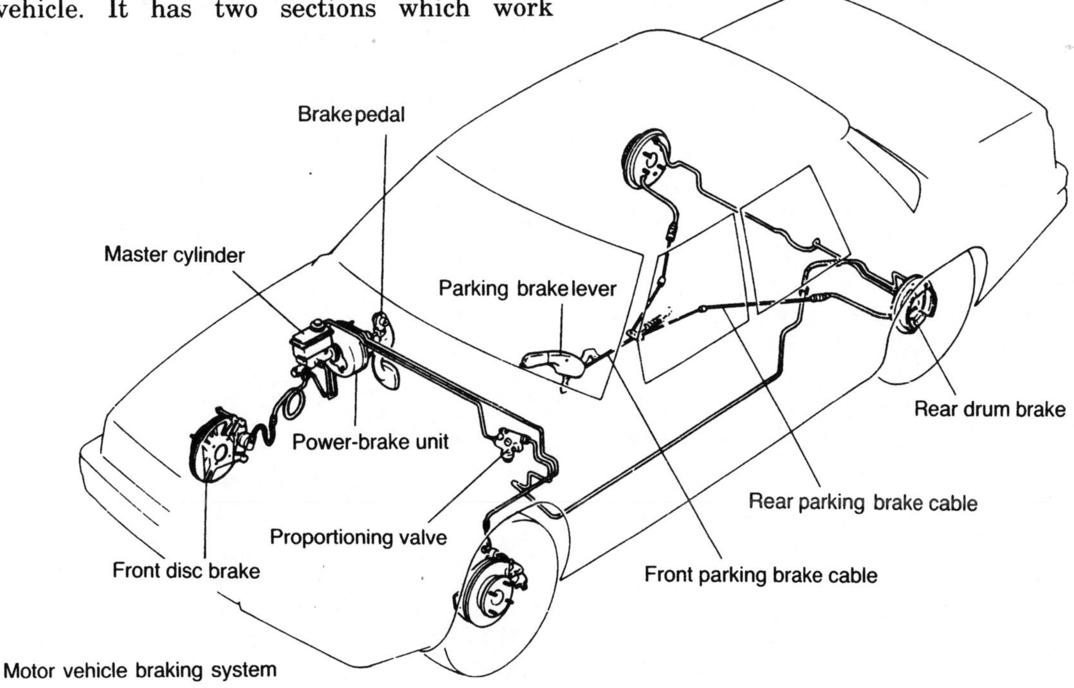

Motor vehicle braking system

LOCATION AND FUNCTION OF MAJOR COMPONENTS

The correct servicing of a braking system is extremely important for safety reasons, and being able to correctly locate the component parts and state their function is essential for:

- understanding the operation of the braking system;
- testing, diagnosing and repairing the braking system.

The **brake pedal** is generally of a pendulum type, which pivots on a bracket mounted under the dash panel above the steering column. It is connected to the brake booster by a push-rod. The brake pedal increases the effort applied by the driver.

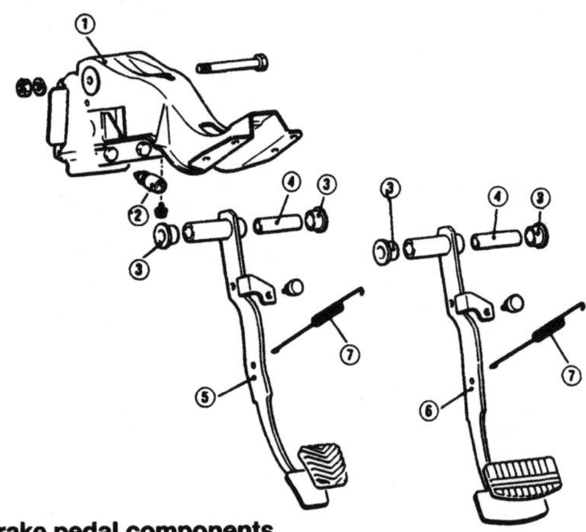

Brake pedal components

1. Pedal support member
2. Stop lamp switch
3. Bushing
4. Spacer
5. Brake pedal (manual trans.)
6. Brake pedal (auto trans.)
7. Return spring

Source: Mitsubishi Magna TM series manual

The **brake booster** is mounted on the under-bonnet firewall between the brake pedal and the master cylinder. It is also connected to the inlet manifold of the engine by a hose. The brake booster multiplies the pedal effort in the range from very light to extremely heavy braking.

The degree of braking is governed by the pressure difference between engine vacuum and atmospheric pressure in the booster housing. This is controlled through the brake pedal by the driver's foot. The booster's effort is transferred to the master cylinder by a push-rod.

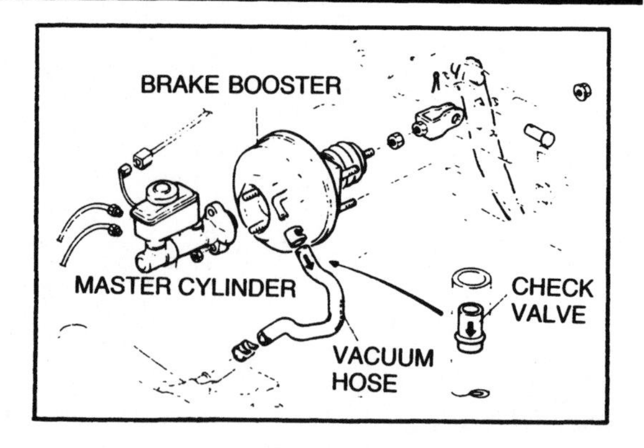

The **master cylinder** is mounted on the brake booster and connected to the brake assemblies by pipes and hydraulic hoses. The common master cylinder is a dual type, which really is two master cylinders in one. The front half operates two brake assemblies while the rear half operates the other two brake assemblies.

Included in the master cylinder is a reservoir which contains the brake fluid for the braking system and a combination valve comprising a pressure differential valve/switch and proportioning valve.

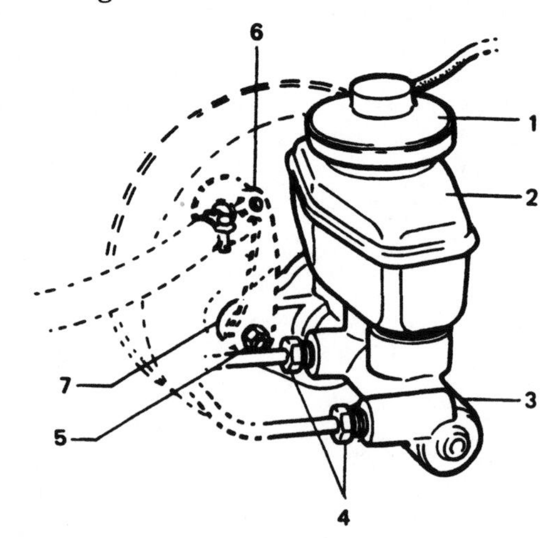

1. Reservoir cap
2. Reservoir
3. Master cylinder
4. Flare nuts
5. Attaching nuts
6. Check valve bracket
7. Gasket

Brake master cylinder

The master cylinder changes the effort from the push-rod of the brake booster into hydraulic pressure. This pressure is transmitted via the pipes and hoses to the brake assemblies, where it applies the brakes.

The **pressure differential valve/switch** used in a dual braking system is generally an integral part of the master cylinder. Some early model vehicles used a separate type connected between the pipes to the front and rear brake assemblies.

An electrical wire connects the pressure differential switch to a warning light mounted in the instrument panel. The pressure differential switch warns the driver of a brake failure by turning on the brake warning light.

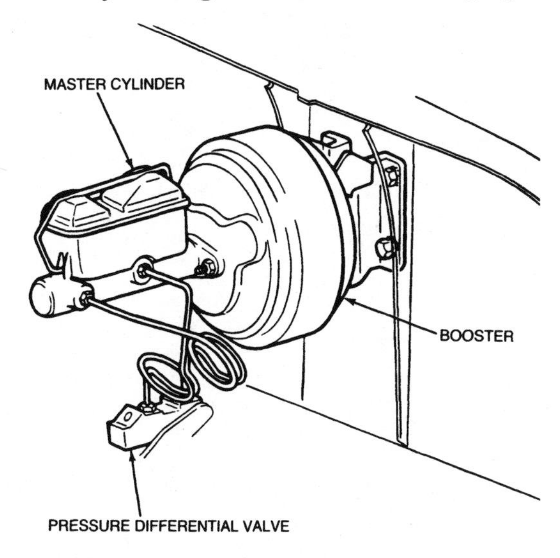

The **proportioning valve** is used in a dual braking system which has the two front brake assemblies operating independently from the two rear brake assemblies. Generally, it is an integral part of the master cylinder.

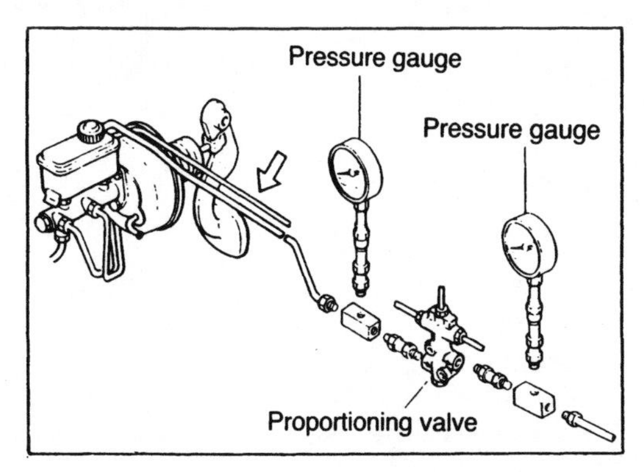

The proportioning valve is connected between the master cylinder outlets or the brake pipes to the front and rear brake assemblies. It is designed to increase braking efficiency by reducing the tendency of the rear wheels to skid under heavy braking.

The **brake pipes** are made from double-wall, welded steel tubing (bundy tubing) and their ends are flared to provide a leak-free seal at the connection points. A flare nut is used at each connection point to join the end of the brake pipe firmly to another component.

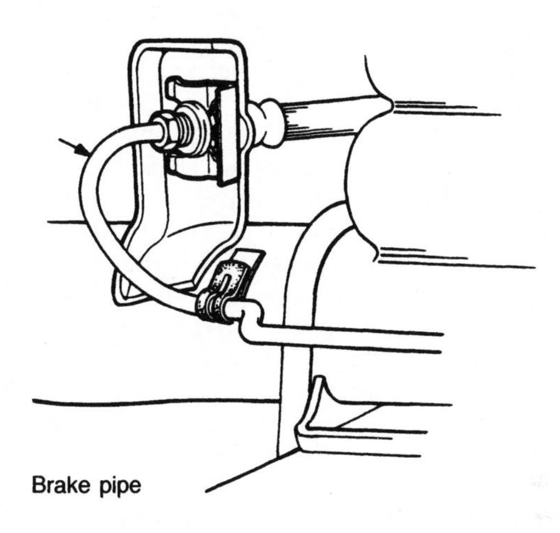

Various types of clamps and clips hold the brake pipe against the body panels or chassis of the vehicle. The brake pipes transmit the brake fluid pressure produced in the master cylinder to the brake assemblies.

The **brake hoses** are made from special reinforced rubber hose (hydraulic hose) with metal fittings clamped at each end. The metal fittings allow each hose to be securely joined to a pipe and component. The brake hose not only has the same function as a brake pipe, but also provides a flexible line between the brake pipe clamped to the body and a brake assembly or Tee piece on the axle housing.

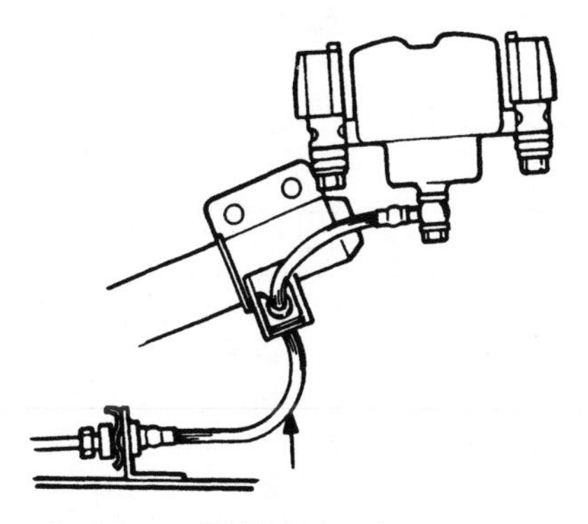

The **brake assembly** may either be a drum or a disc type fitted to the two wheels on the same axle or to all four wheels. Generally, disc brake assemblies are fitted to the front wheels and drum brake assemblies to the rear wheels. They consist of a rotating section which is attached to the wheel, and a non-rotating section that is attached to the steering knuckle support or the end of the axle housing.

The non-rotating section changes the hydraulic pressure from the master cylinder into a force which moves the pads or linings against their disc or drum. Their function is to slow and/or stop a moving vehicle and hold the vehicle stationary when it has been parked. **Caution: the brake assemblies convert the energy of a moving vehicle into heat and after heavy braking will be extremely hot.**

The **park brake** consists of a hand lever located within reach of the driver, a set of cables and levers which connect the hand lever to the brake assemblies on one axle (generally the rear axle), and a special mechanism in the brake assemblies which applies or releases the brakes independently of the hydraulic section.

When the park brake is applied it must continue to hold the vehicle stationary until it has been released. In the released position the park brake must not retard the motion of the vehicle.

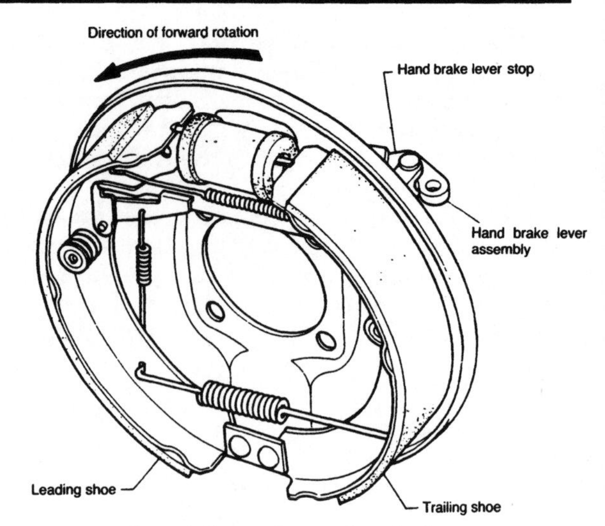

Non-rotating section — drum brake

Source: Nissan Australia

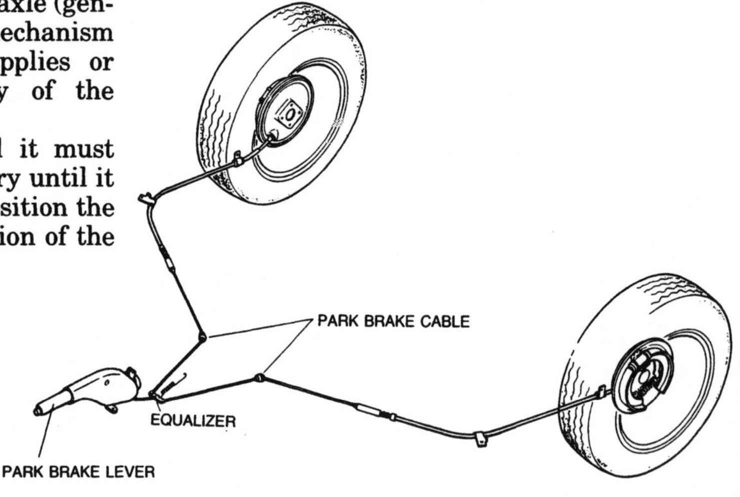

Park brake layout

IDENTIFICATION OF MINOR COMPONENTS

It is necessary to be able to identify the various components related to the different types of brake assemblies as most minor services and adjustments are performed on these units.

The four types of brake assemblies that have been used are:

1 adjustable shoe;
2 self-adjusting shoe;
3 shoe assemblies with park brake linkages (both 1 and 2);
4 disc and pads.

As most of the brake shoe assemblies have similar components, the adjustable shoe type will be fully described while only components which are different in the other shoe assemblies will be described.

Adjustable shoe brake assembly

The **backing plate** is a pressed steel disc with a small rolled-over edge which is bolted to the stub axle or the end of the axle housing. It forms the base of the shoe brake assembly.

The **brake shoe** is made from two pressed steel sections which are interlocked and welded. One of the sections is formed in an arch and drilled at even intervals along each edge for the lining rivets. This is called the table.

Note: Those shoes fitted with bonded linings do not need rivet holes.

The other section is cut to suit the curve of the table and drilled at various places to provide mounting points for the other components. This is called the web.

Two brake shoes are used in each brake shoe assembly and they provide rigid surfaces for the brake linings.

The **brake lining** is the friction surface of the brake shoe. It is made of a mixture of materials which include asbestos, soft metals and resins. It is moulded to the curve of the shoe table and is about 6 mm in thickness. The brake lining may be riveted or stuck (bonded) onto the brake shoe.

To rivet the brake lining to the shoe table, the lining must be drilled to suit the hole pattern of the shoe and counter-sunk to the rivet dimensions. The rivet holes in both parts are aligned and a rivet is inserted into each hole then peened over on the underside of the shoe table.

To bond the brake lining onto the shoe table, a special resin is placed on the mating surfaces. The surfaces are clamped together and then heat-cured under high temperatures and pressures.

The brake lining is the softer material of the two rubbing surfaces, but can transfer heat quickly to reduce wear.

The **wheel cylinder** is cast from iron or aluminium alloy to form a small, thin-walled cylinder with mounti g lugs. A single action wheel cylinder is closed at one end and houses a spring, cup, piston and boot. A double action wheel cylinder houses a spring, two cups, two pistons and two boots.

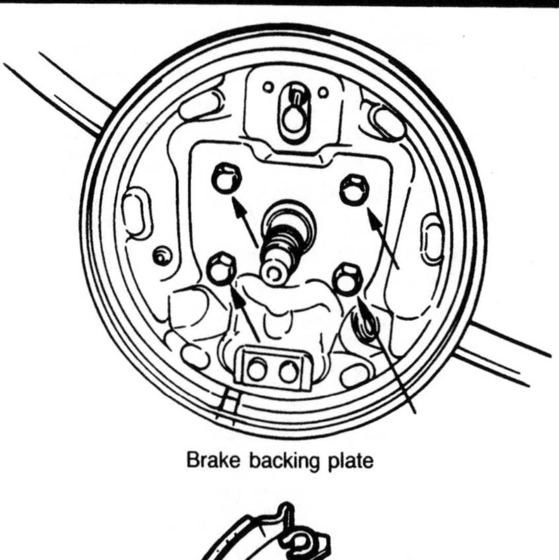

Brake backing plate

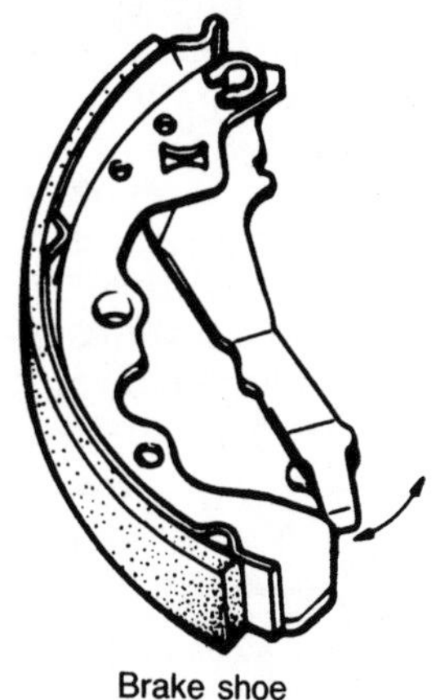

Brake shoe

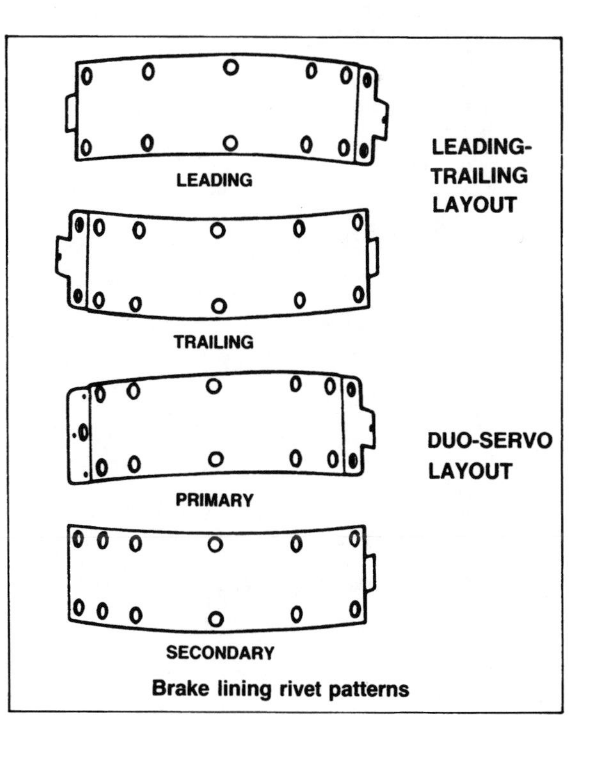

Brake lining rivet patterns

Note: Some wheel cylinders have an expander between each cup and their spring.

The wheel cylinder is bolted to the backing plate and located between the ends of the brake shoes. When the hydraulic pressure from the master cylinder is applied, the wheel cylinder changes the pressure on its pistons to a force at the end of each brake shoe. This applies the brakes.

The **return spring** is an extension spring with specially formed ends which clip in a hole drilled in each web of the brake shoe or onto the anchor pin. The *duo-servo* brake shoe layout has two return springs. These are colour-coded to ensure that each tension spring is connected to the correct shoe.

The *leading and trailing* brake shoe layout has one return spring, which is located as close as possible to the wheel cylinder to ensure that the wheel cylinder pistons are retracted and the brake shoes are returned to their stops when the brake pedal has been released. Similarly, it retracts the brake shoes when the park brake has been released.

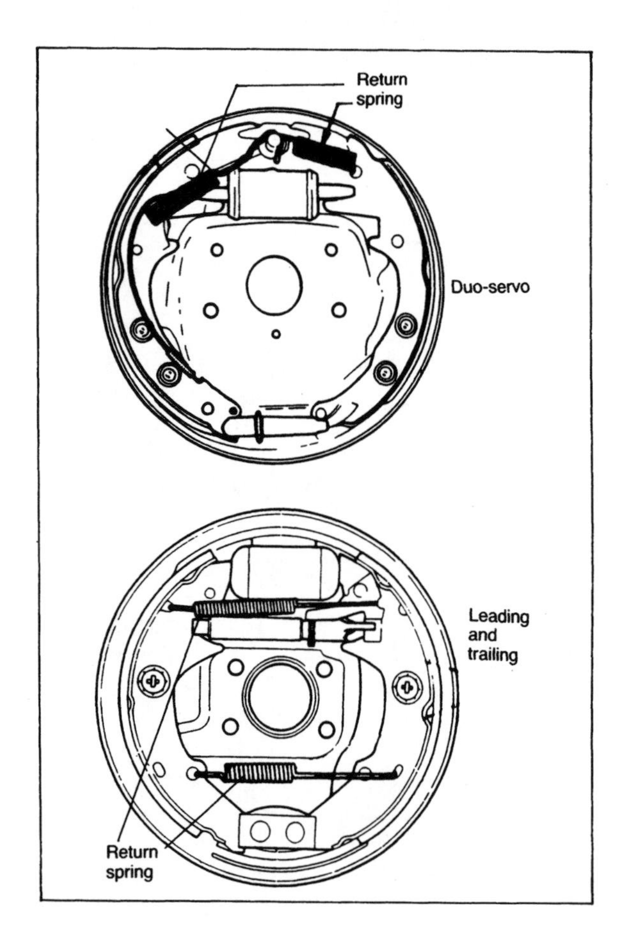

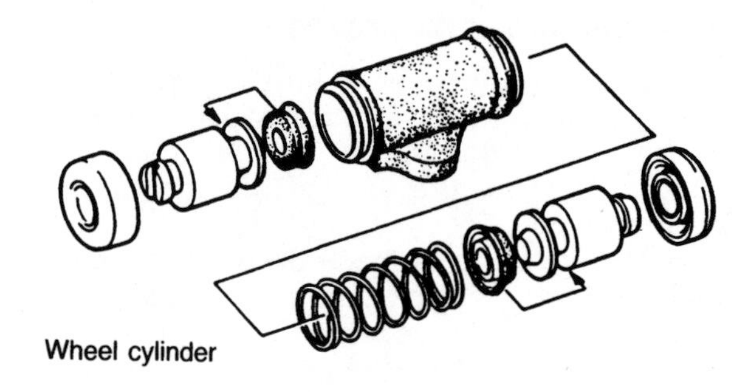

The **anchor spring** is an extension spring which is connected at the anchor end of the brake shoes in a leading and trailing brake shoe layout. It holds the shoes against the anchor.

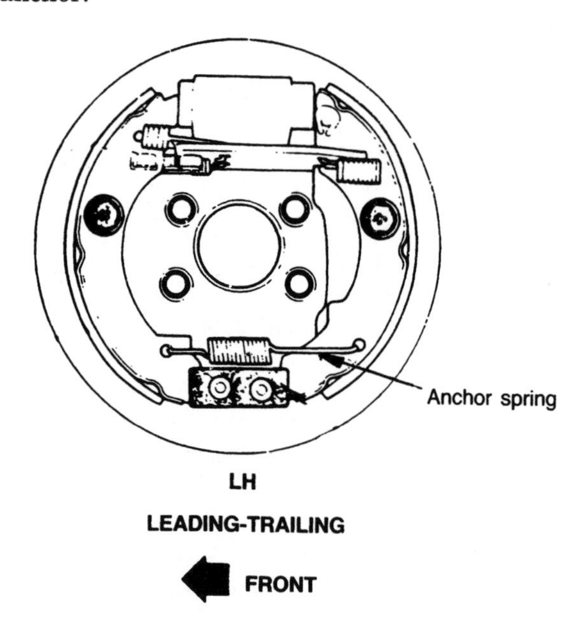

The **shoe retainer** consists of pin, compression spring and plate. The spring and plate may be replaced with a spring clip. Each shoe has a retainer. The pin has a flat head at one end and a flat tip at the other end. It is inserted in a hole in both the backing plate and the brake shoe web.

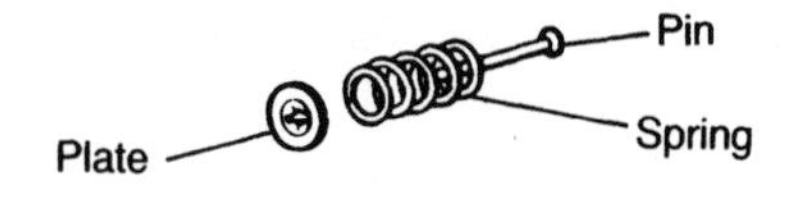

The plate is slotted to allow the flat end of the pin to pass through before it is turned to its locking position. The plate compresses the spring against the web of the brake shoe.

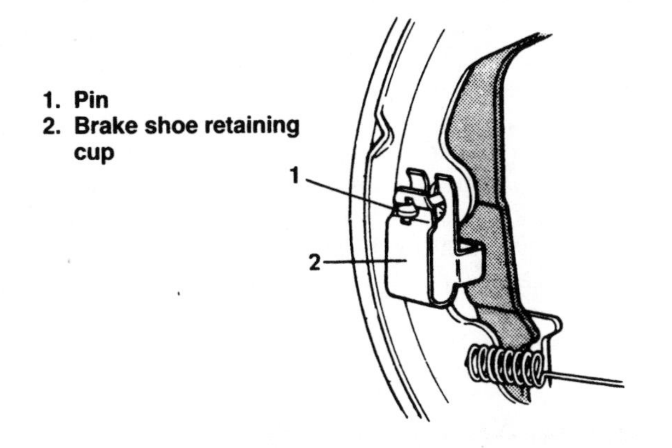

The spring clip which replaces the compression spring and plate is a 'C' shape with a hole at one end and a slot at the other end.

The shoe retainer holds the brake shoe against the backing plate, but allows the shoe to slide a small distance across it.

The **adjusting mechanisms** may be one of the following types.

- wedge;
- cam;
- adjustable stop;
- expandable link.

The common expandable link type consists of:

a an adjusting screw which has a star wheel and short spigot at one end. It screws into the pivot nut;

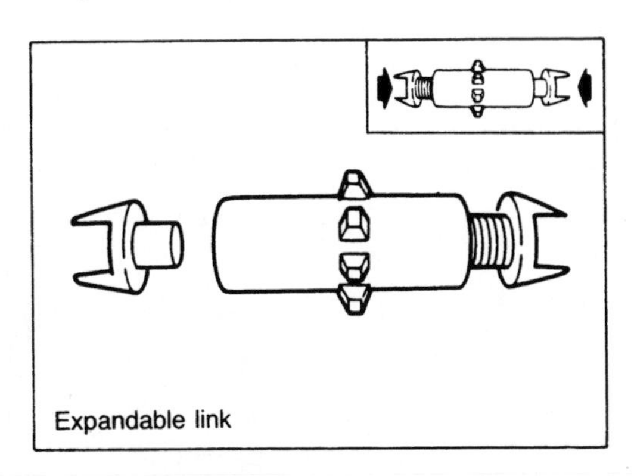

Expandable link

b a pivot nut, which is a long sleeve with a yoke at one end and an internal thread at the other end. Its yoke fits into a slot at the end of the brake shoe web;

c a socket, which is a short sleeve with a yoke at one end and hollow at the other end. It slides over the spigot on the adjusting screw and fits into a slot at the end of the brake shoe web.

Note: Some expandable links have a wave washer between the socket and the star wheel.

The adjusting mechanism allows the clearance between the brake shoe and brake drum to be maintained as wear occurs at the brake lining.

Self-adjusting shoe assembly

The components of a self-adjusting shoe assembly are similar to those of the adjustable shoe assembly, with the exception of the adjusting mechanism which has more components.

The **self-adjusting mechanism** is generally of an expanding link type which uses a screw, cam or wedge to automatically change the length of the link.

The common screw type self adjusting mechanism consists of:

a an expandable link similar to the type described above;

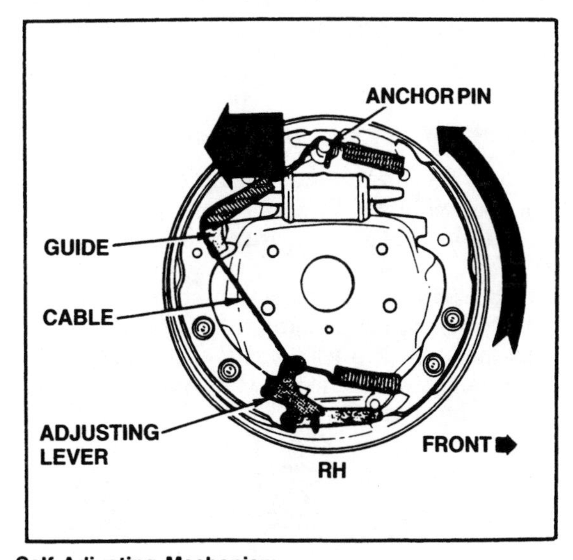

Self Adjusting Mechanism

b a spring steel lever which pivots on the brake shoe web and engages with the notches on the star wheel;

c an extension spring which attaches to the lever and the brake shoe;

d a strut or cable which attaches to the lever and the park brake cable or the anchor pin respectively.

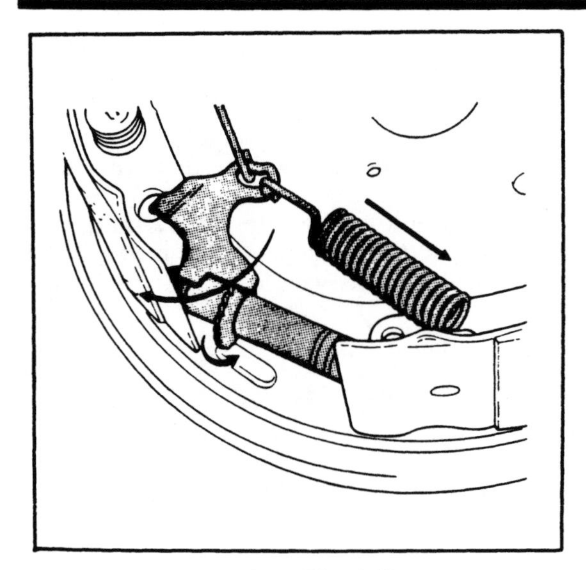

Movement of Lever Over Wheel Star

The self-adjusting mechanism automatically maintains the clearance between brake shoe and brake drum as wear occurs at the brake lining. On some systems adjustment occurs during handbrake operation, however, the most common type adjusts the brake during foot brake operation while the vehicle is reversing.

Shoe assemblies with park brake linkages

The park brake linkages on the two brake shoe assemblies that have been described are similar. The park brake linkage consists of:

a a pressed steel lever, which is connected to the park brake cable at one end and free to move on a pivot at the other end;
b a pivot pin attached to the web of one brake shoe by a clip;

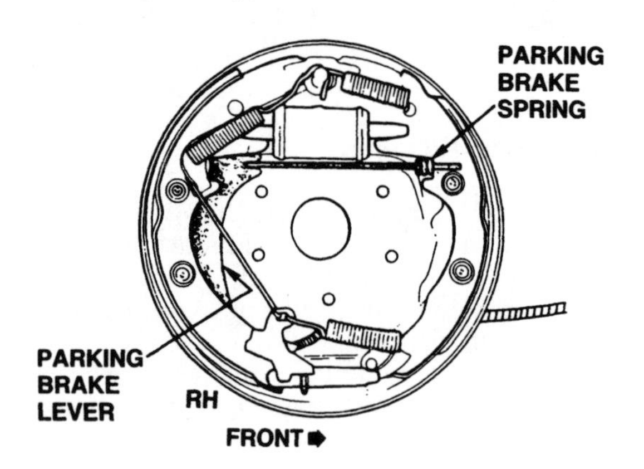

c a strut located between the other brake shoe web and the lever near its pivot;
d a wave washer located at the pivot to reduce rattling noise.

Note: The strut is replaced by the expanding link in the leading and trailing brake shoe assembly.

The two brake assemblies on the same axle that include the park brake are fitted with park brake linkages. These linkages force the brake shoes against the brake drums when the park brake is applied, and allow the brake shoes to return to their stops when the park brake has been released.

Pads and disc brake assembly

The three types of disc brake assemblies are the:

1 fixed head caliper;
2 floating head caliper;
3 swinging head caliper.

The caliper is the assembly that fits over the disc and is mounted on the steering knuckle support or the axle housing. It houses the hydraulic pistons and two pads.

The fixed head caliper has opposed pistons; one or more on each side of the disc. These pistons force the non-metal pads against the sides of the disc when the hydraulic pressure is applied.

The floating head caliper has a piston on one side of the disc which forces one pad onto the disc and causes the head to slide towards the disc to apply the other pad, when the hydraulic pressure is applied.

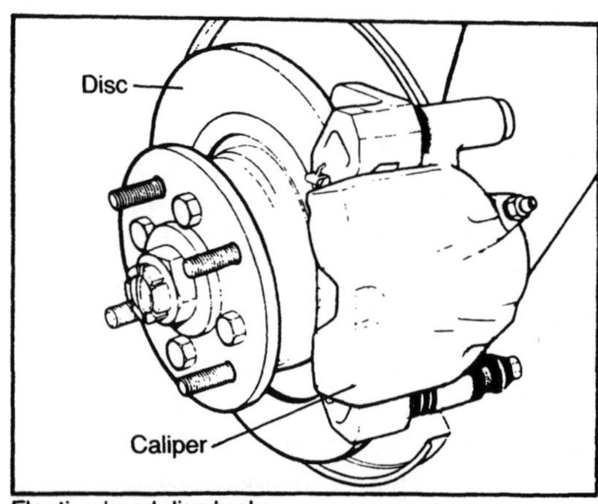

Floating head disc brake
Source: Mitsubishi Colt RB/RC series manual

The swinging head caliper has a piston on one side of the disc which forces one pad onto the disc and causes the head to swing around a pivot to apply the other pad, when the hydraulic pressure is applied.

Since the components of all these calipers are similar, the following components are related to the floating head caliper.

The **anchor bracket** is cast from steel or aluminium alloy. It bolts to the steering knuckle support or the axle housing and provides the mounting point for the two slide pins.

The **slide pins** are made of steel, ground to size and hardened. They are grooved at each end for the seals and locking pins. While each slide pin is locked into the head with a split pin, it is free to slide in the anchor bracket.

The **head** (housing) is cast from steel or aluminium alloy to form a 'U' shape with a cylinder on one side and a set of pad location lugs on the other side. The casting provides four mounting lugs for the slide pins, a piston seal groove in the cylinder bore, a bleeder nipple and a threaded port for the brake hose. It houses the piston and the two pads.

The **piston** is made from steel, ground to size and chrome-plated for high wear resistance. It is externally grooved at its open end for the dust seal. The piston fits into the

Floating head caliper

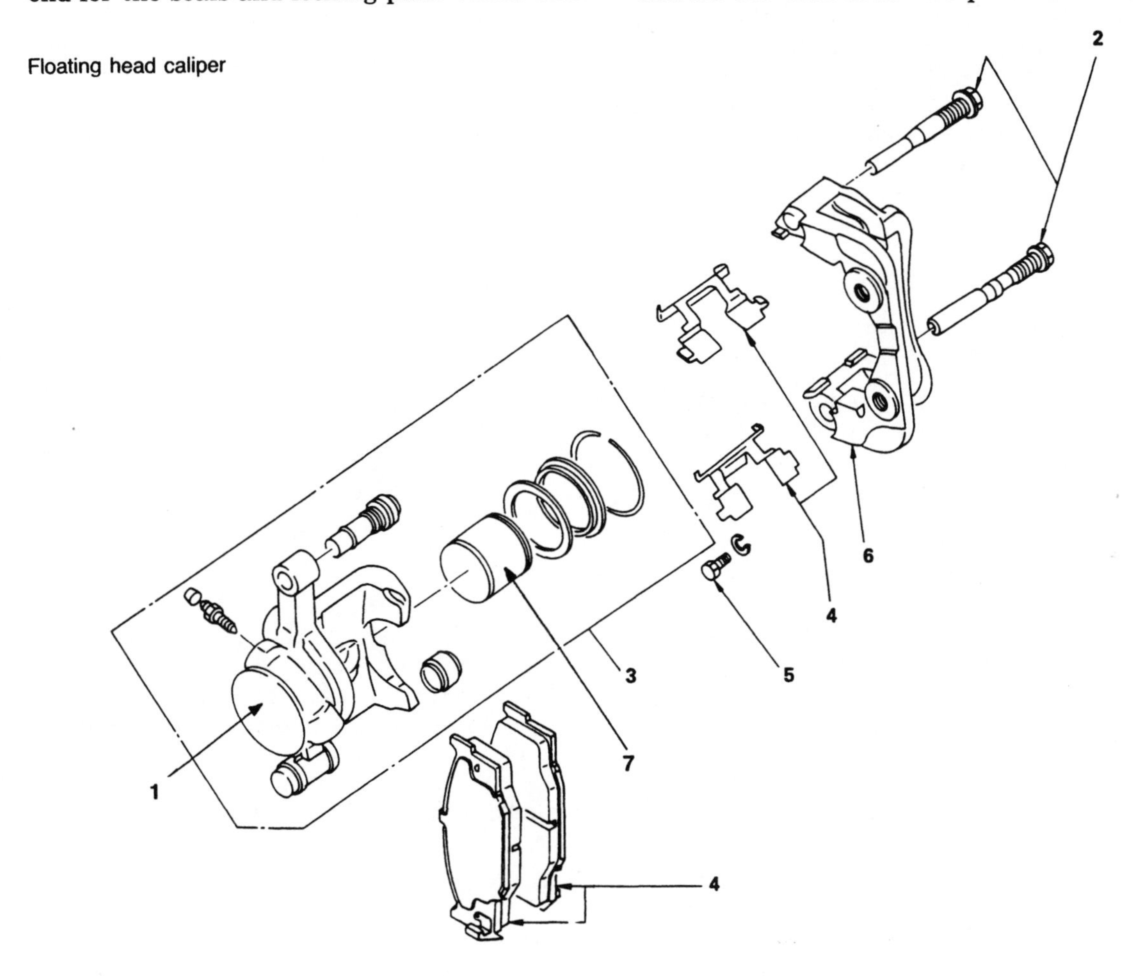

1. Head
2. Slide pin
3. Caliper assembly
4. Pad assembly with shim and clips
5. Bracket attaching bolt
6. Support bracket
7. Piston

piston seal in the head and provides the mounting point for the inner pad.

When the brakes are applied, the piston changes the hydraulic pressure from the master cylinder into a force which pushes the inner pad and pulls the outer pad against the disc.

The **pads** are formed by bonding (sticking) friction surfaces to steel backing plates. These friction surfaces are made of a mixture of materials which include asbestos, soft metals and resins. The backing plates are shaped to suit the curve of the disc, and spring steel mounting clips are riveted to their outer surfaces. The friction material of the pads is the softer material of the two rubbing surfaces, but can transfer heat quickly to reduce wear. *Note*: The park brake can be included as a part of a disc brake assembly in the form of a small drum brake inside the disc or lever mechanism attached to the piston and the caliper housing.

ADJUSTMENT

The three brake system adjustments which normally should be carried out on regular servicing of a vehicle are:

1 brake pedal clearance;
2 brake shoe clearance;
3 park brake lever travel.

These three adjustments are necessary for the adjustable brake shoe assemblies, but the brake shoe clearance adjustment should not be necessary for the self-adjusting brake shoe assemblies.

The pads to disc clearance in a disc brake assembly is not adjustable.

Important: read the brake section of the workshop manual for specification and procedures.

The following adjustment procedures will provide an insight into tools and techniques required to service a brake system.

Brake pedal clearance

Clearance must exist between the brake pedal push-rod and the brake booster control valve to ensure that the brake will fully release. To adjust this clearance:

- Locate the push-rod between the pedal and the booster.
- Using an open end spanner of the correct size, loosen the push-rod lock nut.
- Using a steel ruler, measure the 'free' travel of the pedal. Place one end of the ruler on the firewall and the other end near the pedal. Apply a light pedal force with your fingers until resistance is felt. Measure and note this distance.

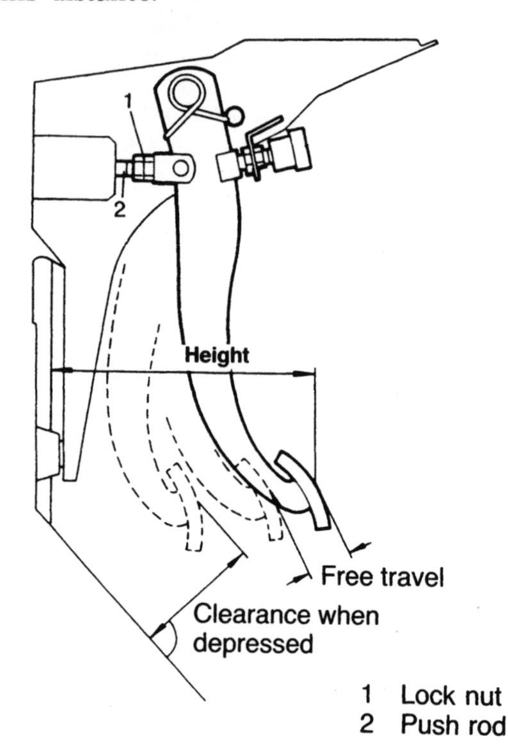

- Adjust the 'free' travel of the pedal to 3 to 5 mm. To reduce the travel, turn the push-rod in a direction that will increase its length. To increase the travel, turn the push-rod in a direction that will reduce its length.
- Tighten the push-rod lock nut.
- Check the 'free' travel of the pedal.

Brake shoe clearance

Clearance between the brake linings and the brake drums must exist to ensure that the wheels will rotate freely when the brakes are released. This clearance is very small (0.1 to 0.5 mm). Two methods can be used to adjust the brake shoe clearance. The first method is by measuring and the second is by feeling the drag.

To adjust the clearance by measuring:

- Loosen the wheel nuts.

- Raise the vehicle so all wheels may be rotated by hand.

Caution: observe all safety procedures.

- Release the park brake and place the gearbox in neutral.
- Using the correct tool, disconnect the park brake cable at its lever or equaliser.
- Remove the wheels and the brake drums.
- Measure the diameter formed by the brake shoes.
- Measure the inside diameter of the brake drum.
- Subtract the brake shoe diameter from the brake drum diameter to obtain the clearance.

Note: A special caliper is available to obtain these measurements.

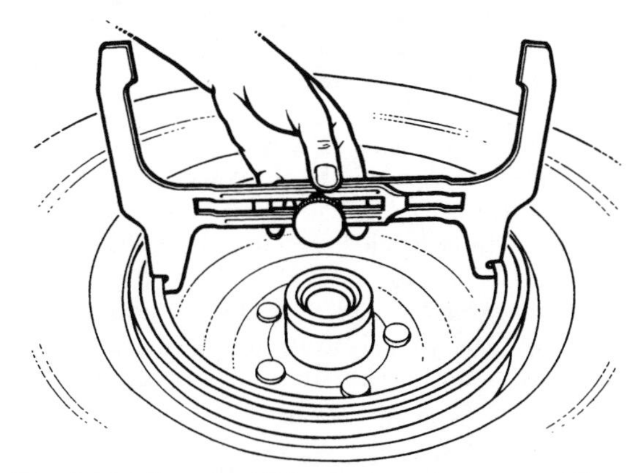

To establish the correct lining-to-drum clearance, set the lower part of the gauge for the drum diameter.

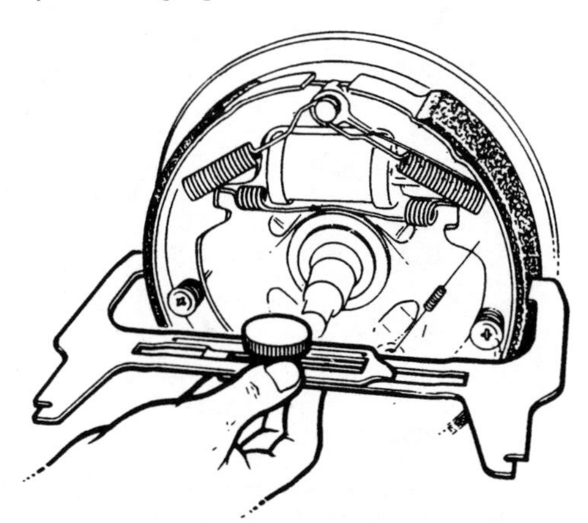

Place the upper, or outside caliper part of the gauge across the newly-installed linings. Adjust the star wheel to expand the shoes to meet the gauge to correct clearance.

- Hold the self-adjusting lever away from the star wheel.
- Using a suitable brake adjusting tool, turn the star wheel in the direction that will expand the shoe diameter.
- Repeat the last five steps until the clearance is in the range of 0.1–0.3 mm.
- Repeat the above procedure on the other brake shoe assemblies.
- Replace the brake drums and wheels.
- Ensure that all wheels turn freely.
- Connect the park brake cable to its lever or equaliser.
- Adjust the park brake lever travel.

To adjust the clearance by feeling the drag:

- Raise the vehicle so all wheels may be rotated by hand.

Caution: observe all safety procedures.

- Release the park brake and place the gearbox in neutral.
- Using the correct tool, disconnect the park brake cable at its lever or equaliser.
- Select the correct brake adjusting tool.
- Remove the dust cover from the backing plate at one of the rear brake assemblies.

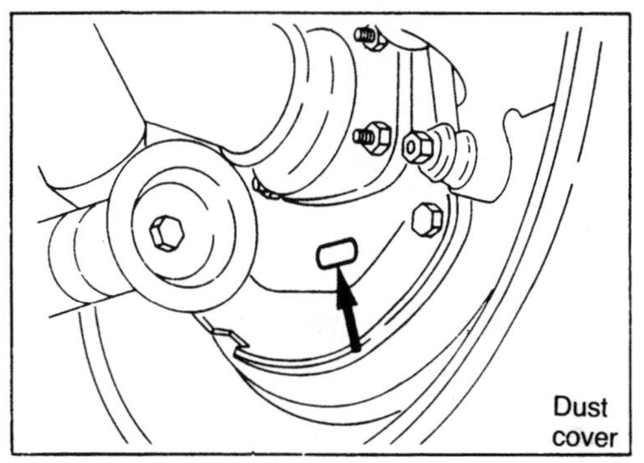

- Insert the adjusting tool through the slot until it contacts the star wheel on the expanding link.
- Move the outer end of the tool upwards to increase the length of the expanding link. This will reduce the shoe to drum clearance.
- While turning the wheel by hand, continue to expand the link until heavy drag is felt.
- Move the outer end of the tool downwards to decrease the length of the expanding link. This will increase the shoe to drum clearance.
- While turning the wheel by hand, continue to contract the link until no drag is felt, and the wheel turns freely.
- Refit the dust cover into the backing plate slot.
- Repeat the above procedure on the other brake shoe assemblies.

- Connect the park brake cable to its lever or equaliser.
- Adjust the park brake lever travel.

Park brake lever travel

To adjust the park brake travel:
- Adjust the brake shoe clearance.
- Apply the park brake lever two notches.
- Using the correct spanners, loosen and screw the lock nut (C) away from the equaliser several turns.
- Turn the adjusting nut (D) in the direction that will apply tension to the cable.

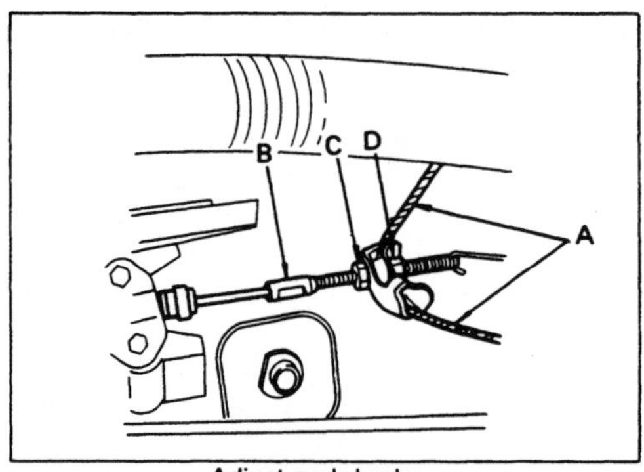

Adjust park brake

- Continue to turn the adjusting nut until a slight drag is felt when the wheels are rotated by hand.
- Using the correct spanners, tighten the lock nut.
- Release the park brake lever.
- Check the rotation of the wheels. They must turn freely (no drag).

Note: When the wheels do not turn freely, repeat the above procedure BUT reduce the tension on the cable.

- Check the park brake lever travel. Apply the foot brake. Count the notches as the park brake is fully applied. Readjust the park brake when the number of notches is not between four and seven.

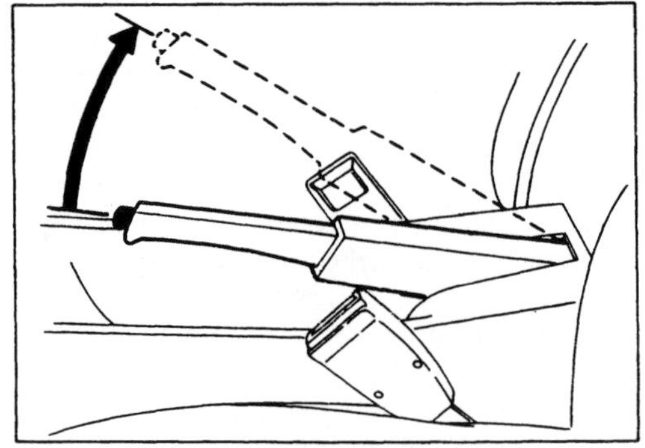
Check lever travel

- Release both brakes.
- Check the rotation of the wheels. They must turn freely (no drag).
- Apply the park brake.
- Return the vehicle to floor level.

Caution: observe all safety procedures.

REMOVE AND REPLACE BRAKE DRUM

The reasons for removing a brake drum are to:

a inspect the brake lining wear;
b remove the brake dust;
c repair the components in the brake assembly.

To remove a brake drum:
- Using a suitable wheel brace, loosen the wheel nuts.
- Raise the vehicle so all wheels may be rotated by hand.

Caution: observe all safety procedures.
- Remove the wheel nuts, hub cap and wheel.
- Remove the brake drum securing device, when fitted.

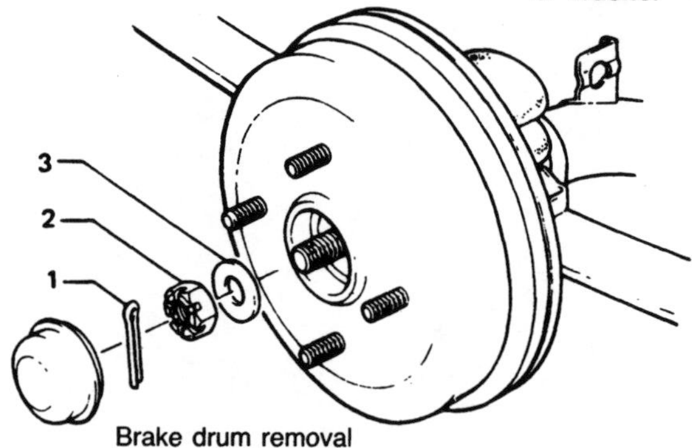

Brake drum removal

- Pull evenly on each side of the brake drum until it has cleared the wheel studs and brake linings. When the drum jams evenly on both shoes, it will be necessary to increase the brake shoe clearance.

Note: On some occasions it may be necessary to hit the brake drum with a soft faced hammer to release it from the axle flange.

Caution: the dust in the brake drum is a health hazard. Carefully follow the correct cleaning procedure.

To replace a brake drum:

- Ensure the mating surfaces of the brake drum and axle flange are clean.
- Align the brake drum holes with the studs on the axle flange. When a screw type securing device is fitted, align the holes.
- Slide the brake drum evenly over the brake linings and wheel studs until it contacts the axle flange.

Note: If the drum jams before it is against the axle flange, remove it and try again.

- Refit the drum securing device, where necessary.
- Refit the wheel, hub cap and wheel nuts. Lightly tighten the wheel nuts with the wheel brace.
- Adjust the foot and park brakes.
- Return the vehicle to floor level.

Caution: observe all safety procedures.

- Tighten the wheel nuts in the correct sequence and to the correct tension.

REMOVE AND REPLACE BRAKE SHOES

The reasons for removing the brake shoes are to:

a renew the brake linings.
b repair or renew the wheel cylinders.
c renew the park brake cable.

The procedure for removing and replacing a pair of brake shoes is similar for the various brake shoe layouts.

To remove the brake shoes from a leading and trailing brake shoe layout:

- Remove the brake drum. Carefully follow the procedure described above.
- Install a clamp on the wheel cylinder. This will retain the pistons and seals in the cylinder.
- Disconnect the return spring from one of the shoe webs.
- Remove the shoe retainers.
  - — Depress the spring clip or plate and turn the pin a quarter of a turn.
  - — Slowly release the spring clip or plate.
  - — Remove the spring clip or plate and spring from the pin.
  - — Withdraw the pin from the rear of the backing plate.

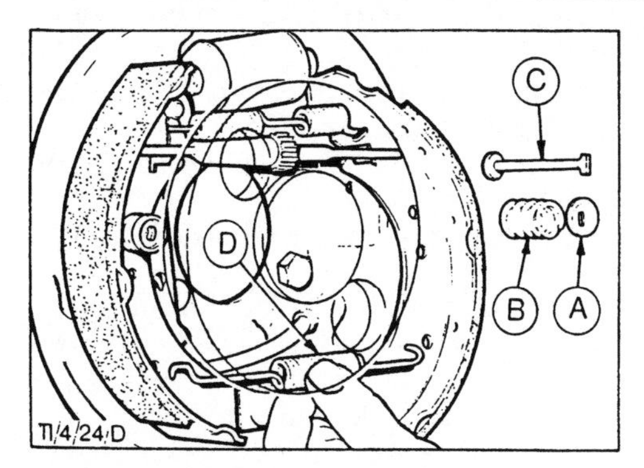

Front shoe securing pin assembly.
A — Plate. B — Spring. C — Pin.

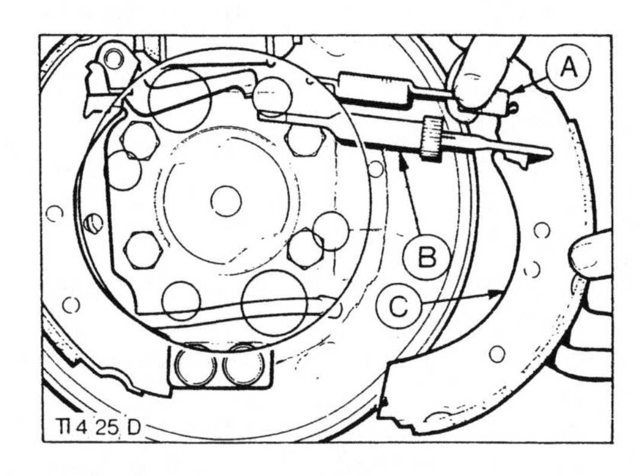

Unhooking spring from front brake shoe.
A — Spring. B — Adjusting lever. C — Front brake shoe.

- Remove the shoe that does not have the park brake lever attached.
  - — Carefully pull the shoe away from the cylinder until it clears the boot and the park brake strut.
  - — Slide it down the anchor block to a position that will allow the anchor spring to be disconnected.
  - — Withdraw the shoe.

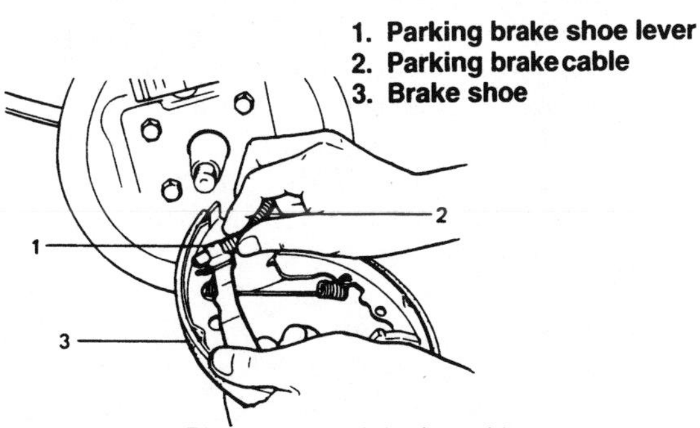

Disconnect park brake cable

- Disconnect the park brake cable from the lever.
  - — Using a pair of pliers, hold the spring away from the lever.
  - — Slide the end of the cable past the hook on the lever.
- Remove the shoe with the park brake lever attached.
- Remove the park brake lever from the shoe.
  - — Disconnect the adjusting lever spring.
  - — Remove the adjusting lever and expandable link from the shoe.
  - — Note the length and the position of the lining on the brake shoe.
  - — Note the side of the web to which the pivot is fitted.
  - — Using a suitable pair of circlip pliers, remove the pivot clip.
  - — Withdraw the pivot and lever from the brake shoe web.

Note: before the brake shoes have been replaced, the wheel cylinders must be inspected for leakage and repaired or renewed when necessary.

To replace the brake shoes on a leading and trailing brake shoe layout:

- Apply a small amount of lubrication on the pivot pin.
- Select the correct shoe. The length and position of the lining on the shoe must be the same as noted when the shoes were being removed.
- Fit the pivot into the web so the lever is on the side that was noted during its removal.
- Install the pivot clip. Using a pair of pliers, clamp the clip firmly in its groove.
- Screw the star wheel of the expandable link in the direction that will reduce its length to the minimum.
- Fit the expandable link to the park brake lever and brake shoe.
- Install the adjusting lever and its spring in the shoe.
- Connect the park brake cable to the lever.
  - — Using a pair of pliers, clamp the spring away from the end of the cable.
  - — Slide the end of the cable over the hook on the end of the lever.
- Position the shoe on the backing plate so one end is against the wheel cylinder piston and the other end is against the anchor block.
- Install the shoe retainer.
  - — Insert the pin through the backing plate and the shoe web.
  - — Slide the spring clip or the spring and plate over the pin.
  - — Depress the clip or plate so the flat end of the pin passes through it.
  - — Turn the pin a quarter of a turn.
  - — Release the clip or the plate.
  - — Ensure the pin is locked by the clip or the plate.

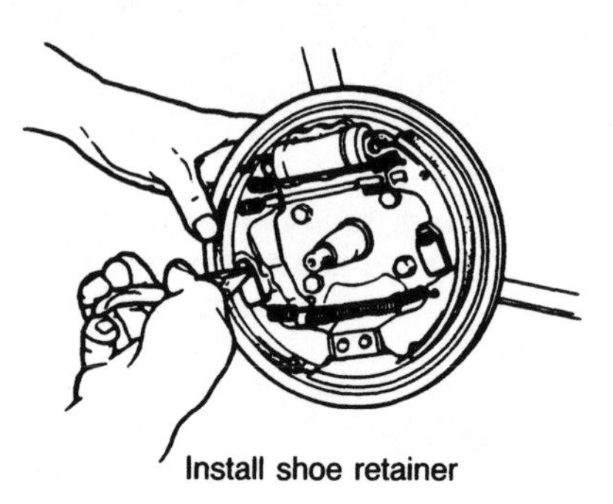

Install shoe retainer

- Fit the shoe return and anchor springs to the web of the shoe that has already been fitted.
- Install the remaining shoe.
  - — Ensure that the shoe is facing in the correct direction.
  - — Connect the shoe to the anchor spring.
  - — Slide the end of the shoe onto the anchor block.
  - — Pull the other end of the shoe towards the wheel cylinder.
  - — Engage the expandable link end into the slot in the shoe web.
  - — Push the shoe firmly against the wheel cylinder piston.
- Install the shoe retainer. Use the same method as described above.
- Connect the shoe return spring to the shoe.
  - — Ensure the spring is attached correctly to the other shoe.
  - — Hold the shoe against the wheel cylinder piston.
  - — Using a suitable pair of pliers, extend the spring far enough to allow its end to hook into the web of the shoe.
- Remove the cylinder clamp.
- Centralise the shoes on the backing plate.
  - — Starting at the wheel cylinder end of the shoes, push the shoes across the backing plate until the linings are the same distance from its edge.
  - — Carefully, slide the shoes onto the anchor block until each end of the linings is the same distance from the backing plate edge.

Remove cylinder clamp

— Attempt to fit the brake drum but do not force it past the shoes.
— Repeat the above steps until the brake drum will slide on and off without touching the brake linings.

- Bleed the hydraulic system. This procedure should only be necessary when the hydraulic system has been repaired.
- Adjust the brake shoe to brake drum clearance. Use one of the methods previously described.

Note: When the brake drum cannot be fitted or there is too much drag, the adjusting lever will have to be lifted away from the star wheel to allow it to be turned in a direction that will decrease the length of the expandable link.

- Replace the brake drum. Use the procedure described above.

CLEANING

The cleaning of the brake shoe assembly and the brake drum will ensure maximum friction surface life and increase the braking efficiency of the vehicle. The removal of the brake dust from the brake drums and brake shoe assemblies must be done using extreme care, as it is a health hazard. The asbestos dust can cause eye irritations and respiratory problems.

Caution: do not remove the brake dust with a jet of compressed air.

The three methods that can be used to remove the dust are:

1 wiping the components with a damp rag and a bristle brush;
2 washing the components with a jet of water or a steam cleaner;
3 extracting the dust from the components with an industrial vacuum cleaner.

Before the cleaning procedure starts, cover your eyes, nose and mouth with a pair of goggles and a respirator that will fit snugly on your face.

To remove the brake dust from a brake drum with an industrial vacuum cleaner:

- Place a sheet of paper or waste rag onto a flat surface, such as a bench or the floor.
- Tap the rim of the drum on the rag to remove the excess dust.
- Fold the rag and place it in waste bin.
- Using the vacuum cleaner, brush the internal surfaces of the drum.
- Using a clean piece of rag, wipe the friction surface of the drum.

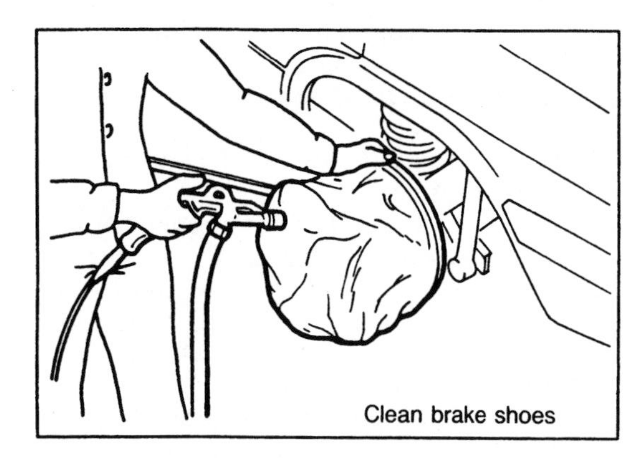
Clean brake shoes

To remove the brake dust from the brake shoe assembly with an industrial vacuum cleaner and a wire or bristle brush:

- Using the vacuum cleaner, brush the surfaces of the components.
- Using the wire or bristle brush, loosen the dust behind the components.
- Remove the loose dust with the vacuum cleaner.
- Check the level indicator on the vacuum cleaner. Dispose of the dust bag when necessary.

REMOVE AND REPLACE CALIPER AND DISC PADS

The reasons for removing a caliper are to:

a renew the brake pads;
b remove the disc;
c renew the piston and seal.

The removal and the replacement of the caliper and disc pads does vary slightly from the type previously mentioned.

Note: Before proceeding, carefully read the brake section of the appropriate workshop manual.

To remove the floating head caliper from a disc assembly:

- Lower the fluid level in the brake master cylinder reservoir.
  - — Raise the bonnet and place a cover over the mudguard.
  - — Remove the top of the master cylinder reservoir.
  - — Using a syringe, extract about half of the fluid.
  - — Dispose of this fluid by placing it into a waste oil container.
- Using a suitable wheel brace, loosen the wheel nuts.
- Raise the vehicle so the wheel may be rotated by hand.

Caution: observe all safety procedures.

- Remove the wheel nuts, hub cap and wheel.
- Using a pair of pliers, remove the locking wire from the caliper mounting bolts. Some mounting bolts may have lock tabs or spring washers.
- Push the brake pads away from the disc.
- Using the correct ring spanner, remove the caliper mounting bolts.

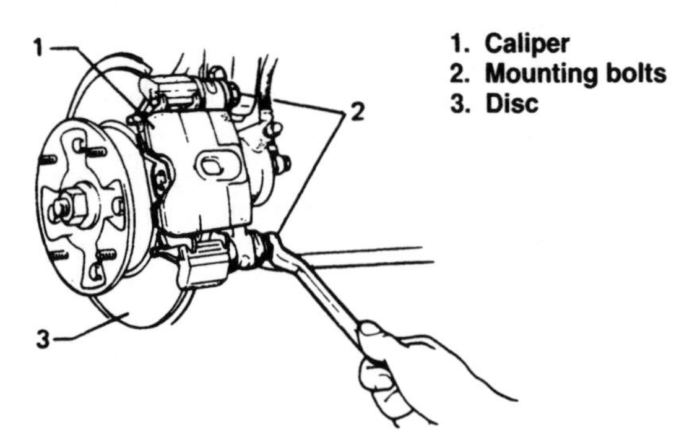

- Ensure the caliper is free to move away from its mounting points. There may be shield screws or brake pipe brackets preventing the caliper from moving.
- Pull the caliper away from the disc.
- Support the caliper on a piece of wire attached to the suspension.

To remove the pads from a floating head caliper:

- Press the outer pad towards the inner pad until the locating dimples have released.
- Pull the outer pad from the caliper head.

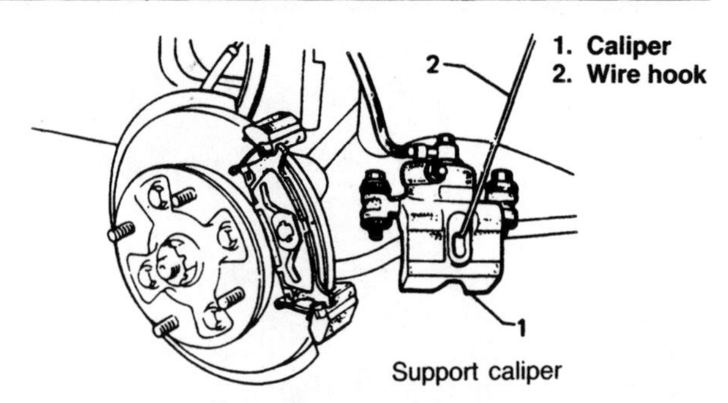

Support caliper

- Using a suitable tool placed between the inner pad and the piston, pry the pad from the piston until the spring clip is free.

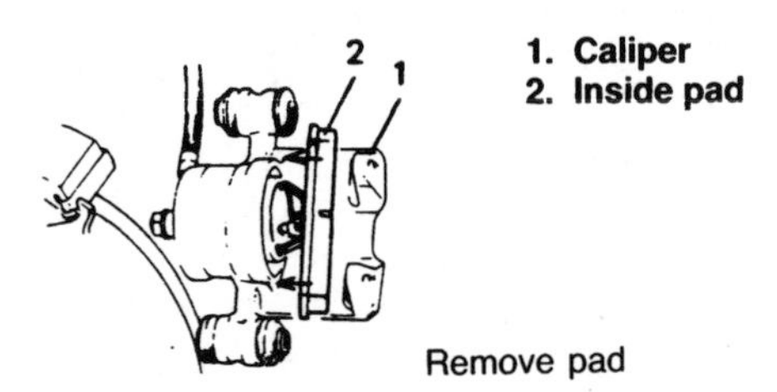

Remove pad

- Pull the inner pad from the anchor bracket.

To replace the pads to a floating head caliper:

- Push the piston as far as it will travel into its bore.
- Position the inner pad into the anchor bracket with its clip towards the piston.
- Push the inner pad until its plate is against the piston.
- Position the outer pad in the head with its spring clips facing the lugs.
- Using a suitable tool, lever each clip away from the plate to allow the pad to be pressed into place.
- Ensure the locating dimples on the pad are in the lug holes.

To replace the floating head caliper to a disc assembly:

- Disconnect the caliper from the supporting wire.
- Position the caliper so the edge of the disc is facing towards the gap between the pads.
- Carefully, slide the caliper over the disc.
- Install the mounting bolts. Renew the lock tabs or spring washers when necessary.
- Using a tension wrench, tighten the bolts to the specified torque.
- Using a pair of pliers, install new lock wire to the mounting bolts.
- Refit the wheel, hub cap and wheel nuts. Lightly tighten the wheel nuts with the wheel brace.
- Return the vehicle to floor level.

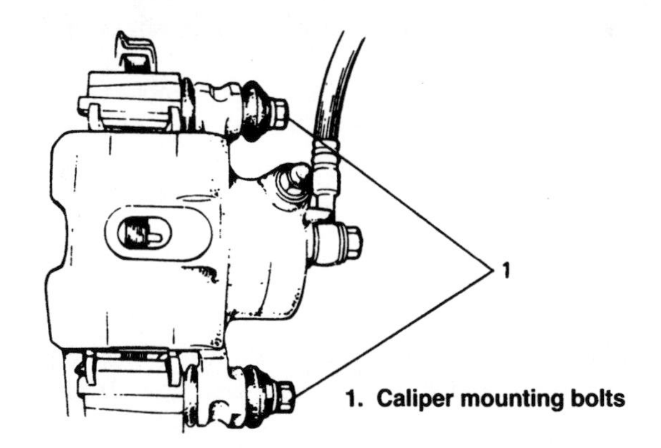

Caution: observe all safety procedures.
- Tighten the wheel nuts in the correct sequence and to the correct tension.
- Top up the master cylinder reservoir with fresh brake fluid. Add fluid to the reservoir until it is three-quarters full.

REMOVE AND REPLACE DISC

The reasons for removing a disc are to:
a machine or renew the disc;
b repack the wheel bearings with grease;
c repair the suspension.
The removing and replacing procedure will vary slightly between different makes and models of vehicles and whether or not the discs are fitted to the driving axle.

To remove a disc fitted to a non-driving axle with tapered roller bearings:
- Using a suitable wheel brace, loosen the wheel nuts.
- Raise the vehicle so the wheel may be rotated by hand.

Caution: observe all safety procedures.
- Remove the wheel nuts, hub cap and wheel.
- Remove the caliper.
- Remove the hub grease cap.
- Using a pair of side cutting pliers, extract the split pin from the axle nut.
- Using the correct size spanner, unscrew the axle nut and place it in the hub cap.
- Hold the hub and slide the washer and outer wheel bearing off the stub axle.
- Place the washer and bearing in the hub cap.
- Grip the hub with both hands and pull it off the stub axle.

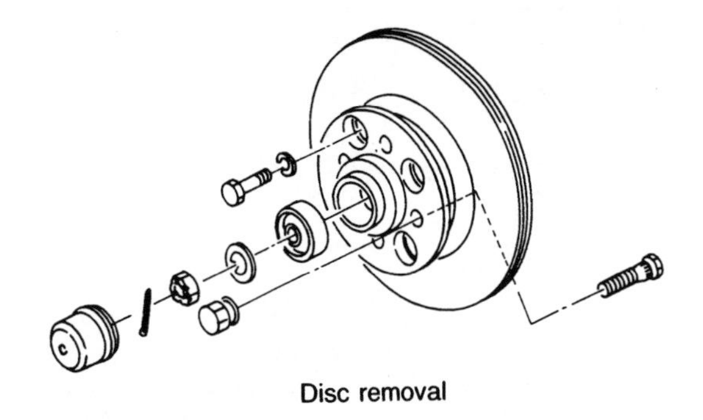
Disc removal

Note: Sometimes the inner wheel bearing and grease seal will remain on the axle. Remove them and place them in the hub cap.
- Place the hub and disc on a clean surface.

To replace a disc to a non-driving axle with tapered roller bearings:
- Clean and repack the wheel bearings with grease. See chapter 8, 'Seals and bearings', for the required procedure.
- Install the hub to the stub axle.
  - — Carefully, slide the hub over the stub axle while ensuring the seal does not contact the axle thread.
  - — Push the hub firmly into place.
  - — Install the outer bearing, washer and nut on the stub axle.

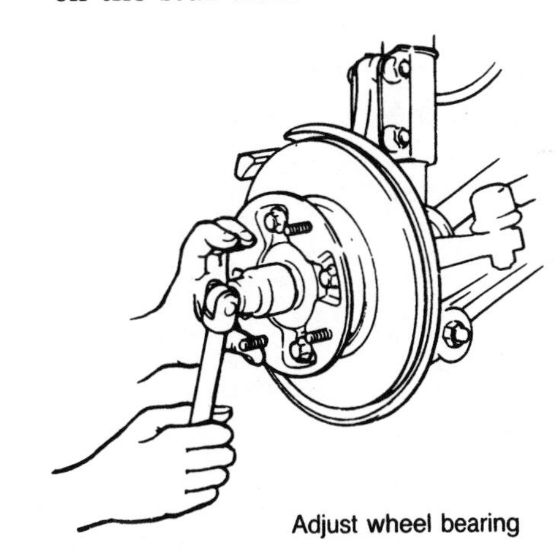
Adjust wheel bearing

  - — Adjust the wheel bearings (see ch. 8).
  - — Install a new split pin.
  - — Replace the hub cap.
- Replace the caliper.
- Refit the wheel, hub cap and wheel nuts. Lightly tighten the wheel nuts with the wheel brace.
- Return the vehicle to floor level.

Caution: observe all safety procedures.
- Tighten the wheel nuts in the correct sequence and to the correct tension.

SETTING PRESSURE DIFFERENTIAL SWITCH

The types of pressure differential switches are:
a self-setting;
b manual-set.
As soon as a brake failure occurs, the hydraulic system causes the pressure differential switch to close and 'turn on' the warning light. When the hydraulic system has been repaired and bled, the switch must open to 'turn off' the warning light. The self-setting switch does this automatically, but the manual-set switch has to be reset before the brakes are bled. Generally, those switches included in the master cylinder are of the self-setting type and those fitted as a separate unit are manual-set.
Note: When specification about the switch cannot be obtained, treat it as a manual-set type.

To reset a manual-set pressure differential switch:

- Disconnect the electrical wire from the terminal at the switch.
- Using the correct ring spanner or deep socket, loosen the switch.
- Unscrew the switch from its housing.

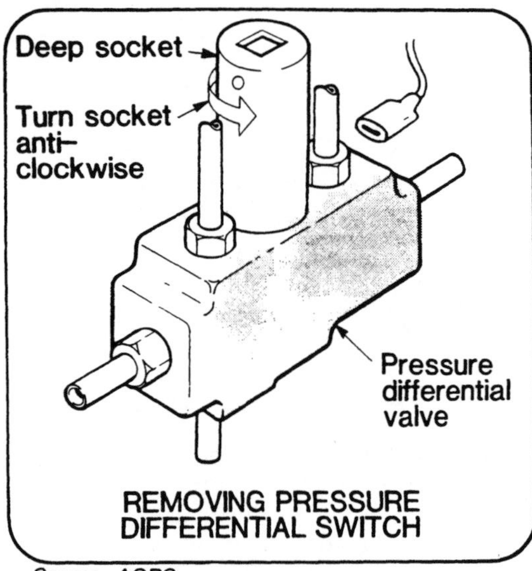

REMOVING PRESSURE DIFFERENTIAL SWITCH

Source: AGPS

- Bleed the brakes. See the next section for the correct procedure.
- Screw the switch into its housing.
- Using the correct ring spanner or deep socket and a tension wrench, tighten the switch to the correct torque.
- Connect the electrical wire to the terminal at the switch.
- Turn on the ignition and check that the warning light remains 'off'.

BRAKE BLEEDING

Brake bleeding is the name given to the procedure for removing the air from the hydraulic system after a repair or a new component has been fitted to the brake system.

When air is trapped in the hydraulic system, the brake pedal feels springy (spongy) as the brakes are applied. In some cases, the pedal will travel its full distance. On the dual brake system when only one section is faulty, the pedal will travel about half its distance.
Note: The warning light would be glowing to warn the driver of a brake problem.

These effects are due to the air being compressed. In an air-free system, the brake fluid is considered to be incompressible.

The three common methods of brake bleeding are:
1 manual bleeding;
2 pressure bleeding;
3 surge bleeding.

Manual bleeding requires two people to carry out the procedure. One person must pump the brake pedal and top up the master cylinder reservoir while the other person opens and closes the bleeding nipples at the brake assemblies.

Pressure bleeding requires a special brake fluid pump and one person to carry out the procedure. The special pump is attached to the master cylinder reservoir to pressurise the hydraulic system while the person opens and closes the bleeding nipples at the brake assemblies.

Surge bleeding is a combination of the other two methods. It requires two people and a special brake fluid pump. The special pump is attached to the master cylinder reservoir to pressurise the hydraulic system while one person pumps the brake pedal and the other person opens and closes the bleeding nipples at the brake assemblies.

Regardless of the brake bleeding method, the two important tasks that must be performed before starting to bleed the hydraulic system are:
1 adjust the brake shoe assemblies;

2 drain all the old brake fluid from the hydraulic system as it may be contaminated with moisture.

Note: Moisture in the brake fluid reduces its boiling point to a limit where it will boil under heavy braking. This will cause loss of brakes.

To bleed the hydraulic system using the manual method:

- Raise the bonnet and cover the mudguard.
- Top up the master cylinder with clean brake fluid.

Caution: spilt brake fluid must be wiped off the paint work with a clean, damp cloth.

- Raise the vehicle to a suitable working height.

Caution: observe all safety procedures.

- Adjust the brakes.
- Locate and remove each dust cap from the bleeding nipples at the brake assemblies.
- Using a suitable single hexagonal ring spanner, loosen each bleeding nipple but ensure it stays seated.
- Select a bleeding hose that will fit snugly on the end of the bleeding nipples.
- Remove the pressure differential switch.
- Starting at the brake assembly furthermost from the master cylinder, place the ring spanner on the nipple and connect the bleeding hose.

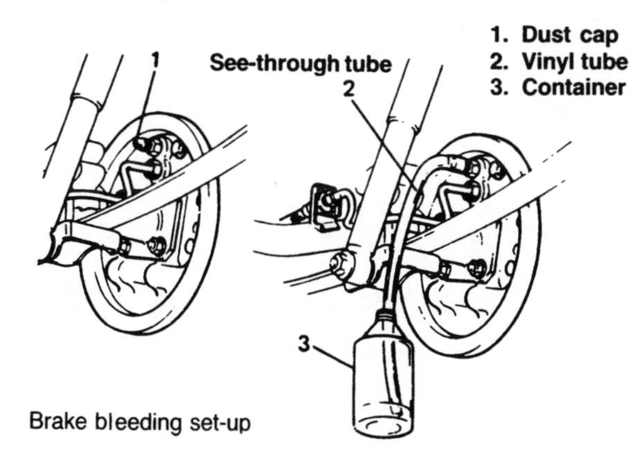

Brake bleeding set-up

- Place the other end of the bleeding hose into clean brake fluid held in a small glass container.
- Open the bleeding nipple about three-quarters of a turn.
- Instruct the second person to slowly depress the brake pedal.
- Observe the stream of fluid as it flows into the container.
  — Bubbles indicate that there is still air trapped in the system.

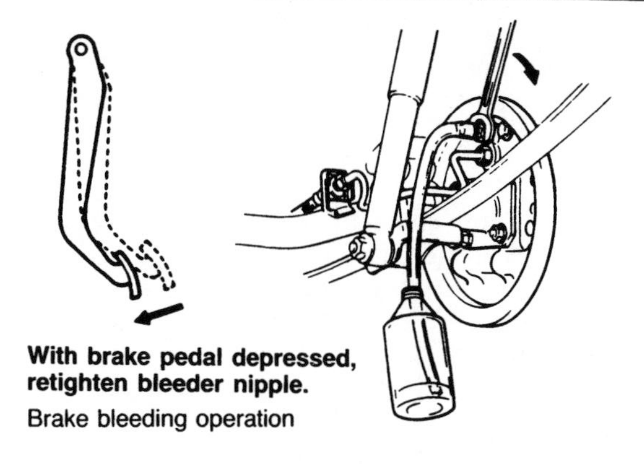

Brake bleeding operation

  — An unbroken stream of fluid indicates the air has been removed.

- Close the bleeding nipple.
- Instruct the second person to slowly release the brake pedal and top up the master cylinder.
- Repeat the last five steps until all the air has been removed from that brake assembly.
- Repeat the last seven steps at the other brake assemblies.

Note: Some brake system have a cross layout, which means one section of the master cylinder operates the left front and right rear brake assemblies and the other section operates the right front and left rear brake assemblies.

- Ensure that the bleeding nipples are tight but not overtight.
- Apply the brakes with a heavy force and check for brake fluid seepage at the bleeding nipples. Should a seepage occur, loosen and retighten the offending nipple.
- Refit the dust covers to the bleeding nipples.
- Return the vehicle to floor level.

Caution: observe all safety procedures.

- Check the fluid level in the master cylinder reservoir.

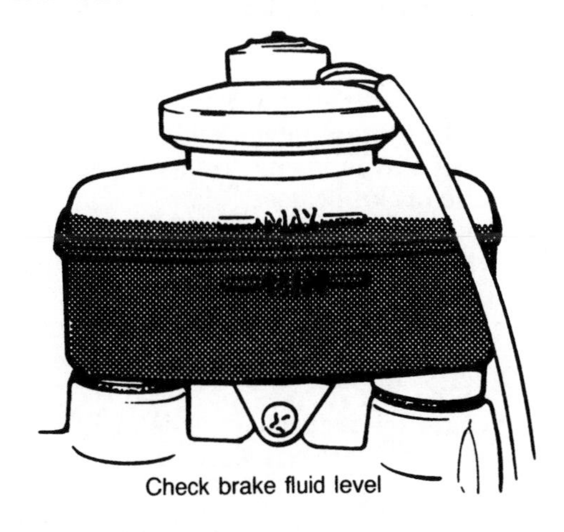

Check brake fluid level

- Refit the pressure differential switch.
- Road test the vehicle.

Important

It is possible, on some drum brake equipped vehicles, that an unsatisfactory (spongy) pedal may be present after the bleeding process has been completed. When this occurs, proceed with the following steps:

- Raise the vehicle to a suitable working height.
- Remove all brake drums.
- Adjust the brake shoes to a position that will give minimum volume in the wheel cylinders. See the diagram.
- Clamp the brake shoes in their full unadjusted position.
  —Use G clamps or similar devices, but protect the brake linings with thin pieces of wood or thick cardboard.
- Bleed the brakes using the method previously described.
- Remove the clamps from the brake shoes.
- Refit the brake drums.
- Readjust the brakes.

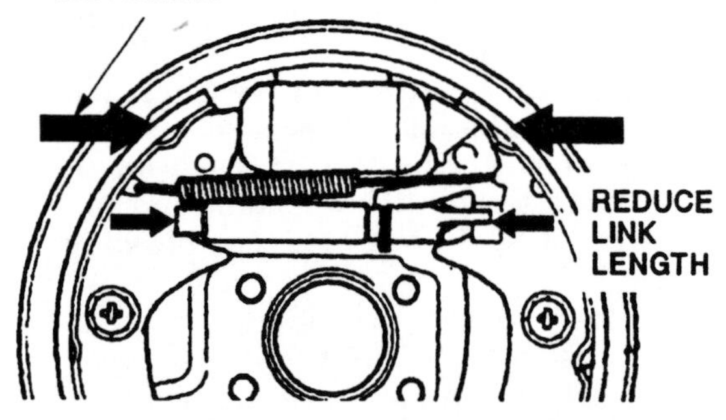

BACK-OFF BRAKE SHOE ADJUSTER

The exceptions to this procedure are related to early brake system layouts which are fitted with some of the following components.

a Hydraulic brake booster.
  - To be bled first.

b Two wheel cylinders in the brake shoe assemblies.
  - For series-connected wheel cylinders, bleed the one connected to the brake pipe first.
  - For parallel-connected wheel cylinders, bleed the lowest one first.

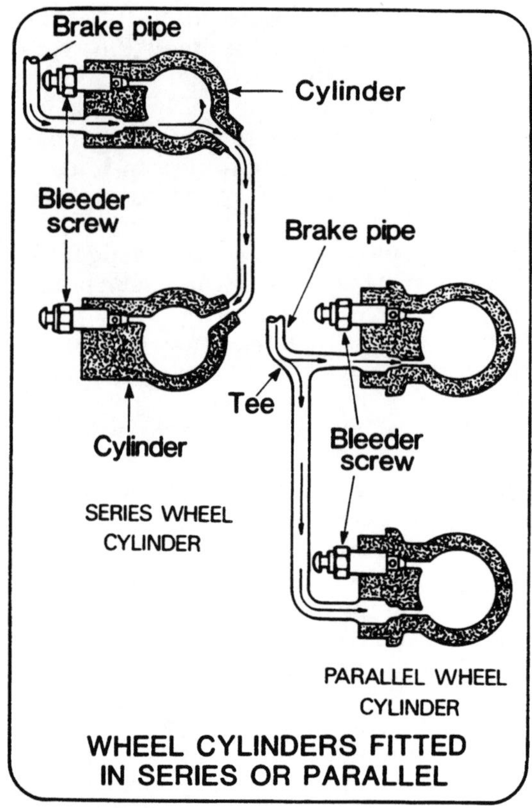

WHEEL CYLINDERS FITTED IN SERIES OR PARALLEL

Source: AGPS

c Load compensating valve.
  - Bleed the front wheel assemblies first.

REMOVE AND REPLACE MASTER CYLINDER

The reasons for removing a master cylinder are to:

a repair or renew the brake booster;
b repair or renew the master cylinder.

To remove a dual master cylinder:

- Apply the park brake and place a manual transmission in reverse gear, or an automatic transmission in 'P'.
- Attach a 'NO BRAKES' sign to the steering wheel.
- Disconnect the earth terminal at the battery.
- Place a protective cover over the mudguard.
- Using the correct flare nut spanner, loosen and unscrew all the brake pipes from the master cylinder.

- Using a suitable ring spanner, loosen and remove the mounting nuts or bolts.

1. Reservoir cap
2. Reservoir
3. Master cylinder
4. Flare nuts
5. Attaching nuts

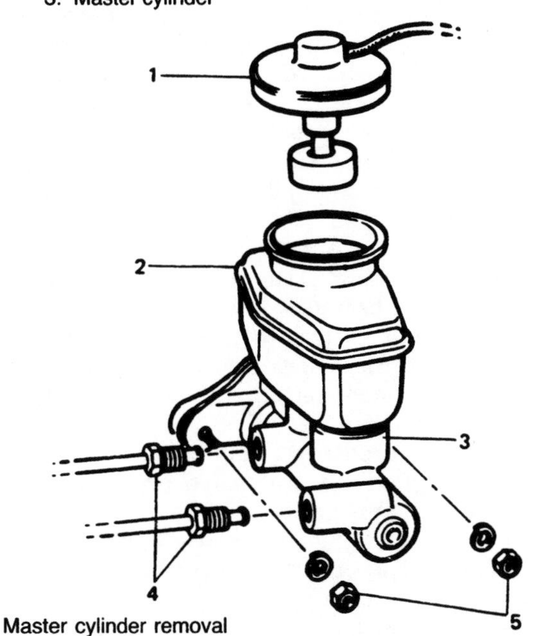

Master cylinder removal

- Carefully, pull the master cylinder away from the brake booster. Take note of any shims or gaskets that may be stuck on the mating surfaces.
- Empty the master cylinder reservoir.

To replace a dual type master cylinder:

- Bleed the master cylinder.
  - — Attach a bleeding pipe to each outlet so its open end is in the reservoir.
  - — Grip the master cylinder lightly between the soft jaws of a vice.
  - — Using clean brake fluid, top up the reservoir.
  - — Using a long, slender, round-end tool, push the master cylinder piston to the end of its stroke.
  - — Release the piston.
  - — Check the fluid level in the reservoir and top up if necessary.
  - — Repeat the last three steps until all the air has been removed.
- Remove the bleeding pipes and plug the outlets.
- Replace the reservoir cap.
- Clean the master cylinder flange and the mating surface on the brake booster.
- Refit or renew the shims or gasket, of the recommended type and thickness, when necessary.

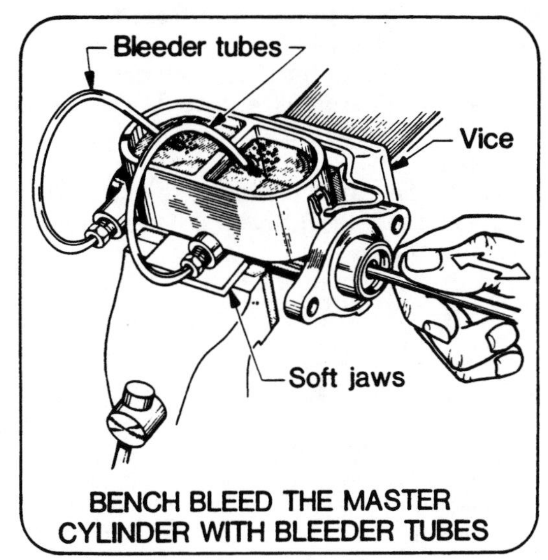

BENCH BLEED THE MASTER CYLINDER WITH BLEEDER TUBES

Source: AGPS

- Carefully, attach the master cylinder to the booster but leave the mounting nuts or bolts loose.
- Remove the plugs and screw the flare nuts on the brake pipes into the outlets but leave them finger tight.
- Using the correct sized socket on a tension wrench, tighten the mounting nuts or bolts to the correct torque.
- Wrap a piece of rag around each pipe just below the flare nut.
- Instruct a second person to slowly depress the brake pedal.

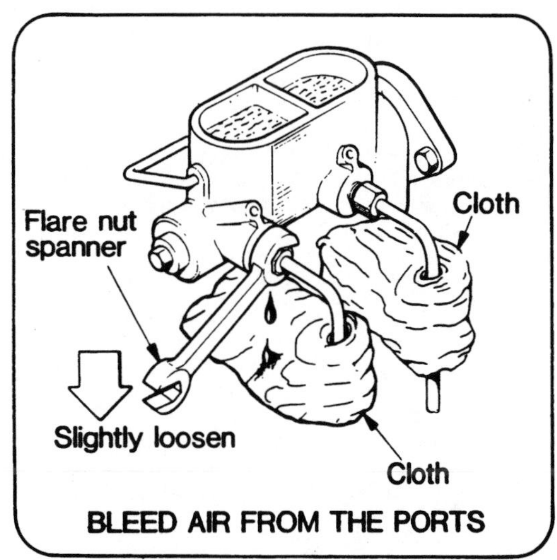

BLEED AIR FROM THE PORTS

Source: AGPS

- Using the flare nut spanner, tighten the flare nuts while the pedal is being depressed.
- Remove the rag from around the pipe.

- Clean any spilt brake fluid from the paint work with a wet rag.
- Top up the brake fluid in the reservoir.
- Check the operation of the brake pedal. Should the pedal feel spongy, bleed the brakes.
- Connect the earth terminal to the battery and remove the 'NO BRAKES' sign from the steering wheel.
- Road test the vehicle.

REMOVE AND REPLACE BRAKE BOOSTER

The reason for removing a brake booster is to repair or renew it.

To remove a brake booster:

- Apply the park brake and place a manual transmission in reverse gear, or a automatic transmission in 'p'.
- Attach a 'NO BRAKES' sign to the steering wheel.
- Disconnect the earth terminal at the battery.
- Place a protective cover over the mudguard.
- Remove the master cylinder. Carefully follow the instructions given in the last section.
- Cap the brake pipes after they have been disconnected to prevent dust from entering.
- Disconnect the hose that connects the booster to the engine manifold at the check valve and plug it.
- Remove the clevis pin from the push-rod at the brake pedal. The clevis pin may be retained with a spring clip or a split pin.
- Using a suitable open end spanner, remove the mounting nuts from the booster mounting studs. These may be located on the inside of the firewall under the dash.
- Withdraw the booster from the firewall.
- Remove the seal (booster to firewall).

Before the brake booster is installed, it is important to check the setting of the output push-rod. The output push-rod must protrude a set distance past the brake booster front shell. When the output push-rod protrudes too far past the shell, the brakes may lock on or burn out because the master cylinder piston will not return to its stop. When the output push-rod does not protrude far enough past the shell the brake pedal will have excessive travel.

The two methods that may be used to check the output push-rod protrusion are:

1 a bridge no-go gauge;
2 a pair of vernier calipers and a steel rule.

To check the output push-rod setting with a bridge no-go gauge:

- Clean the surfaces on the shell either side of the output push-rod.
- Place the gauge so its middle section is above the end of the output push-rod.
- Carefully, look for clearance between the gauge and shell or the gauge and output push-rod. No clearance must exist.
- Increase the output push-rod length when clearance exists between the gauge and output push-rod.

1. Vacuum hose
2. Booster
3. Push rod clevis
4. Dash panel
5. Attaching nut
6. Master cylinder
7. Gasket
8. Splitpin
9. Clevis pin
10. Gasket
11. Attachingnut

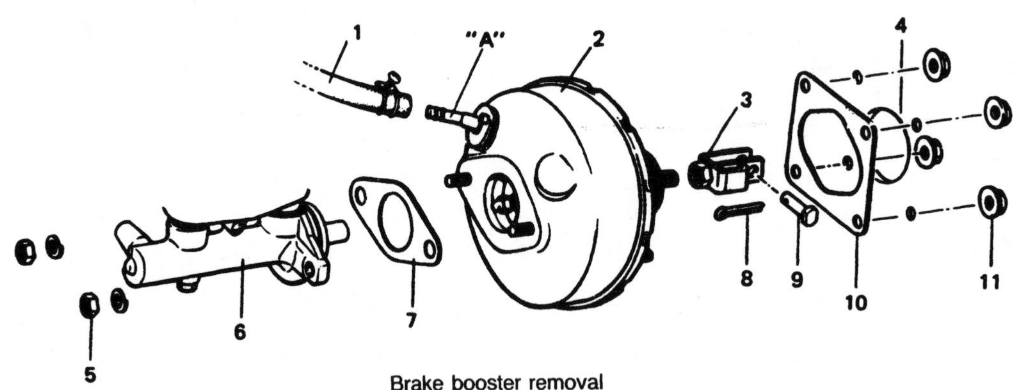

Brake booster removal

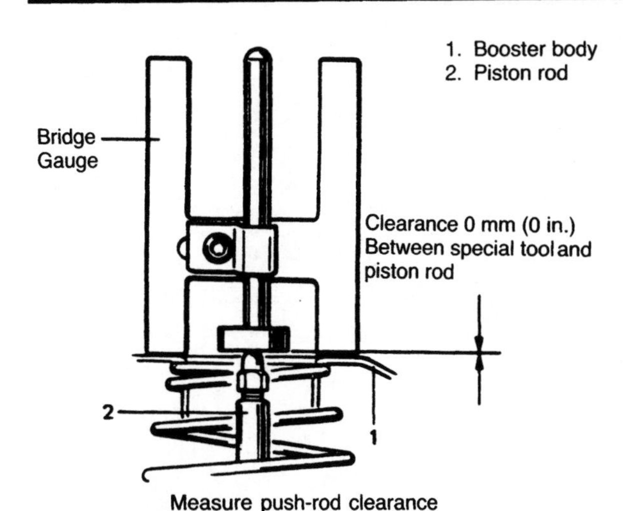

Measure push-rod clearance

- Decrease the output push-rod length when clearance exists between the gauge and shell.

To check the output push-rod setting with a pair of vernier calipers and a steel rule:

- Clean the surfaces on the shell on either side of the output push-rod.
- Place the steel rule so its edge is against the shell and its side is against the output push-rod.
- Using the depth gauge on the vernier calipers, measure the distance from the output push-rod end and the top edge of the steel rule.
- Record the measurement.
- Using the vernier calipers, measure the width of the steel rule.
- Record the measurement.
- Subtract the first measurement from the second measurement.
- Compare the result to the manufacturer's specifications.
- Adjust the output push-rod length to obtain the correct protrusion.

To adjust the output push-rod length:

- Select two spanners that will fit the lock nut and push-rod tip.
- Loosen the lock nut.
- Turn the tip a small amount in the required direction.
- Hold the tip firm and tighten the lock nut.
- Check the output push-rod protrusion.
- Repeat the last four steps until the clearance is zero.

To replace a brake booster:

- Inspect the booster-to-firewall seal and renew or replace it.

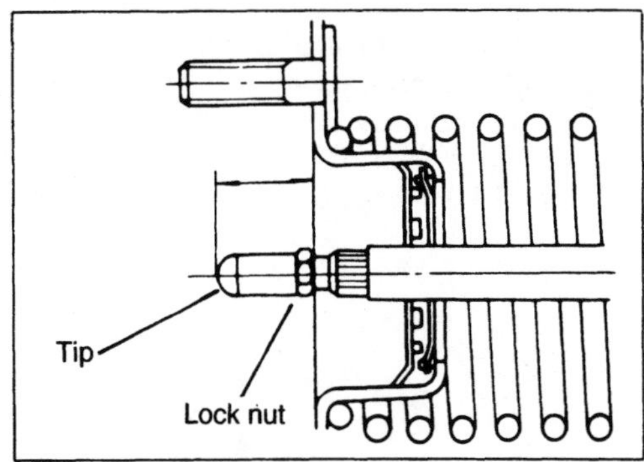

Adjust push-rod length

- Attach the booster-to-firewall seal to the booster.
- Insert the brake pedal push-rod end of the booster through the firewall.
- Refit the mounting nuts to the mounting studs and tighten them to the correct torque.
- Install the clevis pin into the push-rod at the brake pedal. Use a new split pin where necessary.
- Unplug and connect the hose to the check valve.
- Install the master cylinder.
- Check the brake pedal clearance (free travel).
- Connect the earth terminal to the battery and remove the 'NO BRAKES' sign from the steering wheel.
- Road test the vehicle.

CHAPTER 14 REVISION

1 Where is the master cylinder located?
2 What is the master cylinder's function?
3 Name the two components between which the brake booster is located.
4 Why is a brake booster installed in the brake system?
5 Name the two sections in a brake system that must work independently from one another.
6 How is the brake warning light 'turned on'?
7 What component reduces rear wheel skidding under heavy braking?
8 Where is a pressure differential switch located?
9 Name the component that increases the effort from the driver's foot.
10 How is the brake fluid pressure transmitted to the brake assemblies?

11 List two types of brake assemblies.
12 Why do brake assemblies become hot when the brakes are applied?
13 State the reason for a park brake.
14 Name the two brake shoe assemblies.

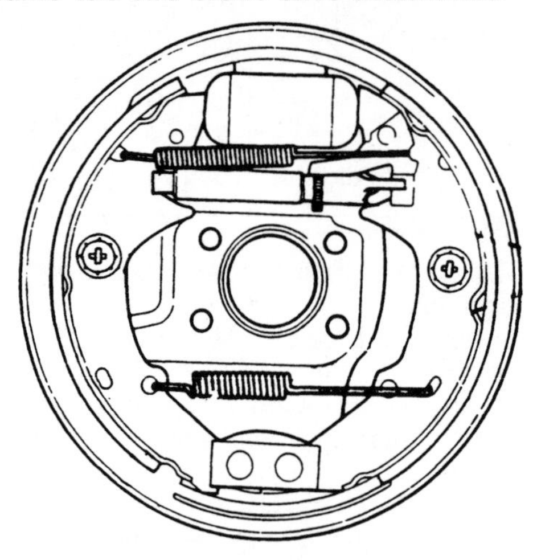

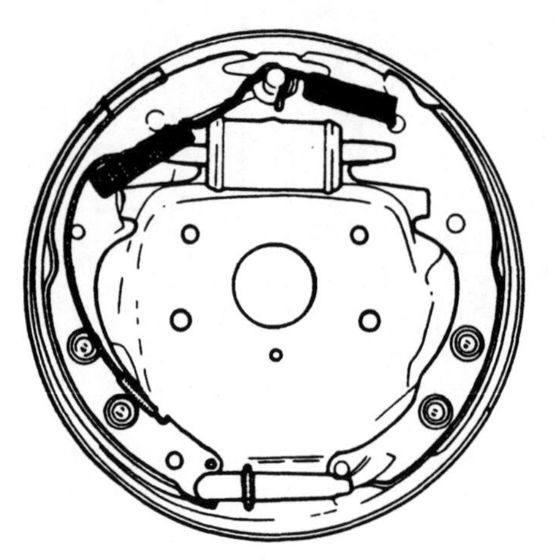

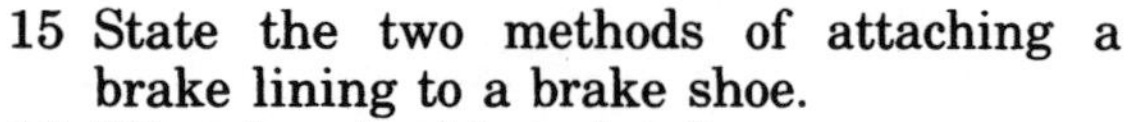

15 State the two methods of attaching a brake lining to a brake shoe.
16 What is a backing plate?
17 Name the component that changes brake fluid pressure into a force.
18 What holds the brake shoes against the backing plate?
19 List the types of adjusting mechanisms.
20 How is the self-adjusting mechanism operated?
21 Name four components found in a park brake linkage.

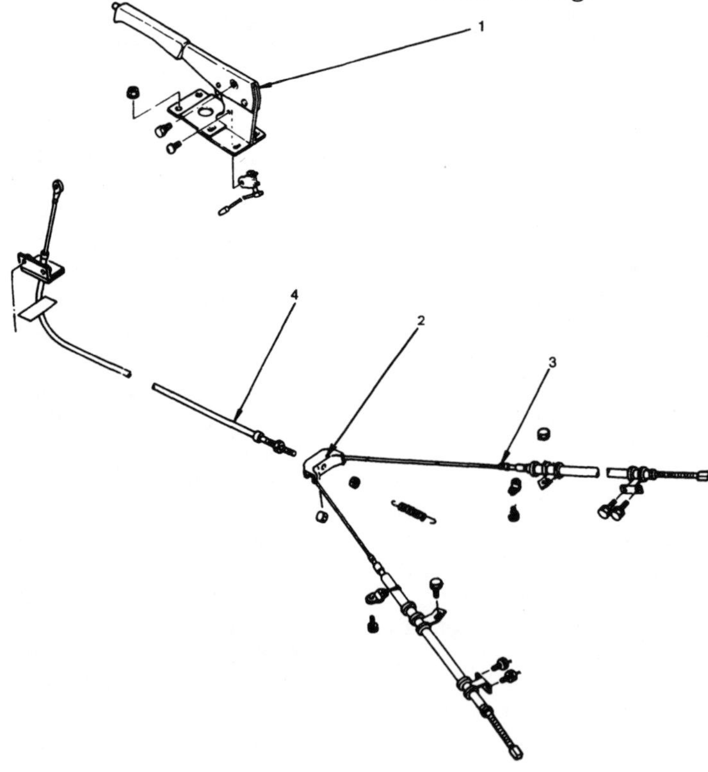

22 Name the disc brake assembly components.

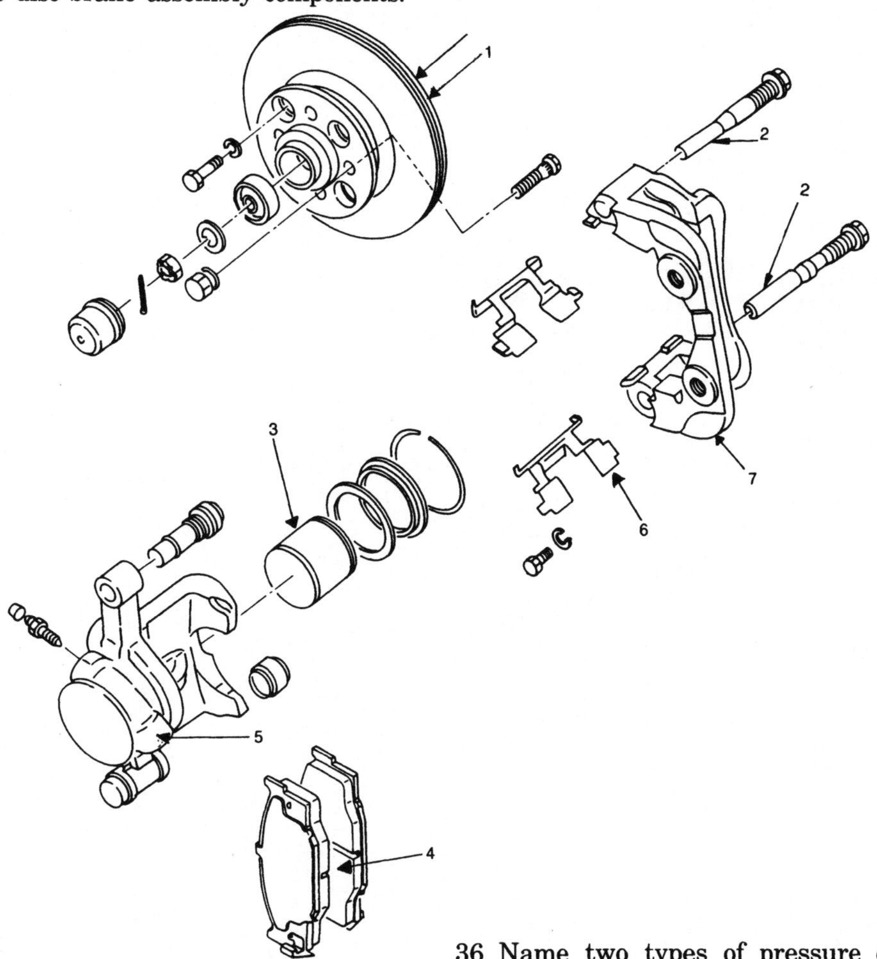

23 In a floating head caliper, how are the pads brought into contact with the disc?
24 What are the three brake system adjustments?
25 Which brake assembly is not adjustable?
26 What is 'free' travel?
27 How is brake shoe clearance checked?
28 How far should a park brake lever travel after it has been adjusted?
29 List the reasons for removing a brake drum.
30 Why is it necessary to clamp the wheel cylinder when the brake shoes are being removed?
31 What is the purpose of centralising the brake shoes when they are being replaced?
32 Why is brake dust a health hazard?
33 When is it necessary to remove a caliper?
34 How are the pads located in the caliper head?
35 When is it necessary to remove a disc?
36 Name two types of pressure differential switches.
37 What is the name of the procedure for removing air from the hydraulic system of the brakes?
38 Why must all the air be removed from the hydraulic system?
39 When the brakes are to be bled, what is so important about the brake shoe adjustment?
40 What indicates that all the air has been removed from the particular brake assembly that is being bled?
41 When is it necessary to remove a master cylinder?
42 What does a spongy brake pedal indicate?
43 Before replacing a brake booster, why is it necessary to check the setting of the output push-rod?
44 After the brake booster has been replaced, why is it necessary to check the 'free' travel of the brake pedal?

15

WHEELS AND TYRES

INTRODUCTION

Wheels and tyres have improved, along with so many other automotive products throughout the years. New materials, new technology and new production methods have all contributed to provide safer driving, better road holding properties during all weather and road conditions, and a degree of comfort which has not previously been experienced by the driver.

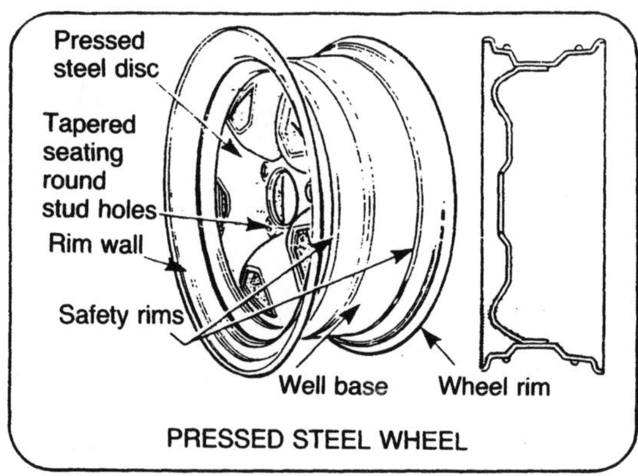

PRESSED STEEL WHEEL

Source: AGPS

WHEELS

The functions of the wheels are to:
1 support the weight of the vehicle;
2 transmit the driving and braking torques;
3 resist side thrusts due to cornering;
4 be easily balanced statically and dynamically;
5 be easily removed and fitted;
6 be easily cleaned.

CONSTRUCTION

Pressed steel wheel

The most common type of wheels on cars is the pressed steel disc type. It consists of pressed or stamped steel discs which are either welded or riveted to the outer circular rims. The rims have a drop centre or well which allows the tyre to be removed and fitted easily. They also have safety humps to which the beads of the tyres are fitted, the safety humps ensuring that the tyre beads do not move sideways when the tyre is under-inflated. The rim also has a hole in which the valve assembly is fitted.

The steel disc has a hole in the centre, to allow the wheel to fit over the hub assembly. Four or five tapered holes are spaced evenly around the disc, in which tapered mounting nuts centralise the wheel on to the hub assembly. There are also slots or holes cut into the disc, near the rim, their purpose being to allow circulating air to cool the brake drum or brake disc assembly.

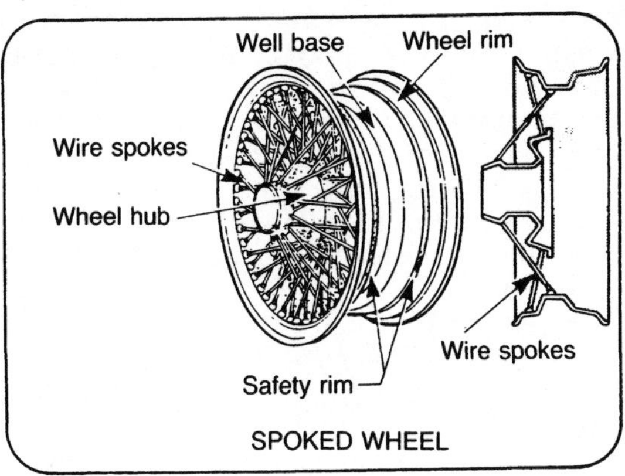

SPOKED WHEEL

Source: AGPS

Spoked type wheel

This is mainly used on some sports cars and motor cycles. It has the advantages of being strong, light-weight, and having one central locknut which allows the wheel to be changed quickly.

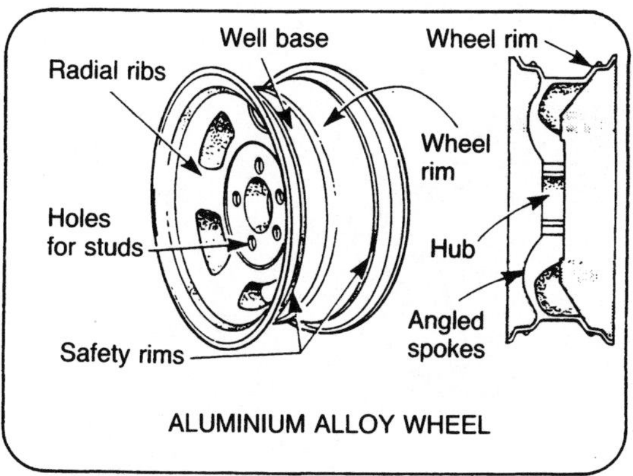

ALUMINIUM ALLOY WHEEL

Source: AGPS

Alloy wheel

This type of wheel is manufactured by a casting process, using aluminium and magnesium alloys. The light-weight construction is strong and the alloys, being good conductors of heat, disperse the heat generated by the tyres and brakes quicker than a steel wheel.

The alloy wheel was originally sold mainly as an 'after market product', however, more and more production cars are now being fitted with alloy wheels as standard equipment.

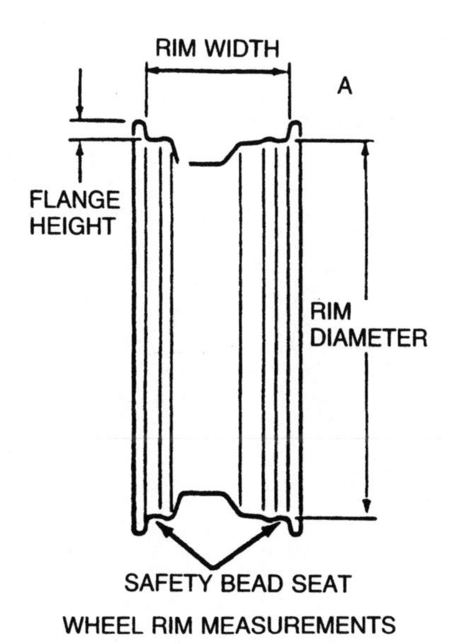

WHEEL RIM MEASUREMENTS

WHEEL SIZE

A wheel is identified by three measurements. The identification markings are usually stamped on the wheel rim. A typical marking of **6JJ×14** refers to:

6 = 6 inch wheel rim width
JJ = flange height
14 = 14 inch wheel rim diameter

Note: The flange height is identified by letters (J, K, JJ, etc.). These letters refer to specific heights, for example — a 'K' rim flange is 0.77 inch high.

TYRES

The functions of tyres are to:

1 Provide a cushion of air between the road and the wheels, to absorb shocks due to irregularities in the road surface. Tyres are also an important part of a vehicle's suspension system;
2 provide the road grip or adhesion between the wheels and the road for efficient, safe traction during acceleration, braking and cornering.

Construction

The most common types of tyres on the modern motor vehicle are the radial ply tyre and the cross ply (conventional).

The **radial ply** tyre can have one or more layers/plies of nylon, polyester or rayon textile casing cords, extending from bead to bead at approximately 90° to the centre line of the tyre. A belt, consisting of two or more plies of textile fibreglass or steel cord, runs around the tyre under the tread rubber. This belt stabilises the casing and gives this tyre good road handling characteristics.

The **cross ply** (diagonal ply) tyre consists of two or more layers/plies of textile casing cords, running diagonally from bead to bead at an angle of 35° to the tyre centre line. Alternate plies are at opposite angles, giving the tyre strength and assisting in steering stability. Nylon and polyester are the most commonly used materials.

These tyres have been known as 'conventional' tyres for many years.

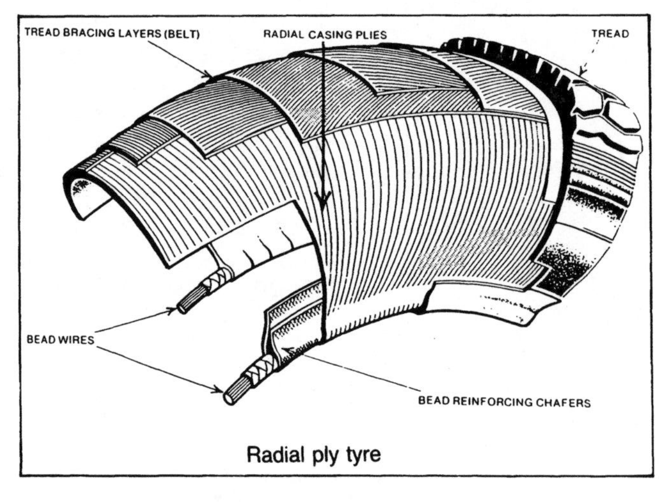

Radial ply tyre

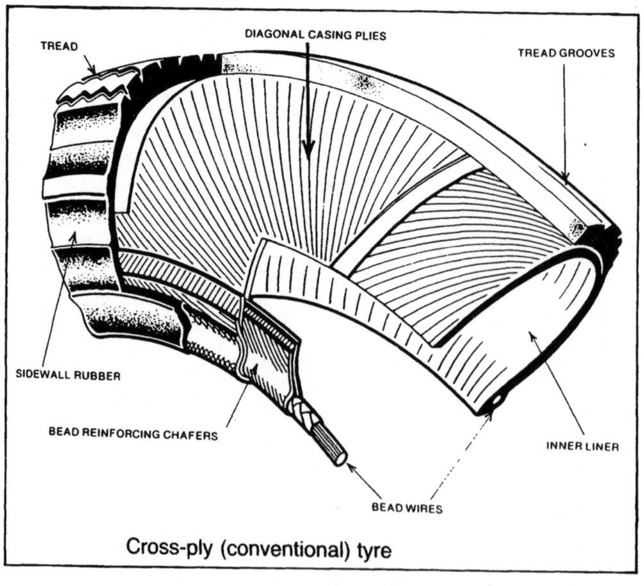

Cross-ply (conventional) tyre

Valves and valve caps

It is good practice to renew valve assemblies when new tubeless tyres are fitted, as valves deteriorate after a period of time and can leak, causing tyre damage due to underinflation.

Valve cores should be checked regularly for leaks, and properly sealing valve caps must be fitted to keep out dust and dirt which can cause valves to leak.

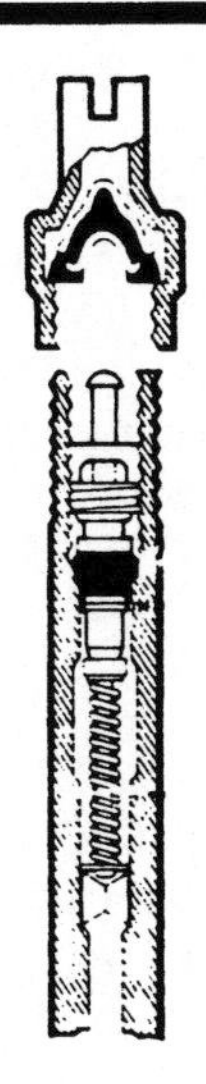

Valve core and cap

Tyre pressures

Pressure should be checked regularly and the air pressures maintained as recommended in the wheels and tyres specification placard, or in the car owner's manual. Air pressures should be checked when the tyres are cold (before the vehicle has run any appreciable distance).

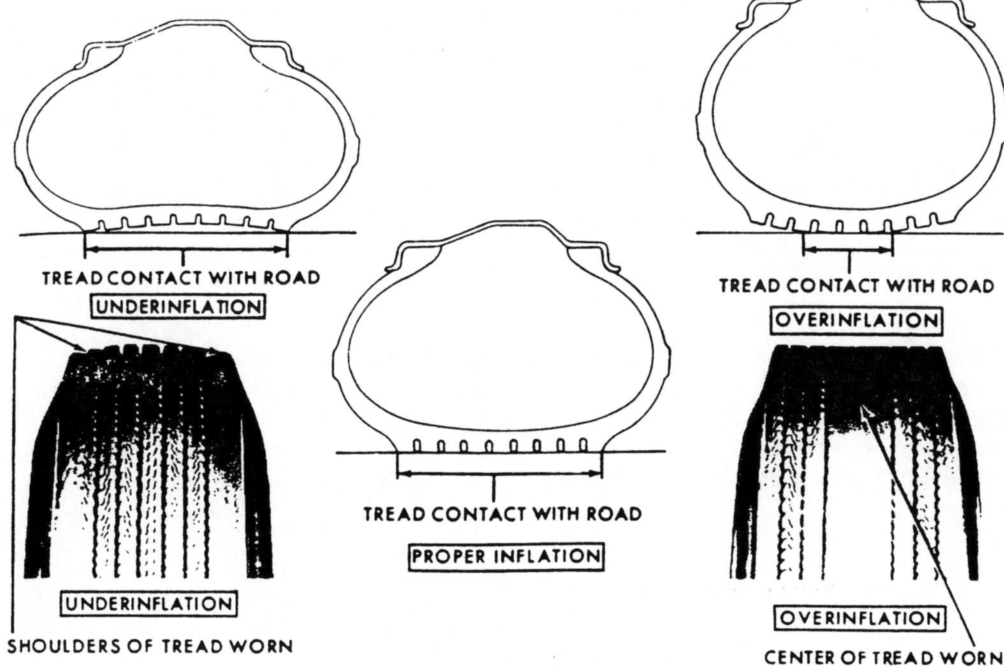

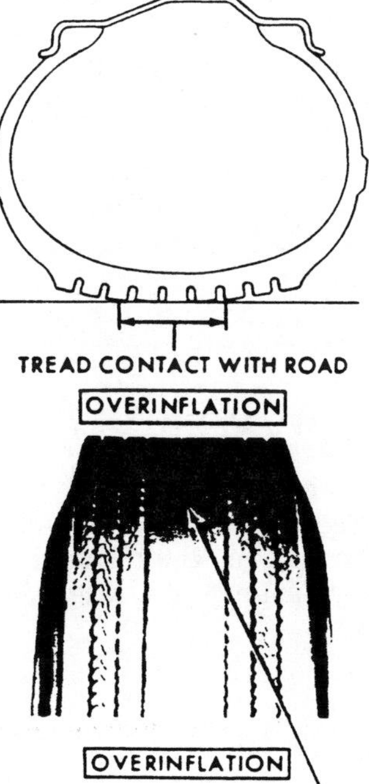

Tyre pressures

Tyre codes

The codes on the sidewalls of tyres provide information on the tyre size, construction, performance rating and load carrying capacity of the tyre. There are many variations to the system of marking the tyre sizes. The following are examples and explanations of the most common types of size markings currently being used in Australia.

1 **Numeric**. Example: 6.95 S 14
- 6.95 refers to the section width in inches
- S refers to speed category
- 14 refers to the wheel rim diameter in inches

Note: Speed category is based on the following: the tyre being in good condition, fitted correctly, at recommended air pressure, and specified for the vehicle.
The speed rating categories:

Speed Category Symbol	L	M	N	P	Q	R	S	T	U	H	V
Max. speed for which tyre is rated (km/h)	120	130	140	150	160	170	180	190	200	210	Over

2 **Alpha numeric**. Example: BR 78 S 14
- B refers to the load carrying capacity
- R denotes a radial ply tyre
- 78 refers to the profile ratio of the tyre
- S refers to speed category
- 14 refers to the wheel rim diameter in inches.

Note: Load carrying capacity markings. There are various markings on tyres to denote their maximum load carrying capacity. Some American tyres have alphabetic markings, **A, B, C, D,** which refer to the ply rating of the tyre (ply rating (PR) is an index of the tyre strength).
- **A** Two ply rating
- **B** Four ply rating
- **C** Six ply rating
- **D** Eight ply rating

Most tyres have their maximum loading marked in pounds or kilograms on their sidewalls.

LOAD INDICES AND EQUIVALENT MAXIMUM LOADS

Load Index	Maximum kg	Load (lbs)	Load Index	Maximum kg	Load (lbs)
66	300	(661)	83	487	(1074)
67	307	(667)	84	500	(1102)
68	315	(694)	85	515	(1135)
69	325	(716)	86	530	(1168)
70	335	(739)	87	545	(1202)
71	345	(761)	88	560	(1235)
72	355	(783)	89	580	(1279)
73	365	(805)	90	600	(1323)
74	375	(827)	91	615	(1356)
75	387	(853)	92	630	(1389)
76	400	(882)	93	650	(1433)
77	412	(908)	94	670	(1477)
78	425	(937)	95	690	(1521)
79	437	(963)	96	710	(1565)
80	450	(992)	97	730	(1609)
81	462	(1019)	98	750	(1653)
82	475	(1047)	99	775	(1709)

Other load markings are by a service index. For example, refer to the chart. The first two digits are the load index, which indicates the maximum load carrying capacity of the tyre according to the load index table.

Load Index	*Maximum load* kg (lbs)
80	462 (992)

Radial: refers to the construction type of the tyre. All radial tyres are marked with the letter **R**.

Profile ratio: refers to the percentage height to width ratio. For example, a tyre having a height of 175 mm and a width of 175 mm. As the height is the same as the width then the profile ratio would be 100 per cent. With reference to a tyre marking of BR 78 S 14, the profile ratio of 78 refers to the height of the tyre being 78 per cent of the width.

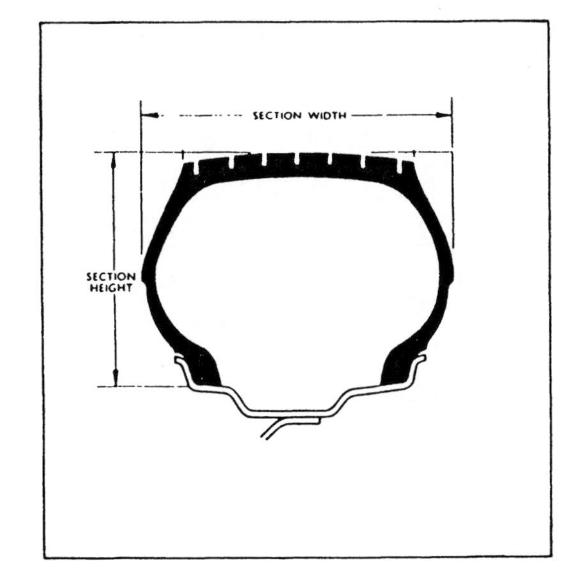

3 **Metric**. Example: 185/70 HR 14
- 185 refers to the section width in millimetres
- 70 refers to the profile ratio
- H refers to the speed category
- R denotes a radial ply tyre
- 14 refers to the wheel rim diameter in inches.

4 **P series metric**. Example: P 165/78 SR 13
- P refers to the tyre being suitable for passenger carrying vehicles
- 165 refers to the section width in millimetres
- 78 refers to the profile ratio
- S refers to the speed category
- R denotes a radial ply tyre
- 13 refers to the wheel rim diameter in inches

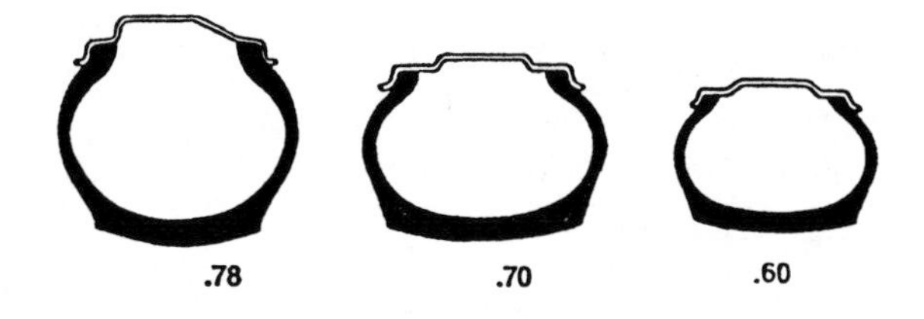

TYRE APPLICATION

Most tyre manufacturing companies produce a tyre application chart, which specifies the tyre size and construction type suitable for a specific wheel size and design. Examples of tyre application charts are shown.

Size	Name	Pattern	Design Rim Size	Section Width (ins.)	Overall Dia. (ins.)	Loaded Radius (ins.)	R.P.M.	(R.P.K.)
5.95L10	AIR-RIDE	C64	4JJ	5.63	19.7	9.0	1070	(665)
5.50L12	AIR-RIDE	C64	4JJ	5.40	21.8	10.2	965	(600)
6.00L12	AIR-RIDE	C64	4JJ	5.80	22.5	10.5	934	(580)
6.20L12	AIR-RIDE	C64	4½JJ	6.16	21.6	10.0	975	(606)
5.60-13	AIR-RIDE	C64	4JJ	5.80	23.8	11.2	880	(547)
5.90-13	AIR-RIDE	C64	4JJ	5.84	24.3	11.3	865	(537)
6.00-13	AIR-RIDE	C64	4½JJ	6.12	23.8	11.0	884	(549)
6.15L13	AIR-RIDE	C64	4½JJ	6.14	22.9	10.6	920	(572)
650-13	AIR-RIDE	C64	4½JJ	6.70	25.0	11.6	840	(522)
6.70-13	AIR-RIDE	C64	5JJ	6.98	25.7	12.0	816	(507)
7.00-13	AIR-RIDE	C64	5JJ	7.08	25.7	11.7	818	(508)
7.25S13	AIR-RIDE	C64	5JJ	7.40	25.9	11.9	811	(504)

ROADWORTHY

It is important to refer to the wheels and tyres selection placard on the vehicle, or in the service manual, before fitting new or replacement tyres and/or wheels. Fit only the specified tyre and wheel combinations to the vehicle, and maintain the air pressures as indicated in the wheel and tyre selection placard.

RECOMMENDED TYRE & RIM SIZES
& INFLATION PRESSURES kPa (P.S.I.) COLD

TYRE SIZE DESIGNATION	RIM CODE	NORMAL LOAD FRONT	REAR	MAX LOAD FRONT	REAR
CR78S14	5½JJ	200	200	210	250
P185/75SR14	6JJ	(28)	(28)	(30)	(36)
ER78S14					
ER70H14	5½JJ	180	180	200	220
P195/75SR14	6JJ	(26)	(26)	(28)	(32)
P215/65HR14					
P205/65HR15	7JJ				

The tyres fitted to this vehicle shall have a maximum load rating not less than 535 kg, or a load index of 87 and a speed category not less than 5.

FOR CONSISTENT HIGH SPEED OPERATION, COLD INFLATION PRESSURES MUST BE INCREASED BY 30 kPa (4PSI). FOR TRAILER TOWING AND ADDITIONAL TYRE CARE ADVICE REFER OWNERS MANUAL

84DA-1532-BA

The fitting of unspecified wheel and tyre combinations could result in the following:

1 The vehicle being unroadworthy.
2 The vehicle's insurance cover being void.
3 The tyres and wheels' warranty being void.
4 The person responsible for the fitting of the tyres and/or wheels being liable for prosecution.

Tyres with a higher load rating than specified on the wheels and tyres selection placard, or in-service manual, can be fitted to the vehicle. Refer to chapter nine for further information on wear limits and acceptability of wheels and tyres.

REMOVAL AND REPLACEMENT

A common type of truck wheel assembly consists of a flanged wheel, a detachable rim, and a split lock ring.

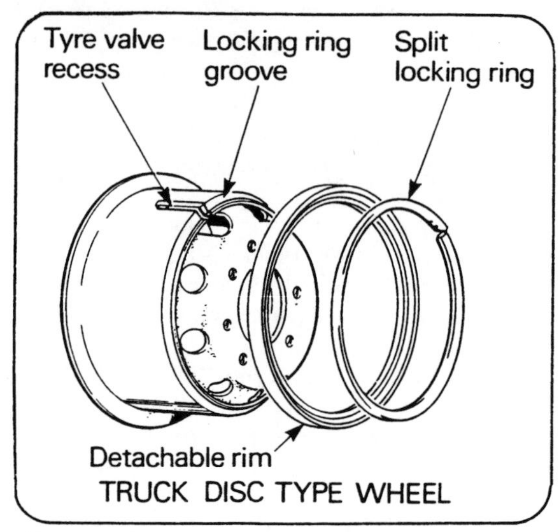

TRUCK DISC TYPE WHEEL

Source: AGPS

Truck tyre removal

1 Place the wheel, detachable ring upwards on a solid block of wood.
2 Remove the valve core to ensure that the tube is completely deflated.
3 Using a tyre lever or a 'bead breaking tool' move the bead of the tyre away from the detachable rim and split lock ring.

TRUCK TYRE BEAD-BREAKING TOOL

Source AGPS

4 Place a tyre lever in the slot at one end of the detachable split ring and prise it out of the wheel groove.
5 Gradually move the tyre lever around the wheel, at the same time removing the split lock ring and the detachable rim.

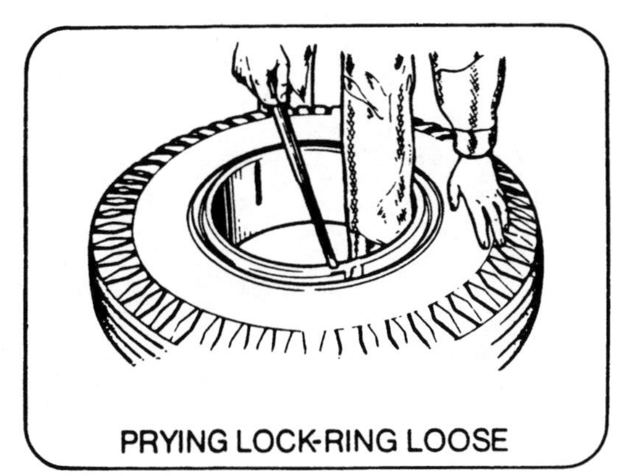

PRYING LOCK-RING LOOSE

Source: AGPS

6 When the detachable rim and split ring are removed, reverse the wheel on the block of wood.
7 The tyre can now be removed from the wheel. Be careful, do not damage the valve stem.

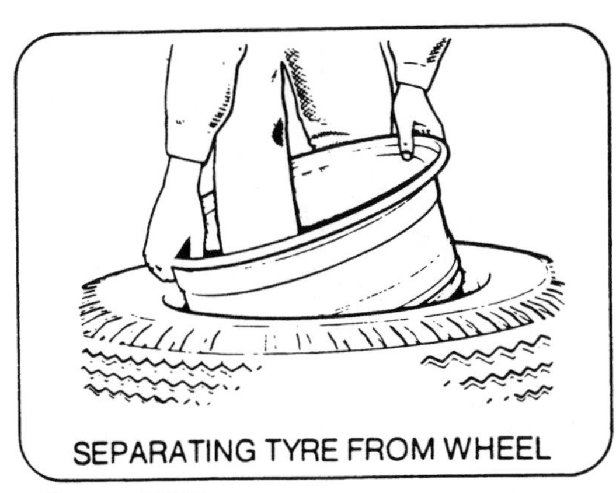

SEPARATING TYRE FROM WHEEL

Source: AGPS

8 Remove the tube protector flap and the tube from the tyre.

Truck tyre refitting

1 Slightly inflate the tube and insert it into the tyre. Refit the tube protector flap.
2 Place the wheel on the wooden block, place the tyre on the wheel and guide the valve stem into the wheel.

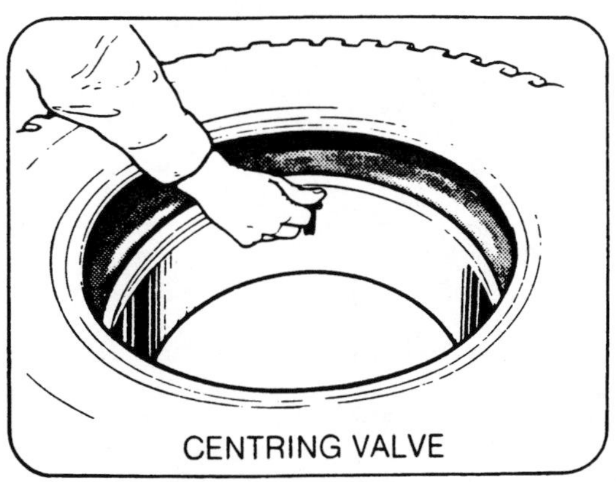

CENTRING VALVE

Source: AGPS

3 Fit the tyre onto the wheel.
4 Refit the detachable rim and split lock ring, by inserting the end of the split ring into the wheel groove opposite the valve stem, and by working around the wheel gradually forcing the split ring into the groove, finally tapping it into position.

WALKING LOCK-RING INTO GROOVE

Source: AGPS

5 Place the wheel and tyre assembly into an approved tyre inflation safety cage. Inflate the tyre to a low pressure (30 kPa) and check that the detachable split ring is seated correctly.

Caution: Extreme care must be taken during this procedure. Make sure the detachable split ring is seated correctly before continuing to inflate the tyre.

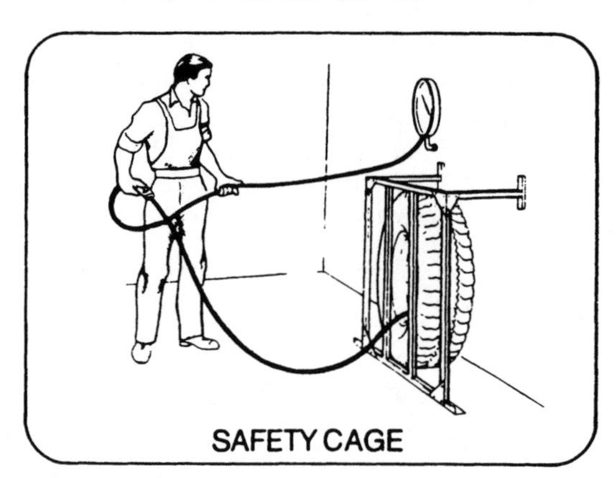
SAFETY CAGE

Source: AGPS

6 When the detachable split ring is seated properly, inflate the tyre to the recommended air pressure.
7 Check the valve for leaks before removing the tyre from the approved tyre inflation safety cage.

Tyre removal and replacement (passenger car)

It is recommended that an approved tyre changing machine be used when removing and fitting passenger car tyres. A common type of tyre changing machine (page 72), uses compressed air power to assist in bead breaking, levering the tyre from the wheel and refitting the tyre to the wheel.

If an approved tyre changing machine is not available, tyres can be removed and replaced by using tyre levers and a lot of care, to ensure that the tyre is not damaged during the removal and the refitting operation.

Removal

1 Remove the valve core to deflate the tyre.
2 Using the bead breaking tool, force the beads of the tyre away from the rim on both sides of the wheel.
3 When the beads on both sides of the wheel are 'broken', lubricate both sides of the tyre at the beads, with an approved 'tyre gel'. Then place the wheel on the ground and insert a tyre lever between the tyre bead and the wheel rim, at the valve side of the wheel.
4 Lever the tyre bead over the wheel rim, insert another tyre lever alongside the first one, and lever the tyre bead over the wheel rim.
5 Progressively move around the wheel levering the tyre bead from the wheel, until one side of the tyre is removed from the wheel rim.
6 The inner tube, if fitted, can be removed at this point.

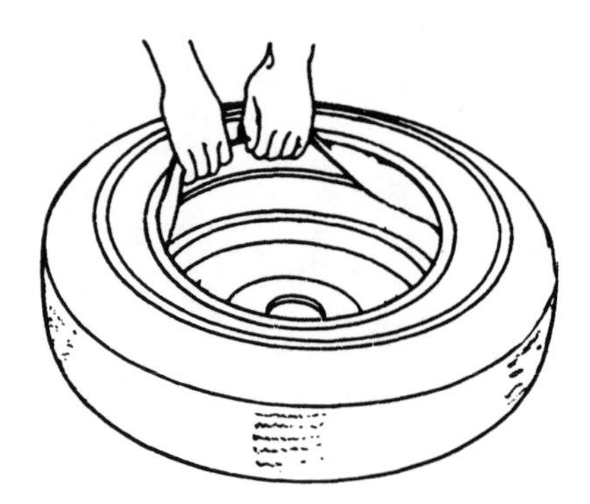
INNER TUBE REMOVAL

7 Stand the tyre upright with the tread on the floor and the wheel pushed down into the tyre. Insert a tyre lever at the top of the wheel and pry off the tyre from the wheel.

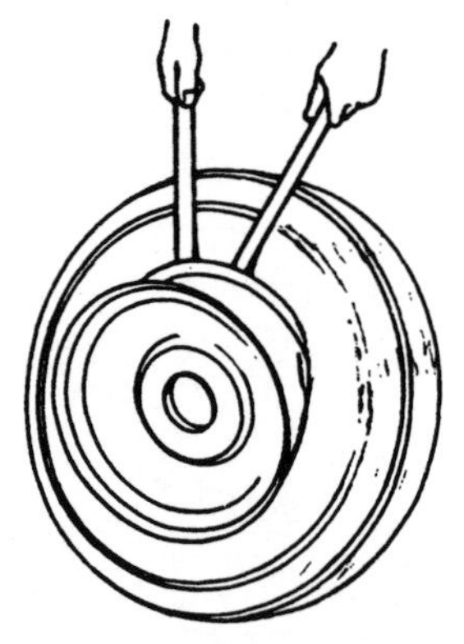
TYRE REMOVAL FROM WHEEL

8 Thoroughly clean the rims of the wheel with a wire brush or steel wool, to ensure that a tubeless tyre will seal correctly. A coating of vegetable oil or special tyre lubricant on the beads of the tyre will assist in the fitting of the tyre onto the wheel.

Refitting

1 Place the tyre over the wheel and carefully lever the bead of the tyre over the rim of the wheel.

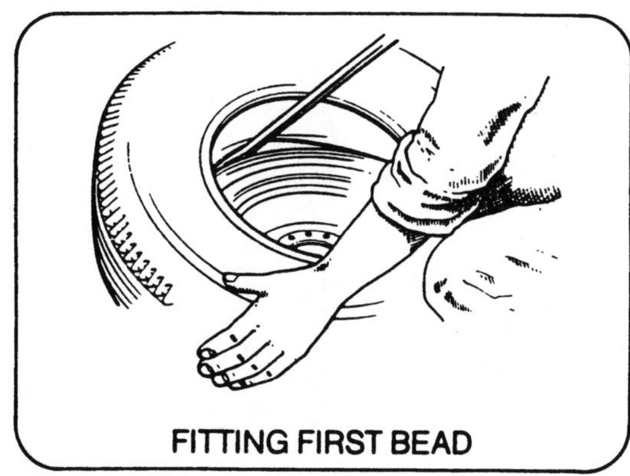

FITTING FIRST BEAD

Source: AGPS

2 Refit the inner tube, if required, at this point.
3 Lever the other bead of the tyre over the rim of the wheel.
4 Refit the valve core and, if an inner tube is fitted, inflate the tyre to the recommended air pressure.
5 For a tubeless tyre, difficulty may be experienced in inflating the tyre, and sealing the tyre beads onto the rim.
6 A special tubeless tyre mounting band may be required to force the bead onto the rims, to seal the tyre to the rim.

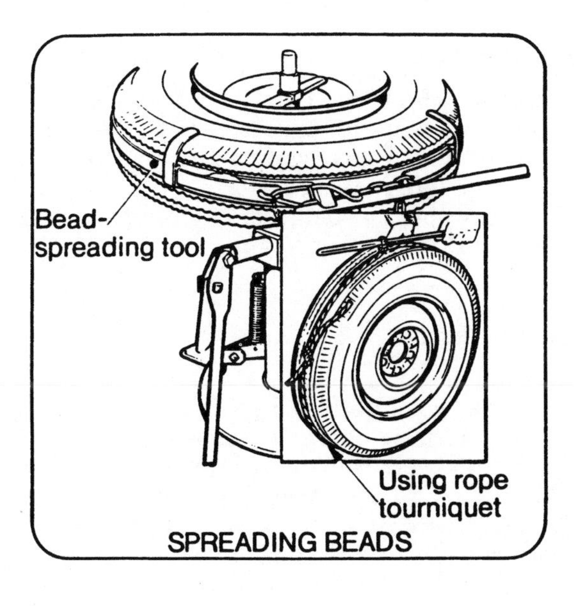

SPREADING BEADS

7 Inflate the tyre to a low pressure, remove the mounting band and inflate the tyre to the recommended pressure.
8 Check the valve and the tyre for air leaks by using soapy water.

TUBE REPAIR

Procedure for a common method of repairing a punctured tube:

1 Fit the valve into the tube and inflate the tube until the walls of the tube are round but not bulging.
2 Immerse the tube in a tank of water, or pour water over the tube to locate the puncture. The hole can be found by the air bubbles coming from the water.
3 Dry the tube thoroughly and buff the area around the hole in tube by using a wire brush or other suitable tool. Ensure that all the dust is removed from the area.
4 Apply a coating of tube repair cement around the previously buffed area. Remove the protective backing from the recommended type and size of tube patch. Place the patch over the punctured area when the cement is 'tacky'. Using a small roller tool, press the patch into the 'tacky' tube repair cement.
5 Retest the tube for any further leaks by immersing it in water.

TYRE REPAIR

1 Remove the tyre from the wheel and inspect it carefully inside and out.
2 Inspect all cuts, snags, and punctures for separation between the plies. Reject for repair any tyres having ply separation, damaged beads and/or extensive cuts.

3 Only punctures up to 6.5 mm (¼ inch), located in the central tread area, may be repaired.

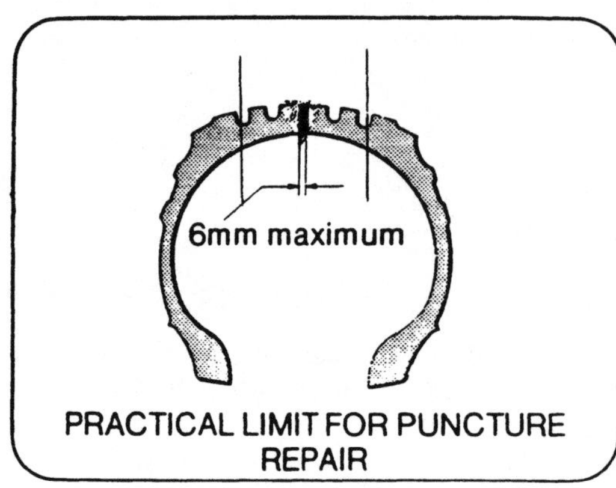

PRACTICAL LIMIT FOR PUNCTURE REPAIR

Source: AGPS

4 Thoroughly scrub the inside area of the tyre around the puncture with a solvent to remove any lubricant.
5 Buff the area around the puncture by using a buffing disc or wire brush and ensure that all the dust is removed from the inside of the tyre.
6 Apply a coating of tubeless tyre repair cement around the previously buffed puncture area.
7 Remove the protective backing from the recommended type and size of tyre patch.
8 When the repair cement is 'tacky', apply the patch to the inside of the tyre. By using a roller, press and roll the patch to ensure that it fits securely to the inside of the tyre.

Note: This repair method does not apply to steel-belted radial ply tyres. These tyres need to be plugged to stop moisture entering the hole to prevent the steel from rusting.

WHEEL BALANCE

Out-of-balance wheels and tyres cause steering vibrations, wheel wobble and irregular tyre wear. When a spinning wheel is in balance, the centrifugal forces acting around the wheel have no noticeable effects on the wheel, because the wheel is balanced and the forces are balanced. When a spinning wheel is out of balance (meaning one part of the wheel is heavier than other parts), the forces acting on it are different around the wheel. This causes wheel wobble, resulting in irregular tyre wear and steering problems.

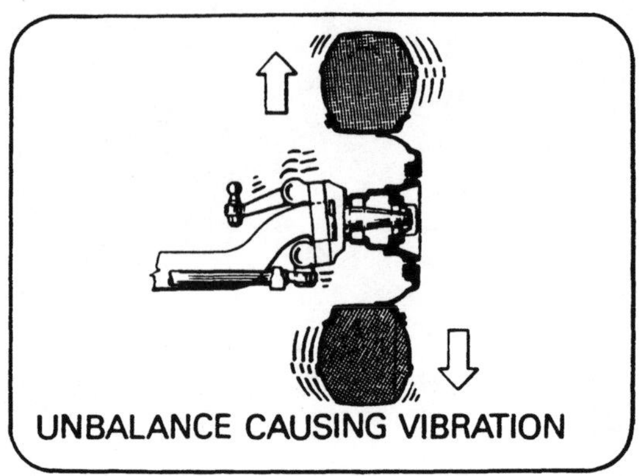

UNBALANCE CAUSING VIBRATION

Source: AGPS

There are two main types of out-of-balance wheels, namely static and dynamic.

Pre-checks

1 Before checking a wheel for static or dynamic balance, ensure that the wheel is clean, free of any mud or dirt and that the tyre air pressure is correct.
2 Ensure that the balancer is on a clean level floor.

Static balance

Static balance is when the mass of the wheel and tyre are uniformly distributed around the centre. A static balanced wheel, when placed on a hub, will remain in any position it is placed. A wheel which is out of static balance, when placed on a hub, will rotate until the heaviest part of the wheel is nearest the ground.

CHECKING STATIC BALANCE

A common type of static wheel balancer consists of a balance head containing a circular bubble gauge, which pivots on a central post attached to a base plate. A foot-operated lever engages the pivot with the balance head. The pedal is activated

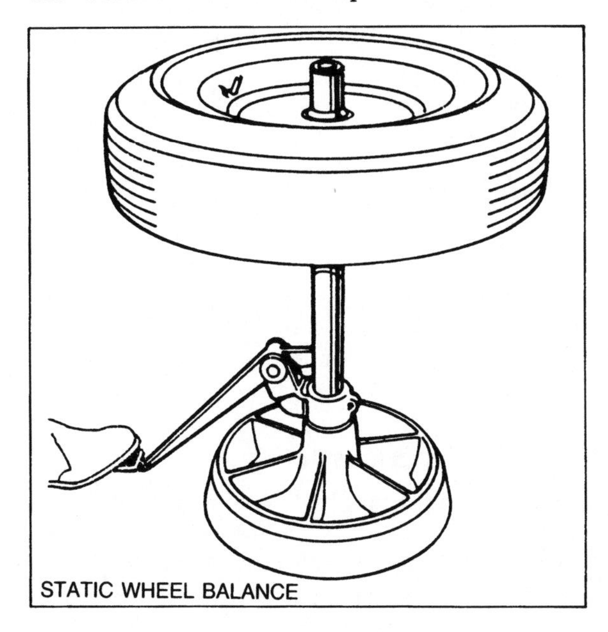

STATIC WHEEL BALANCE

during the actual balance check, and the pivot is protected when the pedal lever is disengaged.

PROCEDURE

1 Place the wheel on the balance head and depress the foot lever. This engages the balance head with the pivot.
2 If the wheel is in static balance, the bubble will be in the centre of the small circle on the bubble gauge.
3 If the wheel is out of static balance the bubble will not be in the centre of the small circle on the bubble gauge. Add balance weights to the wheel until the bubble is in the centre of the small circle on the bubble gauge.

Note: A wheel out of static balance will cause the wheels to have a hopping effect whilst travelling along a road.

Dynamic balance

A wheel can be in static balance and out of dynamic balance. For example, if a wheel has a heavy spot on the inside of the tyre at the position of 12 o'clock and it has a similar heavy spot on the outside of the tyre at the position of 6 o'clock, the wheel would be in static balance. However, when this wheel is spinning, centrifugal force would be acting on the heavy spot at 12 o'clock, pushing the wheel inward and on the heavy spot at 6 o'clock pushing the wheel outward. This would cause the wheel to wobble from the side to side.

CHECKING DYNAMIC BALANCE

There are various types of dynamic balance machines. Most modern types are electronically controlled, and can be programmed for wheel balance accuracy to within one gram, however a 'round off' balance of five grams is acceptable on most wheels.

DYNAMIC WHEEL BALANCER

PROCEDURE

The procedure for checking the dynamic balance of wheels depends on the specific dynamic balance machine being used. Most of these machines correct static and dynamic balance in one operation. It is important to follow the operating instructions as specified by the manufacturer of the wheel balancing machine being used. The drawing shows one type of wheel balancing machine.

Note: A wheel out of dynamic balance will have a weaving or zigzag effect whilst travelling along the road.

TYRE MATCHING

All vehicles should have five matching wheels and tyres of a size and type as specified in the vehicle's wheels and tyres specification placard, or service manual. **Do not** mix tyres of different sizes, profile ratios or casing construction. Pairs of tyres of the same construction but with different profile ratios can be fitted to a vehicle, providing the lower profile tyres are fitted to the rear.

It is not recommended that radial tyres be mixed with bias ply (conventional) tyres on the same axle. The reason for this is that radial tyres have different road holding characteristics to conventional tyres. Mixing the two types may cause steering and handling problems.

With reference to the following diagram, the forces acting on tyres due to cornering are different on conventional tyres to the same forces acting on radial tyres.

A radial tyre, having flexible walls, bends easily in the direction of the force, allowing the tread to remain on the road surface.

The walls on a bias ply tyre are not as flexible as the walls on a radial ply tyre. Thus on

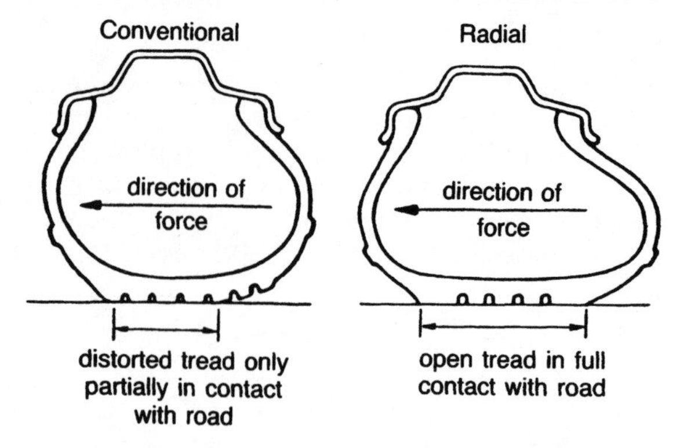

Distortion of a conventional and a radial ply tyre, when the car is on a curve

concerning, the tread on the conventional tyre tends to lift the tyre from the road surface, reducing traction and possibly causing sideways skidding.

Precautions on tyre matching

The following chart will assist you in tyre matching, by reading across the chart for front wheel matching and down the chart for rear wheel matching.

PASSENGER CAR TYRE MATCHING

FRONT WHEELS	REAR WHEELS: 70 & 80+ Series Bias	70 Series Bias	78 Series Bias-Belted	70 Series Bias-Belted	60 Series Bias-Belted	78 Series Radial	70 Series Radial	60 Series Radial
78 & 80+ Series Bias	Yes	Permitted	Permitted	Permitted	Permitted	Permitted	Permitted	Permitted
70 Series Bias	*	Yes	*	Permitted	Permitted	No	Permitted	Permitted
78 Series Bias-Belted	*	Permitted	Yes	Permitted	Permitted	Permitted	Permitted	Permitted
70 Series Bias-Belted	No	*	*	Yes	Permitted	No	Permitted	Permitted
60 Series Bias-Belted	No	No	No	No	Yes	No	No	Permitted
78 Series Radial	No	No	No	No	No	Yes	Permitted	Permitted
70 Series Radial	No	No	No	No	No	*	Yes	Permitted
60 Series Radial	No	No	No	No	No	No	No	Yes

CHAPTER REVIEW

1 What is the function of the wheels on a motor vehicle?
2 What is the function of the drop centre or well on a wheel?
3 Explain the purpose of the safety hump on a wheel.
4 For what reasons are slots or holes cut in a wheel disc?
5 Name three common types of wheels.
6 State the advantages of a spoked wheel.
7 State the advantages of an alloy wheel.
8 What are the three measurements used to identify the size of a wheel?
9 Explain the 'JJ' marking on a wheel.
10 State the function of a tyre.
11 Explain the main construction differences of a radial ply tyre and a cross ply tyre.
12 Name the two most common materials used in the manufacturing of a cross ply tyre.
13 Why should valve assemblies be replaced when new tubeless tyres are fitted?
14 What type of tyre tread wear would indicate an under-inflated tyre?
15 Explain the meanings of the following tyre codes:
a) 6.95 L 14
b) BR 78 S 13
c) 185/70 HR 14
d) P 165/78 SR 13
16 What is the meaning of ply rating in reference to a tyre?
17 What is the tyre and wheel specification placard?
18 Explain the possible results if an unspecified wheel and tyre combination is fitted to a vehicle.
19 What is the function of a bead breaking tool?
20 Why should a tyre lubricant be used when fitting a tyre to a wheel?
21 Explain the procedure for repairing a punctured tube.
22 Explain the procedure for repairing a tubeless tyre.
23 Explain centrifugal force.
24 Name the two types of wheel balance.
25 Explain a simple method of checking a wheel's static balance.
26 Explain the effect of a static out-of-balance wheel.
27 Is it recommended to fit tyres having different profile ratios on the same axle?
28 Explain the meaning of the term profile ratio, with reference to a tyre.
29 Why should radial ply and cross ply tyres never be fitted on the same axle?
30 What combination is permissible when fitting a pair of cross ply tyres and a pair of radial ply tyres to a vehicle?

16

STEERING AND SUSPENSION

INTRODUCTION

If all road surfaces were as smooth as a billard table the task of controlling ride comfort for the occupants and maintaining steering control would be very easy. Unfortunately, our road systems are far from perfect and unless the vehicle has adequate body strength, an efficient suspension system and steering system capable of negotiating many obstacles the vehicle occupants will suffer a ride which is both uncomfortable and dangerous.

The steering system is designed to permit the driver to guide the vehicle along a roadway and around corners without causing excessive wear on the tyres. Generally, this system is attached to the front wheels of the vehicle.

The suspension is designed to dampen the road bumps before they are transferred to the chassis of the vehicle. Each road wheel is fitted with a suspension system. The same type of suspension system must be used on the same axle, for example, both front wheels must be fitted with the same system.

FRAMES

The vehicle body can be constructed in one of two basic methods. Either method, when applied to a passenger car, produces a vehicle which can be similar in external appearances but very different in the manner in which the body strength is obtained. These two methods are the:

1 base-frame or chassis.
2 mono-construction.

A **chassis** is a frame to which all other components are attached, and it is currently used on heavy cars and commercial vehicles. It consists of two 'U' shaped steel members held parallel to one another by a number of cross rails. The joint between each rail and side member is reinforced to provide strength and rigidity. Additional reinforcement is placed at each suspension connection point to enable the frame (chassis) to carry the design load but permit a degree of flexibility along its length.

The final shape will vary slightly depending on the vehicle design, but each chassis will

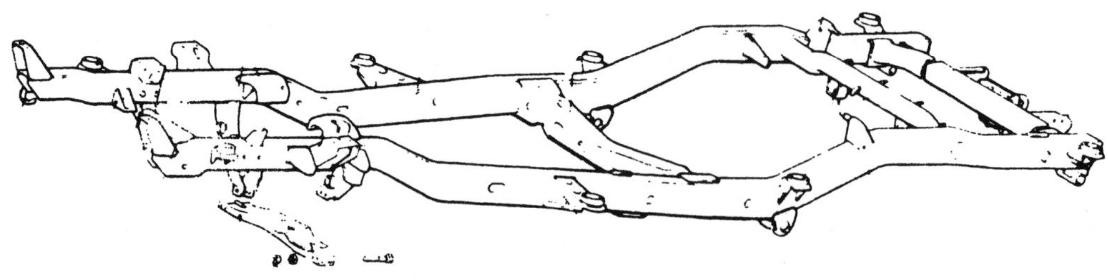

CHASSIS
Source: Mitsubishi Pajero manual 1983

be similar in appearance to the one depicted in the diagram.

This type of construction provides excellent strength and a base for different body styles. For example, a chassis originally designed to accept a sedan body may, with modifications, be suitable for a utility body. The body is attached to the side members through rubberised mountings.

Mono-construction refers to the method of combining the frame and the body into a single unit. In this type of construction there are no frame rails but the body strength and rigidity is maintained by welding all the structural panels and members together. Reinforced sections for load carrying are welded directly into the structure, but a number of manufacturers prefer to include a sub-frame in the design. These sub-frames are of particular value when the basic body design is intended to have a long model run over many years. By designing a front and rear sub-frame in the vehicle a manufacturer can introduce design changes in suspension and/or engine-transmission combinations to suit different countries without variation to the basic body floor pan or engine bay body members. This feature is useful when a vehicle is supplied with options of four, six and eight-cylinder engines, automatic or manual transmission and a variety of suspension packages.

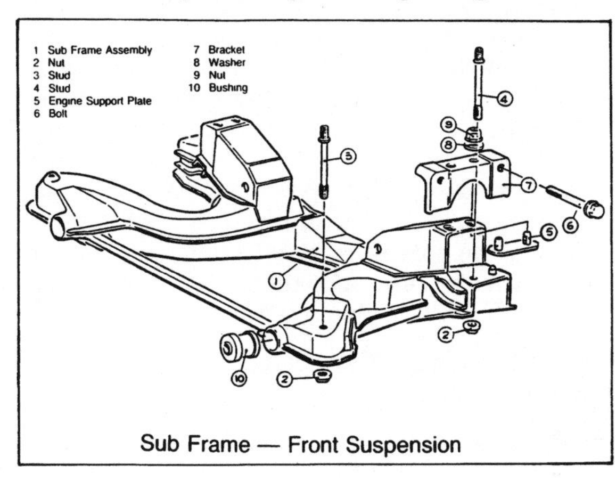

Sub Frame — Front Suspension

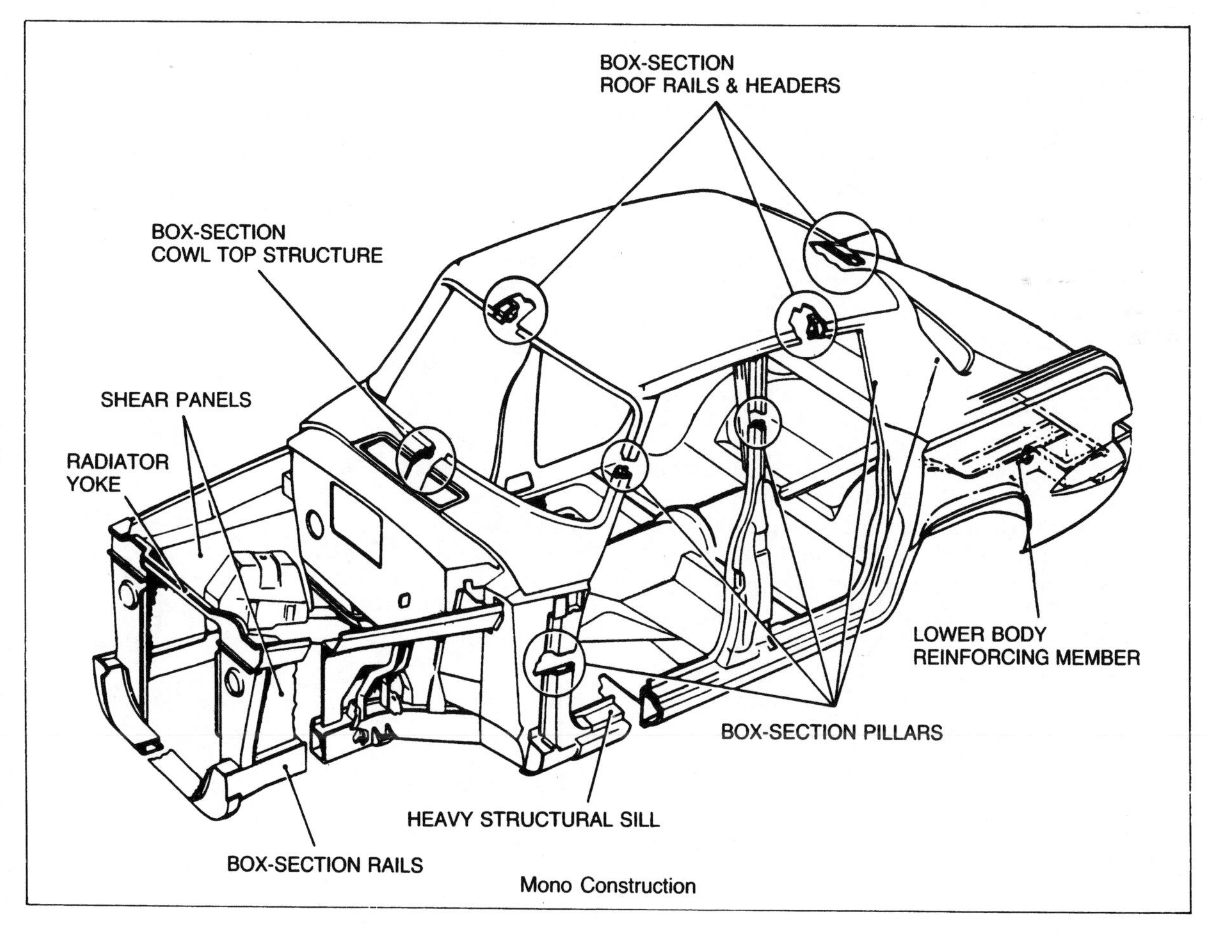

Mono Construction

SPRING TYPES

The spring types used in motor vehicles are:
- coil;
- leaf;
- hydrolastic;
- torsion bar.

Coil springs are the most common spring type used in the modern motor car. They are compact, robust and virtually maintenance-free. Two types of motor vehicle coil springs are:

1 Conventional. This type has a constant spring rate. For example, if a 20 kg load caused the spring length to decrease by 10 mm, then a 40 kg load would cause the spring length to decease by 20 mm.

Note: All coils in the spring have the same diameter and thickness.

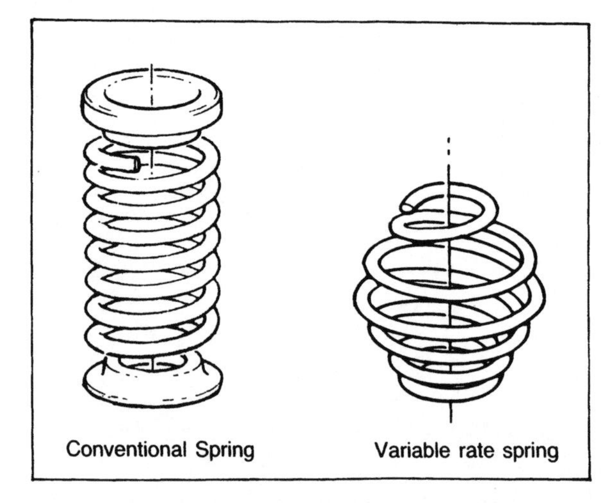

2 Variable rate. This type has different diameter and thickness coils which causes the spring rate to vary with the applied load. For example, if a 20 kg load caused the spring length to decrease by 10 mm, it does not mean that a 40 kg load will cause the spring length to decrease by 20 mm.

Both types of coil springs are located in the suspension between a wishbone and a special reinforced section of the frame.

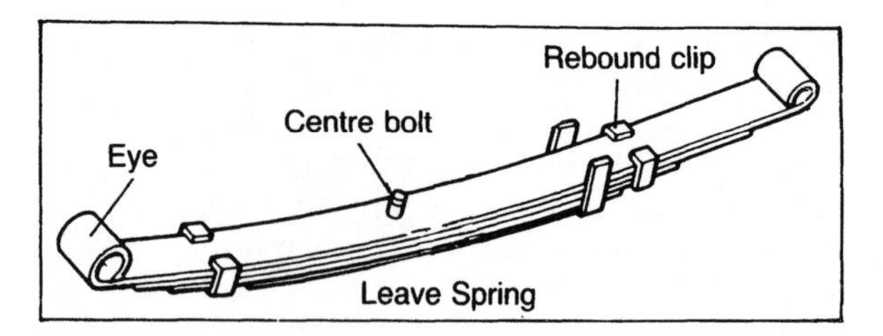

Leaf springs consist of a number of strips of spring steel called leaves held together by centre bolt and clips. One leaf, generally the top leaf, is fitted with an eye at each end to form a main leaf. The eyes in the main leaf are normally fitted with rubber bushes to allow the spring to be attached to the frame with shackle bolts. The other leaves, which decrease in length as they are positioned further from the main leaf, are held in alignment with the main leaf by rebound clips.

The complete assembly is located on the axle at the centre bolt and clamped in position by two 'U' bolts.

Hydrolastic springs were mainly used in a limited range of British cars. This type of suspension system consists of four combined rubber block and hydraulic damper units. The front and rear units, on each side of the vehicle, are connected together by hydraulic pipes and hoses. The system is designed to keep the vehicle height constant when passing over bumps and potholes.

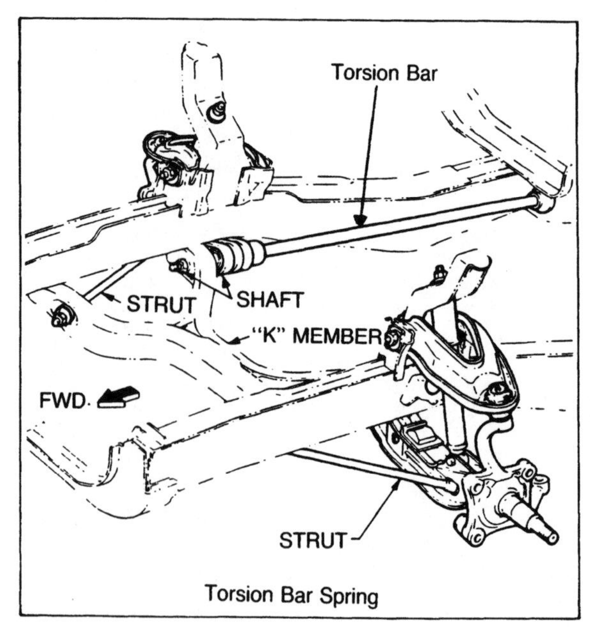

Torsion Bar Spring

Torsion bar springs are spring steel rods which have a spline or a polygon shape formed at each end. One end of the rod is connected to a wishbone on the suspension and the other end is connected to the frame via a mounting block. In this type of spring, the tension on the torsion bar can be adjusted to maintain the correct body ride height.

SUSPENSION TYPES (FRONT)

There are a number of front suspension types, including the following:

- single and twin 'I' beam axle;
- independent lateral arm;
- independent trailing arm;
- MacPherson strut.

Single beam axles consist of a steel beam forged to an 'I' shape with a cylindrical hole cast in each outer end to locate and position the king pins. Flat drilled surfaces are formed at an equal distance from each end to provide mounting pads for leaf springs. The springs are located on these pads by their centre bolts and secured by 'U' bolts.
Note: The king pin connects the stub axle to the 'I' beam.

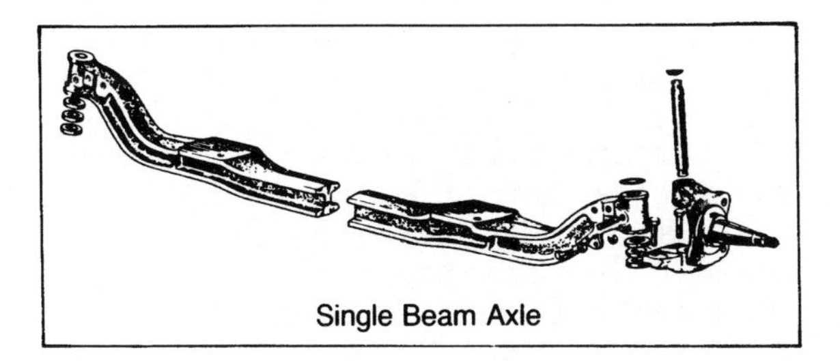

Single Beam Axle

Twin 'I' beam axles have been used on a limited range of light commercial vehicles. This type of axle, used in conjunction with coil springs, achieves the desired effect of having each front wheel independently sprung, and retains the strength of a beam axle.

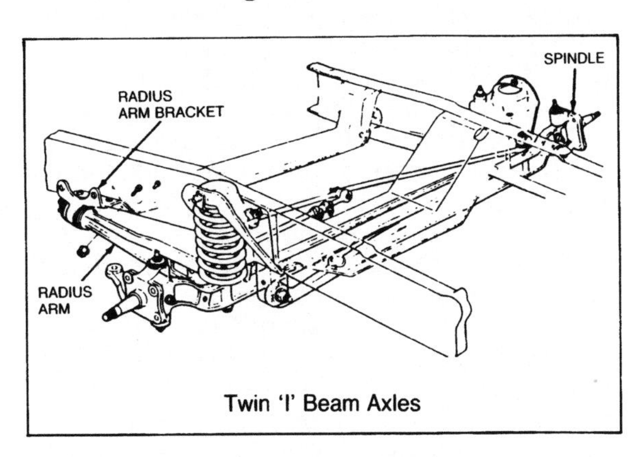

Twin 'I' Beam Axles

Independent lateral arm suspension systems are called 'independent' because each front wheel is sprung independently of the other. The suspension arms are attached at 90° to the frame. The diagram shows one type of independent lateral arm front suspension system, which consists of a coil spring, upper and lower control arms (wishbones), a stub axle and a shock absorber located inside the coil spring.

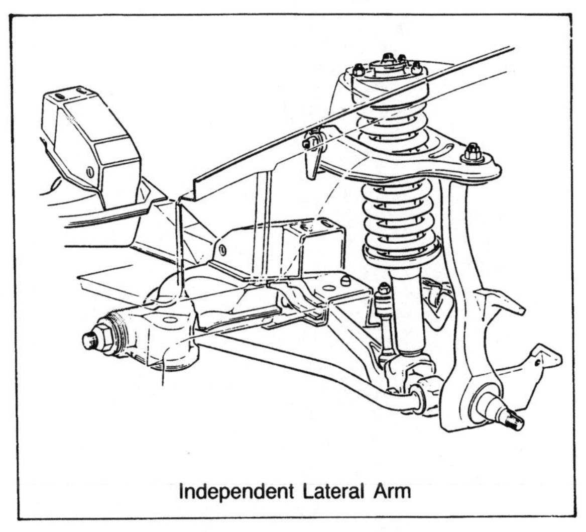

Independent Lateral Arm

Source: Mitsubishi Sigma GJ series manual

Independent trailing arm suspensions generally contain two torsion bars which provide independent springing for the trailing arms, on which the wheel spindles are mounted. The suspension arms are attached parallel to the frame so the wheel spindle mounting point 'trails' behind the arm pivot.

MacPherson strut suspension units consist of a telescopic shock absorber and a coil spring assembly. The wheel spindle and the shock absorber outer casing are either bolted or welded together to form a single unit. The upper end of the assembly is attached to a strengthened section of the frame by a flange containing a rubber mounted thrust bearing. The wheel spindle and a ball joint are located at the lower end of the strut. The ball joint allows the assembly to turn and connects the strut lower section to a lateral suspension arm which is attached by a pivot to the frame. Forward movement of the lateral arm is controlled by either a sway bar or strut rods. This type of suspension unit is popular in many small cars, because it is compact and simple in design.

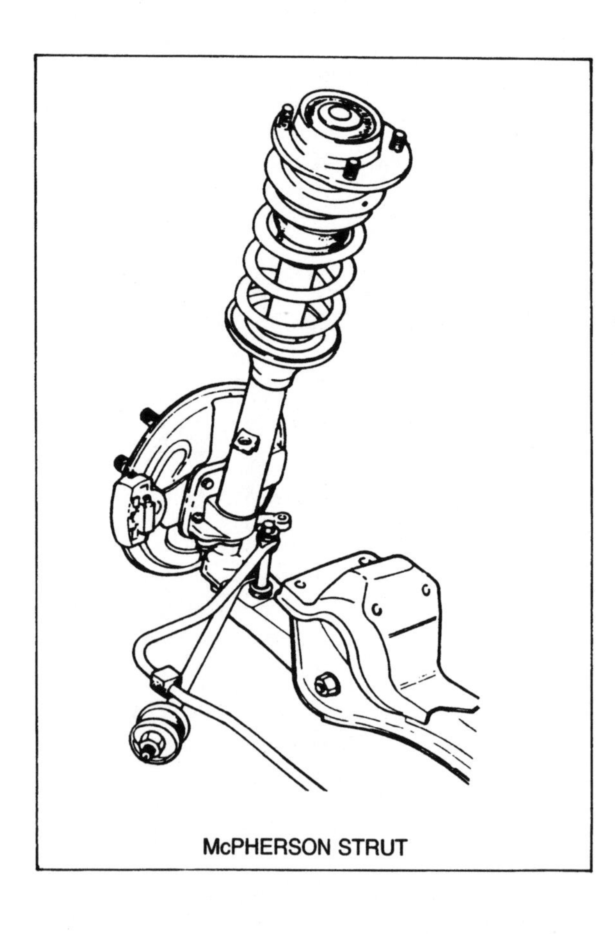

McPHERSON STRUT

SUSPENSION TYPES (REAR)

Rear suspension development, for mass production vehicles, has lagged behind the changes in front suspension design. The combination, which has continued for several decades, of semi-elliptic springs, a 'live' rear axle and hydraulic shock absorbers is inexpensive to manufacture and provides remarkably good insulation of the passenger compartment from rear axle and road noise. Despite these advantages, the arrangement does suffer from two deficiencies: the axle is not positively located and the weight of the axle housing is not carried by the springs (unsprung weight).

The axle location problem is particularly noticeable when the vehicle is fitted with long, 'soft', semi-elliptic leaf springs. When the vehicle is subjected to heavy acceleration or braking forces it is possible for the forward section of the spring to flex into a wave or flat 'S' shape. The curves in the spring reduce the distance between the front spring eye mounting and the axle location point. The axle is then pulled slightly forward, on the spring under load, and does not maintain precise location at 90° to the vehicle centre-line. The rear axle then steers the rear of the vehicle on a different angle to the front and directional stability is affected.

Independent suspension systems reduce unsprung weight to a minimum. However, the increase in rear suspension and axle complexity has been rejected by a number of manufacturers who have retained the 'live' rear axle but have achieved positive axle location and improved ride control. Because of this success the 'live' rear axle is likely to remain in a number of front engine, rear wheel drive cars for several years into the future.

The introduction of front wheel drive vehicles has allowed the removal of the heavy axle housing from the rear suspension. This change reduces the unsprung weight problem, completely removes concerns related to axle or drive shaft deflection angles during suspension movement and allows the designer to introduce new suspension systems.

When studying the suspension systems in this section look carefully at how axle location is maintained for both lateral and twisting movements and how independent suspension units reduce unsprung weight.

Solid (live) axle suspension

Semi-elliptic leaf springs used with a solid or 'live' rear axle housing are mounted longitudinally and are clamped in position by 'U' bolts. The forward end of each spring is attached to the body by a rubber bush and steel pin. The rear attachment consists of a swinging shackle containing rubber bushes. The shackle allows for changes in spring length caused by loading or suspension movement. The springs can consist of a single leaf, the more common four or five leaves or, when used in panel vans/utilities, a second spring fitted below the main unit and only effective under heavy loads.

Two double acting hydraulic shock absorbers are usually fitted to the spring saddle plate.

Improvements which have been fitted to some leaf spring equipped vehicles include sway bars, Panhard rod for lateral control, upper and low control rods which prevent axle housing movement. These types of improvements are found on performance variants of popular sedans.

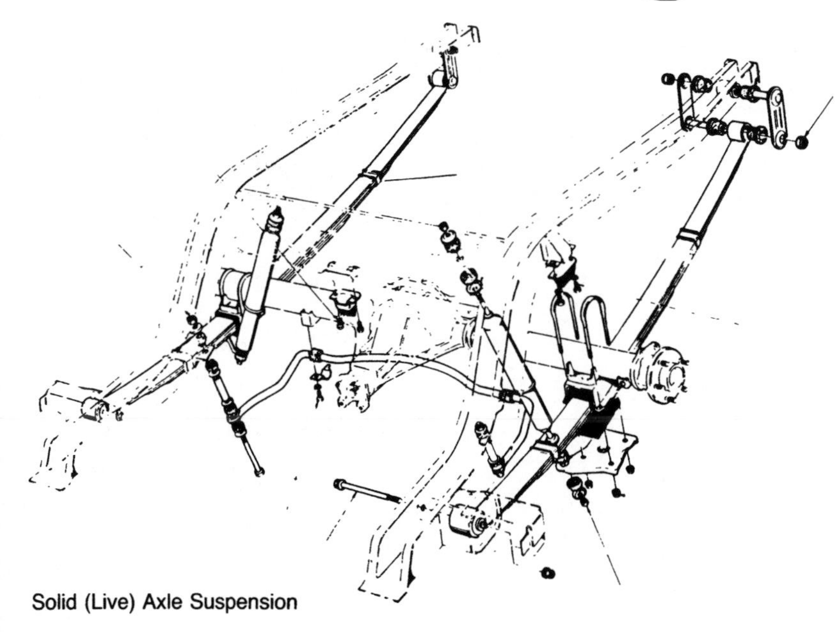

Solid (Live) Axle Suspension

Four link systems combine a 'live' rear axle housing with a pair of coil springs. The axle housing is located by two longitudinally mounted lower arms and two angled upper arms. Rubber bushes are fitted at both ends of the locating arms to prevent sound transfer. The spring deflection problem found on leaf spring systems is eliminated and acceleration and braking forces are effectively controlled.

Ride control can be improved by using variable rate springs and a sway bar. Two double acting shock absorbers complete the standard system.

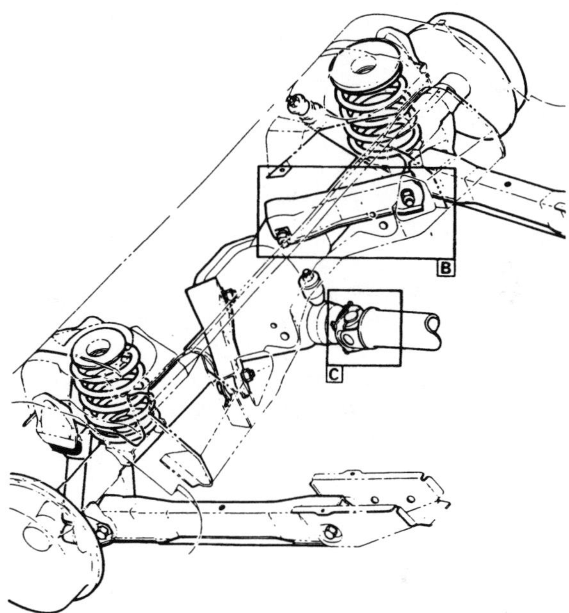

Four Link Suspension

Five link systems are basically a four link system improved by the addition of a Panhard rod and revised upper arm mounting angles. The Panhard rod improves the lateral stability of the axle although suspension movement does still cause slight lateral movement.

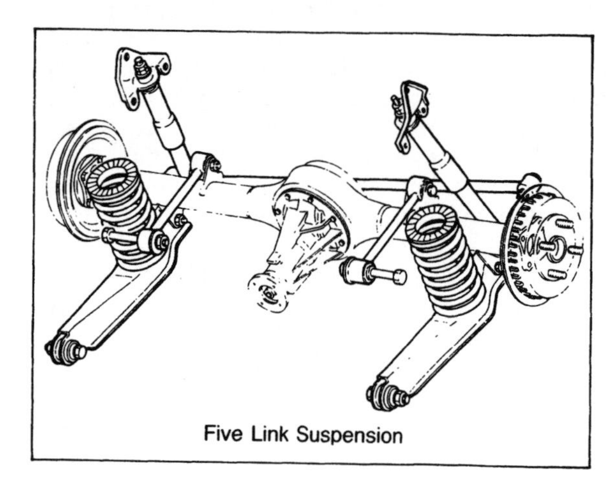

Five Link Suspension

Six link (Watts) systems have two upper and two lower trailing arms which attach the 'live' axle housing to the body. The arms are of unequal length and axle movement in response to braking, acceleration forces and suspension deflection are precisely controlled. To prevent lateral movement an additional two arms (Watts linkage) are angled from the body to a pivot lever on the rear of the axle housing. Each of the arms contains rubber bushes for sound isolation.

The addition of shock absorbers, a sway bar and variable rate coil springs completes the typical installation of this type of system.

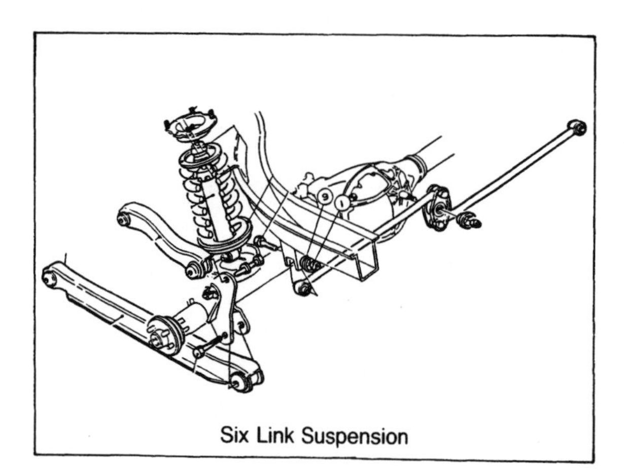

Six Link Suspension

Independent rear suspensions

The design improvements in solid axle suspension systems solved most of the problems, however, the axle housing remained as unsprung weight. Independent rear suspensions provide an answer to this problem by attaching the final drive housing to the body floor. The heavy final drive unit is now carried by the vehicle springs and is classified as part of the sprung weight. This relocation introduced complexity to the rear axles, as the shaft must now move with suspension deflection and maintain effective drive angles without vibration. The axle shafts must therefore contain slip joints and constant velocity joints to compensate for the change in drive distance and angle.

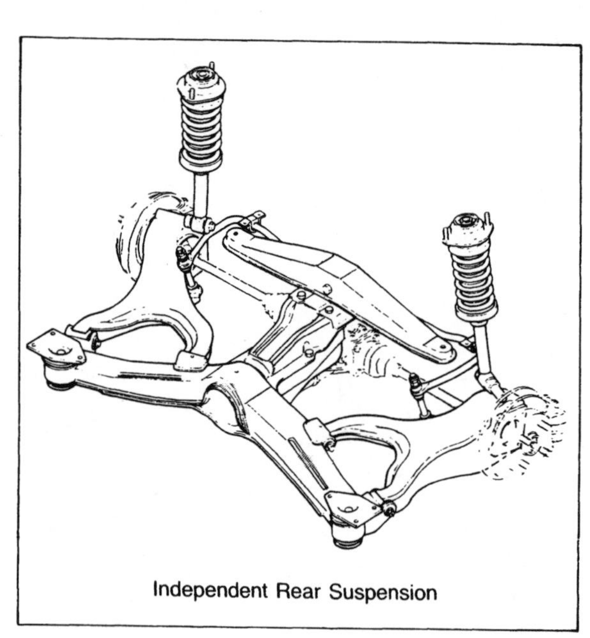

Independent Rear Suspension

Separate coil springs and shock absorbers are the most popular method of springing, however, many vehicles are fitted with an integral spring/shock absorber unit similar to a MacPherson front suspension unit. A sway bar is normally included as a standard component.

Torsion bars have also been used as the major spring in different models and, while not as popular in current vehicles, the concept has evolved into the rear suspension of many front wheel drive cars as a torsion beam or axle.

Transverse beam trailing link suspensions are very popular on front wheel drive vehicles. Generally, this arrangement consists of a transverse torsion beam mounted into the vehicle with a trailing arm wheel carrier fitted to each end of the beam. The spring, beam mounting points and shock absorber locations vary but the concept of a transverse torsion beam is common.

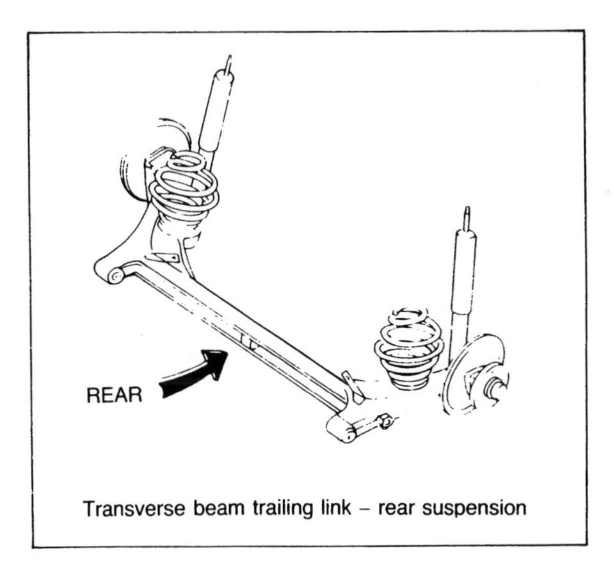

Transverse beam trailing link – rear suspension

Lateral/trailing link suspensions consist of a telescopic strut to which an axle stub is bolted. The strut is located by two transverse lower links and fore and aft movement is prevented by a longitudinal link between the lower strut and the body.

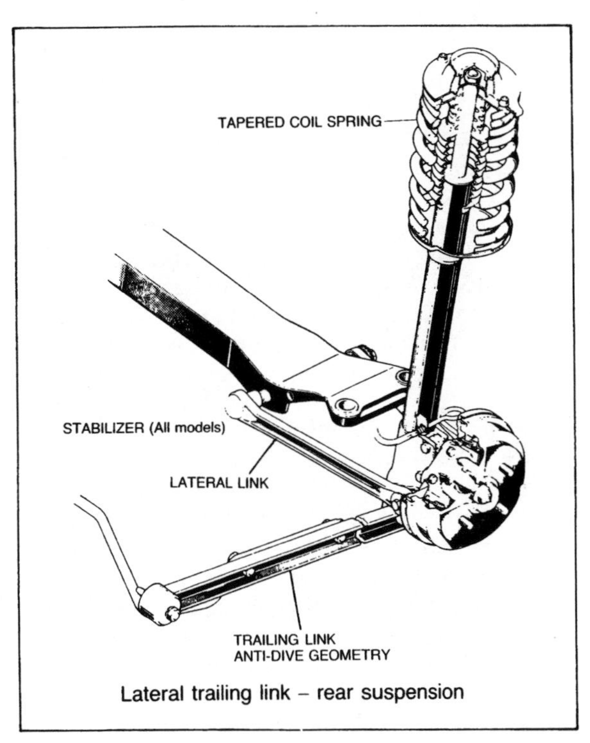

Lateral trailing link – rear suspension

Trailing arm suspensions have the lower arms pivoted at 90° to the vehicle centre line. The rear section of the arm contains a hub and bearings through which the axle shaft passes. When the suspension rises and falls the 90° attaching points ensure the wheel moves without alignment or camber change.

Semi-trailing arm suspensions are widely used and very similar in appearance to the trailing arm suspension. The major variation is the angle of the lower arm attaching points. The pivot points are set at a sharp angle to the vehicle centre line and movement of the arm causes changes to the alignment and camber of the wheel as it rises and falls. The change in angles can be calculated by the designer and vehicle handling can be improved depending on where the lower arm is pivoted.

Swinging half axles vary in design, however, one version used on production vehicles consisted of coil springs mounted on a solid housing. The housing was split through the middle and the two halves could move on each other controlled by a pivot arm and large compensating spring. Large camber variation could be expected as the wheel moved up or down through its travel. The design did not find widespread acceptance.

Many new rear suspension designs are emerging. These designs offer significant improvements in ride control and handling characteristics. Several advanced types now provide automatic toe-in control when the vehicle is cornering or when the suspension deflects as it encounters road bumps. The release of four wheel steering systems featuring either mechanical or electronic control is an indication of the attention being focused on the rear suspension area. Now manufacturers have offered vehicles for sale it is only a question of time before significant changes will occur to the rear suspension systems of many vehicles.

Shock absorbers

As previously described, each road wheel is equipped with a suspension system containing one type of spring. This spring stores energy each time it is distorted by the road wheel contacting a bump. As the energy is released by the spring returning to its original shape, the wheel travels past its normal position and the spring distorts in the opposite direction. This back and forth movement (bouncing), as the spring stores and releases the energy, will continue until the energy is absorbed by the stiffness of the spring and tyre.

Bouncing not only causes discomfort to the driver and passengers but is extremely dangerous because it seriously affects the steering and braking performance of the vehicle. To overcome this problem, each spring is fitted with a spring damper (shock absorber) which changes the mechanical energy to heat.

The shock absorber is mounted between the moving suspension arm and the body or frame. The mounting points of the shock absorber are fitted with rubber insulators or bushes to reduce noise. The main types of insulators or bushings are the axial compressed type and the radial compressed type.

The axial compressed type consists of a bushing located either side of a hole, in the mounting plate, which are slightly compressed between two special washers by a nut located on the end of the shock absorber shaft.

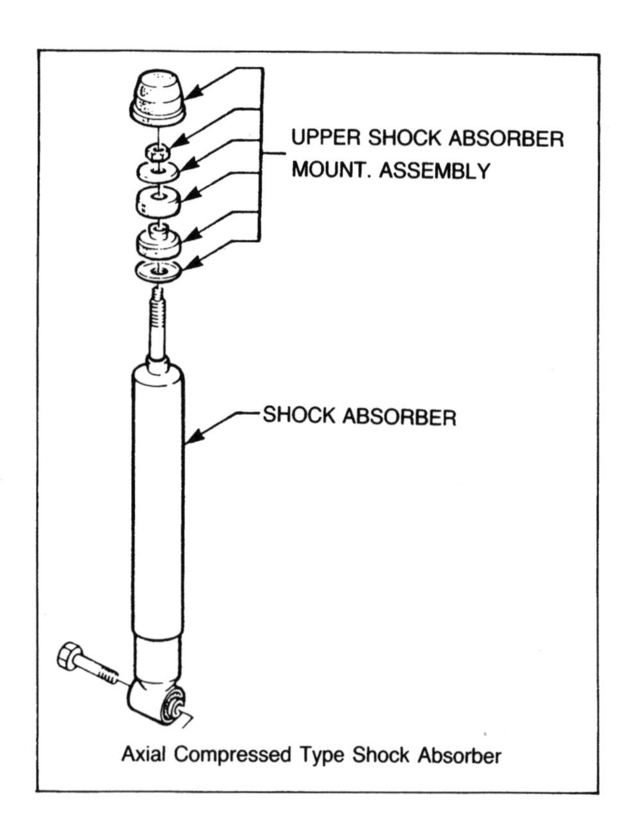

Axial Compressed Type Shock Absorber

The radial compressed type consists of a bushing pressed into a mounting eye with a central sleeve or bar providing the connection point between the shock absorber and the mounting plate.

Of all the various types of shock absorbers used in past years, the telescopic type has been accepted for universal applications.

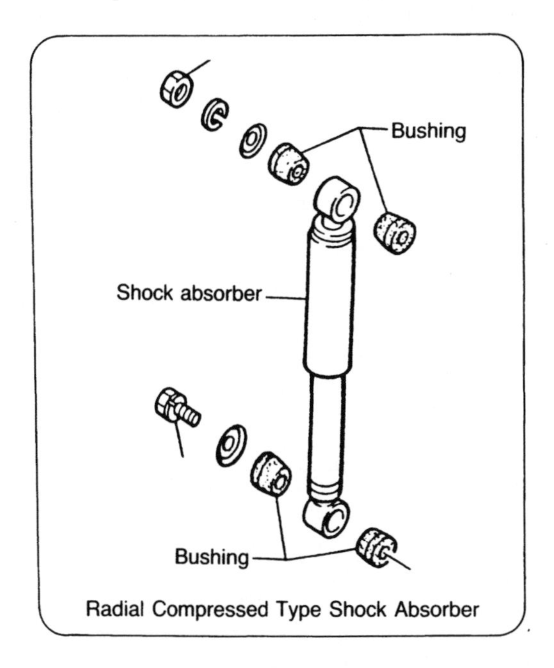

Radial Compressed Type Shock Absorber

Telescopic shock absorbers tend to be similar in appearance but their internal construction may vary considerably. These units can be:

- single acting. Offers resistance in only one direction.
- double acting. Offers resistance in both directions, and can be calibrated to provide the greatest resistance during the rebound movement.

The basic working fluid inside the shock absorber is hydraulic oil, however, some modern units are designed with an internal section containing pressurised gas which improves the shock absorber performance.

A vehicle is equipped with shock absorbers which are adequate for a recommended maximum load, however, an additional load such as a caravan can cause the rear suspension to 'sag'. This problem of a temporarily increased load can be overcome by installing a set of 'after market' shock absorbers containing an air 'bag' or pump-up section. When inflated to working pressure the shock absorber will assist the rear springs to restore the vehicle ride height. The hydraulic action is the same as a conventional shock absorber, and working pressure in the air section is adjusted through a valve located at the rear of vehicle.

Caution: the working air pressure must not exceed the manufacturer's specifications.

SHOCK ABSORBER SERVICE

The operation of a shock absorber can be tested with a shock absorber testing machine, however, most mechanics follow the simple test procedure recommended by the roadworthy regulations. This procedure is:

1. Bounce the vehicle through its full suspension travel.
2. Release the vehicle to allow it to bounce freely.
3. Count the oscillations (up and down movements) the shock absorber takes to bring the vehicle to rest.
4. Compare the number of oscillations with the roadworthy regulations. More than 1 1/2 oscillations indicates a fault.
5. Inspect the shock absorber for:
   - oil leaks;
   - road damage;
   - cracked or broken mountings;
   - worn or loose bushings.

SHOCK ABSORBER REMOVAL

Removal of a shock absorber from a vehicle is a simple task and can generally be achieved without removing any other suspension components or the road wheel. The problem most likely to occur is the tendency for the shock absorber shaft to turn

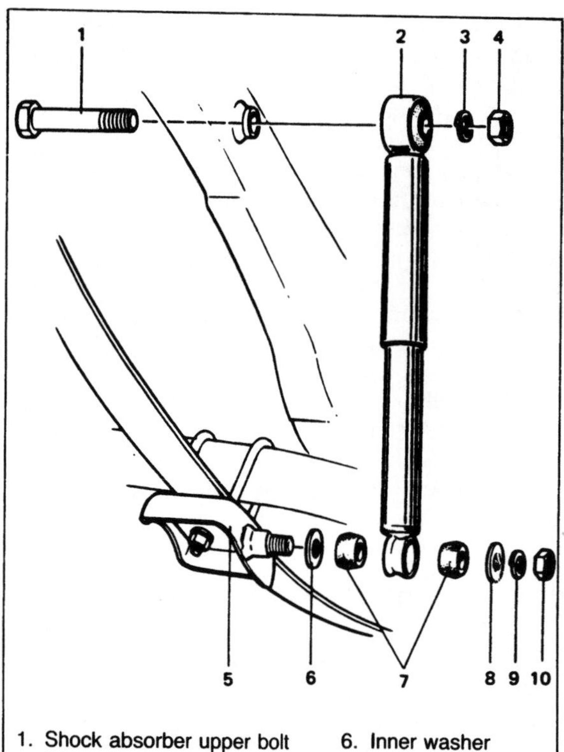

Shock Absorber Removal

with the nut. For a shock absorber fitted with a protection shield over its shaft, grip the shield while removing the nut. For a shock absorber with an exposed shaft, hold the shaft with a spanner located on the two flats provided at the end of the shaft while removing the nut.

Caution: do not grip the polished shaft with any tool because a damaged surface will lead to rapid seal failure.

BENCH TESTING A SHOCK ABSORBER

When a shock absorber has passed an inspection, it can be bench tested. The procedure is:

1. Grip the lower mounting bracket firmly in a vice fitted with soft jaws.
2. Pull the shaft to its fully extended position.
3. Push the shaft to its fully retracted position.
4. Repeat steps 2 and 3 several times.
5. Observe the shock absorber's operation during each stroke.
   - Its action must be smooth.
   - Resistance must be felt in both directions for a double action type or in only one direction for a single action type.
   - Sudden loss of resistance or noise indicates a fault.

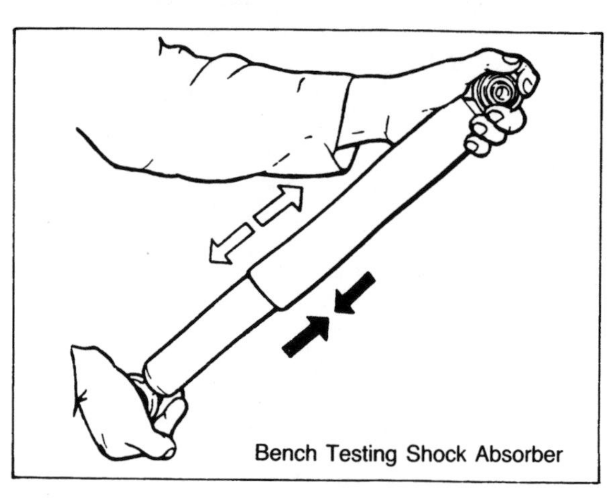
Bench Testing Shock Absorber

SHOCK ABSORBER REPLACEMENT

The replacement of a shock absorber is basically the reversal of the removal process. Some important points to observe are:

1. Ensure that the step on each rubber insulator is correctly located in the mounting hole.
2. Tighten the eye type mounting bushing when the vehicle is in its normal position on the floor (weight on its tyres).

Note: When the mounting bolts are torqued with the bushing hanging at an angle to the suspension, it is possible that the bushing will be torn or damaged by the twisting effect.

3. Ensure that the axial compressed type are not over-tightened.

NEW SHOCK ABSORBER INSTALLATION

New shock absorbers should be 'stroked' using the same method as the testing procedure. This will ensure that the unit is operating smoothly.

Note: Any variation in resistance should stabilise after a few strokes.

AIR SHOCK ABSORBER REMOVAL AND REPLACEMENT

The removal and replacement of 'air' shock absorbers is similar to the procedure for standard shock absorbers, except:

1. Deflate the shock absorber and disconnect its air line before removal.
2. Connect the air line and inflate the shock absorber to at least its minimum operating pressure after replacement.

3 Test the line and fittings for leaks.
—Inflate the system with air to a pressure of 700 kPa.
—Apply soapy water to the ends of the fittings and check for bubbles. Bubbling indicates that an air leak is present.
4 Inflate the system with air to the correct operating pressure.

Note: Typical operating pressures for a system of this type are:
- Maximum pressure, when loaded, is 600 kPa.
- Maximum pressure, when unloaded, is 450 kPa.
- Lowest pressure permissible is 70 kPa.

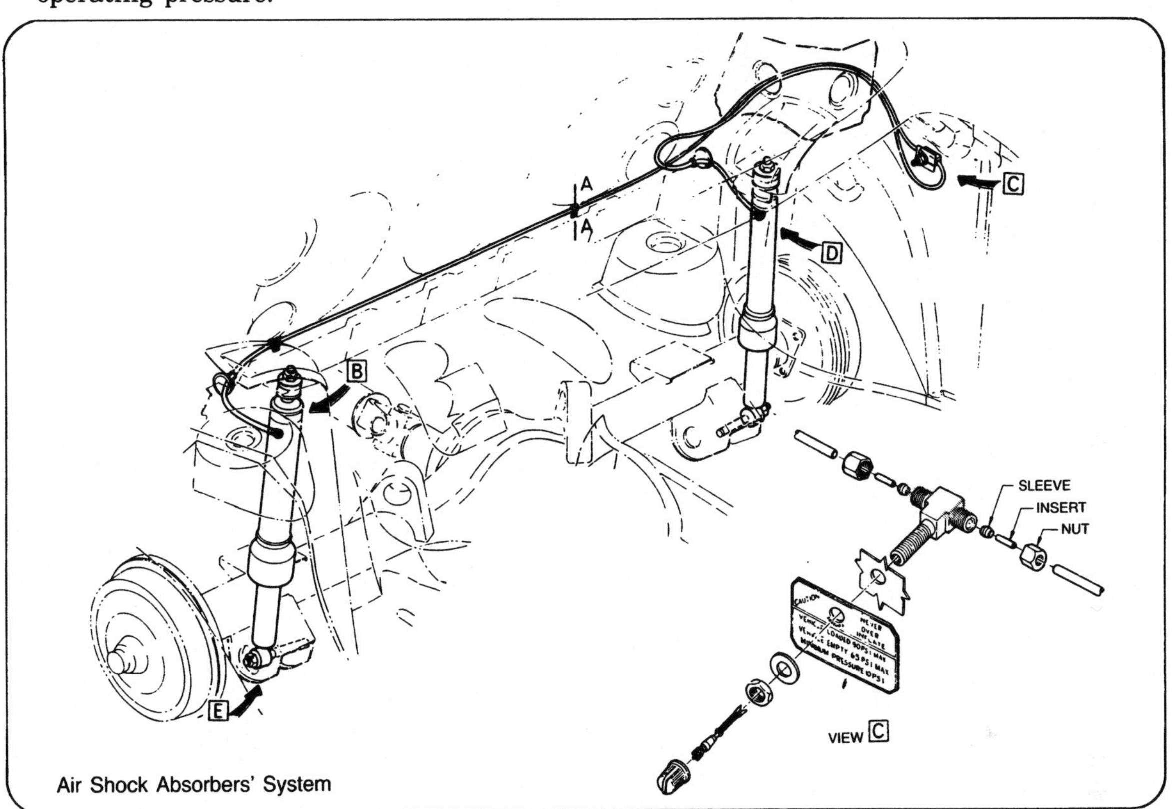

Air Shock Absorbers' System

STEERING

A typical steering system consists of a:
- steering wheel;
- steering column;
- steering box;
- steering pump (power steering option only);
- pitman arm;
- drag link;
- idler arm;
- tie rods/tie rod ends;
- steering arms.(see next page for diagram).

Steering wheels vary in size, rim thickness and material. Most wheels are manufactured on a steel centre with steel radial spokes and a steel rim. The centre is splined to the steering column shaft and firmly attached by a single nut. It is the covering material which makes the wheels different in appearance.

The wheels, made with hard plastic material, usually have a horn button centrally mounted on the hub but are of drop centre design to provide protection to the driver's chest in an accident.

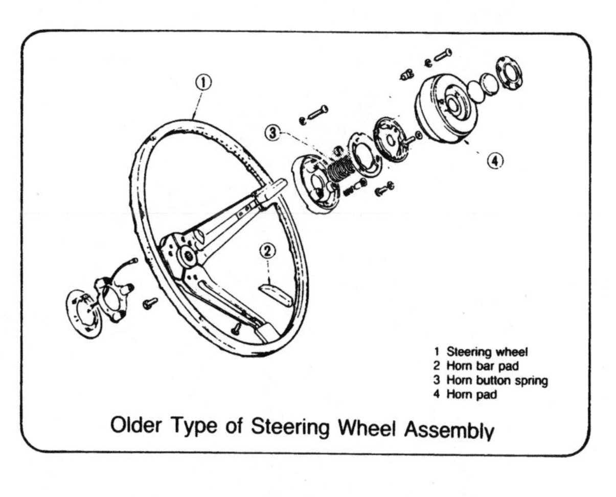

Older Type of Steering Wheel Assembly

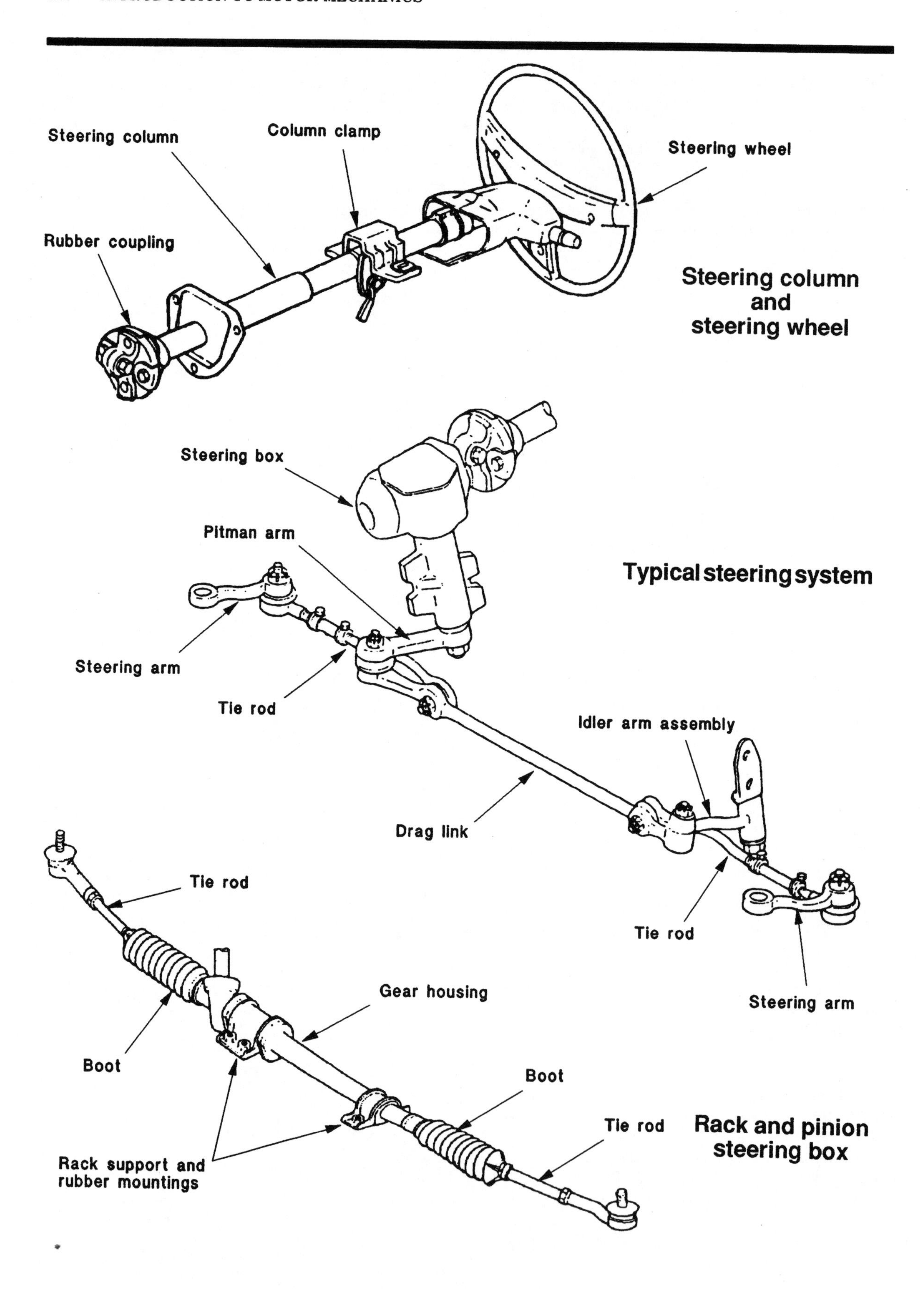

Steering column
Column clamp
Steering wheel
Rubber coupling
Steering column and steering wheel
Steering box
Pitman arm
Typical steering system
Steering arm
Tie rod
Idler arm assembly
Drag link
Tie rod
Tie rod
Steering arm
Gear housing
Boot
Boot
Tie rod
Rack and pinion steering box
Rack support and rubber mountings

Modern steering wheels use a 'soft' feel material which offers excellent grip, and provide driver protection with an extensively padded centre.

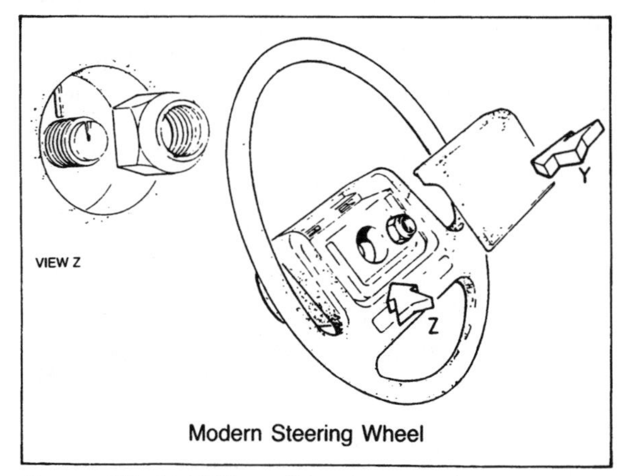

Modern Steering Wheel

Source: Mitsubishi Sigma GJ series manual

A recent development which has not yet spread to many models is the addition of a crash protection air bag fitted into the wheel centre. Check the manufacturer's instructions prior to removing steering wheels fitted with these air bags.

Steering columns in early design vehicles were rigidly attached to, and formed part of, the steering box. This type is no longer used and all modern vehicles are fitted with collapsible steering columns consisting of:

- a steel inner shaft which transmits the turning effort applied at the wheel to the steering box. The shaft is usually connected to the steering box by a universal joint or fabric coupling;

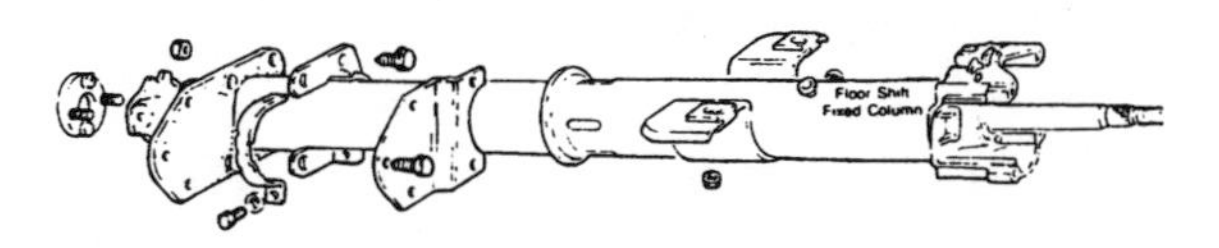
Steering Column

- an outer casing which supports the inner shaft on bearings and is attached to the vehicle by mountings located under the instrument panel and at the engine firewall.

Both column sections (inner and outer) are designed to collapse under severe impact.

Steering boxes are available in many different designs, of which the three most popular are the recirculating ball, the rack and pinion, and the hourglass worm and roller types. When the vehicle is fitted with an engine driven pump to supply hydraulic assistance to the steering box the steering is described as 'power assisted'. Regardless of type, all steering boxes convert the rotational movement of the steering wheel into a lateral movement capable of moving the steering linkage. The steering box must be firmly attached to the vehicle frame and located in a position which allows direct coupling with the steering column.

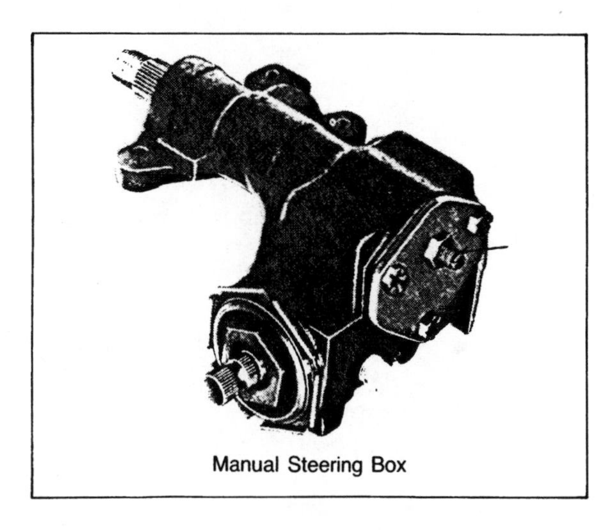
Manual Steering Box

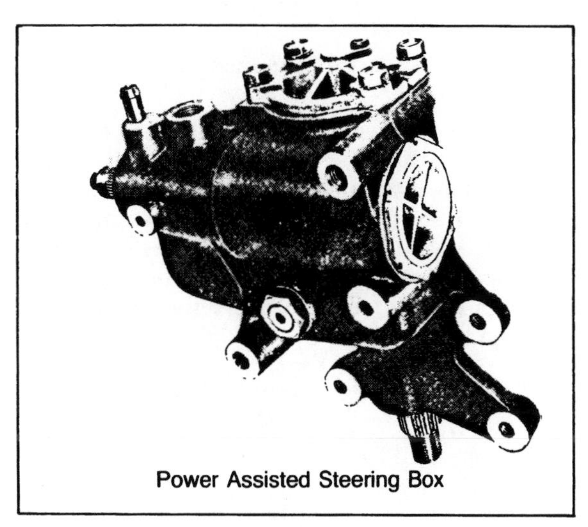
Power Assisted Steering Box

Steering pumps are belt driven from the engine crankshaft pulley and provides hydraulic fluid under pressure to the steering box. The pump is connected to the steering box by two flexible hoses, and when the engine is operating the pressurised fluid reduces the driver effort required to turn the steering wheel.
Note: Rack and pinion steering systems do not contain a pitman arm, drag link or idler.

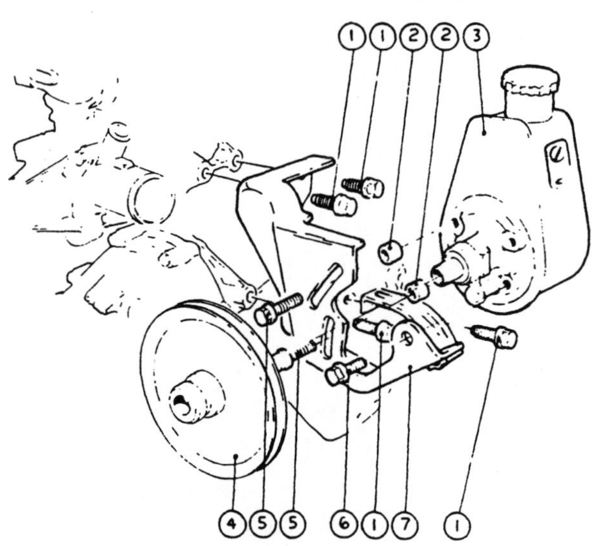

1. Attaching Bolt-Mounting Bracket
2. Spacer
3. Power Steering Pump
4. Pulley
5. Attaching Bolt-Pump Inboard
6. Attaching Bolt-Pump Outboard
7. Mounting Bracket

Steering Pump Assembly

Pitman arms are drop forged and are secured to the steering box sector shaft by a nut and lockwasher. Movement between the arm and shaft is prevented by a number of splines. When the steering wheel is turned the arm will move in a lateral direction, providing both movement and support for the steering linkage.

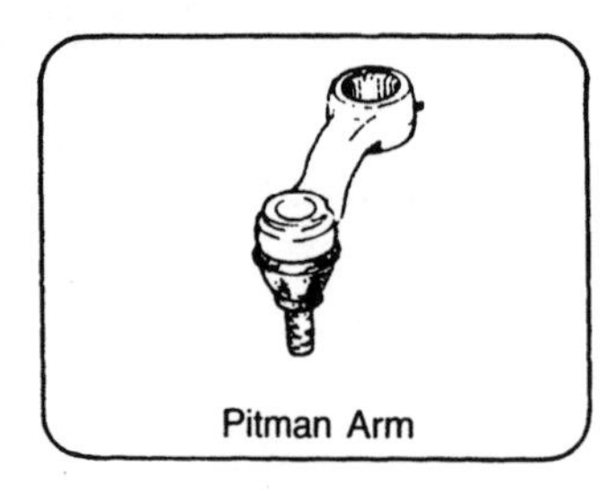

Pitman Arm

Drag links may be known by several other names, depending on the vehicle's country of origin or individual preference of the vehicle manufacturer. Possibly the name which best conveys the function of this part is 'steering connecting link assembly'. A vehicle has steering components on both left and right sides. The drag link connects the two sides so both sections move in unision. The normal connection points are between the pitman arm and the idler arm assembly.

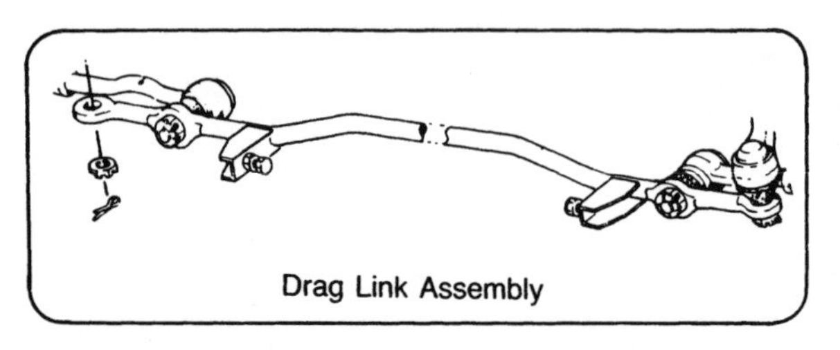

Drag Link Assembly

Idler arm assemblies consist of a fixed pin mounted to the frame rail, and a movable arm which guides and supports the passenger side steering linkage.

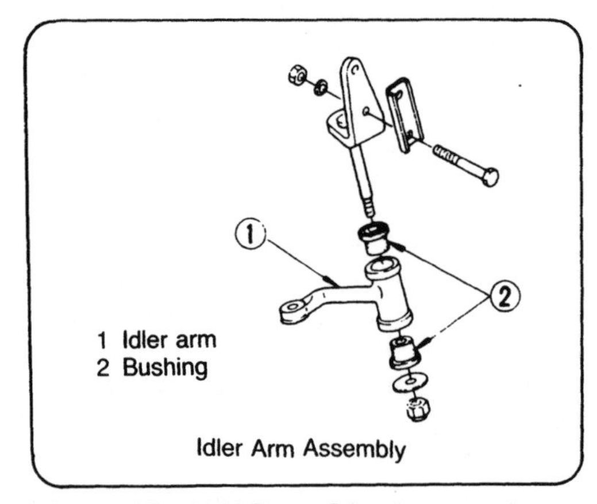

Idler Arm Assembly

Source: Mitsubishi Sigma GJ series manual

Tie rods and tie rod ends are the links between the drag link and the steering arms. The tie rods are fitted with small ball joints (tie rod ends) at each end to compensate for the vertical movement of the steering arms as the road wheels strike bumps in the road surface. Most tie rods are constructed with a threaded sleeve which, when rotated, will alter the length of the tie rod. This feature is used to adjust an alignment angle known as 'toe-in'.

A similar arrangement is used on rack and pinion steering systems. The major difference is that the inner tie rod ends are linked directly to the steering box through ball sockets in the end of the steering rack.

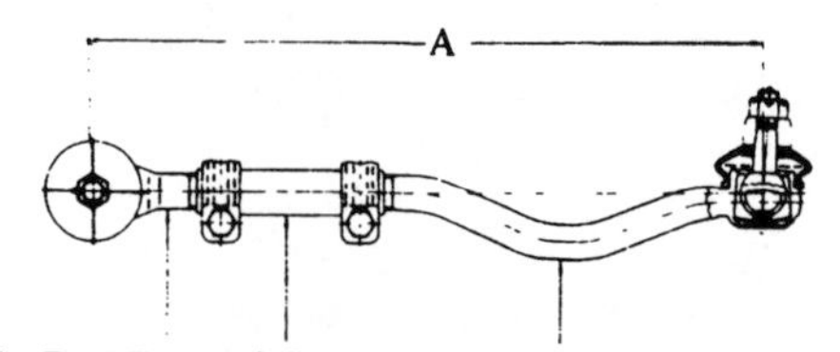

Source: Mitsubishi Sigma GJ series manual

Steering arms are the final connection between the steering system and the road wheel hub and spindle. The arms are bolted onto, or can be an integral part of, the wheel spindle. The arm is angled slightly inward from the wheel and the outer tie rod end is located in a tapered hole near the end of the arm.

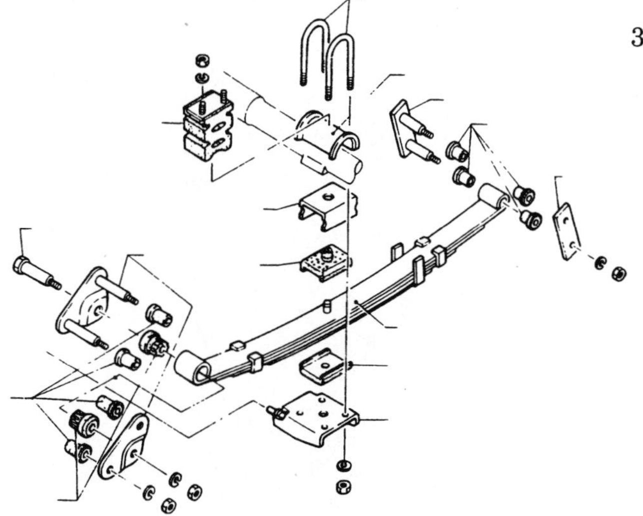

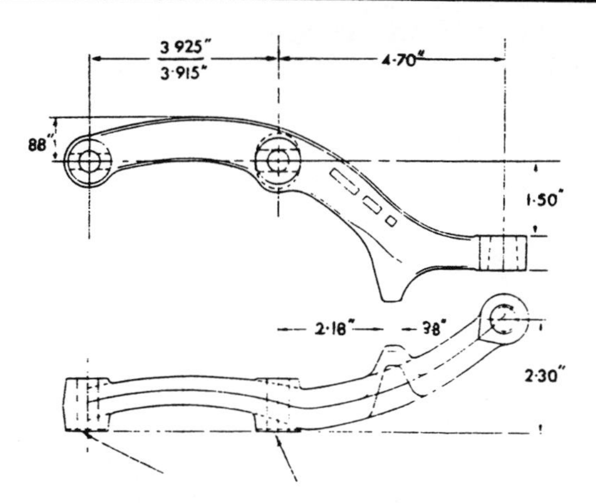

CHAPTER 16 REVISION

1 Describe three differences between mono-construction vehicles and separate chassis vehicles.
2 List two advantages and two disadvantages of mono-construction and chassis vehicles.
3 Identify the spring types as shown in the following diagrams.

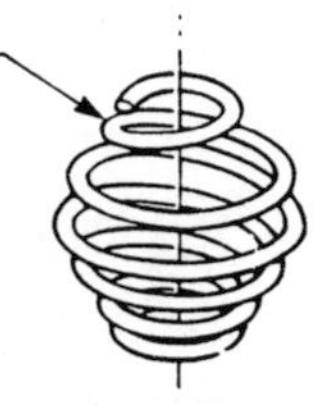

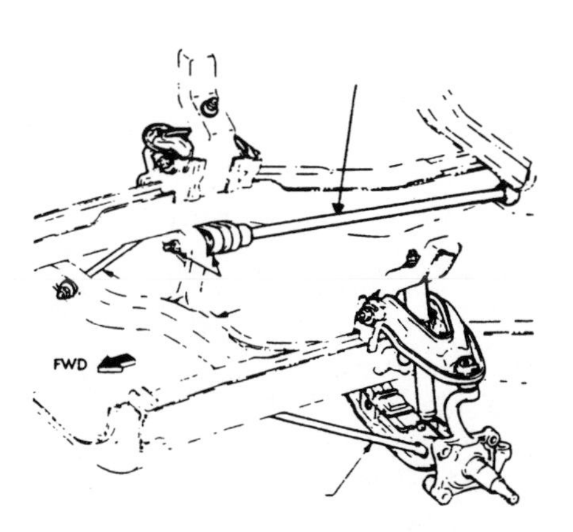

Name the front suspension units as shown in the following diagrams.

5 Identify the rear suspension units as shown in the following diagrams.

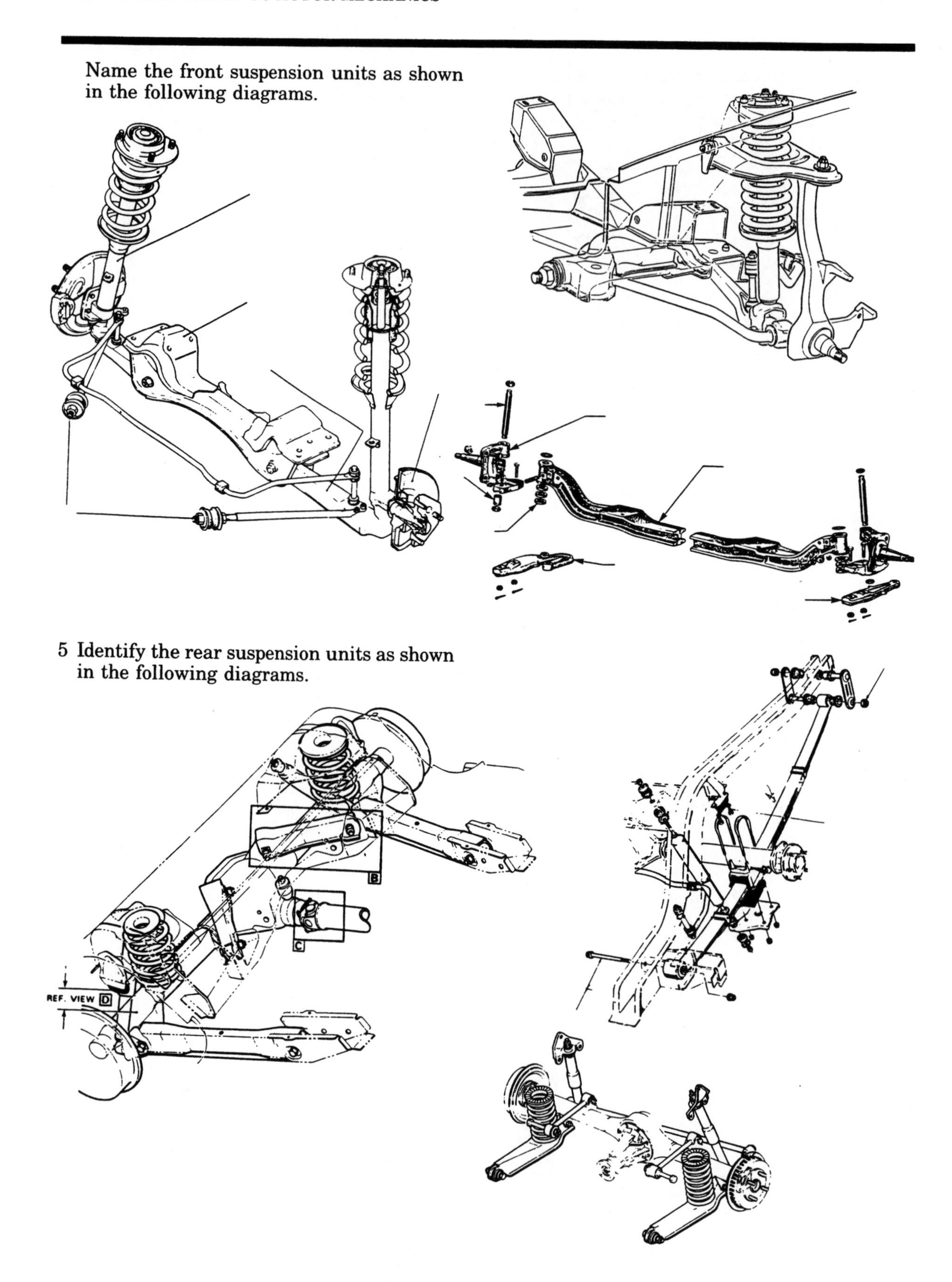

6 Describe the function of the suspension unit on a vehicle.

7 Identify and list the suspension components on the following diagrams.

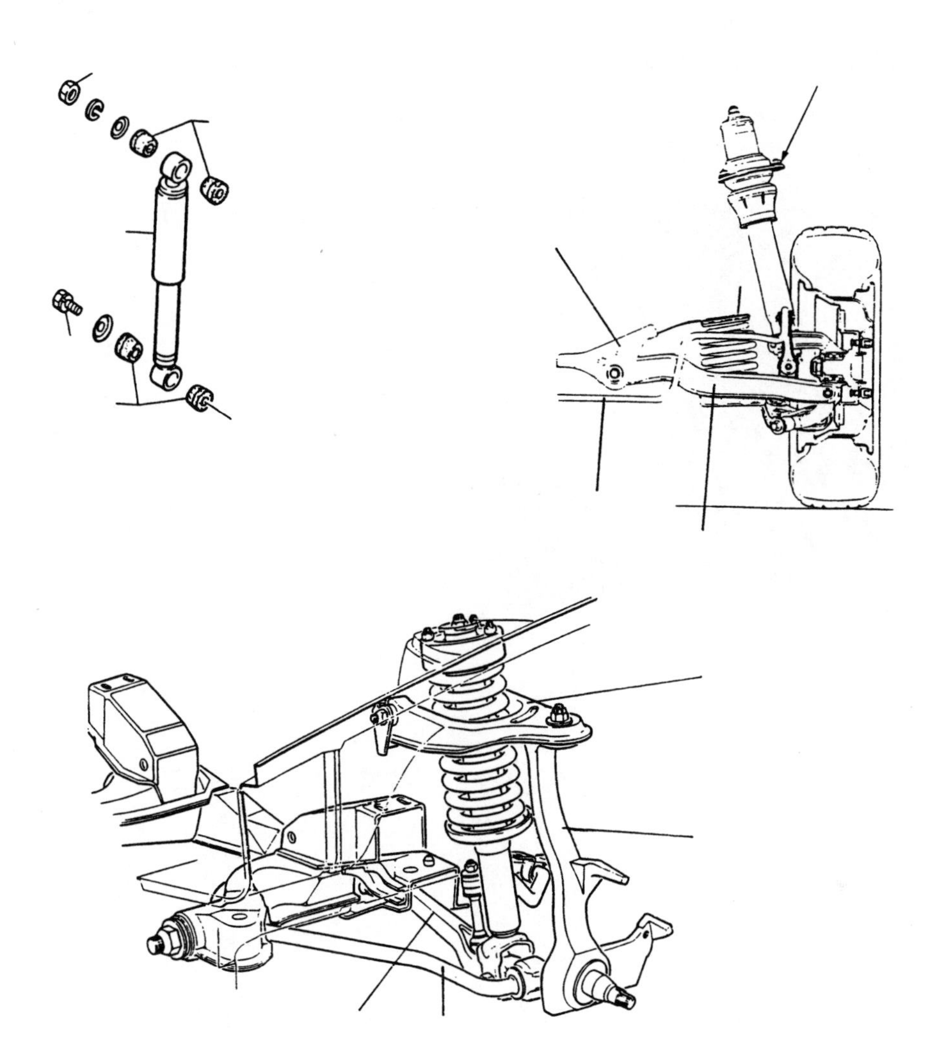

8 State the function of a shock absorber.
9 Explain procedures and precautions when removing and refitting a shock absorber.
10 Describe the test procedures required on a shock absorber.
11 Label the steering components as shown in the following diagram.
12 Describe the function of the steering components as shown in the diagram.

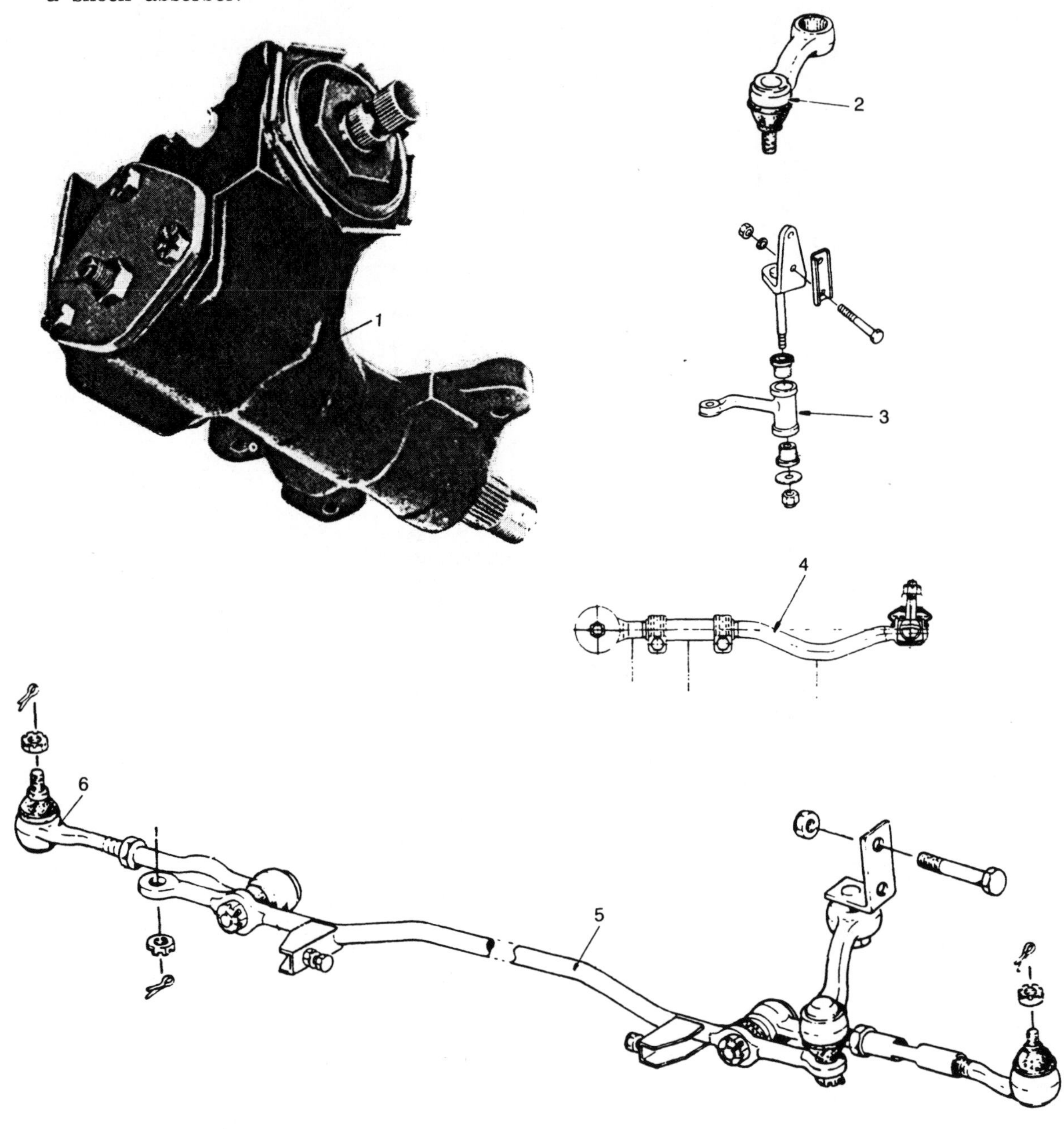

17

DRIVELINES AND FINAL DRIVES

For many years the motor vehicle was designed with a front mounted engine and transmission and power was applied to the road surface by a rear mounted final drive unit driving the back wheels. There were a few exceptions, mainly French and English designs, but the rear wheel drive continued as the dominant design until the 1980s. Major design changes in recent years have resulted in mass production of several different drive arrangements. The most common of these arrangements are described below.

TWO WHEEL DRIVE

The engine and transmission are mounted at the front. A driveshaft (propeller shaft or tailshaft) connects the transmission to a rear-mounted axle housing which contains the final drive gears, differential and drive axles.

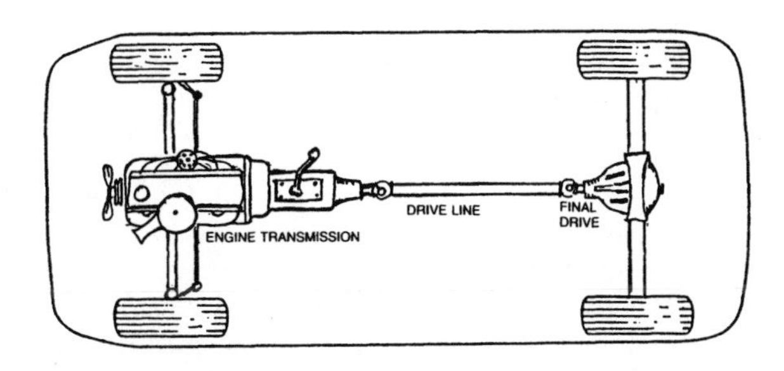

This drive arrangement is similar to that above, but the rear housing is mounted directly to the body and contains the final drive gears and the differential but not the drive axles. The axles are mounted externally, contain flexible joints to allow for movement of the independent suspension, and connect to the rear hubs and wheels.

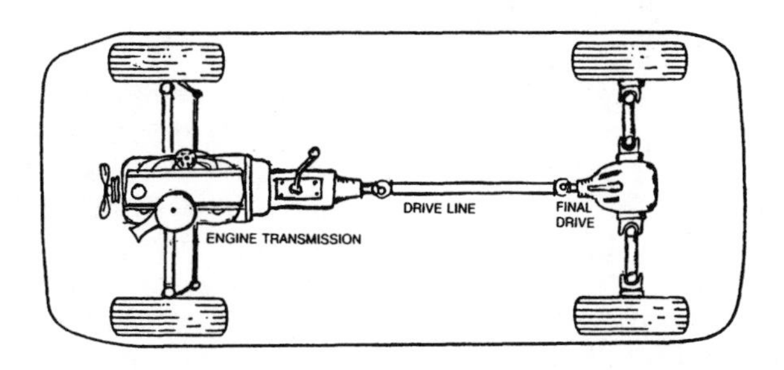

The engine is turned 90° in the frame and is mounted transversely across the engine bay. This mounting arrangement is popularly referred to as an 'east-west' engine mounting. A special type of transmission known as a transaxle is fitted to the rear of the engine. The transaxle contains the gearbox, final drive gears and the differential. The drive axles are mounted externally and connect to the front hubs and wheels. The rear wheels are not driven and cars of this type are known as front wheel drive vehicles.

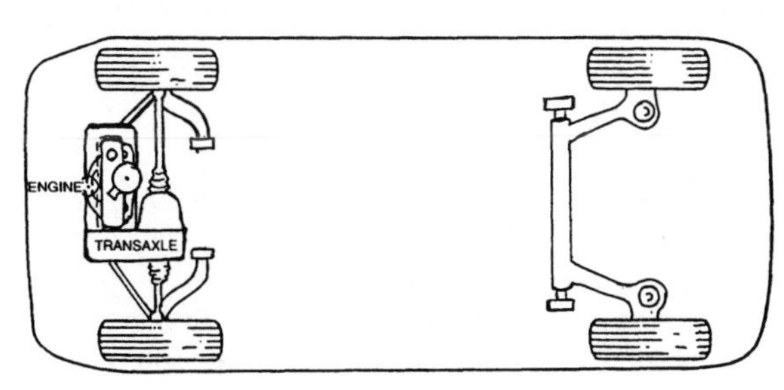

The engine and transaxle are mounted at the rear of the vehicle with the transaxle being fitted to, but mounted forward of, the engine. The rear wheels are connected to the transaxle by externally mounted axles. This concept was very popular in German and Italian cars but is tending to disappear in favour of modern designs.

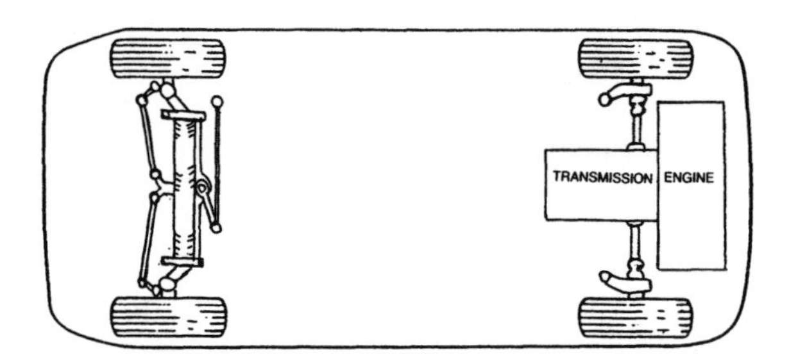

FOUR WHEEL DRIVE

The engine and transmission are mounted at the front of the vehicle. Attached to the rear of the transmission is a transfer case which provides output drives to both a front and rear drive shaft. The front drive shaft connects to a front axle housing which contains final drive gears, a differential and drive axles. The outer ends of the axle housing and the drive axles are designed with joints so the front wheels can be steered. The rear drive shaft connects to a rear axle housing which also contains final drive gears, a differential and drive axles. The transfer case allows the driver to select high or low ratio and two or four wheel drive.

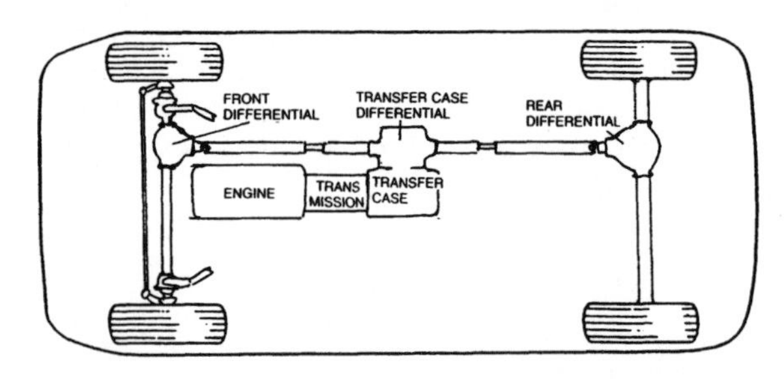

The vehicle is fitted with an east-west front mounted engine and a special transaxle containing a transfer unit which provides constant four wheel drive. The transfer unit is connected to the rear final drive housing by a drive shaft. Power flow from the transmission is passed through a centre (or third) differential which splits the power 50/50: to the front wheels through final drive gears and a differential contained in the transaxle; and to the rear wheels through the transfer unit and drive shaft to a rear mounted housing which contains the rear final drive and differential. This type of vehicle is designed for operation on a highway in four wheel drive. If the vehicle is to be used off-highway in loose dirt or mud the driver can 'lock' the third power sharing differential so power will be maintained to all four wheels even if one wheel set becomes bogged.

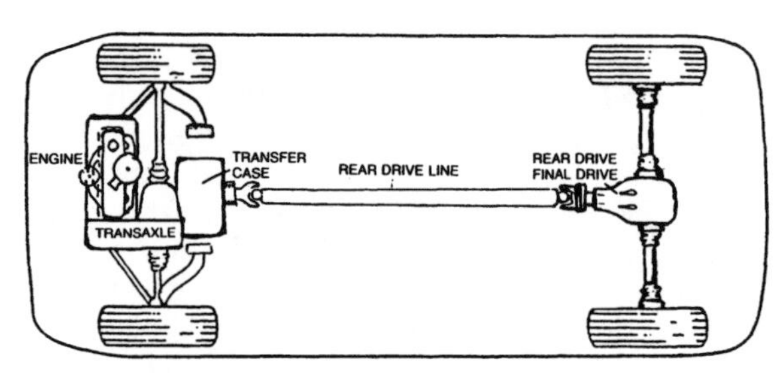

Many other engine/transmission/driveshaft/final drive combinations have been designed and introduced into vehicles but the concepts have tended to be restricted to a limited model range. The constant four wheel drive concept is gaining acceptance in high performance road cars and the next development is likely to be a combination of mid-mounted engines and transmissions and constant four wheel drive for specialised sports cars.

DRIVESHAFTS

The driveshaft is used to connect a front-mounted transmission to a rear-mounted final drive assembly. A basic driveshaft contains two universal joints (normally Hookes type) and has provision for a slip yoke or joint. The slip yoke compensates for suspension movement as wheels and housings move up or down in response to bumps in the road.

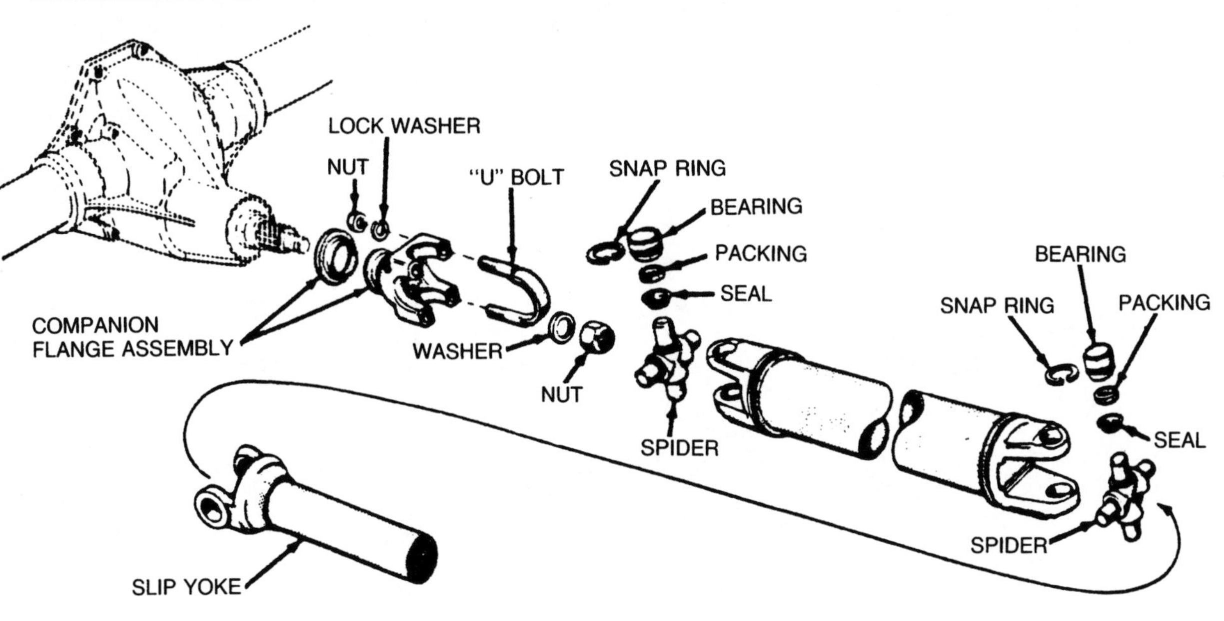

Drive Shaft Disassembled — Removable Carrier Axle only

The driveshaft can also be of two-piece design and be fitted with two Hookes joints, a centre bearing and a third universal joint of either constant velocity (C.V.) or Hookes type.

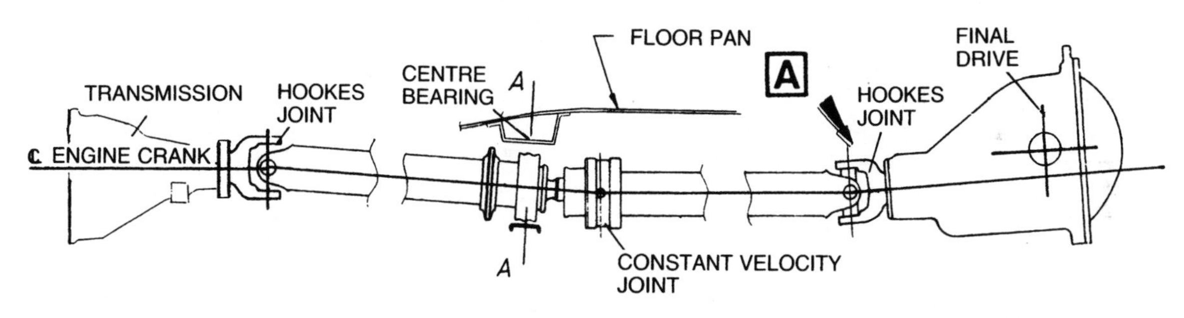

Two-Piece Drive Line

A few manufacturers have produced a different type of driveshaft where the rotating shaft is enclosed in a tube. This concept was never very popular as use of this arrangement, known as a 'torque tube drive', usually required extensive modification of the suspension mountings and/or the transmission mountings.

UNIVERSAL JOINTS

Many of the driveshafts and drive axles fitted to both front and rear wheel drives are required to operate through varying drive angles. The change in angle can be as a result of steering and suspension movement, however, it is necessary to provide flexible joints in the shafts to prevent unacceptable vibrations and/or unit damage. Many different types are available but most passenger cars utilise one or both of the following types.

Hookes joint

The Hookes joint is a simple yet rugged design. When used on driveshafts the joints are normally fitted in pairs to minimise shaft vibration. If the drive angle is increased, such as in front wheel drive axles and steering hubs, or if a smoother transfer of power is required, it is necessary to discard the Hookes joint in favour of a C.V. joint.

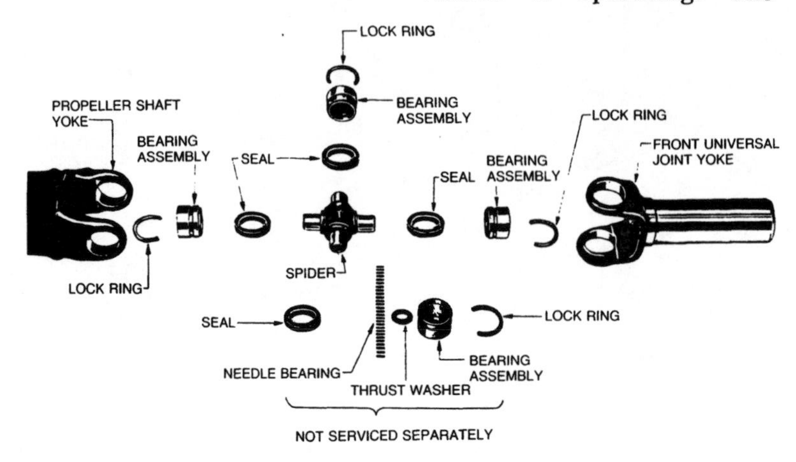

Hookes Joint Assembly – Exploded View

Constant velocity joints

C.V. joints can be fitted at any location requiring the use of a flexible joint. They are usually more expensive than a Hookes joint, therefore a combination of Hookes and C.V. joints are very common on driveshafts. The use of a single C.V. joint and two Hookes joints on a driveshaft ensures that minimum vibration levels will be present when the vehicle is operating. The C.V. joint is particularly suitable at the steering pivot points of front wheel drive axles where large drive deflection angles occur as the steering is turned from lock to lock. They are also fitted to the drive axles of rear wheel drive vehicles with fully independent rear suspension.

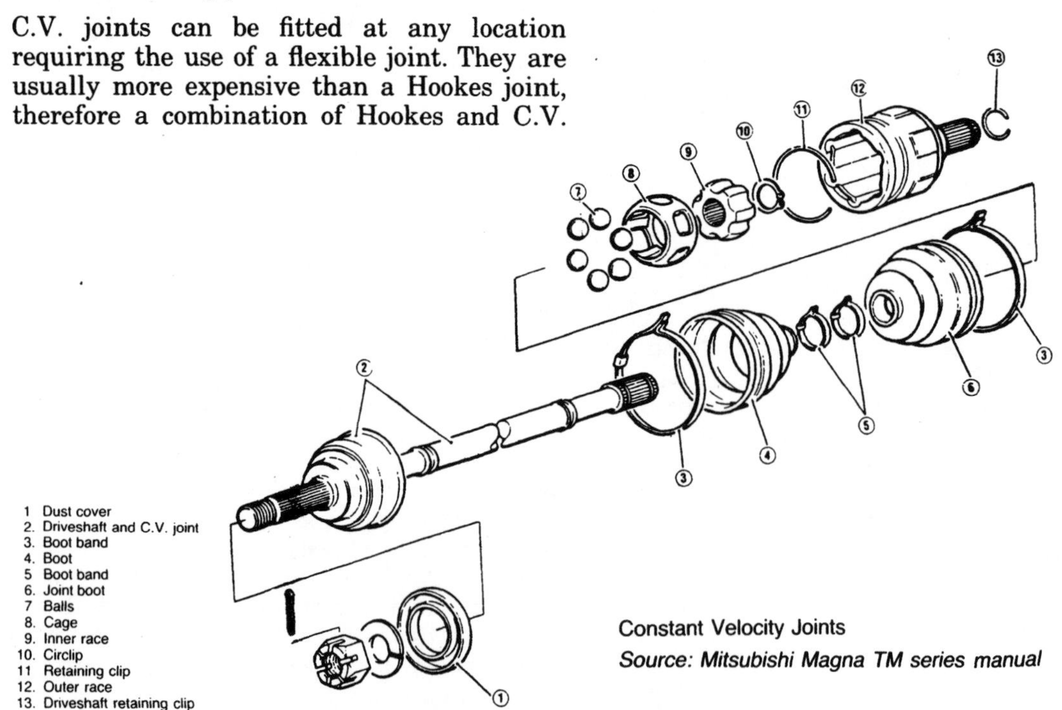

Constant Velocity Joints
Source: Mitsubishi Magna TM series manual

FINAL DRIVE

Final drive assemblies (rear, front or four wheel drive) have two major internal sections. A popular term used to describe the final drive assembly is the 'diff'. Despite popular usage, the term is technically incorrect. The 'diff', or differential, is one of two major sections, the second of which is the final drive gears or crownwheel and pinion.

Differential gear action is essential for any drive axle. When turning a corner the inside wheel has less distance to travel than the outside wheel. If both drive wheels were rigidly fixed to the same drive axle the vehicle would be extremely difficult to control and tyre wear would be rapid. The function of the differential is to allow one drive wheel to rotate at a different speed to the opposite drive wheel yet still provide power to both wheels.

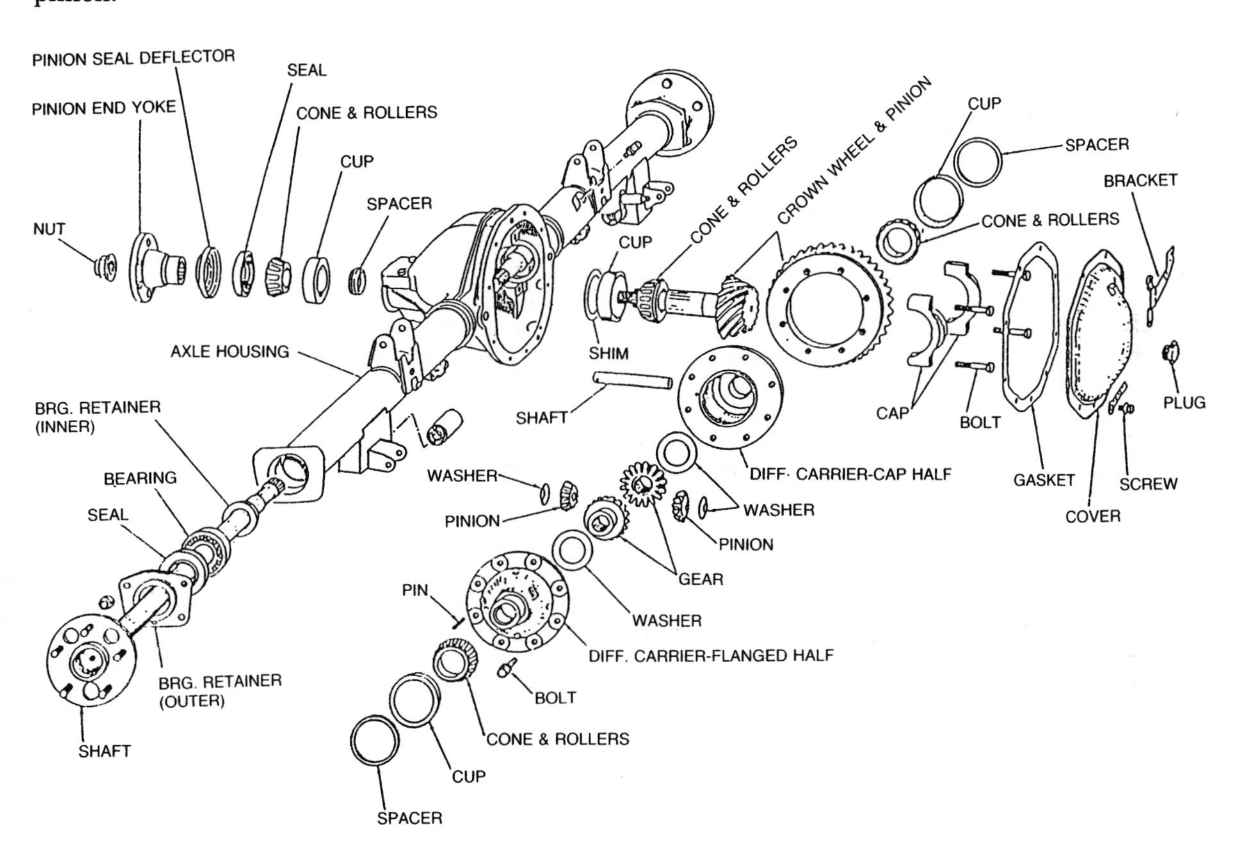

Final Drive Assembly

Source: Nissan Australia

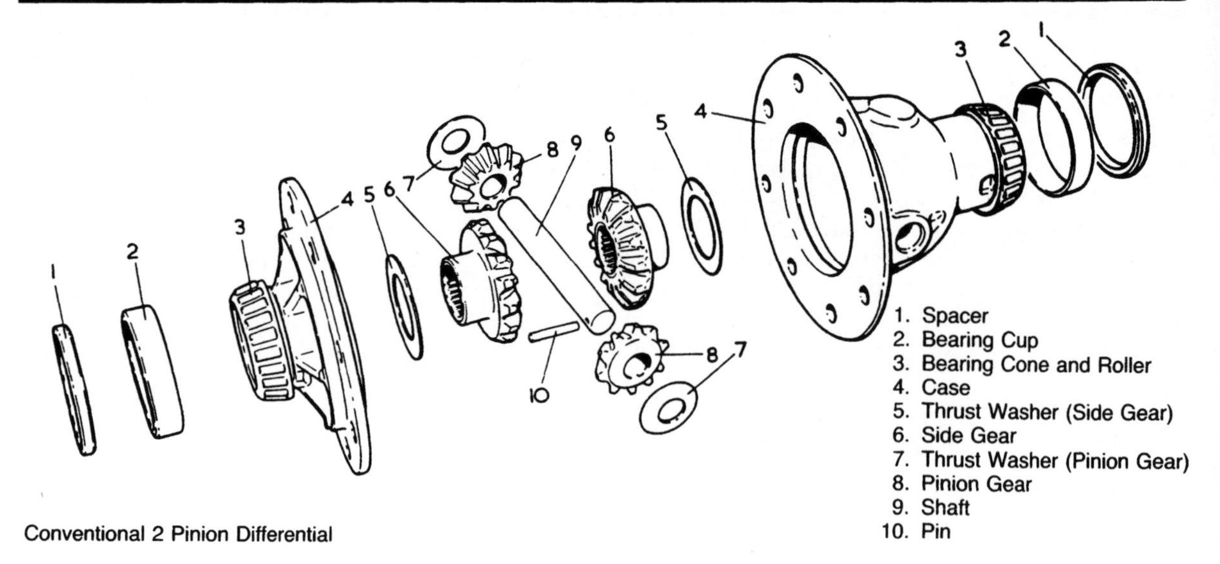

Conventional 2 Pinion Differential

The differential assembly consists of two axle side gears which are splined to the drive axles. Meshed with the side gears are two (heavy duty units may have four) planetary pinions. A large pin passes through the pinions and is locked into a casing which encloses the differential gear set. The differential casing is bolted onto the crownwheel, and the drive line from the final drive pinion to the drive axles is now complete.

Variations of the differential gear set, incorporating clutches or cones fitted between the axle side gears and the differential casing, are available. These units are known as L.S.D. (limited slip differentials) and improve vehicle drivability in loose dirt or boggy conditions.

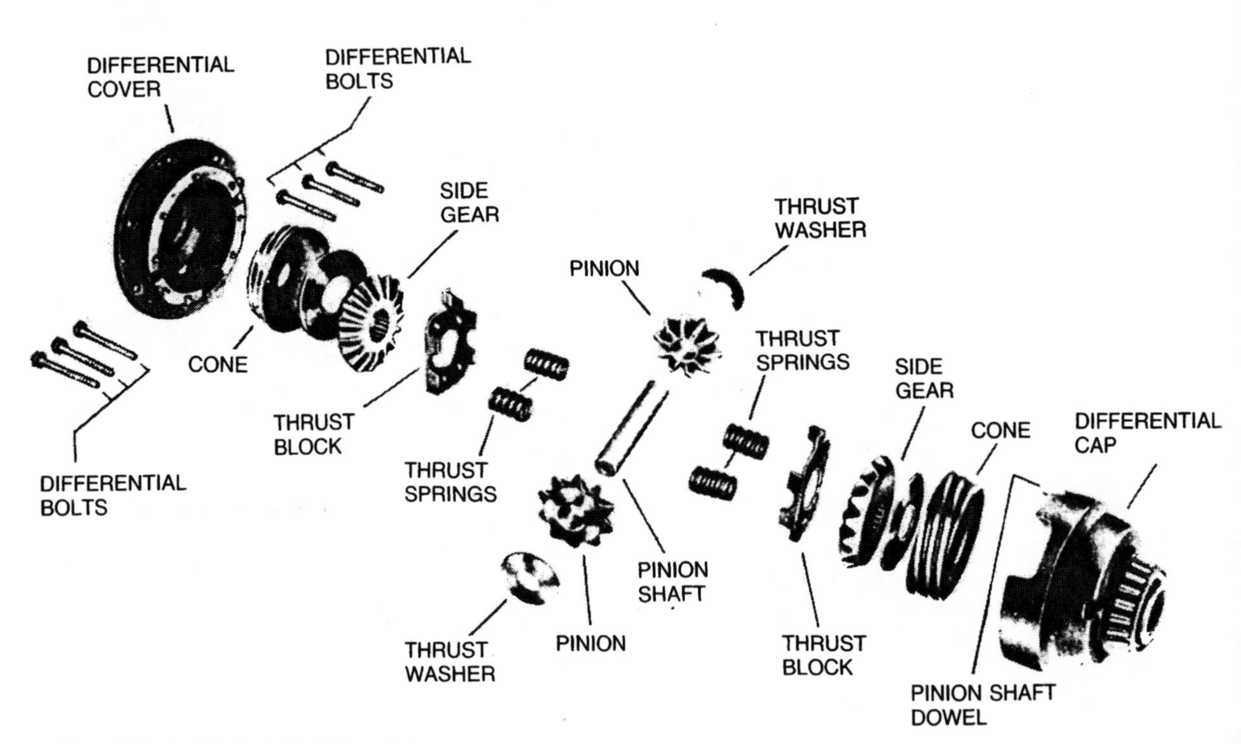

Integral Carrier Limited Slip Differential

Final drive gearing normally consists of a crownwheel and pinion. To achieve a reasonably flat passenger compartment floor pan it is desirable to have the drive shaft mounted as low as possible under the car. This generally means the pinion engages with the crownwheel below the crownwheel centre line. The name applied to gear sets which mesh in this type of alignment is a 'hypoid' gear set. The final drive gears have two functions: to turn the rotation of the drive shaft through 90° providing drive for the axles, and to provide a fixed gear ratio reduction for the vehicle.

Front wheel drive vehicles incorporate the final drive gears inside the transaxle, therefore the gear design can be simplified as the function of the gears is simply to provide a fixed gear ratio reduction.

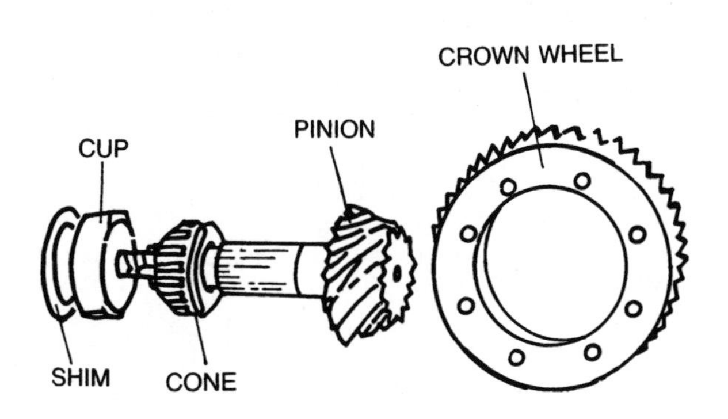

1. Gear case
2. Speedometer drive gear
3. Side bearing inner race
4. Thrust washer
5. Side gear
6. Thrust washer
7. Pinion gear
8. Pinion shaft
9. Knock-pin
10. Ring gear

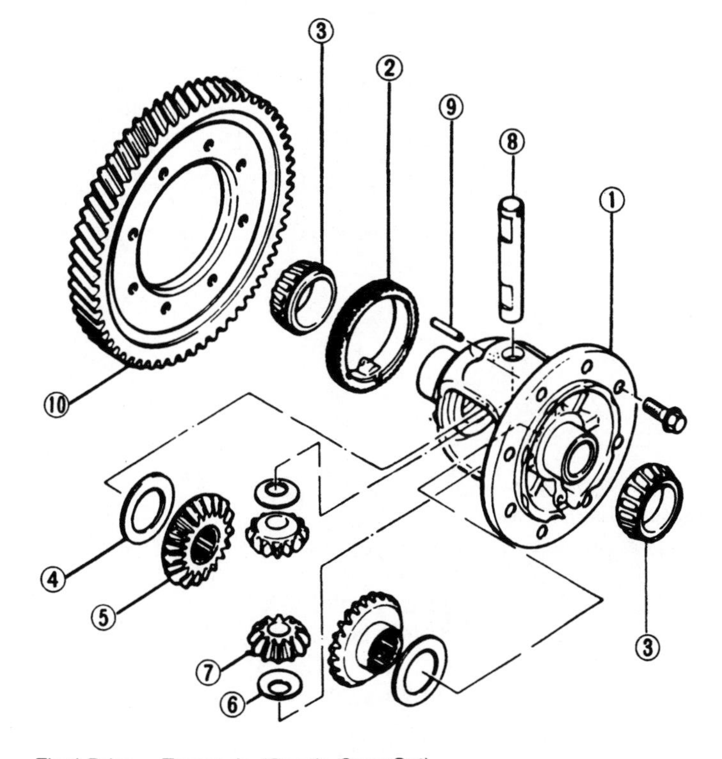

Final Drive – Transaxle (Simple Gear Set)

A further variation in final drives has occurred with at least one mass production automatic transaxle deleting conventional gears and using a planetary gear system to provide the final drive fixed ratio reduction.

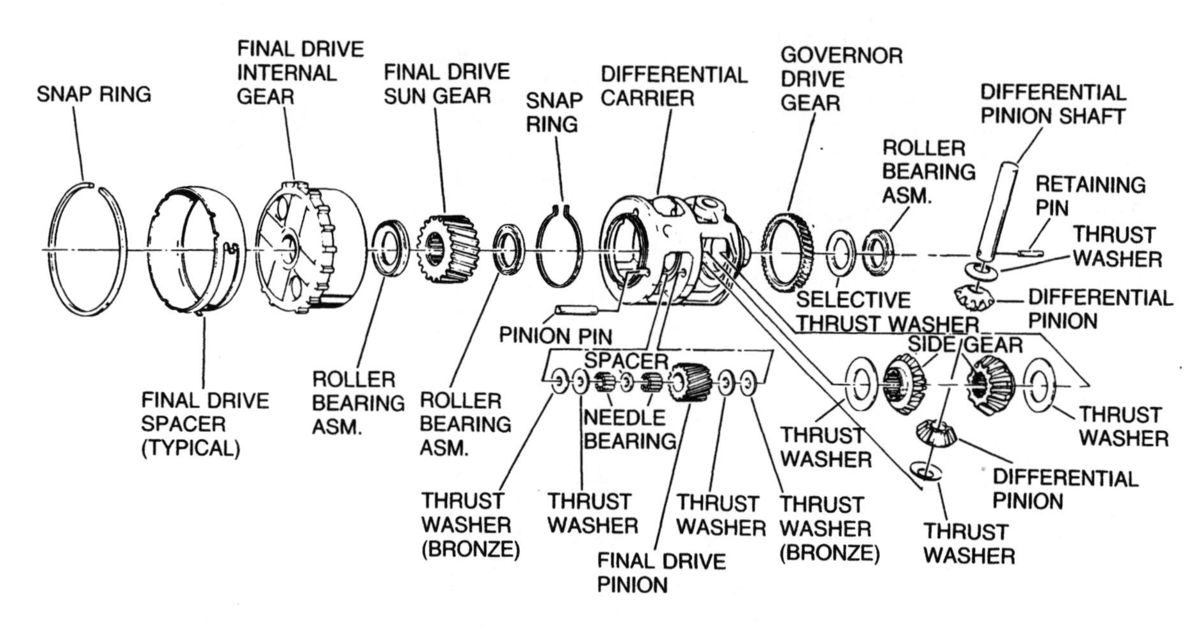

Final Drive – Transaxle (Planetary Gear Set)

CHAPTER 17 REVISION

1 Sketch four different engine/transmission and drive combinations.
2 What is the purpose of a slip yoke when fitted to a driveshaft?
3 Why are Hookes joints fitted in pairs?
4 Name the two major sections contained within a final drive assembly.
5 What is the function of a crownwheel and pinion?
6 What is the function of the differential?

1S

AIR CONDITIONING

INTRODUCTION

The world is now aware of the harmful effects of ozone depleting substances such as the refrigerant gases (CFCs) used in automotive air conditioning systems. The ozone layer, located in the stratosphere, protects the earth from harmful ultra-violet rays. CFCs rising into the ozone layer attack and destroy the ozone molecules causing a thinning of the layer.

Manufacturers world-wide are designing air conditioning systems which use alternative refrigerants. These types of refrigerant gases will not harm the ozone layer. Unfortunately, the new refrigerants will not be able to be used in current systems. The problem of ozone depletion caused by the earlier systems will remain for many years. Therefore, it is essential that refrigerants are reclaimed from vehicle air conditioning systems and are not discharged into the atmosphere.

Every serviceperson maintaining or repairing automotive air conditioning systems must comply with Environmental Protection Authority (EPA) regulations. Unlawful discharging of refrigerant gas into the atmosphere will incur a severe financial penalty.

This chapter relates to the basic operation and safety requirements of an automotive air conditioning system and is not intended to meet the licensing and accreditation requirements of regulatory authorities. Do not proceed to repair or service these systems until further studies have been undertaken.

SAFETY PRECAUTIONS WHEN HANDLING REFRIGERANT GASES

Skin and eye contact

Automotive air conditioning systems use a refrigerant gas. The evaporation temperature of refrigerants can be as low as −30°C therefore, the possibility of liquid refrigerant striking the eyes or the skin must be avoided. It is easy to avoid personal hazards. ALWAYS WEAR GOGGLES AND GLOVES when working on air conditioning systems.

Refrigerant, when inside the air conditioning system, is normally under pressure and will be stored as a liquid. If accidental discharge of liquid refrigerant occurs from a leaking hose, fitting or component, the pressure is immediately reduced to atmospheric pressure. When this occurs, the liquid will change into a gas. To achieve this change, the refrigerant must attract large quantitites of heat from its surroundings. If the refrigerant was to spray onto a person at this time it would obtain the required heat from the person's body. Imagine the consequences if the refrigerant was sprayed into your eyes or onto exposed skin. The refrigerant would absorb heat from your eyes or skin which could result in blindness or frostbite.

If an accidental contact does occur, it is essential to quickly splash large quantities of cool water on the affected areas. Do not rub the eyes. The water will minimise the damage to the eyes or skin.

Note: Seek professional medical attention if refrigerant contacts the eyes.

Refrigerant contact with fire

Refrigerant gas, such as R12, is normally non-poisonous. But if the gas is exposed to an open flame or very hot metal, a poisonous gas will be formed. Inhalation of this gas may result in severe illness. Continued inhalation of small quantities of the gas over a period of time can result in a toxic condition.

Refrigerant storage

Refrigerant gases are generally considered to be safe, but the cylinders are under significant pressure, therefore, care should be observed when cylinders are stored:

- The cylinders should not be stored in direct sunlight or near a heat source such as steam cleaners or welding areas.
- Do not drop or handle the cylinder roughly.
- Cylinders, when not mounted on a servicing station, must be stored in an upright position. The valve must be closed and the valve outlet nut fitted.
- Refrigerant gases should not be mixed, therefore, cylinders must be dedicated to one type of refrigerant.
- Space must be left within the cylinder for expansion of the refrigerant therefore do not fill any cylinder completely.

BASIC PRINCIPLES OF REFRIGERATION

Movement or transfer of heat

Most of us have experienced the transfer of heat during the summer when we have sat in the sun near a pool. The sun warms our bodies (heat is transferred from the sun to our bodies), we then dive into the water (where heat is transferred from our body to the surrounding water) and we feel cool. Heat always moves from a warmer substance to a cooler substance.

Degree of heat

Temperatures can range from absolute zero (-273°C) to very high temperatures in the thousands of degrees. All objects are assumed to contain some heat and therefore can be placed at some point on the temperature scale. For practical purposes, we will look at a range of temperatures between -30°C and 50°C. Heat always transfers from an object higher on the temperature scale to objects lower on the temperature scale, therefore, when two objects are brought into contact, heat will pass to the colder object. The illustration shows a container of water heated to a temperature above room temperature. When the heating source is removed the heat in the warmer water will move to the cooler surrounding air until the water and the air are at the same temperature.

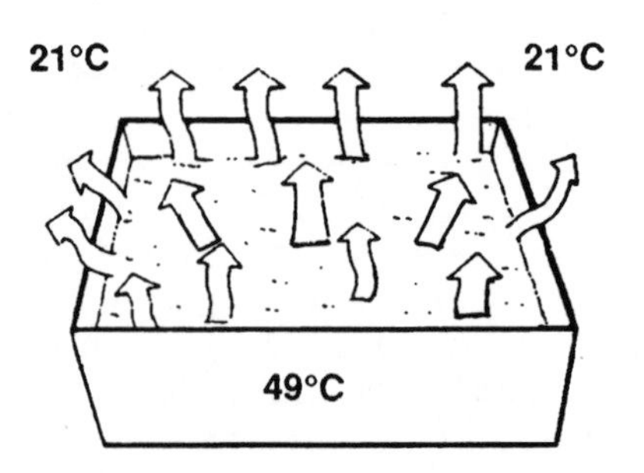

Heat travels from hot to cold.

Quantity of heat

The quantity of heat stored in an object depends on its size, temperature and material. Heat is measured in two ways: intensity or temperature of the object, and the amount or quantity of heat.

Change of state

It is generally accepted that objects will be found in one of three states: solid, liquid or gas. The addition or removal of sufficient heat from an object will cause it to change state. Heat will change a solid into a liquid (ice into water) and with continued heating, it will change a liquid into a gas (water into steam). The reverse will occur if heat is removed from an object, gas will change to a liquid, then with further cooling the liquid will change into a solid.

- Liquids absorb heat when changed from a liquid to a gas.
- Fluids give off heat when changed from a gas to a liquid.

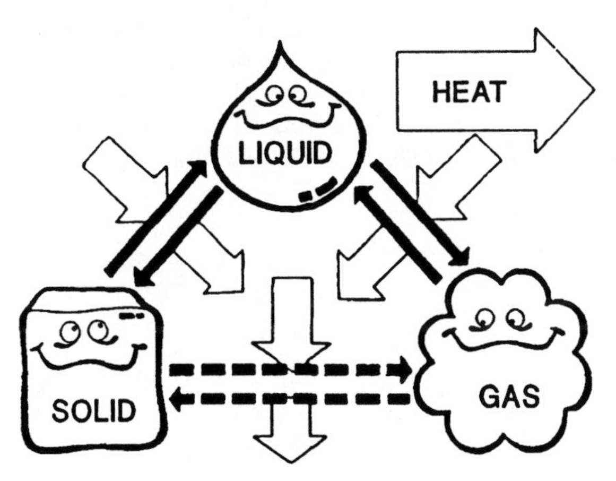

Three states of matter.

Latent heat

To this point we have accepted that a temperature rise will occur in an object as heat is applied to it. However, as it changes state, heat can continue to be applied without any apparent change in temperature. Consider a container of water subjected to heat. The water temperature will rise until the water starts to boil (100°C). Heat will continue to be applied but the temperature of the water will not rise while boiling takes place. If the temperature of the steam was measured and found to be 100°C, it would appear that the heat had been 'lost'. Remember, intensity (temperature) and quantity of heat are two different measurements. During the change of state the steam absorbed the additional heat input and it is contained within the steam. This lost or hidden heat is known as the **latent heat of vaporisation**.

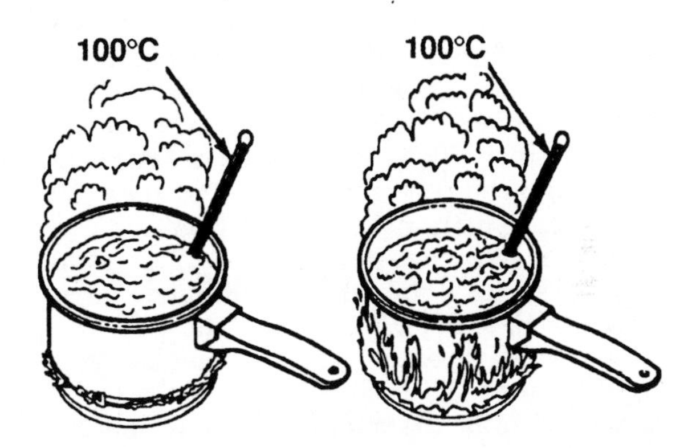

Heat of boiling water never exceeds 100°C.

The heat source is then removed and the steam is allowed to cool. As the steam cools, it is giving up the heat which had been contained within it. The temperature of the steam will not fall until all the stored heat has been released to the surrounding air and the steam has condensed back into water. After the steam has changed to water any further heat loss will cause a drop in temperature of the water. The heat given off during the change of state is known as the **latent heat of condensation**.

An important rule can be stated from our knowledge of latent heat. For a substance to change state from a liquid to a gas, heat must be absorbed by the gas. When a substance changes state from a gas to a liquid heat must be given off by the gas.

Basic air conditioning cycle

A simplified air conditioner can be constructed as shown in the basic air conditioning cycle diagram. The system is arranged with a container (heat exchanger) on each side of a wall. The containers are linked together with pipes, a pump is fitted to circulate the refrigerant and cooling air is blown over the container on the shady side of the wall. Refrigerant is placed inside the containers and the system is sealed from the outside air.

Heat from the sun warms the walls of the container and the liquid refrigerant starts to change state (liquid to gas). As the change of state occurs, the gas takes heat from the container walls. The gas (carrying heat taken from the container walls) passes through the pipe to the shady side of the wall and into the second container. Cool air passing over the second container allows heat to pass from the gas through the container walls and into the atmosphere. The removal of heat from the gas (latent heat of condensation) allows the gas to change state to a liquid. The liquid is then pumped back to the sunny side of the wall and the process is repeated. The continuous cycle of operation of the system allows heat to be transferred from the sunny side container to the shady side of the wall. The container on the sunny side will remain cool.

EFFECT OF PRESSURE ON REFRIGERANT

Before continuing our study into vehicle components, it is necessary to consider the refrigerant behaviour under different conditions.

It was stated previously that the evaporation (boiling) temperature of the refrigerant was in the region of –30°C. This would tend to indicate that the refrigerant would always be in a gas state because our normal day temperature is greatly in excess of the refrigerant boiling point. Increasing the pressure on the gas will cause it to give up its latent heat and change to a liquid. This will still occur at temperatures of 50°C provided the pressure is sufficiently high. Therefore, liquid refrigerant can be present in air conditioning components that are warm to touch.

It was also stated that heat will only move from hot to cold. Consider the high temperature of the air outside a car on a sunny day, yet warm air from the passenger compartment must be passed out to an atmosphere which may be at a higher temperature. This problem is overcome by using the engine driven compressor to raise the temperature of the gas as it is compressed. The refrigerant gas will then be at a higher temperature than the atmosphere and will give up heat through the condenser. As the heat is given off the refrigerant gas will change state back to a liquid.

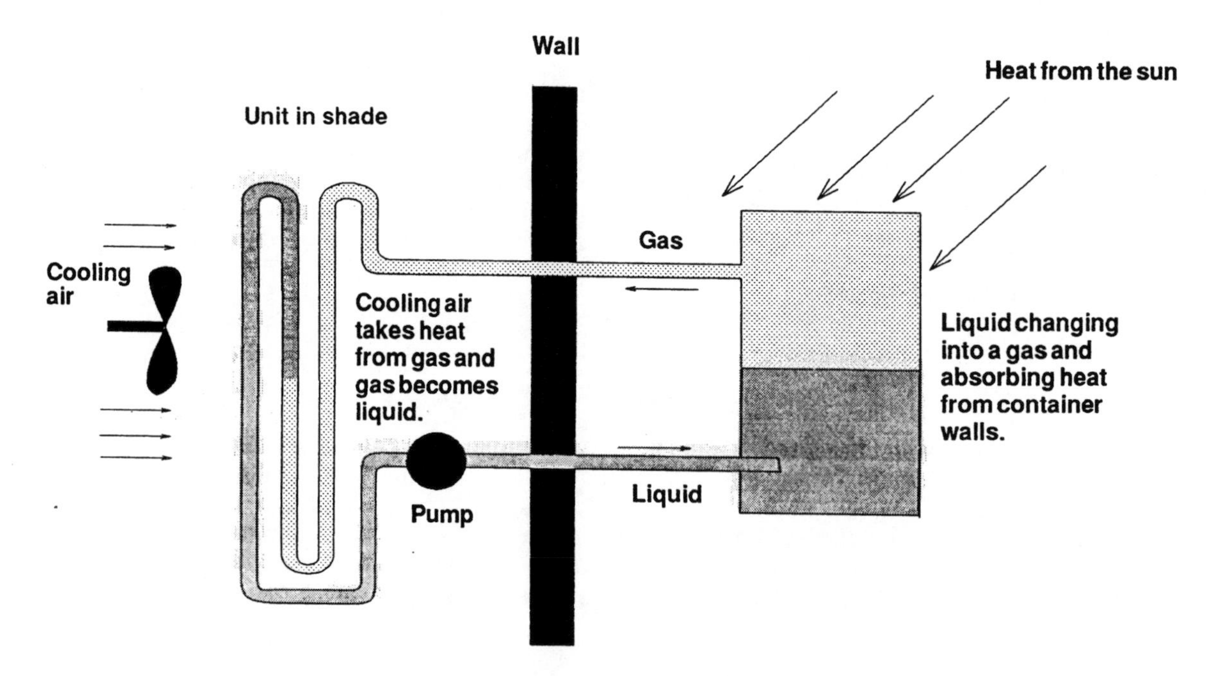

Basic air conditioning cycle.

BASIC AIR CONDITIONING CYCLE FITTED TO A CAR

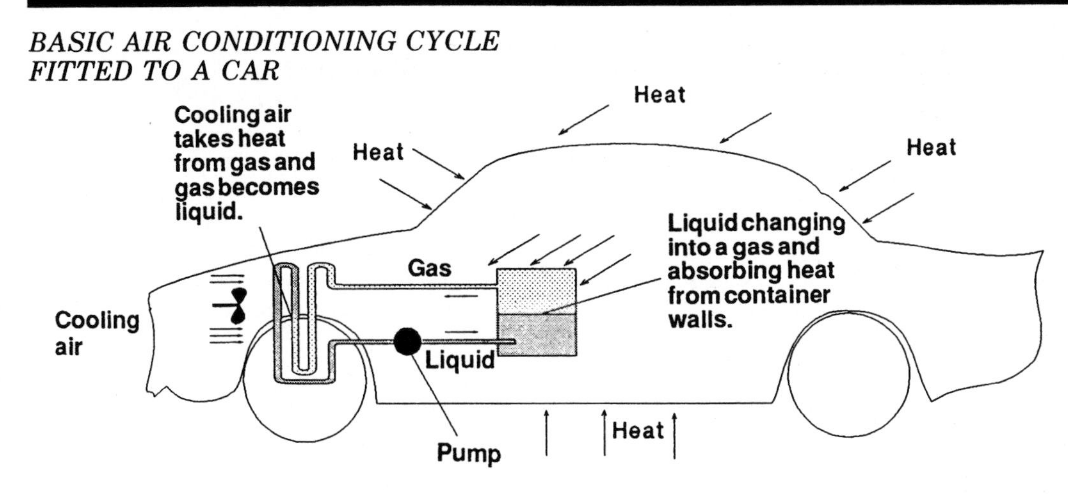

Heat can be removed from the interior of the car by using the basic air conditioner cycle.

The basic air conditioner described previously could be fitted into a car as shown in the diagram. The engine compartment bulkhead would take the place of the wall. Heat enters the passenger compartment through the glass, floor and from the engine bay. Heat is also given off by the car occupants. The heat load which has built up in the passenger compartment can then be transferred into the engine compartment and given off to the atmosphere.

The system shown would not be suitable for a car in its current form. However, by redesigning the components, placing them in suitable locations and adding a couple of controls the system's efficiency can be increased to an adequate performance level.

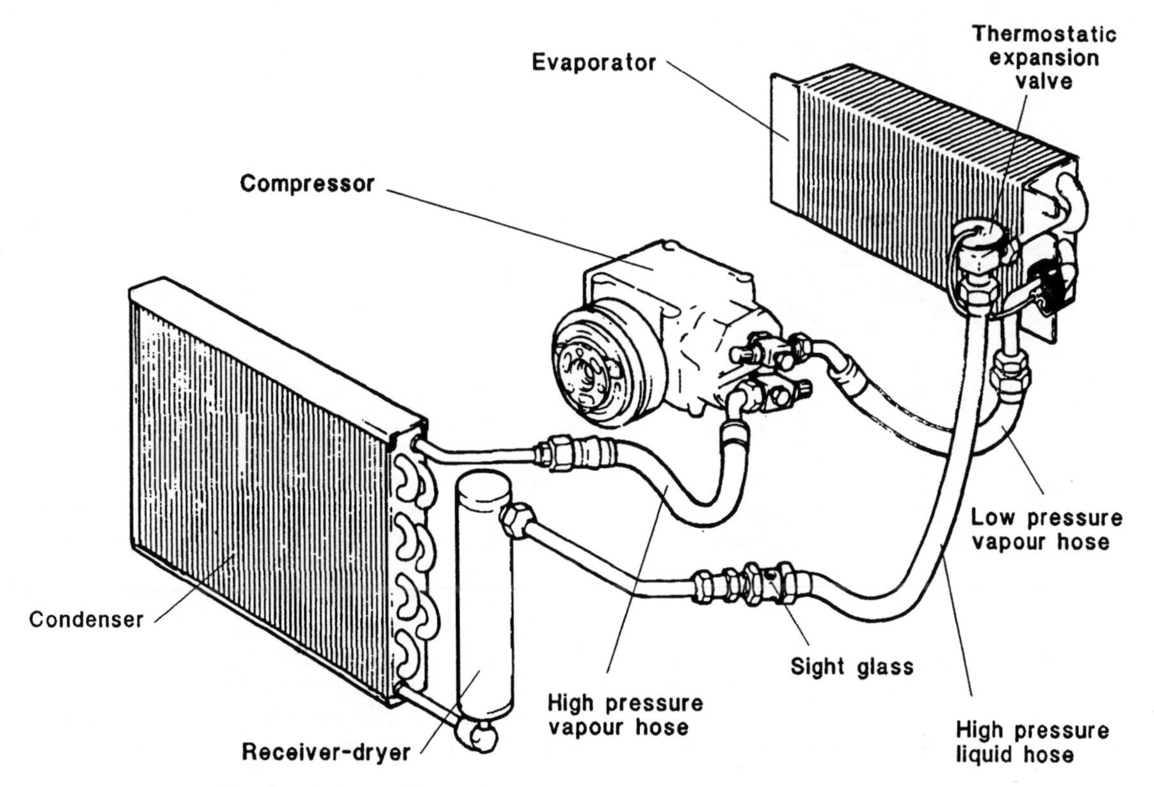

Basic automotive air conditioning system components.

AIR CONDITIONER SYSTEM

A basic vehicle air conditioning system requires five major components:
1 compressor;
2 condenser;
3 receiver-drier;
4 thermostatic expansion valve;
5 evaporator.

1 COMPRESSOR

This is the pump of the system; it raises the pressure (and the temperature) of the gas in the system. The compressor is designed to operate on refrigerant vapour only. There is a small quantity of refrigerant oil held in vapour form which will pass through the pumping section of the compressor but entry of liquid, either oil or refrigerant, will damage the compressor.

There are several different types of compressors used on passenger vehicles. All are belt driven from the engine and have electromagnetic clutches fitted to the drive pulley. The clutch will engage or disengage the drive pulley from the compressor input shaft, as required by the design of the system controls.

The internal construction of the compressor varies greatly between different types. Three major compressor types will be encountered:

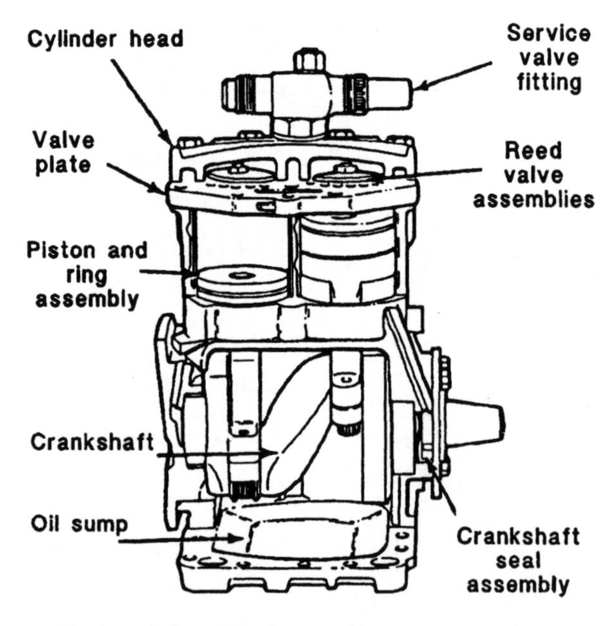

Twin piston reciprocating compressor.

a) The two cylinder, inline, reciprocating piston compressor may be mounted vertically or horizontally. The crankshaft, driven by the pulley, causes connecting rod and piston aseemblies to move up and down the cylinder bores. A valve plate assembly, containing inlet and outlet reed valves is clamped to the top of the cylinder bores by a cylinder head. Compressor oil is contained in the lower section to provide lubrication.

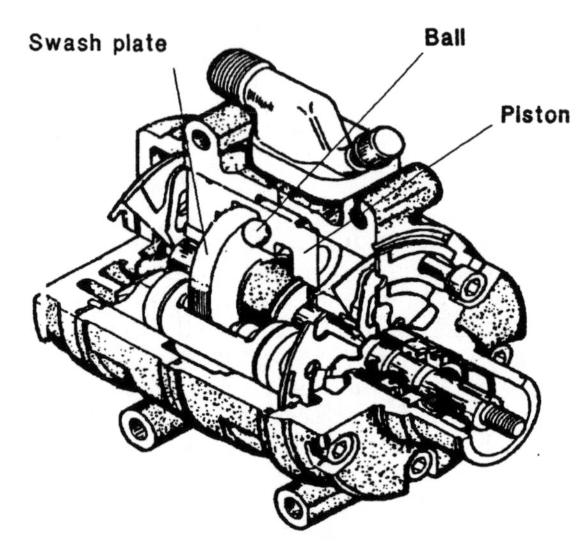

Swash plate type compressor.

b) The 'swash' plate type which has an angled plate driven by the input shaft and pulley. There are two major types using this swash plate construction. Type 1 has the swash plate mounted centrally, to which are fitted three pairs of double acting pistons. The compressor can be considered to be two halves. The front half contains three front facing cylinders and pistons. The rear half also contains three cylinders and pistons, but they face to the rear. Each end of the compressor has a reed valve plate but interconnecting passages link all the pistons to the inlet and outlet service fittings. Type 2 has an end mounted swash plate to which five pistons are mounted each with its own cylinder bore. A valve plate is mounted at one end and the pistons have common inlet and outlet fittings.

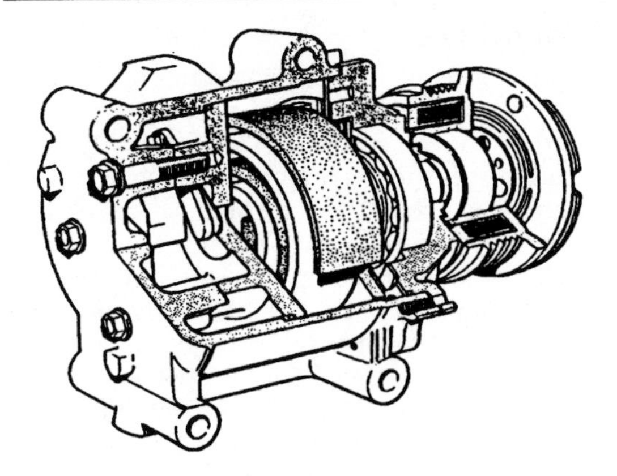

Twin scroll type compressor.

c) The twin scroll type. One spiral remains fixed to the compressor body. The second spiral is placed within the first spiral and rotated by the input shaft. This rotation produces an orbiting action and compression occurs in two pairs of compression recesses.

2 CONDENSER

The condenser, at first sight, resembles a radiator. A closer inspection shows that it consists of a tube bent into a serpentine shape. The shape is enclosed in a series of cooling fins. One end of the tube has an inlet fitting and the opposite end has the outlet fitting. The condenser is mounted at the front of the car forward of the cooling system radiator to ensure maximum cooling air will pass over it. Air passing over the fins removes heat from the refrigerant gas and the refrigerant will condense back to a liquid within the condenser tube.

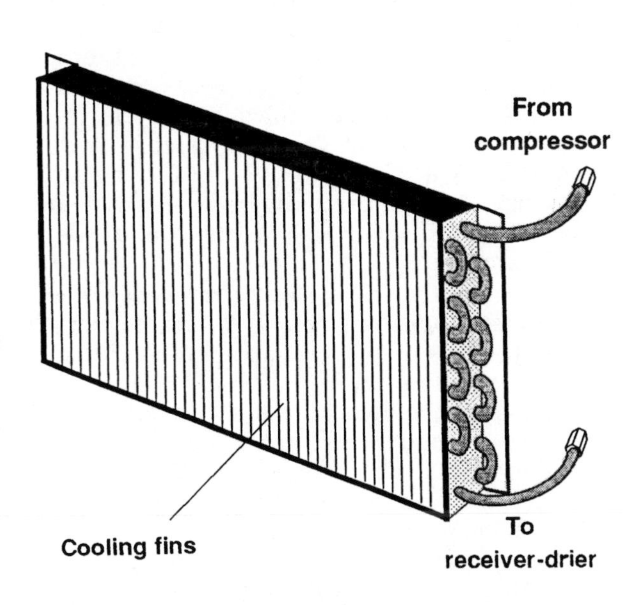

The condensor is mounted in front of the radiator.

3 RECEIVER-DRIER

The receiver-drier performs two functions. It acts as a storage tank for liquid refrigerant. A filter and a desiccant (dehydrating agent) are fitted inside the receiver-drier. The filter prevents the entry of foreign particles which could damage the moving parts of the system. The dehydrator has a limited ability to absorb moisture from within the system. A number of receiver-driers also contain a sight glass through which refrigerant condition (oil streaking, low charge) can be observed. The receiver-drier is mounted on the return line from the condenser.

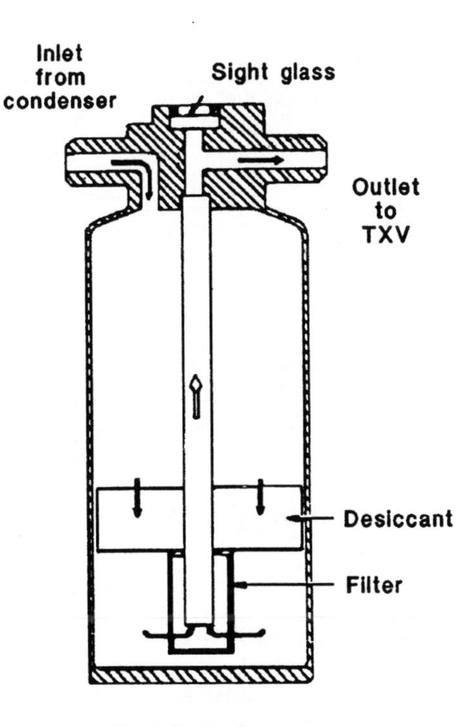

Receiver-dryer

4 THERMOSTATIC EXPANSION VALVE (TXV)

The TXV controls the rate of flow of the refrigerant within the system. The basic refrigeration cycle explained earlier did not have a means of controlling the flow rate through the system therefore the cooling output could not be controlled. The valve action is simple; an orifice is opened or closed to control the rate at which refrigerant is supplied to the evaporator. The method by which the valve moves is slightly more complex. The TXV has a diaphragm chamber mounted at one end. The underside of the diaphragm is linked to the valve operating pin so diaphragm movement will open and close the valve orifice. The upper side of the diaphragm is connected through a tube to a heat sensing bulb. The heat sensing bulb is attached to the outlet pipe of the evaporator. The temperature of the evaporator outlet pipe can therefore be used to control the rate at which liquid refrigerant is admitted to the evaporator inlet. More details of this action are contained in the TXV control explanation.

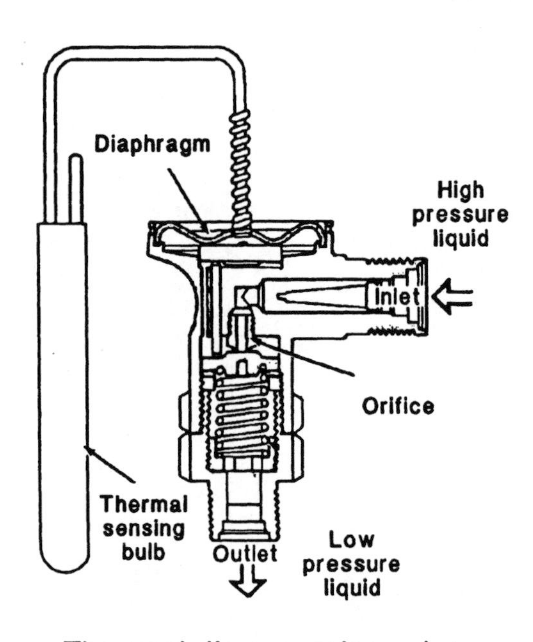

Thermostatic expansion valve.

5 EVAPORATOR

The evaporator is a smaller version of the condenser. It may have several serpentine tubes encased by many thin fins. The fins are to provide maximum heat transfer surface within a confined space. The units are usually mounted in a casing under the dash cowling. Warm air from the passenger compartment is blown by an electric fan over the fins of the evaporator. The refrigerant within the evaporator will accept the heat transfer from the warm air and change state into a gas. The latent heat contained within the gas will be passed through the outlet pipe to the compressor. The warm air passing over the evaporator will give up its heat and emerge back into the passenger compartment as cool air. The warm air may also have contained a significant quantity of water vapour. As the air is cooled by passing over the evaporator fins the water vapour will condense back to water and pass out of the car through drain tubes. The air passing back to the driver and passengers is dehumidified in addition to being reduced in temperature.

Air conditioning system operation

The air conditioning system consists of a low pressure section and a high pressure section. The high pressure section begins at the discharge side of the compressor and includes the condensor, the receiver-drier and all connecting piping up to the inlet part of the thermostatic expansion valve. All of these components and pipes are subjected to what is often called 'high side' pressure. The low pressure (or low side) starts at the outlet of the thermostatic expansion valve, through the evaporator and the suction (return) pipe to the intake side of the compressor.

- Operation of the system commences when the air conditioning on/off switch is selected by the driver to operate the air conditioner (engine running).
- The on/off switch allows a current to flow to the compressor drive pulley electromagnetic clutch. The clutch engages and the compressor will operate.
- The compressor will raise the pressure and temperature of the refrigerant gas. The high pressure gas flows through the compressor outlet and is forced through piping to the condenser.

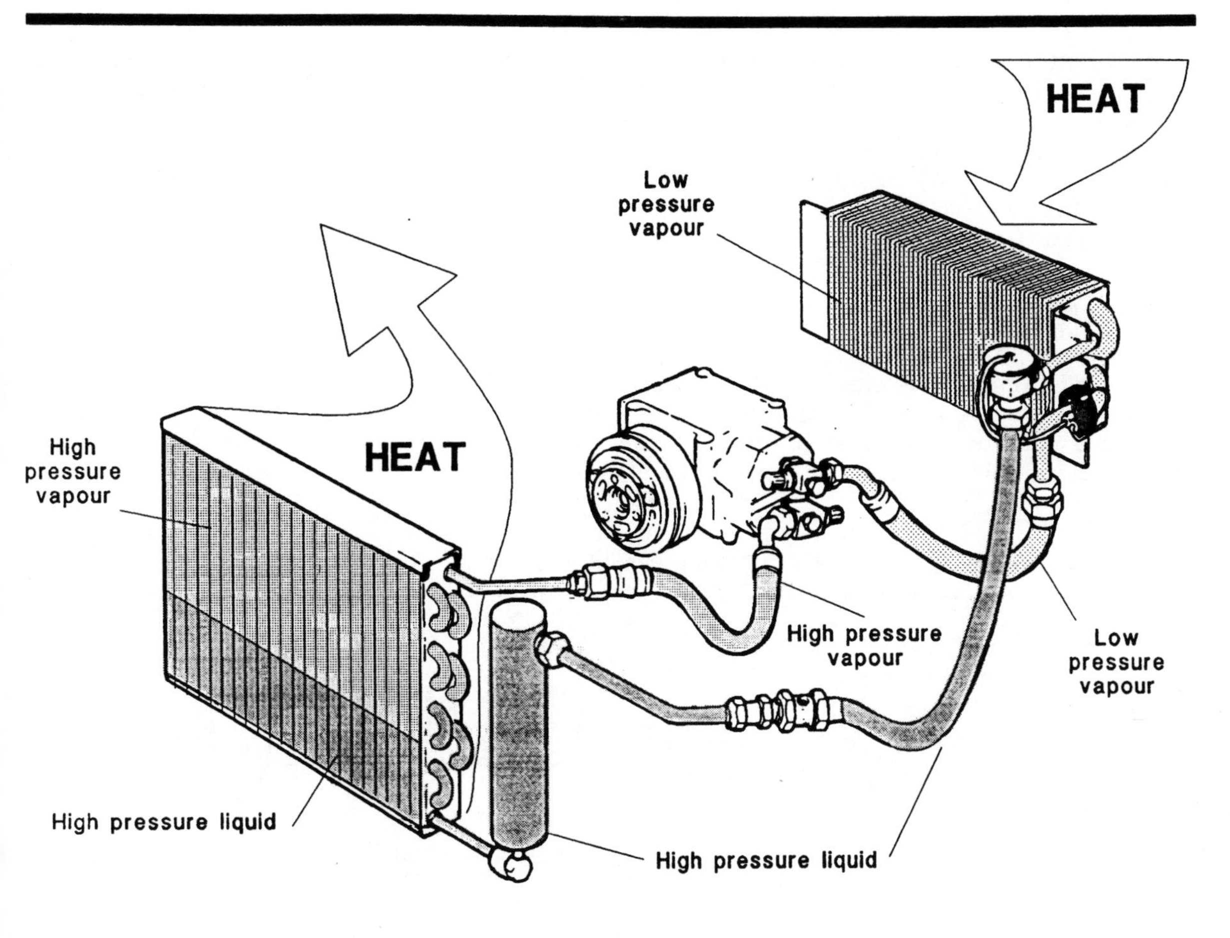

High pressure and low pressure parts of the system.

- As the car moves down the road, external air will be forced through the condensor fins. The external air is cooler than the gas inside the condensor and heat is removed from the gas. As heat is removed the hot gas will change state into a warm liquid. The warm liquid refrigerant will exit the condensor and flow through piping to the inlet of the receiver-drier.
- The receiver-drier acts as a storage reservoir, filters out particles and absorbs small quantities of moisture which could damage the mechanical components of the system. The filtered and demoisturised liquid refrigerant then passes out of the receiver-drier to the inlet of the TXV (thermostatic expansion valve).
- The thermostatic expansion valve is the most important control device in the system, it separates the high and low pressure sides and must be capable of controlling the rate of flow of the refrigerant within the system.
- As the refrigerant passes out of the TXV it enters the low pressure area present in the tubing of the evaporator. A blower fan ensures the warm passenger compartment air is continually blown over the external fins of the evaporator. The heat from the warm air will cause the refrigerant to change state to a gas. The refrigerant gas containing the heat from the passenger compartment flows out of the evaporator and through the suction pipe to the compressor where the cycle will be repeated.

Air conditioning systems of the simple type described in this chapter require a number of control devices to operate effectively. Control devices, such as TXVs, low and high pressure cut-out switches and compressor control thermostatic switches will be studied in later years of your course.

Climate control systems

These systems are a combination of the normal air conditioning systems and the vehicle heating system. By a combination of casings and blend doors it is possible to 'mix' air so the ideal comfort setting is maintained regardless of the outside conditions. The driver can set the comfort level required and the system will provide the same temperature on either hot or cold days. These systems have become quite complicated and may include a micro-processor and sensors. The passenger compartment, ambient, evaporator and engine temperatures are all monitored. A number of units have solar load detectors to measure the strength of direct sunlight. This information allows the micro-processor to select the combination of outlet ducts, fan speeds, heating and cooling blends to ensure the required temperature and dehumidification levels are maintained.

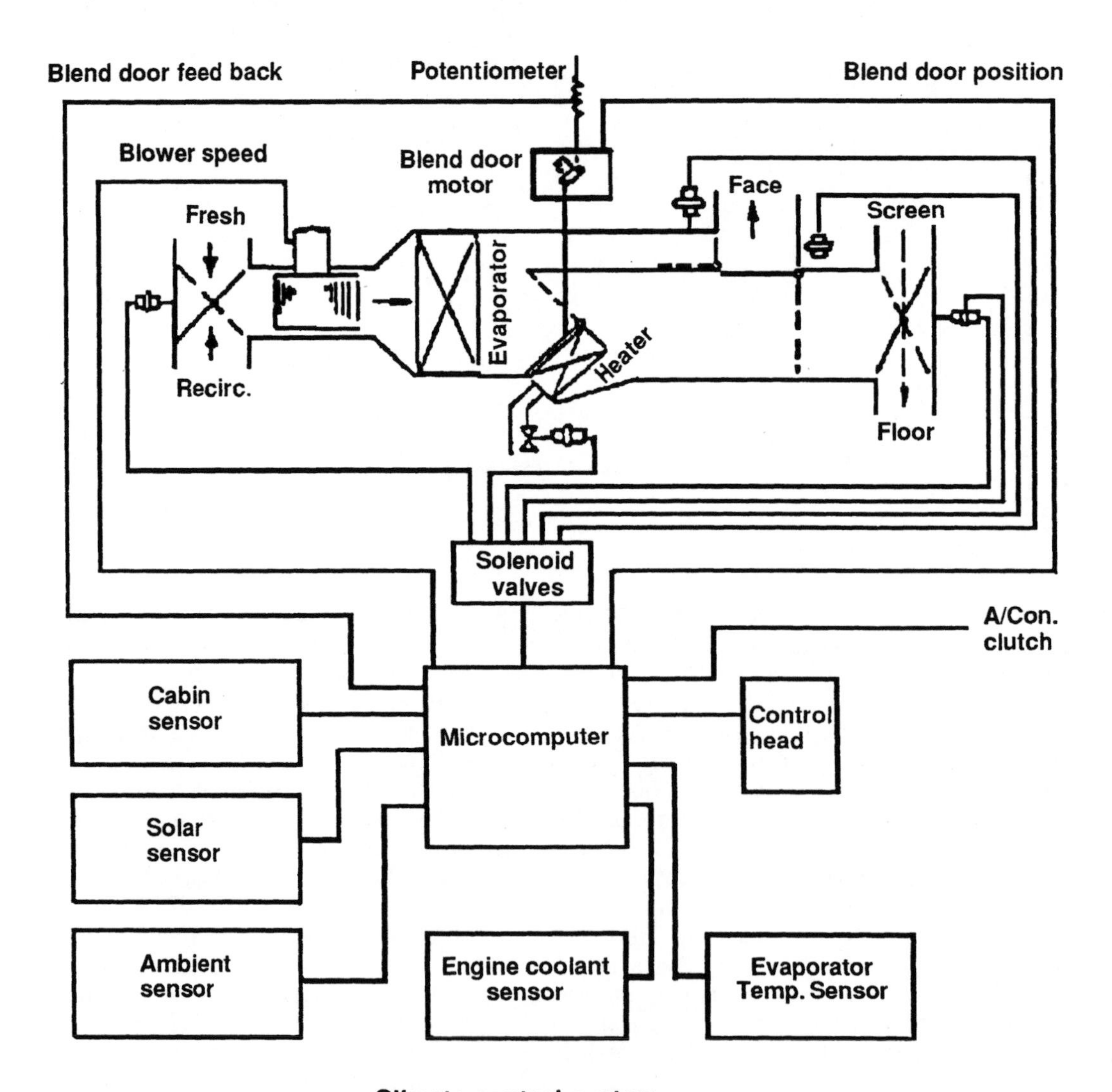

Climate control system.

CHAPTER 18 REVISION

1 Why is it essential to recover R12 from air conditioning systems?
2 Why should goggles and gloves be worn when working on automotive air conditioning systems?
3 List three precautions which should be observed when handling R12 cylinders.
4 Write a short note to explain the term 'latent heat of vaporisation'.
5 List the major components of a basic automotive air conditioning system.
6 Write a function description for each of the components listed in point 5.

GLOSSARY

Accelerator a pedal that is directly or indirectly connected to a throttle butterfly or fuel pump rack (diesel)

Acetylene the fuel gas used in cutting and welding operations

Advance (spark) the number of degrees that a spark occurs before the piston reaches top dead centre (T.D.C.)

Air/fuel ratio ratio of the mass of air and fuel which is induced into the engine cylinder

Air gap (spark plugs) distance between the centre and earth electrode

Air pollution contaminatation of the atmosphere by vehicle emissions

Align to put various parts of assembly into their correct operating positions

Alignment state of adjustment or setting related to the front or rear suspension of a vehicle. Also applies to the aiming of the headlamps

Alloy a mixture of two or more metals

Alpha-numeric combination of letters and figures used in marking of tyres

Alternator a belt driven device that produces electricity when the engine is running

Ammeter instrument used to measure the rate of current flow in amperes

Ampere unit of electric current produced by one volt acting through a resistance of one ohm. Commonly referred to as an amp

Ampere hour capacity a method used to rate batteries related to a current flow over a time period

Anti-clockwise (counter-clockwise) direction of rotation related to the left-hand movement of a shaft, gear, wheel, axle or disc. For example, grip the top of a wheel and then move your hand towards the left

Anti-freeze an additive mixed with engine coolant to lower its freezing point

Anti-friction bearing a bearing that relies on a rolling action to reduce friction

Aspect ratio ratio between the height and width of a tyre

Atmospheric pressure a measurement of the mass of air. Generally, 101 kPa at sea level

Axial direction parallel to a shaft

Axle a shaft that transfers drive from the differential to the road wheel

Ball joint a flexible joint used in steering linkage and suspension systems

Ballast resistor resistance used to control current flow in an ignition system

Battery rating a standard measurement of a battery's ability to supply electrical energy under specified conditions

B.D.C. bottom dead centre related to the crankshaft position when the piston has moved to a maximum distance from the cylinder head

Bleeding a process that describes the removal of air from a sealed system, such as hydraulic brake system and diesel fuel injector system

Blend door a device in a climate control system which controls the air flow and air temperature

Blow-by escape of exhaust gases past piston rings

Carburettor a device used to mix petrol and air in correct proportions

Carburising flame a welding flame in which there is an excess of acetylene

Catalytic converter an exhaust component (looks like a muffler) that reduces the toxic gases before they are emitted to the atmosphere

Charcoal canister fuel vapour storage device connected to the engine inlet manifold and fuel system components (carburettor and fuel tank)

Chassis backbone (frame) of a vehicle on which all other components are mounted

Choke mixture enrichment device located in the top of a carburettor

Clockwise direction of rotation related to the right-hand movement of a shaft, gear, wheel, axle or disc. For example, grip the top of a wheel and then move your hand towards the right

Cluster (counter) gear series of gears cut on the same casting that are rotated by the input shaft of a transmission
Clutch device used to connect or disconnect two rotating parts
Clutch housing (bell housing) stationary metal casting that surrounds a clutch device
Coil an ignition component used to increase battery voltage to the high voltage needed at a spark plug
Cold cranking rating (battery) cranking amps that can be maintained over a period of thirty seconds, at −18°C, while the cell voltage is above 1.2 volts, for example 7.2 volts for a 12 volt battery
Combustion burning of air/fuel in a combustion chamber
Combustion chamber volume (space) remaining when a piston is at its closest position (top dead centre) to the cylinder head
Compound two or more elements mixed together, for example two different metals, several gases, pieces of data
Compression reduction in length or volume of an object, gas or substance caused by applying at least two opposite forces, for example squeezing a coil spring placed between your hands
Compression ratio ratio formed when the total cylinder volume, with a piston at B.D.C., is divided by the combustion chamber volume
Compression rings piston rings that are closest to the top of a piston (generally, the first and second rings on a piston)
Compression stroke piston's movement that reduces the volume of (compresses) air/fuel mixture or air induced into the cylinder
Compressor the component in an air conditioning system which raises the pressure and temperature of the refrigerant gas
Concentric two or more round parts that share the same centre, for example two gears on the same shaft or a wheel on an axle
Condensation action of vapour turning to liquid
Condenser a component of an air conditioning system which allows the pressured gas to be cooled and change state to a liquid
Conduction transfer of heat between objects when they are in contact. The heat flows to coldest object
Conductor material through which current (amps) can flow
Constant mesh gears two or more gears that are always in mesh
Constant velocity universal joint transfers drive from one shaft to another without speed fluctuations
Consumable material that is used during a process, for example petrol in a running engine, electrode in arc welding
Convection transfer of heat from one object to another through the surrounding air (the objects are not in contact with one another)
Coolant liquid used in a cooling system
Cooling system system, generally using air or water, that removes excess heat from a running engine
Corrode chemical action that eats away the surface of an object
Counterbalance weight attached to a specific point on a rotating part to make it balanced
Counter-clockwise see anti-clockwise
Current movement of electrons (amps) through a conductor
Cycle series of events that repeat in the same definite order
Damper device that protects moving parts from excessive vibrations or oscillations
Degree (circle) 1/360th part of a circle
Degree (temperature) 1/100th part of a Celsius scale (thermometer)
Dehydrate remove moisture (water) from an object (dry out)
Desiccant a drying agent used in the receiver drier of an air conditioning system, its function is to absorb and hold moisture, usually silica gel
Die formed by cutting a shape into a block of metal
Die casting forming an object by forcing molten material into a die
Diff slang for final drive and differential unit
Diode electronic device that allows current (amps) to flow freely in one direction
Dipstick indicates the level of oil in the sump of an engine
Direct current current (amps) that flows in only one direction around an electrical circuit
Direct drive one part is rotated by another rotating part at the same speed
Discharge (battery) current (amps) flowing from a battery
Displacement (piston displacement) volume, measured in litres, of a cylinder when a piston moves from T.D.C. to B.D.C. This does not include the combustion chamber
Dowel pin steel pin fitted to one part so that it can be aligned with another part
Drop forged pounding a red hot piece of metal into a given shape by using a drop hammer
Dry friction resistance to movement between two unlubricated parts
Dynamic balance alignment of the centre of mass of an object with a place on its centre of rotation, for example balancing of tyre
Eccentric two or more round parts that do not share the same centre but rotate about the same point, for example cam lobe on a camshaft
E.E.C. see evaporative emission control

E.G.R. see exhaust gas recirculation
Electrochemical chemical reaction that produces electricity, for example a lead acid battery
Electrode (spark plug) centre rod and metal strip welded to the body of a spark plug are called electrodes. They are subjected to high temperatures and voltages
Electrode (welding) fluxed coated metal rod which melts during an arc welding process
Electrolyte mixture of sulphuric acid and water used in a battery
Electromagnet magnet produced when a current flows through a coil of wire wound onto a steel bar
Electronic low voltage and low current devices or circuits, for example diodes, transistors, integrated circuits (I.C.s), computers
E.M.F. electromotive force
Emissions harmful gases that are discharged from a motor vehicle into the atmosphere, for example carbon monoxide, hydrocarbons and oxides of nitrogen
End play axial (lengthwise) movement between two parts
Energy capacity for doing work
Engine displacement displacement of one cylinder (see displacement) multiplied by the number of cylinders in an engine. For example, if one cylinder has 0.5 L displacement and the engine has 8 cylinders, the engine displacement is 4 (0.5 × 8) L
Engine mounting thick rubberised pads fitted between the engine/transmission and the chassis
E.P. (extreme pressure) lubricant special additives mixed with oil so that it can lubricate hypoid gear teeth operating under very heavy loads conditions, for example final drive oil (diff. oil)
E.P.A. Environmental Protection Authority
Evaporation action of liquid turning to vapour
Evaporative control system a system that prevents hydrocarbons (petrol vapours) from escaping into the atmosphere (reduces hydrocarbon emissions)
Evaporator a component of an air conditioning system which is fitted inside the passenger compartment, its function is to change the liquid into a gas and control the temperature within the vehicle
Exhaust gas recirculation a system that directs a small amount of exhaust gas into the inlet manifold to lower combustion temperature (reduces oxides of nitrogen emissions)
Exhaust stroke piston's movement that forces the burnt gases into the exhaust system
Exhaust system system that transfers the burnt gases from the engine to the atmosphere at the rear of the vehicle (pipes, manifold, muffler and catalytic converter)
Exhaust valve part that opens and closes the exhaust port in a combustion chamber to control the flow of burnt gases
Ferrous metal metal containing iron
Field magnetic force lines which pass through the air around a magnet
Filament fine wire inside a light globe that produces light when it glows, as a current is passed though it
Filter removes foreign particles from air, oil, water or petrol
Firing order sequence of firing (spark at spark plug) each cylinder of a multi-cylinder engine
Force push or pull on a body that changes its shape, speed or movement
Forge similar to drop forging but a light hammer is used
Frame see chassis
Freon-12 see R-12
Frequency number of processes repeated in a given time, for example a number of cycles in a second
Friction act of rubbing one object against another which results in the production of heat
Friction force resistance on an object as it moves across the surface of another object. It depends on the type of surfaces, weight of object and clamping force
Friction bearing thin section tube or disc placed between a moving part (shaft or side of gear) and a stationary (housing) part
Fuel hydrocarbon or alcohol substance that is mixed with air to produce a highly combustible mixture
Fulcrum point about which a lever turns, for example pin that supports a brake pedal
Fuse device that melts when the current flow in an electrical circuit exceeds a safe value. It protects the other electrical parts in the circuit from higher than normal current flow
Fusible link short piece of special wire placed in the main electrical wire close to the battery. It melts when the main wire shorts or is overloaded (excessively high current flow)
Fusion two metals flow together when they reach their melting points, for example welding
Gas a substance that can fill the whole volume of a closed container. It flows easily, is compressible, expands rapidly when heated, and contracts rapidly when cooled
Gasket special substance placed between two mating surfaces to form a seal
Gassing boiling battery electrolyte which emits hydrogen gas during battery charging
Gear disc or cone shaped object that has teeth cut at regular intervals around its circumference
Gear ratio number of turns that the driver gear (gear rotated by its shaft) rotates while the driven gear (gear rotated by driver gear) completes one turn. For example two turns of

the driver gear to one turn of the driven gear produces a gear ratio of 2:1

Grid (street directory) inferred, evenly spaced vertical and horizontal lines that criss-cross a map

Grind to remove metal from an object by using a rough abrasive moving surface

Ground (electrical) chassis and body of a vehicle to which one battery terminal is connected, thus forming a common return current path for all electrical circuits

Gudgeon pin hollow metal pin that connects the piston to the connecting (con) rod. Other names used are piston and wrist pins

Halogen bulb quartz lightbulb filled with a halogen gas such as iodine or bromine

Hand brake wheel brake assemblies on one axle or driveline operated by the driver through a mechanical linkage. Prevents a parked vehicle from moving. Sometimes called a park brake

Harmonic balancer reduces the vibrations in a crankshaft by converting twisting energy from the power pulses into heat. Sometimes called a vibration damper

Headlight converts electrical energy into light that is strong enough to allow the vehicle to be driven at night

Heat engine machine that changes heat energy into mechanical energy which can be used propel a vehicle

Heat range (spark plug) means of classifying spark plugs by their operating temperature. The heat range is determined by the length and width of a spark plug's centre electrode insulator

Heat sink device that prevents electric parts from overheating by transferring the heat they produce to the atmosphere

Heat treatment (metal) method of changing the characteristics of a metal by heating it to a specific temperature

Helical spiral shape such as a coil spring or a screw thread

Helical gear teeth cut on a gear which are at an angle to its centre line

Herringbone gear 'V' shaped teeth formed on a gear or two helical gears, with exact opposite angle teeth, pressed together to form a 'V'

High tension refers to the very high voltage produced in an ignition coil and delivered to the spark plugs

Hone to remove metal from an object by using a fine abrasive moving surface

Hotchkiss drive transmission connected through an open tailshaft to the final drive and the driving force is transferred from the axle housing to the chassis by leaf springs or link arms

Hot spot small area whose temperature is considerably higher than its surroundings

Hydraulic brakes brakes operated by hydraulic pressure. The master cylinder provides operating pressure, which is transmitted through steel pipes to wheel cylinders or pistons and applies the brake shoes or disc pads

Hydraulic lifter valve lifter, which uses hydraulic pressure from the engine's lubrication system. It automatically adjusts to any variation in valve stem length

Hydrocarbons a combination of hydrogen and carbon atoms. All petroleum based products consist of hydrocarbons

Hydrometer an instrument using a float to determine the specific gravity of a liquid

Ignition system designed to produce a spark within the cylinders of an engine to ignite the mixture of petrol and air

Independent suspension a suspension system which allows each wheel to move up and down without undue influence on other wheels

Intake manifold a series of tubes connecting the carburettor with the intake valves in the cylinder head

Internal combustion engine an engine which burns fuel internally (inside) to develop its power

Jumper leads heavy duty electrical cables, having connector/clips at each end, which are used to temporarily connect a charged battery to a discharged battery for starting an engine

Jet a small hole or orifice in a carburettor used to control the flow of petrol

Kingpin a steel pin which connects the stub axle to the beam axle. The stub axle pivots around the king pin

Kilopascal 1000 pascals (Pa)

Latent heat amount of heat required to change a substance from one state of matter to another without any change in its temperature

Lead (petrol) petrol containing tetraethyl lead as an anti-knock additive

Leaf spring a suspension spring made up of several pieces of flat spring steel

Lean mixture (petrol) a fuel mixture, with an excessive amount of air in relation to petrol.

Level plug a plug which can be unscrewed to check the oil level in transmission and final drive assemblies

Leverage increasing force by using one or more levers

Live axle an axle, having a wheel attached to it and directly driving that wheel

L.P.G. liquefied petroleum gas

Lubrication reducing friction between two surfaces by coating them with grease or oil

MacPherson strut a front suspension system, incorporating a long telescopic strut, which allows the wheels to move up and down and pivot around the strut

Mag wheels light-weight wheels made from magnesium. The term 'mag' is often applied to alloy wheels

Master cylinder a component part of the hydraulic braking system, which provides hydraulic pressure

Micrometer a precision measuring instrument

MIG welding metal inert gas, a process of electric welding using a coil of wire and an inert shielding gas, usually argon

Molecule the smallest portion of matter which can be divided and yet still retain all the properties of the original matter

Monoblock all cylinders in an engine cast as one unit

Muffler (exhaust) a component through which all the exhaust gases flow to allow expansion and reduce the pressure of the gases

Newton unit of force. A force of 1 Newton accelerates a mass of 1 kg at the rate of 1 m/sec^2

Newton metre a measurement of torque. When a force of 1 Newton is applied to a lever at 1 m from the centre of a bolt, the bolt is subjected to a torque of 1 Nm

O.E.M. original equipment manufacturer

Octane rating a classification of a petrol's ability to resist detonation. Obtained by comparison with a standard reference fuel

Ohmmeter an instrument for measuring the resistance in an electrical circuit or unit (in ohms)

Oil classification — SG oil recommended for most modern petrol engines. Can be used where SC, SD, SE, SF are specified

Oil cooler a device used to remove excess heat from the engine and/or the transmission oils

Oil pump a component used to force oil under pressure to various parts of the engine

Oil sump a casing bolted to the base of the engine, which holds the oil used for engine lubrication

Oscillating action a swinging action, similar to the pendulum of a clock

Oscilloscope a test instrument which displays the performance of ignition and other electrical components action on a screen

Otto cycle the four stroke cycle of an engine consisting of intake, compression, power and exhaust

Oversteer the tendency of a vehicle, when cornering, to turn more sharply than the driver intended

Oxidising flame a welding blowpipe flame which contains an excess of oxygen

Ozone layer is located in the stratosphere, it protects the earth from harmful ultra violet rays

Parallel circuit an electrical circuit, which provides several paths for current to flow through and/or around a resistance unit

P.C.V. positive crankcase ventilation

Pinging rattling sound from the engine, usually heard during acceleration. Can be caused by ignition timing and/or wrong grade of fuel being used

Pitman arm is the lever connecting the steering box cross shaft to the steering linkage

Potentiometer a variable resistor having three connections, its function is to vary the amount of resistance applied to a circuit

Printed circuit an electrical circuit consists of electrically conductive lines printed onto a board, thus eliminating the use of wires

Ply rating (tyres) an indication of the tyre's load-carrying capacity

Polarity (battery) identifies the battery terminals (positive and negative)

Power booster (brakes) a component which increases the brake pedal effort by utilising engine vacuum

Preignition fuel mixture being ignited before the correct time

Preloading (bearings) is a mild loading applied by adjustment to anti-friction bearings to place the bearings in their operating position

Quad cam refers to an engine having four camshafts

R-12 (refrigerant) manufactured gas that is used in air-conditioning units

Radial (direction) the line at right angle (perpendicular) to a shaft

Radius the measurement by a straight line from the centre to the circumference of a circle

Ratio (gears) the fixed relationship between gears in mesh. For example, if the driving gear turns three revolutions to the driven gear's one revolution, then the ratio is 3:1

Receiver drier a component of an air conditioning system, its function is to absorb moisture and be a storage point for the refrigerant gas

Reciprocating a back and forth, up and down motion, same as piston movement in a cylinder

Regulator (gas) adjustable device to control pressure

Relay (electrical) electro-magnetic operated switch, which causes contacts to close or open to control the current flow in another circuit

Resonator small muffler fitted into an exhaust system to provide additional silencing

Retard (ignition timing) to set the ignition timing later than specified

Rocker arm pivots on the rocker shaft. When one end of the rocker arm is pushed up by a pushrod the other end pushes down and opens a valve

Rotary engine (Wankel) this engine does not use reciprocating pistons but consists of central rotors turning inside a housing. Each rotor is of triangular shape and provides three piston faces. The engine is normally of two-rotor design but single and triple rotor types have been built

R12 the type of refrigerant gas commonly used in automotive air conditioning systems is known as R12

Runout refers to the surface or edge of a rotating object. When the object is not 'running' true to its centre line the outer edge or surface will move in and out or rise and fall. Runout can be measured at right angles to the centreline or along the length of the object

Safety rim a wheel rim which has a rolled ridge beside the lip on each side of the rim. These ridges are designed to keep the tyre bead from moving into the drop centre of the rim in the event of a tyre failure

Seat has several variations, for example car seat upon which the vehicle occupants sit, engine valve seat upon which the engine valve seals, pump valve seats, seated injector plunger. The word 'seat' will occur in almost any application where one part is required to rest against or seal against another part

Series circuit the circuit has a single path for the current to flow. Resistances in the circuit are wired into the circuit so that any current in the circuit must pass through the first resistance before reaching the second resistance, etc.

Series-parallel circuit the circuit has a section (series) containing resistances through which all current must flow but also has a junction point where the circuit divides into two or more paths (parallel) in which the current can flow

Shock absorber fitted to each suspension unit to stop the vehicle bouncing after bumps. The most common type is an oil-filled telescopic tube fitted with valves which control both spring compression and rebound (double acting)

Sliding gear transmission gears which, when a gear change occurs, slide along a splined shaft until the gear teeth mesh with another gear

Snap ring a pre-tensioned ring (which is split for fitting purposes) fitted into grooves in shafts or housing recesses for the purpose of holding bearings, gears, shafts etc. in position

Solid state a term used to describe modern electronic devices which have tended to supersede earlier electromechanical designs. For example, vibrating contact voltage regulators are replaced by solid state regulators which contain diodes, transistors etc. and have no moving mechanical parts

Spark current flow across an air gap between two electrodes

Specific gravity the ratio of the mass of a given volume of a substance when compared to the mass of an equal volume of water

Spongy pedal (brakes) the brake pedal, when applied, does not have a normal resistance to application but feels spongy or springy. Generally this condition indicates air is present in the hydraulic system

Spot weld two electrodes clamp the work pieces together and a large current flows from one electrode through the work pieces to the second electrode, effectively fusing the metal pieces together

Spur gear a gear with straight teeth machined parallel to the mounting shaft

Stabiliser bar a shaped spring steel bar mounted across the vehicle frame (or subframe) and attached to both left and right suspension members. The purpose of the stabiliser is to reduce body roll during cornering, and it may be fitted to front, rear or both suspensions

Static balance the uniform distribution of the mass of a tyre and wheel rim about the wheel centre. A statically unbalanced wheel, when fitted to a freely turning hub, will stop rotating with the heavy spot at the bottom. This unbalanced spot, when the wheel is rotated at speed, will cause the wheel to bounce up and down

Steering arms are the final connection between the steering linkage and the steering spindle. The arms may be separate pieces bolted to the spindles or may be an integral part of the spindle forging

Steering geometry the various angles (camber, caster, toe-in, toe-out on turns and steering axis inclination) built into the vehicle suspension systems are collectively referred to as steering geometry

Synchromesh device fitted to manual transmissions to assist the driver in achieving smooth gear changes. Early design transmissions were not fitted with synchromesh units on each gear and the driver was required to use the accelerator and clutch to match gear speeds for 'crunch free' selection of gears. The only gear not fitted with a synchromesh unit, in modern transmissions, is reverse gear. Reverse should only be selected from neutral with the vehicle stationary so synchromesh is not required

Tap this term may refer to the tool (cutting tool with flutes and serrated edges) or to the action performed by the tool (cut threads in a hole)

Tap and die set a toolset of taps for internal thread cutting and dies for external thread (bolt) cutting

Tension a stretching force applied to an object. For example, when a nut is tightened onto a cylinder head stud the stud is placed under tension

Thrust bearing a bearing designed to accept forces parallel to the bearing mounting shaft

Timing the relationship of one moving component to another moving component, for example the timing of the spark at the spark plug in relation to piston position in the cylinder bore. This timing is usually achieved by comparing the crankshaft position marks when illuminated by a timing light flash triggered from the spark at the plug

Tolerance the variation from the exact specification allowed by the manufacturer

Torque turning or twisting force

Torque convertor a hydraulic coupling fitted between the engine and an automatic transmission. The unit contains an impellor, turbine and a stator

Torque multiplication when the engine torque is increased by the application of gears or a torque convertor

Transaxle the name applied to a unit when the transmission, final drive and differential components are combined into a single casing. This arrangement is ideal for front wheel drive vehicles

Transfer case this unit applies to four wheel drive vehicles and is used to supply the driving force to both front and rear axles. The unit can be fitted to the rear or the transmission or may be located in a separate casing and driven by a shaft from the transmission output

Valve clearance sometimes referred to in American publications as valve lash. This clearance is the space which exists between the valve stem and the operating mechanism when the camshaft follower is resting on the base circle of the camshaft. An engine fitted with hydraulic camshaft followers may operate with zero valve clearance

Viscosity the tendency of an oil to resist flowing

Viscosity index relates to the change of viscosity of an oil at low and high test temperatures

Volt unit of electrical force that will cause a current flow of one ampere through a resistance of one ohm

Voltage (E.M.F.) number of volts that an electrical supply (battery or alternator) must provide to an electrical circuit to maintain a given current flow (amps)

Voltmeter instrument used to measure voltage

INDEX

ACKNOWLEDGMENTS

Without the generous assistance of the organisations listed below, the cost of this book would have been prohibitive to students. The publishers and authors gratefully acknowledge the kindness of these organisations in giving permission for reproduction of the many illustrations.

Telecom
Universal Business Directories
McPherson's Limited
Sykes-Pickavant Ltd (illustrated products are available from Tridon Pty Ltd)
Ford Motor Company of Australia
Myttons Ltd
Litchfield Distributors Pty Ltd
Siddons Industries
Patons Brake Replacements
Repco Ltd
Repco Auto Parts
Stihl Pty Ltd
Australian Government Publishing Service. Material is Commonwealth of Australia copyright and is reproduced by permission.
Caterpillar Inc.
Nissan Australia
Mitsubishi
General Motors-Holden
Mitutoyo Precision Measuring Instruments
Torrington Bearings
Champion Spark Plugs
Perkins Engines Ltd
Chrysler Valiant
Datsun
Pitman Publishing
McGraw-Hill
Dunlop Industrial Sales
Mazda
Lisle
Honda
F.M.C.
C.I.G.
Oklahoma State Department of Vocational and Technical Education/Curriculum and Instructional Materials Center, Vocational Trade and Industrial Education, *Auto Mechanics* Vol. 1, p. AM-33-C. Copyright. Used by permission.
Thiessen/Dales, *Automotive Principles and Service* (2nd edn) © 1980. Reprinted by permission of Prentice-Hall Inc., Englewood Cliffs, N.J.
Deere & Co., *Fundamentals of Service* and *Tractor Operators' Manual*. Copyright. Used by permission. All rights reserved.